Synthesis of Subsonic Airplane Design

Synthesis of Subsonic Airplane Design

An introduction to the preliminary design of subsonic general aviation and transport aircraft, with emphasis on layout, aerodynamic design, propulsion and performance

Egbert Torenbeek

with a foreword by
H. Wittenberg

1982

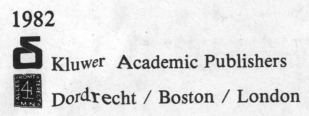

Kluwer Academic Publishers

Dordrecht / Boston / London

Library of Congress Cataloging in Publication Data

Torenbeek, Egbert.
Synthesis of Subsonic Airplane Design.

Includes bibliographical references and index.
  1. Airplanes–Design and Construction.
  I. Title.
     TL671.2.T67   1982   629.134'1   82–12469
ISBN 978-90-481-8273-2

CIP

Joint edition published by
Delft University Press,
Mijnbouwplein 11,
2628 RT Delft, The Netherlands
and by
Kluwer Academic Publishers
P.O. Box 17
3300 AA Dordrecht, The Netherlands.

Sold and distributed in the U.S.A. and Canada by
Kluwer Academic Publishers
101 Philip Drive
Norwell, MA 02061
U.S.A.

In all other countries, sold and distributed by
Kluwer Academic Publishers Group,
P.O. Box 322
3300 AH Dordrecht
The Netherlands.

Reprinted 1984,1985,1987,1988,1990,1993, 1995, 1996.

Contents

Foreword

Since the education of aeronautical engineers at Delft University of Technology started in 1940 under the inspiring leadership of Professor H.J. van der Maas, much emphasis has been placed on the design of aircraft as part of the student's curriculum. Not only is aircraft design an optional subject for thesis work, but every aeronautical student has to carry out a preliminary airplane design in the course of his study. The main purpose of this preliminary design work is to enable the student to synthesize the knowledge obtained separately in courses on aerodynamics, aircraft performances, stability and control, aircraft structures, etc.

The student's exercises in preliminary design have been directed through the years by a number of staff members of the Department of Aerospace Engineering in Delft. The author of this book, Mr. E. Torenbeek, has made a large contribution to this part of the study programme for many years. Not only has he acquired vast experience in teaching airplane design at university level, but he has also been deeply involved in design-oriented research, e.g. developing rational design methods and systematizing design information. I am very pleased that this wealth of experience, methods and data is now presented in this book.

In the last twenty years of university education for engineers much attention has been devoted to the fundamental sciences such as mathematics and physics. Recent years have seen a revival of the interest in "design" and a number of general textbooks have now been published on this subject. However, very few modern textbooks on the science and the art of aircraft design, are available. It is my sincere hope that Mr. Torenbeek's book will contribute to a renewed interest in airplane design in many parts of the aeronautical world, both inside and outside universities.

In view of the immense increase of knowledge in the aeronautical sciences and engineering since the Second World War, it seems a formidable task, requiring much courage on the author's part, to write a textbook on airplane design. It is well-nigh impossible to deal with all problems of airplane design at the same depth and undoubtedly personal choice has to prevail in many areas with regard to the material to be presented. In my view, Mr. Torenbeek has made an excellent choice of his subjects, preserving a careful balance between the presentation of a design manual and a general textbook on airplane design. This volume will therefore be a most worthwhile guide to everybody who in the course of

his professional training or career, is interested in the initial design phase of airplane projects, an activity which is very important for shaping the future of aviation.

Delft University of Technology   H. Wittenberg

August 1975   Professor of Aerospace Engineering

Author's preface

This textbook is intended to offer readers with a professional interest in airplane design a general survey of the layout design process. It contains a large amount of data and numerous methods which will be useful for carrying out the initial design calculations associated with the dimensioning of all major airplane parts. To a certain extent it has the character of a design manual, but considerable attention is also devoted to qualitative background information.

Several of the design methodologies and procedures presented have already appeared in the literature on the subject, while others have been developed recently by the author. They have been chosen on the basis of two criteria: they are not overdependent upon the state of the art and they give reliable results with a minimum of information. Most of the procedures have been extensively tested and considerably improved during the decade for which the author was responsible for students' design courses and projects in the Department of Aerospace Engineering of the Delft University of Technology. Emphasis is laid on conventional subsonic airplane designs in the civil category, i.e., broadly speaking the airplane types to which the American FAR Parts 23 and 25 and equivalent BCAR requirements apply (light and transport-type aircraft). Although many of the aspects to be discussed are equally relevant to V/STOL and military aircraft, other complicating factors are involved in the design of these types, resulting in a radically different approach to the design process. The large variety of design specifications and configurations in these categories prohibits a general treatment.

The author makes no apology for the fact that his approach to airplane design may be biased by a university environment, probably not the ideal one in which to carry out design studies. The teaching of design in the aeronautical departments of universities and institutes of technology has, unfortunately, not kept pace with developments in industrial design practice. Aircraft design and development have become a matter of large investments, even in the case of relatively small projects. The manhours required have increased considerably in recent years and the time is almost past when a single designer could consider himself the spiritual father of a new type.

In contrast with the increased sophistication to be observed in industrial design very few regular design courses at technological universities and institutes have been able to survive the process of continuous curriculum evaluation and revision.

Although experienced designers in the industry may possibly be the only authors qualified to write an authoritative textbook on airplane design, they are usually not in a position to devote enough of their time to a task which is not felt to be in the direct interest of their employers. The reader may therefore conclude that the present book will be most useful for teaching and study purposes and for people who need a general introduction to the vast field of initial aircraft design and development. Nevertheless, some of the procedures and data presented will certainly be of some assistance to design departments in industry.

A knowledge of the principles of applied aerodynamics, airplane structures, performance, stability, control and propulsion is required to derive the utmost from this book. Its usefulness for degree design courses will therefore be greatest in later stages of the course. In the presentation of the individual subjects the need to balance design considerations is frequently stressed. This is particularly the case in the second chapter, where the initial choice of the general arrangement is discussed, the basis adopted being a synthesis of many considerations of widely differing character. The main body of the book is devoted to the rationale behind layout design and although estimation methods for lift, drag, geometry, etc. are considered essential parts of the design process, they have been brought together in a separate set of appendices with a limited amount of text. Considerable attention is devoted in all the chapters to the impact of airworthiness requirements on design and to subjects that have been covered only very briefly by other authors. Particular emphasis is laid on the interior layout of the fuselage (Chapter 3), a survey of the present and future potentials of aircraft engines (Chapter 4), systematic design studies based on performance requirements (Chapter 5), and weight estimation methods (Chapter 8). The complex interaction of wing location, center of gravity range, and horizontal tailplane design is treated in Chapters 8 and 9. The consistent collection of prediction methods for lift, drag and pitching moment estimation will, it is thought, be useful as a general survey and as a tool for wing design (Chapter 7) and performance calculations (Chapter 11). A large collection of statistical data, illustrations and diagrams is added to this presentation, which aims at providing the individual student/ designer or the small design team with reliable guidelines. For industrial applications some of the methods may have to be refined and/or extended.
A large and systematic list of references to literature is presented, which will help the reader to find more information on the subjects specifically dealt with and on other related subjects. As he glances through these references the reader's attention may be drawn to a particular subject that interests him, possibly stimulating him to add another innovation to the design synthesis of his project and thereby contribute to the overall quality of aircraft design technology.

ACKNOWLEDGEMENTS

As is the case in the preparation of most technical books, the author of this volume is indebted to many persons who have aided in its completion.
For many years Professor H. Wittenberg has been the promotor of courses in preliminary aircraft design, and the idea of writing this book came about as a direct consequence of his activities. The author wishes to express his appreciation to him for his general support, for his critical revision of the text and for his willingness to write the foreword.
I am indebted to Mr. G.H. Berenschot, who has given general and technical assistance by collecting information and data, preparing many figures and tables, compiling the index and revising the text in detail. His perseverance, friendship and the moral support he has given me for many years have been particularly invaluable.
I would like to express my appreciation to Professor J.H.D. Blom, chief aerodynamicist, and Ir. P.F.H. Clignett, preliminary design engineer, both of Fokker-VFW International

as well as to Ir. C.H. Reede, head of the scientific department of the Royal Dutch Airlines (KLM), and to my colleagues Ir. F.W.J. van Deventer and Dr. Th. van Holten, who provided valuable and detailed suggestions together with actual text after reading parts of the book for their technical content. In addition, many students have used forerunners of the present text during their studies and their useful feedback has resulted in many improvements.

Thanks are due to the Department of Aeronautics and Space Engineering of Delft University of Technology for granting permission to prepare and publish this book, and for providing the necessary typing and duplication facilities. Many members of this Department and of the Photographic Office of the Central Library have given professional help in producing the illustrative material. I am also indebted to the Delft University Press, and in particular to Ir. P.A.M. Maas and Mrs. L.M. ter Horst-Ten Wolde for their support, encouragement and assistance in editing the publication.

The author wishes to express his gratitude to Messrs. J. van Hattum, J.W. Watson M.A., and D.R. Welsh M.A., who have made admirable contributions to the translation of the Dutch text and to improving the readability of the book. I am extremely indebted to Mrs. C.G. van Niel-Wilderink, for the excellence with which she performed the formidable task of typing not only the manuscript, but also the final copy for photoprinting. I would like to thank my uncle, the Rev. E. Torenbeek, for his painstaking efforts in checking the typographic accuracy of the copy, and mr. P.K.M. De Swert for preparing the final layout and for his methodical checking.

The following individuals, companies and organizations have kindly provided data and drawings.

Advisory Group for Aerospace Research and Development: Figs. 5-24, 10-1, 10-2, 10-3, F-17 and F-22;

American Institute of Aeronautics and Astronautics: Figs. 2-1, 4-39, 7-20, 6-1, 9-11, 10-5, F-12 and G-25;

Aeronautical Research Council: Figs. 2-24, G-16, G-17 and K-6;

Aérospatiale: Figs. 2-3, 2-10 and 7-25;

Airbus Industries: Figs. 11-5 and 11-9;

Aircraft Engineering: Figs. 3-1, 3-11, 3-21, 3-26, 6-16, 7-25, 10-17, 10-18, 12-3 and 12-5;

Alata Internazionale: Fig. 2-3;

The Architectural Press Ltd.: Fig. 3-6;

Avco Lycoming: Fig. 4-14;

Aviation Magazine: Fig. 2-8;

Avions Marcel Dassault-Bréguet Aviation: Figs. 3-3, 3-11 and 10-20;

Boeing Aircraft Company: Figs. 2-11, 2-15, 3-11, 5-18, 7-25 and 10-6;

Mr. S.F.J. Butler: Fig. F-22;

Canadair Limited: Fig. 3-21;

Canadian Aeronautics and Space Institute: Fig. 3-27;

Centre de Documentation de l'Armement: Figs. 2-5 and 10-21;

The De Havilland Aircraft of Canada Ltd.: Fig. 3-11;

Detroit Diesel Allison Division of General Motors Corporation: Fig. 4-46;

Dowty Group: Fig. 6-27;

Engineering Sciences Data Unit Ltd.: Figs. F-23, G-26 and K-9;

Flight Control Division of U.S. Air Force Flight Dynamics Laboratory: Figs. A-2, A-3, E-3 and E-9;

Fokker-VFW International: Fig. 11-14;

Flight International: Figs. 2-2, 2-10, 2-11, 3-8 and 7-30;
Hamilton Standard Division of United Technologies Corporation: Fig. 6-5;
Hartzell Propeller Inc.: Fig. 6-7;
Hawker Siddeley Aviation: Figs. 2-20, 3-21, 12-2 and 12-4;
International Civil Aviation Organisation: Figs. 10-4 and 10-5;
De Ingenieur: Fig. 5-22;
Institute of Aerospace Sciences: Fig. 6-24;
Lockheed Aircraft Corporation: Fig. 3-17;
McDonnell Douglas Corp.: Figs. 3-14, 6-2, 7-25;
McGraw-Hill Book Cy.: Fig. 6-19;
Messerschmitt-Bölkow-Blohm: Fig. 2-7;
(British) Ministry of Aviation: Fig. 10-6;
National Aeronautics and Space Administration: Figs. 2-25, 6-5, 7-5b, 7-22, 9-8, E-2,
E-5, E-6, G-11, G-12, G-19 and G-22;
National Research Council of Canada: Fig. 3-27;
Pergamon Press: Fig. 5-11;
Mr. D.H. Perry: Fig. K-6;
Piper Aircraft Corporation: Fig. 4-10;
Polytechnisch Tijdschrift: Figs. 2-10 and 7-25f;
Pratt and Whitney Aircraft Division of United Technologies Corp.: Fig. 4-46;
The Royal Aeronautical Society: Figs. 1-1, 2-14, F-6, F-22 and K-8;
Rolls Royce 1971 Ltd., Derby Engine Division: Figs. 4-15, 4-19, 4-20, 4-37, 4-46 and
6-25;
Dr. W. Schneider; Figs. C-2 and C-3;
Ir. G.J. Schott Jr.: Fig. 5-22;
Society of Aeronautical Weight Engineers: Figs. 11-7 and D-1;
Society of Automotive Engineers, Inc.: Figs. 4-7, 4-8, 4-9, 6-2, 6-7, 6-23, 6-27, 7-5
and 11-13;
Mr. W.C. Swan: Fig. F-17;
Turboméca Bordes: Fig. 4-46;
Vereinigte Flugtechnische Werke-Fokker Gmbh: Fig. 3-11;
Mr. R.E. Wallace: Fig. 5-24;
John Wiley and Sons Ltd.: Fig. F-8.

Finally, I would like to thank my wife, Nel, for her unstinting help without which this
book would hardly have been possible.

September 1975   Ir. E. Torenbeek
Delft University of Technology

UNITS

In accordance with the convention used in publications such as Jane's All the World's
Aircraft and Flight International, all data and most of the figures have been given in
the technical unit system, both in British and metric units. Hence, lb and kg refer to
pound and kilogram forces, respectively. An exception is made in Appendix J, where sea
level data of the Standard Atmosphere have been given both in the technical and SI
systems.

Preface to the student Edition

Textbooks on the rapidly advancing subject of aircraft design tend to become
obsolete within a few years. In spite of this the first edition has proved its
value up to the present time, as a reference source for design efforts and
publications in many places all over the world.
It therefore pleases me that the publishers have decided to launch this new edition,
aimed at an expansion of the market into the university classroom, thereby making
the book affordable by many more individuals.
This has given me the opportunity to further refine some of the methods and
formulations, mainly on the basis of suggestions and comments of attentive students.
In spite of the reduction in size, the contents are not abbriviated.

Delft, May 1981
E. Torenbeek

Chapter 1. General aspects of aircraft configuration development

SUMMARY

It is shown that there is an interaction between the development work for a new design and the various factors determining the need for a new type of aircraft. Preliminary aircraft design is an essential part of this development phase; its aim is to obtain the information required in order to decide whether the concept will be technically feasible and possess satisfactory economic possibilities.
Attention is paid to the impact of the design requirements laid down in the initial specification and the airworthiness regulations. General observations aimed at illustrating the aircraft design and optimization process are presented.

1-1. INTRODUCTION

In the pioneering era of civil aviation the aircraft designer had only a very limited choice. There was practically only one category of powerplant at his disposal, namely the - nearly always aircooled - piston engine, which gave very limited power. Either there were no aerodynamic aids to augment the lift of the wing at low speeds, or those that did exist were, for various reasons, seldom used, with the result that wing loadings were kept low and high speeds were consequently unobtainable. Low wing loading favored the biplane layout which, with its high parasitic drag, formed another obstacle to high speeds. Flight was rarely above 10,000 feet (3,000 meters), since there were no pressurized cabins. During this epoch aircraft design was generally the work of one or a very few designers in each factory and the scope of development work for each new aircraft type was limited. In the twenties it was possible to design and produce a new aircraft for delivery to the customer within half a year, one of the reasons being that series were relatively small. This enabled Anthony Fokker and his staff to build fourteen entirely different commercial designs during the relatively short period of eighteen years (1918-1936).

But the nature of aircraft project design has undergone a radical change since the Second World War. Development of the jet engine and subsequently the turbofan, now supplying a thrust up to about 50,000 lb (22,500 kg), has greatly widened the choice of powerplants. Transport aircraft now cruise at altitudes of 30,000 to 40,000 feet (9,000 to 12,000 m) at speeds not far below that of sound. The takeoff and landing speeds of the largest aircraft have, of course, risen steadily and runways have consequently become longer and longer, but means to call a halt to this trend have meanwhile proved technically feasible.

Air transport has passed through an era of unprecedented growth. During the period from 1950 till 1970 the average yearly increase in the passenger-miles flown reached 14%, a growth figure which was only exceeded by the sale of plastics. The increase in transport productivity (payload times speed) of the largest transport aircraft has been equally impressive. In addition, modern aircraft have to satisfy an ever increasing number of severe safety regulations, while economic requirements, resulting from intense competition, have steadily become more exacting, with the result that the development and construction of new types of aircraft - even relatively small ones - demand a very high capital outlay and entail considerable financial risk. An aircraft industry nowadays is generally unable to produce an entirely new type of aircraft oftener than once every 12 years (roughly), quite apart from the question of whether there is any need for more rapid replacement of existing types. An exception to this "rule" must be made for the giants in the aircraft industry, e.g. Boeing and McDonnell Douglas, who are able to bring out one or two additional new designs during the same period. The sheer size and long leadtime of new projects have led various firms to share the risks by cooperation, while in Europe they have resulted in international joint ventures.

Although new concepts have been and will continue to be proposed from time to time by talented designers, the time is past when a chief designer could be regarded as the spiritual father of a new type of aircraft. A possible exception to this might be made in the case of private aircraft and small transports. Preliminary design departments nowadays have staffs numbering some dozens to several hundreds of highly trained technicians and engineers and computing facilities have increased immensely, while some preliminary design teams even have wind tunnels permanently at their disposal. More manhours are now being invested in the project design phase than were formerly spent on the entire detail design. Work in the design department has developed into a professional occupation, carried out in teams, with regular consultations between specialists in various disciplines.

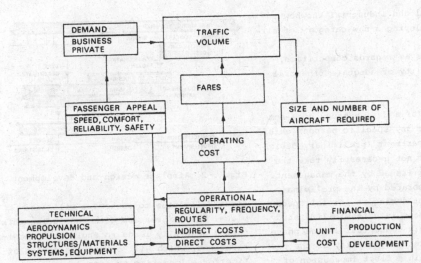

Fig. 1-1. Factors contributing to the growth of air transport, according to Morgan (Ref. 1-26)

This does not imply that the man in overall charge of preliminary design has himself to be a specialist. As will be shown further on, he must be able to take a wide interest in and have a sound insight into a great many disciplines related to design as a whole.

1.2. AIRCRAFT DESIGN AND DEVELOPMENT

Although the technical aspects of aircraft design form the subject of this book, it will be appreciated that this is not an activity carried on in remote offices by specialists generating designs of any kind that may occur to their imaginations. There is close interaction between the development work for a new aircraft type and the other factors which together determine the growth of and/or changes in aeronautical activities. These interrelationships being different for the various fields of aeronautics (passenger and cargo transport, business aviation, tourism, flying instruction, etc.), we will deal with the matter by quoting a single example here.
The development of new airliners has always been stimulated mainly by the growth of the traffic volume and the improvement of technical and operational standards. Fig. 1-1 is a scheme for allocating the various fac-

tors contributing to the growth of air transport. Growth in air traffic stems from reduction of fares, improved quality of the aircraft (speed, comfort), increased business activity and growth of private incomes, aircraft capacity growth, increasing number of routes, increasing frequency on existing routes, and greater utilization of aircraft and ground facilities. The contribution of research and development to this process, indicated in the lower left-hand corner of the diagram, is unique because this block shows only output lines. Although the diagram is obviously a simplification of the real situation and, at the same time, must not be considered as a control system*, it does indicate that aircraft development is a primary cause of growth. As a corollary to this, it is necessary in launching a new development program to appreciate the interacting effects of the "aeronautical environment" in advance and thereby ensure that there will be no conflict with the (future) needs of operators, passengers and the general public. Moreover a number of industrial constraints set limits to the feasibility of new projects, namely:
a. the available project development organization and production capacity;

*For example, the role of the government's aeronautical activities has been omitted.

3

b. the technical and industrial knowhow required for developing a new category of aircraft;

c. the prospects as regards competition;

d. the availability of adequate financial backing.

The initiative for a design study does not always stem from any specific person (chief designer) or department (preliminary design office) and does not necessarily take the form of an order issued by the management. The idea is elaborated by the preliminary design department during an initial speculative design phase in a feasibility study. The object of this conceptual design phase is to investigate the viability of the project and to obtain a first impression of its most important characteristics.

If the results seem encouraging both from a technical point of view and as regards the market prospects, a decision may be taken to develop the design further in order to initiate a new-design aircraft development program (Fig. 1-2). Comparisons will be made with some alternatives, preferably on a systematic basis. The design that has scores the highest rating will be elaborated in greater detail in the preliminary design phase. A characteristic of this phase is that modifications are made continuously until a decision can be taken to "freeze" the configuration*, and this marks the end of the preliminary design phase. If the market is considered likely to receive the design favorably and finance for the project is assured, the management may give permission for further development (go-ahead approval). The subsequent phases of detail design, construction and testing will lead to the granting of a Certificate of Airworthiness and some time later the

*The expression "configuration" as used in this chapter refers to the general layout, the external shape, dimensions and other relevant characteristics. It is not intended to indicate the actual airplane configuration as characterized by the position of the flaps, landing gear, etc.

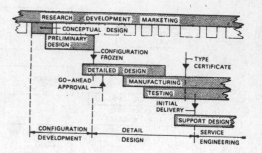

Fig. 1-2. Airplane design and development

first deliveries to customers will take place. The information gathered during this period generally leads to engineering modifications, which will occupy the design office for a long time to come.

After the first production series has left the factory, the company will continue to develop its product. These developments may take the form of an increase in the airplane's transport capacity (stretching), installation of an improved type of engine, improvement of performance by introducing aerodynamic refinements, such as "cleaning up" the aircraft, etc. Successful aircraft generally go through a process of growth which offers the customer a choice of a number of variants, each suitable for a specific transport assignment, and this considerably strengthens the company's ability to face up to competition. In Fig. 1-2 these activities have been arranged in three groups: the configuration development* phase the detail design phase and the service engineering phase. The first two of these have been taken as separate phases, since the decisions taken during the first stage are still partly based on the statistical probabilities that specific technical aims will be achieved and the actual construction is only defined in broad general terms, whereas in the detail design phase the aircraft is designed "down to the last rivet" and the detailed production schedule is laid down. During this period the number of

*Frequently referred to as "preliminary design engineering".

4

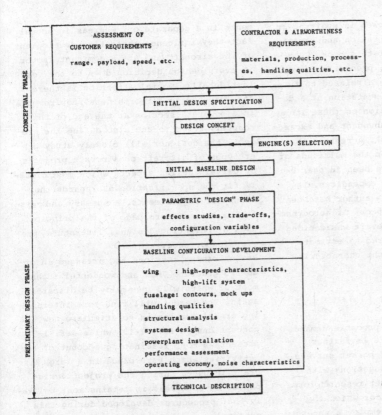

| ASSESSMENT OF CUSTOMER REQUIREMENTS
range, payload, speed, etc. | CONTRACTOR & AIRWORTHINESS REQUIREMENTS
materials, production, processes, handling qualities, etc. |

INITIAL DESIGN SPECIFICATION

DESIGN CONCEPT

ENGINE(S) SELECTION

INITIAL BASELINE DESIGN

PARAMETRIC "DESIGN" PHASE
effects studies, trade-offs,
configuration variables

BASELINE CONFIGURATION DEVELOPMENT

wing   : high-speed characteristics,
            high-lift system
fuselage: contours, mock ups
handling qualities
structural analysis
systems design
powerplant installation
performance assessment
operating economy, noise characteristics

TECHNICAL DESCRIPTION

Fig. 1-3. Configuration design and development of a high-subsonic transport aircraft.

documents and drawings will increase rapidly and the development costs will show a nearly proportional rise.

1.3. CONFIGURATION DEVELOPMENT

The principal aim in this phase of design (Fig. 1-3) is to obtain the information required in order to decide whether the concept will be technically feasible and have satisfactory economic possibilities. In contrast to the detail design phase, neither the actual construction process nor the detailed production schedule plays a dominant part here.

An important aspect of the entire development of a new type of aircraft is that it takes place in a succession of design cycles. In the course of each of these cycles the aircraft is designed in its entirety and investigation is carried out

into all the main groups and airframe systems and equipment to a similar degree of detail. The extent of this detailing steadily increases as the design cycles succeed each other, until finally the entire aircraft is defined in every detail. On the basis of the terminology given in Figs. 1-2 and 1-3, the subsequent basic design stages might be designated as follows:
a. conceptual design;
b. initial baseline design;
c. baseline configuration development;
d. detail design.

These design stages might also be referred to as the speculative design, the feasible design, the best conceivable design and the final hardware design cycles. A number of aspects of the first three of these design cycles will be further elaborated and discussed in the course of this book. The conceptual design will be the subject of the second chapter. Chapters 3 through 10 discuss the procedure followed in evolving an initial baseline configuration and constitute the main portion of this book. Chapters 11 and 12, finally, present a survey of several items which are related to further elaboration and presentation of a design. Since the information on these items available elsewhere is abundant and exceeds the scope of our textbook, a representative choice has been made from the multitude of subjects, while some have been further elaborated in the form of appendices. In view of the fact that the author has been mainly concerned with aspects of aerodynamic design and performance, it was decided to present methods required to estimate some aerodynamic and flight characteristics.

1.3.1. The design concept

In the initial phase the probable demand for a new type of aircraft is further specified with the aid of market surveys, further inquiries and discussions with potential customers. Market research forms a specialized discipline for which the large aircraft companies employ a separate department or office, while in smaller factories the work is often done by the design team. In either case, it is essential that the designer or the design team is closely involved since there is no sense in starting a design before the nature of the design requirements has been studied from all angles and a clear picture has emerged on which to base the general design philosophy.

The market survey will lead to an initial specification which will mainly define the transport performance - payload and maximum range, often also the cruising speed - as well as the most relevant field and climb performance, cabin arrangement, airframe services and equipment, etc. A decision will also have to be made as to which set of airworthiness and operational requirements the design will have to comply with. Fig. 1-3 oversimplifies the situation by presenting the airworthiness require-

ments in a separate box whereas in actual fact they influence the entire development of the aircraft and are a dominant factor in every design decision down to the very last detail. Special attention is therefore devoted to airworthiness requirements in a separate section at the end of this chapter. Prior to and also during the project, the designer will closely study other types of aircraft or aircraft projects or certain design aspects which most closely fit the specification as regards the transport requirements. A summary and critical assessment are made of the principal data of competing designs, literature, previous experience, etc.

Brainstorming sessions are arranged in order to generate "new and wonderful ideas", most of which will generally be discarded again. The reader will find some interesting creative attempts to stimulate new conceptions in Ref. 1-13, while Ref. 1-35 presents a most fascinating account of the activities which take place in a large aircraft industry during the project design phase. Nevertheless it remains most unlikely that procedures developed during this extremely speculative first design phase will offer any certainty of leading to a successful new conception. There is no real substitute for the originality of the engineer who is capable of forcing a breakthrough with a unique brainwave such as the monocoque construction, the sweptback wing, jet engines at the rear of the fuselage (à la Caravelle), area ruling and other inspired innovations, sometimes generated in an environment where project design is engaged in only incidentally, if at all. Design concepts are therefore being developed continuously, while only very few actually result in a preliminary design and subsequent development program (Fig. 1-2). The conception phase will result in preliminary layout sketches of the kind shown in Fig. 1-4, including a summary of the principal characteristics and basic design philosophy on which subsequent design stages will be founded. Although hardly any thought has been given at this stage to

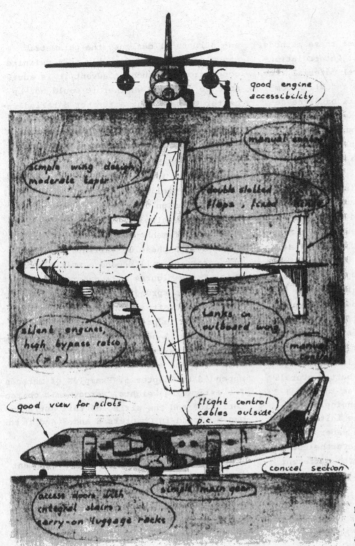

good engine
accessibility

manual control

simple wing design,
moderate taper

double slotted
flaps, fixed (?)

silent engines,
high bypass ratio
(≥ 5)

tanks in
outboard wing

manual

good view for pilots

flight control
cables outside
p.c.

conical section

access doors with
integral stairs;
carry-on luggage racks

single main gear

Fig. 1-4. Initial design concept of an ultra-short-haul
airliner (30 passengers)

details, a complete aircraft has already
been put on paper and the designer may have
a strong feeling that it could be the answer.

1.3.2. Initial configuration design and
configuration variations

Since design is not a deterministic process,
particularly in the early stages in which
the conception is realized, various solutions to attain the desired goal will present themselves. If it proves impossible to

weigh up the pros and cons and arrive a
realistic answer, based on the intuition
and experience of the designer(s), comparative studies will have to be undertaken.
Since it will generally not be found feasible to transform all likely configurations into fully developed projects, a parametric design phase, as shown in Fig.
1-3, is often decided upon. This first entails the development of an initial baseline design (or point design), using relatively easily applied sizing methods, provided these are available. In this book the

reader will find a number of these methods, based partly on theoretical interrelationships, partly on statistical material (see, for example, Chapters 5, 6 and 7). A complete layout is made of this baseline design, showing three views and some principal cross-sections. The next step is to check to what extent the characteristics and performance of the design will meet the design requirements.

Changes are now made in this baseline design, preferably in a systematic manner, following predetermined and clearly defined working rules. There will thus emerge a family of designs which are easily comparable with each other as well as with the baseline design. The object of this exercise is twofold: first to improve the design where it does not meet the requirements, and second to investigate the most likely possibilities and see whether other variants may prove a better proposal. It may also show that changes in the design requirements would yield a better overall balance. Although the diagram in Fig. 1-3 suggests that the type of engine has already been chosen before the initial baseline design has been put on paper, the parametric design phase may nevertheless include studies of variants with different types of engines and even a different number of engines per aircraft. When a number of variants are studied, a systematic approach is essential in order to obtain a sound basis for comparison. Although the absolute accuracy of the methods used should be as high as possible, the main objective is to differentiate between the designs. Final judgement at the end of this phase will result in a baseline configuration which, subject to approval by all concerned, will be chosen for further development. It can be presumed that this design, after detailed engineering, will probably meet the initial specification while, in addition, it will be the best conceivable design. On the other hand, in the absence of complete certainty as to specific aerodynamic characteristics of the consequences of variations in weight as between differ-

ent structural designs, the parametric study may not lead directly to a definite conclusion. In such an advent it is advisable to consider whether it would not be better to carry out a further detailed investigation of, say, two alternative configurations before an irrevocable choice is made. This point is treated more fully in Ref. 1-38.

To give some idea of the magnitude of this phase of a preliminary design, the initial baseline design of a small transport will require something in the order of several thousands of manhours, whereas the subsequent design phase of variants and parametric studies will demand a multiple of this. These figures will naturally show a considerable spread and will depend on the type of aircraft and on the extent to which the company is prepared to pursue the investigation. A number of examples of parametric studies are given in this book (Section 5.5) and also in the references appended to Chapter 5. Examples of methods for estimating weights, aerodynamic characteristics and the sizing of tail surfaces can be found in Chapters 8 and 9 and in the appendices.

1.3.3. Baseline configuration development

During this phase the baseline design is further developed to a depth of detail which can be regarded as meaningful. Various sections of the design department will be called in to contribute to the aerodynamic design, the stressing of the main structure, design of the airframe systems and equipment, etc. At the earliest possible stage a start is made with tests in the wind tunnel, while the external lines are determined and mockups showing the internal layout of the fuselage, cable runs and installations, are built and all the remaining tasks involved in arriving at a complete definition of the project are carried out.

During the development of the baseline configuration, errors will be observed which have been made during the phases already described and which are usually caused by lack of data. Correction of these errors will entail corrections in the design which will have repercussions for all the disciplines involved.

Coordination during this phase will be the responsibility of the preliminary design department, for this team will be most familiar with the project and is consequently best able to visualise the consequences of the corrections. One of the most frequent jobs during this phase is the setting-up of a weight control program, particularly for those weight components which have been estimated solely on the basis of statistical material. These rectifying programs and corrections may in some cases be covered by feedbacks in Fig. 1-3, but specific mention of them there has been omitted to preserve the clarity of the illustration.

As soon as the project can be regarded as sufficiently mature and any doubts regarding its essential characteristics have been removed, the project manager may take the decision to freeze the configuration and this means the end of the preliminary design phase. The characteristics of the design are summarized in a technical description which serves as a basis for discussions with potential customers.

Some idea of the scope of a configuration development program can be obtained from Ref. 1-36, which gives the following information concerning the Lockheed L-1011: "In the two years of the configuration development, over two million man hours were expended to investigate various configurations and approaches to determine the optimum design. More than 10,000 hours of testing have been completed in seven different wind tunnels to establish the most efficient overall configuration".

1.3.4. The preliminary design department

When the development of a new type of aircraft is to be undertaken, the general practice nowadays is to form a project group, containing not only preliminary design engineers but also experts in other disciplines, such as:
- aerodynamicists who are directly concerned with the design of the external shape,
- structural engineers, dealing with preliminary research into the overall structural layout and carrying out the dimensioning and optimisation,
- production experts and experts in the materials field who investigate what types of production methods should be adopted,
- service experts, to ensure easy maintenance and overhaul,
- weight engineers, whose job is to deal with the prediction and control of the weight (distribution) and moments of inertia,
- engineers to design the flight control system and analyse flying qualities,
- designers of airframe systems and equipment, and
- financial and economic experts who are not only able to estimate the first and operating costs of the aircraft, but also keep a close check on the financing of the entire design project.

Since this book does not deal primarily with the organizational aspect of project development, but rather with the technical aspects of airplane design, we will specify only the various tasks of the preliminary design team. Unlike the other departments of the design office, this team is permanently engaged in project work and its work mainly consists of the following activities:
- market analysis and the drawing-up of initial specifications for new types of aircraft in close cooperation with the sales department;
- devising various solutions to a given design problem;
- evaluation of different design proposals using preliminary design methods in order that decisions are taken on the basis of a sound assessment of the pros and cons;
- setting up and coordinating detail research oriented on aerodynamical, structural and other problem areas. These tasks

9

may be of a general character, such as the development of design methods for estimating drag, weights, etc., or project-oriented, such as the aerodynamic design of a flap system;
- discussions with potential customers and (future) subcontractors for main components such as engines, landing gear, airframe services, avionics, etc.;
- assisting the sales department by supplying technical data;
- making product development studies, aimed at increasing the utility of existing aircraft.

1.4. THE INITIAL SPECIFICATION

There is certainly no need to prove that sufficient material on a subject such as "market analysis aimed at the development of new aircraft" exists to warrant the publication of a separate volume. The present paragraph will of necessity have to be restricted to a few general observations with civil aviation as their main background. The example used will be an initial design specification for a hypothetical short-haul airliner for 180 passengers in the all-tourist layout, referred to as "Project M-184". A design evaluation of this project can be found in Ref. 1-64; it was intended as a highly simplified example for the purpose of illustrating the design process in a series of lectures. An apology is due for the fact that most of the considerations which follow in the present section apply to this particular design, intended for introduction into service around 1980.

In civil aviation the specification of a new aircraft type is generally drawn up by the manufacturer. Airlines are usually more content to evaluate projects offered to them for use on their own route network, though in a few cases they themselves have taken the initiative and written the specification which they felt was required. The designer will, however, realize that

a project can only be justified when it is likely to find a worldwide market. A specification issued by an airline will only be interesting provided it also appeals to competitors. It should also be realized that operators do not necessarily possess the best insight into the technical capabilities and knowhow of the airplane manufacturer. Nevertheless, there are examples of successful aircraft which have been designed to an operator's specification or a specification written with the customer's active cooperation (Viscount, Tristar and DC-10). All the same, the responsibility for the specification and the resulting project will still rest squarely on the shoulders of the aircraft manufacturer. This procedure is quite different from the case of military aircraft, where the specification will nearly always be issued by the customer: the armed forces.
The term "initial specification", as opposed to the more detailed type specification of a design, is used here to emphasize that there is an interaction between the technical design work and the development of the design requirements as a result of market analysis, engine development and various assessments during the development phase.

1.4.1. The need for a new type of aircraft

The following are some good reasons for initiating a new aircraft design:
a. Existing aircraft are becoming either technically or economically obsolescent, and a new type may do the job better. New standards for equipment, maintenance, operational use, noise suppression, passenger comfort, etc. may make renovation of the operator's fleet desirable.
b. Certain developments in traffic patterns have created a need for new types of air transport. For example, the growth of traffic may, as explained in Section 1.2, result in a new class of (larger) transport aircraft, or new travel habits (home to work and back) may open up the possibility of a new class of commuter aircraft. Air transport may fulfil the needs of developing countries, where the infrastruc-

* NUMBER OF PASSENGERS IN AN ALL-TOURIST LAYOUT (SEAT PITCH 34 IN., .87 M): 180 OR MORE. CORRESPONDING DESIGN PAYLOAD: 20,000 KG (44,100 LB). AN UNDERFLOOR FREIGHTHOLD VOLUME OF AT LEAST 50 M^3 (1,762 CU.FT) WILL BE REQUIRED. STANDARD SIZE BELLY CONTAINERS ARE PREFERRED.

* RANGE, WITH ABOVE MENTIONED PAYLOAD: 2,200 KM (1,200 NM) IN A HIGH-SPEED CRUISE, AT A DOMESTIC RESERVES. MAXIMUM RANGE (REDUCED PAYLOAD): 3,200 KM (1,726 NM) AT LONG-RANGE CRUISE TECHNIQUE.

* MAX. CRUISING SPEED AT 9,150 M (30,000 FT) AL-TITUDE: M = .82. DESIGN LIMITS: M_{MO} = .85, V_{MO} = 704 KMH (380 KTS) EAS.

* FIELD LENGTH REQUIRED FOR TAKEOFF AND LANDING, ACCORDING TO AIRWORTHINESS RULES: 1,800 M (5,900 FT) AT SEA LEVEL, ISA + 20 $^{\circ}$C (95 $^{\circ}$F), AT MAXIMUM (CERTIFICATED) TAKEOFF WEIGHT. RUNWAY LOADING: LCN = 30, RIGID PAVEMENT, 18 CM (7 IN.) THICKNESS.

* REGULATIONS: FAR PARTS 25, 36 AND 121. THE NOISE CHARACTERISTICS MUST SHOW AN IMPROVEMENT RELATIVE TO THE 1969 VERSION OF FAR PART 36 OF 10 EPNdB.

Fig. 1-5. Initial specification of a hypothetical short-haul airliner for introduction into service around 1980

ture is inadequate for surface transport.
c. A new type of aircraft is built and tested in order to give added impetus to an important new technical development, such as a V/STOL demonstrator prototype. Since experimental aircraft nearly always lead to a financial loss, at least in the first stages, there will have to be government funding, e.g. in the form of a development contract.

Manufacturers should be wary of aiming at filling the "gap in the market". That gap may well have remained unfilled for the simple reason that the need for an airplane of the kind was insufficient. Another danger which should be warned against is the adoption of a particular technical novelty which in itself may be a very clever achievement but is unlikely to contribute to profitable operation of the aircraft. Nevertheless, the design office will be continually involved in studies aimed at determining the potentialities of new technical developments and innovations. Any new type designed will have to be marketed in accordance with a properly thought-out time schedule. It is important to remember that if it is offered too early the production rate will increase too slow-ly, resulting in a productivity loss on investments which the company has put into the project. A launching delayed too long may be equally disadvantageous, either because the market has meanwhile been saturated by competitors' products or because the production line has to expand too fast and excessive manpower has to be (temporarily) hired and additional investments made.

The initial specification shown in Fig. 1-5 was drawn up for an airliner intended to augment and replace the current class of high-subsonic short-haul passenger transports: the BAC 1-11, McDonnell Douglas DC-9 and Boeing 737, and to some extent also aircraft designed for medium ranges: the Hawker Siddeley Trident, Aérospatiale Caravelle and Boeing 727. The category considered does not include smaller aircraft such as the Fokker F-28 or the VFW-614. The aircraft mentioned above are powered by low-bypass turbofans and have a capacity of 80-120 passengers (short-haul) or 120-180 (medium-haul). The need for a new type stems from the following considerations:
a. The increased traffic volume requires larger-capacity aircraft.
b. The new standard of passenger comfort

11

introduced by the wide-body jets will undoubtedly be extended to short-haul traffic.

c. Reduction of noise production will be a
prerequisite in the eighties.

d. Technology improvements in the fields of
high-speed and low-speed wing aerodynamics,
new structural materials (composite structures), lightweight avionics and improved
flight control systems may be considered
for application in this new aircraft category.

In view of the large volume of short-haul
traffic the market seems to offer scope
for a new aircraft with smaller capacity
as compared to the Airbus A-300, for example.

1.4.2. Transport capacity

When a new specification is being drawn up,
the first step will have to be a forecast
of the traffic and the transport demand
over the route sector concerned during the
period under review. A technique commonly
used here is a statistical analysis of the
yearly growth percentage of the total distance covered by passengers in terms of
passenger-miles (passenger-kilometers). On
the basis of an extrapolation of this
growth percentage, the total transport demand for the period considered may be estimated. Assumptions will next have to be
made regarding the frequency of the flights,
the average load factor and yearly utilization and from these the desired productivity (number of passengers times block
speed) can be deduced. A rule of the thumb
sometimes used states that the most favorable time between successive flights over
a particular route is about equal to the
time taken to fly the route. Hence, if the
block speed increases, the frequency of the
service should also be stepped up. The following are some other aspects to be considered:

a. For a large capacity aircraft the operating costs per aircraft-mile will be high,
but those for a seat-mile will be low,
since certain costs do not rise proportion-

ally to the size of the aircraft, e.g. total salaries of the flight crew and the
cost of avionics and certain services, and
will therefore decline with each additional seat.

b. A comparatively small aircraft will
show a low cost per aircraft-mile and its
critical load* will be smaller than that
of a large aircraft. This does not necessarily apply to the critical load factor
(critical load/maximum load).

Generally speaking, large aircraft are
best suited to routes with high traffic
density, provided the frequency of operations is compatible with the market requirements.

In drawing up the specification for the M-
184 project (Fig. 1-5), the following conclusions were arrived at:

a. During the 1960-1970 period short-haul
traffic grew considerably, both in the
United States and in Europe. A yearly
growth of 15 percent, resulting in a
doubling in five years, was no exception
and the growth was even more marked during
the 1965-1970 period. Charter traffic in
fact underwent an explosive expansion during that same period, with growth percent-
ages as high as 25 to 30. Factors which
contributed to this growth were: regular
tariff decreases, a rising level of pros-
perity, and the greatly improved comfort
of jet aircraft compared with other means
of transport.

b. A gradual decrease in the yearly growth
can be expected for the period 1975-1985
as a result of a slackening-off or decline
in the economy, a certain measure of satu-
ration of the transport market, and unavoidable increases in tariffs. The latter
are a result of the rapidly increasing
costs of fuel and the measures which have
to be taken to meet the certification requirements regarding noise levels. Assuming
an annual growth of 10 percent for the
years 1973-1980 the total yearly production

*The number of passengers required to pay
the cost of the flight.

on short routes will have to rise to 195 percent of the 1973 value, while during the first three years after the airplane's introduction the traffic demand will rise to about 250 percent.

c. On very busy routes the Airbus A-300 and possibly also the Trijets McDonnell Douglas DC-10 and Lockheed 1011 will take over a large share of the short/medium-haul traffic. On routes where the growth will be less progressive, however, the jump in capacity from current short-haul aircraft to the A-300 will probably be too great and there will be an opening for aircraft with a capacity some 80 to 100 percent greater than that of the DC-9, provided it offers good possibilities for further growth.

d. For the M-184 a capacity of at least 180 passengers has been chosen for an all-tourist layout with a possible later "stretch" to about 250 passengers, while the cargo holds require a total volume of at least 1800 cu. ft (50 m^3). Compared to that of current airliners the passenger accommodation must show an improvement in the level of comfort, but this need not necessarily be achieved by the use of two aisles. A very close watch will have to be kept on the economical consequences of an increased level of comfort.

1.4.3. Design cruising speed and range

The speed factor has constituted an outstanding contribution to the development of aviation; the aircraft has proved to be the only means of transport in which increased speed does not necessarily lead to an increase in fuel consumption. Although a fast means of transport will be attractive to the passenger, the air transport companies in particular rate the speed element highly because, broadly speaking, it means that more trips can be made per day and production is increased. It is not only the cruising speed, however, that is important; equally vital is the time devoted to taxying, takeoff, climb, descent, approach and landing, which means that the block speed is a better yardstick than the cruising speed. Any new type of short-haul aircraft will have to possess a considerably higher cruising speed than the one it is intended to replace, in order to save the time needed for an extra flight.

In the case of smaller general aviation aircraft the value of speed mainly depends on how the aircraft is used. A top executive whose working hours are assumed to be extremely valuable will be prepared to pay considerably more for speed than the owner of a small utility aircraft which is used for tourism or in regions with an underdeveloped infrastructure where reasonable surface transport is lacking.

In drawing up the specification for the M-184 project (Fig. 1-5), it has been assumed that the design cruising speed must not be less than that of existing aircraft. In the high-subsonic speed bracket, however, any increase in speed will considerably influence the external shape (angle of sweep, airfoil shape and thickness), generally resulting in an empty weight increment, extra development costs and increased fuel consumption. The extent to which the economic advantages of the higher block speeds will outweigh these losses cannot be predicted offhand; this would have to be ascertained by a tradeoff study, which could also take into account the possibilities of recent developments in high-speed wing aerodynamics.

In the case considered here a design Mach number of .82 in high-speed cruise has been chosen on the basis of conventional section shapes, while the possible gain resulting from the use of an advanced wing shape may be either the use of a thicker airfoil - and hence a lighter structure -, a larger wing span, or a higher economical cruising speed.

As regards the choice of the design range of the M-184 it was concluded from a survey of route distributions that a peak occurs for traffic on ranges of about 280 nm (500 km), e.g. Los Angeles - San Fran-

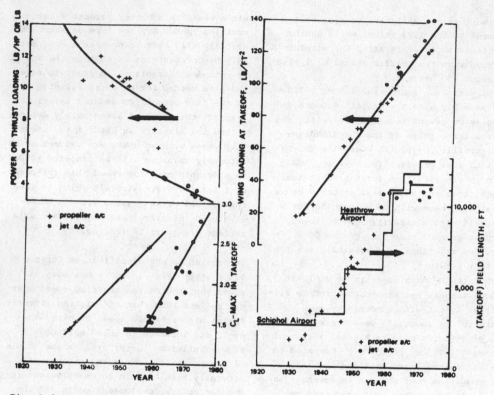

Fig. 1-6. Trends in the takeoff performance of civil aircraft (only trend-setting types have been plotted)

cisco. Another peak, although less pronounced, is observed around 500 nm (900 km). An aircraft designed to fly ranges between 110 and 1,200 nm (200 and 2,200 km) will cover 87 percent of the traffic market. Although a decrease to 600 nm (1,100 km) in the range for maximum payload may lead to a slight improvement in the direct operating costs at short ranges, 25 percent of the short-haul routes are longer and a considerable number of operators would not choose the aircraft. A design range of 1,200 nm (2,200 km) at high-speed cruise was decided for the M-184. In view of the specified field performance there may be an opening for a version with increased all-up weight and fuel capacity to suit operators who require a longer range version and put less emphasis on low-speed performance.

1.4.4. Low-speed characteristics and field performance

Two starting-points may be used for specifying the runway length for takeoff and landing:

a. The aircraft is optimized for cruising flight. The shape and dimensions of the wing, as well as the cruising altitude, are so chosen that the fuel consumption is a minimum for the design range flown at the design cruising speed. The thrust of the engines will be based either on the required climb performance or on the design cruising speed requirement. The takeoff and landing performance will now become more or less derived values and can be influenced only to a limited extent by the design of the flaps and the wheel brakes. The continuous growth in aircraft weight and consequent increase in wing loading (Fig. 1-6) have resulted in increased takeoff distances which have demanded a steady lengthening of the runways, in some cases and certainly at the principal international airports to as much as about 13,000 feet (4,000 m). The approach speeds for the landing have risen to 160 to 170 kts (300 to 315 km/h), although the landing distance is not critical for most long-range transports.

Any further continuation of this trend would only be justified if adaptation of the aircraft to existing runways led to a considerable increase in operating costs and, moreover, the lengthening of the runways was environmentally acceptable. If we also take into account the 1969 requirements regarding noise production (FAR 36) and a possible tightening up of these in the future, it would not appear very likely that future generations of transport aircraft will require any appreciable lengthening of runways which are now being used by aircraft like the DC-8, Boeing 707 and Boeing 747.

b. The runway performance of the new design will be adapted to the airports from which the future customer is now operating the aircraft that the new product will have to replace. For a new short-haul aircraft this means that the runway length should not exceed that used for the category to which the DC-9, BAC 1-11 and Boeing 737 belong and that the design of the landing gear should be adapted to the strength of these runways. Any increase in operating costs resulting from these requirements should be carefully watched and realistic data should be available when it comes to discussing the tradeoff between shorter runways and cost increase. The design study will therefore have to include an investigation into the effect of field requirements on the design characteristics, direct operating costs and noise characteristics.

With a specified runway length for the M-184 project (Fig. 1-5) of not more than 6,000 ft (1,800 m) at sea level (standard atmosphere), it is anticipated that the majority of potential customers will be able to operate the aircraft from the runways now being used, provided that the runway Load Classification Number at Maximum Takeoff Weight does not exceed 30 on a rigid pavement 7 inches (18 cm) thick*.

1.4.5. Other requirements

a. The engine constitutes an important factor in the reduction of the operating costs and its choice should be carefully matched to the aircraft. In the case of transport aircraft the design range is particularly important, while the noise level has to satisfy exacting requirements if restrictions are not to be applied to the use of the aircraft. Fuel consumption has to be carefully watched.

b. Much attention should be paid to an optimum cabin arrangement to enable the operator to use different layouts. In general the distance between fixed partitions at the front and rear of the cabin should be as great as possible.

c. Equipment and instruments. The specification will state the amount of NAV/COM equipment to be carried and its degree of duplication. This will result from discussions with customers and will be based on the mode of operation of the aircraft (VFR and/or IFR flights category of landings), and a distinction will generally be made between standard and optional equipment.

d. Construction, inspection and maintenance. Apart from the airworthiness requirements (Section 1.6) the specification will generally also feature special requirements such as a fail-safe or safe-life design philosophy and a service life of the structure, expressed in terms of the maximum number of flight hours or flight cycles or both. The manufacturing and production processes, etc., may also be subject to special requirements which can have far-

*cf. Chapter 10.2.1.

15

reaching effects when certain structural parts or even main structures are adopted from types already in use. A case in point is the Boeing 707, 727 and 737 family of aircraft all of which have almost identical fuselage cross-sections.

e. Airframe services and noise level. The principal design requirements to be met by the air-conditioning and pressurization system are related to the air supply, temperature and degree of humidity, cabin pressure differential, etc. Noise levels, both internal and external, are also decided upon. Requirements may also be written into the specification with respect to the electrical, hydraulic and pneumatic systems, anti-icing equipment and possibly also the Auxiliary Power Unit (APU).

1.5. THE "CONTINUOUS THREAD" RUNNING THROUGH THE DESIGN PROCESS

1.5.1. The iterative character of design

The creation of an airplane configuration cannot be laid down in a universal, detailed procedure. However, some general characteristics of the design process may be amplified with the help of Fig. 1-7,

which shows the principal phases schematically. This diagram deals with technical and computational elements and could apply equally well to the design of other technical products, unlike Fig. 1-3 which refers specifically to an aircraft development.

An essential element of the design process is that it is always made up of iterations. After a trial configuration has been subjected to a first analysis of its characteristics (weights, mass distribution, performance, flying qualities, economy, etc.), it will be seen either that it does not meet all the requirements, or that it does comply with them but improvements in some respects are possible. Only after a number of configuration changes have been incorporated will the designer be able to determine whether the final configuration satisfies the requirements in every respect and may also be regarded as the best conceivable design, bearing in mind the inevitable uncertainties which are peculiar to the preliminary design phase. The convergence test has been incorporated in the diagram to indicate that a situation may arise in which, despite all the improvements made in the design, no configuration can be found which entirely meets all requirements simultaneously.

The reason may be that certain requirements in the specification and other constraints have proved to be contradictory or too extreme, taking into account the state of the art, or that the basic conception has not been chosen properly. For example, the designer may be confronted with a situation which, to ensure that the engines selected will supply the power required to keep the aircraft in the air after engine failure, would necessitate leaving a large part of the payload back at the airport. The convergence test in Fig. 1-7 is therefore a general indication showing whether the attempts to improve the design have brought it closer to the requirements or not.

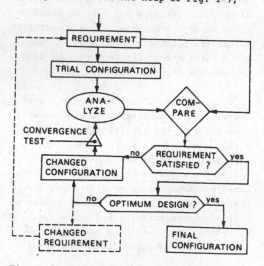

Fig. 1-7. General design procedure

1.5.2. Searching for the optimum

The search for the best conceivable design
may be illustrated by a hypothetical case
in which the quality of the design is
judged on the basis of a single numerical
criterium, referred to as the "merit function" or "objective function". In the case
of transport aircraft this may be the Direct Operating Costs (DOC, see Section
11-8) at the design range but it may also
be the Maximum Takeoff Weight.
In Fig. 1-8 it has been assumed that vari-

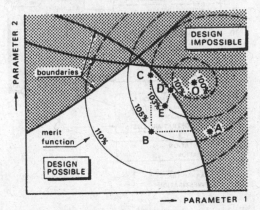

Fig. 1-8. Graphical representation of design optimization

ations in the design are limited to two
independently variable parameters, such as
the wing loading and the thrust loading.
Each point on the diagram represents a
fully defined design, the merit function
of which can in principle be determined.
In the case considered in Fig. 1-7 it is
expressed as a percentage of the minimum
value which can be obtained. The requirements of the specification have been incorporated in this overall picture as follows:
a. Requirements which sharply define the
transport capacity and/or other aspects of
performance should be used as a basis for
the aircraft's general arrangement, layout
and shape, serving as uniquely defined, explicit conditions.
b. Requirements which are put in the form

of constraints in the sense of minima for
flying speeds, maxima for the runway
length, etc. appear as boundaries of the
area within which the design parameters
may be chosen.
In the shaded area we find those combinations of the design variables for which it
will not be possible to satisfy certain
requirements laid down in the form of constraints. In the example it can be seen
that the trial configuration (point A) lies
in the region of unacceptable combinations.
However, all requirements can be met by
changing just one parameter, although at
the cost of a less favorable merit function
(point B). If the second parameter is now
also changed (point C), we find that the
merit function has been improved. Point D,
where one of the limitations is tangent to
the line of the constant merit function,
indicates the combination with which all
requirements can be met and, at the same
time, the most favorable assessment obtained. Nevertheless, the designer may decide to choose point E for the final configuration since in general the positions
of the design boundaries are still subject
to some doubts and design E offers a certain margin which considerably reduces the
risk of crossing the borderline.
Although the "absolute optimum" (point O)
is of no immediate importance in the case
under review, it may still be useful to explore this design in somewhat greater detail, because the difference between the
merit functions of designs D and O is an
indication of the price that has to be
paid for a requirement which makes it impossible to achieve the theoretical absolute optimum. If this disadvantage were to
prove serious, the incorporation of more
advanced techniques might be considered,
enabling the designer to approach the optimum more closely. Although it is true
that these will generally lead to an increase in development costs and influence
the overall evaluation, the result might
well be a saving in operating costs and/or
an increase in productivity.
Another approach might be a certain relax-

17

ation of the critical requirement in the specification, assuming this relaxation is likely to be less harmful than radical adaptation of the aircraft to the extreme requirement. In such a case a special version of the design, suitable for a particular type of operation, may be worth considering.

The example in Fig. 1-7 is essentially a highly simplified picture of the actual world of the aircraft designer. The method outlined is nothing more than a tool to arrive at a better justification for decisions. An aircraft is never evaluated on the basis of a single quantifiable criterion, while the number of variable parameters will always be much larger than two. For instance, increasing the fuselage diameter will generally lead to greater comfort and increase the passenger appeal with a possible increase in yield, but it will also increase the empty weight and the drag and hence the operating costs. It will be almost impossible to find a way out of this dilemma without relying upon the sound judgement of the designer.

1.5.3. A suggested scheme for preliminary design

In the above we have made no reference to how a designer lays out a trial configuration and how he introduces changes. When the design problem lends itself to quantification so that the survey given in Fig. 1-8 can be calculated, the problem may be tackled by means of a computer program. Routine calculation can be done very quickly, but the designer will have to monitor the program. This procedure is gaining popularity in most large companies which undertake costly projects. In Great Britain the Royal Aircraft Establishment also has a facility of this kind at its disposal. In many other cases, however, it will be desirable to lay out a trial configuration using relatively simple procedures and statistical/analytical relationships which approach the optimum reasonably closely. Further investigation may then be limited to the introduction of relatively minor changes which do not affect the design very drastically, provided the original

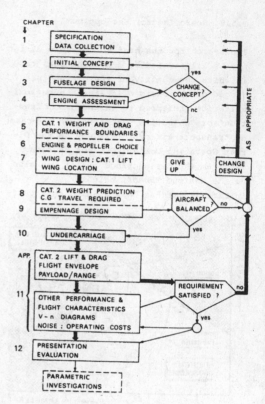

Fig. 1-9. Survey of the initial baseline configuration design

concept was right.

This book has been composed in such a way that the reader will be able to use this simplified procedure with the help of the methods presented in the text. Not only will he find the necessary formulas and relationships, but his attention will also be drawn to considerations which precede decisions. No attempt has been made to streamline the design process as such, although the sequence of the chapters does show some affinity to the continuous thread that runs through the design process for a conventional subsonic transport aircraft. This is illustrated in Fig. 1-9, subject to the following reservations:
a. The diagram, although representative in character, does not possess universal validity. Designers do not always consciously work according to a set program.

b. During the design process assumptions will repeatedly be made which will later have to be verified and, if necessary, corrected until the results agree with the assumptions. To preserve the clarity of Fig. 1-9 the number of such iterations ("feedbacks") has been drastically limited.

c. Some of the procedures indicated may not be required for a particular design. It will, for instance, be up to the designer to decide whether he will make use of the systematics to determine the wing loading and engine thrust (power) discussed in Chapter 5.

A diagram comparable to Fig. 1-9 but compiled for use in a computer program will be found in Section 5.5.2. (Fig. 5-18).

1.6. IMPACT OF CIVIL AIRWORTHINESS REQUIREMENTS AND OPERATIONAL RULES

1.6.1. General

Airworthiness Requirements, Operational Rules and other regulations are framed by national governments and imposed on airplane manufacturers and operators in order to guarantee the general public a certain level of safety. These rules have a far-reaching influence on the design of the structure, systems, installations, performance and flying qualities of aircraft. To begin with, the designer will have to make the correct choice of the airworthiness code to which the airplane will be designed, particularly when an international market is envisaged. He must realize that rules differ from country to country and that distinctions between various airplane categories and types of operation have to be observed. The purpose of this section is to give some insight into the most relevant criteria on which he can base his choice. Emphasis will be placed on the following rules:

a. FEDERAL AVIATION REGULATIONS (FAR), issued by the Federal Aviation Administration (FAA), an office of the Department of Transportation of the United States of America;

b. BRITISH CIVIL AIRWORTHINESS REQUIREMENTS (BCAR), issued by the Civil Aviation Authority (CAA) of Great Britain.

Several other countries have drawn up similar rules, but there are always more or less troublesome differences, leading to confusion and extra costs when an attempt is made to satisfy different rules simultaneously. In view of the considerable economic impact of these requirements, efforts have been made and will continue to be made to arrive at greater uniformity.

In the past the International Civil Aviation Organization (ICAO) has attempted to promote international requirements. For example, a well-known requirement was ICAO Circular 58-AN/53: "Provisional Acceptable Means of Compliance, Aeroplane Performance", dated 1959. These have not been generally accepted and only one type of aircraft, the Fokker F-27 (Fig. 1-10a), was certificated accord-

a. A "large" aircraft: the Fokker F-27 Friendship (Maximum Takeoff Weight 45,000 lb)

b. A "light" aircraft: the Scottish Aviation Jetstream (Maximum Takeoff Weight 12,499 lb = 5,670 kg)

Fig. 1-10. Examples of large and light aircraft

ing to these performance rules, after adoption of the code by the Dutch Civil Aviation Authority (RLD).

In this section only those items will be reviewed which may affect the design of the

COUNTRY (BUREAU)		UNITED STATES OF AMERICA (F.A.A.)			GREAT BRITAIN (C.A.A.)	
GROUP	AIRPLANES PERFORMANCE GROUP	SMALL		GENERALLY LARGE	LIGHT	LARGE
	MAX. TAKEOFF WEIGHT	≤ 12,500 LB	≤ 12,500 LB		C, D, E, ≤ 12,500 LB	A >12,500 LB
CATEGORY		NORMAL, UTILITY ACROBATIC AND AGRICULTURAL (RESTR.)	NORMAL	TRANSPORT	NON-, SEMI- AND AEROBATIC AGRICULTURAL	NON-AEROBATIC
CLASS	NUMBER OF ENGINES	ONE OR MORE	TWO OR MORE	TWO OR MORE	ONE OR MORE	TWO OR MORE
	TYPE OF ENGINE	ALL TYPES*	PROPELLER ENGINES ONLY	ALL TYPES*	ALL TYPES*	ALL TYPES*
MINIMUM CREW	FLIGHT	ONE OR MORE	TWO	TWO OR MORE	ONE OR MORE	TWO OR MORE
	CABIN ATTENDANTS	NONE	<20 PASS.: NONE ≥20 PASS.: ONE	<10 PASS.: NONE ≥10 PASS.: ONE OR MORE	-	-
MAX. NUMBER OF OCCUPANTS		10	11 THRU 23	NOT RESTRICTED	-	NOT RESTRICTED
MAX. OPERATING ALTITUDE		25,000 FT	25,000 FT	NOT RESTRICTED	NOT RESTRICTED	NOT RESTRICTED
MAX. DESIGN DIVING SPEED			NOT RESTRICTED		300 KTS/M = .6	NOT RESTRICTED
APPLICABILITY		*reciprocating,turbofops, -jets and -fans				
AIRWORTHINESS STANDARDS AIRPLANES " " ENGINES " " PROPELLERS		FAR PART 23 " " 33 " " 35	SFAR PART 23 FAR " 33 " " 35	FAR PART 25 " " 33 " " 35	BCAR SECTION K " " C " " C	BCAR SECTION D " " C " " C
NOISE STANDARDS		FAR PART 36	PROP. DRIVEN: APPENDIX F	" " 36	-	-
GENERAL OPERATING AND FLIGHT RULES		FAR PART 91	FAR PART 91	FAR PART 91	LAID DOWN IN AIR NAVIGATION REGULATIONS	
OPERATIONS	DOMESTIC, FLAG AND SUPPLEMENTAL COMM. OPERATORS OF LARGE AIRCRAFT	-	-	FAR PART 121		
	AIR TRAVEL CLUBS USING LARGE AIRCRAFT	-		FAR PART 123		
	AIR TAXI AND COMM. OPERATORS		FAR PART 135	-		
	AGRICULTURAL AIRCRAFT	FAR PART 137				

Table 1-1. Classification of aircraft categories in the American and British airworthiness requirements

aircraft in the preliminary design stage. Obviously, this summary is not a substitute for study and consultation of the relevant airworthiness requirements. Designers are also advised to take due notice of the FAA Advisory Notes, which are intended as explanatory information to prevent misinterpretation of the regulations.

Airworthiness requirements sometimes do not cover new developments in civil aviation, while in other cases changes in the regulations are to be expected in the future. All rules are subject to continuous revision and the authorities should be consulted in connection with any particular problem areas that are not covered by current legislation. In many cases the design must incorporate provisions for (retro-)fitting changes which may be required by the regulations during development and production and sometimes even after the start of service.

A choice must be made of the group, category or class of aircraft to which the design will belong. The upper part of Table 1-1 shows the division of aircraft into groups while the lower part lists the appropriate American and British requirements. The most relevant point to note in this table is that civil aircraft are classified as "light" (U.K.) or "small" (U.S.) when their Maximum (certificated) Takeoff Weight is less than 12,500 lb (5,700 or 5,760 kg). An important class of light aircraft is formed by the feeder liners and twin-engined business (executive) aircraft of 12,499 lb Maximum Takeoff Weight, an example of which is the Scottish Aviation Jetstream (Fig. 1-10b). "Large aircraft" - for the purpose of airworthiness standards - have a Maximum (certificated) Takeoff Weight of more than 12,500 lb (5,700 or 5,760 kg). The division into categories in Table 1-1, as used with respect to certification, indicates a grouping of aircraft

SUB CHAPTER	CONTENTS	PART
		1
A	DEFINITIONS AND ABBREVIATIONS	
B	PROCEDURAL RULES: General Rule-making Procedures	11
	Enforcement Procedures	13
	Nondiscrimination in Federally Assisted Programs of the FAA	15
C	AIRCRAFT: Certification Procedures: type, production, airworthiness certification, delegation options, production approval	21
	Airworthiness Standards : Airframes: Normal, Utility, Acrobatic Category airplanes	23
	Transport Category airplanes	25
	Normal Category rotorcraft	27
	Transport Category rotorcraft	29
	Manned Free Balloons	31
	Aircraft Engines	33
	Propellers	35
	Noise Standards: Aircraft type Certification	36
	Technical Standard Orders Authorizations for Materials, Parts, Appliances	37
	Airworthiness Directives	39
	Maintenance, Preventive Maintenance, Rebuilding, and Alteration	43
	Identification and Registration Marking	45
	Aircraft Registration	47
	Recording of Aircraft Titles and Security Documents	49
D	AIRMEN: Beyond the scope of this book	61 – 67
E	AIRSPACE: Designation of Federal Airways, Controlled Airspace and Reporting Points	71
	Special Use Airspace	73
	Establishment of Jet Routes	75
	Objects Affecting Navigable Airspace	77
F	AIR TRAFFIC AND GENERAL OPERATING RULES:	
	General Operating and Flight Rules	91
	Special Air Traffic Rules and Airport Traffic Patterns	93
	IFR Altitudes	95
	Standard Instrument Approach Procedures	97
	Security Control of Air Traffic	99
	Moored Balloons, Kites, Transport of Dangerous Articles, Parachute Jumping, etc.	101 – 107
G	CERTIFICATION AND OPERATIONS: AIR CARRIERS, AIR TRAVEL CLUBS, AND OPERATORS FOR COMPENSATION OR HIRE:	
	Certification and Operations: Air Carriers and Commercial Operators of Large Aircraft	121
	Air Travel Clubs using Large Airplanes	123
	of Scheduled Air Carriers with Helicopters	127
		129
	Operations of Foreign Air Carriers	133
	Rotorcraft External-Load Operations	135
	Air Taxi Operators and Commercial Operators	137
	Agricultural Aircraft Operations	141 – 149
H	SCHOOLS AND OTHER CERTIFICATED AGENCIES: Beyond the scope of this book	
I	AIRPORTS: Of interest only:	
	Notice of Construction, Alteration, Activation and Deactivation of Airports	157
J	AIR NAVIGATION FACILITIES: Beyond the scope of this book	171
		181 – 189
K	ADMINISTRATIVE REGULATIONS: Beyond the scope of this book	

Table 1-2. Subdivision of FAR requirements based upon intended use or operating limitations.

1.6.2. Federal Aviation Regulations

The FAA Regulations are divided into Subchapters, each containing one group of subjects, and these are further subdivided into Parts, as exemplified by Table 1-2. For practical purposes, the FAA issues these Parts in a Volume system, each volume containing one or more Parts. Subchapters A, C, F and G are of particular interest to the designer. It is useful to note the following distinction:

a. Certification Rules and Procedures (Subchapter C), relating to the airworthiness aspects of aircraft, irrespective of the manner in which they are operated. For example: rules are established for defining and measuring the landing distance, but no criteria for deciding whether the aircraft can be used on the particular airfields considered are given here.

b. Operating and Flight Rules (Subchapters F and G), specifying conditions to comply with certain types of operations, relating the takeoff and landing distances required to the available runway lengths, etc. Although these rules have no direct consequence for the airworthiness of a particular type of aircraft, the designer must appreciate the interaction between the Operational and Flight Rules in order to design the aircraft so that it fulfils its task. Some examples are given below.

1. Part 121 contains criteria for minimum fuel supply (Pars. 121.639 to 121.647), particularly the reserve fuel required for holding, diversion, etc. of transport aircraft. These rules have a considerable impact on payload-range characteristics and hence on operating economy (cf. Section 11.8).

2. In Paragraphs 121.185 and 121.195 it is stated that upon landing at a destination airport each transport aircraft must come to a full stop within 60% of the effective length of the runway from a point 50 ft (15.24 m) above the runway.

3. Part 135, applicable to Air Taxi Operators and Commercial Operators of small aircraft, defines a category of "small aircraft" different from the one mentioned previously. This refers to a class of airplanes, operated under an individual exemption and authorization of the Civil Aeronautics Board (CAB) or under the exemption authority of the Economic Regulations of Part 298 (Ref. 1-86). In this particular context a "small transport aircraft" means a multi-engined aircraft having a maximum passenger capacity of 30 seats or less or a maximum payload of 7,500 lb (3,400 kg) or under. The operational requirements for this particular category are greatly simplified in relation to the transport category. The British Short SD 3-30 feeder lin-

er is an example of an aircraft designed to these particular American regulations.

4. Aircraft with more than 9 passenger seats have to be operated with at least two pilots. When category II operations (cf. FAR Part 97) or operations in IFR conditions are conducted, a second-in-command pilot is also required. Aircraft with 8 passenger seats or less may be operated with a pilot's seat occupied by a passenger if appropriate measures are taken to ensure that the passenger cannot interfere with the pilot's actions. A flight attendant is required when 20 seats or more are installed

1.6.3. British Civil Airworthiness Requirements

The British Regulations are published in sections which are comparable to the FAR Parts or combinations of Parts. Operating Rules are laid down in separate Air Navigation Regulations, which will not be reviewed here. The subdivision of requirements according to subjects is compared with the American equivalent in Table 1-3.

SECTION	REQUIREMENT	AMERICAN EQUIVALENT
A	PROCEDURE	SUBCHAPTER B
C	ENGINE AND PROPELLER	PARTS 33 AND 35
D	AEROPLANE*	PART 25
E	GLIDER	PART 21 + GLIDER CRITERIA HANDBOOK
G	ROTORCRAFT	PARTS 27 AND 29
J	ELECTRICAL	PARTS 37 AND 41
K	LIGHT AEROPLANE	PART 23, SFAR PART 23 (PART 135, App.A)
L	LICENCING	SUBCHAPTER D
R	RADIO	PARTS 37 AND 41

* LARGE AEROPLANES

Table 1-3. Subdivision of BCAR requirements

To the preliminary design engineer of British aircraft the following subdivision into performance groups is the most significant:

Group A - aircraft that, following a power-unit failure, are not forced to land.

Group D - aircraft that, following an engine failure, are not forced to land after takeoff, during initial climb or when flying ·on instruments has started.

Group C - aircraft whose performance is not specified with regard to engine failure.

Group E - aircraft for which the extent of performance scheduling is limited (Maximum Takeoff Weight below 6,000 lb or 2,730 kg).

1.6.4. Airworthiness standards and design

The American airworthiness standards FAR Parts 23 and 25 and the BCAR Sections D and K contain several subsections relating to very similar subjects. The FAR is subdivided into the following Subparts:

A - general (and definitions)
B - flight
C - structures
D - design and construction
E - powerplant (installation)
F - equipment (installation)
G - operating limitations (and information)

SUB-PART B. The level of safety intended by the airworthiness regulations will only be achieved by relating the characteristics of the aircraft to those of the airport, the surroundings (obstacles) and the route. The requirements of this sub-part lay down absolute performance minima, as well as the methodology needed to define and measure flight characteristics.

SUB-PART C defines the loads on the structure, the safety factors, and the minimum strength which shall be provided in the airplane as a whole and in its components. They are directly related to the primary structure.

SUB-PARTS E AND F. The powerplant and equipment installations requirement must ensure safe operation within the airplane structure during all appropriate phases of the flight.

SUB-PART·G. Certain limiting values are to be established, warning notices (placard speeds) displayed, and instructions made available to the flight crew.

The subdivision into "small" or "light" aircraft on the one hand and "large" aircraft on the other hand is very similar to the subdivision into the non-transport category and the transport category, except that the transport category is not restricted in Maximum Takeoff Weight. Some aircraft lighter than 12,500 lb (5,670 kg) have been certificated under FAR Part 25. The differences in airworthiness standards for transport and non-transport categories are significant. The main reason for this is that transport aircraft are intended to be operated by airlines, carrying fare-paying passengers, while the other category is primarily intended for private use by individuals and companies in general aviation. The airworthiness authorities assume that, unlike the private owner, the average passenger knows little about flying and relies upon the government to ensure the highest degree of safety in every phase of the flight.

However, to reduce the effects on safety of different certification policies for the two categories, the private aircraft is normally restricted in its operations, especially above congested areas. Besides, a steady improvement in the safety of operation is aimed at for all categories. A small category of light aircraft may carry up to 15-18 passengers and for this category it has been agreed that the airworthiness standards of FAR Part 23 are too low. The arbitrariness of this subdivision may be demonstrated by Fig. 1-10b, which shows a "light" aircraft and Fig. 1-10a, which shows a "large" aircraft with comparable operational characteristics. Consequently, the FAA has developed a special set of regulations, Special FAR Part 23 (SFAR 23), with more stringent performance requirements, particularly after engine failure (Ref. 1-78). In order to take advantage of the future growth capacities of an airframe design, a manufacturer may decide to satisfy certain FAR 25 requirements from the outset, even though the airplane is intended primarily for FAR 23 certification.

SPECIFIC REQUIREMENT	AIRWORTHINESS STANDARD	PART 25	SFAR PART 23	PART 23
GENERAL	. weight limitation	no	≤ 12,500 lb	≤ 12,500 lb
	. min. number of engines	two	two	one
	. seating capacity	no restriction	more than 10 occupants	up to 9 passengers
PERFORMANCE	. engine failure req.'s in takeoff	yes	no	no
	. accelerate-stop	complete stop	limited	no
	. landing	detailed	limited	> 6000 lb: limited ≤ 6000 lb: no
	. wet runway	yes	no	no
	. climb capability after engine failure	throughout flight	takeoff, landing	multi-engine: limited en-route
FLIGHT CHARACTERISTICS	. lateral c.g. shift	included	no	no
	. minimum control speed	related to liftoff speed and stall speed	related to stalling speed at MTOW	
	. spin characteristics	no	limited	complete
	. maneuver load factor margin in cruise	avoid buffet onset	no	no
STRUCTURAL DESIGN, CONSTRUCTION	. maneuver and gust load envelope	yes	yes	limited for single engine
	. fatigue evaluation	fail-safe, safe life fatigue evaluation of major parts	for pressure cabin, wing and associated structure	
	. fail safe / safe life	specified throughout	for wing and carry-through structure	
	. bird-proof windshield	yes	no	no
	. limit descent velocity for landing gear loads	10 fps	dependent on landing wing loading, but ≤ 10 fps.	
	. max. cabin pressure alt. after system failure	15,000 ft	no	no
	. special emergency provisions for pax.	yes	yes	no
	. ice protection prov.	yes	limited	no
	. restarting capability of engines	yes	yes	no
SYSTEMS, EQUIPMENT	. powerplants and related systems	complete independence	complete independence	limited independence
	. system redundancy	throughout	essential functions duplicated	no
	. equipment for adverse weather flight	yes	yes	no

Table 1-4. Differences in FAR airworthiness standards for small and transport category airplanes

The most relevant differences between the transport and non-transport categories have been listed in Table 1-4. It should also be noted that transport category aircraft require an approved Flight Manual (FM), containing all the necessary information on performance, operating limitations and procedures, in both normal and emergency situations. Similar documentation for the non-transport category is relatively limited.

Although Table 1-4 applies to the American regulations, most of the differences are valid for British aircraft as well. Some exceptions are:

1. Section K of the BCAR limits the design dive speed to 300 kts (556 km/h) EAS or M = .6 for take-off weights up to 12,500 lb (5,670 kg).

2. The BCAR Section D is limited to aircraft with a Maximum Takeoff Weight above 12,500 lb (5,670 kg), but may apply to non-transport category aircraft as well.

3. British light aircraft are divided into performance groups and weight groups, American small aircraft into weight groups only, with some exceptions for single-engine aircraft.

4. Birdproof windshields are required for both categories of British aircraft, in contrast to American practice, where only transport-category airplanes must be provided with them.

5. British rules are more stringent and detailed with respect to wet runway landing performance certification.

6. Special requirements for British agricultural aircraft are included in Section K, while American agricultural aircraft are certificated under restricted operations of FAR Parts 21 and 22.

In conclusion, the similarities between the British and American airworthiness standards far outweigh the differences, resulting in practically the same overall standard of safety, especially for the transport category. The American rules are generally more precise and the designer can apply them more easily without having to consult the authorities. On the other hand, the British rules are more flexible to accommodate new developments, avoiding special regulations. They constitute a basis for the assessment of the airworthiness of a new type of aircraft, the overall assessment being made on an engineering basis.

It will be clear that several problems have to be solved when certification is intended to comply with both the American and British regulations. For example, Ref. 1-80 shows that in the case of the Beechcraft 99, certificated for several years under FAR Part 23, many performance penalties were imposed when a British certification was applied for. After an extensive revision of the interpretation of the requirements, the performance figures were less divergent. On the other hand, the Short Skyvan is certificated according to both regulations and exhibits noticeable differences in performance, with the British version on the conservative side.

There are many important details of airworthiness which cannot be examined in this chapter. A number of them will be discussed in the appropriate place where they bear directly on the design. This introduction is merely intended to prove that airworthiness rules and requirements form a most important source of information for the designer and as such should belong to his daily inventory and mental toolkit.

1.7. CONCLUSION

We end this chapter with a summary of some of the more characteristic tasks of the staff of a preliminary design department.

a. During the development of a preliminary design and the coordination of the configuration development phase, the designer will come into contact with a number of disciplines related to aeronautical engineering: aerodynamics, flight mechanics, propulsion, the science of materials and structures, operational analysis, statistics and optimization. The designer should also know how aircraft are certificated, how flights are carried out under widely differing conditions, and how aircraft are operated. It follows that he should have a wide and up-to-date knowledge, spread over a large number of disciplines, in a profession which is characterized by its dynamic development. He should also be able to give proper attention to details.

b. Typical of almost every design is the

use of iterations. It starts with a trial configuration which will then be analyzed and altered after comparison with the requirements. The entire cycle will then start afresh, until the result shows either that the design is not feasible or that it is reasonably well defined and may in fact be further developed with some confidence. The designer should have the courage to put something on paper to break the chicken-and-egg conundrum. He will have to carry out many calculations and record the results in a clear and well-ordered manner, so that others may be able to follow the procedure. In spite of much apparently meaningless work, he has to remain motivated in order to do a professional job.

c. Particularly during the initial phase the designer should be able to anticipate on the later development and experimental results. The organisation of the project should nevertheless leave room to clear up any vital problem areas as early as possible, for instance by carrying out wind tunnel tests. As the designer's experience grows, this sense of anticipating will come to him more easily.

d. The design department must be able to deal statistically with the ever increasing flow of information on new developments and the outcome of research and make it reproducible. A well adapted documentation and library of data will be essential for permanent use, but should also be augmented for each new design (see Fig. 1-9, first box). The latest edition of Jane's All the World's Aircraft is invaluable, though the same can be said of many aviation journals and magazines. Even so, it is becoming in- creasingly difficult to evaluate the multitude of publications on the quality and reliability of their contents.

e. There will be permanent discussions with experts in other disciplines as well as with (prospective) customers, subcontractors and suppliers. Designers are generally very active in attending aviation symposia and conferences. Teamwork will be the order of the day, particulary where large projects are concerned.

f. In view of the long development period required for a modern aircraft, it will be necessary for the designer to do some crystal-gazing from time to time. A new type will only be successful if it does the job better than the obsolescent type it is intended to replace and preferably better than the designs competing with it for the same slice of the market. A careful balance should be struck between the need for technical innovations on the one hand and the desire to avoid excessive financial risk on the other. Decision-making has sometimes to be based on vague and only broadly defined considerations; for this reason the experience of the design organisation is essential in making proper decisions. The designer will have to possess a faculty for judgement and a feeling for what can and what cannot be done.

In spite of the heavy demands on the designer's capacities in a modern preliminary design department, his work will still be fascinating, because it brings new challenges and offers opportunities for innovations which may have a great influence on the success of the final product.

Chapter 2. The general arrangement

SUMMARY

A sound choice of the general arrangement of a new aircraft design should be based on a proper investigation into and interpretation of the transport function and a translation of the most pertinent requirements into a suitable positioning of the major parts in relation to each other. The result of this synthetic exercise is of decisive importance to the success of the aircraft to be built. However, no clear-cut design procedure can be followed and the task of devising the configuration is therefore a highly challenging one to the resourceful designer.

Considerations, arguments and some background information are presented here in order to provide the reader with a reasonably complete picture of the possibilities. The differences between a high wing and a low wing layout, and the location of the engines either on the wing or fuselage or elsewhere, are discussed on the basis of various cases from actual practice. Examples of unconventional layouts and many references to relevant literature are given to stimulate further study and may possibly generate ideas for new conceptions.

The study of possible configurations should result in one or more sketches of feasible layouts. They serve as a basis for more detailed design efforts, to be discussed in later chapters, and they can therefore be regarded as a first design phase.

2.1. INTRODUCTION

Before a general arrangement drawing of a new design can be put on paper, a choice will have to be made as to the relative location of the main components: wings, fuselage, engines, tail surfaces and landing gear.

A specific configuration is often inspired by a trend or line of evolution which may have its origin somewhere in the past. It may be that previous experience with aircraft in a similar category has established a tradition which cannot be easily discarded. But even when a company tackles an entirely new type, it is generally found that designers fall back on research work done years before by the company's research department or aeronautical laboratories. One example is the Boeing 707 - or its imme-

diate predecessor, the KC-135 Stratotanker - in which certain design features can be traced back as far as the 1945 Stratocruiser design, which itself was developed from the B-29 Superfortress (Fig. 2-1). At first sight the final version shows practically no similarity to any aircraft the Boeing company had previously built. Even so, the 367-60 and 367-64 preliminary designs have much in common with the Stratocruiser Model 377, particularly as regards the fuselage, while an obvious similarity also exists with the B-47 with respect to the location of the engines (Ref. 2-2). Although the Model 707 pioneered the new era of long-range high-subsonic transport aircraft with jet propulsion, its general shape still had its origin in previous designs. It follows that a sound evaluation of practical solutions incorporated in ex-

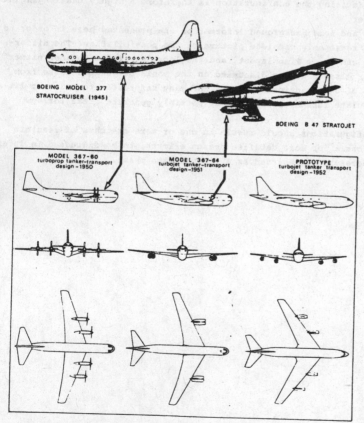

BOEING MODEL 377
STRATOCRUISER (1945)

BOEING B 47 STRATOJET

MODEL 367-60
turboprop tanker-transport
design -1950

MODEL 367-64
turbojet tanker-transport
design-1951

PROTOTYPE
turbojet tanker transport
design -1952

Fig. 2-1. Similarity between various designs by Boeing (Ref. 2-2)

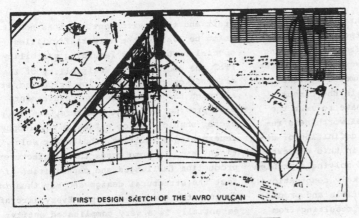

FIRST DESIGN SKETCH OF THE AVRO VULCAN

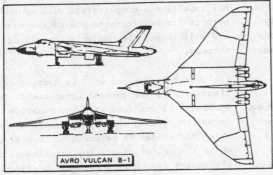

AVRO VULCAN B-1

Fig. 2-2. Development of the external shape during the design of the Avro Vulcan bomber. (Ref.: Flight, 31 Jan. 1950)

isting successful designs should be the first step in the conceptual phase.

A successful first choice of the configuration does not mean that no major changes will be required as development proceeds. This is illustrated in Fig. 2-2 in which an early design sketch of the AVRO Model 698 is compared with the final layout of the B-1 Vulcan bomber (approximately 1945-1948). Though these versions exhibit considerable differences, a gradual evolution of the initial baseline configuration took place in the course of the project development (Ref. 2-7). If this design is compared with that of the Handley Page Victor and the Vickers Valiant, which were both based on the same specification, it can be concluded that various solutions, each with its own particular merits, are possible. This will be discussed in more detail in Section 2.3. Unfortunately, for various

reasons few examples of design evolutions have been published, and it is therefore difficult to draw general conclusions from which recommendations can be deduced. The list of references includes one publication, Ref. 2-6, which is particularly interesting in this connection since it presents some very unusual arrangements dating from the introduction of jet propulsion.

The general arrangement adopted can, in fact, only be properly justified once the design has been finalized. A satisfactory comparison of two different solutions for the same specification will not always be possible, as many design details add up to determine the characteristics of an aircraft and the design considerations published by the manufacturers are as a rule insufficiently detailed.
Competition forces manufacturers to explore new solutions, which is one of the

reasons why competition has the long-term effect of advancing the technology. Excessively large departures from the existing state of the art may, however, lead to the taking of unwarranted commercial risks. Another restraining factor is the circumstance that all designs have to meet the existing or anticipated airworthiness requirements. Hawthorne's[*] definition of design may be aptly quoted in this context: "Design is the process of solving a problem by bringing together unlikely combinations of known principles, materials and processes". A typical example resulting from such a procedure is the Sud-Aviation Caravelle (Fig. 2-3), which successfully pioneered the location of the engine pods at the tail of the fuselage. The spiritual father of this design was Pierre Satre.

It is scarcely possible to give hard and fast rules for arriving at a sound configuration. Some relevant considerations will be presented in the sections which follow but these should be interpreted with caution, as it sometimes happens that even small dimensional differences between the designs may lead to completely different conclusions. Sketches that are reasonably accurate with respect to dimensions are indispensable in the design stage. Without a correct representation of the relative size of the major components, the design drawing might perhaps result in a good artist's impression of the designer's ideas but it is likely to be useless as a basis for further engineering. Engine dimensions, especially in the case of high bypass ratio engines, are often of particular importance in view of their relation to duct sizes, landing gear height, etc. Certain dimensions needed for these drawings may be deduced from data of similar aircraft, preferably using parameters such as wing loading, aspect ratio, relative airfoil thickness, etc. The statistical data presented in other chapters of this book may also be used as a source of information.

During the configuration study the designer should have a clear picture in his mind of the operational requirements of the air-

[*] Engineering Laboratory, Cambridge University

craft and the environment, such as how it is to be loaded, the airport facilities, special requirements regarding visibility from the cockpit, the desirability of the aircraft carrying a very low price-tag, etc. Although the general design requirements will provide important pointers, the designer should develop a "design philisophy", determining priorities, indicating solutions, etc. In some cases the manufacturer's production facilities and capabilities affect the structural design and may thus influence the general concept. Every aircraft essentially is a very complicated entity; all superfluous complication will be costly both to the manufacturer and to the user and will lessen the design's chances of success.

The next few sections will be devoted to discussion of the general arrangement. This chapter will not deal with the fuselage layout, including the use of tailbooms, for which the reader is referred to Chapter 3, while the center of gravity limitations of the design and their influence on the general configuration will be discussed in Chapter 8. When engine location is dealt with (Para. 2.3), it will be assumed that the number and type of engines have already been decided. If this is not so, the reader should turn to Section 6.2 for more information.

2.2. HIGH, LOW OR MID WING?

The vertical location of the wing relative to the fuselage must be considered first. Fig. 2-4 shows three layouts of aircraft design projects in different categories. It will be obvious that the wing location relative to the fuselage is to a very large extent determined by the operational requirements. Although the aerodynamic and structural differences are not without importance, they can only be deciding factors when the choice between high, low and mid wing is not dictated by considerations of maximum operational flexibility.

30

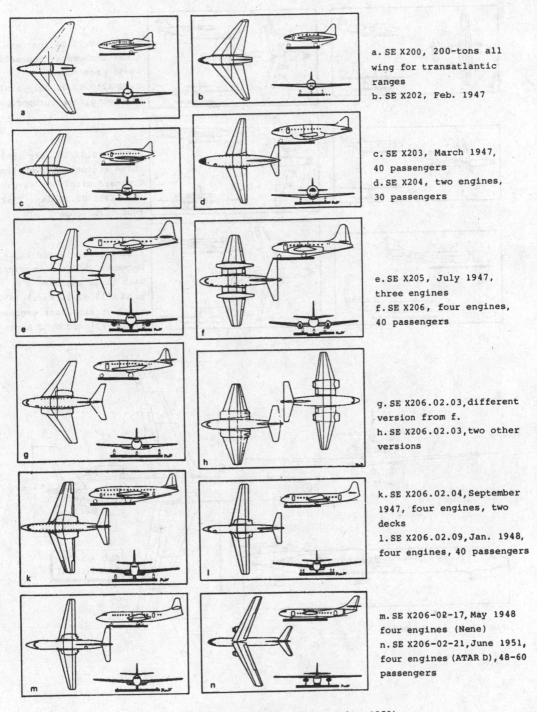

a. SE X200, 200-tons all wing for transatlantic ranges
b. SE X202, Feb. 1947

c. SE X203, March 1947, 40 passengers
d. SE X204, two engines, 30 passengers

e. SE X205, July 1947, three engines
f. SE X206, four engines, 40 passengers

g. SE X206.02.03, different version from f.
h. SE X206.02.03, two other versions

k. SE X206.02.04, September 1947, four engines, two decks
l. SE X206.02.09, Jan. 1948, four engines, 40 passengers

m. SE X206-02-17, May 1948 four engines (Nene)
n. SE X206-02-21, June 1951, four engines (ATAR D), 48-60 passengers

Fig. 2-3. Design projects by Sud-Aviation (Ref. ALATA, Febr. 1959)

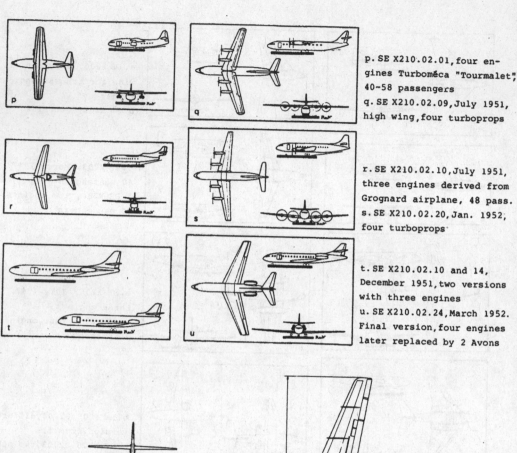

p. SE X210.02.01, four engines Turboméca "Tourmalet", 40-58 passengers
q. SE X210.02.09, July 1951, high wing, four turboprops

r. SE X210.02.10, July 1951, three engines derived from Grognard airplane, 48 pass.
s. SE X210.02.20, Jan. 1952, four turboprops

t. SE X210.02.10 and 14, December 1951, two versions with three engines
u. SE X210.02.24, March 1952. Final version, four engines later replaced by 2 Avons

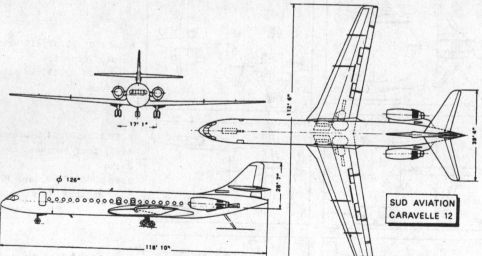

SUD AVIATION
CARAVELLE 12

112' 6"
39' 4"
28' 7"
17' 1"
ø 126"
118' 10"

Fig. 2-3. (continued)

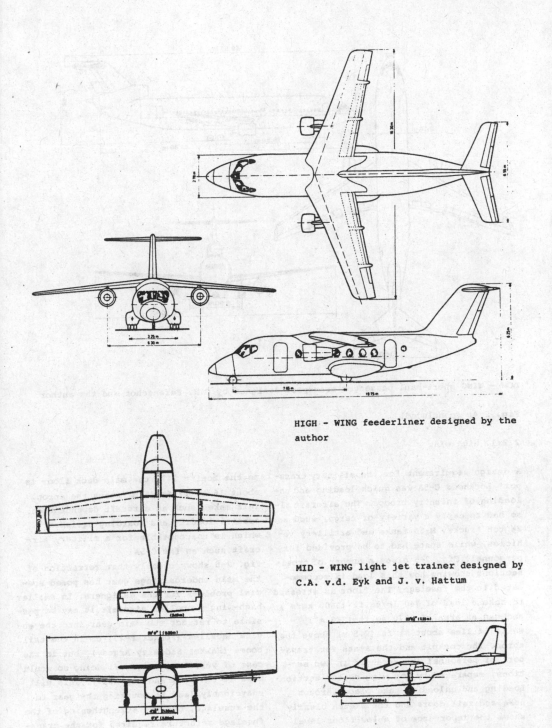

HIGH - WING feederliner designed by the author

MID - WING light jet trainer designed by C.A. v.d. Eyk and J. v. Hattum

Fig. 2-4. Examples of high-wing, mid-wing and low-wing layouts (preliminary designs)

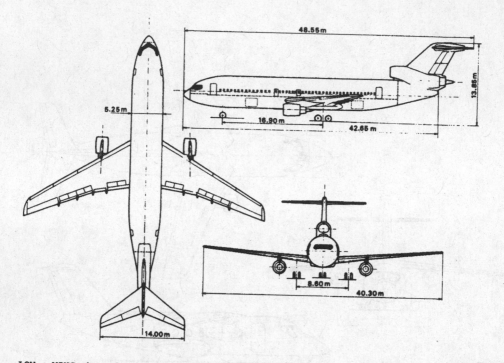

LOW - WING short-haul passenger transport designed by G.H. Berenschot and the author

Fig. 2-4. (concluded)

2.2.1. High wing

A design requirement for the military transport Lockheed C-5A was quick loading and unloading of infantry troops. The aircraft also had to carry a variety of cargo, such as 2½ ton trucks, M-60 tanks and artillery vehicles, while space had to be provided for personnel. Fig. 2-5 gives a number of cross-sections showing how this load is accommodated in the fuselage. The floor is stressed to take a load of 740 lb/sq.ft (3600 kg/m²) and has an area of 2370 sq.ft (220 m²), while it lies about 8¼ ft (2.5 m) above the apron. The cockpit and the seats for transport of personnel are arranged in two sections, separated by the wing center-section. Loading and unloading take place through nose and tail doors and the sketch clearly shows the importance of a low floor level for this very large aircraft. In the case of a low wing aircraft of comparable size, such

as the Boeing 747, the main deck floor is about 16 to 17 ft (5 m) above the apron. This makes such an aircraft dependent on special loading and boarding equipment, which is unacceptable for a military aircraft such as the C-5A.

Fig. 2-5 shows clearly that retraction of the main undercarriage gear has posed special problems for the designers. In smaller high-wing propeller aircraft it may be possible to retract the main gear into the engine nacelles (Fokker F-27) or in the tail booms (Hawker Siddeley Argosy), but in the case of very large aircraft doing so would make it too tall and too heavy. This will unavoidably lead to mounting the gear to the fuselage, but the strengthening of the fuselage structure required for the transmission of the landing impact loads will result in a weight increase. This is only

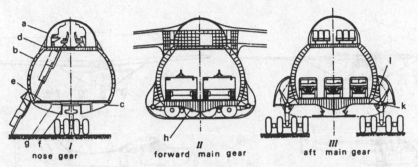

nose gear   forward main gear   aft main gear

a: upper lobe; b: central lobe; c: lower lobe; d: upper deck; e: main deck; f,g: longitudinal supports; h: main fuselage frames; k: main gear shock strut; l: external mounting
frame

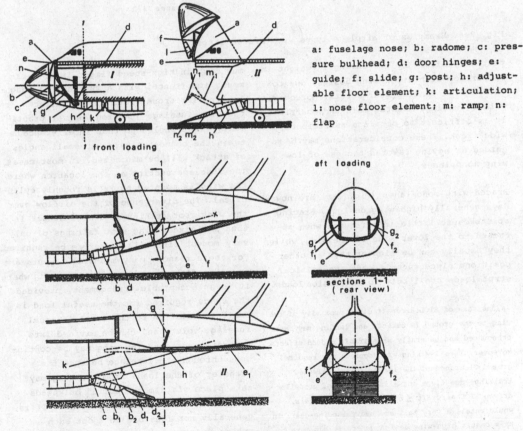

a: fuselage nose; b: radome; c: pressure bulkhead; d: door hinges; e:
guide; f: slide; g: post; h: adjustable floor element; k: articulation;
l: nose floor element; m: ramp; n:
flap

a: load-carrying structure; b: adjustable floor element; c: articulation; d,k: flap; e:
central loading door; f_1, f_2: lateral doors; g_1, g_2: screwjacks; h: levers

Fig. 2-5. Loading provisions and undercarriages supports for the Lockheed C-5A (Ref.:
DOC-AIR-ESPACE No 113 - Nov. 1968)

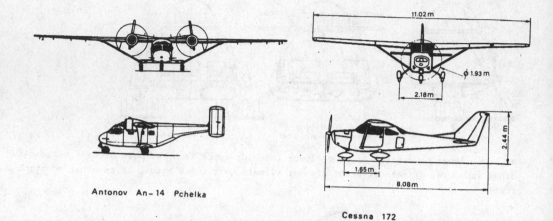

Antonov An-14 Pchelka

Cessna 172

Fig. 2-6. Examples of airplane types with braced wings

partially offset by the saving in weight in comparison with a low-wing design, due to the shorter landing gear struts. Moreover, with a fuselage-mounted main undercarriage it is difficult to obtain a sufficiently wide track. These considerations may be regarded as having favored the use of low-wing monoplanes.

Braced-wing monoplanes (Fig. 2-6) are nowadays generally high-wing designs. Bracing struts cause little interference when attached to the lower side of the wing, while they usually can be lighter than in other positions since in this case the critical strut loads are likely to be tension loads.

In the case of STOL aircraft close proximity of the wing to the ground in takeoff and landing may cause pronounced and generally undesirable ground effects. Moreover, if a low wing was adopted, the required ground clearance of the large, fully deflected trailing-edge flaps and - in the case of propeller-driven STOL aircraft - of the large propellers, would entail a very tall and heavy landing gear. In this case a high-wing design generally has more to recommend it.

2.2.2. Mid wing

This layout is generally chosen when mini-

mum drag in high-speed flight is of paramount importance. With a fuselage of roughly circular cross-section, the surfaces at the wing-fuselage junction meet at practically right angles so that interference between the boundary layers at small angles of attack will be minimized. In most cases the fuselage section at the location where the wing is mounted to it is roughly cylindrical. The divergence of the airflow over the wing root at high angles of attack is thus minimized. Wing root fairings of only very modest size will therefore be required. For these reasons many mid-wing layouts are found in fighter and trainer aircraft, where it is an acceptable arrangement, provided the space required for the useful load is small in relation to the total internal fuselage volume and can be divided into separate sections. The wing may be continuous through the fuselage because the transfer of the loads from the wing may take place via almost "solid" bulkheads to which each winghalf is attached. It is generally not feasible to adopt such a scheme for transport aircraft, and very few mid-wing monoplanes are to be found in this category. However, it is worth noting that on large transport aircraft the wing arrangement adopted approaches the mid-wing position, the reason being that the cabin

floor, which is located just above the wing center section, is positioned relatively high in the fuselage cross section.
Another exception is the HFB 320 Hansa (Fig. 2-7), which features a negative angle

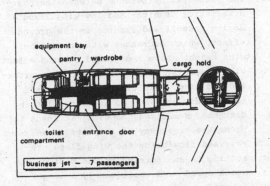

Fig. 2-7. Fuselage layout of the HFB Hansa jet

of sweep, with the engines at the tail balancing the cabin ahead of the wing center section. However, with this layout it is difficult to avoid considerable shift of the center of gravity for different loading conditions unless serious loading restrictions are accepted.[*] The swept-forward wing presents certain aeroelastic problems which are difficult to solve without the use of tip tanks. Although the manufacturer claims that the aircraft possesses low drag characteristics, maximum cabin height for a given fuselage diameter, and a good view for all the occupants, it remains doubtful whether these outweigh the disadvantages.

2.2.3. Low wing

The low wing position frequently offers many advantages. It is true that light aircraft still account for a fair number of high-wing monoplanes, but this may be more a matter of company tradition than an obvious technical advantage. The low cargo floor height is of benefit to small freighters designed for operating into secondary air-

[*]See Section 8.5.4

ports and from airfields and airstrips where special loading equipment is not available. In the case of most passenger aircraft, the height of the cabin floor above the ground is of lesser importance as use can be made of steps of loading bridges.

Efficient use of the underfloor space in the fuselage for the stowage of cargo is possible only if the fuselage is at a suitable height above the apron. Without resorting to a tall undercarriage, this is more easily achieved in a low-wing design. The generally larger fuselage height above ground level on a low-wing configuration may also offer advantages when, after a fuselage stretch, the tail angle available is still sufficiently large to allow for optimum rotation during the takeoff, without creating an unacceptable geometrical pitch angle limitation (Fig. 2-8).

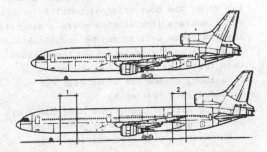

Fig. 2-8. A stretched version of the Lockheed Tristar is envisaged in the design stage. The fuselage will be stretched by 3.56 m (1) and 4.57 m (2) in front of and behind the wing resp. The undercarriage is sufficiently long to allow rotation over 12° after stretching (Ref.: Aviation Mag. No. 550, 30 Nov. 1970)

2.2.4. Effect of the wing location on the general arrangement

a. Interior arrangement.
On a high-wing aircraft the fuselage section below the floor is generally flattened in order to reduce the undercarriage height and to keep the floor at the desired level

above the ground, truckbed height being generally 4 to 4½ ft (1.20 to 1.40 m). A flat fuselage belly leaves little or no room to carry underfloor cargo and this may necessitate a longer cargo hold in the cabin, as compared to aircraft with a circular fuselage cross-section. This in turn may lengthen the fuselage, particularly when, in addition, most equipment and services will have to be located above the floor. A large center of gravity travel may result. On low-wing aircraft the landing gear may be retracted into (propeller) engine nacelles or into the fuselage just behind the center-section of a swept-back wing. Retraction of the main undercarriage between the main wing spars is more easily achieved with a non-stressed lightly loaded skin than with a stressed-skin structure. In small high-wing transports the aisle is sometimes sunk in relation to the rest of the cabin floor to provide adequate standing room. The most critical point is at the wing-fuselage intersection where a slightly lowered cabin ceiling may be unavoidable. It may be worthwhile investigating whether this space can be used for stowage or a cloakroom, or whether lavatories can be extended into this area. Low-wing aircraft sometimes have the wing protruding slightly below the fuselage, necessitating extensive fairings (e.g. Hawker Siddeley 125). In some small touring aircraft the front seats may be mounted directly on the wing box, thus saving space.

b. Safety.
The low wing and possibly the engines will form a large energy-absorbing mass during a forced landing, although they also present potential fire hazards upon contact with the ground. The wing generally contains fuel and the tanks are likely to be damaged, particularly if they are of the integral type. If the impact is not too heavy, damage and fire risk in a high-wing aircraft may be limited. When an aircraft is forced down on water, the fuselage of a high-wing monoplane will be submerged; provisions must therefore be made for escape through the cabin roof. It should be noted, however, that not all aircraft have to be certificated for overwater operation.

c. Performance and flying qualities.
The principal difference between the characteristics of high- and low-wing layouts during takeoff and landing is the ground effect, which decreases with increasing wing height above the ground. Ground effect will generally cause a reduction in vortex-induced drag, resulting in a decreased takeoff distance and an increased landing distance. Sometimes, however, it leads to premature breakaway of the airflow and even reversed flow below the wing flaps, resulting in an increase in the Minimum Unstick Speed[*] and consequently a longer takeoff run.
Probably more important is the decrease in downwash at the horizontal tail, leading to a nose-down pitching moment. This will require greater elevator deflection for the takeoff rotation and the landing flare-out and this may be a determining factor for the elevator power required. The proximity of the ground may have an opposite effect, causing the aircraft "to land itself". This means that after a properly executed final approach little or no elevator movement is required for the flare-out. This can be the case when the wing is placed in such a low position that ground effect causes a marked lift increment, while the nose-down pitching moment mentioned above is approximately compensated by a nose-up pitching moment due to the wing lift. Though this characteristic may in itself be advantageous, it is practically impossible to design the aircraft from the outset in such a way as to achieve it.
With respect to maximum lift and minimum drag, there are admittedly differences between the high and low wing locations, but these may be minimized by proper use of fillets and fairings (Fig. 2-9). Even so, the high wing is superior in this respect to the low wing, particularly where induced

[*]See Appendix K, Section K-2

38

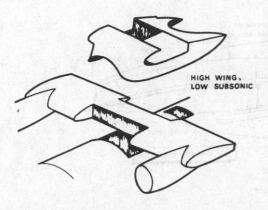

HIGH WING,
LOW SUBSONIC

a. fillets mainly on lower wing surface

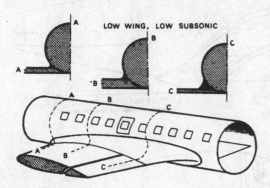

LOW WING, LOW SUBSONIC

b. fillets mainly on upper wing surface

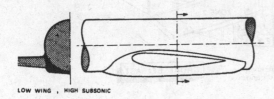

LOW WING , HIGH SUBSONIC

c. fillets act to obey the area rule and to
house the main landing gear

Fig. 2-9. Several types of fillets to reduce
unfavorable wing-fuselage interference

drag at high lift is concerned. The potential differences in damping the Dutch roll
may be largely suppressed by good design,
particularly proper choice of the wing dihedral angle and fin area. Negative wing
dihedral, desirable on swept-back wings,
can easily be incorporated in a high-wing
design without resorting to a tall undercarriage. However, both configurations may
possess comparable flying qualities, except
for fast maneuvers in aerobatics, which are
favored by the mid- and low-wing layout.
High-wing aircraft will generally require
roughly 20 percent more vertical tail area
than low wing types.

d. Structural aspects.
The Lockheed C-5A has already been cited as
an example of the difficulties encountered
when designing the undercarriage of a high-
wing aircraft. Although the weight penalty
in the fuselage structure is partly offset
by a lighter wing and a shorter and lighter
nosewheel gear, on balance the high-wing
layout will be at a disadvantage with respect to empty weight and complication of
the structural design.

2.3. LOCATION OF THE ENGINES

2.3.1. Propeller aircraft

Aircraft powered by piston engines are generally seen in two layouts: the single
tractor engine type with the powerplant in
the fuselage nose and the twin tractor type
with both engines fitted to the wing. New
aircraft types with four piston engines are
not being built any more since any rating
over, say, 500 hp is produced more efficiently by the turboprop engine, and piston
engines in that class are practically obsolete. Configurations are occasionally
observed which differ from the generally
accepted solutions described above, but in
such cases the choice must have been influenced by special considerations, such as
the desirability of creating a high thrustline (amphibians) or avoiding asymmetry in

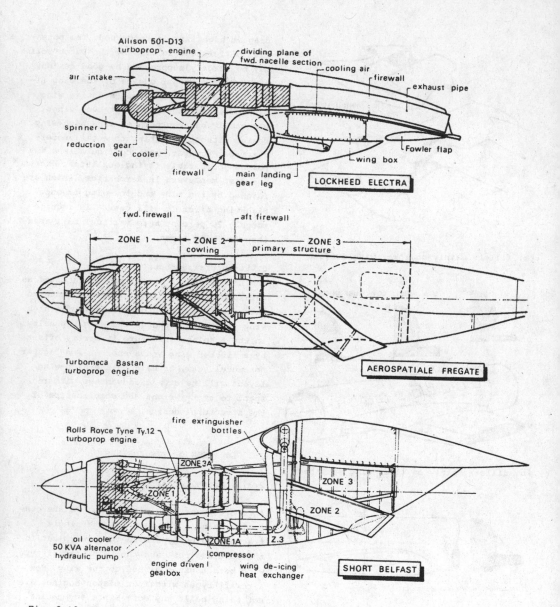

Fig. 2-10. Three types of turbopropeller engine installation

the event of engine failure by using one tractor and one pusher engine in the plane of symmetry (e.g. Cessna Skymaster).

Positioning the propeller engines in front of the wing generally results in the most attractive configuration from the aerodynamic and structural point of view. The propeller slipstream of operating engines generally has a favorable effect on the stall and increases the wing lift, in particular when trailing-edge flaps are extended, thus forming a kind of built-in safeguard against stalling. On the other hand, an engine failure may cause considerable windmilling drag before the propeller is feathered and while the flow over the wing is still disturbed. The yawing and rolling moments induced by engine failure present control problems and downgrade the flight performance, in particular when the engine fails in the takeoff. Variation of the engine power will change the downwash behind the wing and this is of particular influence on the stabilizing contribution of the tail surfaces (Section 2.4.2).

The location and installation of the engines in nacelles mounted to the wing leading edge is illustrated in Fig. 2-10 for several aircraft types. As will be shown in Chapter 6, the engine configuration and the propeller design have considerable influence on this. In the case of a high-wing layout there is generally more freedom with respect to the vertical position of the engines relative to the airfoil as compared to low-wing aircraft, since propeller clearance over the ground is relatively easily provided.

When turboprop engines are used, an engine nacelle which is placed low relative to the wing is to be favored, both for its light supporting structure and for effective discharge of the hot gases, requiring only a short exhaust pipe. On a low-wing aircraft designers are often forced to adopt a relatively high position for the engine nacelles in order to ensure sufficient propeller to ground clearance. This may lead to unfavorable interference effects between the nacelle and the wing, causing premature breakaway of the airflow and additional induced drag.

2.3.2. Jet-propelled transport aircraft

When the jet engine became an acceptable prime mover for both transport and large military aircraft (about 1947-1950), the traditional piston engine layout was discarded and a new configuration was sought which would suit the specific characteristics and demands of the jet engine. Smaller jet aircraft were generally designed for military purposes and had a single engine in the fuselage; in the case of transport

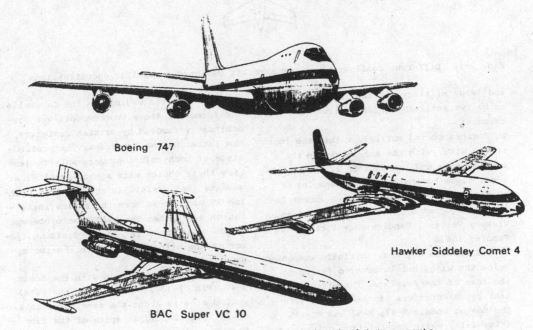

Boeing 747

Hawker Siddeley Comet 4

BAC Super VC 10

Fig. 2-11. Examples of powerplant installation on subsonic jet transports

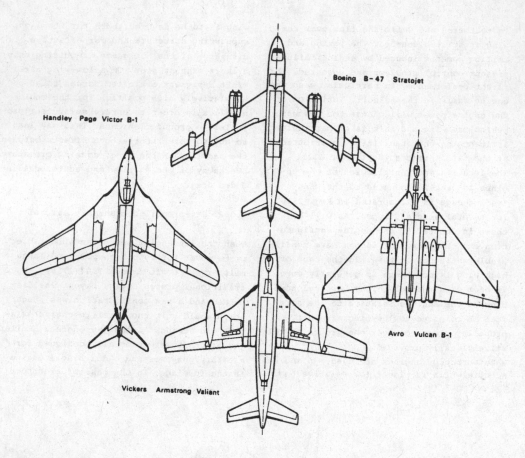

Fig. 2-12. Different configurations for aircraft designed to similar specifications

and large military aircraft there appeared to be two entirely different lines of thought:

a. Engines buried entirely within the root of the wing, with the air intake in the leading edge and the exhaust at the trailing edge, close to the fuselage. Examples of such an arrangement may be seen in the De Havilland Comet (Fig. 2-11), Avro Vulcan, Vickers Valiant, Handley Page Victor and Tupolev 104.

b. Pod-mounted engines, initially suspended below the wing, but later also fitted to the rear of the fuselage. The first important representatives of this school were the Boeing models B-47, B-52 and KC-135 (Fig. 2-1), while Sud-Aviation originated

the rear-mounted engines in the Caravelle. The former of these two concepts was predominantly favored by British designers, the latter by the Americans. The protagonists of both solutions were able to justify their choice with sound technical arguments (Ref. 2-11). In the author's opinion it was not so much the engine installation but rather the aerodynamic concept of the wing which formed the deciding factor in the difference between the two approaches.

Comparing the Avro Vulcan with the Boeing B-47 (Fig. 2-12), we find that the total wetted area is about the same for both aircraft (Fig. 2-13), in spite of the fact that the wing area of the Vulcan is nearly

	BOEING B-47	AVRO VULCAN
GROSS WING AREA ~ ft²(m²)	1430 (133)	3446 (320)
TOTAL WETTED AREA ~ ft²(m²)	11300 (1050)	9600 (885)
SPAN ~ ft (m)	116 (35A)	99 (30.2)
MAX. WING LOADING ~ lb/ft²(kg/m²)	140 (690)	435 (212)
MAX. SPAN LOADING ~ lb/ft (kg/m)	1750 (2590)	1520 (2250)
ASPECT RATIO	9.43	2.84
C_{D_o} (ESTIMATED)	.0198	.0069
$\sqrt{\pi Ae}$ (e=OSWALD FACTOR)	.0425 (.8)	.125 (.9)
L/D_{max} ; $C_{L_{opt}}$	17.25 ; .682	17.0 ; .235

Fig. 2-13. Similarity in max. lift/drag ratios for two widely different configurations

three times that of the B-47 (wing area*). In contrast to this, there is the difference between the aspect ratios**, namely a figure in excess of 3. However, both aircraft have nearly the same span loading (weight/span). The remarkable conclusion can be drawn that for a given dynamic pressure both the profile and vortex-induced drag*** will be roughly equal for these aircraft. Although the comparison shown in Fig. 2-13 is based on estimated values, it clearly shows that it is possible to achieve a comparable range performance with both wing layouts. There are, all the same, considerable differences between the two types:

a. The maximum lift/drag ratio occurs at C_L = .235 for the Vulcan and at C_L = .68 for the B-47. When cruising at high altitude the Vulcan had more freedom to maneuver without experiencing serious buffeting due to compressibility.

b. The structural height at the wing root of the Vulcan, namely about 6 feet 8 inches (2 meters), proved ample to house the engines internally; in the case of the B-47 the height available was only about 26 inches (.66 m).

* The gross wing area used in Fig. 2-13 is as defined in Appendix A Section A-3.1.

** Aspect ratio = span²/area = span/geometric mean chord. See Appendix A.

*** Definitions in Section 11.2.

In a sense this means that the design philosophy with regard to the engine installation was mainly decided by the shape of the wing. Although this example is predominantly of historical interest, it shows that there is a close connection with decisions in other fields.

The protagonists of podded engines attached to the wings by means of pylons use the following arguments to support their views:

a. Separately spaced engines are well placed from the safety point of view. In the event of fire in one of the pods the likelihood that fire will spread to the fuel in the wing is limited. In fact this was the main argument for the choice of the B-47 configuration.

b. The short intake and exhaust ducts enable the engine to run under optimal conditions.

c. The mass of the engines and the pylons lead to a reduction in the bending moment of the inner wing, thus lightening part of the wing structure. When they are located ahead of the flexural axis, they constitute a mass balance against flutter.

d. The engines can be made easily accessible at the cost of very little increase in structure weight since the pods do not form part of the stressed structure. Access to engines buried in the wing roots has to be provided by detachable skin panels at a location where the wing is highly stressed.

e. The engine pylons appear to have a favorable effect on the airflow at large angles of attack and tend to counteract pitch-up of sweptback wings (Fig. 2-14). The pylons act in a manner similar to

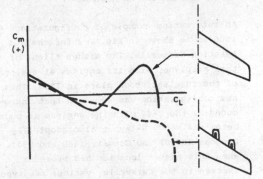

Fig. 2-14. Effect of wing-mounted pods on longitudinal stability (Ref. 2-11)

fences, which are often used on "clean" wings.
Against this the protagonists of completely buried
engines list the following arguments:

a. The extra drag resulting from the buried engine
installation is only a few percent as against about
15% of the total drag in the case of a configuration similar to the B-47. Incidentally, the current
generation of turbofans show a value of about 8 to
10%.

b. As a result of the low wing loading and low
value of C_L in cruising flight, maneuvering is
possible without compressibility problems such
as buffeting.

c. The pitch-up problem of swept wings is less
significant for low aspect ratio wings.

d. As a result of the low wing loading, the low-
speed performance will be better.

e. The relatively low aspect ratio wing box structure will lead to greater stiffness and aeroelasticity will be less of a problem.

Many of these arguments are only valid up to a
point, and in particular the progress in engine
technology towards high bypass ratio engines, together with the development of more efficient high-
lift devices in 1950-1970, has settled the case in
favor of high wing loadings and pod-mounted engines.
This does not mean that buried engines will not return to favor again in the future. For example, the
application of laminar flow control by suction of
the boundary layer to reduce drag might eventually
lead to a totally different design approach, such
as a combination of low wing loading and engines
of relatively low thrust in cruising flight, integrated into the wing or fuselage.

An interesting example of configuration
studies is shown in Fig. 2-3 for the Sud
Aviation Caravelle. The maiden flight of
this airplane, with its engines at the rear
of the fuselage, took place in 1955. Thus a
new configuration was added to that introduced by the B-47 with the engines in pods
below the wing, a layout also adopted by
Douglas (DC-8) and Convair (880 and 990).
When this engine location had proved a
success in the Caravelle, various new types
were designed to practically the same formula: the BAC 1-11, Vickers VC.10, Hawker
Siddeley Trident, Douglas DC-9, Boeing 727,

Fokker F-28 and all executive turbojet aircraft. Though this layout has the obvious
advantages of a "clean" wing, low door sill
height and little asymmetric thrust after
engine failure, it also has a large center
of gravity variation with variation in
loading condition and it has to be carefully designed to avoid the superstall
problem. Therefore, after 1965, a new trend
towards engines on the wings occurred
(Boeing 737, Lockheed L-1011, Douglas DC-10,
Dassault Mercure).

It would serve little purpose to express a
general verdict in favor of either of these
two configurations. Each specification, as
well as every new type of engine, will require renewed study to support a particular
choice and the outcome can only be properly
assessed when various configurations have
been designed according to the same ground
rules. This procedure was followed by the
Boeing Company for the development of their
Model 737 (Fig. 2-15), when two competing

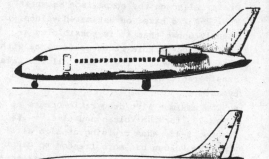

Fig. 2-15. Two configurations of the Boeing
Model 737, resulting from a competition of
two design teams

design teams were put on the same exercise.
Factors which should be investigated in the
case of a transport aircraft are indicated
in Fig. 2-16. The principal differences between the Caravelle and the Boeing 737 engine location will be briefly discussed
with reference to that list.

• EMPTY WEIGHT	• FLYING QUALITIES
• ENGINE MAINTENANCE	stalling
• FOREIGN OBJECTS INGESTION	engine-out control
• SYSTEMS	go around
fuel	cruise dutch roll
anti-icing	• NOISE
air conditioning	• PERFORMANCE
• CARGO LOADABILITY	drag
	max. lift
	second segment climb

Fig. 2-16. Engine location factors (Ref.: ATA Engng and Maint. Conf., Oct. 27, 1964)

a. Empty weight.
The following factors have to be considered:
- A wing structure weight saving is possible with wing-mounted engines due to the mass relief effect on the bending moment on the inner wing.
- Engines placed too far outboard increase the landing impact loads and necessitate a large vertical tailplane.
- Engines at the rear of the fuselage require local "beef-up" and lead to loss of useful space in the tail, resulting in added structure weight and a larger fuselage for the same payload.
- Differences in weight of the tail surfaces depend on various factors which do not permit a general conclusion.
Summing up, we may say without too much emphasis that the empty weight of a Caravelle-type layout will typically be 2 to 4% more than that of a comparable design with the engines on the wing.

b. Engine maintenance.
Although the size of the aircraft comes into it, engines below the wings are generally better accessible from the ground than in any other layout.

c. Flexibility of loading.
This depends primarily on the location of the load relative to the center of gravity of the empty aircraft, a subject that is treated more fully in Chapter 9. Both configurations may be designed for good loading characteristics, although a greater

c.g. travel must be catered for in the case of rear-mounted engines. At full payload the download on the tail, with consequent loss in the lift to drag ratio, will be considerable with engines mounted to the rear fuselage. Besides, the layout with wing-mounted engines will have a larger underfloor cargo-hold behind the wing, which is generally more easily used.

d. Performance.
Regarding drag in cruising flight there is little to choose between the two layouts, assuming that both have been well designed aerodynamically. Douglas, however, claims that in the case of the DC-9 the drag of the wing plus nacelles at high subsonic speeds is reduced as a result of favorable aerodynamic interference, (Ref. 2-30). Generally speaking, a layout with wing-mounted engines will lead to an increase in induced drag and a slight reduction in the drag-critical Mach number. The drag resulting from the asymmetrical flight condition following engine failure, rapidly increases with the lateral distance of the failed engine to the aircraft centerline and will, therefore, be greater with engines mounted on the wing. The protagonists of the Caravelle layout claim that their clean wing gives a gain of 20% in maximum lift. The Boeing Cy. does not agree, basing its opinion on test data and arguing that in the case of the clean wing the useful lift is reduced by gadgetry to ensure a favorable pitching behaviour at the stall. And indeed, looking at maximum C_L values for a number of aircraft with different layouts, it is not possible to discern a clear-cut tendency either way.

e. Flying qualities.
Engines mounted to the rear of the fuselage are often combined with a tailplane on top of the fin (T-tail). This particular layout has a potential problem in the high incidence range, namely the "deep stall"[*]

*Also referred to as "superstall" or "locked-in stall"; see Section 2.4.2

(Section 2.4.2). In the case of engines on the wings, the yawing moment resulting from engine failure will be more pronounced.

f. Mounting of a central engine.
Three-engined aircraft generally have one engine mounted centrally at the rear of the fuselage. The problem which will have to be faced here is whether to bury this engine in the fuselage, which will require a fairly long and curved inlet (Boeing 727, Hawker Siddeley Trident, Lockheed L-1011) with consequent loss of intake efficiency and extra weight. Alternatively, the engine can be installed in a pod on top of the fuselage, but in that case the vertical tail surface forms an obstruction. Fig. 2-17 depicts some possible solutions, all

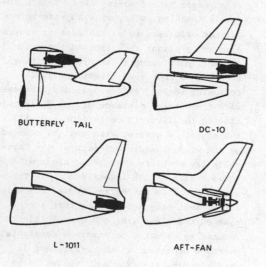

BUTTERFLY TAIL   DC-10

L-1011   AFT-FAN

Fig. 2-17. Installation of the central engine on a three-engined jet aircraft

of which present particular design problems. The thing to do here is to optimize the chosen solution in such a way that the disadvantages will be limited. Ref. 2-23 shows that a purely objective comparison of two solutions is very difficult. Manufacturers' data for structure weight, fuel consumption and economy for both the L-1011 and the DC-10 are used to show that both solutions are best.

2.3.3. Single-engined subsonic jet aircraft

Aircraft of this type have the engine mounted inside the fuselage and the intake and exhaust ducts often present a problem. The inlet duct has to supply a constant flow of air at different operational engine settings and in different flight conditions. Flow distortion and turbulence at the compressor face must remain within the limits

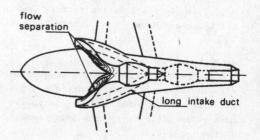

flow separation

long intake duct

Fig. 2-18. Intake problem on a stubby fuselage with side intakes

laid down by the engine manufacturer. Pronounced curvature in the inlet duct should therefore be avoided. This is not very easy to comply with in the case of a fuselage which is relatively wide at the location of the air inlets unless a long inlet duct is acceptable (Fig. 2-18). The latter is generally undesirable for reasons of space or balancing. It also costs weight and results in inlet pressure loss.
At different angles of attack variations in the direction of the incoming air should

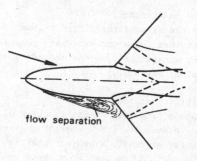

flow separation

Fig. 2-19. Asymmetric intake condition in a sideslip

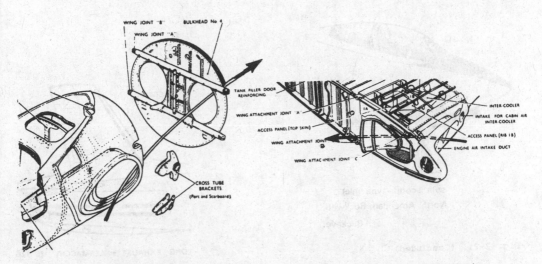

Fig. 2-20. Structural arrangement of wing-root air intake (De Havilland Vampire)

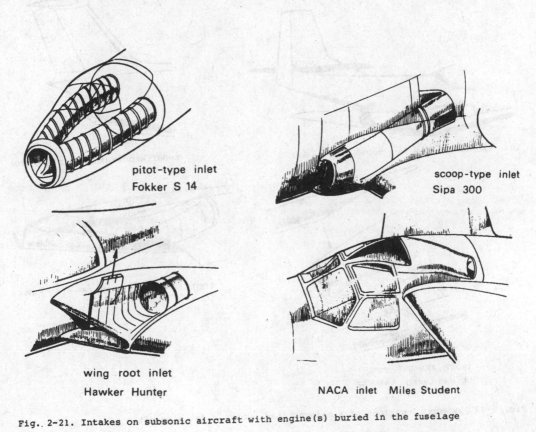

pitot-type inlet
Fokker S 14

scoop-type inlet
Sipa 300

wing root inlet
Hawker Hunter

NACA inlet Miles Student

Fig. 2-21. Intakes on subsonic aircraft with engine(s) buried in the fuselage

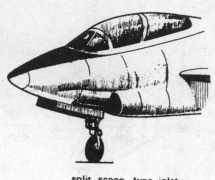

split scoop-type inlet
North American Rockwell
Buckeye

Fig. 2-21. (Concluded)

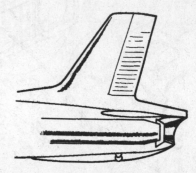

LONG EXHAUST — AERMACCHI MB 326

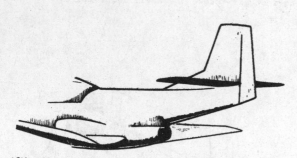

LOW EXHAUST, SINGLE TAIL BOOM — SIPA 300

SHORTENED EXHAUST — L 29 DELFIN

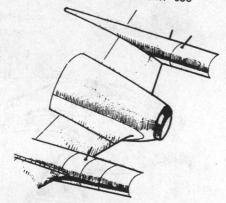

SHORT EXHAUST PIPE
DE HAVILLAND VAMPIRE

SPLIT EXHAUST — HAWKER SEA HAWK

Fig. 2-22. Exhaust locations of single-engined subsonic jet aircraft

not be excessive. The wake of the partly stalled wing must not enter the inlet duct, which means that the leading edge of the wing would be an unsuitable location unless special measures were taken. When split intakes are used, a sideslipping condition will cause dissimilar flow patterns, which may lead to unstable flow and even to air oscillating instead of entering the duct as shown in Fig. 2-19.

When the engine is installed in the fuselage, the designer has to decide whether it is desirable to continue the wing structure through the fuselage without interruption. On a highly maneuverable aircraft, designed for high normal load factors, such a continuous structure is very attractive. It will then depend upon the relative proportions of the inlet duct and the thickness of the wing whether it is feasible to lead the inlet ducts through the spar webs. Fig. 2-20 shows that this was possible in the case of the De Havilland Vampire, but in other cases it may prove desirable to lead the inlet either over or under the wing. Fig. 2-21 shows different types of inlet which will be briefly discussed here.

The pitot type (a) provides the engine with undisturbed airflow for all flight conditions. It requires a long inlet duct, which generally will have to be divided at the level of the cockpit, and intake efficiency is low. This type is now rarely used on subsonic aircraft.

An intake in the wing-root (b) is difficult to realize as the intake opening must be able to supply the required airflow at different intake velocities and also cope with changes in the angle of attack and angle of sideslip. At the same time the local airfoil shape must not be modified more than is strictly necessary.

Side inlets on either side of the fuselage form scoops and thus cause additional drag. To keep this drag low, the airscoops must not be kept too short and must be well faired. A divertor is needed to prevent the fuselage boundary layer from entering the duct but this also adds to the drag. The inlet opening should be located sufficient-

ly far ahead of the wing in order to avoid interference with the wing and excessive variations in the intake conditions.

An air inlet on top of the fuselage has sometimes been used in experimental aircraft and was adopted for the Miles Student. The opening has to be raised sufficiently far above the fuselage to avoid boundary layer and wake ingestion at large angles of incidence.

A split inlet at the bottom of the fuselage may be regarded in some ways as a compromise between the pitot inlet and side inlets. When measures are taken to avoid the ingestion of debris during takeoff and taxying, this layout may be particularly attractive for mid-wing and high-wing aircraft.

The exhaust nozzles should be so positioned and directed that the (hot) jet efflux will not impinge on the structure. At subsonic speeds in a parallel flow, the expanding gases of a pure jet may be assumed to expand within a cone with half the top angle equal to 6 degrees. Exhaust nozzles are manufactured from stainless steel sheet and are fairly heavy; on pure jet engines they will weigh from 1 to 1.5% of the engine weight per foot of length (3 to 5% per meter). The weight will be even greater in the case of bypass engines. Moreover, exhaust nozzles cause a thrust loss of about .3% per foot (1% per meter). They should therefore be kept as short as possible. Some examples shown in Fig. 2-22 will be discussed.

When the exhaust nozzle is located in the rear end of the fuselage, it is possible to keep the efflux away from the aircraft without having to take any special precautions. A single tail boom is sometimes adopted in order to shorten the exhaust. Another solution consists of a split exhaust with two openings on either side of the fuselage. Unfortunately, both configurations lead to structural problems, while complicated fairings must be used around the exhausts. Another way to shorten the length of the ex-

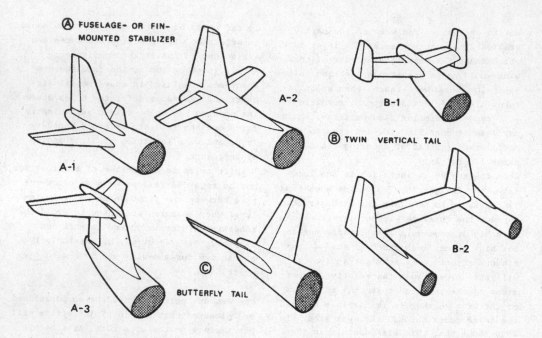

Fig. 2-23. Classification of tailplane configurations

haust pipe is to use two tail booms. This has the added advantage that it provides excellent accessibility to the engine.

2.4. ARRANGEMENT OF THE TAILPLANE

The design of the tail surfaces probably depends more on the general arrangement and the detail layout of the aircraft than any other major part. Because of their location, their effectiveness is influenced by the wing and the operation of the engines, particularly in the case of propeller-driven aircraft. The way in which the empennage is mounted to the fuselage, or possibly to tail booms, affects the structural layout of the tail surfaces and that of the fuselage. General instructions applicable in the preliminary design stage are therefore very difficult to lay down.

2.4.1. Classification of tail surface configurations

Examples are given in Fig. 2-23 of the principal configurations seen in practice. Although there are many intermediate solutions, these will not be discussed here.

Group A: A single fin with the stabilizer mounted either on the fuselage or on the fin represents the most common current layout. It also ensures structural simplicity and stiffness, although in the case of the T-tail (A-3) attention must be devoted to preventing tailplane flutter. Aerodynamic considerations leading to this choice are discussed in Section 2.4.2.

Group B: The considerable height of a large fin will cause a rolling moment due to rudder deflection as a consequence of the large distance of the fin aerodynamic center from the longitudinal axis of the aircraft. If this is considered to be objectionable, a twin fin may be well worth investigating as a means of minimizing this effect. When twin tail booms are used (group B-2), such a layout is the fairly obvious choice.

Group C: The V- or butterfly tail is often
adopted for sailplanes, with the object of
avoiding damage to the tail when landing on
overgrown terrain. The V-tail is sometimes
also used on powered aircraft, e.g. the
Fouga Magister where it served to keep the
tail surface clear of the jet efflux of the
engines, without having to resort to a T-
tail. Another classical example is the
Beechcraft Bonanza. The V-tail has never
become popular, mainly because the moving
surfaces have to serve both as rudders
(differential deflection) and as elevators
(simultaneous deflection), which leads to
a complication in the control system design.

2.4.2. The location of tail surfaces

a. Jet efflux effects.
The tail surfaces must never be in the jet
efflux. Assuming that the efflux of a pure
jet spreads out conewise with half the top
angle equal to 6 degrees, this defines a
region which may be regarded as "out of
bounds" so far as the tail surfaces are
concerned. If necessary, the centerline of
the jet efflux may be diverted a few degrees in any desired direction. Another
possibility is to apply a moderate dihedral
to the horizontal tailplane. It is advisable to have as great a distance as possible
between the noise generating regions and
the tail surfaces, since otherwise the very
high intensity of the engine noise may cause
acoustic fatigue in the relatively flat skin
panels of the tail. Any special measures to
prevent this will entail a weight penalty.
A jet efflux close to the stabilizer will
affect the direction of the airflow and diminish its stabilizing contribution due to
the jet pumping effect.

b. Slipstream effects.
In symmetrical flight, the lift distribution
of the wing with deflected flaps depends on
the engine speed. The same applies to the
downwash and the local velocity distribution
at the tail. When the airspeed and the angle
of attack are changed, the stabilizer moves
in a vertical direction relative to the

slipstream, which causes variations in the
longitudinal stability. These depend partly
on the location of the stabilizer, measured
in the vertical direction. Fig. 2-24 shows

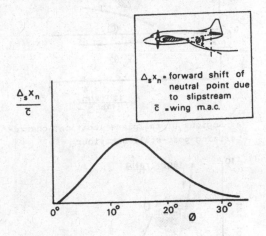

Fig. 2-24. Forward shift of the neutral
point due to slipstream (Ref.: A R C R & M
2701)

that loss of static stability is small with
the stabilizer placed very high or very low,
but this cannot always be realised in practice. As the power to weight ratio and the
maximum lift coefficient increase, the slipstream effects will also become more pronounced and generally the tail size will
have to be increased.
In flight with one engine inoperative there
will be a yawing moment which has to be
counteracted mainly by rudder deflection.
There will also be a non-symmetrical lift
distribution over the wing and this will
cause a sidewash at the fin, effectively
resulting in an increase in the yawing moment. This condition of flight provides a
criterion for the size of the fin and rudder
in the case where the engines are mounted
on the wing.

c. Stability and control in the stall and
post-stall condition.
Although in normal operating conditions a
wing stall is avoided by applying adequate
safety margins relative to the minimum fly-

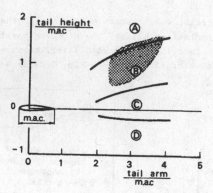

a. Regions of tailplane location, characterizing post-stall behaviour

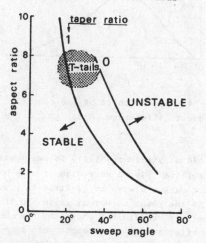

b. Stability boundary for the wing alone

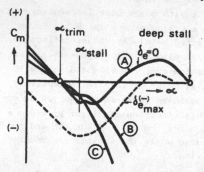

c. Post stall stability for the complete airplane

Fig. 2-25. Static stability at high angles of attack as affected by wing shape and tailplane location (Ref.: NASA TM-X-26)

able speed, a stall may be encountered occasionally. The stall speeds must be demonstrated during certification testing as they form the baseline for most performance figures in takeoff and landing. Safe recovery from a stall is therefore a requirement. The longitudinal flight characteristics are affected primarily by the "stiffness in pitch", represented by the slope of the C_m-α curve (Fig. 2-25c). A negative slope corresponds to positive static stability, while the trimmed condition is equivalent to $C_m = 0$, to be obtained by elevator and/or stabilizer deflection. The wing and horizontal tailplane are the main contributing components, the tailplane location being of prime importance. Fig. 2-25b shows combinations of wing sweep angles and aspect ratios which ensure a stable wing pitching moment slope at the stall. The stable region marked is only approximate and may be influenced by airfoil variation, wing twist, boundary layer fences, engine pylons and leading edge high-lift devices. The boundary of the stable region, as derived from windtunnel tests, indicates a reason why highly swept wings generally are of low aspect ratio. A slightly unstable wing pitch-up may be acceptable, provided the horizontal stabilizer is sufficiently effective. The effect of the vertical location of the stabilizer is illustrated in Fig. 2-25c for several cases, defined in Fig. 2-25a. In region A, which covers most T-tails, instability at large angles of incidence is generally preceded by a less pronounced instability at the stall. In region B the stabilizer only enters the wake of the wing when the latter becomes unstable. Region C does not show these phenomena at low speeds, but pitch-up may occur on maneuvering flight at high subsonic speeds. Region D is a location which may be regarded as satisfactory for all angles of incidence. This arrangement may sometimes be possible in the case of high-wing aircraft, but attention should be paid to the location of the wake, particularly when flaps are deflected.

Most tailplanes designed for normal opera-

ting conditions will be sufficiently effective to provide stability at high angles of
attack as well. However, if the wing wake
is augmented by the wake of a wide fuselage
and pod-mounted engines on either side of
the rear fuselage tail, conditions may exist such that a T-tail aircraft encounters
extended regions of post-stall instability.
At very high angles of attack the tailplane
contribution to longitudinal stability will
be reduced to 10-20% of its normal value.
At angles of attack between 30 and 40 degrees the tailplane itself stalls and the
slope of the C_m-α curve is once more reversed into a stable one. At $C_m = 0$ the aircraft is trimmed in a "deep stall". In that
condition the pitching moment due to elevator deflection may be insufficient to restore the normal attitude and the airplane
is locked in this condition. A very fast
descent at low forward speed is unavoidable
and recovery from it is very doubtful. There
are various methods of curing such unacceptable behavior, e.g. increasing the tailplane span and modifying the wing shape.
For added safety a stick shaker can be installed to warn the pilot at a preset angle
of attack, while a stick pusher is frequently used to force the steering column
forward when the stalling angle of attack
is approached.
Adoption of a T-tail does not necessarily
face the designer with disadvantages only.

Fig. 2-26 shows that placing one tail surface at the tip of another leads to an increase of about 50 percent in the aerodynamic aspect ratio, so that the stabilizer
may increase the lift curve slope of the
fin by roughly 15 percent. A similar improvement in the effectiveness of the horizontal stabilizer may be obtained by the
use of two fins at the tips. Another point
is that the downwash at moderate angles of
incidence decreases with increased verti-
cality of the stabilizer, which in the case
of a T-tail may sometimes justify reducing
the area. The same effect is achieved by
placing the stabilizer on top of a sweptback fin, thus increasing its moment arm.

d. Recovery from spins.

In the case of aircraft designed for aerobatics (e.g. trainers), recovery from a
spin must be possible. In small aircraft
this involves use of the rudder, which must
therefore be effective even at very large
angles of incidence. It will be seen from
Fig. 2-27 that the indicated location of

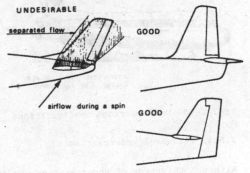

Fig. 2-27. Effectiveness of the rudder
during a spin

the stabilizer will cause the greater part
of the rudder to be shielded. Some layouts
for avoiding this are indicated. V-tails
and fins at the tips of the stabilizer are
favorable in this respect.

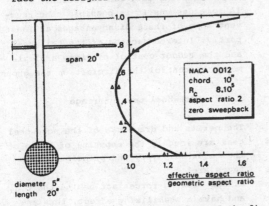

Fig. 2-26. Effective aspect ratio of the fin
in combination with a horizontal tailplane
(Ref.: NACA TN 2907)

2.5. ARRANGEMENT OF THE UNDERCARRIAGE

Various configurations for the undercarriage have been adopted in the past, but many of them were designed for special purposes. Only three of these need be discussed in the present context. The discussion is further amplified by Fig. 2-28 and Refs. 2-35 through 2-38.

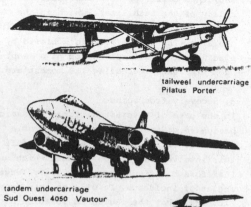

tailweel undercarriage
Pilatus Porter

tandem undercarriage
Sud Ouest 4050 Vautour

nosewheel undercarriage
Fokker F 28 Fellowship

Fig. 2-28. Undercarriage configurations

2.5.1. Tailwheel undercarriage

Although this type of undercarriage was in general use during the first three decades of aviation, it must now be regarded as obsolete for most designs. Its advantages should nevertheless be mentioned:
a. The tailwheel is small, light and simple to design.
b. The location of the main gear legs makes attachment to the wings an easy matter.
c. A three-point landing can be carried out by bringing the aircraft to a stalled condition. The aerodynamic drag will provide a retarding force, which is particularly

needed when the airfield is unsuitable for full application of brakes (e.g. wet grass).
d. When brakes are applied the vertical load on the main gear will increase, thereby reducing the risk of skidding.

The reason why the tailwheel undercarriage has been almost completely superseded by the nosewheel or tricycle gear is that it also possesses the following drawbacks:
a. Violent braking tends to tip the aircraft onto its nose.
b. The braking force acts ahead of the center of gravity and thus has a destabilising effect when the aircraft is moving at an angle of yaw relative to its track. This may cause a ground loop.
c. In a two-point landing a tail-down moment will be created by the impact force on the main landing gear, resulting in an increase in lift which makes the aircraft bounce.
d. The attitude of the wing makes taxying difficult in a strong wind.
e. In the case of transport aircraft the inclined cabin floor will be uncomfortable for the passengers and inconvenient for loading and unloading.
f. In the tail-down attitude the inclination of the fuselage will limit the pilot's view over the nose of the aircraft.
g. During the initial takeoff run drag is high until the tail can be raised.
In some designs it is possible to circumvent some of these disadvantages at least partly. Interconnection of the tailwheel and the rudder control provides a simple means to control the aircraft on the ground.

2.5.2. Nosewheel undercarriage

The merits and drawbacks of the nosewheel gear are roughly the opposite of those of the tailwheel type. The principal advantages are:
a. The braking forces act behind the c.g. and have a stabilizing effect, thus enabling the pilot to make full use of the brakes.
b. With the aircraft on the ground, the

fuselage and consequently the cabin floor are practically level.

c. The pilot's view is good.

d. The nosewheel is a safeguard against the aircraft turning over and so protects the propeller(s) when used.

e. During the initial part of the takeoff the drag is low.

f. In a two-point landing the main gear creates a nosedown pitching moment.

The steady increase in landing speeds of modern aircraft has accentuated these advantages, so that they carry more weight than the following disadvantages:

a. The nose unit must take 20 to 30% of the aircraft's weight in a steady braked condition and it is therefore relatively heavy.

b. The landing gear will probably have to be fitted at a location where special structural provisions will be required. In the case of a retractable nosegear on light aircraft it may also prove difficult to find stowage space inside the external contours of the aircraft.

Although there is still a measure of choice during the preliminary design stage, this constitutes one of the most difficult problems to be solved.

Summing up, we may state that the nosewheel undercarriage has gained favor because it greatly facilitates the landing maneuver and enables the brakes to be used more efficiently.

2.5.3. Tandem undercarriage

Here the main wheels are arranged practically in the plane of symmetry of the aircraft and the front and rear wheels absorb landing impact forces of the same magnitude. Use of the tandem gear is justified when much emphasis has to be placed on the following advantages:

a. Both main legs are placed at nearly equal distances ahead of and behind the center of gravity, thus locally creating space for payload close to it.

b. The wheels may be retracted inside the fuselage without interrupting the wing

structure. The increase if any in fuselage weight will depend on other factors.

Against these we have to set the following disadvantages:

a. Outrigger wheels will be required to stabilize the aircraft on the ground and these may increase the all-up weight by approximately 1%. However, by using two pairs of main legs instead of single ones, a certain amount of track may be obtained, resulting in a reduction of the load on the outriggers (Boeing B-52).

b. The pilot must carefully maintain the proper touchdown attitude in order to avoid overstraining the gear. Care has also to be taken to limit the angle of bank during the landing to avoid overstraining the outriggers. It may sometimes be possible to locate the rear legs close to the center of gravity of the aircraft, and so reduce this disadvantage, but that also means losing the opportunity to have an unobstructed space.

c. A large tail-download is required to rotate the aircraft. It will therefore be desirable to chose the attitude of the aircraft at rest so that it will fly itself off, but this may lead either to an increase in drag during the takeoff roll or to a high liftoff speed.

Generally speaking, the arguments against the tandem gear are of such a nature that its adoption should only be considered when no other solution meets the case.

2.6. SOME UNCONVENTIONAL AIRCRAFT CONFIGURATIONS

The characteristics of different general arrangements discussed in the preceding sections mainly apply to the classic airplane layout for which a clear distinction can be made between lifting, non-lifting and stabilizing major components. It was assumed that the tailplane was mounted at the rear of the aircraft. The payload is carried inside the fuselage while the fuel is mainly stored in the wings and, if necessary, the fuselage. The fuselage is basically designed for optimum transport and rapid loading and unloading, and contributes little to the lift.

A radical departure from the classic layout is the integrated configuration of which the flying wing is the purest representative. The wing is designed to produce lift as well as to contain the entire payload while it also provides stability and control. Less radical is the tailless aircraft which does have a fuselage but no horizontal tail surfaces. A third unusual layout is the tail-first or canard.

When the choice of one of these types is considered, there have to be obvious points which indicate that materially better performance, a considerably lighter structure or improved flying qualities will be achieved. For example, the flying wing layout would most probably be considered only in the case of a sailplane or a long-range aircraft, both of which make full use of the potential improvement in lift drag ratio. Practical experience with this type indicates that a new design will require extensive research before a reliable product can be put on the market. Many examples illustrating this point are known in the history of aviation.

2.6.1. The flying wing

During the period around the Second World War, several designers in various countries regarded the flying wing as the ideal layout, promising large reductions in drag and structure weight. They included A. Lippisch and the Horten brothers in Germany, J.K. Northrop in the U.S.A. and G.H. Lee in Great Britain. Round about 1965 Lee attempted to draw attention to a flying wing design for a short-haul airliner (Fig. 2-30, Ref. 2-45).

Since a pure flying wing possesses no fuselage and no horizontal tail surface, it may be possible to achieve a very low zero-lift drag coefficient. This may be of the order of .008 to .011 as compared to .015 to .020 for conventional aircraft. The maximum lift/drag ratio being inversely proportional to the square root of this figure, a theoretical improvement of about 40% may be obtained for a given aspect ratio (Fig. 2-29). Assuming similar fuel weights, takeoff weights and cruising speeds, the same improvement applies to the range. Alternatively, this gain may be taken in the form of a reduction in fuel consumption, engine power and takeoff weight for a specified payload and range.

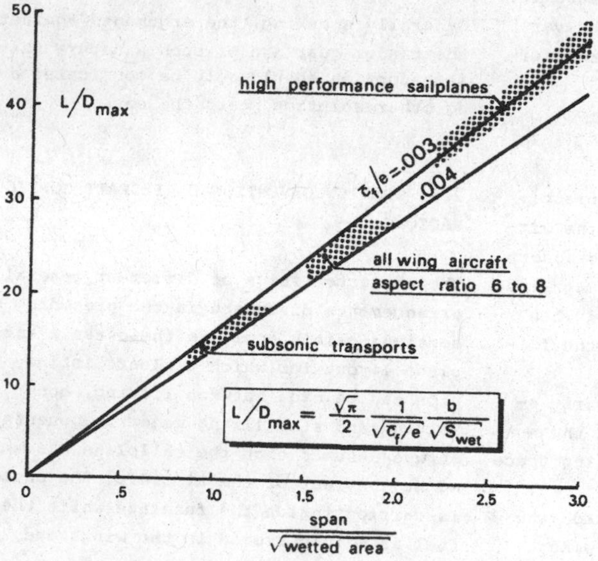

$$L/D_{max} = \frac{\sqrt{\pi}}{2} \frac{1}{\sqrt{\bar{c}_f/e}} \frac{b}{\sqrt{S_{wet}}}$$

b = wing span
$\bar{c}_f$ = mean skin friction coefficient, based on S_{wet}
e = Oswald's span efficiency factor
S_{wet} = total airplane wetted area
D = drag
L = lift

Fig. 2-29. Maximum lift/drag ratio at subsonic speeds

The empty weight of the flying wing could be less, mainly as a result of the favorable mass distribution within the wing, which reduces the bending moment at the root. Supposing the mass to be distributed along the span in a similar manner to the lift, it would even be theoretically possible to reduce the bending moment to zero in 1-g flight. Hence the bending moment will predominate in the landing and the torsion moments in flight and a large part of the wing structure will be designed on the basis of stiffness requirements. By and large, a reduction in structure weight relative to the conventional layout is likely to be possible. To ensure stability in a trimmed condition, the following requirements must be fulfilled:

a. The aerodynamic pitching moment at zero lift and zero control deflection must be positive (i.e. nose-up). This condition can be met by using a special wing section with a bent-up trailing edge or a sweptback wing with washout at the tip, or aileron deflection. Both measures tend to increase the vortex-induced drag.

b. The center of gravity must be ahead of the aerodynamic center, but this condition is difficult to fulfil in the case of a straight wing, as it implies that the entire load must be concentrated in the forward part of the wing. With a sweptback wing there is less trouble, first because the aerodynamic center is situated further back and second because more space is available in the plane of symmetry ahead of it.

A high aspect ratio sweptback wing is longitudinally unstable at large angles of incidence (Fig. 2-25b). In the case of a flying wing this instability cannot be corrected by means of a horizontal tailplane and a high aspect ratio is therefore detrimental to stability. Consequently, part of the aerodynamic gain is lost and in the case of Lee's flying wing (Fig. 2-30) its aerodynamic superiority over the conventional layout has largely disappeared.

A low aspect ratio wing enables the design-

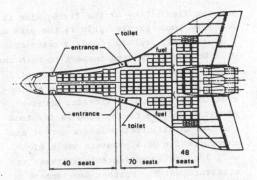

Fig. 2-30. Proposal for a short-range allwing aircraft

er to get a sufficiently thick wing to accommodate the load to be carried. In addition to comparing a conventional aircraft and a flying wing on the basis of equal aspect ratios, we can also do so on that of equal volumes. With a given payload density an optimum design will generally be one in which the space within the external contours is fully utilized. For equal volumes both configurations have roughly the same wetted area and the flying wing can only gain through a greater span and the use of buried engines, creating less drag. Another drawback of the flying wing is that it is incapable of achieving a high maximum lift coefficient. Effective flaps at the trailing edge cause a nose-down pitching moment which cannot be trimmed. A low wing loading is not just a secondary effect here but an absolute must. However, high load factors in turbulent air will be the inevitable result; these will be objectionable less from the structural viewpoint than from that of the occupant's comfort and the pilot's workload. The flying wing can be made longitudinally stable but its response to control surfaces deflections and bumps will always be accompanied by a poorly damped phugoid and an oscillatory short period motion, both annoying characteristics to the pilot, although this might be improved by some form of artificial stability augmentation.

It should finally be pointed out that the

loading flexibility of the flying wing is not very good, particularly in the case of a low-density payload. Loading restrictions will be necessary with respect to both the longitudinal and the lateral position, which is an undesirable factor in the operation of transport aircraft. Moreover, the shape of the flying wing is far from that of an efficient pressure vessel and incorporation of a pressure cabin might well lead to a considerable increase in structure weight. Further development to increase the payload is not feasible as the stretch potential of the flying wing is almost nil.

Summing up the case for the flying wing, we may say that this configuration is potentially capable of reaching a high lift to drag ratio with a low structure weight but the flying and operational characteristics are troublemakers. Since the control function is integrated in the wing, there will be additional trim drag. As a passenger transport aircraft the flying wing does not appear to be a suitable proposition but it may be considered for special purposes, such as sailplanes, long-distance reconnaissance or very large, special-purpose cargo aircraft.

2.6.2. Tailless aircraft

Although the flying wing and the tailless aircraft share the characteristic of not possessing a horizontal tailplane, the latter type has a conventional fuselage which carries a large part of the load. The tail of the fuselage is relatively short and carries only a vertical tail surface. The tailless aircraft is generally designed for supersonic speeds and utilises a slender delta wing. The movable parts at the trailing edge act as elevators when deflected in the same direction and as ailerons when deflected in different directions. Like the flying wing the tailless aircraft is unable to carry effective landing flaps and sufficient lift for landing is obtained by choosing a low aspect ratio wing of large

area, resulting in large approach angles. Some of the disadvantages mentioned in connection with the flying wing also apply to this type, although they generally weigh less heavily on account of the lower aspect ratio and larger mean chord.

Since the tailless delta is of a much less radical nature than the flying wing, various successful aircraft have been built to this formula and have reached series production. The best known are: Avro Vulcan (almost a flying wing, Fig. 2-12), Convair B-58, Convair F-102, Lockheed YF-12A, Douglas F-4D, Dassault Mirage, SAAB Draken and the BAC-SUD Concorde. These aircraft all operate in the transonic or supersonic speed region, the tailless delta being one of the best configurations for supersonic cruise. Ref. 2-44 gives general information concerning the design of such aircraft.

2.6.3. Tail-first or canard aircraft

Canard aircraft have attracted the interest of designers from time to time on account of several particular characteristics. After all, the Wright brothers' aircraft was a canard and it appears an attractive idea to place the longitudinal control surface in front of the wing and out of the wing downwash to where it can never be in its wake. The aircraft's equilibrium is preserved by means of an upward force on the forward plane, which contributes to the lift in a positive sense. In contrast to the conventional layout, this will increase maximum lift and reduce the trim drag, a characteristic which is of particular advantage in the case of high-speed, highly maneuverable aircraft.

A distinction can be made between a long-coupled and a close-coupled canard (Fig. 2-31). In the former category emphasis is laid on the reduction of drag in cruising flight, which is obtained by placing the forward plane far ahead of the wing, thereby reducing the mutual interference. Dynamic stability may be assured by keeping the area of the forward plane below 10% of that of the mainplane. Design problems re-

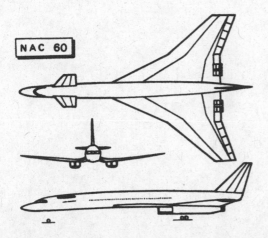

a. Long-coupled canard (North American design for an SST, 1964)

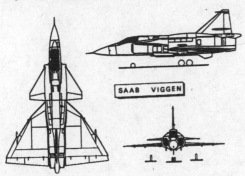

b. Short-coupled canard

Fig. 2-31. Tail-first airplane configurations

lating to this configuration are:

a. To achieve an acceptable range for the center of gravity, the forward plane has to be capable of producing a higher maximum lift coefficient than the main wing. Generally speaking, this can only be achieved when the main wing possesses a low aspect ratio. The forward plane has to be provided with a sophisticated flap system.

b. The trailing vortices of the forward plane affect the flow over the wing and will set up a rolling moment in a sideslip. The vortices may also strike the fin.

In the case of the short-coupled canard the mutual interference between the two planes is deliberately used to achieve a high maximum lift. This effect is obtained at large angles of attack on surfaces of low aspect ratio with sharply sweptback leading edges. The large drag which now occurs will only be acceptable for aircraft with sufficiently powerful engines.

In short, the canard layout appears to be suitable for transonic or supersonic and highly maneuverable aircraft, in the latter case if sufficient thrust reserve is available.

Chapter 3. Fuselage design

SUMMARY

This chapter starts with an introductory section dealing with the design requirements, the possibility of achieving an optimum external shape and suggestions for a design procedure.
The second section presents detailed instructions for the design of the passenger cabin, stressing the desirability of achieving efficient arrangement. This is important in order to ensure that, for a given level of passenger comfort, the fuselage makes the maximum possible contribution to the operation of the aircraft.
Some attention is given to freight aircraft where the choice of specified density, the use of containers and pallets, and the loading and unloading provisions are of considerable influence on the design. The final sections contain directives relating to the design of the flight deck and the external shape.

3.1. INTRODUCTION

3.1.1. Function and design requirements

The preliminary general arrangement of the aircraft is closely tied up with the fuselage, the main dimensions of which should be laid down in some detail. In fact, the fuselage represents such an important item in the total concept that its design might well be started before the overall configuration is settled.

The main characteristics of the fuselage are as follows:

a. It constitutes the shell containing the payload which must be carried a certain distance at a specified speed. It must permit rapid loading before the flight and rapid unloading after it. The fuselage structure also offers protection against climatic factors (cold, low pressure, a very high wind velocity) and against external noise, provided suitable measures have been taken.

b. The fuselage is the most suitable part for housing the cockpit, the most functional location generally being in the nose.

c. The fuselage may be regarded as the central structural member to which the other main parts are joined (wings, tail unit and in some cases the engines) on the one hand, and as the link between the payload and the aircraft on the other. In some aircraft a number of these duties are assigned to tail booms.

d. Most of the aircraft systems are generally housed in the fuselage, which sometimes also carries the engines, fuel and/or the retractable undercarriage.

Although the installation of aircraft systems will not be dealt with in this chapter, the reader may refer to Fig. 3-1 which shows how the Auxiliary Power Unit (A.P.U.) and the air-conditioning equipment can be installed in the fuselage.

Many of the requirements laid down in relation to the fuselage limit the designer's range of choice. The list below - though far from complete - enumerates the factors which should be given serious attention as they affect most designs.

a. The drag of the fuselage should be low, since it represents 20 to 40% of the zero-lift drag. At a given dynamic pressure the

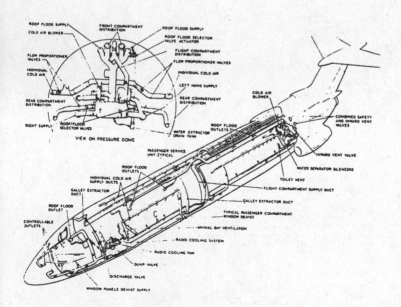

Fig. 3-1. The air conditioning system of the Hawker Siddeley Trident

drag is mainly determined by the shape and the wetted area. If we choose a fuselage diameter 10% larger than strictly necessary, the direct result will be a 1.5-3% increase in total drag. This will mean a higher fuel consumption or decreased range, increased takeoff weight, hence another drag increment, etc. This "snowball effect" in weight growth depends on the type of operation for which the aircraft is to be used, and this, in turn, determines how much effort should be made to achieve minimum drag. In the case of a freight aircraft designed for low speeds and modest yearly utilization, such as the Short "Skyvan", a good aerodynamic shape has been sacrificed to easy loading by means of a readily accessible rear loading door.

b. The structure must be sufficiently strong, rigid and light, possess a fixed useful life and be easy to inspect and maintain. In order to avoid fatigue failure of the pressure cabin, a relatively low stress level should be chosen for the skin, e.g. 12,000 p.s.i. (850 kg/cm^2) which is about 30% of the $\sigma_{.2}$ - limit for Al 2024-T3. Pressure cabins have a circular cross-section, or a cross-section built up of segments of a circle.

c. Operating costs are influenced by the effect of the fuselage design on fuel consumption and by manufacturing costs. Generally speaking, we gain by keeping the fuselage as small and compact as possible within acceptable limits. On the other hand, it must be remembered that the design and dimensions of the fuselage are decisive factors with regard to the aircraft's earning capacity. In aiming for a compact design, the designer should never go so far that potential customers will reject the aircraft because it lacks comfort as a result of cramped accommodation.

d. The fuselage does not merely serve to carry the empennage, but also affects the tailplane configuration. It will generally contribute a destabilizing effect to the aerodynamic moments in pitch and yaw which is approximately proportional to its volume, while the stabilizing contribution of the

tail surfaces is mainly dependent on the length of the fuselage tail.

3.1.2. Drag and optimization of the external shape

Surprising though it may be, the fuselage of transport aircraft - a category which is particularly suited for optimization - are seldom ideally streamlined in shape. Amongst subsonic aircraft, the Lockheed Constellation and Airspeed Ambassador were the last to be developed and a more recent example, although in a different category, is the HFB Hansa. We may ask to what extent aerodynamic optimization of the fuselage's external shape is both desirable and possible. Apart from the question of optimization in a broader sense, which constantly occupies the designer - what is the best arrangement for the seats, where is the best location for the freight hold, etc. - the dominant questions to be answered in the preliminary design stage are:

a. Should the aim be to achieve the ideal streamline shape with minimum drag, or is a cylindrical mid-section to be preferred?

b. Should a long, slender shape be adopted or would a short, squat fuselage be better?

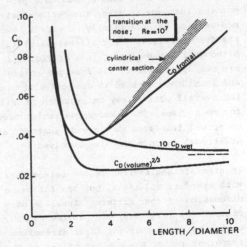

Fig. 3-2. Drag coefficient of streamline bodies of revolution at low speeds (Curves calculated with Section F-3.4.)

Fig. 3-2 illustrates the influence of the fuselage slenderness ratio λ_f (length/diameter) on drag, by showing the drag coefficient of a fuselage related to several reference areas. These are: the frontal area, the wetted area and the (volume)$^{2/3}$. The figures refer to a fully streamlined body of revolution for slenderness ratios higher than 4. The same coefficients are also given for a fuselage with a cylindrical section. It is assumed here that for these values of λ_f the drag coefficient, based on the wetted area, is essentially equal to that of the pure streamline body. We may now draw the following conclusions from the figure.

a. The drag coefficient, based on the wetted area, approaches the flat plate value for very slender shapes. When λ_f is low, there is a considerable pressure drag.

b. The coefficient based on the frontal area shows a pronounced minimum at $\lambda_f = 2.5$ to 3. When a cylindrical mid-section is used, the drag rises considerably, particularly in the case of high values of λ_f.

c. On the basis of (volume)$^{2/3}$ the drag coefficient shows a shallow minimum for $\lambda_f = 4$ to 6, but rises only slightly for higher values of the slenderness ratio. Slenderness ratios of less than 3 lead to a pronounced increase in drag. The cylindrical mid-section has practically no effect on the drag coefficient for all values of λ_f. Although the numerical values of Fig. 3-2 are not applicable to all fuselage shapes and should not be used in drag calculations, the overall picture may be regarded as valid for most cases. The slenderness ratios used on actual fuselages show a wide scatter as additional factors are also involved.

In aircraft engineering we can seldom work with standard solutions, but the following discussion of four different fuselage configurations (Fig. 3-3) may provide an indication as to whether the ideal streamline shape, or rather the minimum of one of the curves of Fig. 3-2, might have been the designer's aim. In the case of transport aircraft (Fig. 3-3a), the space allotted to the load takes up to between 60 and 70% of the fuselage volume. The shape of the fuselage here is derived from an efficient arrangement of passengers or freight.

A cylindrical mid-section is used for the following reasons:

a. Structural design and manufacture are considerably simplified.

b. It is possible to obtain an efficient internal layout with little loss of space.

c. The flexibility of the seating arrangement is improved.

d. Further development by increasing the length of the fuselage (stretching) is facilitated.

As the length of the fuselage increases, the areas of the tail surfaces will be reduced, but this is only true up to a point. Slender fuselages in large aircraft with a fineness ratio of 12 to 15, may well involve stiffness problems. The analytical approach used in Ref. 3-3 indicates that it is not so much the fuselage drag but more particularly the weight which is the deciding factor where the optimum shape is concerned. This is confirmed by the small variation in the drag coefficient based on (volume)$^{2/3}$ with the slenderness ratio (Fig. 3-2). For slender fuselages there is also a favorable influence of the Reynolds number on friction drag.

In the case of passenger as well as freight aircraft, the possibilities of varying the shape of the fuselage are limited by practical considerations relating to the load. The fuselage should be designed "from the inside outwards", and the skin should envelop the load in such a way that the wetted area is minimum, thus avoiding breakaway of the airflow as far as possible.

In the case of freight aircraft (Fig. 3-3b), loading and unloading in the longitudinal direction will be the aim. A door in the nose is unsuitable for relatively small freighters as the cockpit would have to be extended to an unreasonable extent on top of the fuselage. Nor is it an easy matter to design a freight door in the tail where stresses are introduced by the tail unit

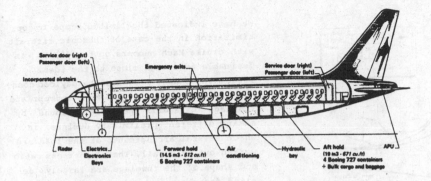

a. Fuselage with relatively large payload volume and efficient internal arrangement (Dassault Mercure)

b. Freighter aircraft (Nord 2508) with tail booms and short fuselage afterbody

c. Jet trainer (Fouga Magister) with small useful load volume

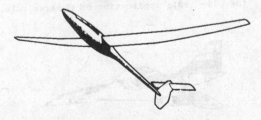

d. Sailplane (Sigma Glider) with fuselage forebody and afterbody designed to different aerodynamic criteria

Fig. 3-3. Categories of fuselage shapes

and by pressurization. In such a case the use of tail booms may be considered (Section 3.3.3), although other solutions are also possible. The length of the fuselage is chosen as short as aerodynamic considerations permit, resulting in a short, squat body. Although this may not approach the optimum streamline shape, it may come close to the minimum drag value for a given frontal area, as shown in Fig. 3-2 (e.g. H.S. Argosy $\lambda_f = 4.85$; IAI Arava: $\lambda_f = 3.75$) Fig. 3-4 shows that the afterbody slenderness ratio may be quite small without giving rise to a large increase in drag. It should, however, be remembered that the flow induced by the wing lift will generally alter this picture in an unfavorable sense. The tail booms, including the engine nacelles of the Arava, have a slenderness ratio of 14, which shows that these parts have been designed with the object of keeping the wetted area to a minimum.

Trainers and small touring aircraft (Fig. 3-3c) carry a relatively small useful load in relation to the size of the fuselage and a certain measure of freedom is present in choosing the disposition of the occupants, the engine or engines, equipment and possibly the fuel. The length of the fuselage tail will mainly be decided as a function of the operation of the tail surfaces. When two occupants are seated side by side a slenderness ratio of about 6 is a common

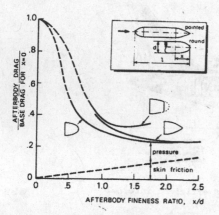

Fig. 3-4. Afterbody drag of a fuselage tail, when added to a cylindrical shape

We have indicated the limited scope for optimization in the case of subsonic aircraft with cruise Mach numbers up to .85. It is desirable to design close to the ideal streamline shape, provided the payload does not dictate otherwise. The most appropriate fuselage shape will therefore be found by making different preliminary designs in which several arrangements of the load are worked out in detail. The slenderness ratio and shape of the fuselage are largely derived factors.

At flying speeds above M = .85, fuselages having conventional slenderness ratios will generally create wave drag. As this chapter will not deal with

value; in a tandem arrangement this figure should be at least 8. An interesting exception is the Sipa Minijet dating from 1952 (Fig. 3-5a). This had a fineness ratio of 3.3, which obviously showed that the aim was to achieve minimum drag for a given frontal area by using tail booms to carry the empennage.

In the design of sailplanes (Fig. 3-3d) great emphasis is laid on minimum drag. For the forward part of the fuselage, there is a tendency to choose a shape which results in minimum drag for a given cross-sectional area of the cockpit. From Fig. 3-2 it can be seen that in such a case λ_f = 2.5 to 3 would lead to a low drag figure. Contrary to the examples already discussed, the assumption that the boundary layer is fully turbulent does not apply to gliders. When a favorable shape is chosen, the body shape may be compared to a laminar airfoil section revolved about an axis and the boundary layer will become turbulent some distance behind the nose. In that case the optimum slenderness ratio would be 3 to 4. The length of the tail boom should be decided by the moment arm of the tail surfaces. The extremely slender boom will have a small wetted area as well as a low drag coefficient. Attention should also be paid to the required stiffness and weight.

a. Sipa Minijet (1952); fuselage length/diameter = 3.3

b. Boeing design for a transonic airliner (M= .95- .98); application of the area rule

c. Aérospatiale/BAC Concorde supersonic passenger transport; fuselage length/diameter ∿ 20

Fig. 3-5. Examples of special fuselage shapes

the aerodynamic problems occurring at transonic and supersonic speeds we shall confine ourselves to discussing a few brief examples. Boeing's design for a transonic transport aircraft (Fig. 3-5b) shows how a satisfactory drag at M = .95 to .98 can be obtained by applying Whitcomb's area rule, combined with supercritical aerofoils. The central portion of the fuselage has a kind of waist, which is unlikely to present many problems with regard to interior planning in the case of aircraft designed for a relatively small payload. Detailed studies, however, will be required to show to what extent a gain in cruising speed of 10 to 15% outweighs the practical drawbacks of this fuselage shape.

Flight at supersonic speeds demands a very slender fuselage in order to keep the wave-drag down to an acceptable value. For the Aérospatiale/BAC Concorde, the Tupolev 144 and the Boeing 2707 SST project, slenderness ratios of about 20, 18.5 and 18 to 25, respectively were chosen. Incidentally, it is interesting to note that both the Concorde (Fig. 3-5c) and the Tu-144 have a cylindrical mid-section.

3.1.3. A design procedure for fuselages with cylindrical mid-section

The following design procedure, derived from Ref. 3-1, is applicable to fuselages of transport aircraft with a cylindrical mid-section. It applies particularly to the large wide-body category transports. In the case of smaller transports catering for, say, less than 120 passengers, a somewhat simplified procedure may be used by assuming that the passenger cabin is almost entirely cylindrical. Some of the steps to be discussed below are also applicable to cargo aircraft, but in many cases the location of the freight doors will be the deciding factor in the design (Section 3.3.3).

a. Choose the number of seats abreast in a cross-section and/or the dimensions of the cargo load, selecting a section which will most likely determine the diameter of the central part of the fuselage. The remaining principal dimensions of the fuselage will be largely dependent on this parameter.
b. Design the shape of the cross-section on the basis of certain predetermined rules

(seat dimensions, seat pitch, safety provisions, etc.). If more than 150 to 200 passengers are to be accommodated the use of two aisles should be considered; if more than 500 passengers are carried, it becomes feasible to think of using two decks. Inside contour points are laid down which limit the internal dimensions of the cabin and the freight hold. Around this a contour is drawn as tightly as possible. This will generally be circular in shape, but may be built up of circular segments (double bubble, flattened belly).
c. The external shape may be determined by assuming the minimum thickness of the fuselage walls (skin, formers, upholstering, etc.).
d. It will now be possible to draw in the planview of the fuselage nose and tail, including the cockpit. These are parts which do not possess a cylindrical contour. Since they generally contain fewer passengers per unit of volume than the cylindrical part, they should be kept as short as possible. Some guidelines can be found in Section 3.5.1.
e. The capacity of the fuselage nose and tail portions is subtracted from the total payload. For the remainder, essentially the major part of the payload, a prismatic portion is chosen.
f. On the basis of the plans of side and front views, the following details may be decided:
- The main dimensions of the cockpit.
- The dimensions and location of doors, windows and emergency exits, spaces required for embarcation and disembarcation or evacuation in case of emergency.
- The tail of the fuselage. This must be planned in such a way that it will not create an unacceptable geometrical limit to rotation in takeoff and landing.
- The indication of spaces for the wing centre-section, attachment frames of the engines, retraction of the landing gear, pressurization and air-conditioning, electrical and electronic systems, etc., insofar as they are present.
- The presence of adequate space below the

cabin floor for cargo and passenger baggage. If this is not available, consideration may be given to altering the shape of the cross-section (e.g. double bubble), increasing the diameter, raising the cabin floor, changing the arrangement of the seats, etc.

g. If the first assumption does not lead to a satisfactory design, or if various possible solutions are to be evaluated, the procedure must be started again at a. and repeated as often as necessary. It will generally be necessary to check whether alternative planning schemes are possible within the fuselage as it stands, e.g. different seating arrangements in various classes, combiplane*, freighter version , etc.

The work scheduled above will lead to a provisional design when a number of data, such as those regarding the wing root location, are not yet available. In particular the position of the center of gravity may necessitate a fundamental re-arrangement of the payload at a later stage. The preliminary design of the fuselage is made almost entirely on the drawing board since very few analytical studies are available in the existing literature. Provided the outcome of these investigations is limited to the influence of the fuselage shape on empty weight and drag, the results sometimes prove reliable, though they are rarely accurate. Apart from this it is important to consider such factors as the comfort of the occupants, easy access for the passengers, embarcation and servicing and the influence which the shape of the fuselage has on the general configuration of the aircraft. These aspects all have an important bearing on operating economy, although their influence cannot be evaluated quantitatively. Operating economy is of such vital importance that manufacturers generally build one or more mockups of the fuselage before deciding on the

*an airplane with combined transportation of passengers and cargo on the main deck.

final design. A separate group is responsible for the details of the interior design, but a useful starting point for this should be established in the preliminary design.

3.2. THE FUSELAGE OF AIRLINERS AND GENERAL AVIATION AIRCRAFT

3.2.1. Importance of comfort and payload density

Although passengers judge the accommodation in an aircraft according to many yardsticks, there are a number of minimum requirements which should be met. Comfort is mainly dependent on the following factors:

a. The design and arrangement of the seats. This applies particularly to the adjustability of the seat and the legroom available.

b. The general aesthetic impression created by the interior, especially the suggestion of spaciousness within the limited dimen - sions of the cabin.

c. The room available for the passenger to move about in the cabin.

d. The climate in the cabin: temperature, moisture, freedom from draughts and the provision of an adjustable supply of air. It is important to keep the rate of pressure variations during climb and descent within acceptable limits.

e. Noise in the cabin, or more specifically Speech Interference Level (SIL) and the presence of resonances.

f. Accelerations, mainly normal to the flight path but also in the direction of roll during braking. Apart from external factors such as the weather, comfort is largely influenced by wing design and the flexibility of the fuselage structure.

g. The aircraft's attitude during climb and descent.

h. The duration of the trip.

i. The number and accessibility of lavatories, washrooms, lounges (if provided) and suchlike amenities.

j. Stewardess service, in-flight entertainment, meal service, snacks, etc.

The aircraft designer has, at least to
some extent, some direct influence with
respect to these factors: the space in the
fuselage and the influence of wing design
on accelerations (wing loading, aspect ratio, angle of sweep). Other factors, such
as air-conditioning and pressurization,
sound proofing etc. will be dealt with at
a later stage of the design. The designer
has no direct say in the in-flight services
provided for the passengers, although he
should allow for the weight and locations
of toilet facilities, pantries and cloakrooms.

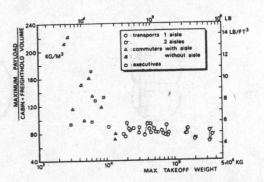

Fig. 3-7. Equivalent payload density

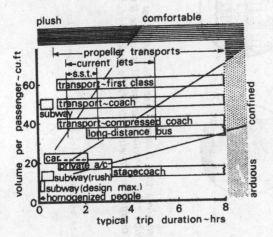

Fig. 3-6. Volume per passenger and trip duration (Ref.: The Architectural Review)

Fig. 3-6 illustrates the space available in
an aircraft as compared with a number of
other vehicles. It shows the relationship
between the available volume per passenger
and the average trip duration with respect
to comfort. In the case of aircraft, a distinction is made between different classes
of fares. Any attempt to increase the level
of comfort by choosing a large cabin volume,
will result in a growth in fuselage dimensions which will have a considerable effect
on operating costs. International agreements
and competition generally make it impossible
to offset this by increasing fares. On
profitable routes where there is keen competition, however, an increase in space will

pay dividends. Statistical data (Fig. 3-7)
clearly show that for takeoff weights in
excess of 25,000 lb (11,000 kg) approximately, the "density" of the load varies
little with the size of the aircraft. For
most of the current fleet of transport aircraft the average density is 5.0 to 5.5
lb/cu. ft (80 to 90 kg/m^3) while it comes
to about 4.5 lb/cu. ft (70 kg/m^3) in the
case of the new category of large wide-
body jets (Boeing 747, Douglas DC-10, Lockheed L-1011, Airbus A-300 B). In the case
of small short-haul aircraft with an all-
up weight of up to 25,000 lb ($\sim$ 11,000 kg)
approximately, the design specifications
vary so much that the load densities lie
between 5.0 to 12.5 lb/cu. ft (80 - 200
kg/m^3) even rising to 14 lb/cu. ft (220
kg/m^3) in the Britten Norman BN-2A.
In the above an average has been taken for
the payload density and no allowance has
been made for a distinction between the
weights of passengers, luggage and cargo.
This factor will be discussed in more detail in Section 3.2.5, when dealing with
the dimensions of the cargo holds.

3.2.2. Cabin design

a. Cross-section.
The first step is to decide upon the number
of seats to be placed abreast in a cross-
section. Fig. 3-8 shows several cross-sections investigated by the McDonnell Douglas
Aircraft Corp., in connection with the de-

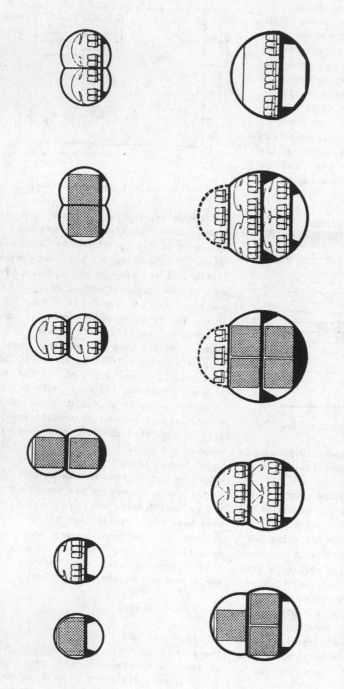

Some of the basic cross-sectional alternatives that Douglas has considered for the very large aircraft project, with a present-day cross section to provide scale. Prime requirement is for the 8 ft x 8 ft container to be carried efficiently. The bottom-row (middle) designs could alternatively be made circular, dispensing with the upper (dotted) lobe.

Fig. 3-8. Fuselage configuration studies by Douglas (Ref.: The Aeroplane, Aug. 4, 1966)

sign of a very large transport which had also to be operated as a freighter. When studying large aircraft of this kind various configurations are projected.

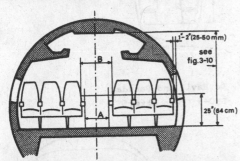

NUMBER OF SEATS	A minimum	B minimum
10 OR LESS	12 in. ,305 mm	15 in. ,381 mm
11 THROUGH 19	12 in. ,305 mm	20 in. ,508 mm
20 OR MORE	15 in. ,381 mm	20 in. ,508 mm

Fig. 3-9. Minimum aisle width for passenger transports (Ref.: FAR 25.815 and BCAR D4-3 par. 5.2.6)

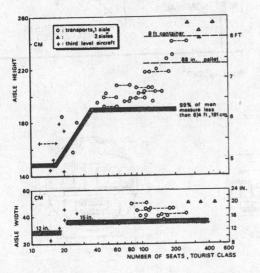

Fig. 3-10. Statistical data on dimensions for the aisle

Details such as the dimensions of the seats (Section 3.2.3) and the aisle(s) (Figs. 3-9 and 3-10) are entered in the cross-section.

FAR 25.817 limits the number of seats on each side of the aisle to three, so if more than six passengers are planned in a cross-section the designer will have to allow for two aisles. The minimum permissible width of the aisle in transport aircraft is laid down in FAR 25.817 (Fig. 3-9). All passengers must be able to move their heads freely without touching the cabin walls. This requires a free space with a radius of at least 8 to 10 inches (.20 to .25 m), measured from the eyes. The cabin wall can then be drawn accordingly. In the case of pressure cabins the cross-section will generally be a circle, or it may be built up from segments with different radii (see examples in Fig. 3-11). If no luggage can be carried under the cabin floor, the fuselage belly contour may be flattened. This is sometimes done in the case of high wing monoplanes (e.g. the Fokker F-27) and has the advantage that the undercarriage may be shortened. The external fuselage diameter can be found from the internal dimensions by adding about 4 inches (10 cm) for the thickness of the cabin wall. Remarkably enough statistics show that aircraft size has hardly any effect on the wall thickness of pressure cabins, though wide variations do, of course, exist.

In aircraft where the pressure cabin is limited to the cockpit, or in unpressurized fuselages, a rounded rectangular, elliptical or oval cross-section is a common choice. With the internal dimensions specified, this generally leads to a minimum frontal area. In this category of aircraft we may assume the wall thickness to be 2% of the fuselage width plus approximately 1 inch (25 mm).

The result of this design procedure may be compared with Fig. 3-12 which is based on existing aircraft and shows the fuselage width as a function of "total seat width" in the cross-section.

b. Location of seats and dimensions of the cabin.

In order to increase the flexibility of the cabin interior, the seats are mounted on

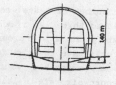

Dassault "Falcon" 10

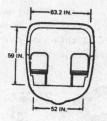

De Havilland Canada
DHC-6 "Twin Otter"

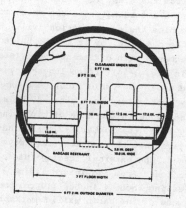

De Havilland Canada DHC-7

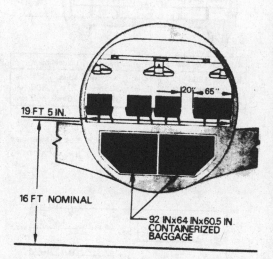

Boeing 747

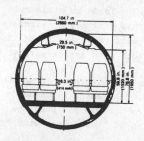

VFW-Fokker 614

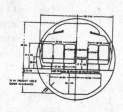

BAC-1 11

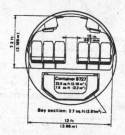

Dassault Bréguet
Mercure

Fig. 3-11. Examples of some typical fuselage cross-sections of transport aircraft

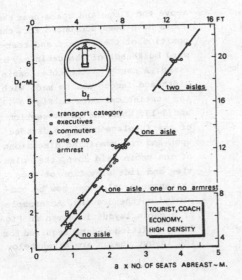

Fig. 3-12. Fuselage width vs. "total seat width"

rails sunk into the floor. Standard seat rails allow the seat pitch (i.e. the longitudinal distance between corresponding points on the two nearest seats in a row) to be adjusted by increments of one inch.

Seat pitch is generally associated with the class of service. At present, however, the terminology in the comfort standards is somewhat confused due to the historical developments in passenger service.

Up to about 1950 the level of seating comfort on normal journeys could be compared to the current first class standard which offers a seat pitch of 40 inches. As a result of pressure by airlines which charged lower fares for special flights ("air coach"), the tourist class was introduced around 1952. This differed from first class not only in that seats were more closely spaced (pitch about 38 inches), but in some cases the number of seats abreast was also increased. Another factor was the type of meal served. Around 1959 the official tourist class was replaced by the economy class with a typical seat pitch of 34 inches. Nowadays the expressions "tourist", "coach" and "economy" are used without any explanation of the exact distinction between them. There is also "high density" seating

with a pitch down to 29 or 30 inches. To avoid this somewhat unattractive term, some companies prefer to use the expression "economy class" in order to avoid adverse passenger reactions.

As a guideline the following figures may be used for typical seat pitch values:
first class   : 38 to 40 inches
tourist/coach/economy: 34 to 36 inches
high density/economy: 30 to 32 inches.
Since comfort is not only a matter of seat pitch, the choice of pitch may also be influenced by the trip duration and the width of the seat. Another important point is the maximum number of seats abreast: passengers tend to dislike three seats in a row. This can be improved by choosing a greater pitch or by using a wider center seat.

If there is a wall or partition in front of a row of seats some space should be left to allow sufficient leg room and permit limited adjustment of the seat back. A distance of about 40 inches between the seat backrest and the partition should be adequate. Extra space is also required at the emergency exits (see Section 3.2.4).

The cabin floor should preferably be kept level in the normal cruise attitude, although this may not always be possible nor necessary in a small aircraft. It is particularly important to have a level floor in large aircraft where food and drinks are served from carts. The floor should be sufficiently strong to support the maximum number of passengers in a high density layout. The permissible floor loading should generally be at least 75 to 100 lb/sq. ft ($350 - 500$ kg/m^2), but 200 lb/sq. ft (1000 kg/m^2) is required when the floor has to carry freight. The thickness of a cantilever floor will be about 5% of the fuselage diameter. The total floor area can be found from the fuselage design drawing. Statistically average values are 6.5 sq. ft ($.6$ m^2) per passenger for normal aircraft and 7.5 sq.ft ($.7$ m^2) for wide-body jets, both in an all-tourist configuration.

When the seats have been arranged and lavatories, pantries, wardrobe(s), freight holds

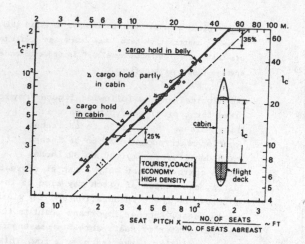

Fig. 3-13. Statistical correlation at the cabin length

above the floor and space near the doors have been accounted for, the position of the forward and rearward bulkheads of the cabin may be fixed. A check on the total cabin width and length can be made with the statistical data in Fig. 3-12 and 3-13. When the cross-section of the fuselage has been decided upon and the remaining dimensions of the cabin laid down, the plan view and side elevation of the passenger section can now be completed on the drawing. An example of a cabin layout is given in Fig. 3-14. Additional data required for the layout design are given below.

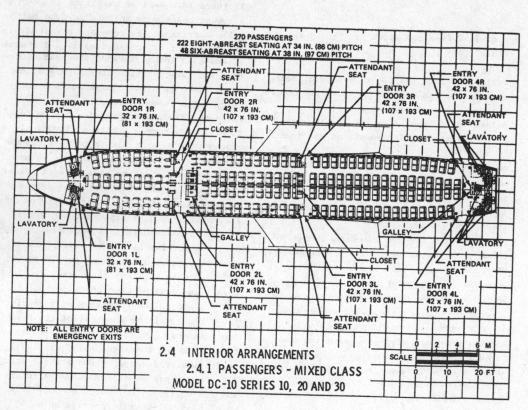

Fig. 3-14. Cabin arrangement of the McDonnell Douglas DC-10

3.2.3.Passenger seats

Although the preliminary design will be based on a certain type of seat, due allowance should be made for the fact that airlines tend to lay down their own specifications for the cabin furnishings. The passenger seats on which data are given in Fig. 3-15 and Table 3-1 are representative

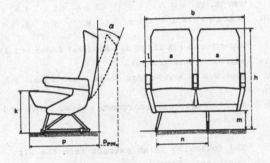

Fig. 3-15.Definitions of seat dimensions

SYMBOL[*]	UNIT	SEAT CLASSIFICATION		
		DE LUXE	NORMAL	ECONOMY
a	inch	20(18½-21)	17(16½-17½)	16.5(16-17)
	cm	50(47-53)	43.5(42,5-45)	42(40,5-43,5)
b_2[**]	inch	47(46-48½)	40(39-41)	39(38-40)
	cm	120(117-123)	102(100-105)	99(47-102)
b_3[**]	inch	-	60(59-63)	57
	cm	-	152(150-160)	145
l	inch	2¾	2¼	2
	cm	7	5.5	5
h	inch	42(41-44)	42(41-44)	39(36-41)
	cm	107(104-112)	107(104-112)	99(92-104)
k	inch	17	17¾	17¾
	cm	43	45	45
m	inch	7¾	8¼	8¼
	cm	20	22	22
n	inch	usually	32	(24-34)
	cm		81	(61-86)
p/p_{max}	inch	28/40	27/37½	26/35½
	cm	71/102	69/95	66/90
α/α_{max}	deg	15/45	15/38	15/38

*definitions in Fig. 3-15

**the index denotes the number of seats per block

NOTES

1. The data represent normal values and are not standard values. A statistical range is indicated in brackets.
2. In wide-body aircraft, seats are used in the tourist/coach class with a=19" (48 cm), b_3=66" (168 cm), h=43" (109 cm). In high-density arrangments the "normal" type seat is used.
3. In third-level aircraft it is customary to install seats with only one or no armrest, with typical dimensions: width 16½" (42 cm), h=35" (89 cm), p=26" (66 cm).

Table 3-1.Seat dimensions (Ref.: Seat manufacturers brochures, Flight Int., July 8, 1965)

of the types in normal use at the time of writing. They are classified as follows:
de luxe type: seat pitch 37-42 inches
normal type : seat pitch 32-36 inches
economy type: seat pitch 28-31 inches
The de luxe type is used in the first class section, while economy-type seats are fitted in the high-density class. Apart from this there is no clear-cut relationship between the comfort classes and the classification given above. Seats may also be listed according to the width between the armrests (distance "a" in Fig. 3-15), as follows:
de luxe: a = 19 inches
normal : a = 17 inches
narrow : a = 16 inches
A seat width of 19 inches (48 cm) corresponds approximately to first class in most passenger aircraft, but is also used in the tourist class of wide-body jets. In the latter type of aircraft it is also possible to use a layout with normal seat width, thus enabling one more seat to be added to a cross-section. Ref. 3-9 employs a somewhat more detailed distinction according to the standards used in international and domestic transport. The seats used in long-range aircraft are often finished more luxuriously, with a resultant effect on weight.
The above applies to aircraft used for normal transport routes. The following data apply to other types:
a. Small passenger aircraft for low density traffic: third level, commuters, feederliners. These only make short flights and this

justifies the use of simply designed seats without armrests. The seat pitch would be 30 - 36 inches (76 - 81 cm).

b. Business aircraft are generally furnished in lavish style and there is a considerable variation in seat dimensions. One finds seats with a width of 24 inches (60 cm) across the armrests and a pitch of 34 - 36 inches (86 - 92 cm). When the aircraft is used for passenger transport the pitch is reduced to 30 inches (76 cm) and a bench seat for three passengers may be provided at the rear of the cabin.

c. Private aircraft generally have no aisle between the seats and it is not so much the seat width as the cabin width which matters. Taking the average shoulder width of an occupant as 20 inches (51 cm) and allowing for 2 inches (5 cm) clearance on either side, the minimum internal width will be 21 inches (61 cm) for a single tandem arrangement and 46 inches (117 cm) for side-by-side seating. A narrower cabin will give most occupants a cramped feeling.

SEAT CLASSIFICATION	MEDIUM/ LONG HAUL		SHORT HAUL	
	LB	KG	LB	KG
de luxe –single	47	21.3	40	18.1
–double	70	31.8	60	27.2
normal –single	30	13.6	22	10.0
–double	56	25.4	42	19.0
–triple	78	35.4	64	29.0
economy –single	24	10.9	20	9.1
–double	47	21.3	39	17.7
–triple	66	29.9	60	27.2
commuter–single	-	-	17	7.7
–double	-	-	29	13.2
lightweight seats	-	-	14	6.4
attendants' seats	18	8.2	14	6.4

executive seats,
single - VIP   : 50 lb (22.7 kg)
      - normal : 40 lb (18.1 kg)
      - small a/c: 32 lb (14.5 kg)
ejection seats - trainers: 150 lb (installed)

Table 3-2. Typical seat weights for civil aircraft

Seats and seat mounts are designed for passenger weights of 170 lb (77 kg). Normal loads must be absorbed during flight and on the ground, but the loads occurring during an emergency constitute a more critical case. These are laid down in the airworthiness regulations as follows:

	for-wards	rear-wards	up-wards	down-wards	side-ways
FAR 25.561	9 g	-	2 g	4.5 g	1.5 g
BCAR D3-8	9 g	1.5 g	4.5 g	4 g	2.25 g

According to FAR 25.785 an additional factor of 1.33 applies to seat fittings.

Some data on seat weights can be found in Table 3-2.

3.2.4. Passenger emergency exits, doors and windows

The following is an extract from the airworthiness requirements which contain the most relevant points for a preliminary design. This extract has no legal validity and the actual requirements should be consulted for more detailed particulars.

a. Passenger aircraft, to be certificated according to FAR 25 and BCAR Section D (see FAR 25.807 through 813 and BCAR Section D para. 5 of chapter D4-3).

Emergency exits are generally grouped in four classes, the particulars of which are given in Fig. 3-16 and Table 3-3. Types I and II are located at cabin floor level, unless type II is placed above the wing. Types III and IV are located above the wing. Apart from these, FAR 25.807 also describes ventral emergency exits and escapes through the tail cone.

The minimum number of exits required is shown in Table 3-4. Aircraft carrying more passengers than are given in this table must comply with special conditions. The FAR requirements demand exits of type A, not less than 42 inches wide and 72 inches high (107 x 183 cm^2). The reader should refer to FAR 25.807 for details regarding location, accessibility, escape chutes, etc. for this class of aircraft.

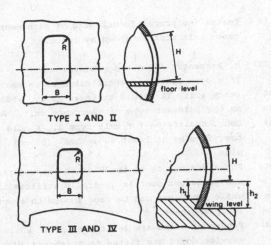

TYPE I AND II

floor level

TYPE III AND IV

wing level

Fig. 3-16.Classification of emergency exits

EMERGENCY EXIT CLASSIFICATION AND LOCATION		B (min) inches (mm)	H (min) inches (mm)	R (max) inches (mm)	MAX. HEIGHT OF STEP	
					inside (h_1) inches (mm)	outside (h_2) inches (mm)
I	FLOOR LEVEL	24 (610)	48 (1219)	$\frac{1}{3}$ B	-	-
II	FLOOR LEVEL	20 (508)	44 (1118)	$\frac{1}{3}$ B	10 (254)	17 (432)
	ABOVE WING					
III	ABOVE WING	20 (508)	36 (915)	$\frac{1}{3}$ B	20 (508)	27 (686)
IV	ABOVE WING	19 (483)	26 (661)	$\frac{1}{3}$ B	29 (737)	36 (914)

NOTE
dimensions defined in Fig. 3-16, according to
FAR 25.807.

Table 3-3.Classification of emergency exits

When it is impossible to place exits above
the wing, as in the case of high wing aircraft, an exit having at least the dimensions of type III should replace each exit
III and IV as shown in Table 3-4.
When these requirements are difficult to
fulfil, the exceptions listed in FAR 25
should be consulted. Additional requirements apply to aircraft which are to be
certificated for making emergency descents
on water (FAR 25.807 para. d). There should
be unobstructed access to emergency exits;

SEATING CAPACITY (EXCL. CABIN STAFF)	NUMBER OF EXITS REQUIRED ON EACH SIDE OF THE FUSELAGE			
	TYPE I	TYPE II	TYPE III	TYPE IV
1 through 10	-	-	-	1
11 through 19	-	-	1	-
20 through 39	-	1	-	1
40 through 59	1	-	1	-
60 through 79	1	-	1	-
80 through 109	1	-	1	1
110 through 139	2	-	1	-
140 through 179	2	-	2	-

NOTE
1. BCAR requirements are slightly different for
1-10 passengers; for this case an emergency exit
of type III is required on both sides of the
fuselage. Two exits of type I and II are required
for a seating capacity of 180 up to 219.
2. The relevant rules should be consulted where
passenger seats exceed this number and for
special regulations.
3. Exits need not be at locations diametrically
opposite each other. They should be located in
accordance with the passenger seating distribution.
4. Two exits of type IV may be used instead of each
type III exit.
5. The classification of emergency exits is defined
in Table 3-3 and Fig. 3-16.

Table 3-4.Minimum number of passenger emergency exits according to the FAR Part 25
requirements

the width is laid down in the BCAR require-
ments as 20 inches (51 cm) for types I and
II.
Reference 3-4 recommends the following
standards:
Type I   : 36 inches.
Type II   : 20 inches.
Type III and IV: 18 inches.

These distances determine the seat pitch
next to emergency exits and will therefore
affect the total cabin length.

b. Aircraft to be certificated according to FAR Part 23.

The following is quoted from FAR Part 23.807

"Number and location: Emergency exits must be located to allow escape in any probable crash attitude. The airplane must have at least the following emergency exits:
(1) For all airplanes, except airplanes with all engines mounted on the approximate centerline of the fuselage that have a seating capacity of five or less, at least one emergency exit on the opposite side of the cabin from the main door specified in para. 23.783.
(2) Reserved.
(3) If the pilot compartment is separated from the cabin by a door that is likely to block the pilot's escape in a minor crash, there must be an exit in the pilot's compartment. The number of exits required by subparagraph (1) of this paragraph must then be separately determined for the passenger compartment, using the seating capacity of that compartment.

Type and operation: Emergency exits must be movable windows, panels, or external doors, that provide a clear and unobstructed opening large enough to admit a 19-by-26 inch (483 x 660 mm) ellipse. In addition, each emergency exit must
(1) be readily accessible, requiring no exceptional agility to be used in emergencies,
(2) have a method of opening that is simple and obvious,
(3) be arranged and marked for easy location and operation, even in darkness,
(4) have reasonable provisions against jamming by fuselage deformation, and
(5) in the case of acrobatic category airplanes, allow each occupant to bail out quickly with parachutes at any speed between V_{S_o}* and V_D .

*V_{S_o} = stalling speed, flaps down;
V_D = design diving speed. .

Tests: The proper functioning of each emergency exit must be shown by tests."

c. Passenger doors and windows.
If a door is to be certificated as an emergency exit, it should at least be as wide as the relevant type of emergency exit. A door, qualifying for exit type A, should therefore be at least 42 inches (107 cm) wide.
For a maximum of up to 70 or 80 passengers, one passenger door is generally sufficient, while two doors can be used for up to about 200 passengers.
Passenger doors are located to port while service doors are fitted to starboard. Widebody jets are an exception as these can be boarded from both sides.
Doors should preferably be 6 ft high and 3 ft wide (1.80 x .90 m^2) but these dimensions are difficult to achieve in the case of smaller aircraft.
The window pitch is not always decided by the seat pitch, but frequently by the optimum distance between the fuselage formers. An average figure is 20 inches (.50 m) for the former and window pitch. In pressure cabins the windows are circular, rectangular with rounded corners, elliptical or oval in shape. The top of the window is roughly at the passenger's eye level.
In the case of smaller aircraft the main frames will have to be installed at the wing attachment points and these will influence the location of doors and windows. Access to passenger doors, service doors and panels should be unobstructed from both the outside and the inside. For example, the space between the wing and the fuselage-mounted engines should permit sufficient freedom for maneuvring food carts and cargo loaders.

3.2.5. Cargo holds

The design specification does not always stipulate the amount of cargo to be carried. Airlines may have radically different requirements, depending on the kind of traffic they carry, so the best way is to conduct an inquiry among potential customers. If there is no time for this, the method given below may provide a quick answer. This is based on the following assumptions:

1. Volume-limited and structure-limited payload are equal.
2. Weight of a passenger is 170 lb (77 kg), see BCAR Section D, ch. D 3.1 para. 3.4.
3. Luggage weight is 35 lb (16 kg) per passenger on short-haul flights and 40 lb (18 kg) on long-haul flights.
4. Loading efficiency is 85%, i.e. 15% of the space is lost.
5. Average density of cargo is 10 lb/cu.ft (160 kg/m^3) and of luggage 12.5 lb/cu.ft (200 kg/m^3).

Ignoring storage losses at the freight doors, the following expression can be derived:

Freight hold volume = .118 cu.ft per lb (.0074 m^3 per kg) of max. payload minus 20.8 cu.ft (.59 m^3) per passenger.

Alternatively, this expression can be used to obtain the volume-limited payload[*]:

Max. payload = 8.5 lb per cu.ft (136 kg per m^3) of freight hold volume plus 177 lb (80 kg) per passenger.

This yardstick may be used in the preliminary design stage but it cannot be applied to all aircraft. There is sometimes a space limit to the load, while the structure is strong enough to carry greater loading weights. In other cases, a limit is set by the difference between the Maximum Zero Fuel Weight and the Operational Empty Weight. This is generally an undesirable condition on civil aircraft. Belly freight holds should have an effective height of at least 20 inches (50 cm) but a height of more than 35 inches (90 cm) is to be preferred, particularly when it is necessary for staff to work in the hold. This condition cannot be

*This term is explained in Section 8.2.2.

satisfied for fuselage diameters of less than about 10 ft (3 m), so either a double-bubble fuselage cross-section will have to be adopted or the freight holds must be located above the floor.

To control the center of gravity travel it might be of advantage to keep the underfloor holds both ahead and behind the wing. Small twin-engined aircraft sometimes have a luggage hold in the (fuselage) nose ahead of the cockpit or in the engine nacelles. In pressurized airliners the freight and baggage holds are pressurized as well, though the temperature may be lower than in the cabin. They must be easily accessible by means of hatches or be located close to a door. When determining the volume required, allowance should be made for possible loss of space near the hatches.

In the case of very large aircraft, it is recommended that freight holds be designed to take the universal containers used in other wide-body jets; the relevant dimensions are given in Fig. 3-20.

3.2.6 Services

Although the airliners belonging to the IATA have come to certain arrangements regarding the service to be offered to the passenger, individual companies have varying ideas about this. Before starting the design of the cabin, the outcome of a separate study devoted to this subject should be obtained and be incorporated in the specification. An example of one of these studies is provided by Ref. 3-9.

a. Pantries, lavatories and wardrobes. The number and dimensions of the above facilities are shown in Table 3-5. The data are derived from standard type specifications and do not necessarily apply to individual users. Some flexibility in layout design should be incorporated.

Location: For aesthetic reasons toilets should preferably be located so that they are not directly visible from the pantry. They should be easily accessible, and when

Table 3-5 with column structure:

Aircraft type:	N_{pass}	Range N M	galleys number	galleys l x b (inch)	toilets number	toilets l x b (inch)	pass toilet	wardrobes number	wardrobes l x b (inch)
Aérospatiale N-262 Frégate	29	400	1	23 x 20	1	41 x 28	29	1	40 x 24
Grumman Gulfstream I	19	2100	1	34 x 25	1	67 x 37	19	1	36 x 32
Hawker Siddeley 748 srs 200	44	1000	1	37 x 14	1	53 x 35	44	-	
Fokker-VFW F.27 Friendship srs 200	48	1100	1	43 x 35	1	47 x 46	48	1	31 x 16
De Havilland Canada DHC-7	44	800	1	26 x 24	1	46 x 30	44	1	26 x 24
Lockheed L-188 Electra	95	2300	2	46 x 26	4	46 x 41	24	2	46 x 34
HFB 320 Hansajet	7	1000	1	24 x 24	1	30 x 26	7	1	24 x 15
Hawker Siddeley HS-125 srs 400	8	1450	-		1	35 x 28	8	1	24 x 12
Dassault Falcon 20.F	10	1500	1	27 x 18	1	44 x 30	10	1	51 x 25
Dassault Falcon 30/Mystère 40	34	750	-		1	41 x 31	34	-	
VFW-Fokker 614	40	700	1	35 x 28	1	55 x 32	40		65 x 40
Fokker VFW-F.28 Mk 1000	60	1025	1	44 x 25	1	58 x 25	60	1	25 x 21
BAC-111 srs 200/400	74	900	2	49 x 22	2	65 x 35	37	1	49 x 22
Mc Donnell Douglas DC-9 srs 10/20	80	1100	1	48 x 33	2	48 x 48	40	2	48 x 21
Boeing 737 srs 200	115	1800	1	55 x 43	2	43 x 34	58	1	55 x 43
Aérospatiale Caravelle 12	118	1000	1	51 x 43	2	55 x 43	59	2	24 x 17
Dassault Mercure	140	800	-		2	47 x 34	70	2	49 x 16
Boeing 727 series 200	163	1150	2	51 x 32	3	43 x 39	55	-	
Europlane	191	1400	3	42 x 42	4	42 x 42	48	1	52 x 26
A-300 B/4	295	1600	3		5	59 x 35	59	-	
Lockheed L-1011	330	2700	1	20 x 13.5 ft^2 under floor galley	7	45 x 36	47	-	head racks
Mc Donnell Douglas DC-10	380	3000	1	49 x 32	9	40 x 40	42	2	6.3 x 1.8 ft^2
BAC-VC-10	135	4200	1	79 x 47	5	47 x 41	27	2	42 x 24
Boeing 707-320 B	189	5000		48 x 34	4	40 x 37	48	1	79 x 43
Mc Donnell Douglas DC-8 srs 63	251	4000	2		5	42 x 42	50	4	34 x 20
Boeing 747	490	5000	4	6.6 x 2.1 ft^2	12	40 x 40	41	2	5.9 x 2.3 ft^2

NOTES:

N_{pass} = maximum number of passengers, tourist class, approx. 34" seat pitch

Range at about N_{pass} x 205 lb payload and including normal fuel reserves

Dimensions are approximate average length x width; toilets are not always rectangular.

Table 3-5.: Number and dimensions of galleys, lavatories and wardrobes of some airliners

the cabin arrangement includes a separate first-class section it is desirable to provide toilet facilities in that part too. Toilets are generally not movable since they form an integral part of the aircraft structure and require special provisions. Only limited flexibility is available with regard to the location and arrangement of galleys. It is advisable to locate these facilities at the forward and/or rearward end of the cabin, thus allowing for different cabin layouts. It is permitted to locate them in the plane of the propellers. In the case of wide-body jets, space may be saved by placing the pantry below the floor.

When servicing, loading and unloading the aircraft between flights, the following operations are performed:
- replenish potable water,
- remove left-over food, drinks and waste

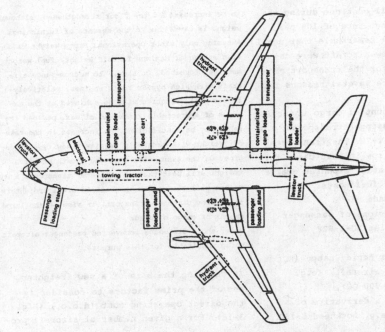

Fig. 3-17. Ground handling of the Lockheed L-1011 Tristar

from the pantries and take on fres supplies,
- service toilets,
- clean cabin,
- unload and load passengers freight and
  luggage.
The trucks, stairs, carts, loaders, etc.
needed for these operations should not ob-
struct each other, which means that careful planning of door locations and service
points is required, particularly in the case
of large aircraft. An example is given in
Fig. 3-17.

b. Cabin systems.
Among the facilities which should be available in every passenger aircraft are a public address system, lighting, cooling air
(operated by the passenger), hatracks over
the seats and space for hand luggage. A
supplementary oxygen supply is compulsory
when cruising height is above 25,000 ft
(7620 meters), cf. FAR 91.32.

c. Cabin staff.
The minimum number of flight attendants is
specified by the airworthiness regulations

(e.g. FAR 91.215); the actual number of cabin staff is fixed by the company. The following data from Ref. 3-9 give the average
number of passengers per member of cabin
staff:

	first class	mixed	tourist class
International sched- uled flights	16	21	31
U.S.A. domestic flights	20*	29	36
Other domestic flights	21	-	39

At least one folding chair is placed at
each exit for members of the cabin staff.
This should permit a good view into the
passenger cabin.
*revised number

3.3. THE FUSELAGE OF CARGO AIRCRAFT

3.3.1. The case for the civil freighter

During the sixties the transport of air
freight has shown a very rapid annual growth
of the order of 19% in terms of ton-miles
carried per year. According to ICAO projec-

tions this growth will continue during the seventies at an average rate of 16% per year. It is therefore remarkable that until recently only very few aircraft were designed specifically for the transport of air freight. There are several reasons for this.

a. A considerable amount of cargo is carried in the bellies of passenger transports, e.g. approximately 60% in 1970. The transportation costs are quite low in that case, for the extra direct operating expenses emerge mainly in the form of fuel costs.

b. Extensive use is made of
- special freight versions of passenger transports (e.g. Douglas DC-8-62F, Boeing 707/320 C),
- Quick Change (QC) or Rapid Change (RC) versions of passenger aircraft, (e.g. DC-9-30RC, Boeing 727-200 QC),
- civil freighters as a derivative of military freighters (e.g. Lockheed C-130 and L100/L200),
- obsolete passenger transports, converted into freighters (Douglas DC-6, Lockheed L-1049).

Provided the growth of air freight continues on the same lines in the coming years, an expanding market can be expected for new freighter aircraft. The following arguments favor such a development.

a. Most converted passenger transports have loading doors in the side of the fuselage. In view of the increasing popularity of 8 ft x 8 ft containers, very large doors might be required in the future, while most passenger cabins, except on the widebody jets, are not suitable for these sizes. It may prove prohibitive to design the cabins of short-to-medium haul aircraft especially for container transport in view of the cost penalty involved.

b. The average density of freight is considerably higher than the payload density of a passenger cabin. The difference is likely to increase in the future, as passenger comfort will be improved, while freight densities are tending to increase. Consequently, the passenger transport converted into a freighter will have a payload that is 1.5 to two times as large. The Max Zero Fuel Weight* will have

*This term is explained in Section 8.2.2.

82

to be increased and the floor strengthened, although weight is saved due to the absence of furnishings. Assuming only minor Operational Empty Weight reduction and equal Maximum Takeoff Weight, fuel weight must be decreased. On short- to medium-range aircraft with high bypass ratio engines, relatively little fuel reduction will be achieved at the expense of appreciable range and at max. payload the range may be insufficient. An increase in the takeoff weight will generally require a new type of engine. On the assumption that the growth of air freight will beat that of passenger transport, a new market will emerge for specialized civil freighters,
- for long distance transport, in view of the large amount of cargo offered;
- for short distances, converted passenger aircraft being unsuitable for this purpose.

In choosing the size of a new freighter, one of the prime factors to consider is the direct operating cost (d.o.c.) (Fig. 3-18). For a given number of aircraft pro-

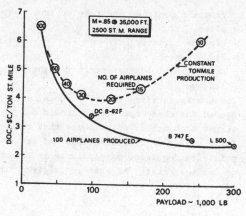

Fig. 3-18. Effect of freighter size on direct operating costs

duced, the d.o.c. will decrease with airplane size. The reasons are mainly the lower fuel cost per lb of air freight, the decreasing cost of flying staff, systems and maintenance. On the other hand, it will be extremely difficult, if not impossible, to sell 100 aircraft of the size of the L-500*.

*A civil version of the C-5A (project).

Therefore, assuming a market share in the form of a constant ton-mile production, the result will be an optimum size. Incidentally, in the example presented here, the specialized freighter cannot compete with freighter versions of passenger transports.

3.3.2. Payload density and volume of the freight hold

The following factors may play a part in deciding whether to ship goods by air or by surface transport:
a. Fast transport may prevent decay or depreciation. Examples are: foodstuffs, fresh vegetables, fruit, cut flowers, certain types of animals. Some highly valuable goods and expensive instruments are ideal items for air cargo.
b. Rapid distribution of an article may increase its sales potential (e.g. newspapers) or enhance a service (airmail).
c. Air transport may lead to a reduction in storage space and capital investment in spares. It is more advantageous to ship certain goods by sea, the ship sometimes also serving as storage space. These are generally goods with a high specific density.
d. Transport of spare parts, modified products or new models may be important in the case of a hold-up in a production line. Goods whose value depends on fashion will generally be sent by air.
e. Packing costs for air freight are sometimes considerably lower than those for

surface transport. Less damage is incurred during loading and unloading, particularly when containers are used. This results in lower insurance rates.
f. Isolated regions will be difficult to reach owing to time-consuming surface transport. In such cases air transport will be the only means of meeting their requirements with regard to medical supplies and perishable foods, etc.

From histograms giving dimensions and densities (examples in Refs. 3-11 to 3-14, see also Fig. 3-19) it can be seen that freight presents a wide variety of characteristics. Processing on the airfield as well as in the aircraft demands more equipment and manpower and is considerably more costly than the transport of passengers.
It is possible to imagine the optimum case when the freight hold is completely filled, while at the same time the payload is maximum. However, when loads of a typically high density are carried, the hold will be only partly filled, with the result that the unused empty space will increase the drag and weight. With low density loads, however, there will be less than the permissible weight in freight and the aircraft will in some way be excessively strong. A proper mean can only be found on the basis of detailed data concerning the dimensions and weights of the goods to be carried and these should be supplied by potential customers. In some cases a few preliminary layouts are made, using fuselages suited to

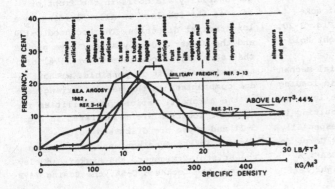

Fig. 3-19. Histograms of specific density of freight

different load densities in order to see
which choice will maximize the revenue. Although most histograms show a peak at 10 to
12 lb/cu.ft (190 kg/m^3), goods of higher
density are offered at widely varying frequency. Ref. 3-12 recommends that the volume of the hold should be determined by
means of the following data which apply to
bulk transportation of goods.

$$\text{Net volume} = \frac{\text{density reserve}}{\text{av. density of freight offered}} \times$$

$$\times \frac{\text{max. payload} \times \text{average freight percentage}}{\text{stowage factor}}$$

The average freight percentage, comparable
to the average load factor for passenger
transports, may be put at .65. The reserve
density magnitude 1.20 to 1.30 is required
to allow for freight having a lower density
than the average.

The stowage factor represents the percentage of usable space in relation to the net
volume. This allows for space losses for
storage nets, clearance between cargo and
structure, room for inspection, etc. Depending on the fuselage shape and the particular nature of the freight, this factor
may vary between .70 and .85. Combining
these data, the result will be a space-
limited payload in the case of average and
low densities. For a density of 15 - 25
per cent higher than the average, the maximum payload capacity will be obtained.

3.3.3. Loading systems

The use of pallets and containers has progressively increased in recent years. The
characteristics and dimensions of some
standard sizes, as presented in Fig. 3-20,
form the starting-point for freight hold
design.

Modern civil airliners have special mechanical loading systems for the rapid loading
and unloading of standard size pallets and
containers. These systems are resulting in
increased aircraft utilization, especially
on short-haul routes. The Douglas Corp. developed the 463L system for the USAF, a
complete system for handling freight both

on the ground and in the aircraft itself.
It is being used in the Lockheed C-130,
C-133, C-135, C-141 and C-5A aircraft. The
system utilizes 88 x 108 inch pallets and
trailers with a platform, 20 ft wide and
48 ft long, provided with rollers and adjustable to heights between 40 inches and
156 inches in order to fit various aircraft
floor levels. Roller conveyor strips are
mounted on the fuselage floor and the rails
used to guide the pallets can be adjusted
to various standard sizes. Similar to this
is the Rolamat system used in the Hawker
Siddeley Argosy (Fig. 3-21), which is designed to increase the loading capacity to
4,000 lb per minute. Latching points in the
floor are present to secure the load by
means of nets and ties. When these nets are
designed for normal loads only, a strong
net is required in front of the freight
hold to catch the load in case of a 9g
deceleration. Frequently, however, containers and pallet nets are designed to cope
with this load and the catching net is not
required.

In the case of combined transportation of
passengers and freight on the same floor,
it is advisable to locate the freight hold
in front of the passenger cabin. A passageway between the cockpit and the passenger
cabin is then required.

3.3.4. Accessibility of the freight hold

Although many passenger aircraft which have
been converted into freighters have side
doors, a pure freighter should have better
accessibility via doors in the front or
rear of the fuselage to allow loading in a
longitudinal direction. The Lockheed C-130
and the Hawker Siddeley Argosy have proved
that a readily accessible floor level of
approximately 4 ft is possible, without undue compromises in the general arrangement
of the aircraft. Various possibilities for
the door location are illustrated in Fig.
3-21 and these are discussed below.
a. Door in the fuselage nose, as used on
the Bristol Freighter, the Hawker Siddeley
Argosy, the Lockheed C-5A, the Boeing 747F

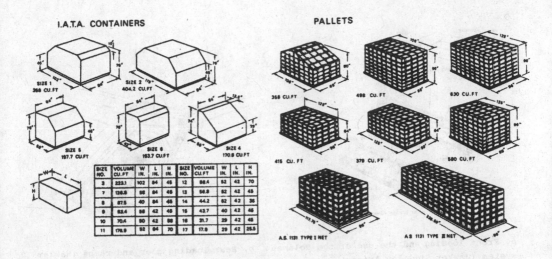

I.A.T.A. CONTAINERS

SIZE 1 358 CU.FT
SIZE 2 404.2 CU.FT
SIZE 5 197.7 CU.FT
SIZE 6 163.7 CU.FT
SIZE 4 170.8 CU.FT

SIZE NO.	VOLUME CU.FT	W IN.	L IN.	H IN.	SIZE NO.	VOLUME CU.FT	W IN.	L IN.	H IN.
3	223.1	102	84	45	12	88.4	52	42	70
7	128.5	88	84	45	13	56.9	52	42	43
8	87.5	40	84	45	14	44.2	52	42	36
9	83.4	86	42	45	15	43.7	40	42	45
10	70.4	90	43	88	16	31.7	29	42	45
11	178.9	52	84	70	17	17.9	29	42	25.5

PALLETS

358 CU.FT
498 CU.FT
630 CU.FT
415 CU.FT
379 CU.FT
580 CU.FT
A.S. 1131 TYPE I NET
A.S. 1131 TYPE II NET

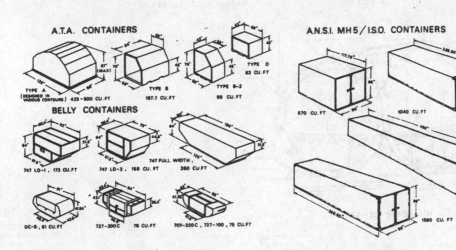

A.T.A. CONTAINERS

TYPE A (DESIGNED IN VARIOUS CONTOURS) 425-500 CU.FT
TYPE B 197.7 CU.FT
TYPE B-2 99 CU.FT
TYPE D 63 CU.FT

BELLY CONTAINERS

747 LD-1, 173 CU.FT
747 LD-3, 158 CU.FT
747 FULL WIDTH, 350 CU.FT
DC-8, 81 CU.FT
727-200C, 76 CU.FT
707-320C, 727-100, 78 CU.FT

A.N.S.I. MH5 / I.S.O. CONTAINERS

570 CU.FT
1040 CU.FT
2090 CU.FT
1560 CU.FT

pallet weights (nets included)
88" x 108": 220-230 lb (100-104 kg)
88" x 125": 255-270 lb (116-122 kg)
96" x 125": 285 lb (129 kg)

container (8'x8'x10') weight: 1000 lb (454 kg)

Fig. 3-20. Some standard pallets and containers

S.A.E. CONTAINERS

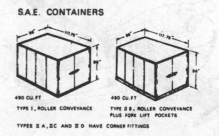

490 CU.FT
TYPE I, ROLLER CONVEYANCE

490 CU.FT
TYPE II B, ROLLER CONVEYANCE PLUS FORK LIFT POCKETS

TYPES II A, II C AND II D HAVE CORNER FITTINGS

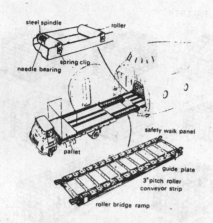

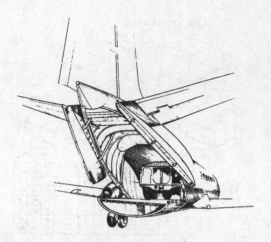

a. Front loading and the use of the Rolamat system (Hawker Siddeley Argosy)

b. Rear loading door and ramps (Hawker Siddeley Andover)

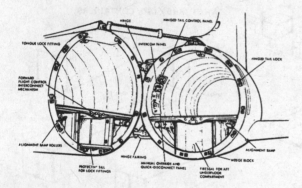

c. Swing-tail (Canadair CL-44D)

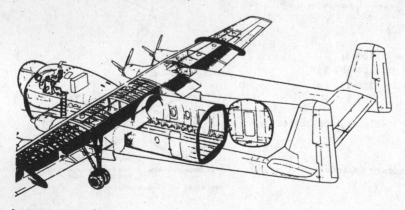

d. Rear loading door on aircraft with tailbooms

Fig. 3-21. Accessibility and loading of freighter aircraft

and the Aviation Traders Carvair. The major problem is to avoid the considerable drag caused by the high cockpit, which is difficult to avoid on relatively small aircraft.
b. Door in the fuselage tail, as used on the Short Skyvan, Transall C-160, Hawker Siddeley Andover, De Havilland Caribou and Buffalo, Short Belfast, Breguet 941, Lockheed C-130, C-141 and C-5A. For easy access, especially in small freighters, it is essential to camber the fuselage tail upwards, thus creating an aerodynamic problem (Section 3.5.1). The fuselage weight penalty is of the order of 6 to 10%, depending on the structural details. The zero-lift drag increment is of a similar order of magnitude, but may equally well be much higher. The door size is relatively large and it may become difficult to seal the pressure cabin.
c. Tail boom layout, in combination with a door in the rear part of a stubby fuselage. This configuration, occasionally seen on freighter aircraft (H.S. Argosy, Noratlas, IAI Arava, Fairchild C-82 and C-119), offers a readily accessible freight hold and permits the use of a beaver tail for dropping purposes, if required. The high aerodynamic drag is a disadvantage.
d. Swing-tail, a layout proposed by Folland in a freighter project as far back as 1922. To date, the swing-tail has been implemented only on the Canadair CL-44, at the expense of a penalty of some 1,000 lb (450 kg) structure weight relative to a side door, i.e. about 6½% of the fuselage structure weight. From an aerodynamic standpoint, however, the swing-tail is ideal and the structural complexity may be outweighed by a considerable reduction in fuel comsumption.
e. A swinging fuselage nose (including the flight deck) creates considerable difficulties in carrying through cables, wires, plumbing, etc. The weight penalty is of the order of 12% of the fuselage structure weight. Its use may be considered in very special cases (e.g. the Guppy family).

The maximum floor loading of freighters for an evenly distributed load must be at least:

- civil   : 125-300 lb/sq.ft (600-1,500 kg/m^2)
- military: 225-1,200 lb/sq.ft (1,100-6,000 kg/m^2)

Design criteria for local loads are:
- civil   : 3,500-9,000 lb (1,600-4,000 kg)
- military: 3,000-10,000 lb (1,300-4,500 kg)

Door sizes must be adapted to the type of freight to be loaded and unloaded, and in the case of loading in the longitudinal direction a clearance of at least one inch (2.5 cm) must be present on both sides. The freight hold ceiling must be at least 6 inches (15 cm) above the freight for ease of loading. The need for a passageway through the loaded freight hold depends on the type of freight carried. Inspection during the flight is not always necessary in the case of containerized freight. For preference the freight hold should be prismatic in shape; steps in the floor are not acceptable, except in very special cases. A separate door for the cockpit crew is necessary. Several windows are usually incorporated in the fuselage walls. For civil freighters it may be useful to consider a passenger version (convertible freighters) and in that case more windows and passenger doors are required.

3.4. FLIGHT DECK DESIGN

3.4.1. Location of the pilot's seat and the flight controls

On light aircraft the cockpit may, to some extent, be arranged in line with the particular design requirements. This applies particularly to the location of foot pedals in the vertical direction, as this factor affects the cross-sectional height and hence the fuselage frontal and wetted areas. The pedals must be placed below the level of the seat bottom to avoid tiring the pilot.
Instructions for the location of the seat and stick controls are presented in Fig.3-22.

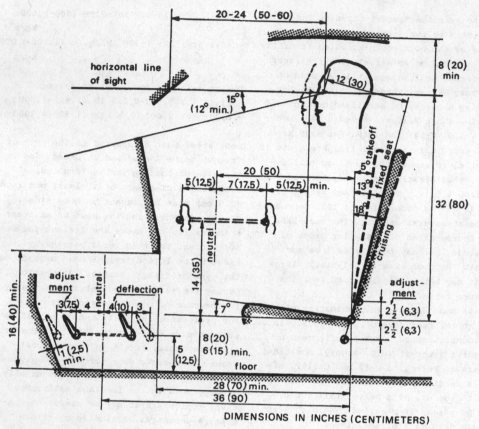

DIMENSIONS IN INCHES (CENTIMETERS)

NOTES:

1. Distance between foot pedal centerlines: 8 inches (20 cm) minimum, 12 inches (30 cm) maximum.

2. The indicated floor is a reference line; the actual floor need not be horizontal. Only the local height of the foot pedal relative to the floor is of importance.

3. For many light aircraft the seatback has a fixed position. The recommended setting relative to the vertical is 13°.

REFERENCES:

1. F. Maccabee: "Light aircraft design handbook". Loughborough University of Technology, Feb. 1969.

2. Draft ISO recommendation No. 1558, 1973.

3. Mil. Standard MS 33574.

Fig. 3-22. Recommended dimensions for the cockpit of a light aircraft with stick control

For a control-wheel layout it is advisable to use the data for transport aircraft. Generally speaking, the outside view from the cockpit is only obstructed by the wing and no special measures are necessary. The downward view forward is determined by the instrument panel, the glare shield, the fuselage nose or the engine cowling. Part of the cockpit roof of light aircraft can be of a light-alloy construction to improve stiffness and strength and to provide protection against sunlight.

Particularly on transport aircraft, more is required than merely the convenient location of the flight controls and instruments. The position of the pilot relative to the cockpit windows, and the window shape, are equally important. Pilots of varying body dimensions must feel at ease in the cockpit and be able to take up a position from which a clear outside view is possible. A design aid which is usually employed here is the reference eye point. This is a fixed point chosen by the designer in the aircraft, which serves as a reference for defining both the outside view and the seat position. It is defined as follows:
(1) The reference eye point must be located not less than five inches aft of the rearmost extremity of the primary longitudinal control column when the control is in its most rearward position (i.e. against the elevator up stops).
(2) The reference eye point must be located between two vertical longitudinal planes which are one inch to either side of the seat centerline.
(3) Any person from 5'4'' (1.63 m) to 6'3'' (1.91 m) tall, sitting in the seat must be able to adjust the seat with the seat back in the upright position, so as to locate the midpoint between his eyes at the reference eye position. With the seat belts fastened, he must also be able to operate the aircraft controls with lap strap and shoulder harness fastened.
In the proposed para. FAR 25.777 dated Jan. 12, 1971, a requirement is laid down with respect to the seat adjustment relative to

a position of the seat bottom located 31½ inches below the reference eye point. The Society of Automotive Engineers (SAE) and the International Standardization Organization (ISO) have made recommendations (e.g. Ref. 3-24) for the standardization of other dimensions of the cockpit. A condensed version of the various proposals is presented in Fig. 3-23, which may serve as a starting-point for the cockpit design or mock-up design.
On most transport aircraft the crew seat position can be adjusted horizontally and vertically, while the seat back is reclinable. In some cases seat rails extend far back and/or allow sideways displacement to facilitate easy access/egress or to permit crew members to take up a position where other controls can be handled or panels read off. The seatback in its upright position is used for takeoff and landing; on short trips this position is generally not changed. The cruise position is used when the autopilot is operative.

3.4.2. Visibility from the cockpit

During VFR flights the pilot must have a clear view of such a part of the air space that he has adequate information to control the flight path and avoid collisions with other aircraft or obstacles. For design purposes this general requirement can be evaluated in the form of minimum angles of vision during cruising flight, takeoff, landing and taxying.

a. Horizontal flight (Fig. 3-24).
To define clear areas of vision, binocular vision and azimuthal movement of the head and eyes are assumed to take place about a radius, the center of which is the central axis. The areas of vision are measured from the eye position with the airplane longitudinal axis horizontal. For example, in level flight, with the pilot looking straight ahead from the reference eye position, clear vision must be possible up to 17° downward and 20° upward. The complete envelope of the clear areas of vision is given

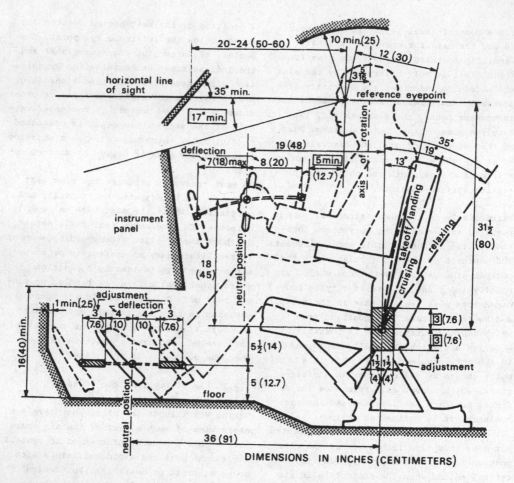

DIMENSIONS IN INCHES (CENTIMETERS)

NOTES:

1. Distance between the centerlines of both seats: see Table 3-6.

2. Distance between the centerlines of the foot pedals: 14 inches (35 cm).

3. Most dimensions can be chosen within wide ranges, except the framed ones: these are specified in the rule proposed in FAR 25.772.

4. The indicated floor is a reference line; frequently a footrest is used.

REFERENCES:

1. FAR 25.772 (proposed), dated Jan. 12, 1971.

2. Draft ISO recommendation 1558, 1973.

3. Mil. Standard MS 33576.

Fig. 3-23. Recommended flight deck dimensions for transport aircraft with wheel controls.

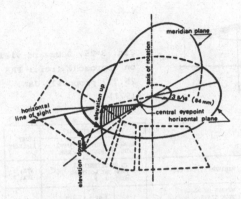

a. Definition of the pilot's view

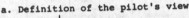
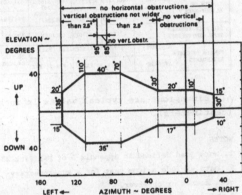

b. Minimum required clear areas of vision

Fig. 3-24. Vision from the pilot's (port) seat in horizontal flight (Ref.: FAR Part 25.777 proposal, Jan. 1971)

in Fig. 3-24b, where areas are also indicated where no obstructions may impair the pilot's vision. This determines the location of windshield posts, instruments and other cockpit equipment. Areas are also indicated where windshield posts of limited width are considered to be acceptable.

b. Visibility during approach (transport category).
In the case of modern transport aircraft, considerable variations can be observed in the airplane attitude during low-speed flight. These are caused by great differences in wing aspect ratio, angle of sweep and type of high-lift devices. Accordingly,

standards must be evolved for this category of aircraft to ensure clear areas of vision during the approach. The angle of view forward and downward must be sufficient to allow the pilot to see the approach and/or touchdown zone lights over a distance equal to the distance covered in 3 seconds at the landing speed when the aircraft is

(1) on a $2\frac{1}{2}^{\circ}$ glide slope,

(2) at a decision height which places the lowest part of the aircraft at a height of 100 feet above the touchdown zone (see Fig. 3-25),

(3) yawing $\pm 10^{\circ}$,

(4) making an approach with 1,200 feet Runway Visual Range, and

(5) loaded to the most critical weight and center of gravity location.

In the British requirement BCAR Appendix No. 2 to Chapter D4-2 some additional stipulations are made:
(1) When taxying, the pilot should be able to see the ground at a maximum of 130 ft from the airplane, but preferably this distance should be 50 ft or less.
(2) When climbing, the pilot should be able to see at least 10° below the horizon and preferably $15-20^{\circ}$ below it.
(3) When landing, the pilot should be able to see below the horizontal when the airplane is in the tail-down attitude.
Another desirable feature is that during taxying the pilot should be able to see the wingtip on his side of the airplane.

When all these requirements have to be incorporated in cockpit design, the designer of high-speed aircraft with a pressurized fuselage may run into considerable trouble. Unacceptable deformation of the fuselage contour, high drag penalties and unacceptable noise levels may be the result. Therefore most transport aircraft do not completely meet all requirements, but nevertheless these should be used as a starting-point for crew compartment design.

3.4.3. Flight deck dimensions and layout

The minimum number of flight crew is based on the total work load, consisting of the

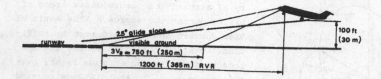

Fig. 3-25. Downward view on approach. (ref.: FAR 25.777 proposal, Jan. 1971)

following activities:

(1) Flight path control.

(2) Collision avoidance.

(3) Navigation.

(4) Maintaining contact with Air Traffic Control Centers.

(5) Operation and supervision of systems.

(6) Taking decisions concerning the execution of the flight.

The total work load is affected by the duration of the flight, the degree of automation and complication of the systems and the operational limitations. Accordingly, the data in Fig. 3-26 and Table 3-6 relating

	TRANSPORT AIRCRAFT			LIGHT AIRCRAFT
	LONG HAUL	MEDIUM HAUL	SHORT HAUL	
MINIMUM FLIGHT CREW	TO BE DETERMINED FROM THE WORKLOAD[1], MINIMUM: 2			VFR: 1 IFR: 2
NUMBER OF FLIGHT DECK SEATS	4	3 or 4	2 or 3[3]	2
LENGTH OF FLIGHT DECK[4] MINIMUM	140(355)	125(317)	LOW-SUBS. 90(228)	63(160)
AVERAGE	150(380)	130(330)	HIGH-SUBS. 105(267)	70(178)
DISTANCE BETWEEN SEAT CENTERLINES	42(107)	42(107)	40(102)	30(76)[5]
NUMBER OF CABIN ATTENDANTS MINIMUM	1 PER 50 PASSENGERS (PAX)			1 FOR 20 PAX OR MORE
AVERAGE	1 PER 30 PAX		1 PER 35 PAX	

All dimensions are typical values in inches (centimeters).

NOTES

1. Work load defined in Appendix D of FAR Part 25.

2. Data exclude a jump seat for a supernumerary crew member or observer.

3. According to an old rule a flight engineer must be present for a max. takeoff weight above 80,000 lb (36,300 kg).

4. Definition in Fig. 3-26. Space for electronics rack included on transport aircraft, excluded on light aircraft.

5. This figure varies widely; it is affected primarily by the external fuselage width.

References

FAR Part 23.1523 and 25.1523; Appendix D to Part 25;

FAR Operating Rules 91.211; 91.213; 91.215 and 121.385 through 121.391.

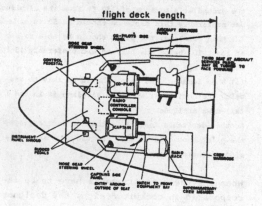

Fig. 3-26. Crew cabin layout for a medium-range transport

to the number of seats are statistical figures and obviously not standard requirements.

Transport aircraft must have duplicated flight controls and must be operated by at least two pilots. Short-to medium-range aircraft can be operated by two crew members. However, long-range and several medium-range aircraft require a third crew member because of the generally long duration of

Table 3-6. Statistical data on number of crew members and flight deck dimensions

the flights and the complexity of the systems installed. A control panel and a seat are then provided for a flight engineer or third pilot (systems operator). Fig. 3-26 shows that a fairly large flight deck is required in long-range airplanes. Adequate space must be provided so that crew members can stow their baggage, coats, etc. in or adjacent to the flight deck.

The flight deck accommodation in general aviation aircraft is generally limited, so that the length of the flight deck is not more than 5 ft (1.5 m) for small touring aircraft, and up to about 6 ft (1.8 m) for business aircraft.

3.4.4. Emergency exits for crew members (Ref. FAR 25.805 and BCAR Section D para. 5.2.1 of chapter D 4-3)

The following requirements are quoted from the FAR-regulations:
"Except for airplanes with a passenger capacity of 20 or less, in which the proximity of passenger emergency exits to the flight crew area offers a convenient and readily accessible means of evacuation for the flight crew, the following apply:
(a) There must be either one exit on each side of the airplane or a top hatch, in the flight crew area.
(b) Each exit must be of sufficient size and must be located so as to allow rapid evacuation of the crew. An exit size and shape of other than at least 19 by 20 inches (482 x 508 mm) unobstructed rectangular opening may be used only if exit utility is satisfactorily shown, by a typical flight crewmember, to the Administrator".

3.5. SOME REMARKS CONCERNING THE EXTERNAL SHAPE

3.5.1. Fuselages with a cylindrical mid-section

The following applies to the fuselage nose, i.e. the non-cylindrical front part of the fuselage.
a. A frequently used value for the length/diameter ratio is 1.5 to 2.0. A lower value may be used on freighters provided that this

lightens the door and door support structure to such an extent that it outweighs the extra drag.
b. All passenger transports and many high-speed general aviation aircraft have a radar installation, for which a reflector must be planned in the nose section.
c. It may be advantageous to locate the nose gear in front of the forward pressure bulkhead: in that case the wheelbay has no pressure walls.
d. On small aircraft the fuselage nose can be used to contain Nav/Com equipment and/or luggage. In the case of piston engines this may lead to a forward location of the center of gravity and the wing can be so located that the propellers are in front of the cockpit. The accessibility of such a nose bay on the ground is generally quite satisfactory.

The following hints are pertinent to the fuselage tail, i.e. the non-cylindrical rear part.
a. To avoid large regions of boundary layer separation and the associated drag increments, the tail length is usually 2.5 to 3 times the diameter of the cylindrical section. For a tail boom configuration a slenderness ratio of 1.2 to 1.5 may be acceptable, provided that the weight of the fuselage and door structure can be reduced.
b. For ease of production, part of the fuselage tail may be conical; half the top angle of this cone should be 10° to 11°, or at most 12°. The transition between the cone and the cylinder ought to be smooth with sufficiently large radii of curvature.
c. Tail cross-sections may approximate circles or standing ellipses in shape. Beavertails have unfavorable drag characteristics and should be avoided on civil aircraft.
d. During takeoff and landing the fuselage tail must clear the ground under normal operating conditions.
e. There is usually plenty of space in the fuselage tail to contain the A.P.U. and/or the air conditioning system, provided that the position of the center of gravity will permit this location. If a central engine

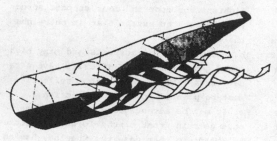

a. Schematic drawing of flow separation and vortex shedding from a rear-loading fuselage (Ref.: NCR Aeron. Report LR-395)

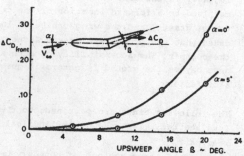

b. Drag increment vs. upsweep angle (Ref. 3-26)

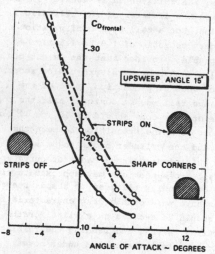

c. Effect of cross-sectional shape on drag (Ref. 3-27)

Fig. 3-27. Flow phenomena around cambered rear fuselages

is present, the minimum tail length may be determined by the allowable curvature of the intake duct. This situation may be improved by locating the engine fairly high.

Fillets
Where the wing is connected to the fuselage, too much divergence in the airflow must be avoided. Some form of filleting is required but its exact shape must be determined by means of windtunnel experiments. Some examples of fillets are shown in Fig. 2-9.

Cambered fuselage tail
The rear part of the fuselage is often slightly upswept in order to obtain the required rotation angle during takeoff or landing. The drag resulting from this slight camber is negligible. However, on freight aircraft with a rear loading door the fuselage must be swept up over a considerable angle, especially on small freighters like the De Havilland Caribou and Buffalo. Adverse interference may occur in the flow fields induced by the wing (downwash), the wheel fairings and the rear fuselage. The formation of vortices below the rear part of the fuselage is shown in Fig. 3-27a. These vortices are unstable and can cause lateral oscillation, especially at low speeds, high power, and high flap deflection angles. A considerable drag penalty in cruising flight is also caused by a large fuselage camber (Fig. 3-27b). Sharp corners on the lower part of the fuselage may relieve the problem by generating stable vortices, inducing upwash below the fuselage and thereby creating attached flow. Measurements (Fig. 3-27c) have shown that the drag penalty can be limited to reasonable values. References 3-26 and 3-27 give more detailed descriptions of the aerodynamic phenomena involved.

3.5.2. Fuselages for relatively small useful loads

Several considerations outlined in the previous section also apply to this category, together with the following:

a. If an engine is mounted in the fuselage nose, the required downward view of the pilot (Fig. 3-22) and the required propeller-to-ground clearance determine the vertical level of the engine.

b. Allow sufficient width in the fuselage around the rudder pedals and for the pilot's shoulders and elbows.

c. Avoid sharp changes of cross-sectional area, as well as discontinuities in the radius of curvature in the longitudinal direction. The fuselage should not be tapered in the region where the wing is attached, for this will entail the use of large fillets.

d. The fuselage tail length is determined by the tailplane moment arm required. A reasonable value for the distance between the wing and horizontal tailplane quarter-chord points* is 2.5 to 3 times the wing MAC*.

e. The details of the external lines are affected by the type of structure. A design sketch of the structural concept should be made at an early stage. In the case of welded frames, the fuselage sides will be flat and not curved like the panels used on semi-monocoque fuselages.

f. For ease of production, a substantial part of the fuselage should have single curvature.

g. In the case of a jet engine (or engines) buried in the fuselage, attention must be paid to the possibility of removing the engine(s) for major overhauls and to ensuring their accessibility for inspection.

*Definitions in Appendix A Section A-3.3.

Chapter 4. An appreciation of subsonic engine technology

SUMMARY

This survey presents some background information which is required when an engine has to be chosen for a new subsonic aircraft design.

The first chapter compares piston and turbo-engines and their principal applications. The second chapter presents a survey of the characteristics, possible applications and performance of piston engines with a power rating of up to about 500 h.p. Various engine configurations are discussed and a generalized method is given for estimating the take-off power and weight of the piston engine.

Single flow, bypass turbojet engines and turboprop engines are compared, taking a division between the gas generator and the propulsive device as the point of departure. An explanation of the significance of the thermal efficiency and propulsive efficiency, as well as specific fuel consumption, specific thrust and power rating, is also given.

The influence of the compressor pressure ratio, turbine inlet temperature and bypass ratio on engine performance, general configuration, weight, drag and external engine noise is discussed on the basis of generalized data. Some attention is paid to possible future developments.

NOMENCLATURE

A_e — exhaust nozzle area

BHP — Brake Horse Power (P_{br})

BMEP — Brake Mean Effective Power

BPR — By Pass Ratio

C_T — specific fuel consumption of turbojet engine

C_P — specific fuel consumption of propeller engine

c_p — specific heat of engine air at constant pressure

D — drag

FPR — Fan Pressure Ratio (= ε_f)

G — gas generator function

g — acceleration due to gravity

H — heating value of fuel

h — enthalpy; altitude

IMEP — Indicated Mean Effective Pressure

ISA — ICAO Standard Atmosphere

K — ratio for estimating engine weight

k — constant of proportionality

M_o — flight Mach number

METO — Maximum Except Take-Off (power)

$\dot{m}$ — mass flow per unit time

N_{cyl} — number of cylinders per engine

n — engine rotation speed; exponent of V in polytropic process; exponent of V_{cyl}

OPR — Overall Pressure Ratio (= ε_c)

P — power

P_{br} — brake horsepower

$P_{g_{is}}$ — convertible energy, generated by gasifier

P_{to} — static takeoff power

p — (static) pressure

P_o — ambient (static) pressure

P_t — total (stagnation) pressure

R — gas constant; ratio for estimating piston engine takeoff power

rpm — revolutions per minute

SFC — Specific Fuel Consumption (C_T or C_P)

shp — shaft horsepower

T — (static) temperature; thrust

T_o — ambient temperature

T_s — standard ambient temperature

T_t — total (stagnation) temperature

TET,TIT — Turbine Entry (Intake) total Temperature (T_{t_4})

V — specific volume of a charge (piston engine)

V_o — flight speed

V_{cyl} — total swept cylinder volume per engine

v — velocity of fully expanded exhaust flow

W_e — dry engine weight

$\dot{W}$ — weight flow per unit time (no index: engine air)

γ — ratio of specific heats of air

δ_m — relative static pressure = static pressure/ambient pressure

Δ — increment

ε — compression ratio (piston engine); pressure ratio (gasturbine engine)

η — efficiency

λ — bypass ratio

ρ — atmospheric density

σ — relative (atmospheric) density

ϕ — non-dimensional TET

INDICES

B — combustion chamber

c — (high pressure) compressor

cr — cruising flight

cyl — cylinder(s)

d — intake duct

e — engine (installation); exhaust opening

F — fuel

f — fan

g — gas generator; gearing

i — fuel injection

id — ideal (thrust definition)

j — jet

m — manifold inlet

mech — mechanical transmission (gearbox)

n — rotational speed

p — propulsive; pressure

prop — propeller

r — ram effect

s — supercharging

sn — standard net (thrust definition)

t — turbine

tf — combination turbine-fan

to — takeoff

tot — overall (efficiency definition)

4.1. INTRODUCTORY COMPARISON OF ENGINE TYPES

Engine types suitable for use on subsonic aircraft are:
- piston engines
- turboprop engines | propeller-driven aircraft
- single-flow jet engines (straight jet engines)
- bypass jet engines (turbofans) | jet-propelled aircraft

As far as subsonic aircraft are concerned, rocket engines and ramjets can only be regarded as suitable for particular applications, e.g. when additional thrust is required for a short period of time. These engines will not be discussed in this chapter, but the interested reader can find relevant information in the various textbooks mentioned in the list of references.

The type of engine suitable for a particular aircraft design is mainly determined by the following considerations:

a. Flight envelope.
The range of normal flying speeds at which the aircraft will operate (the flight envelope). The propeller operates at a high propulsive efficiency up to M = .5 to .6, after which compressibility phenomena at the tips will cause a considerable loss in efficiency. For higher speeds only the jet engine may be regarded as a suitable means of propulsion.

b. Fuel consumption.
Fig. 4-1 shows the quantity of fuel used per hour by some representative examples in the categories mentioned above, the figures being for cruising flight at a given thrust which is equal to the drag. In the case of jet engines, this specific fuel consumption is referred to as Thrust Specific Fuel Consumption (TSFC, c_T). For propeller aircraft, it should be compared with $c_p V / \eta_{prop}$, where V is the flying speed, c_p the specific fuel consumption related to the shaft horsepower and η_{prop} the effi-

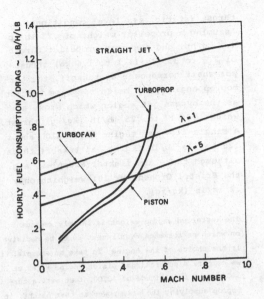

Fig. 4-1. Engine comparison based on fuel consumption to overcome drag

ciency of the propeller*. Fig. 4-1 shows that up to about M = .4 to .5 the piston engine has the lowest fuel consumption. Generally speaking, the turboprop engine has a slightly higher fuel consumption than the piston engine, but it burns kerosine which is cheaper than gasoline. Over the entire speed range the single-flow jet engine is the thirstiest type, while at high subsonic speeds the bypass engine is the most economical with regard to fuel consumption.

c. Engine weight.
Piston engines generally weigh about 1.1 to 1.75 lb (.5 to .8 kg) per takeoff shaft horsepower, while for turboprop engines the equivalent figure is between .35 and .55 lb/hp (.15 to .25 kg/hp). In order to arrive at a fair comparison with jet engines, the specific weight may be expressed as the weight of the engine with installed propeller per unit of propeller

*The terms used will be explained in Section 4.3.7.

thrust (static, sea level conditions). Assuming a propeller weight of .22 lb/hp (.1 kg/hp) and a static propeller thrust of 2.5 to 3.5 lb (1.1 to 1.6 kg) thrust per shaft horsepower at takeoff, the turboprop engine will be in the same bracket as the bypass jet engine which shows a weight of .17 to .25 lb/lb (kg/kg) thrust. A single-flow jet engine will weigh about .25 to .35 lb/lb (kg/kg) at takeoff thrust, although there are lighter examples, e.g. the Bristol Orpheus Mk 101 weighing only .2 lb/lb (kg/kg).

The installed engine weight is largely dependent on which performance requirement is to be decisive in the choice of the engine. To take an example: if we compare a supercharged piston engine rated at 340 shp at an altitude of 20,000 feet with a turboprop supplying the same power at that height, it is clear that their specific weights, based on this condition, are much closer as compared with the takeoff condition. This is because the turboprop has a rating of 550 shp at sea level as compared with 450 bhp supplied by the piston engine. Whether or not the excess takeoff power of the turboprop can be used efficiently will depend on the type of aircraft. In this particular case, however, the piston engine will still weigh between 2 and 2.5 times more than the turboprop. Broadly speaking, this means that by choosing the turboprop it will be possible to carry more useful load for the same takeoff weight.

It is not possible to give a generally valid conclusion, but Table 4-1 shows that the piston engine is at a disadvantage in weight when compared to the other types. Aircraft with turbojet engines differ very little with regard to the powerplant installation weight.

d. Dimensions.

When the engines are installed in separate nacelles, their dimensions become important with regard to parasite drag. Here only a comparison between piston engines and turboprop engines will be meaningful. The following comparison shows the dimensions of the nacelles of two propeller-driven aircraft with engines of comparable rating. Cessna 414: two Continental TSIO 520-J engines of 310 takeoff/bhp each, nacelle width: 3 ft 4 in. (1 m); height 2 ft (.6 m); length: 13 ft (3.9 m); frontal area: .016 sq.ft (.0015 m^2) per hp.
Government Aircraft Factories N-22 two Allison 250-B17 of 400 takeoff bhp each, nacelle width and height: 2 ft 2 in. (.65 m); length: 8 f (2.4 m); frontal area: .009 sq.ft (.00083 m^2) per hp.
Here, too, the piston engine is seen to be at a disadvantage, although in the case of the Cessna 414 the nacelle contains a small luggage/baggage hold.

e. First cost.

At a first cost of some $25 to $50 per hp the piston engine is the cheapest type of powerplant. The price of a turboprop amounts to approximately $60 to $100 per shp (price levels 1974). Turbojet engines cost about $20 to $40 per lb takeoff thrust,

Type of aircraft	MTOW lb	Number and type of engines	Powerplant weight % of Basic Empty Weight
Convair 340	47,000	2 Pratt and Whitney R-2800 of 2400 BHP each	31.0
Fokker F-27 Mk.500	45,000	2 Rolls-Royce Dart R.Da7 of 2100 ESHP each	20.0
Sud Aviation Caravelle VIR	110,230	2 Rolls-Royce Avon 533R of 12,600 lb each	14.6
VFW 614	41,000	2 Bristol Siddeley M45H of 7,480 lb each	14.9

NOTES

1. Basic Empty Weight is defined in Section 8.2

2. Thrust and power refer to sea leavel static (dry) takeoff rating

Table 4-1. Weight of powerplant installation as a fraction of the Basic Empty Weight

dependent on size.

f. Engine overhaul.
Time between two major overhauls (Time Between Overhaul, TBO), will be about 1,500 to 2,000 hours in the case of a normally aspirated piston engine; for supercharged types this time amounts to 1,000 to 1,500 hours. In the case of a good turboprop engine the time between two revisions may be anything up to 3,000 to 4,000 hours.

g. Engine noise and vibration.
Engine noise and vibration, which largely result from the recoprocating movement of the pistons, are the principal drawback of the piston engine, in spite of the attention paid to balancing and noise suppression. The turboprop engine makes less machinery noise, but the propeller noise is predominant. For the occupants of the aircraft the jet engine is the most silent, but the observer on the ground will regard the jet aircraft as the most annoying of the three, particularly during the takeoff and approach phases. As a result of the low exhaust velocity of the engine gases the bypass engine is much quieter than the single-flow jet engine, provided suitable measures are taken to suppress the noise of the fan. The external noise of the propeller engine can only be reduced drastically by adopting a slowly rotating propeller (low tip speed).

h. Passenger appeal.
It is generally accepted that jet aircraft possess more passenger appeal than propeller aircraft.

A final assessment of the factors enumerated above can only be made by means of detailed design studies with proper attention to the installation of the engines in the aircraft and their influence on the general layout of the aircraft. It is, however, possible to draw some general conclusions:
1. For high-speed aircraft only jet propulsion can be considered. The relatively simple and less expensive single-flow engine is suitable today for a limited category of aircraft where price is the dominant factor and the higher fuel consumption is regarded as less important.
2. For low-speed aircraft with a power of more than, say, 500 hp per engine, the propeller turbine will be the best choice.
3. When power ratings of less than about 500 hp are used, the piston engine will be a competitor of the propeller turbine, due to the relatively low cost of the power unit. In the case of aircraft designed for high utilization (small passenger aircraft, some business aircraft, commuters), the cost of the powerplant will be spread out over a large number of flying hours and the propeller turbine will be at an advantage because of the low maintenance costs and cheaper fuel.
4. Small aircraft for private use (sports and touring aircraft, club trainers), are nearly always fitted with piston engines.

The various engine types will be discussed more fully in the following sections. Methods for calculating engine performance will not be presented here since these can be found in the publications mentioned in the list of references and in Appendix H, nor will an attempt be made to explain how engines are designed. The discussion will be confined to some of the most important characteristics with which the aircraft designer should be familiar when he has to choose an engine for his design. Some background information is supplied in connection with the future technological development of engines.

4.2. CURRENT RECIPROCATING ENGINES

4.2.1. Some characteristics of the four stroke engine

A short description is given here of the cycle of the four stroke Otto-type engine, explaining the p-V diagram presented in Fig. 4-2a.

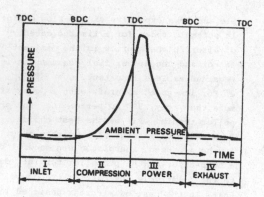

a. Cylinder pressure vs time
TDC-Top Dead Center
BDC-Bottom Dead Center

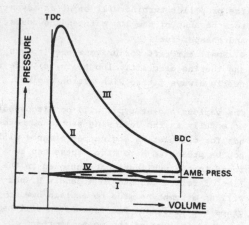

·b. Indicator diagram

Fig. 4-2. Cylinder pressure of a four
stroke engine

a. During the induction stroke the inlet
valve is open and the gas mixture (carburetor) or the air (injection) is drawn
from the inlet manifold into the cylinder.
b. During the compression stroke the gas
mixture is compressed, following the polytropic relationship pv^n = constant, in
which n is mainly dependent on the compo-
sition of the gas mixture (n ∿ 1.2 to 1.35).
Apart from the inlet pressure, the final
pressure is determined by the compression

ratio*, that is, the ratio between the
volumes contained in the cylinder above
the piston crown at the beginning and at
the end of the compression stroke.
The compressed gas mixture is ignited after being compressed and combustion takes
place at a practically constant volume.
Combustion temperature and gas pressure
are very high since the mixture ratio is
nearly stoichiometric, with the result
that the thermal efficiency of the cycle
is relatively high. Thermal efficiency for
a given mixture ratio is a function of the
compression ratio and, contrary to the jet
engine, is not affected by the rotation
speed of the engine.
c. During the power stroke the burning and
expanding charge transmits power to the
piston. This power is transferred to the
propeller by means of the connecting rod
and the crankshaft. The process of expansion again follows the polytropic relationship pv^n = constant.
d. After the exhaust valve is opened, the
gas is expelled into the exhaust manifold
during the exhaust stroke at approximately
constant volume.

The total energy imparted to the piston may
be determined by adding the work done
during these processes algebraically and
multiplying the sum by the number of power
strokes per unit of time (n/2) and the number of cylinders per engine. The power determined in this way can be derived from
an indicator diagram (Fig. 4-2b) by inte-
grating the pressure versus the volume
swept by the piston, resulting in the
Indicated Mean Effective Power. This power
is often associated with the Indicated Mean
Effective Pressure (IMEP**), defined as
follows:

*Note that the compression ratio is a ra-
tio of volumes and not pressures
**In (4-1) and (4-2) the total cylinder
volume in cu.inch must be divided by
396,000 to obtain BMEP in psi, when the
engine speed is in rpm

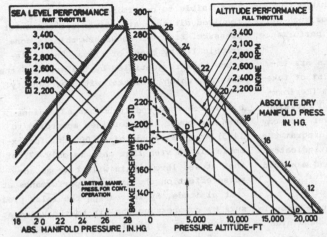

NOTE:

To find the actual horsepower from altitude, rpm, manifold pressure and air inlet temperature:

1. Locate A on full throttle altitude curve for given rpm and manifold pressure

2. Locate B on sea level curve for rpm and manifold pressure and transfer to C

3. Connect A and C by a straight line and read horsepower at given altitude: D

4. Modify horsepower at D for variation of air inlet temperature T from standard altitude temperature T_s by the formula:

Actual hp = hp at D x $\sqrt{\dfrac{T_s}{T}}$

where T and T_s are absolute temperatures

Fig. 4-3. Example of a standard power diagram of a normally aspirated engine

$$IMEP = \frac{\text{Indicated Mean Effective Power}}{\text{Cylinder Volume x (n/2)}} \quad (4-1)$$

The Brake Horse Power (BHP) is the product of the Indicated Mean Effective Power and the mechanical efficiency. Allowing for power losses due to friction and driving accessories, this efficiency declines with increasing engine speed.

The specific engine performance is often expressed as Brake Mean Effective Pressure (BMEP*):

$$BMEP = \frac{\text{Brake Horse Power}}{\text{Cylinder Volume x (n/2)}} \quad (4-2)$$

The standard engine diagram, an example of which is given in Fig. 4-3, shows BHP as a function of the manifold pressure, rpm and altitude in the standard atmosphere, generally with a mixture control which leads to maximum power. Sometimes, however, the takeoff power of continuous power is given for a rich mixture. Corrections are given for non-standard atmospheric conditions. Engines can be fed with air of roughly at-

*In (4-1) and (4-2) the total cylinder volume in cu.inch must be divided by 396,000 to obtain BMEP in psi, when the engine speed is in rpm

mospheric pressure (normally aspirated) or with air compressed by a blower (blown, supercharged or turbocharged engines), resulting in standard diagrams of different shape.

a. Normally aspirated engines.
The full throttle inlet pressure differs from atmospheric pressure by only about 2% to 5%. On the left hand section of the diagram the power at sea level is given as a function of the (inlet) Manifold Absolute Pressure, MAP, and the rpm. Power limits may be dictated by:
- the maximum permissible rpm,
- the manifold pressure corresponding to the full throttle condition (throttle valve fully open),
- an inlet pressure limit in the case of a continuously running engine.
The following engine ratings are distinguished:
a. Takeoff Power*, which is the maximum power permitted for a limited period of time during the takeoff of the aircraft, e.g. 1 to 5 minutes.

*Takeoff power is sometimes called "Rated Power" and the maximum continuous rating is known as METO power (Maximum Except Take Off)

103

b. Normal Rated Power*, the maximum continuous power which may be used for such parts of the flight as maximum climb performance and maximum level speed.

c. Cruising flight, where limits are imposed for fast cruising, e.g. 75% of takeoff power at 90% of maximum rpm (Performance Cruise), and for economical cruising flight, e.g. 65% of takeoff power (Economy Cruise). In the standard power diagram the permissible region is sometimes indicated in the form of limits for rpm and manifold pressure.

The effect of altitude in the full-throttle condition is shown on the right hand side of the standard power diagram (Fig. 4-3). Since the power is directly proportional to the mass of the charge per unit of time, the brake horsepower will diminish for a given rpm and fully opened throttle with increasing altitude as a result of decreasing density of the air. An expression often used to calculate this relationship reads:

$$\frac{\text{BHP at altitude}}{\text{BHP at sea level}} = (1+c)\ \sigma - c \qquad (4-3)$$

The factor c is often taken as .132, although this number is based on limited experimental evidence with several radial engines.

When determining the BHP for part throttle working conditions at increasing altitude, we can make use of the characteristic that for a given rpm and inlet pressure, power will increase slightly as a result of:
- a decrease in air temperature, which increases the mass of air aspirated by the engine,
- a decrease of back pressure in the exhaust manifold, which leads to better filling of the cylinders.

It has been observed that the relationship between power and density is also linear in nature. By using this characteristic it

*Takeoff power is sometimes called "Rated Power" and the maximum continuous rating is known as METO power (Maximum Except Take Off)

is possible to determine the BHP for the required altitude at a given rpm and inlet pressure. Fig. 4-3 shows how this is done graphically by means of linear interpolation; this is justified by the scale of the diagram, which is linear with the density.

b. Supercharged and turbocharged engines. When the power is determined according to (4-3) at an altitude of 20,000 ft, for example, it is seen that this is only 47% of that at sea level. This will have an adverse effect on the aircraft performance at that altitude. By increasing the inlet pressure by compressing the air to the value at sea level, it is possible to regain this loss of power. In these "blown" or supercharged engines (Fig. 4-4), one generally

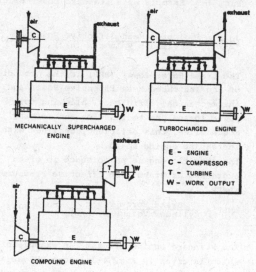

Fig. 4-4. Methods of boosting piston engine performance

uses a centrifugal compressor which is driven by either the crankshaft or by a turbine placed in the exhaust gases.
Mechanical superchargers are driven through gears, sometimes with multiple gear ratios. This means, however, that power is used from the crankshaft (about 6 to 10%), thus reducing the output and slightly increasing the fuel consumption in relation to the work output of the engine.

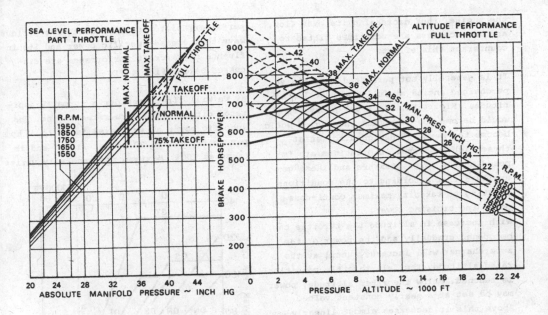

Fig. 4-5. Performance diagram of a supercharged engine

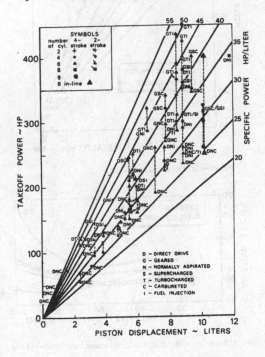

Fig. 4-6. Takeoff power of reciprocating engines vs total piston displacement (Ref. Jane's All the World's Aircraft 1972-'73)

The turbo-supercharger not only increases the inlet pressure but the back pressure in the exhaust as well, with the result that during the working stroke more work has to be done and the cylinders will not be completely filled. Part of the energy present in the exhaust gases, which is of the order of one-third of the energy supplied by the fuel, is absorbed by the turbine and the exhaust thrust will diminish. Incidentally, this thrust is small in any case and is generally disregarded when considering engine performance, unless special devices, like nozzles, are incorporated to utilize some of the energy present.

Compound engines may have different configurations. The one shown in Fig. 4-4 is mechanically supercharged and utilizes the energy of the exhaust gases to drive a turbine, thus contributing directly to the effective work output. This system may be considered as intermediate between the piston engine and the gas turbine. Due to its complexity, however, the compound engine has never been very successful. Superchargers will heat the air admitted to

105

the engine, thus decreasing its mass flow, and intercoolers are sometimes fitted to counteract this effect.

It is generally not permitted to use a supercharged engine at full power at low altitudes (Fig. 4-5), since the pressures would become too high, and the engine would become too heavy if designed to resist these high pressures. Limitations are imposed on the inlet pressure and these generally differ according to the conditions relating to takeoff, maximum continuous power and cruising power.
With increase in altitude the throttle can be opened gradually and the power of the supercharger will increase, until at the rated altitude, the full-throttle condition is reached. Below the rated altitude, power may be set at a nearly constant value. Above this it decreases almost linearly with density, as in the normally aspirated engine.

c. Engine control and fuel consumption in cruising flight.
A lean mixture will basically result in low fuel consumption. This is seldom used, however, particularly in the case of engines with high compression ratios, where there is a risk of detonation. The engine manufacturer will generally supply information regarding the relationship between the (specific) fuel consumption, rpm and power, and sometimes also for various mixture ratios. By means of these data and the relationship between speed and the shaft power required to propel the aircraft, it will be possible to determine the most favorable cruising condition where the fuel consumption per mile is a minimum.

4.2.2. Engine design and its influence on flight performance

The takeoff power of various engines in the 50 to 450 hp power bracket has been plotted against piston displacement in Fig. 4-6. Specific BHP is observed to lie between about .40 to .80 hp per cu.in. (25 to 50 hp

per liter) piston displacement. The various ways to increase specific power and its influence on aircraft performance are discussed below (Fig. 4-7).

a. We will first consider a normally aspirated direct drive carbureted engine. The specific power at sea level is .4 to .5 hp per cu.in. (25 to 30 hp per liter) and is taken as 100%. Maximum rpm generally varies

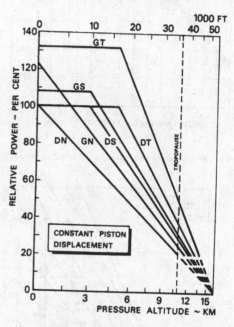

a. Power-altitude relationship

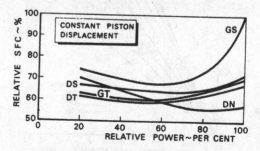

b. Effect on specific fuel consumption, METO-rating

Fig. 4-7. Reciprocating engine design improvements (Note: symbols defined in Fig. 4-6. Ref. 4-16)

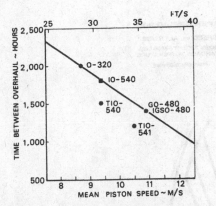

Fig. 4-8. The effect of piston speed on TBO
(Ref. SAE Paper 710381)

Continental Tiara engine it is even possible to obtain 4,000 - 4,400 rpm. For a given BMEP and cylinder volume, the brake horsepower increases with rpm, as demonstrated by (4-2). Gearing will cost a few per cent of the engine power and raise the weight and cost by approximately 12% and 50% respectively (Ref. 4-13). Specific powers of .53 to .56 hp/cu.in.(33 to 35 hp/liter) are obtained from geared engines.

The following limiting factors should be taken into consideration:
- Engine vibration results from the torsional fluctuations in the crankshaft.
- High piston speeds have an adverse influence on the time between overhauls (Fig. 4-8).

between 2500 and 2800. When a higher value is taken, either the tip speed of the propeller will become too high or the propeller diameter will be too small, resulting in a low efficiency.

b. Higher engine speeds are possible with the use of reduction gearing between the crankshaft and the propeller, raising the rpm to 3,200 or 3,400. In the case of the

c. Effect of cylinder geometry.

With a given piston speed and total swept cylinder volume (piston displacement), the power developed may be increased by:
- choosing a relatively large cylinder bore with respect to the stroke;
- reducing the piston displacement by increasing the number of cylinders (N_{cyl}), which leads to higher rpm. It can be shown

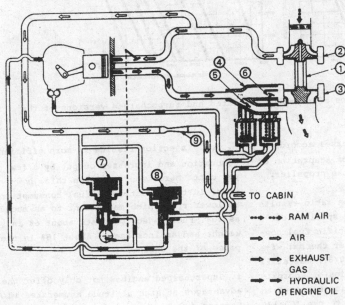

TO CABIN

RAM AIR

AIR

EXHAUST GAS

HYDRAULIC OR ENGINE OIL

1 turbocharger
2 compressor
3 turbine
4 dual butterfly valve assembly
5 wastegate valve
6 diverter valve
7 variable absolute pressure controller
8 fixed absolute pressure controller
9 sonic venturi

Fig. 4-9. Turbocharging engine and cabin system

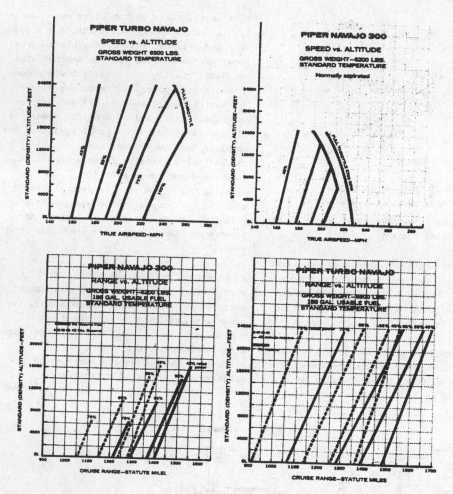

Fig. 4-10. Flight performance of normally aspirated and turbocharged versions of the Piper Navajo

(Ref. 4-13) that power increases according to $(N_{cyl})^{1/3}$, but it requires adaptation of the reduction gear driving the propeller.

d. An increase in compression ratio results in some increase in BMEP and specific output and reduction of the specific fuel consumption as a result of better thermal efficiency. Low octane fuel (80/87) may still be used with compression ratios of approximately 7.5, but when the ratio is raised to 8 or 10 it will be necessary to use 100/130 octane fuel.

e. Fuel injection results in more efficient combustion and increases output by a few per cent. More important than this, however, is the reduction in fuel consumption at lower rpm, which may amount to as much as 10%. Fuel injection costs about 6% in weight and an increase of about 18% in the price of the engine.

f. Supercharged engines not only offer the advantages at high altitude enumerated in Section 4.2.2., but also the secondary advantage that they are able to supply pres-

108

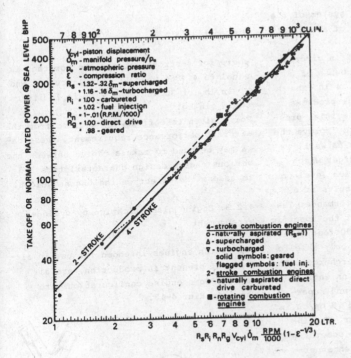

Fig. 4-11. Estimation of takeoff horsepower of reciprocating engines at SL/ISA

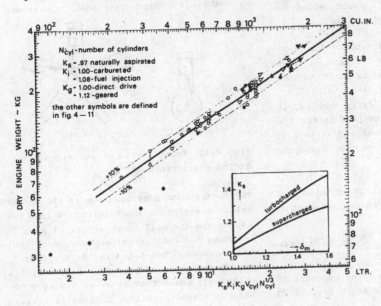

Fig. 4-12. Estimation of dry weight of reciprocating engines

sure to the air conditioning system of the cabin. A diagrammatic layout of a typical system is shown in Fig. 4-9.

An example of the improvement in aircraft performance resulting from the use of a supercharger is given in Fig. 4-10. The supercharger used in this case produces practically no increase in the inlet pressure at sea level. In order to improve the takeoff and climb performance as well, it is possible to use a supercharger which raises the inlet pressure at sea level to, say, 1.5 atmospheres. This results in a marked improvement in the performance. The example in Fig. 4-7 shows that the specific power amounts to 110% at 20,000 feet (6,150 m) with this engine, which is only able to supply 47% output without supercharger. The specific takeoff power of piston engines with propeller reduction gear and supercharger may lie between .7 and .8 hp/cu.in. (45-50 hp/liter) or even higher. The increase in engine weight may be something like 18% for a crankshaft-driven supercharger and 30% for a turbo-supercharger, but will depend on the desired increase in performance. In order to avoid detonation, the compression ratio should be low.

f. In the case of large outputs (about 3000 hp) of the type used in the past, a further increase may be achieved by adopting the compound principle (Fig. 4-4). Although this type of engine is fairly complex, it is not impossible that engines in the lower power brackets may be developed in this direction to meet competition from turboprop engines.

A semi-empirical correlation of takeoff power and engine design characteristics is given in Fig. 4-11:

$$P_{to} = const \times R_S R_i R_n R_g V_{cyl} \delta_m \frac{RPM}{1000} (1 - \varepsilon^{-1/3}) \quad (4-4)$$

where for 4-stroke engines the constant is equal to .41 when V_{cyl} is in cu.in., or 25 when V_{cyl} is in liters.
Engine weight may be estimated as follows (Fig. 4-12):

$$W_e = const. (K_s K_i K_g V_{cyl} N_{cyl}^{1/3}) .6582 \quad (4-5)$$

where for 4-stroke engines the constant is equal to 4.4866 when V_{cyl} is in cu.in. and W_e in lb, or 30.465 when V_{cyl} is in liters and W_e in kg.
For a given takeoff power, deduced from the aircraft performance requirement, these data may be used to make a choice between the various engine design characteristics and to assess the effect on the engine weight.

4.2.3. Engine classification by cylinder arrangement

All piston engines intended for use in aircraft at present in production, are air cooled. Some engine configurations are shown in Fig. 4-13.

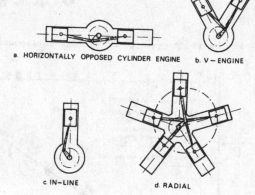

a. HORIZONTALLY OPPOSED CYLINDER ENGINE b. V — ENGINE

c. IN-LINE d. RADIAL

Fig. 4-13. Frontal view of reciprocating engine configurations

The most common configuration is the opposed cylinder engine as shown in Fig. 4-13a. Examples are the well known series of Avco Lycoming and Continental engines, e.g. the Avco Lycoming IGSO-540 (Fig. 4-14). This configuration leads to a flat engine with a relatively small frontal area, having good balance characteristics. Cooling is satisfactory with up to four cylinders on either side.

A type which is no longer produced is the

Fig. 4-14. Avco Lycoming IGSO-540-A series; 380 horsepower: ¼ left rear view

"Vee" engine (Fig. 4-13b), which is narrower than the horizontally opposed cylinder arrangement and has a smaller frontal area. In-line engines (Fig. 4-13c) have all cylinders placed one behind the other with their axes parallel, resulting in a very small frontal area. When more than four cylinders are required, air cooling will present a problem, particularly when the engine is used in a pusher installation. The in-line arrangement has frequently been used in the past for water cooled engines. When the engine is installed in nacelles on the wings, both the drag and the adverse influence on the lift will be minimum when a vertical in-line engine is adopted. With a tractor engine in the nose of the aircraft, however, the frontal area is generally dictated by the cross-section of the passenger cabin. In the case of side-by-side seating the only advantage of the in-line engine is that it improves the view of the occupants of the forward seats. When two seats are arranged in tandem, the in-line engine is well suited for the narrow fuselage.

The radial engine (Fig. 4-13d) possesses a large frontal area, resulting in high drag, but the excellent cooling makes it very suitable for aircraft which have to fly at low speeds for long periods (helicopters, agricultural aircraft). Radial engines with two rows of 7-9 cylinders have been built and used for airplane propulsion in large numbers, e.g. in the Lockheed Constellation and Douglas DC-6. They are now only built in the East European countries.

4.2.4. Two stroke and Rotary Combustion engines

a. The two stroke engine is only offered in low power outputs and is suitable for power-assisted sailplanes and unmanned aircraft.

111

It has the advantages of extreme simplicity and low cost, and low specific weight in terms of dry weight per shaft horsepower. The unfavorable reputation of the two-stroke engine is largely caused by its inefficient combustion, high fuel consumption, irregular ignition and difficulties in starting up. These disadvantages may be cured by using fuel injection, pre-compression of the air admitted to the engine and the use of high rpm, but these measures will to some extent eliminate the essential advantages.

b. The Rotary Combustion (RC) or Wankel engine is still in the experimental stage as far as its role in aviation is concerned. Up to 1972 the only operational applications were the Curtiss Wright RC-2, installed in the Lockheed Q-Star - the quiet Lockheed observation aircraft - and some powered sailplanes. Functionally speaking, the RC engine is in some respects comparable with the two stroke engine: it possesses no inlet or exhaust valves and the engine does not act for half the time as an air compressor, as in the case of the four stroke engine. It is, therefore, possible to obtain more power per unit of cylinder volume which makes the RC engine lighter and more compact for a given output. This is confirmed in practice in the case of the RC-2/60, which has an output of 200 hp with a cylinder volume of only some 120 cu.in. (two liters), running at 5,000 rpm. A more recent project is the German RFB Fantrainer A WI-2, a two-seat training aircraft powered by a 300 hp Wankel four-disk RC engine. It features a ducted propeller of small diameter to cater for the high rpm.
A second important advantage is that the rotary disk does not follow a reciprocating movement. This makes it possible to transmit the disk movement directly to the shaft, resulting in less noise and vibration. It also leads to lighter balancing, a lighter propeller, engine mounting and installation, together with a more compact unit. Development work still being carried out

is mainly directed at the sealing between the rotary disk and the cylinder wall and at obtaining the most favorable combustion characteristics. In view of the high rpm it will be necessary to choose a greater propeller reduction ratio than would be used in the case of more conventional engines.

Fig. 4-11 shows the performance of some two-stroke engines and of the Wright RC-2/60 as a function of the same parameter as used for four stroke engines. It is striking to note that both types show an output only 10 to 15% higher than the four stroke engine. The improvement in the performance of the Wankel type engine is mainly due to the considerably higher rpm, which can be obtained without an excessive penalty for proper balancing. The Tiara four stroke engines made by Continental (Ref. 4-17) are interesting in this respect since they use a new method to reduce torsional vibration during engine starting (Vibratory Torque Control). This makes it possible to run these engines at relatively high rpm (up to 4,400), and these engines also have a high specific power output, running at low noise levels.

4.3. BASIC PROPERTIES OF AIRCRAFT GAS TURBINES FOR SUBSONIC SPEEDS

In spite of the fact that at first sight gas turbine engines show considerable mutual differences in general configuration (Fig. 4-15), it is possible to compare them on a common basis with respect to their thermodynamic properties.
The majority of the gas turbine power plants designed for use in civil aviation can be classified according to one of the following categories:
- straight jet (simple jet) or single-flow turbojet engines,
- bypass engines (turbofans) or double-flow engines,
- gas turbines driving a propeller or rotor: turboprop and turboshaft engines, re-

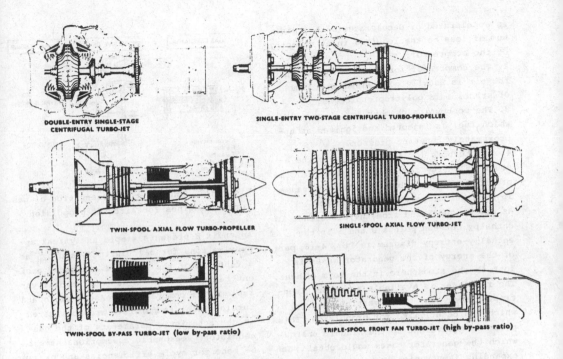

DOUBLE-ENTRY SINGLE-STAGE CENTRIFUGAL TURBO-JET

SINGLE-ENTRY TWO-STAGE CENTRIFUGAL TURBO-PROPELLER

TWIN-SPOOL AXIAL FLOW TURBO-PROPELLER

SINGLE-SPOOL AXIAL FLOW TURBO-JET

TWIN-SPOOL BY-PASS TURBO-JET (low by-pass ratio)

TRIPLE-SPOOL FRONT FAN TURBO-JET (high by-pass ratio)

Fig. 4-15. Aircraft gas turbine engine configurations

spectively. The engine category which only supplies power from the shaft (turboshaft engines) will here be regarded as belonging to the turboprop category. Engines with afterburning are mainly installed in aircraft designed for supersonic cruising speeds and military aircraft which must be capable of short bursts of speed. These types will not be dealt with here.

In comparing gas turbine engines it is useful to split them up as follows:
1. the gas producer or gasifier (gas generator) which is the source of gases with high energy,
2. the propulsive device which is the means to transform the energy of the gases into useful propulsive power (thrust times speed).

4.3.1. The gas producer

The gas producer is the essential element of the engine in which a continuous self-

sustaining process takes place in the production of hot gases according to the Brayton cycle (Fig. 4-16). The component cycles take place in the following main elements:

1. The entry duct in which atmospheric air

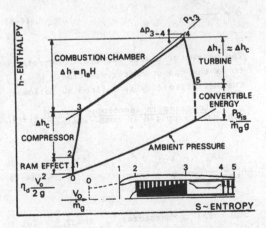

Fig. 4-16. Gas generator thermodynamic cycle

is accelerated or decelerated with a minimum of loss to the required inlet velocity of the compressor.

2. The compressor (or compressors) in which the air is usually compressed in a number of stages in a polytropic process.

3. The combustion chamber (or chambers) in which fuel is injected and ignited at approximately constant pressure.

4. The turbine (or turbines) where the amount of energy required to drive the compressor(s) is absorbed from the hot gases in a number of polytropic processes.

As may be concluded from the difference in enthalpy between points 0 and 5 in the enthalpy-entropy diagram in Fig. 4-16, part of the energy of the generator gases is lost in the atmosphere in the form of heat. The remainder is available as useful isentropic power ($P_{g_{is}}$, convertible energy), which can be utilized for propulsion. This power is equivalent to the kinetic energy which the generator gases would obtain when expanding isentropically to atmospheric pressure.

The thermodynamic performance of the gas generator may be characterized by the specific convertible energy and the thermal efficiency. The specific convertible energy, $P_{g_{is}}/\dot{m}_g$, is written in non-dimensional form:

$$G = \frac{P_{g_{is}}}{\dot{m}_g c_p T_o} \qquad (4-6)$$

The parameter G will henceforth be referred to as the gas generator function.

Thermal efficiency is defined as follows:

$$\eta_{th} = \frac{\text{increase in isentropic energy}}{\text{heat supplied in comb. chamber(s)}} \quad (4-7)$$

or

$$\eta_{th} = \frac{P_{g_{is}} - \frac{1}{2} \dot{m}_g V_o^2}{\dot{m}_g c_p (T_{t_4} - T_{t_3})} \qquad (4-8)$$

The total temperatures T_{t_3} and T_{t_4} refer to the entry to the combustion chambers and that of the turbine (Fig. 4-17).

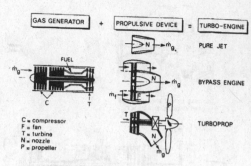

Fig. 4-17. Schematic representation of gas turbine engines for aircraft propulsion

Appendix H presents simple analytical approximations for calculating G and η_{th} as a function of the pressure altitude, ambient temperature and M_o as operational variables. The compressor pressure ratio and turbine inlet temperature are considered as characteristic parameters of the gas generator. Reasonable assumptions have to be made for cycle efficiencies and pressure losses. The thermodynamic calculations which form the basis of these expressions can be found in various publications, e.g. Refs. 4-30 and 4-39.

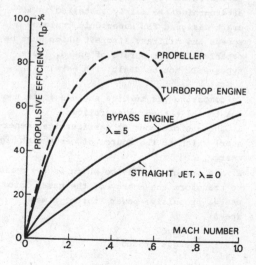

Fig. 4-18. Propulsive efficiency of subsonic turbo-engines

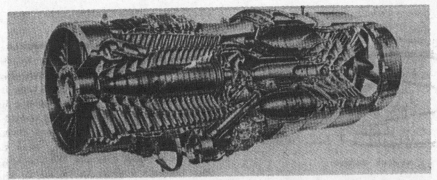

a. Rolls-Royce Avon - straight jet engine

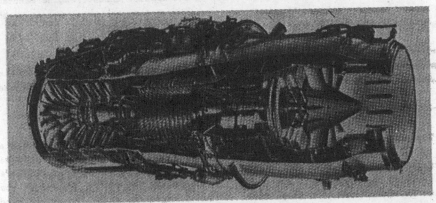

b. Rolls-Royce Spey - low bypass engine ($\lambda = 1$)

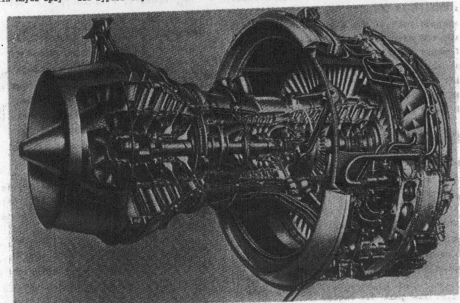

c. Rolls-Royce RB-211 - high bypass engine ($\lambda = 5$)

Fig. 4-19. Examples of turbojet engine configurations (data can be found in Table 6-1)

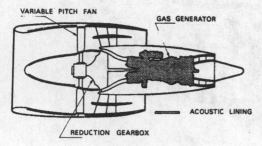

VARIABLE PITCH FAN
GAS GENERATOR
ACOUSTIC LINING
REDUCTION GEARBOX

fan tip speed 1,050 fps (320 m/sec)
exhaust velocity generator 900 fps (275 m/sec)
exhaust velocity fan 600 fps (180 m/sec)

d. Rolls-Royce/Bristol Siddeley M-45S (project)-
very high bypass ratio ($\lambda = 10.5$); geared fan; variable incidence rotor blades

Fig. 4-19. (concluded)

4.3.2. The propulsive device

The propulsive device is that part of the
engine in which the convertible energy of
the gases is transformed into propulsive
power. This part is shown diagrammatically
in Fig. 4-17 where it may be regarded as a
separate part of the engine, as is the case
with the aft-fan (Fig. 4-15). In the case
of other engine configurations, however,
this is a more fictitious distinction, as
will be discussed later on.
Gas turbines can be compared on the basis
of their propulsive efficiency η_p,

$$\eta_p = \frac{\text{useful propulsive power}}{\text{increase in isentropic energy}} \qquad (4-9)$$

or:

$$\eta_p = \frac{T\,V_o}{P_{g_{is}} - \frac{1}{2}\,\dot{m}_g\,V_o^{\,2}} \qquad (4-10)$$

Fig. 4-18 shows the propulsive efficiency
of various types of engines under representative and comparable conditions.

4.3.3. The pure jet engine

The propulsive device of a straight jet
(example in Fig. 4-19a) is the nozzle in

which the exhaust gases are accelerated and
expelled rearwards (Fig. 4-17). For subsonic aircraft applications a converging
nozzle is usually adopted. The flow in the
exit opening of the nozzle in the design
condition (high rpm) of modern turbojets: the nozzle is said to be choked.
The local static pressure (p_e) will be a-
bove the ambient pressure and the engine
thus delivers a pressure thrust in addition
to the impulse thrust. Engine manufacturers
usually define the total thrust as the
standard net thrust:

$$T_{sn} = \dot{m}_g\,(v_e - V_o) + A_e(p_e - p_o) \qquad (4-11)$$

In order to obtain a simple expression for
the propulsive efficiency, we may use the
ideal thrust instead:

$$T_{id} = \dot{m}_g\,(v_g - V_o) \qquad (4-12)$$

The velocity v_g is reached some distance
behind the nozzle opening when the gases
have expanded to atmospheric pressure. If,
to simplify the matter, we assume that complete expansion is an isentropic process,
the propulsive efficiency will be:

$$\eta_p = \frac{\dot{m}_g\,(v_g - V_o)\,V_o}{\frac{1}{2}\,\dot{m}_g\,v_g^{\,2} - \frac{1}{2}\,\dot{m}_g\,V_o^{\,2}} = \frac{2}{1 + v_g/V_o} \qquad (4-13)$$

This expression may be recognized as the
familiar Froude efficiency. At low speeds
(takeoff and climb), $v_g \gg V_o$ and η_p will
be low; at high subsonic speeds it will be
of the order of .50 (Fig. 4-18).

4.3.4. The turbofan engine

Although the principle of the bypass engine (examples are given in Fig. 4-19b, c
and d) had been patented as early as 1937
and an engine of this type was built and
tested in 1946*, large-scale application
was only possible when designers were able
to build efficient compressors with high
pressure ratios and turbines capable of
withstanding high gas temperatures could

*The Metropolitan Vickers F 2/3 aft-fan
engine

116

be designed. The pure jet engine was the
only turbojet engine available for transport aircraft up to around 1960, but in
later years it has been entirely replaced
by the bypass engine as far as civil aviation is concerned.

The propulsive device of the bypass engine
(Fig. 4-17) consists of the following elements:

1. a low pressure turbine which absorbs the
energy from the generator gases and transmits it through a shaft or directly to
2. a low pressure compressor (fan) which
compresses cold bypass air,
3. jet nozzles, both for the generator exhaust (hot flow) and for the bypass air
(cold flow). A modest gain in thrust can
be obtained for bypass engines by mixing
the hot and cold flows in a special mixing
device.

Since in the turbofan the convertible energy is
spread over a greater quantity of air than in the
case of the pure jet engine, the mean exhaust velocity will be lower and, according to (4-13), the
propulsive efficiency higher (Fig. 4-18). However,
the transmission of energy from the hot to the
cold flow will entail losses, so that an improvement will only be achieved when the gas generator
specific output is sufficiently high. It can be
shown that useful application of the bypass principle is dependent on the fundamental condition
that:

$$G > \frac{\gamma - 1}{2} M_o^2 \frac{\eta_d}{\eta_{tf}^2} \qquad (4-14)$$

Here η_d is the isentropic inlet efficiency of the
fan (cf. Fig. 4-16) and η_{tf} is the efficiency of
the transmission of energy from the hot to the cold
flow ($\eta_{tf} = \eta_t \times \eta_f$). With modern gas generators
this condition is easily met at subsonic speeds,
but at su ersonic speeds this will not always be
the case.

The most important parameter of the bypass
engine is the ratio of the mass flows per
unit time through the fan ($\dot{m}_f$) and through
the gas generator ($\dot{m}_g$); the bypass ratio is

$$\lambda = \frac{\dot{m}_f}{\dot{m}_g} \qquad (4-15)$$

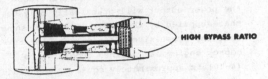

MEDIUM BYPASS RATIO

HIGH BYPASS RATIO

Fig. 4-20. Configuration differences of
low/medium and high bypass engines

Obviously, for straight jet engines $\lambda = 0$.
The following distinction is often made
(Fig. 4-20)
a. Engines with a low bypass ratio ($\lambda <$
about 2): low to medium bypass engines. The
bypass air is completely ducted around the
core engine and is sometimes also mixed
with the hot gases.
b. Engines with a high bypass ratio ($\lambda >$
about 2). Engines of this type are generally fitted with a relatively short fan cowling, though the desirability of installing
noise suppression material (acoustic lining) favors the use of a longer duct.

The standard net thrust of a bypass engine
is defined as:

$$T_{sn} = \sum_{g,f} \left| \dot{m} (v_e - V_o) + A_e (p_e - p_o) \right| \qquad (4-16)$$

in which the mass flows and exhaust conditions refer to those of the gas generator
(g) and fan air (f).

4.3.5. The turboprop engine

In this type of engine the propulsive device (Fig. 4-17) consists of:
1. the power turbine,
2. the propeller shaft plus reduction gear,
3. the propeller,
4. the exhaust nozzle for the engine gases.
Most of the energy derived from the gases
is used to drive the propeller. The propulsive power from the exhaust gases is about
10 to 20% of that of the shaft, with the

117

result that when we deduct the ram drag $(\dot{m}_g \, V_o)$ the net propulsive thrust of the engine gases at high flying speeds will be relatively small (order of magnitude: 5 to 10%). In the case of turboshaft engines the power output will practically only be that supplied by the shaft. It can be shown that the propulsive efficiency of the turboprop engine, as defined according to (4-10) is approximately equal to the product of the turbine efficiency, the mechanical efficiency of the reduction gear and the propeller efficiency:

$$\eta_p = \eta_t \, \eta_{mech} \, \eta_{prop} \qquad (4-17)$$

At flying speeds up to $M_o = .5$ to $.6$ the propulsive efficiency of the turboprop engine will be superior to that of turbojet engines (Fig. 4-18). With increasing speeds, however, critical compressibility phenomena will occur at the propeller tips and propeller efficiency will deteriorate progressively.

The total power of the turboprop engine is often expressed as Equivalent Shaft Horse Power (ESHP):

$$P_{eq} = P_{br} + \frac{T_j \, V_o}{\eta_{prop}} \qquad (4-18)$$

Since in the static condition $V_o = \eta_{prop} = 0$, the second term is indeterminate, and the ESHP is sometimes expressed as follows:

$$P_{eq} = P_{br} + \frac{T_j}{2.5} \quad (T_j \text{ in lb}, P_{eq} \text{ in hp}) \qquad (4-19)$$

Here the thrust of the engine gases is compared to that of a propeller which has a static thrust of 2.5 lb per shaft horsepower.

4.3.6. Overall efficiency, specific fuel consumption and specific thrust (power)

The definitions given above of the thermal efficiency and propulsive efficiency are logical as they enable us to compare different types of engine with the same gas generator on the basis of their propulsive

devices.

The distinction between the gas generator and the propulsive device may, however, be fictitious where the actual engine hardware is concerned. The outer part of the fan - which compresses cold air - forms part of the propulsive device, while the inner part should be regarded as part of the gas generator. Also, the low pressure turbine supplies power to both the fan and the gas generator. The total pressure ratio of the bypass engine, therefore, is the product of the pressure ratios across the fan and across the high-pressure compressor when the fan is mounted ahead of the compressor.

In some publications dealing with the bypass engine the propulsive efficiency is put equal to the Froude efficiency, assuming one (average) exhaust velocity. In that case, however, the thermal efficiency will have to be corrected for losses in the fan and the low-pressure turbine.

The overall or total efficiency is defined as follows:

$$\eta_{tot} = \frac{\text{useful propulsive power}}{\text{heat content of the fuel}} = \frac{T \, V_o}{\dot{W}_F \, H} \qquad (4-20)$$

where H is the heating value of the fuel. The overall efficiency may be split up into combustion efficiency,

$$\eta_B = \frac{\text{heat supplied in combustion chamber(s)}}{\text{heat content of the fuel}}$$
$$(4-21)$$

and the thermal and propulsive efficiencies given by (4-7) and (4-9). Hence the following will apply to the product:

$$\eta_{tot} = \eta_B \, \eta_{th} \, \eta_p \qquad (4-22)$$

The specific fuel consumption is often used to represent the efficiency of the engine. For turbojet engines the Thrust Specific Fuel Consumption (TSFC) is defined as follows:

$$C_T = \frac{\text{fuel consumption per hour}}{\text{net thrust}} \qquad (4-23)$$

while the Power Specific Fuel Consumption (PSFC) is used for turboprop engines:

$$C_P = \frac{\text{fuel consumption per hour}}{\text{shaft (or equivalent) power}} \qquad (4-24)$$

Since fuel consumption is a direct indication of the quantity of heat supplied to the engine, we may deduce from (4-20) and (4-23):

$$C_T = \text{constant} \frac{V_o}{\eta_{tot} H} \qquad (4-25)$$

where C_T is in lb/lb/h (or kg/kg/h) and the conversion constant is equal to 7.820 when V_o is in kts and H in BTU/lb or equal to 8.435 when V_o is in m/s and H in kcal/kg. Knowing that the speed of sound at sea level is equal to 661 kts (340.43 m/s) and assuming a heating value of 18,550 BTU/lb (10,300 kcal/kg), the corrected TSFC ($C_T/\sqrt{\theta}$) can be shown to be given by:

$$\frac{C_T}{\sqrt{\theta}} = .2788 \frac{M_o}{\eta_{tot}} \qquad (4-26)$$

where θ = the relative ambient air temperature. The specific fuel consumption in cruising flight is a figure of merit of the engine which determines the quantity of fuel required to overcome the drag. In the case of the straight turbojet the use of C_T has the advantage that it varies but little with speed. As opposed to this, for turbofan engines the variation with speed will be greater with increasing bypass ratio. Turbofan engines may be regarded as being situated midway between the straight jet engine and the turboprop engine, as the fuel consumption of the latter is related to the power output and also varies little with the speed.

The specific thrust of a jet engine indicates how much thrust the engine develops for a given quantity of air flowing through it per unit of time:

$$\frac{T}{\dot{W}} = \frac{\text{net thrust}}{\text{weight of airflow per unit of time}}$$

$$(4-27)$$

The specific thrust is also equal to the product of the fuel/air ratio divided by the specific fuel consumption. When the specific thrust is high, a relatively small inlet diameter will be needed in order to produce the required thrust and the weight and drag of the powerplant will generally also be low. This conclusion only applies, however, when engines are compared which have a comparable mechanical configuration.

For a single-flow engine it is possible to derive a relation between the specific thrust and the propulsive efficiency, by combining (4-12), (4-13) and (4-27):

$$\frac{T}{\dot{W}} = \frac{2 V_o}{g} \left(\frac{1}{\eta_p} - 1 \right) \qquad \text{for } \lambda = 0 \qquad (4-28)$$

Jet engines with a relatively high propulsive efficiency will therefore have a low specific thrust at a given flying speed. The same conclusion is also valid for turbofan engines. It follows that when an engine is being designed or selected for a specific aircraft, a compromise between contradictory requirements will always be unavoidable. In the case of long-range aircraft a low fuel consumption (i.e. a high value of η_p) will be regarded as essential. When a high momentary performance is desired, however, a high thrust for a limited engine size must be achieved and it will be necessary to choose a high value of the specific thrust.

In the case of turboprop engines, the power may be related to the air mass flow through the engine. The specific power is maximum when G^* is maximum, but this condition does not result in a minimum for the specific fuel consumption, so here again a compromise will have to be sought.

4.3.7. Analysis of the engine cycle

When an off-the-shelf engine is not available for a particular aircraft design, or when parametric studies are being carried out, it may be desirable to make a cycle

*Section 4.3.1

119

analysis of the engine, taking various engine design parameters as variables. Engine manufacturers have computer programs for performing these studies, and the same applies to some aircraft companies. Methods are also presented in the relevant literature (Refs. 4-25, 4-26, 4-29 and 4-50). In Appendix H a summary is given of analytical expressions which, in spite of the simplifications introduced, are suitable for making an initial estimate of engine performance in preliminary aircraft design. This method has been used for the generalized calculations which follow. When engine performance is calculated a distinction can be made between:

1. performance at the design point,
2. off-design performance.

Design point is intended to indicate the working condition at high rpm where the efficiencies of the compressor and turbine are optimum. The design point may be chosen for a representative situation, such as the takeoff, climb or cruise. In this way the engine can be adapted as far as possible to its use in the aircraft to match the most critical performance requirement. In the case of the straight jet engine, the design point is generally assumed as the takeoff, whereas the most critical condition for high bypass engines will often be the cruising flight.

4.4. ASSESSMENT OF TURBOJET ENGINES

The most important design parameters determining engine performance are:

1. Cycle efficiencies and pressure losses
2. Overall Pressure Ratio of the gas generator-compressor (OPR, $\epsilon_c = p_{t_3}/p_{t_2}$)
3. Turbine Entry (Intake) Temperature (TET, TIT, T_{t_4})
4. Bypass ratio (λ)
5. Fan Pressure Ratio (FPR, ϵ_f)

As shown by Fig. 4-21, an improvement in the process efficiencies is an effective method of enhancing engine performance. In the simplified example given, a 5% improve-

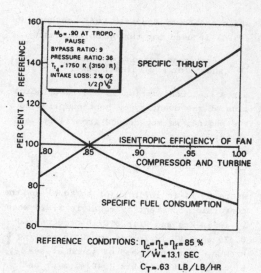

REFERENCE CONDITIONS: $\eta_c = \eta_t = \eta_f = 85\%$
T/W = 13.1 SEC
C_T = .63 LB/LB/HR

Fig. 4-21. Effect of cycle efficiency variation on turbofan performance

ment in all efficiencies (e.g. 85% to 90%) will result in an 11% decrease in fuel consumption and a 16% increase in specific thrust. A high bypass engine will be more sensitive to increase in efficiency than a low bypass engine. It should be noted, however, that this 5% improvement in efficiency is equivalent to a reduction in losses by one-third. The ways in which the engine manufacturer can reduce these losses fall outside the scope of this book.

We shall now explain the influence of the other parameters relating to

1. the fuel consumption in cruising flight,
2. the specific thrust under various conditions and the decrease of thrust with altitude and speed (thrust lapse rates),
3. the weight and drag of the powerplant as installed in the aircraft,
4. the noise production of the engine.

The initial cost, maintenance and service life are very important factors influencing the choice of an engine, but these will only be dealt with incidentally, as insufficient data are available to enable us to determine conclusively how far they are affected by the engine design. The only general rule is that engine price and mainte-

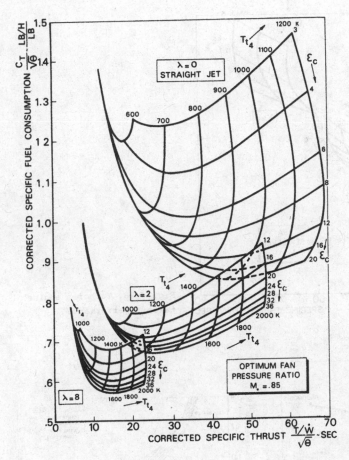

NOTE: This diagram is drawn for the tropopause, but is equally valid for other altitudes, provided T_{t_4} is chosen so that the same value of T_{t_4}/T_0 is obtained

Fig. 4-22. Effect of OPR and TET at cruising conditions

nance cost increase with complexity.

4.4.1. Overall Pressure Ratio

When a cycle analysis is carried out on turbofan engines with varying working pressures and temperatures at the design condition, a convenient representation of the results can be given in the form of Figs. 4-22 and 4-23. The corrected specific thrust and fuel consumption have been used in order to make the result valid for different altitudes. This figure shows that in all configurations an increase in OPR leads to a reduction in the specific fuel consumption in cruising flight.

In order to obtain a high specific thrust, either in cruising flight (Fig. 4-22) or during takeoff (Fig. 4-23), a high value for the OPR is not required at the given TET. In the case of the pure jet engine, for example, with TET = 1,000 to 1,200K in cruising flight, the condition for maximum specific thrust is ε_c = 7 to 9 and in the takeoff between 9 and 11. In the case of turbofan engines these values will be slightly higher, say 12 to 16 at the generally acceptable turbine temperatures (Section 4.4.2). In view of the fact that - within certain limits - an increase of OPR over these optimum values has little influence on the specific thrust, the designer generally aims at higher pressure ratios which lead to a better fuel con-

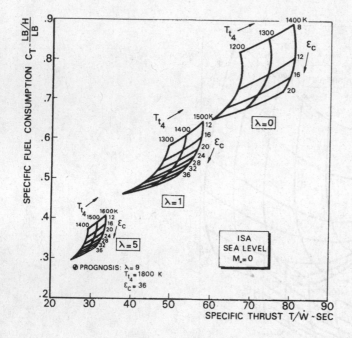

Fig. 4-23. Turbofan engine takeoff performance

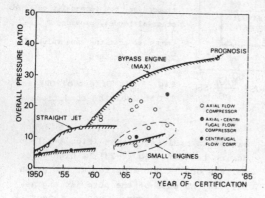

Fig. 4-24. Historic development of the OPR (References: Jane's All the World's Aircraft, various engine manufacturers' data)

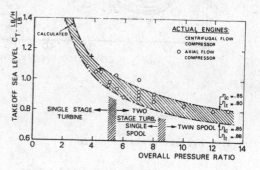

Fig. 4-25. Specific fuel consumption of straight jet engines at sea level

sumption in the design point and also at reduced rpm. Up to about 1953 jet engines were designed with a single stage centrifugal compressor (Fig. 4-15) which gave a maximum OPR of around four. As these engines had a high fuel consumption and large frontal area, they have been superseded by engines with an axial compressor (Figs.

4-15 and 4-19a). There are some recent applications of combined axial and centrifugal compressors on relatively small engines, such as the Garret AiResearch ATF-3, which have quite acceptable pressure ratios and a small frontal area.

Axial compressors make it possible to obtain large pressure ratios and a small frontal area. A value of $\varepsilon_c = 8.5$ is still possible with a single compressor and single spool. When this value is exceeded, the design conditions for the foremost and rearmost compressor stages are so far apart, that it will be necessary to employ two compressors, running at different rpm: twin spool engines.

The highest value for the OPR actually used in straight jet engines is about thirteen. Fig. 4-24 shows how the pressures have risen during the fifties. Fig. 4-25 gives an impression of the influence of OPR on C_T during takeoff. The actual fuel consumption of a number of engines is compared with calculated values.

The tendency towards high pressure ratios has been continued in the turbofan engine up to values of 25 to 30 in the case of the latest generation for large transport aircraft. Fig. 4-26 gives a schematic repre-

A. Twin spool; fan combined with low pressure section of compressor   B. Twin spool; separate fan

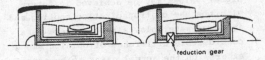

reduction gear

C. Triple spool   D. Twin spool; geared fan

Fig. 4-26. Principle of various fan-compressor configurations of high bypass engines

sentation of some engine configurations in which these high pressure levels can be achieved.
Layout A has the disadvantage that the last stages of the low pressure compressor contribute little to the increase in pressure, as they operate at relatively low tip speeds.

Layout B does not have that disadvantage, but demands a very high pressure ratio from the high pressure part, which entails the use of variable incidence guide vanes. Layout C, employing three shafts, is attractive for various applications, since a relatively small pressure ratio is demanded from each of the compressors (Ref. 4-34). The specific weight is lower than that of the others.
Layout D, the two-shaft engine with geared fan (see also Fig. 4-19d) appears to be particularly suitable for relatively small engines where the power to be transmitted by the reduction gear is limited. The reduction results in a low fan tip speed, while the turbine is still working at a relatively high tip speed, resulting in a compact turbine. For several existing or projected geared fan engines the gear ratio amounts to between .25 and .50.

From Fig. 4-22 it is seen that for an engine with $\lambda = 8$, for example, a pressure ratio increase of up to 36 will only give a modest gain in the specific fuel consumption. At these high pressures the temperature at the rearmost compressor blades becomes so high that the material properties will deteriorate. For a future generation of large engines for long-range transport aircraft $\varepsilon_c = 36$ may possibly prove to be the limit (Fig. 4-24), but there is also a notable tendency towards considerably lower values in order to achieve simplicity coupled with low cost. In the future we may expect a range of values between $\varepsilon_c = 12$ and 40, depending upon the use for which the engine is designed. The author's forecast is as follows:
- small executive aircraft, feederliners: 12 to 14
- short-range transport aircraft: 20 to 25
- long-range transport aircraft: 35 to 40

4.4.2. Turbine Entry Temperature

Returning to Fig. 4-22, the following may be observed:

a. For a given bypass ratio a TET value
which gives a minimum C_T can be found for
every OPR. This is henceforth indicated as
the "optimum" TET.

b. The optimum TET increases with increasing bypass ratio, the reason being that
for high λ the propulsive efficiency is
not so sensitive to increasing TET as it
is in the case of low bypass engines (Fig.
4-27).

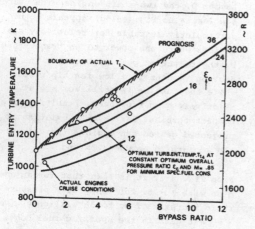

Fig. 4-27. Actual and optimum TET in cruising flight (civil engines)

c. At high bypass ratios, TET may be varied
within fairly wide limits above or below
the optimum value without greatly affecting
C_T. At λ = 8 and ε_c = 28, the TET will be
optimum at 1500 K (2700 R) in the example
presented. However, when the TET is chosen
10% lower (1350 K, 2430 R) it will cost
only about 1% in C_T.

d. For all combinations considered, a high
TET is favorable for obtaining a high specific thrust.

The maximum TET values used in actual engines for cruising conditions (Fig. 4-27)
are generally higher than the "optimum"
values. On the one hand, a high specific
thrust has the advantage of low engine
weight and installed drag, while, on the
other, it will mean that in normal long-
range cruise the engine will not operate
at the maximum permissible cruise TET, but
at a lower value which is closer to the op-

timum for low fuel consumption and long
operating life.

Maximum values for TET are higher during
takeoff than in cruise (Fig. 4-28) as they

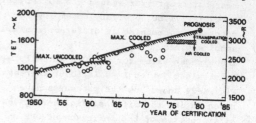

Fig. 4-28. Historic development of maximum
TET during takeoff (References: engine manufacturers' data)

will generally be required only for a limited period of time. In the case of low
bypass engines, the difference will be in
the region of 150 to 200 K (270 to 360 R),
but they will be less (50 to 100 K, 90 to
180 R) for high bypass ratios, since the
size of the engine is determined by the
thrust required in cruising flight.

At present, with the use of uncooled turbine blades, temperatures of about 1200 K
(1260 R) may be regarded as permissible
for short periods. When they are air-
cooled temperatures of up to 1600 to 1650
K (2880 to 2970 R) may be possible. Still
higher values may be achieved in future
with the use of transpiration cooling and
improved materials, although there is a
considerable difference of opinion among
experts on the actual gains to be expected.
Looking at the values used in practice over
the past years (Fig. 4-28), it may be noted
that the average increase comes to about
20 K (36 R) per year. If this trend continues in the future we may expect a generation of large, high bypass engines in
the eighties with a TET of 1800 K (3240 R)
during takeoff and 1750 K (3150 R) during
cruise. The data in Ref. 4-48 indicate that
these temperatures are feasible, with only
moderate quantities of air required for
cooling.

4.4.3. Bypass ratio

The choice of this parameter has such a far reaching effect on the design of the engine and its installatioh, that the optimization of λ cannot be fully dealt with here, in view of the complicated nature of the problem. Accordingly, we shall confine ourselves to summarizing some of the more important aspects.

a. Design of the fan and low pressure turbine.

The power supplied by the gas generator may be divided in various ways over the hot and cold flows. The division of power can be characterized by the work ratio,

$$\text{work ratio} = \frac{\text{power transferred to fan}}{\text{convertible energy}} \quad (4\text{-}29)$$

This ratio is controlled by the pressure ratio of the low pressure turbine and the Fan Pressure Ratio and determines the exhaust velocities imparted to the hot and cold gases. It can be shown (Ref. 4-30) that the propulsive (and overall) efficiency will be maximum when:

$$\frac{v_f}{v_g} = \eta_{tf} = \eta_t \cdot \eta_f \quad \begin{array}{l}(.70 \text{ to } .75 \text{ are} \\ \text{typical values})\end{array} \quad (4\text{-}30)$$

where η_t and η_f are the turbine and fan efficiencies respectively. Fig. 4-29 shows

that for moderate values of λ, the Fan Pressure Ratio (FPR) has no great influence on the specific fuel consumption, at any rate within certain limits. If, say, for λ = 5 we take an FPR of 1.55 instead of the thermodynamic optimum of 1.66, this will result in an increase of only 1% in C_T. The variation in the power absorbed by the fan, however, is closely related to that of the turbine. In view of the limited pressure ratio per stage, the number of turbine stages will have to be adapted for certain combinations of λ and FPR. If we now vary the FPR with a given number of turbine stages, the optimum will be found to lie at another value than in the previous case.

In Fig. 4-30 the optimum FPR is given as a

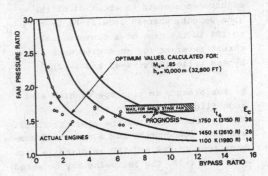

Fig. 4-30. Calculated and actual FPR

function of λ for some combinations of OPR and TET, calculated for M = .80 with the analytical method given in Appendix H. The points plotted in this figure represent values for the takeoff applicable to actual engines; corresponding values for cruising flight are higher (Section 4.4.2). The graph shows that for a given gas generator cycle the optimum value for the FPR decreases with increasing λ. For high bypass engines this effect will be partly compensated by the higher temperature and pressure levels which will generally be chosen in such cases, resulting in a greater specific convertible power. The maximum value for the FPR with a single stage fan will be about 1.60 to 1.70.

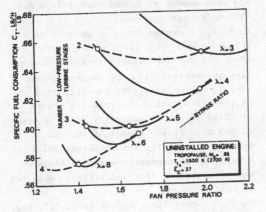

Fig. 4-29. Effect of fan pressure ratio on SFC and number of low-pressure turbine stages

When the bypass ratio is increased at a given rpm, the fan diameter and tip speed will also increase. In the case of many modern engines, part of the fan will work at transonic speeds which, amongst other things, will cause an increase in noise level (sub-para. f). This may be improved by using three engine shafts, thus reducing the rpm of the fan, or reducing fan speed by means of a reduction gear (geared fan, see Fig. 4-19d).

The Dowty-Rotol Company has developed a geared fan with rotor blades which can be adjusted in flight, so that an optimum adjustment may be obtained for various engine settings. Thrust reversal can be achieved by giving the rotor blades a negative pitch. The extra weight of the blade adjustment amounts to about 8% of the engine weight, whereas it will be over 15 to 20% in the case of a conventional thrust reverser for an engine with a high bypass ratio (Refs. 4-47 and 4-52).

b. Performance in high speed, high altitude flight.

The influence of λ on the specific thrust and fuel consumption for a number of combinations of OPR and TET is given in Figs. 4-31 and 4-32. The following may be noted:

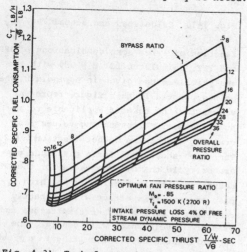

Fig. 4-31. Turbofan performance in cruising flight as affected by OPR and bypass ratio at constant TET

1. In the case of engines without intake and exhaust nozzle losses, C_T will continue to decrease with increasing λ, provided the optimum FPR is chosen for each configuration. A theoretical minimum will be reached with $\lambda \to \infty$, corresponding to a propulsive efficiency $\eta_{tf} = \eta_t \cdot \eta_f$.

2. The specific thrust decreases rapidly with increasing λ, although this effect may be partly compensated by choosing higher values for the TET.

3. When intake and exhaust nozzle losses are accounted for, there will be a bypass ratio for which C_T is minimum. The effect of these losses is greatest when λ is large and will even be accentuated under operational conditions by the power extraction required to drive aircraft systems and bleed air to the de-icing and air conditioning systems. The minimum installed C_T which may be achieved, and the corresponding value for λ, are largely dependent on the magnitude of these losses (Fig. 4-33).

4. Fig. 4-32 shows three combinations (points A, B and C), which are representative of different generations of engines, both recent and present, for transport aircraft. A radical improvement in C_T has been achieved in the present generation of large engines like the Rolls-Royce RB-211, Pratt and Whitney JT-9D and General Electric CF-6 high bypass ratio turbofan engines.

The development of future generations may take different courses.

Point D_1: A bypass ratio of about 9 at a TET of about 1750 K (3150 R) and an OPR of about 36. The improvement in fuel consumption as compared with the current generation is approximately 9% non-installed and 8% installed, while the specific thrust in cruising flight is about 15% less. In view of the high flying speeds of long-haul transport aircraft, a much larger value for λ may lead to excessive engine diameters and installation drag and will probably not be favorable.

Point D_2: Very large bypass ratios of, say, 20 to 30 at a TET of 1600 K (2880 R) and an OPR of 25. This configuration will imply a

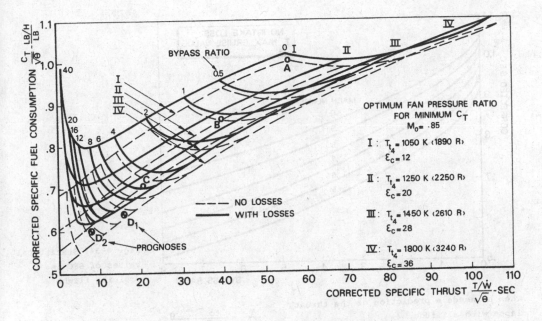

Fig. 4-32. Effect of bypass ratio and intake/exhaust losses on performance in cruising flight

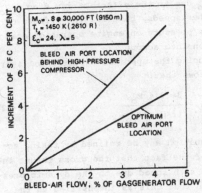

Fig. 4-33. Effect of bleed air takeoff on SFC

very large fan diameter ("prop-fan") and entails the use of a reduction gear. Fuel consumption may be about 15% less at the flying speed chosen for the case in Fig. 4-32, but the gain will be greater as the cruising speed decreases. The drag of these engines increases rapidly with cruising speed.

The engines in this forecast are suitable for different types of aircraft. Type D_1 may be particularly useful for fast long-range transport aircraft, while type D_2 may possibly be used on smaller and less rapid short-haul aircraft which should have good takeoff and landing performance and a very low noise level (sub-para. f).

The effect of altitude on engine performance is dependent on many factors. An approximation for the thrust lapse with altitude is occasionally found in the literature, for example:

$$\frac{\text{thrust at altitude}}{\text{thrust at sea level}} = \sigma^n \quad (n < 1) \qquad (4-30)$$

where both values of the thrust have been defined at the same Mach number and engine rating. Equation 4-30 should be considered as an interpolation method for calculating engine performance at altitudes where the engine manufacturer has not specified the thrust, rather than as a prediction method. The designer should refer to Appendix H or employ any other suitable method available

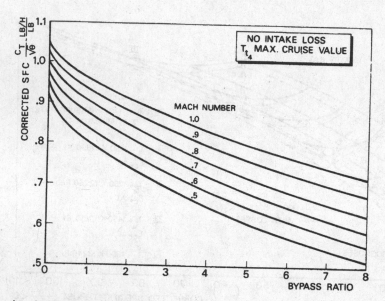

Fig. 4-34. Statistical values of SFC in cruising flight

when he needs a prediction of the thrust lapse with altitude.

Although engine performance can easily be predicted by using the method given in Appendix H, the working pressures and temperatures of the engine should either be known or chosen. When the designer is prepared to accept statistical averages for cruising flight, reasonably accurate figures can be found from Fig. 4-34 for efficient engines with high OPR. As against this, a comparison with Fig. 4-31 shows that for low OPR values the SFC will be considerably higher.

c. Takeoff performance and other flight conditions.

The specific fuel consumption is of less importance for performance in takeoff and climb, and we shall therefore pay rather more attention to the thrust lapse with speed, altitude and engine setting.

If T_{to} represents the static thrust at sea level, and if at the same time we assume that the gross thrust and mass flow through the engine do not change appreciably over a speed range of up to about M = .15, we may write:

$$\frac{T}{T_{to}} = \frac{T_{to} - \dot{m}\,V_o}{T_{to}} \tag{4-31}$$

In this expression T decreases linearly with V_o due to the effect of ram drag $\dot{m}\,V_o$. At higher speeds, however, the effects of dynamic pressure on engine operation must be taken into account.

Introducing the specific thrust into (4-31) it follows that:

$$\frac{T}{T_{to}} = 1 - \frac{34.714\,M_o}{(T/(\dot{W}\sqrt{\theta}))_{to}} \tag{4-32}$$

This equation may be refined by taking account of the fact that the gross thrust increases with speed due to the dynamic pressure and that this is intensified as the bypass ratio increases. Assuming the mass flow through the engine to be constant, we may deduce (Appendix H):

$$\frac{T}{T_{to}} = 1 - \frac{.45\,M_o\,(1+\lambda)}{\sqrt{(1+.75\lambda)}\;G} + (.6 + \frac{.11\,\lambda}{G})\,M_o^{\,2} \tag{4-33}$$

Here G is the gas generator function, defined in Section 4.3.1, which will typically be of the order of .9 to 1.2 during takeoff.

From inspection of (4-32), it will be clear that since the specific thrust decreases

with the bypass ratio (Fig. 4-23), the
takeoff thrust of a high bypass engine
will decrease more rapidly with speed than
is the case with a low bypass engine. Fig.
4-35, which is based on empirical data,

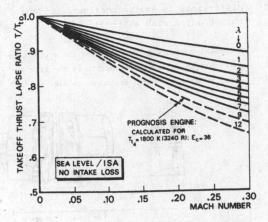

Fig. 4-35. Statistical curves of thrust
lapse during takeoff

may be used for performance calculations
in preliminary design. It shows the variation in thrust, deduced from engines with
λ ranging from 0 to 7. Both curves for
$\lambda = 9$ and 12 apply to a TET of 1800 K (3240
R) and an OPR of 36, and are computed with
4-33.

Some of the consequences of the above are
as follows:
1. If, for a particular aircraft design,
the thrust level is based on a requirement
relating to the runway length or climb
performance at low flying speed, the aircraft which is fitted with low bypass engines will have more thrust in cruising
flight than one with high bypass engines,
since the thrust decreases less rapidly
with speed and altitude.
2. If two aircraft designs, which in other
respects are entirely similar, have the
same maximum cruising speed at a certain
height, while their engines have the same
installed cruise thrust, the high bypass
engine will generally result in better
takeoff and climb performance.

For given aircraft design requirements for cruise
and takeoff, it is possible to find a bypass ratio
where the thrust required in both conditions will
be matched and this might form a basis for choosing
λ. In actual practice there are still other factors
to be considered:
1. Cruising altitude is often a fairly arbitrarily
chosen parameter which may still be varied within
certain limits.
2. The engine manufacturer can influence the engine
thrust lapse to some extent by adaptation of the
compressor design point. When, for instance, the
emphasis lies on the takeoff requirement, the design point of the compressor is chosen accordingly
so that the compressor efficiency will be maximum
during takeoff, which implies that it will be lower
during cruising flight. When, however, the cruising
flight is taken as the sizing condition, the design
point will be chosen with this requirement in mind.
3. The engine manufacturer will design the engine
for use in several types of aircraft. Once the engine configuration has been chosen it is still
possible to influence performance to a limited extent - for example by increasing the TET, which
will increase the quantity of air flowing through
the gas generator, though the engine will work
slightly less efficiently with regard to fuel consumption.

d. Engine weight.
From the decrease in the specific thrust
with λ (Fig. 4-23) one might tend to conclude that both weight and drag due to the
powerplant installation will increase. It
will be seen from Fig. 4-36 that this rea-

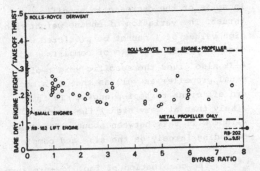

Fig. 4-36. Specific dry weight of turbojet
engines. (References Jane's All the World's
Aircraft, engine manufacturers' data)

soning is not entirely correct, at any rate where the specific engine weight for takeoff conditions is concerned. This may be explained by assuming that the dry engine weight comprises that of the gas generator and the propulsive device*. Assuming the first component proportional to $\dot{W}_g$ and the second proportional to the fan thrust, we may write:

$$W_e = C_1 W_g + C_2 T_f \qquad (4-34)$$

The static fan thrust as a fraction of the total takeoff static thrust can be derived from the analytical expressions in Appendix H as follows:

$$\frac{T_f}{T_{to}} = 1 - \frac{1}{\sqrt{1+\eta_{tf}\,\lambda}} \quad (\text{at } V_o = 0) \qquad (4-35)$$

Elaborating (4-34) further and applying it to actual types of engines, we find $C_1 = 10\,\varepsilon_c^{1/4}$ and $C_2 = .12$. This results in an approximation for the specific dry engine weight**:

$$\frac{W_e}{T_{to}} = \frac{10\,\varepsilon_c^{1/4}}{(T/W)_{to}\,(1+\lambda)} + .12\left(1 - \frac{1}{\sqrt{1+.75\lambda}}\right) (4-36)$$

The second contribution is about 20% higher for geared turbofans and another 20% should be added to it for variable pitch fans. Extra weight for acoustic treatment - if present - should be added (Table 8-8). In general it may be concluded that the specific weight decreases with increasing λ, in spite of the decrease in specific thrust. The variation of engine weight with high values of λ cannot be predicted with certainty. For the sake of comparison Fig. 4-36 also gives the specific weight of a metal propeller as well as that of a modern efficient turboprop engine. It is likely that at very high λ the specific weight will vary between about .15 and .25, depending largely on the size and complex-

*i.e. the combination of fan, low-pressure turbine and nozzle(s), see Fig. 4-17
**A more detailed weight prediction method will be found in AIAA Paper No. 70-669

ity of the engine.

When the specific weight is related to cruising conditions there is no clear tendency to rise or drop with λ. The wide variation observed is mainly caused by differences in engine configuration and thrust lapse with altitude.

A special category of engine where the accent lies particularly on low speed flight, is the lift engine, examples of which are shown in Fig. 4-37.

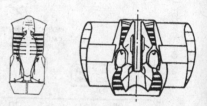

Subject	RB 162-81 LIFT JET	RB 202-36 BOOSTER FAN
Max. T.O. thrust (average perf.) ∿ lb	5992	13700
By-pass ratio	0	9.5
Airflow ∿ lb/sec	85	721
Specific thrust ∿ lb/lb/sec	70.5	19.0
Mean jet velocity ∿ ft/sec	2218	610
Specific fuel consumption ∿ lb/hr/lb	1.152	0.45
Thrust/Weight ratio	14.4	11
Diameter ∿ in	29	80.5
Height ∿ in	54	52.8
Noise at 1500 ft distance ∿ EPNdB	113	89

Fig. 4-37. Principal characteristics of Rolls Royce lift engines

These engines are intended for use on V/STOL aircraft and have found commercial application in the Hawker Siddeley Trident 3, where they only operate as a booster during takeoff. Their main characteristics are:

1. high thrust at low installed weight and small volume,

2. relatively short operational life.
These features permit a specific weight of only .08 to .10 to be achieved. In the Rolls-Royce RB-202 project the objectives include a low velocity of the exhausted engine gases, a low noise level and low specific fuel consumption. As a result of these requirements the general configuration and the engine cycle are very different from those of normal engines designed primarily for high speed propulsion.

e. Installation drag.
This is the drag of the engine mounted in
a nacelle on the aircraft. In accordance
with Fig. 4-38 the following contributions
to the total drag must be taken into account:
1. gas generator nacelle drag,
2. drag of the plug in the hot flow,
3. fan nacelle drag,
4. pylon drag, and
5. powerplant/airframe interference drag.
When the jet effluxes are separated, the
gas generator nacelle will be located in
the fan exhaust flow, which has a higher
velocity than the flight speed. The same
applies to the plug in the hot gases, which,
however, have a still higher velocity. The
extra drag is usually termed the scrubbing
drag* and it decreases with λ since the
velocities of the effluxes decrease. Curve
VIII in Fig. 4-38 has been calculated by
assuming optimum values for the FPR and a
gas generator of constant design. With a
given gas generator an increase in λ will
result in a larger fan diameter, a greater
wetted area of the fan nacelle and the pylon. This increase is approximately proportional to the static thrust. Because of
various counteracting effects, the total
installation drag - except interference
drag - $(D_1+D_2+D_3+D_4)$ does not vary appreciably within a variation of bypass ratios
between 3 and 10, especially when expressed
in relation to the thrust (curve VI in Fig.
4-38).
Although the data in Fig. 4-38 only apply
to a given gas generator, they present a
good impression of the trends. It should
be observed that the frontal area of the
engine is not a sound reference for weight
and drag.

Installation drag increases in proportion
to V_o^2. This increase may be reduced by
lowering λ, though the result will be an

*Due to the interrelationship between
scrubbing drag and the engine operating
conditions, this drag contribution is frequently considered as a thrust loss

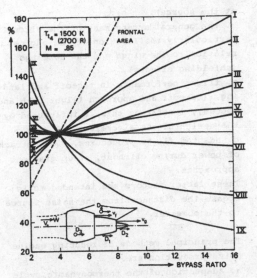

I   Static thrust T_{to}; fan diameter; D_3+D_4
II Basic dry weight W_e
III Specific weight, cruise condition, W_e/T_{cr}
IV Drag of installed engine in cruise condition
    $(D_1+D_2+D_3+D_4=D_e)$
V   Cruise thrust T_{cr}
VI Specific drag, D_e/T_{cr}
VII Specific weight at takeoff W_e/T_{to}
VIII Drag D_1+D_2
IX Specific thrust in cruise T_{cr}/W

Fig. 4-38. Effect of bypass ratio on thrust,
drag and weight for a given gas generator
(Refs. 4-33, 4-36, 4-38 and 4-53)

increase in C_m. In general it can be said
that the optimum value of λ for cruising
performance will decrease with the design
cruising speed.

4.4.4. Engine noise

In view of the fact that very few systematic design rules have as yet been established to reduce engine noise, we shall
have to confine ourselves to a qualitative
summary of the most important factors.
The external noise production of an aircraft will be largely influenced by:
a. the design of the engine and its installation in a nacelle (noise reduction

131

at the source),

b. the general arrangement of the aircraft, particularly the location of the engines in relation to the wings and the fuselage (shielding effects),

c. flight performance in takeoff and landing, (takeoff and approach flight path and speeds), insofar as this is influenced by the design of the aircraft,

d. adapted flight procedures, e.g. cutback of power during climbout, steep gradient approaches.

These latter factors are intended to increase the distance from the noise source to the observer.

The principal methods of reducing engine noise production are:

a. Adaptation of the thermodynamic cycle to reduce the aerodynamic noise, the most important single factor being the bypass ratio.

b. Measures to avoid the creation of machinery noise.

c. Measures to suppress noise that has been generated.

Fig. 4-39 gives a general impression of the variation in different noise sources with bypass ratio. In the straight turbojet the dominant source is the violent jet mixing process aft of the engine, together with the exhaust system. Jet noise suppressors reduce the noise level by some 5 PNdB. The reduced exhaust velocity of low bypass ratio engines lessens the tailpipe and jet noise, but at the same time the fan compressor and turbine noise are not far below this level. Turbomachinery noise frequently becomes dominant at reduced engine rpm. In the case of high bypass engines the fan noise is particularly predominant at all relevant power settings.

Practical measures which will lead to a low noise level are indicated in Fig. 4-40 and enumerated below:

1. Use of low efflux velocities will decrease the jet noise, particularly at low bypass ratios. Since an acceptable specific thrust is desirable, a fairly high TET will be chosen on high bypass engines.

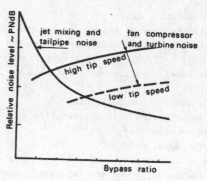

a. Variation of critical noise sources

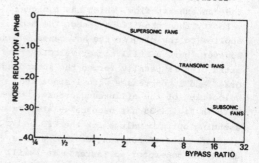

b. Potential noise reduction trend (Ref. 4-47)

Fig. 4-39. Effect of bypass ratio on noise production

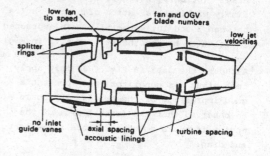

Fig. 4-40. Practical measures to reduce the noise of a turbofan installation

2. Use of jet noise suppressors on low bypass engines.

3. Readjustment of the ratio of the efflux velocities of the cold and hot gases relative to the theoretically optimum

value*. For a given λ, this will bring the combined jet and fan noise to a minimum.

4. Adjustable fan rotor blades, which will have the optimum setting with regard to noise during various flight conditions (variable pitch fans).

5. Low fan tip speed, which is made possible by adopting the three-shaft layout (Fig. 4-19C) or the twin spool layout with reduction gear (Fig. 4-19D).

6. Low pressure ratios of both the fan and the low pressure turbine.

7. Elimination of the inlet stator blades and an increase in the distance between the fan rotor blades and the stators.

8. Optimization of the number of rotor and stator blades of the fan.

9. Use of acoustic lining in the engine housing and/or sound suppressing rings (splitter plates) in the inlet and exhaust. This may reduce the noise by 5 to 10 PNdB.

10. Reduction of machinery noise, particularly turbine and tailpipe noise.

4.4.5. Summary and prognosis for the turbofan engine

a. For each bypass ratio there is a "thermodynamic" optimum for the fan pressure ratio. As λ increases this optimum becomes more marked and it will be more important to choose the optimum Fan Pressure Ratio.

b. The present bypass engine with λ = 5 to 8 shows a considerable improvement in specific fuel consumption as compared with the pure jet engine and engines with a low λ. A further improvement of about 15% will only be possible if λ is chosen equal to 20 to 30, combined with a high Overall Pressure Ratio and Turbine Entry Temperature.

c. Sensitivity to inlet and exhaust losses increases with bypass ratio, while for λ > 10 to 12 there will be a great increase in the total installed drag. These values of λ require another design conception for both the engine (fan reduction gear) and

*i.e. in accordance with the optimum work ratio; see Section 4.4.3 a.

the aircraft, due to the very large engine diameter.

d. The bypass ratio plays an important role in matching the takeoff, climb and cruise performance. When the accent is on low speed performance a high λ will be preferred, but when cruising performance is regarded as the critical requirement, an optimum will have to be found with respect to specific fuel consumption, installed weight and drag. The optimum λ decreases as cruising speed goes up.

e. An increase of λ results in a reduction in noise, provided the tip speed of the fan does not become too high. With bypass ratios larger than about 9 to 10 it will be necessary to adopt a fan reduction gear.

f. From extrapolations of TET and OPR we may expect a generation of engines in the eighties, with the main characteristics summarized in Table 4-2.

Application	Long-haul		short-haul, short runways	
Bypass Ratio	9		20 to 30	
OPR	36		25	
FPR	1.6		1.4 to 1.3*	
Condition	cruise	takeoff	cruise	takeoff
Mach number	.9	0	.7	0
TET	1750K (3150R)	1800K (3240R)	1500K (2700R)	1600K (2880R)
Spec. thrust (sec)	14	27.5	8 to 5.5	18 to 13.5
Corrected SFC $C_T/\sqrt{\theta}$ (h^{-1})	.64	.27	.56 to .52	.19 to .16

*geared fan

Table 4-2. Forecast of principal characteristics for two new generations of turbojet engines (all data apply to the uninstalled engine)

4.4.6. Engine performance in non-standard atmosphere

The thrust is found to be highly dependent on the ratio φ,

$$\phi = \frac{\text{Turbine Entry Temperature}}{\text{Atmospheric Temperature}} \qquad (4\text{-}37)$$

as well as on the temperature correction $\sqrt{\theta}$,

$$\theta = \frac{\text{ambient temperature}}{\text{standard ambient temp., sea level}} \qquad (4\text{-}38)$$

When the influence of the atmospheric temperature is determined on the assumption that the TET remains constant (curve I in Fig. 4-41), it is seen that at an air tem-

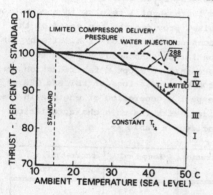

Fig. 4-41. Thrust in non-standard conditions

perature of, say, 95 F (35 C) at sea level the thrust will be 12½% below the value for standard conditions. If ϕ is taken constant, the thrust will be approximately proportional to $1/\sqrt{\theta}$ (curve II). In this case the loss in thrust will be considerably less, but at high atmospheric temperatures the TET will rise too high, with a resultant detrimental effect on the engine's service life.

The thrust of most modern engines is controlled according to curve III, where two regions can be distinguished:

a. A temperature region where the exit pressure of the compressor is limited. As the temperature increases the supply of fuel will be increased in such a way that the thrust will remain practically constant and the TET will also increase.

b. A region where the TET is limited so that the thrust will decrease as the am-

bient temperature rises.

Known as the method of flat rating, this ensures that the engine will deliver a practically constant thrust within a large range of ambient conditions encountered in operational practice.

Another method of reducing thrust losses at high atmospheric temperatures is to use water injection. A mixture of water and methanol is sprayed into the compressor inlet or injected at the combustion chamber inlet. As this evaporates it will cool the air and the mass flow through the engine will increase. The Turbine Entry Temperature is restored by the burning of methanol. With the use of water injection the standard thrust may be maintained up to about 95 F (35 C) and, if required, can be increased by about 3 to 7%. A water injection system is required, but this does not entail any drastic changes in the design of the engine, so that the engine manufacturer will offer it as an optional item.

4.5. ASSESSMENT OF TURBOPROP ENGINES

Contrary to expectations in the fifties, the development and use of turboprop engines in civil aviation has not been so widespread as that of turbojets. This is mainly due to:

a. the relatively great complexity of the combination of a gas, turbine engine with propeller reduction gear and constant speed propeller, as compared with the straight jet engine,

b. the higher flying speeds which can be attained with the jet engine, as compared with the propeller engine.

In the competition with the piston engine, the turboprop was only able to take the lead when technological progress enabled compressor pressure ratios to be raised to about 5 or 8 and Turbine Entry Temperatures of about 1200 to 1300 K (2160 to 2340 R) became feasible. This has resulted in the turboprop unit completely replacing the piston in the 500 to 3000 hp class of civil engines. Where engine first price is an important factor, however, the turboprop unit is still very much at a disadvantage.

4.5.1. Performance

The power developed by the gas generator is distributed partly to the propeller shaft and partly to the engine exhaust gases. It can be shown that the propulsive efficiency and the flying speed are the deciding factors in the optimum distribution of energy. The following optimum jet efflux velocity is found (Ref. 4-7):

$$v_{j_{opt}} = \frac{v_o}{\eta_{prop} \, \eta_t \, \eta_{mech}} \tag{4-39}$$

It is concluded that when the flight speed in the design conditions increases, the optimum gas efflux velocity will also increase. Consequently, when a turboprop engine is designed for relatively high speed operation, the thrust of the exhaust gases will be considerable at low speeds. If, on the contrary, the main emphasis is on high power output at low speeds, a low efflux velocity will be desirable (turboshaft engines).

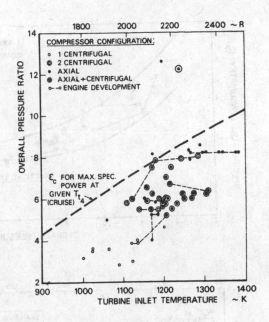

Fig. 4-43. Temperatures and pressure ratios of turboprop engines (data from Ref. 4-27)

When (4-39) is applied to a family of turboprop engines, the specific fuel consumption and power may be computed with the method explained in Appendix H. This has been carried out for a cruising speed of

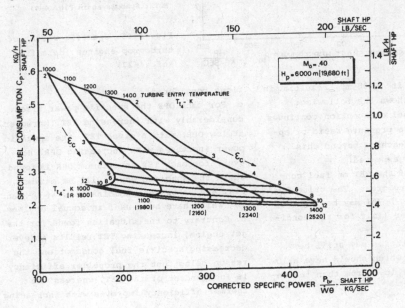

Fig. 4-42. Generalized performance of turboprop engines at cruising conditions

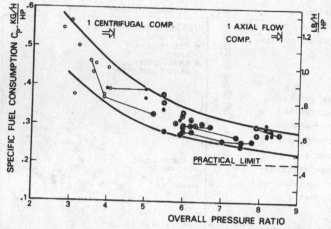

Fig. 4-44. Specific fuel consumption at sea level static conditions

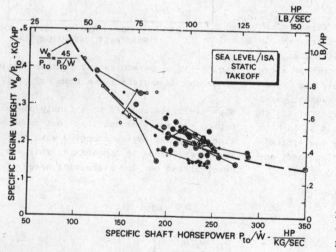

Fig. 4-45. Specific weight of turboprop engines (data from Ref. 4-27)

M_o = .4 at 19,680 ft (6000 m), resulting in Fig. 4-42, which shows the following:

a. The specific fuel consumption continues to decrease until a pressure ratio of approximately 8 is reached; beyond this little gain can be expected.[*]

b. The influence of the TET on fuel consumption is not very great. The value C_p = .45 lb/hp/h (.20 kg/hp/h) may be regarded as a practical lower limit for the conditions considered.

c. For each OPR the specific power rises considerably with increasing TET. Moreover, at low OPR, with a given TET, the specific power increases with increasing OPR. However, for each TET an optimum OPR is found, where the specific power will be maximum. Fig. 4-43 shows the combinations of OPR and TET which have been used in actual engines.

d. Contrary to the situation found for the jet engine, increasing TET results in ever decreasing specific fuel consumption, the reason being that the propeller efficiency is independent of the TET, whereas the thermal efficiency improves with increasing

[*]Greater pressure ratios may still be chosen for engines which should have a low fuel consumption at lower rpm than the design condition.

TET.

Fig. 4-43 shows that for a number of engines the values adopted for the OPR do not depart to any great extent from the calculated optimum. This leads us to think that further development of turboprop engines will not prove spectacular as far as specific fuel consumption and specific power are concerned. There are, however, still possibilities of achieving a low fuel consumption for engines which have to run at low power during very long flights, e.g. regenerative engines, for which the reader should consult Ref. 4-6.

The effect of the OPR on specific fuel consumption is shown statistically in Fig. 4-44.

4.5.2. Weight and drag

The weight of the turboprop engine is largely decided by the mass of air flowing through it and amounts to about 45 lb per lb/s of airflow. Therefore:

$$\frac{W_e}{P_{to}} = \frac{45}{(P/\dot{W})_{to}} \quad \text{(lb/hp or kg/hp)} \qquad (4-40)$$

From this expression, which is compared with actual examples in Fig. 4-45, it is concluded that a high specific power is important for low weight. To this we may add that the frontal area of the engine is also proportional to $\dot{W}$, so for the nacelle drag per hp it is possible to derive a similar relationship, as in (4-40).

4.5.3. Turboprop engine configurations

Older types of turboprop engines are fitted with a single centrifugal compressor. (See Fig. 4-46). Since OPR values of about 4 have been achieved, these engines show a specific fuel consumption of at least .65 to .75 lb/hp/h (.30 to .35 kg/hp/h). For this reason the air is often pre-compressed by means of an axial compressor and with this combination it will be possible to reach pressure ratios of the order of 10. These pressure ratios may also be obtained with a single axial compressor or with two centrifugal compressors in series. Some engines are fitted with two axial compressors on separate shafts and this makes a high pressure ratio possible (e.g. Rolls-Royce Tyne, OPR = 13.5, see Fig. 4-46b).

Many turboprop engines are fitted with a free turbine. This is a low pressure turbine which is not coupled mechanically to the gas generator, but drives the propeller through a separate shaft. With these engines it is often possible for the pilot to adjust the power in certain phases of the flight by selecting the propeller pitch (β-control). This demands a special device which governs the quantity of fuel supplied to the engine and controls the power output in such a way that the propeller rpm remains constant (cf. Section 6.3.3.).

The following features are offered by the free turbine:

a. When output has to be rapidly increased in a flight phase with low power, it will only be necessary to increase the rpm of the gas generator. The high-inertia propeller already revolves at the required rpm, thus making it possible to obtain a quick response.

b. In flight conditions which require control of the aircraft relative to a given glide path the use of β-control will give improved speed stability, particularly in the low speed regime.

c. Due to the aerodynamic coupling with the gas generator, the free turbine runs close to the optimum rpm under different working conditions, resulting in a high efficiency.

d. In the case of engine failure, the freely revolving propeller will only have to drive the free turbine and propeller drag immediately after the failure will be small.

There are various possible engine configurations which can be adapted to feature a free turbine, without resorting to the co-axial layout. Examples are the Allison 250 (Fig. 4-46c) and the Pratt and Whitney PT6A,

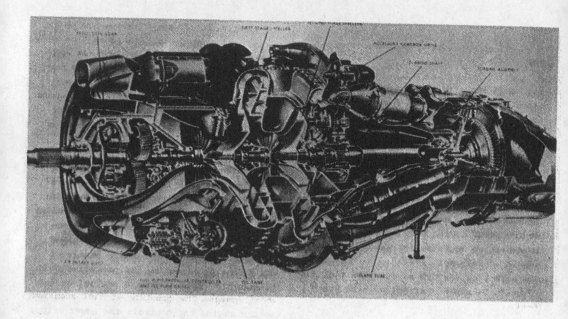

a. Rolls-Royce DART-Mk.510

b. Rolls-Royce TYNE 12

Fig. 4-46. Examples of
turboprop engine configurations (data can be found in
Table 6-2)

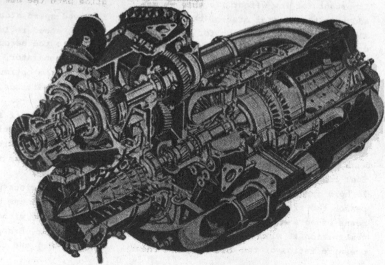

c. Allison Model 250

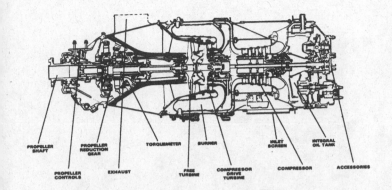

PROPELLER SHAFT · PROPELLER REDUCTION GEAR · TORQUEMETER · BURNER · INLET SCREEN · INTEGRAL OIL TANK · PROPELLER CONTROLS · EXHAUST · FREE TURBINE · COMPRESSOR DRIVE TURBINE · COMPRESSOR · ACCESSORIES

d. Pratt and Whitney PT-6A

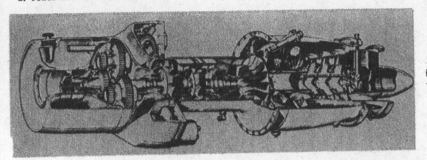

e. Turboméca Astazou XIV

Fig. 4-46. Examples of turboprop engine configurations. (data can be found in Table 6-2)

(Fig. 4-46d), which work on the "reverse flow" principle with reversal of the flow within the engine. The gases are ejected either in front or in the centre of the engine at the sides. The advantage lies in the very compact layout, demanding little space to accommodate the engine. This may be compared with the Turboméca Astazou (Fig. 4-46e) with the jet pipe at the rear of the engine. However, with the jet pipes at the sides it is not possible to take full advantage of the thrust of the engine gases, unless one or more curved jet pipes are used. This will lead to an increase in drag and a loss of thrust.

Chapter 5. Design for performance

SUMMARY

After laying down the broad outlines of the general arrangement and finalizing the design of the fuselage, the designer's next step will be to decide on the type of engine to be installed and the size of the wing. Both have a direct effect on performance and operating costs.

Basic relationships are presented for the performance of systematic design studies. Initial weight and drag estimation methods are given and boundary values are derived for aircraft parameters such as wing loading, thrust or power loading and aspect ratio. Some examples of the use of parametric studies and synthesis diagrams are discussed. Finally performance aspects related to noise reduction are dealt with on a qualitative basis.

Symbols

A	- wing aspect ratio, $A = b^2/S$	P	- power
$\bar{a}$	- mean deceleration or acceleration	P_{to}	- total (equivalent) takeoff horse-power of all engines
$a(a_o)$	- speed of sound (at sea level ISA)	$p(p_o)$	- static atmospheric pressure (at sea level ISA)
BFL	- Balanced Field Length		
b	- span or width; no index: wing span	R	- range
C	- rate of climb	Re	- Reynolds number, $Re = Vl/\nu$
C_o	- coefficient in generalized expression for L/D at V_2	r	- correction factor or ratio for drag estimation
C_D	- airplane drag coefficient, $C_D = D/\frac{1}{2}\rho V^2 S$	S	- distance; area (no index: wing area)
C_{D_i}	- induced or lift-dependent drag coefficient	SEP	- Specific Excess Power
C_{D_o}	- zero-lift drag coefficient	SLS	- Sea Level, Static condition
		T	- thrust; atmospheric temperature ($T_o = T$ at sea level ISA)
C_F	- mean airplane zero-lift drag coefficient, based on wetted area	T_j	- total jet thrust (turboprop engines only)
C_L	- lift coefficient, $C_L = L/\frac{1}{2}\rho V^2 S$		
$C_{L_{max}}$	- maximum lift coefficient, $C_{L_{max}} = W/\frac{1}{2}\rho V_S^2 S$	T_{to}	- total static takeoff thrust, all engines operating, uninstalled engines
C_p	- specific fuel consumption, propeller engines	$\bar{T}$	- mean thrust in takeoff run
C_T	- specific fuel consumption, turbo-jet engines	t	- time; section thickness
		Δt	- equivalent inertia time
c	- chord length	V	- speed
c_F	- friction coefficient of boundary layer	V_a	- approach speed
		V_R	- rotation speed
D	- drag	V_S	- stalling speed; takeoff config-
d_1, d_2, d_3	- factors defining the zero-lift drag coefficient		uration: V_{S_1}; landing configuration: V_{S_o}
E	- coefficient in generalized expression for L/D at V_2		
EPNdB	- equivalent, perceived noise, dB	V_{td}	- touchdown speed
e	- Oswald's induced drag factor, $1/e = \pi A(dC_{D_i}/dC_L^2)$	V_x	- engine failure speed
		V_1	- speed at the decision point
		V_2	- takeoff safety speed
		V_3	- initial climbout speed, all engines operating
f	- field length factor for takeoff (f_{to}) or landing (f_{land})	W	- weight
g	- acceleration due to gravity	W_e	- operational empty weight; weight empty equipped
h	- altitude; height; screen height in takeoff (h_{to}) or landing (h_{50})	W_{eng}	- dry weight of engines
		W_{fix}	- fixed part of W_e (i.e. not determined by W_{to})
ISA	- ICAO Standard Atmosphere		
k_W	- ratio of takeoff weight to actual weight	W_{to}	- (maximum) takeoff weight
		W_{var}	- variable part of W_e (i.e. determined by W_{to})
L	- lift		
l	- length	γ	- angle of climb or descent; ratio of specific heats
M	- Mach number		
N_e	- number of engines	$\bar{\gamma}$	- equivalent angle of climb or descent
n	- load factor ($n = L/W$); exponent		

- of relative density

γ_2 — second segment climb angle
Δ — increment
δ — relative static pressure; $\delta = p/p_o$
ζ_n — nacelle frontal area/cylinder volume per engine
η — efficiency
θ — relative atmospheric temperature; $\theta = T/T_o$
Λ — sweep angle
λ — bypass ratio
μ — friction coefficient; μ = retardation force/normal force
μ' — equivalent friction coefficient, including aerodynamic forces
ν — coefficient of kinematic viscosity
$\rho(\rho_o)$ — air density (at sea level ISA)
σ — relative density; $\sigma = \rho/\rho_o$
ϕ — power/engine frontal area
ψ — specific thrust (jet thrust/airflow); specific power (bhp/cylinder volume per engine)

Subscripts

air — air maneuver in takeoff or landing
comp — compressibility effects
cr — cruising flight
f — fuselage; flap; fuel
g — ground (run)
i — start of cruising flight
land — landing
LOF — lift-off
MD — minimum drag condition
n — engine nacelle(s)
p — payload
res — reserve fuel
run — takeoff or landing run
stop — deceleration phase during aborted takeoff or landing
t — tailplane (empennage)
thr — thrust reverser(s)
to — takeoff
trip — trip fuel (fuel burned)
uc — undercarriage
w — wing
c/4 — quarter-chord line
o — sea level conditions

5.1. INTRODUCTION

After deciding on the general arrangement (Ch. 2), incorporating this in initial design sketches, and finalizing the layout drawings of the fuselage (Ch. 3), the designer's next step will be to decide on the type of engine to be installed and the size of the wing. At this stage of the design the specified mission and flight performance will play an important part. In Chapter 4 it was concluded that engine performance is affected by many parameters, such as cycle temperatures and pressure ratios, bypass ratio, etc. The most important properties of the wing are wing sections and area, aspect ratio and high-lift devices. In a well-balanced design the various parameters and shape factors are combined so as to minimize both the initial costs and operating costs, while meeting all performance requirements.

In view of the limited availability of engines, the type of engine to be installed is occasionally chosen on the basis of limited performance studies. In such a case the wing design procedure is considerably simplified but the result may be unsatisfactory if the number of suitable engines is too restricted. If several engine types are available, a systematic study is required to find optimum combinations of wing design and types of engine. In this Chapter the basic relationships for the performance of such studies will be developed and elementary examples will be given. A complete evaluation of the effect of the design choices on the complete aircraft is virtually equivalent to carrying out as many design studies as there are parametric variations and many details must therefore remain undiscussed.

Simplified illustrations are presented to cover the more complex computational system and the reader may refine all relationships insofar as he thinks fit. The procedure is divided into three major parts:

a. Initial estimation of the empty weight, fuel weight, all-up weight and drag polars

(Sections 5.2 and 5.3).

b. Derivation of boundary values of the design parameters, based on "reversed" performance calculations (Section 5.4).

c. Some examples of performance optimizations (Section 5.5).

The actual choice of the type of engine and wing shape is based not only on performance calculations but also on other factors to be considered in Chapters 6 and 7.

In recent years, low noise production has become a requirement of major importance in aircraft design. Although the methods for achieving a low-noise design have not yet been settled, some aspects will be mentioned in this chapter.

It is emphasized that most of the methods presented here are intended as illustrative examples, applicable to an early stage of the design; they should be refined and improved as soon as more data become available. Chapter 8 and the appendices to this book contain more detailed data and many references to the relevant literature.

5.2. INITIAL WEIGHT PREDICTION

5.2.1. Stages in the estimation of airplane weight

In the early days of aviation, the design engineer was responsible for the design, stress analysis and weight control for the components of the aircraft. As design problems became more complex, specialized fields of engineering were developed. One of these fields is weight determination, control and coordination. Today, weight prediction and control are part and parcel of every phase in the design and development of every type of aircraft.

To be effective, weight prediction and control must be carried out during the early stages of the preliminary design, before a design configuration becomes "frozen". As soon as weight figures are distributed to the various departments of a design office as a starting point for further design evaluation, they will strongly resist any

attempt to alter design weights at a later stage. The weight data issued by the preliminary design department serve as goals for other engineering departments.

Weight prediction is necessary not only for stress and performance computations, but also for design optimization, as reflected in the following categories of estimating methodology:

1. Pre-configuration selection methods.
2. Configuration selection methods.
3. Post-configuration selection methods.

The general requirements to be satisfied by estimating methods are given below.

a. Pre-configuration selection methods.
"Weight guesstimates" are used during the period in the design where the mission requirements are practically the only objectives which are definitely known. The airplane size and structural arrangement cannot be finalized until systematic studies have been made to determine the optimum design. The prediction methods to be used should be elementary in nature, and the use of statistics is appropriate when it produces rapid answers. Some examples are given in Section 5.2.2. and in References 5-6 to 5-10.

b. Configuration selection methods.
The methods normally employed are semi-empirical and seek to account for weight variations due to changes in major design parameters. Such methods are valuable in parametric studies aimed at finding out which combination of parameters yields the best compromise of all the input variables. Not only must the methods used provide accurate answers in an absolute sense, but the predicted effect of the variation of each parameter should be equally correct. The computation methodology is iterative in nature, i.e. starting values for the design weights must be assumed or based on previous estimates, and after completion of the weight breakdown the new values must be used to start a further computation cycle, continuing in this way until the process has con-

verged to within the required accuracy. Examples of these types of methods will be presented in Chapter 8.

c. Post-configuration selection methods. In this phase a baseline configuration has been selected and it is necessary to analyse the loads and weight in greater detail. Major elements of such a method are a detailed weight breakdown and the use of preliminary stress analysis to determine the amount of structural material required to resist the applied loads and provide adequate stiffness. Weights must be added for joints, cut-outs, splices and other features which complicate the structural arrangement. Design specifications of all systems must be available to estimate their configuration, power requirements and weight. Due to the very complex character of the type of work involved, comprehensive methods covering all conceivable aircraft types are not readily available. However, many examples of rational methods applicable to detail weights can be found in the literature. The Society of Allied Weight Engineers (SAWE) has published many of these and a selection of papers appears in the list of references to Chapter 8.

5.2.2. Examples of weight "guesstimates"

The takeoff weight is the sum of operating empty weight, payload and fuel weight:

$$W_{to} = W_e + W_p + W_f \qquad (5-1)$$

The empty weight can be considered as the sum of a fixed weight and a variable weight:

$$W_e = W_{fix} + W_{var} \qquad (5-2)$$

The actual subdivision can be adapted to the case under consideration. By way of example, if the engine to be used is known at this stage, it will be considered as a fixed weight. Like the fuel load, the variable weight can be considered as a fraction of the takeoff weight, resulting in:

$$W_{to} = \frac{W_p + W_{fix}}{1 - \dfrac{W_{var}}{W_{to}} - \dfrac{W_f}{W_{to}}} \qquad (5-3)$$

This equation will now be evaluated for several airplane categories.

a. Light aircraft with piston engines. One important contribution is the engine weight, which is frequently a known factor. Alternatively, it can be estimated from Fig. 4-12. We then write:

$$W_{to} = \frac{W_p + W_{eng}}{1 - \dfrac{W_{var}}{W_{to}} - \dfrac{W_f}{W_{to}}} \qquad (5-4)$$

The following averages were found from data relating to some 100 light aircraft:

$$\frac{W_{var}}{W_{to}} = \frac{W_e - W_{eng}}{W_{to}} = .45 - \text{fixed gear}$$
$$= .47 - \text{retractable gear}$$
$$= .50 - \text{utility category}$$
$$= .55 - \text{acrobatic category}$$

normal category (5-5)

$$\frac{W_f}{W_{to}} = \text{constant } \frac{R}{1000} \, r_{uc} \, A^{-.5} + .035 \quad (5-6)$$

where the constant is .31 when R is in n.m. or .17 when R in km.
The factor r_{uc} accounts for undercarriage drag (see Section 5.3.2.).

b. Turbojet and turbopropeller aircraft. In principle, cruise fuel can be determined with the well-known Bréguet range equation. However, this is complicated by the fact that extra fuel is required for takeoff, climb, descent, reserves, etc. To avoid lengthy computations an estimate can be made as follows:
Turboprop aircraft: the total fuel quantity is obtained from Fig. 5-1.
Turbojet aircraft: the fuel is split up into trip fuel and reserve fuel. Trip fuel is given by Fig. 5-2, explained in Section 5.4.2., and reserve fuel is given by eq. 5-46 or 5-47.

145

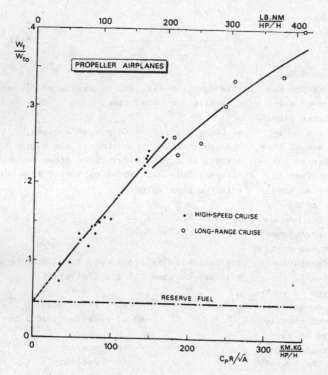

Fig. 5-1. Estimation of fuel weight fraction for turboprop aircraft (small and transport category)

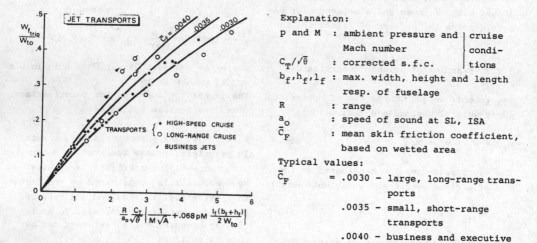

Explanation:

p and M : ambient pressure and Mach number ⎫ cruise conditions

$C_T/\sqrt{\theta}$: corrected s.f.c. ⎭

b_f, h_f, l_f : max. width, height and length resp. of fuselage

R : range

a_o : speed of sound at SL, ISA

$\bar{C}_F$: mean skin friction coefficient, based on wetted area

Typical values:

$\bar{C}_F$ = .0030 - large, long-range transports

.0035 - small, short-range transports

.0040 - business and executive jets

Fig. 5-2. Estimation of trip fuel weight fraction for jet airliners and executive aircraft

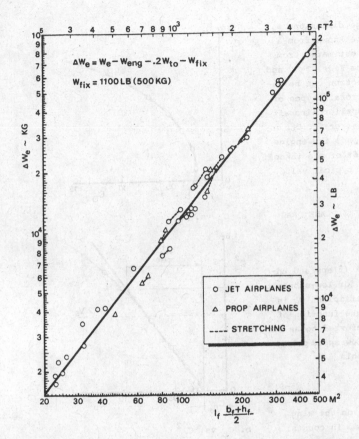

$$\Delta W_e = W_e - W_{eng} - .2W_{to} - W_{fix}$$

$$W_{fix} = 1100 \, LB \, (500 \, KG)$$

○ JET AIRPLANES
△ PROP AIRPLANES
---- STRETCHING

$l_f \frac{b_f + h_f}{2}$

Fig. 5-3. Chart to estimate
the empty weight of transport
and executive airplanes

The empty weight of light aircraft ($W_{to} <$ 12,500 lb, 5670 kg) is roughly 60% of the takeoff weight, hence:

$$W_{to} = \frac{W_p}{.4 - W_f/W_{to}} \qquad (5-7)$$

For transport aircraft with $W_{to} >$ 12,500 lb (5,670 kg), the accuracy of the empty weight prediction can be improved by splitting it up into:
- the weight of the dry engines,
- a fixed weight of approximately 1,100 lb (500 kg),
- a constant fraction of the takeoff weight, mainly associated with wing and undercarriage structure,
- a weight group dependent upon the fuselage size.

A justification for this approach is given in Fig. 5-3. The 20% fraction of W_{to} is derived from data relating to several developed aircraft types, leading to take-off weight growth, without changes in fuselage dimensions. The importance of the fuselage size is obvious not only from the direct contribution of the fuselage structure, but also from the fact that the weight of such items as sound insulation, wall trim, floor covering and the airconditioning system, etc. is closely related to the fuselage dimensions as well.
The result of this approach is:

$$W_{to} = \frac{W_p + W_{eng} + W_{fix} + \Delta W_e}{.8 - \frac{W_f}{W_{to}}} \qquad (5-8)$$

where ΔW_e is derived from Fig. 5-3 and W_{fix} is equal to 1,100 lb (500 kg).

147

At this stage, the fuselage dimensions ℓ_f, b_f and h_f will generally be known from a layout drawing, or may be estimated from statistical data (e.g. from Figs. 3-12 and 3-13). The fuel weight fraction can be estimated from comparable aircraft types or from Fig. 5-1 for turbopropeller aircraft and Fig. 5-2 for turbojet aircraft. The engine weight will be known once the engine is chosen. Otherwise 5 to 6% of the takeoff weight may be assumed as a typical value.

5.3. INITIAL ESTIMATION OF AIRPLANE DRAG

5.3.1. Drag breakdown

In the cruise configuration (flaps and undercarriage retracted) and for low-subsonic flight speeds, the drag coefficient C_D is usually expressed as a unique function of the lift coefficient C_L, referred to as the airplane drag polar. Most low-speed polars are approximated by a parabola,

$$C_D = C_{D_o} + C_{D_i} = C_{D_o} + \frac{C_L^2}{\pi Ae} \qquad (5-9)$$

The drag coefficient, based on the wing area, is frequently expressed in counts; one count = .0001. The symbol $A = b^2/S$ denotes the aspect ratio of a wing with span b and gross planform area S (definitions in Appendix A-2).
When the actual drag polar, as determined from measurements in test flights or a wind tunnel, is compared with eq. 5-9, it is found that in a practical region of C_L values the straight line approximation of C_D vs. C_L^2 is acceptable (Fig. 5-4).
Contrary to the exact polar curve, having a minimum at some small positive value of C_L, eq. 5-9 indicates that C_D is minimum for $C_L = 0$. The zero-lift drag coefficient C_{D_o} is a fictitious quantity for the general case of a non-symmetrical aircraft, the actual C_D for $C_L = 0$ being slightly higher. The induced or lift-dependent drag coefficient can be compared to the theoretical induced drag of a wing with elliptic spanwise lift distribution,

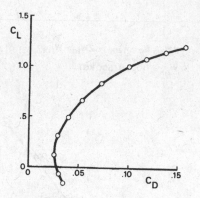

a. C_D vs. C_L

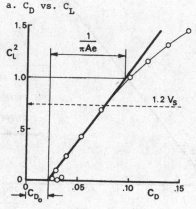

b. C_D vs. C_L^2

Fig. 5-4. Typical low-speed polar curve

$$C_{D_i} = \frac{C_L^2}{\pi A} \qquad (5-10)$$

The Oswald efficiency factor e in eq. 5-9 accounts for the non-ellipticity of the lift distribution, the increase of profile drag of the wing, fuselage, tailplane, nacelles and various interference effects with the angle of attack.

A preliminary stage of drag estimation may be accomplished by adding the individual drag contributions of the various components of the airplane:

$$C_{D_o} = \frac{\sum C_{D_j} S_j}{S} \qquad (5-11)$$

The product $C_{D_j} S_j$ is the drag area of each

component. For wings with thickness/chord ratios of up to 20% and slender fuselages (length/diameter ratio greater than about 4), skin friction drag is predominant and it is customary to deduce the drag coefficient of the components from their wetted area. Table 5-1 shows typical low-speed drag figures for various classes of airplanes in the cruise configuration.

	C_{D_o}	e
high-subsonic jet aircraft	.014 - .020	.75 - .85*
large turbopropeller aircraft	.018 - .024	.80 - .85
twin-engine piston aircraft	.022 - .028	.75 - .80
small single engine aircraft		
retractable gear	.020 - .030	.75 - .80
fixed gear	.025 - .040	.65 - .75
agricultural aircraft:		
– spray system removed	.060	.65 - .75
– spray system installed	.070 - .080	.65 - .75

* The higher the sweep angle, the lower the e-factor

Table 5-1. Drag figures for various aircraft types

5.3.2. Low-speed drag estimation method

The example of a drag prediction method presented in this section, is a very elementary approach, which can be refined if desired. A more detailed drag prediction method is given in Appendix F.

The drag of an aircraft component may be estimated by comparing it to the friction drag of an equivalent flat plate having the same wetted area and length. For this comparison to be valid, the boundary layer must develop in a similar manner for both cases and therefore the Reynolds number, with respect to the length, must be equal, while the transition from laminar to turbulent flow is assumed to occur at the same distance from the nose or leading edge.

A second condition which must be fulfilled for the flat plate analogy to be valid is that there should be no appreciable regions of separation, and the aircraft components should therefore be well-streamlined, moderately cambered and smooth in shape. With regard to sharp corners, rear fuselage upsweep or short stubby fuselage tails, for example, the flat plate analogy yields no realistic answers. In such cases reference should be made to experimental data, a valuable collection of which is to be found in Ref. 5-12.

To account for distributed surface irregularities and roughness drag, the boundary layer is sometimes assumed to be fully turbulent. In this case the friction coefficient according to Prandtl-Schlichting, based on the wetted area, is given by:

$$C_F = \frac{.455}{(\log Re)^{2.58}} \quad (5-12)$$

This equation is depicted in Fig. 5-5 in a normalized form, i.e. the value of C_F for $Re = 10^8$ is taken equal to 1. To account for the thickness of the body, shape factors representing the ratio of actual drag to flat plate drag are given in literature for wing sections and circular streamline bodies (e.g. in Ref. 5-12). For each aircraft the exposed (wetted) area must be calculated, interference effects estimated and extras added to account for protuberances, flap and control surface slots, cockpit windows and the like (Appendix F). A detailed drag estimation is usually a very elaborate exercise, for which most airplane manufacturers have developed their own procedures. For the purpose of sizing the airplane and the engine, a somewhat simplified approach based on statistical data is suggested here. In the present method, the zero-lift drag will be calculated according to the following basic equation:

$$C_{D_o} S = r_{Re} r_{uc} \left[r_t \left\{ (C_D S)_w + (C_D S)_f \right\} + (C_D S)_n \right] \quad (5-13)$$

The various contributions will be explained

149

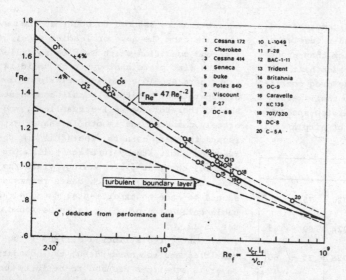

Fig. 5-5. Correction factor
on C_{D_o} for scale effects,
roughness, etc

below. Base* drag is not included but may
be accounted for, if present, by assuming
$\Delta(C_DS) = .13$ times the projection of the
base area on a plane normal to the flow.

a. Wing.
Uncorrected drag area for smooth wings:

$$(C_DS)_w = .0054 \; r_w\{1+3(t/c)\cos^2\Lambda_{.25}\}S \quad (5-14)$$

where r_w = 1.0 for cantilever and 1.1 for
braced wings
t/c = mean thickness/chord ratio
$\Lambda_{.25}$ = sweep angle at the quarter-chord
line
S = gross planform area (Appendix
A-2)

Eq. 5-14 is derived from the flat plate
analogy and a typical thickness correction
for t/c up to .20. The "uncorrected" drag
figure applies to a smooth wing with the
transition region at approximately 10% of
the chord, for a Reynolds number of 12.5
million, based on the geometric mean chord
length.

* Base: a surface, usually at the rear end
of the fuselage, more or less normal to
the flow.

b. Fuselage.
Uncorrected parasite drag for streamline
shapes:

$$(C_DS)_f = .0031 \; r_f l_f \; (b_f + h_f) \quad (5-15)$$

l_f = fuselage length, including propeller
spinner or jet engine outlet, if
present,
b_f, h_f= max. width and height of the major
cross section, including canopy,
r_f = shape factor, i.e. the ratio of actual wetted area to that of a fuselage with elliptical or circular
cross-section and cylindrical mid-
section, for which $r_f = 1.0$,
r_f = 1.30 - rectangular cross-section,
= 1.15 - one side of cross-section
rectangular, other side rounded off,
= $.65 + 1.5 \dfrac{diameter}{length}$ - fully streamlined fuselages without cylindrical
mid-section.

The "uncorrected" drag figure applies to a
fuselage with fully turbulent boundary
layer, for a Reynolds number equal to 100
million, based on l_f. Calculations have
shown that the thickness correction, multiplied by the ratio of gross wetted area to
the cylinder area $.5 \; \pi \; l_f \; (b_f + h_f)$, yields
an almost constant value of approximately
.93 for most practical length/diameter ra-

tios. Great care should be taken when applying eq. 5-15 to aircraft with rear fuselage upsweep, bluff canopies, etc., for which considerably higher drag figures must be expected (cf. Section 3.5.1. and Fig. 3-27).

c. Tailplane.

By analyzing tailplane drag in the same way as the wing drag, it was found that in practical cases the tailplane contribution amounts to roughly 24% of fuselage plus wing drag; hence $r_t = 1.24$ is a good average. On STOL-type aircraft r_t may be as high as 1.30 due to the relatively large tailplane area.

d. Engine installation and nacelles.

The parasite areas will be related to the installed thrust or power, in order to account for variations of nacelle or intake scoop size and shape with engine thrust or power.

TURBOJET ENGINES

The isolated engine pod is taken as the reference case; its wetted area can be related to the engine mass flow, taking due account of differences in pod design for variation in bypass ratio. On the basis of actual pod shapes and wetted areas, an uncorrected friction coefficient of .003 and typical scrubbing and supervelocity drag values for cruising flight, the following expression was found:

$$(C_D S)_n = 1.72 \; r_n \; r_{thr} \left(\frac{5+\lambda}{1+\lambda}\right) \frac{T_{to}}{\psi_{to} P_o} \qquad (5-16)$$

where T_{to} and ψ_{to} refer to standard SLS conditions and

r_{thr} = 1.0 - thrust reversers installed,
     = .82 - no thrust reversers.

The factor r_n is an installation factor to account for pylon and interference drag in the case of podded engines. For buried engines r_n represents drag due to intake scoops, exhaust pipes, etc.

The following figures are suggested as typical:

r_n = 1.50 - all engines podded
   = 1.65 - two engines podded, one buried in fuselage tail (this figure includes internal drag of the central inlet)
   = 1.25 - engines buried in nacelles, attached on the side of the fuselage
   = 1.00 - engines fully buried, intake scoops on fuselage
   = .30 - engines fully buried, wing root intakes

It should be noted that the factor $(5+\lambda)/(1+\lambda)$ in eq. 5-16 has no physical meaning; it has been derived on a statistical basis from the ratio of wetted area to frontal area for actual engines with varying bypass ratios.

TURBOPROP ENGINES

The uncorrected nacelle drag is:

$$(C_D S)_n = .1 \; r_n \; \frac{P_{to}}{\phi_{to}} \qquad (5-17)$$

where P_{to} and ϕ_{to} refer to standard SLS conditions and

r_n = 1.0 - ring-type inlets
   = 1.6 - scoop-type inlets, increasing the frontal area

Eq. 5-17 is based on a typical drag coefficient .065 based on the nacelle frontal area, while for ring-type inlets the engine frontal area is approximately 65% of the nacelle frontal area.

PISTON ENGINES

Wing-mounted:

$$(C_D S)_n = .07 \; \zeta_n \; \frac{P_{to}}{\psi_{to}} \qquad (5-18)$$

where P_{to} and ψ_{to} refer to SLS-ISA conditions and

$$\zeta_n = \frac{\text{nacelle frontal area}}{\text{cylinder volume}}$$

Typical values for ψ_{to} may be obtained from Section 4.2 or from engine manufacturers' data. A favorable effect of size is reflected in the factor ζ_n, which is of the order of .012 to .015 sq.ft per cu.in. (.07-.09 m^2/liter) for engine powers up to

151

500 hp, but may go down to .05 to .07 sq.ft per cu.in. (.03 to .04 m^2/liter) for the 2,000-4,000 hp class.
Fuselage-mounted tractor engines:

$$(C_D S)_n = .015 \ b_f h_f \qquad (5-19)$$

e. Undercarriage.
For undercarriages which are fully retracted within the airplane external lines* a drag penalty is not necessary ($r_{uc} = 1.0$). Other values can be derived by retracing differences in measured performance of aircraft with fixed and retractable gear versions and from published data on drag:
$r_{uc} = 1.35$ - fixed gear, no streamlined
             wheel fairings
    $= 1.25$ - fixed gear, streamlined wheel
             fairings and struts
    $= 1.08$ - main gear retracted in stream-
             lined fairings on the fuselage (like
             C-130 and C-5A)
    $= 1.03$ - main gear retracted in na-
             celles of turbopropeller engines

f. Wing tip tanks.
A typical figure, such as $\Delta(C_D S) = .055$ times the tank's frontal area, should be added to eq. 5-13. To compute the effective aspect ratio, the effective wing span and area should be taken as the distance and area between the tank centerlines (Ref. 5-12).

g. Corrections for Reynolds number and miscellaneous drag.
It can be argued that interference effects, surface irregularities, excrescences, air scoops, aerials, slots, etc. generally affect the boundary layer more on small low-speed aircraft than they do on large high-speed aircraft, due to differences in relative size and emphasis on pure aerodynamic design. In evaluating the present method, the ratio of actual measured to basic zero lift drag was therefore plotted versus the Reynolds number, based on fuselage length (Fig. 5-5).

*i.e. no wheel fairings are required

The following approximation accounts for the effect of the Reynolds number on turbulent skin friction drag and miscellaneous drag items:

$$r_{Re} = \frac{\text{actual zero-lift drag coefficient}^*}{\text{uncorrected drag coefficient}} =$$

$$= 47 \ Re_f^{-.2} \qquad (5-20)$$

where

$$Re_f = \frac{V_{cr} \ l_f}{\nu_{cr}} \qquad (5-21)$$

The subscript "cr" refers to the design cruising altitude and speed. Figure 5-5 shows that the miscellaneous drag contributions amount to 25-30% for light aircraft and 10-15% or less on large transport aircraft. The rms error of the method is 4%, but this figure also allows for inaccuracy in the available data on drag polars.

5.3.3. Compressibility drag

Compressibility effects on drag are generally ignored at Mach numbers below .5. The drag polars of a high-subsonic transport aircraft in Fig. 5-6 illustrate that for

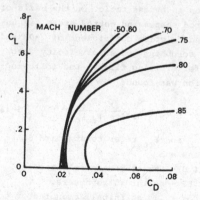

Fig. 5-6. Effect of compressibility on the drag polar

*including excrescences, protuberances, roughness, etc.

low C_L values and Mach numbers up to .70 the effects of compressibility on drag are of secondary nature. Between M = .70 and .80 a steady increase in the drag is observed ("drag creep") and at a critical Mach number of approximately .85 a rapid rise is experienced in both the zero-lift drag and the induced drag. This drag rise is caused by shock waves and boundary layer separation induced by these shock waves. The endeavor to achieve low compressibility drag, or rather a high drag-critical Mach number, is one of the most comprehensive tasks in aerodynamic design. This subject is dealt with more fully in Section 7.2. For the purpose of initial design calculations, it is fair to assume that the aerodynamicists will achieve acceptable drag figures at the design cruising speed, provided that wing sweep and thickness ratio are chosen appropriately. We may therefore assume that:

$\Delta C_{D_{comp}}$ = .0005 - long-range cruise conditions

$\Delta C_{D_{comp}}$ = .0020 - high-speed cruise conditions

5.3.4. Retracing a drag polar from performance figures

In order to compare an estimated drag polar with the drag figures of existing airplanes, drag coefficients may be deduced from performance data as supplied by the airplane manufacturer. To do this for a given type of aircraft, the lift coefficient in cruising flight is calculated first,

$$C_{L_{cr}} = \frac{W/S}{\frac{1}{2}\rho V^2}\bigg]_{cr} \qquad (5-22)$$

Cruising altitude and speed are obtained from the manufacturer's brochure or other appropriate publications. The drag due to lift,

$$C_{D_{i_{cr}}} = \frac{C_{L_{cr}}^2}{\pi Ae} \qquad (5-23)$$

is estimated by assuming a typical value for e (cf. Table 5-1) and taking a condition with a fairly low C_L. As thrust equals drag in cruising flight, we may write:

$$C_{D_{cr}} = \frac{T}{\frac{1}{2}\gamma p M^2 S}\bigg]_{cr} \qquad \text{(jet airplanes)} \quad (5-24)$$

or

$$C_{D_{cr}} = \frac{\eta_p P_{br}}{\frac{1}{2}\rho V^3 S}\bigg]_{cr} \qquad \begin{array}{l}\text{(propeller}\\ \text{airplanes)}\end{array} \quad (5-25)$$

For jet-propelled aircraft, T is the installed thrust, which is lower than the uninstalled thrust due to bleed air, power offtakes and inlet pressure loss. A typical figure may be 4% reduction, but for high bypass engines it may well be as high as 8%. For propeller aircraft η_p represents the combined effect of installed propeller efficiency, extra drag of aircraft components in the slipstream, intake losses, power and bleed air offtakes and cooling drag. Typical figures are:

η_p = .85 - turbopropeller engines (including jet thrust)

= .80 - wing-mounted piston engines

= .78 - tractor piston engine in fuselage nose.

Cruise power or thrust may be obtained from the engine brochure for the appropriate rating, altitude and flight speed. In the case of piston engines, ratings are usually 75% or 65% rated power for performance and economic cruising respectively. Finally, the zero-lift drag is calculated:

$$C_{D_o} = C_{D_{cr}} - C_{D_i}\bigg]_{cr} \qquad (5-26)$$

A more accurate determination can be made by analyzing several flight conditions in a similar way. By plotting the values of C_D vs C_L^2, C_{D_o} and e can be found by a straight-line approximation, in the manner indicated in Fig. 5-4.

5.3.5. Drag in takeoff and landing

Drag in the en route configuration determines the cruise thrust and hourly fuel

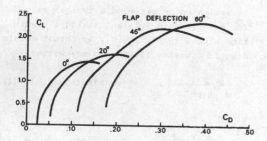

Fig. 5-7. Low-speed polars for a light transport

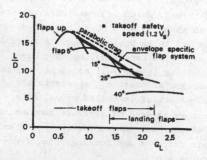

Fig. 5-8. Typical locus of lift/drag ratios for the takeoff safety speed

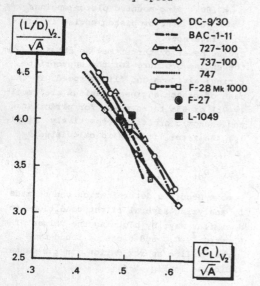

Fig. 5-9. Generalized takeoff lift/drag ratios

consumption. Alternatively, drag in the low-speed configuration (flaps deflected) determines the permissible takeoff weight for a particular flight of a passenger airliner and hence the maximum useful load.[*] On a hot and high airfield, a limitation on takeoff weight may result in a reduced payload or a restricted fuel weight and range. In such a case a drag deterioration of 10% may result in 30% less payload to be carried.

Typical polars for several flap deflections for takeoff and landing are shown in Fig. 5-7. As the available climb gradient,

$$\gamma = \frac{T}{W} - \frac{C_D}{C_L} \qquad (5-27)$$

depends on the lift/drag ratio, the aerodynamic characteristics are sometimes plotted accordingly (Fig. 5-8).[**] For each flap angle, the L/D ratio at the takeoff safety speed V_2 are indicated on the curves. A line connecting these points forms an envelope or locus, which is useful in comparing one flap system with another. Since accurate prediction of the results of a flap design program is obviously impossible, it is useful to compare the V_2 locus on a basis of generalized parameters, by writing the drag polar as follows:

$$\frac{C_L/C_D}{\sqrt{A}} = \left(\frac{C_{D_o}}{C_L} \frac{\sqrt{A}}{} + \frac{C_L}{\pi e \sqrt{A}} \right)^{-1} \qquad (5-28)$$

The values of C_{D_o} and e refer to the configurations with flaps deflected. Fig. 5-9 presents a collection of some published V_2 loci. The following approximation is suggested for preliminary design purposes:

$$C_{D_{V_2}} = C_o + \frac{\left(C_{L_{V_2}} \right)^2}{\pi A E} \qquad (5-29)$$

where C_o = .018, E = .70 slats extended
C_o = .005, E = .61 no slats or slats retracted.

[*] Definition in Section 8.2.1.
[**] A more complete figure is shown on Fig. 11-4

154

Note that this equation does not represent an actual aircraft polar; it refers to the initial climb-out after takeoff. To include drag due to engine failure at low thrust/weight ratios, E may be reduced by approximately 4% for wing-mounted engines and 2% for engines mounted on either side of the fuselage tail.

Although the accuracy of eq. 5-29 for the approach and landing configurations is probably not so good, due to the higher ratio of flight speed to stalling speed, it still remains a useful first-order approximation.

5.4. EVALUATION OF PERFORMANCE REQUIREMENTS

The objective of performance analysis is to predict the performance of an airplane type of given design and geometry. The designer, however, is confronted with the reverse problem: knowing the performance objectives, he must find combinations of design characteristics which will result in a design that satisfies or exceeds all requirements.

The intention of this section is to translate the performance requirements into boundary values for those major aircraft parameters and characteristics, which have a first-order effect on performance:

a. Powerplant.

1. Total SLS takeoff thrust (power) of all engines, uninstalled, usually combined with the takeoff weight in the thrust/weight ratio (T_{to}/W_{to}) or power/weight ratio (P_{to}/W_{to}). Thrust loading or power loading is the reciprocal value of this: W_{to}/T_{to} or W_{to}/P_{to}.
2. The number of engines, N_e.
3. Engine type or configuration. The thrust or power lapse ratios with speed, altitude and ambient temperature $(T/T_{to}$ or $P/P_{to})$, and the specific fuel consumption at cruising conditions are particularly relevant. It is not customary to derive boundary values for these characteristics, as they can-

not be considered as independent variables (cf. Chapter 4, Section 4.4.).

b. Wing.

1. Gross wing area S, usually combined with the all-up weight into the wing loading at takeoff W_{to}/S.
2. Aspect ratio $A = b^2/S$. Instead of A, the span loading W_{to}/b^2 is occasionally used.[*]
3. High-lift devices, in particular the available C_L-max and lift/drag ratios in takeoff and landing configurations.

The wing section shape, the taper ratio and the sweep angle are not considered here; these will be discussed in Chapter 7.

The various performance equations will be presented in a way which permits their ready application. For climb and high-speed performance it is generally convenient to express the engine thrust (power) in terms of the other parameters, for low-speed performance the wing loading limitations are most easily derived explicitly, particularly in the case of a fixed engine and given aircraft weight.

The performance aims may not always be realized. For example, with regard to imaginary solutions for the wing area. Even in the case of a fixed engine it is therefore useful to consider possible engine growth, or alternatively to use the equations to deduce the permissible takeoff weight from the most critical performance item.

5.4.1. High-speed performance

a. Jet-propelled aircraft.

The thrust required to fly at a given speed and predetermined altitude is given by:

$$T = D = \tfrac{1}{2}\gamma p \, M^2 \, C_D S \qquad (5-30)$$

* span loading is sometimes defined as W_{to}/b

155

where T is the installed thrust, as derived from the engine manufacturers' brochure for a given engine, or assumed equal to a certain percentage of the uninstalled thrust when the engine has not yet been sized (cf. Section 5.3.4.).

The drag coefficient, as given by eq. 5-9, is dependent on the lift coefficient,

$$C_L = \frac{W/S}{\frac{1}{2}\gamma p M^2} \qquad (5-31)$$

and eq. 5-30 can be rewritten as follows:

$$\frac{T}{W} = \frac{\frac{1}{2}\gamma M^2 C_{D_O}}{W/pS} + \frac{W}{pS} \frac{1}{\frac{1}{2}\gamma M^2 \pi Ae} \qquad (5-32)$$

The zero-lift drag coefficient may be elaborated in terms of the wing area and the installed thrust. Any drag prediction method available to the designer may be used for this purpose. In this text the semi-statistical method of Section 5.3. is used to derive the following result:

$$\frac{T_{to}}{W_{to}} = \frac{\frac{.7\delta M^2 d_1}{W_{to}/P_O S} + \frac{W_{to}/P_O S}{.7k_W^2 \delta M^2 \pi Ae} + .7\delta M^2 d_2}{T/T_{to} - .7\delta M^2 d_3} \qquad (5-33)$$

where

$$k_W = \frac{W_{to}}{W}$$

$$d_1 = r_{Re} \, r_{uc} \, r_t \, \frac{(C_D S)_w}{S} + \Delta C_{D_{comp}}$$

$$d_2 = r_{Re} \, r_{uc} \, r_t \, \frac{(C_D S)_f \, P_O}{W_{to}} \qquad (5-34)$$

$$d_3 = r_{Re} \, r_{uc} \, \frac{(C_D S)_n \, P_O}{T_{to}}$$

These terms are explained in Sections 5.3.2. and 5.3.3.

In equation 5-33 the three terms in the numerator are associated with wing profile drag, induced drag, fuselage drag and empennage drag. The denominator may be interpreted as an effective lapse ratio of the engines installed in nacelles, including internal and external nacelle or intake scoop drag. The value for T/T_{to} can be ob-

tained from non-dimensional thrust curves or generalized data in Appendix H.

The minimum thrust at a given airspeed and altitude is found when the wing loading is equal to:

$$\frac{W}{S} = \frac{1}{2}\gamma p M^2 \sqrt{d_1 \, \pi Ae} \qquad (5-35)$$

This condition is identical to the condition for minimum drag (or L/D-max) of the wing plus that part of the tailplane contribution which is proportional to the wing drag. The factor d_1 is thus proportional to the wing profile drag coefficient; the order of magnitude is .008 - .010, for aircraft with retractable undercarriage.

b. Propeller aircraft.

The equivalent horsepower required to fly at a given speed and altitude is given by:

$$\eta_p \, P = DV = \frac{1}{2}\rho V^3 \, C_D S \qquad (5-36)$$

The power/weight ratio in the sea level static (SLS) condition is derived in the same way as for jet aircraft:

$$\frac{1}{V} \frac{P_{to}}{W_{to}} = \frac{\frac{1}{2}\rho V^2 \left| \frac{d_1}{W_{to}/S} + \frac{d_2}{P_O} \right| + \frac{2 \, W_{to}/S}{k_W^2 \rho V^2 \pi Ae}}{\eta_p \, P/P_{to} - \frac{1}{2}\rho V^3 d_3} \qquad (5-37)$$

For wing-mounted engines, the factors d_1 and d_2 are defined analogous to eq. 5-34, while

$$d_3 = r_{Re} \, r_{uc} \, \frac{(C_D S)_n}{P_{to}} \qquad (5-38)$$

This effect of compressibility on the drag can generally be ignored. For a tractor engine in the fuselage nose, $d_3 = 0$ and d_1 is equal to:

$$d_1 = r_{Re} \, r_{uc} \left| r_t (C_D S)_w + (C_D S)_n \right| /S \qquad (5-39)$$

The power lapse ratio P/P_{to} must be derived from the engine manufacturer's data. For the installed efficiency η_p typical values may be assumed as presented in Section 5.3.4.

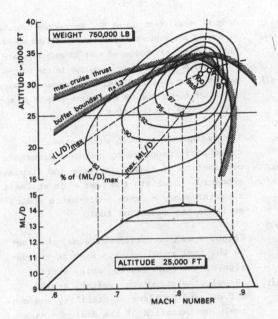

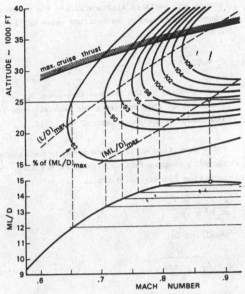

a. With compressibility effects on drag;
100% $(ML/D)^*_{max}$ = 14.8

b. No compressibility effects on drag;
100% $(ML/D)_{max}$ = 14.8

Fig. 5-10. Range performance of a high-subsonic long-range jet transport

5.4.2. Range performance

A specified range or radius of action must be achieved with a given payload or maximum fuel capacity, taking into account fuel reserves for holding and diversion. An indication of the cruise procedure will generally be mentioned: (initial) cruise altitude, long-range or high-speed condition.

a. Jet aircraft.
Although range performance depends upon the cruise procedure, the Bréguet equation is useful as a basis for an initial prediction of cruise fuel. On the conditions of constant angle of attack, airspeed and specific fuel consumption (climb cruise), the range is:

$$R = \frac{V}{C_T} \frac{L}{D} \ell n \frac{W_i}{W_1 - W_f} \qquad (5-40)$$

or:

$$\frac{R}{a_o} = \frac{M L/D}{C_T / \sqrt{\theta}} \ell n \frac{W_i}{W_i - W_f} \qquad (5-41)$$

where W_i is the initial weight and W_f the cruise fuel weight. For a given fuel fraction and engine type, the primary parameter in this equation is M L/D, the range parameter. The operational variables (cruise altitude, Mach number), the wing loading and the drag polar are the primary variables. For medium- and long-range aircraft, the amount of fuel consumed is large and it is necessary to aim at optimum flight conditions.

For a given drag polar and wing loading, lines of constant M L/D can be plotted on a speed-altitude diagram: Fig. 5-10 ; the lower part of this figure is an intersection for one altitude. The inclination of the tangent from the origin to the C_D/C_L-curve represents the condition for maximum M L/D. Considering flight at a specified sub-critical Mach

157

number and variable altitude first, the maximum range is obtained at an altitude where L/D is maximum, hence

$$C_L = \sqrt{C_{D_o} \, \pi Ae} \qquad (5\text{-}42)$$

corresponding to the minimum drag speed M_{MD}.
The locus of this condition for each altitude is indicated in the upper part of Fig. 5-10.

For a specified altitude and variable Mach number, ignoring compressibility effects on the drag, the condition for maximum M L/D is:

$$\frac{d}{dC_L} \left(\sqrt{\frac{W}{S} \frac{2}{\gamma P}} \frac{C_L'}{C_L^2} \right) = 0 \qquad (5\text{-}43)$$

resulting in:

$$C_L = \sqrt{\frac{1}{3} C_{D_o} \, \pi Ae} \qquad (5\text{-}44)$$

corresponding to a Mach number equal to $\sqrt[4]{3} \, M_{MD}$.
The conditions for C_L according to (5-42) and (5-44) are incompatible and no absolute optimum combination of M and altitude can be obtained. This is confirmed in Fig. 5-10, indicating that in the absence of compressibility M L/D continues to increase with altitude.

For high-subsonic speeds, a rapid rise in drag and subsequent deterioration of the range is observed beyond the drag-critical Mach number. Although the complex character of the flow does not allow an analytical treatment of compressibility effects, Fig. 5-10 indicates that a definite condition for maximum M L/D is now present. The locus of C_L for this condition is correspondingly modified and intersects the locus of $(L/D)_{max}$ at the optimum combination of M and altitude.

The condition for maximum specific range (i.e. the distance travelled per pound of fuel consumed) is slightly different from $(M\ L/D)_{max}$ due to the effects of altitude and speed on engine s.f.c. In operational practice the flight speed will always be some 10-20% above M_{MD} in horizontal flight in order to obtain positive speed stability and to avoid buffet during maneuvers. A typical long-range cruise condition results in 98% of the maximum specific range (point A in Fig. 5-10a). In operations where fuel consumption is not a dominant factor, the high-speed cruise at a somewhat lower cruise altitude is significant (point B). The maximum cruise rating of the engines determines the flight speed, and a typical extra drag of some 20 counts is acceptable. An intermediate condition is the cost-economical cruise, resulting in a favorable combination of fuel costs and block time effects on operating costs (Section 11-8).

The designer's problem with respect to range performance is to choose a favorable combination of speed, altitude and airplane geometry, to obtain the best - or at least a satisfactory - range performance and to estimate the amount of fuel.

Flight speed variation has a major effect on the fuel required, but also on the design of the wing (sweep angle, section shape), the structural weight, engine s.f.c., and problems of stability and control. Optimization of the design - Mach number is a very complex study and this parameter is usually specified in a rather arbitrary manner in the design requirements.

Cruising altitude has a direct effect on fuel weight. When the installed engine thrust is based on the cruise condition, the engine size required increases with altitude, as the density decreases, and for a given specific thrust the inlet diameter must increase as well. The weight of engine plus fuel is a minimum for some altitude below the altitude for maximum L/D. This case is elaborated analytically by Küchemann in Ref. 5-4, resulting in an optimum condition for C_L depicted in Fig. 5-11. For long-range aircraft the minimum fuel requirement dominates while on short hauls the amount of fuel consumed is less and the engine weight is the main factor, leading to a relatively low optimum for C_L in cruising flight.

Obviously the matter is more complicated in real life:

— Cruise fuel is only part of the total fuel load.

— Fuel weight and engine weight do not have the same significance from the point of view of achieving minimum operating costs.

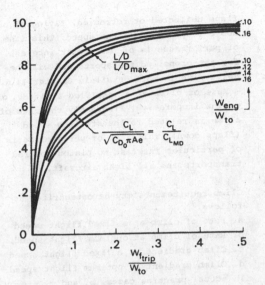

Fig. 5-11. Optimum cruise conditions according to the criterion of Küchemann and Weber (Ref. 5-4)

— For short-haul aircraft the engine thrust is frequently determined by the takeoff field length or an engine-out climb requirement.
— Oxygen system requirements are dependent on cruising altitude (e.g. FAR 121.327-333), which may be a deciding factor.
— Air Traffic Control considerations affect the choice of cruising altitude.

For an initial estimation of the amount of trip fuel required, the fuel consumed for cruising is derived from eq. 5-41 :

$$\frac{W_{f_{cr}}}{W_{to}} = 1 - \exp\left(-\frac{R\,C_T/\sqrt{\theta}}{a_o\,M}\,\frac{C_D}{C_L}\right) \qquad (5-45)$$

where the initial weight is assumed to be approximately equal to W_{to}. Additional fuel is also used during takeoff, climb to and descent from cruise altitude, approach and landing. Fig. 5-2 is based on the following assumptions:
a. The cruise fuel according to eq. 5-45 is the dominant factor in the fuel contribution.
b. For transport aircraft the wing loading

given by eq. 5-35 can be used to find a first-order approximation for the L/D-ratio of the wing.
c. The wetted area of the fuselage is the primary parameter for the fuselage drag area; other drag contributions are assumed proportional to wing plus fuselage drag. Fig. 5-2 can be used, provided the fuselage dimensions are known, for example from a fuselage layout drawing or the data from Figures 3-11 and 3-12. Engine s.f.c. is deduced from the engine manufacturer's brochure or from the data listed in Chapter 4.

Reserve fuel consists of various contributions (cf. Table 11-2). One important item is holding fuel, which is proportional to the airplane minimum drag, hence inversely proportional to $\sqrt{A}$. The following empirical correlation has been found to give a good approximation for transport aircraft:

$$\frac{W_{f_{res}}}{W_{to}} = .18\,\frac{C_T/\sqrt{\theta}}{\sqrt{A}} \qquad \text{(transport a/c)} \qquad (5-46)$$

where $C_T/\sqrt{\theta}$ is the same quantity as used in Fig. 5-2. A fuel reserve for 3/4 hour extra flying time should be allowed for business jets and executive aircraft. This can simply be translated into an equivalent range increment:

$$\Delta R = .75\,V_{cr} \quad \begin{array}{l}\text{(business and}\\ \text{executive jets)}\end{array} \qquad (5-47)$$

Fig. 5-2 is thus valid, provided an equivalent range is used equal to R + ΔR.

If the effects of varying parameters like wing loading and aspect ratio are to be assessed, the aircraft weight distribution must be computed in terms of these parameters. The details of these calculations will not be dealt with here; a simplified example in Section 5.5.3 shows the effects of wing area variation on the weight distribution of a long-range aircraft.

b. Propeller aircraft.
The Bréguet range equation for propeller

aircraft is:

$$R = \frac{\eta_p}{C_p} \frac{L}{D} \ln \frac{W_i}{W_i - W_f} \qquad (5-47)$$

A similar picture to Fig. 5-10 can be drawn up. If s.f.c. and propeller efficiency variations are ignored, the specific range can be shown to be a maximum
- for minimum airplane drag, if the altitude is fixed and flight speed variable,
- for minimum power, if the flight speed is fixed and altitude variable.
As in the case of jet aircraft, these conditions are incompatible and no absolute optimum exists on the basis of flight mechanics. In general, range performance continues to improve with altitude until the available engine power becomes the limiting factor.
Fig. 5-1 can be used for an initial estimation of fuel weight. The wing aspect ratio is the primary factor in obtaining a high L/D ratio and good range performance. It is the only parameter used in this figure to characterize the aerodynamic performance, as no considerable improvement was found when fuselage dimensions were introduced.

5.4.3. Climb performance

Climb performance may be specified in the form of:

a. Operational requirements, derived from desired performance capabilities in normal operating conditions, e.g.
- Rate of climb at sea level, clean configuration, all engines operating.
- Service ceiling altitude for max. rate of climb = 100 ft/min, .5 m/sec), clean configuration, all engines operating or one engine inoperative.

b. Airworthiness requirements, to ensure adequate performance for safety in normal and critical conditions, e.g.
- Minimum climb gradient in various configurations (takeoff, en route, landing), one engine inoperative or all engines operating,

flaps deflected or retracted, flying at or above a specified flight speed. This item of performance is of particular interest for jet-propelled transports and will be dealt with in more detail in Chapter 11.6.
- Rate of climb at a specified altitude, one engine inoperative. Frequently the rate of climb is related to the stalling speed (flaps down). This item of performance is of particular interest to piston-powered transports and all light aircraft.

Climb requirements may be categorized as follows.
a. Rate of climb at a fixed flight speed
b. Rate of climb at optimum flight speed
c. Climb gradient at a fixed flight speed
d. Climb gradient at optimum flight speed
In actual practice cases, b. and c. are by far the most important ones; case a. can be found in airworthiness requirements for transport aircraft with reciprocating engines. Case d. may incidentally occur where a takeoff climb gradient cannot be fulfilled at V_{2min} for transport category aircraft ("overspeed").
A special case of operational climb performance - the time to climb to a given altitude - will not be discussed here as no analytical procedure is available to convert such a requirement into combinations of design variables.
The various cases will first be dealt with as a general performance problem. Some examples and applications will then be presented in order to illustrate the procedure.

A useful general term for specifying climb performance is Specific Excess Power, SEP:

$$SEP = \frac{T-D}{W} V = \frac{V}{g} \frac{dV}{dt} + \frac{dh}{dt} \qquad (5-48)$$

Excess power is available for climbing, accelerating and making turns. In the case of a steady climb at n = L/W $\sim$ 1, SEP is identical to the rate of climb C, provided the angle of climb γ is not too large (approximation: cosγ = 1).
For horizontal flight at n = 1, the SEP is

equivalent to the rate of increase of kinetic energy and is therefore a measure of the time required to accelerate from one speed to another. In a horizontal turn at a specified rate of turn or load factor, the SEP represents the maneuvering and acceleration capability. Note that the dimension of SEP is length per unit time and not power.

a. Jet aircraft.
The thrust required is derived from eq. 5-48:

$$\frac{T}{W} = \frac{SEP}{V} + n \frac{C_D}{C_L} \qquad (5-49)$$

1. In the case of a specified climb gradient for n = 1, steady flight:

$$\gamma = \frac{dh/dt}{V} = \frac{SEP}{V} \qquad (5-50)$$

and eq. 5-49 yields:

$$\frac{T}{W} = \gamma + \frac{C_D}{C_L} \qquad (5-51)$$

The lift and drag coefficients are given by eqs. 5-31 and 5-9. Hence:

$$\frac{T}{W} = \gamma + \frac{\frac{1}{2}\gamma M^2 C_{D_o}}{W/(pS)} + \frac{W/(pS)}{\frac{1}{2}\gamma M^2 \pi Ae} \qquad (5-52)$$

The minimum value for T/W corresponds to the minimum C_D/C_L ratio with

$$C_L = \sqrt{C_{D_o} \pi Ae} \quad \text{at } M = \left\{ \frac{W/(pS)}{\frac{1}{2}\gamma \sqrt{C_{D_o} \pi Ae}} \right\}^{\frac{1}{2}} =$$

$$= M_{(L/D)_{max}} \qquad (5-53)$$

and consequently:

$$\left(\frac{T}{W}\right)_{min} = \gamma + 2\sqrt{C_{D_o}/\pi Ae} \qquad (5-54)$$

2. For a specified rate of climb for dV/dh= 0, n=1 and arbitrary V, eq. 5-49 is modified to:

$$\frac{T}{W} = \frac{SEP/a}{M'} \left(\frac{M'}{M}\right) + n\sqrt{\frac{C_{D_o}}{\pi Ae}} \left\{ \left(\frac{M}{M'}\right)^2 + \left(\frac{M'}{M}\right)^2 \right\} \qquad (5-55)$$

where

$$M' = \left\{ \frac{n \, W/(pS)}{\frac{1}{2}\gamma \sqrt{C_{D_o} \pi Ae}} \right\}^{\frac{1}{2}} = \sqrt{n} \, M_{(L/D)_{max}} \qquad (5-56)$$

If the Mach number is specified, this equation can be used directly to obtain T/W. However, in most cases a condition for M is sought for which the T/W ratio is a minimum. A plot of T/W vs. M, together with a typical thrust lapse curve (Fig. 5-12) in-

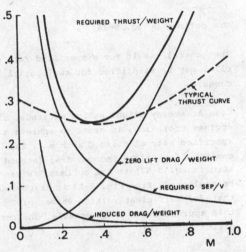

Fig. 5-12. Required thrust in a climb with specified SEP

dicates that an acceptable approximation for T/W is found at the Mach number for which eq. 5-55 has a minimum value. This condition is:

$$\left(\frac{M}{M'}\right)^3 - \left(\frac{M'}{M}\right) = \frac{SEP/a}{2nM'} \sqrt{\frac{\pi Ae}{C_{D_o}}} \qquad (5-57)$$

The solution is presented in graphical form in Fig. 5-13. In general, it is acceptable to calculate C_{D_o} as follows:

$$C_{D_o} = 1.1 \left(d_1 + d_2 \frac{W_{to}}{S \, p_o}\right) \qquad (5-58)$$

where the drag due to powerplant installation is assumed at 10% of C_{D_o} and d_1 and d_2

161

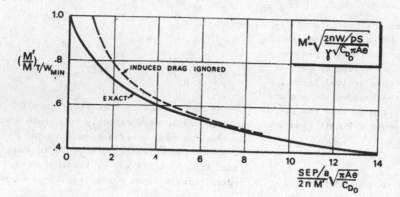

$$M' - \sqrt{\frac{2nW/pS}{\gamma \sqrt{C_{D_o} \pi Ae}}}$$

$\left(\frac{M'}{M}\right)_{T/W_{MIN}}$

INDUCED DRAG IGNORED

EXACT

$$\frac{SEP/a}{2nM'}\sqrt{\frac{\pi Ae}{C_{D_o}}}$$

Fig. 5-13. Condition for the minimum thrust Mach number in a climb with specified SEP (jet aircraft)

are defined by eq. 5-34.

The general result for a specified rate of climb may be simplified for two special cases.

Case A: steady flight at low altitude, all engines operating, in order to achieve a specified rate of climb C at n = 1. The contribution of the induced drag (second term in eq. 5-57) to the optimum T/W can be neglected. From Fig. 5-13 it is obvious that for sufficiently large values of C this approximation is acceptable. The condition for M is:

$$M^3 = \frac{C/a}{\gamma C_{D_o}} \frac{W}{pS} \tag{5-59}$$

Substitution into eq. 5-55 yields:

$$\frac{T}{W} = \frac{3}{2} \gamma^{1/3} \left\{ \frac{\frac{C}{a}\sqrt{C_{D_o}}}{\sqrt{W/(pS)}} \right\}^{2/3} +$$

$$+ \frac{2}{\gamma^{1/3}} \frac{C_{D_o}}{\pi Ae} \left\{ \frac{\sqrt{W/(pS)}}{\frac{C}{a}\sqrt{C_{D_o}}} \right\}^{2/3} \tag{5-60}$$

In most cases this expression is quite accurate, provided C corresponds to low altitude performance with all engines operating. In the derivation of T/W, it was assumed that dV/dt = 0. In operational practice, however, a rate of climb is usually established in a flight with constant EAS or CAS. It can be shown that the acceleration necessary for flying at constant EAS at low altitude is given by:

$$\frac{V}{g}\frac{dV}{dt} = \frac{V}{g}\frac{dV}{dh}\frac{dh}{dt} = .567 \ M^2 C \tag{5-61}$$

This may be translated into an additional engine thrust required for a given C:

$$\frac{\Delta T}{W} = .567 \ \frac{C}{a} \left(\frac{\frac{C}{a} \times \frac{W}{pS}}{\gamma C_{D_o}} \right)^{1/3} \tag{5-62}$$

Instead of a detailed calculation; for which many data must be available, a statistical correlation of T/W and C/$\sqrt{W/S}$ at sea level may be used, as given in Fig. 5-14.

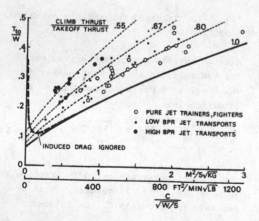

CLIMB THRUST / TAKEOFF THRUST .55   .67   .80   1.0

$\frac{T_{to}}{W}$

INDUCED DRAG IGNORED

o   PURE JET TRAINERS, FIGHTERS
·   LOW BPR JET TRANSPORTS
●   HIGH BPR JET TRANSPORTS

M²/s√KG

400   800 FT²/MIN√LB 1200

$\frac{C}{\sqrt{W/S}}$

Fig. 5-14. Correlation of thrust/weight ratio and maximum rate of climb

Case B: flight at high altitude with low rate of climb, in order to achieve a specified service ceiling. The contribution of

162

SEP in eq. 5-57 can be neglected and the Mach number for minimum T/W is equal to M'. At the service ceiling, C = 100 ft/min (.5 m/sec), hence

$$\frac{SEP}{a_o} = .00147 \qquad (5\text{-}63)$$

and the thrust required at the ceiling is:

$$\frac{T}{W} = 2n\sqrt{\frac{C_{D_o}}{\pi Ae}} + \frac{.00123}{\sqrt{\theta}} \frac{\sqrt[4]{C_{D_o} \pi Ae}}{\sqrt{nW/(pS)}} \qquad (5\text{-}64)$$

where the relative temperature θ and the ambient pressure p refer to the ceiling. The ratio of thrust at the ceiling to static thrust must be used to convert this T/W value into T_{to}/W_{to}.

b. Propeller aircraft.
For n = 1 the available power is equal to the power required to climb plus the power necessary to balance the drag. Assuming steady flight, SEP = dh/dt = C, and eq. 5-48 is modified into:

$$\eta_p \frac{P}{W} = C + \frac{C_D}{C_L} V \qquad (5\text{-}65)$$

If C is defined at a given flight speed, the lift and drag coefficients are known and eq. 5-65 can be used directly.
To find the flight speed for which the power required to climb reaches a minimum, assuming a parabolic drag polar, the P/W ratio for given rate of climb is modified to:

$$\frac{P}{W} = \frac{1}{\eta_p}\left\{ C + \left(\frac{C_{D_o}}{C_L^{3/2}} + \frac{C_L^{1/2}}{\pi Ae}\right)\sqrt{\frac{2W}{\rho S}}\right\} \qquad (5\text{-}66)$$

For a given altitude and engine rating, P/W is affected by η_p and C_L. For a parabolic polar, the value of the term:

$$\frac{C_{D_o}}{C_L^{3/2}} + \frac{C_L^{1/2}}{\pi Ae}$$

is minimum for $C_L = \sqrt{3 C_{D_o} \pi Ae}$. However, propeller efficiency generally improves with increasing airspeed and in practice the most favorable speed is roughly 20% higher than the speed for the minimum power required to

balance the drag. The result is:

$$\frac{P}{aW} = \frac{1}{\eta_p}\left\{ \frac{C}{a} + 2.217 \frac{C_{D_o}^{1/4}}{(\pi Ae)^{3/4}} \sqrt{\frac{W}{pS}}\right\}$$

$$\text{at } M = 1.09 \frac{\sqrt{W/(pS)}}{\sqrt[4]{C_{D_o} \pi Ae}} \qquad (5\text{-}67)$$

The engine power* is generally the maximum continous (equivalent) power for turboprop engines or the rated (METO) power for reciprocating engines, at the ambient conditions for which C is specified.
In the case of a service ceiling, eq. 5-67 can be used by taking C = 100 ft/min (0.5 m/s), hence

$$\frac{C}{a_o} = .00147 \qquad (5.68)$$

while in (5-67) the speed of sound and the static pressure refer to the service ceiling.

c. Applications.
Airworthiness climb requirements are to be found in:
FAR 23.65 and 67
SFAR 23. Amendment 1 ch. 6
FAR 25.65 and 67, 25. 117-119-121
BCAR Chapter D2-4.
A survey of the most pertinent data for transport aircraft will be given in Chapter 11 Section 11.5. It is not intended to deal with all possible requirements in this section. Instead, some examples will be presented to illustrate applications of the formulas already derived and the importance of climb requirements, with particular reference to civil aircraft engine sizing.

Example 1
The rate of climb specified at sea level for a subsonic jet trainer is 4500 ft/min (23 m/s), corresponding to c/a = .0672. The weight is 7000 lb (3170 kg), wing area S = 210 sq.ft (19.5 m²) and

* divide (5-67) by 550 if P is in hp, W in lb and C in ft/sec

A = 5.5, hence $W/(p_o S)$ = .0156 and πAe = 13.8 for
e = .8. Furthermore C_{D_o} = .019.
To use Fig. 5-13, we calculate

$$M' = \left\{ \frac{W/(p_o S)}{\frac{1}{2}\gamma \sqrt{C_{D_o} \pi Ae}} \right\}^{\frac{1}{2}} = .209 \text{ and } \frac{C/a_o}{2M'} \sqrt{\frac{\pi Ae}{C_{D_o}}} = 4.33$$

The figure indicates M'/M = .59; hence M = .35 for
minimum T/W. Using eqs. 5-55 and 5-62, we calculate:

$$\frac{T}{W} = .309 \text{ for steady flight}$$

$$\frac{\Delta T}{W} = 013 \text{ for acceleration (constant EAS)}$$

total T/W = .322

For a thrust lapse rate of .85 at M = .35, the
takeoff SLS thrust/weight ratio must be at least
.38. The approximate equation 5-60 for steady
flight yields T/W = .314, as compared with .309
for the "exact" solution.

Example 2
For a twin-engined subsonic jet passenger transport, the service ceiling with one engine failed
is specified at 15,000 ft (4570 m). The following
data are pertinent to the aircraft.
W = 75,000 lb (34,000 kg); S = 850 sq.ft (79 m^2);
C_{D_o} = .018; A = 8.5 and e = .85, hence πAe = 22.7.
For engine failure, an 8% decrease of e is assumed:
πAe = 20.9. At an altitude of 15,000 ft, W/(pS) =
.072 and θ = .897. According to eq. 5-56,

$$M' = \left\{ \frac{.072}{.7 \sqrt{.018 \times 20.9}} \right\}^{\frac{1}{2}} = .41$$

We now calculate from eq. 5-55:

$$\frac{SEP/a}{2M'} \sqrt{\frac{\pi Ae}{C_{D_o}}} = .0645$$

From Fig. 5-13 we find that M'/M = .975, and therefore the speed for minimum thrust at this altitude
is M = .42. Equation 5-56 yields: T/W = .0624. The
engine thrust lapse at an altitude of 15,000 ft
and with M = .42 is .47. The thrust/weight ratio
required at SLS is: T_{to}/W_{to} = .265. The effect of
flying at constant EAS is neglected in view of the
low rate of climb at the service ceiling.

Example 3
An important airworthiness requirement for civil
aircraft is the so-called second-segment climb gradient, laid down for example in FAR 25.121 (b). It
states that with one engine inoperative, flaps in
takeoff position, landing gear retracted, engines
in the takeoff rating and out of ground effect, a
specified minimum climb gradient is to be obtained
at the takeoff safety speed V_2. This requirement
must be met for all operational ambient conditions
and may, especially on hot and high airfields, limit
the authorized takeoff weight.
In the case of a subsonic passenger transport e-
quipped with 3 engines, for example, the required
climb gradient is 2.7%. The ambient conditions are:
sea level, temperature 95 F (i.e. 35 C, ISA + 20 C).
Other data:
W_{to} = 210,000 lb (95,000 kg), S = 2,060 sq.ft
(192 m^2); V_2 = 1.2 V_s with C_L-max = 2.40 for the
takeoff. Aspect ratio: A = 7.5; Section 5.3.5. is
used to estimate the lift/drag ratio in symmetrical
flight with flaps deflected at $C_L = C_L$-max/(1.2)^2 =
1.67.
We find: C_D/C_L = .094 for a slatted wing, hence
C_D/C_L = .10, assuming an increment of 5% for drag
due to engine failure. The Mach number at V_2 is:

$$M_2 = \frac{V_2}{a} = 1.2 \sqrt{\frac{W/(p S)}{\frac{1}{2}\gamma C_{L_{max}}}} = .2$$

From eq. 5-51 we now find: T/W = .027 + .10 = .127
for two engines operating at M = .2 and 95 F (35 C).
For a thrust lapse rate of .75, including the effect
of non-standard temperature, the total takeoff
thrust (SLS, ISA) must be at least:

$$T_{to}/W_{to} = \frac{N_e}{N_e - 1} \cdot \frac{T/W}{T/T_{to}} = .254$$

For W_{to} = 210,000 lb (95,000 kg), the thrust per
engine must be at least 17,780 lb (8,054 kg).

Example 4
Light aircraft with W < 6,000 lb (2720 kg) must
comply with the requirement FAR 23.67. With takeoff
power at sea level, undercarriage down and flaps in
the takeoff position:

C ≥ 300 ft/min, or c/a ≥ .0045
and
C ≥ 11.5 V_{S_1} ft/min;
where V_{S_1} = equivalent stalling speed (knots).

The first requirement can be substituted directly
into eq. 5-67. The second can be evaluated as follows:

$$\frac{P_{to}}{a_o W_{to}} \geq \frac{1}{\eta_p}\left(\frac{W_{to}}{\rho_o S}\right)^{\frac{1}{2}}\left|\frac{.136}{\left(C_{L_{max}}\right)^{\frac{1}{2}}} + \frac{2.21\, C_{D_o}^{1/4}}{(\pi Ae)^{3/4}}\right| \qquad (5\text{-}69)$$

By way of example, the following data may be applied to a light aircraft: W_{to} = 3,300 lb (1,500 kg); S = 130 sq.ft (12 m^2), hence $W_{to}/(\rho_o S)$ = .0121; $C_{L_{max}}$ = 1.8; πAe = 15.3 and C_{D_o} = .055 with flaps deflected and undercarriage down. Effective propeller efficiency: η_p = .65. Substitution of C/a ≥ .0045 into eq. 5-67 yields: $P_{to}/a_o W_{to}$ ≥ .0305, while eq. 5-69 yields $P_{to}/a_o W_{to}$ ≥ .0408. The latter being the most critical requirement, it is concluded that P_{to} must be at least 275 hp.

d. Design data.

If no better information is available, the following data may be useful in working out climb performance requirements.
Propeller efficiency during climb at sea level:
tractor propeller in fuselage nose,
fixed pitch : η_p = .61 ($\pm$ .052)
constant speed: η_p = .665 ($\pm$ .059)
tractor propellers, wing-mounted,
constant speed: η_p = .73 ($\pm$ .058)
These data were found by application of the present method to the performance data of a large number of aircraft. The second number gives the rms error. All figures include slipstream effects, cooling drag, power off-takes and intake losses.

For the effect of engine failure on drag, 4% may be added to C_{D_o} for the drag of feathering propellers, while the Oswald factor may be reduced by approximately 10% for wing-mounted engines.
The airplane drag polar may be estimated by the method explained in Section 5.3. For the effect of undercarriage extension, it is reasonable to take ΔC_{D_o} = .015 to .020 as a typical value. It should be noted that in most equations for climb performance small variations in C_{D_o} are of minor importance, unless the flaps are deflected. Hence, it may be assumed that the drag due to powerplant installation adds roughly 8% to C_{D_o} on turbine-powered aircraft and 12% with piston engines.

The engine thrust or power lapse rate may be determined from engine manufacturer's brochures. For jet engines the effect of M on the thrust is important, while lapse rates are very sensitive to the bypass ratio as well. Curves of $T/\delta T_{to}$ vs. M may serve the purpose. An example is shown in Fig. 6-3. In the case of turboprop engines the shaft power increases noticeably with M due to the ram effect, unless there is a structural or thermal engine limitation up to some specified altitude. If this is not the case, the following approximation may be used for a given rating and Mach number:

$$\frac{\text{power at altitude}}{\text{power at sea level}} = \sigma^n \qquad (5\text{-}70)$$

where n is generally between .7 and .8. For naturally aspirated piston engines at constant rpm:

$$\frac{\text{full throttle power at altitude}}{\text{full throttle power at sea level}} =$$

$$= 1.132\sigma - .132 \qquad (5\text{-}71)$$

Supercharged engines maintain constant power up to the rated altitude. Above this the power decreases linearly with σ in the same way as in eq. 5-71. With many supercharged engines a cruising power of 65% to 75% of rated power can be maintained up to the cruising altitude.

5.4.4. Stalling and minimum flight speeds

In establishing low speed performance, operational flight speeds must have a specified safety margin relative to the stalling speed, in order to provide the pilot with some measure of freedom to maneuver and in order to avoid stalling due to vertical gusts. The required field length being roughly proportional to the kinetic energy at the screen height, and hence to (velocity)^2, a low stalling speed provides a powerful method of obtaining good field performance. On the other hand, a decrease in stalling speed generally entails a cost penalty as a more sophisticated flap system must be developed or the wing loading decreased, or

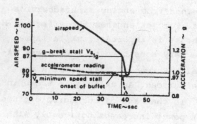

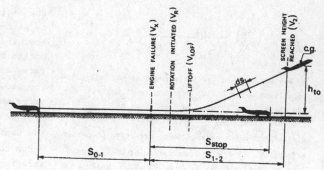

Fig. 5-15. Time history of the airspeed during a stall maneuver

Fig. 5-16. Phases during the takeoff with engine failure

both.

With regard to the definition of stalling speeds, reference is made to Fig. 5-15, which depicts a time history of the airspeed during a stalling maneuver. In principle, several flight procedures can be chosen for this - for example, flight at constant longitudinal deceleration, constant flight path angle (or horizontal flight) or constant normal acceleration. The FAR certification procedure is a stall maneuver at constant dV/dt. Several tests are carried out at different values of dV/dt and the stalling speed is defined by interpolation at $dV/dt = -1$ kt/s.

In approaching the stall, the normal load factor initially remains approximately constant until a break occurs, indicating that wing stalling is progressing rapidly. The corresponding airspeed is referred to as the 1-g stalling speed or g-break stalling speed V_{S-1g}. Immediately after the g-break the airspeed continue V_S to decrease, and the sink speed increases rapidly until the pilot takes corrective action by pitching down the nose, resulting in a positive dV/dt. The minimum airspeed measured in this procedure, V_S, is noticeably lower than V_{S-1g}. The FAR-25 regulations allow V_S to be used in scheduling various reference speeds for definition of the performance, whereas the British requirements do not allow the stalling speed to be less than 94% of V_{S-1g}.

In view of this reasoning, it must be noted that a definition of C_L-max according to

$$C_{L_{max}} = \frac{W/S}{\frac{1}{2}\rho V_S^2} \qquad (5-70)$$

does not result in a "physical" C_L-max as obtained in the wind tunnel, but produces a value which may be some 10 to 20% higher. Alternatively, when calculating the stalling speed from C_L-max obtained by theoretical methods or wind tunnel experiments, it is generally appropriate to assume an apparent C_L-max increment of, say, 13%. All values of C_L-max mentioned in the following sections are corrected in this manner.

Occasionally a limit is imposed on the stalling speed and the corresponding wing loading is limited by:

$$\frac{W}{S} \leqslant \frac{1}{2}\rho V_S^2 C_{L_{max}} \qquad (5-71)$$

For example, in the BCAR requirement Ch. D2-11, an upper limit applies to the stalling speed:
70 mph (112.5 kmh) - group C
60 mph (96.5 kmh) - group D
These values can be substituted into eq. 5-71 to find a limiting wing loading for these classes of aircraft. The reader should also refer to FAR 23.49 (b) for aircraft with a takeoff weight of 6,000 lb (2720 kg) or less.

5.4.5. Takeoff

The following cases are distinguished:
a. All-engine takeoff distance requirements
b. one-engine-out takeoff distance requirements
c. Accelerate-stop distance requirements
For aircraft to be certificated under the FAR Part 23 regulations, no explicit requirements need to be met with regard to the event of engine failure during takeoff. The all-engine takeoff distance applies to airplanes with a takeoff weight of 6,000 lb (2720 kg) or more and is generally defined as the distance required to pass the screen height of 50 ft (15.3 m) at a speed of 1.3 V_S. For a particular class of aircraft operating under the FAR Part 135 operational rules, the FAR Part 23 performance standards are considered to be inadequate. The performance standards laid down in SFAR 23 and NRPM 68-37 are intended as intermediate steps towards improving safety in the operation of small passenger airplanes and air taxis capable of carrying more than 10 persons (Ref. 5-22). Accelerate-stop distances are introduced in these requirements. Under SFAR 23 this is the distance required to accelerate to the critical engine-failure speed V_1 and then to decelerate to 35 knots, while NRPM 68-37 covers the distance needed to come to a full stop. These regulations contain no specifications for a one-engine inoperative takeoff distance.
The takeoff performance of transport category airplanes is a fairly complicated matter, which is dealt with in more detail in Appendix K. It is generally found that the most critical item of performance is determined by the case of a one-engine-out takeoff. The condition that the airplane must be brought to standstill after engine failure at the critical engine failure speed leads to the concept of the Balanced Field Length (BFL), which is usually considered as the most important design criterion as far as field performance of transport category aircraft is concerned. The simplified methods presented in this section are intended to serve as a first approximation of field length for the purpose of sizing the engine thrust or power and wing design. They can be refined if more detailed information is available to the designer; an example is given in Appendix K.

a. All-engines takeoff.
Since the takeoff consists of a takeoff run and an airborne phase, we may write:

$$S_{to} = S_{run} + S_{air} \qquad (5-72)$$

The expression for S_{run} is

$$S_{run} = \frac{1}{2g} \int_{0}^{V_{LOF}^2} \frac{dV^2}{a/g} \qquad (5-73)$$

where the momentary acceleration,

$$a/g = \frac{T}{W_{to}} - \mu - (C_D - \mu C_L) \frac{\frac{1}{2}\rho V^2 S}{W} \qquad (5-74)$$

can be approximated by

$$S_{run} = \frac{V_{LOF}^2/2g}{\bar{T}/W_{to} - \mu'} \qquad (5-75)$$

where $\bar{T}$ is a mean value of the thrust during the takeoff run. Assuming the lift-off speed to be approximately 1.2 V_S and the lift coefficient during the takeoff run to be equal to twice the value for minimum $(C_D - \mu C_L)$, it is found that

$$\bar{T} = \text{thrust at } V_{LOF}/\sqrt{2}$$

$$C_L = \mu\, \pi Ae \qquad (5-76)$$

$$\mu' = \mu + .72\, C_{D_o}/C_{L_{max}}$$

Assuming an air maneuver after lift-off with $C_L = C_{L-LOF}$ = constant and T-D = constant, the following result can be derived from the AGARD Flight Test Manual, Vol. 1:

$$S_{air} = \frac{V_{LOF}^2}{g\sqrt{2}} + \frac{h_{to}}{\gamma_{LOF}} \qquad (5-77)$$

and

$$\frac{V_3}{V_{LOF}} = \sqrt{1 + \gamma_{LOF} \sqrt{2}} \qquad (5-78)$$

where $\gamma_{LOF} = (T-D)/W$ at liftoff and V_3 = velocity at the takeoff height (30 or 50 ft).
The liftoff speed corresponding to a given V_3 is:

$$V_{LOF} = V_3 \left(\frac{1}{1 + \gamma_{LOF} \sqrt{2}}\right)^{\frac{1}{2}} \qquad (5-79)$$

From eqs. 5-75, 5-76 and 5-79 the takeoff distance is now:

$$\frac{S_{to}}{f_{to} h_{to}} = \left(\frac{V_3}{V_S}\right)^2 \frac{W_{to}/S \left\{(\bar{T}/W_{to} - \mu')^{-1} + \sqrt{2}\right\}}{h_{to} \rho g\, C_{L_{max}} (1 + \gamma_{LOF} \sqrt{2})} +$$

$$+ \frac{1}{\gamma_{LOF}} \qquad (5-80)$$

Regula- tions	V_3/V_S	f_{to}	h_{to}
(S)FAR 23	1.3	1.0	50 ft (15.3 m)
FAR 25	1.25 to 1.30 (no require- ment)	1.15	35 ft (10.7 m)

Table 5-2. Characteristic values for the all-engines takeoff according to FAR 23 and 25.

In the absence of better information, the following assumptions and approximations may be made in applying eq. 5-80.
1. In calculating μ' according to eq. 5-76, it is reasonable to assume: $.72\, C_{D_o}/C_{L_{max}} = .010\, C_{L_{max}}$, $\mu = .02$ for concrete and $\mu = .04 - .05$ for short grass.
2. $\gamma_{LOF} = .9 \frac{\bar{T}}{W_{to}} - \frac{.3}{\sqrt{A}}$
3. The mean thrust/weight ratio at mean velocity $V_{LOF}/\sqrt{2}$, allowing for slipstream effects and power offtakes, is as follows:
jet aircraft:

$$\bar{T} = .75 \frac{5 + \lambda}{4 + \lambda} T_{to} \qquad (5-81)$$

aircraft with constant speed propellers:

$$\bar{T} = k_p\, P_{to} \left(\frac{\sigma\, N_e\, D_p^2}{P_{to}}\right)^{1/3} \qquad (5-82)$$

where $P_{to}/(N_e D_p^2)$ is the propeller disc loading; see Fig. 6-9, for example.
$k_p = 5.75$ when $\bar{T}$ is in lb, P_{to} in hp, D_p in ft
$k_p = .321$ when $\bar{T}$ is in kg, P_{to} in kgm/s, D_p in m.
For fixed-pitch propellers the mean thrust is roughly 15-20% below the value given by eq. 5-82.
From eq. 5-77 it follows that for a specified takeoff distance, the wing loading is limited to:

$$\frac{W_{to}}{S} \leqslant \left|\frac{S_{to}}{f_{to}} - \frac{h_{to}}{\gamma_{LOF}}\right| \frac{\rho g\, C_{L_{max}} (1 + \gamma_{LOF} \sqrt{2})}{(V_3/V_S)^2 \left\{(\bar{T}/W_{to} - \mu')^{-1} + \sqrt{2}\right\}} \qquad (5-83)$$

b. Takeoff with engine failure and accelerate-stop distance.
A critical decision speed V_1 is defined so that, with a single engine failure, the total accelerate-stop distance required becomes identical with the total takeoff distance to reach screen height safely. A simple analytical method for determining the BFL in the preliminary design stage, devised by the author and presented in Ref. 5-27, will be summarized below.

As opposed to the usual subdivision (takeoff run to liftoff, transition and climb distance), the continued takeoff is split up into 2 phases (Fig. 5-16):
- Phase 0-1: acceleration from standstill to engine failure speed V_x,
- Phase 1-2: the motion after engine failure, up to the moment of attaining the screen height at takeoff safety speed V_2.
The distance travelled during phase 0-1 is:

$$S_{0-1} = \frac{V_x^2}{2\, \bar{a}_{0-1}} \qquad (5-84)$$

where $\bar{a}_{0-1}$ is calculated in the same way as for the all-engines takeoff.
The energy equation is applied to phase 1-2 (Fig. 5-16), resulting in

$$S_{1-2} = \frac{1}{\bar{\gamma}} \left(\frac{V_2^2 - V_x^2}{2 g} + h_{to}\right) \qquad (5-85)$$

where the equivalent climb gradient $\bar{\gamma}$ is defined as follows:

$$\bar{\gamma} = \frac{1}{W_{to}} \frac{\int_{1}^{2} (T-D_{air}-D_g)\,ds}{S_{1-2}} \qquad (5\text{-}86)$$

The distance required to come to a standstill after engine failure can be represented by:

$$S_{stop} = \frac{V_x^2}{2\,\bar{a}_{stop}} + V_x \Delta t \qquad (5\text{-}87)$$

where Δt is referred to as an equivalent inertia time, affected in principle by the thrust/weight ratio at V_x (Fig. K-6). The condition for balancing the field length is $S_{1-2} = S_{stop}$ and Ref. 5-27 gives the following expression for the critical engine failure speed V_1:

$$\frac{V_1}{V_2} = \left\{ \frac{1+2g\,h_{to}/V_2^2}{1+\bar{\gamma}/(\bar{a}/g)_{stop}} \right\}^{\frac{1}{2}} - \frac{\bar{\gamma}\,g(\Delta t-1)}{V_2} \qquad (5\text{-}88)$$

The condition that $V_1 < V_R$ must be satisfied. To check this, a more detailed analysis of the rotation and flare maneuver is necessary (Appendix K). In the case that $V_1 = V_R$, the field length is generally no longer balanced.

Combination of eqs. 5-84 through 5-88 results in the expression

$$BFL = \frac{V_2^2}{2g\{1+\bar{\gamma}/(\bar{a}/g)_{stop}\}} \left\{ \frac{1}{(\bar{a}/g)_{0-1}} + \frac{1}{(\bar{a}/g)_{stop}} \right\} \times$$

$$\left(1 + \frac{2\,g\,h_{to}}{V_2^2}\right) + \frac{\Delta S_{to}}{\sqrt{\sigma}} \qquad (5\text{-}89)$$

In this expression the inertia distance ΔS_{to} may be assumed equal to 655 ft (200 m) for $\Delta t = 4\frac{1}{2}$ seconds, a value derived for typical combinations of wing and thrust (power) loadings. This result is valid for both propeller and jet aircraft.

To make eq. 5-89 readily applicable for preliminary design, some further simplifications can be introduced.

1. On the basis of several realistic assumptions regarding undercarriage drag, ground effect, etc., it was found (cf. Ref. 5-27) that the following ap-

proximation can be made:

$$\bar{\gamma} = .06 + \Delta\gamma_2 \qquad (5\text{-}90)$$

where $\Delta\gamma_2$ is the difference between the second segment climb gradient γ_2 and the minimum value of γ_2 permitted by the airworthiness regulations.

2. An average value of $\bar{a}_{stop} = .37g$ has been found from application of the method to 15 jet transports, although with optimum brake pressure control, lift dumpers and nosewheel braking, decelerations as high as .45g to .55g can be achieved on dry concrete. For very high decelerations the balancing condition may not be satisfied.

Using these simplifications, we find the following expression:

$$BFL = \frac{.863}{1+2.3\,\Delta\gamma_2} \left(\frac{W_{to}/S}{\rho\,g\,C_{L_2}} + h_{to}\right) \left\{ \frac{1}{\frac{\bar{T}}{W_{to}}-\mu'} + 2.7 \right\} + $$

$$+ \frac{\Delta S_{to}}{\sqrt{\sigma}} \qquad (5\text{-}91)$$

where

$h_{to} = 35$ ft (10.7 m) and $\Delta S_{to} = 655$ ft (200 m)

$\mu' = .010\,C_L + .02$ for flaps in takeoff position

$C_{L_2} = C_L$ at V_2^{max}; normally $V_2 = 1.2\,v_S$, hence $C_{L_2} = .694 \times C_L^{max}$

$\bar{T}$ = mean thrust for the takeoff run, given by eqs. 5-81 and 5-82

$\Delta\gamma_2 = \gamma_2 - \gamma_{2_{min}}$

γ_2 is the second segment climb gradient calculated from eq. 5-51 at airfield altitude, one engine out (cf. example 3 of Section 5.4.3)

$\gamma_{2_{min}} = .024, .027$ or $.030$ for $N_e = 2,3$ or 4 respectively.

For project design, the case of $\Delta\gamma_2 = 0$ presents most interest, as the corresponding weight is limited by the second segment climb requirement and the BFL is a maximum for the particular flap setting, disregarding the case of overspeed (Appendix K). Obviously, when $\Delta\gamma_2 = 0$ is substituted into eq. 5-91, the thrust/weight ratio must be chosen accordingly so that the thrust is sufficient to obtain $\gamma_2 = \gamma_{2_{min}}$.

For a given BFL, eq. 5-91 may be used to find the following limitation to the wing loading:

$$\frac{W_{to}}{S} \leqslant \rho\,g\,C_{L_2} \left\{ \frac{1.159\,(BFL-\Delta S_{to}/\sqrt{\sigma})\,(1+2.3\,\Delta\gamma_2)}{(\bar{T}/W_{to}-\mu')^{-1}+2.7} - h_{to} \right\} \quad (5\text{-}92)$$

5.4.6. Landing

The landing is split up into two phases: the air distance from passing the screen to touchdown and the landing run from touchdown to standstill. The air distance can be derived from conservation of total energy:

$$S_{air} = \frac{1}{\bar{\gamma}} \left(\frac{V_a^2 - V_{td}^2}{2\,g} + h_{land} \right) \qquad (5-93)$$

where $\bar{\gamma}$ is the mean value of $(D-T)/W$. The length of the landing run from touchdown to standstill is:

$$S_{run} = \frac{V_{td}^2}{2\,\bar{a}} \qquad (5-94)$$

where $\bar{a}$ is the mean deceleration, taking into account an equivalent inertia time, similar to the case of the accelerate-stop distance. Expressing the minimum speed in the stall in terms of the wing loading, the density and C_L-max, and adding eqs. 5-93 and 5-94, we may find:

$$\frac{S_{land}}{h_{land}} = \frac{1}{\bar{\gamma}} + \frac{W_{land}/S}{h_{land}\,\rho\,g\,C_{L_{max}}} \times$$

$$\left| \left(\frac{V_{td}}{V_a} \right)^2 \left(\frac{1}{\bar{a}/g} - \frac{1}{\bar{\gamma}} \right) + \frac{1}{\bar{\gamma}} \right| \left(\frac{V_a}{V_{s_o}} \right)^2 \qquad (5-95)$$

The landing distance is affected by variations in the touchdown speed which is largely dependent upon piloting technique. A reasonable estimate can be obtained by assuming that the landing flare is approximately a circular arc, flown with a constant incremental load factor Δn. In tnat case it can be shown that a first-order approximation for the touchdown speed is given by

$$\frac{V_{td}^2}{V_a^2} = 1 - \frac{\bar{\gamma}}{\Delta n} \qquad (5-96)$$

Substitution into eq. (5-95) yields:

$$\frac{S_{land}}{h_{land}} = \frac{1}{\bar{\gamma}} + 1.69 \frac{W_{land}/S}{h_{land}\,\rho\,g\,C_{L_{max}}} \times$$

$$\left| \frac{1}{\bar{a}/g} \left(1 - \frac{\bar{\gamma}}{\Delta n} \right) + \frac{\bar{\gamma}}{\Delta n} \right| \qquad (5-97)$$

The approach speed has been assumed as 1.3 V_{s}, in accordance with most airworthiness regulations.

The following factors have an effect upon landing performance:
a. The mean value of thrust minus drag. The initial value of γ at the threshold is generally .05 for transport category aircraft, corresponding to an approach angle of 3°. The final value is equal to C_D/C_L in ground effect, the thrust being approximately zero at touchdown. It is difficult to calculate an accurate value of $\bar{\gamma}$, but an average of .10 may be taken as reasonable. The screen height is $h_{land} = 50$ ft (15.3 m).
b. The load factor increment during the flare is normally of the order of $\Delta n = .10$, although large variations can be observed, due to differences in atmospheric conditions, piloting technique and airplane response.
c. The average deceleration during the landing run ($\bar{a}$) is affected by the same factors as discussed with regard to the emergency distance in Section K-4 of Appendix K. For design purposes, the so-called arbitrary landing distance is frequently used. This is established on a dry concrete runway from which the required landing field length is found by multiplying by a factor $f_{land} = 5/3$, according to FAR Part 91. In Fig. 5-17 the unfactored landing distance is plotted for actual aircraft and compared with eq. 5-97, in order to estimate typical values for the mean deceleration. For pre-design purposes the following values may be assumed for $\bar{a}/g$:
.30 to .35 light aircraft, simple brakes,
.35 to .45 turboprop aircraft, propeller reverse thrust inoperative,
.40 to .50 jets with ground spoilers, antiskid devices, speed brakes,
.50 to .60 as above, with nosewheel braking.
The limiting wing loading for a given (factored) field length can be approximated by using the mean values for $\bar{\gamma}$ and Δn stated

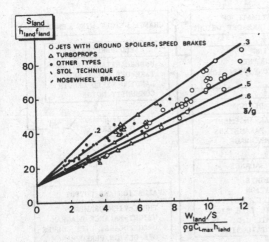

Fig. 5-17. Statistical correlation of the unfactored landing distance

previously:

$$\frac{W_{land}}{S} \leqslant \left(\frac{S_{land}}{f_{land}\,h_{land}} - 10\right) \frac{h_{land}}{\frac{1.52}{\bar{a}/g} + 1.69} \rho g\, C_{L_{max}}$$

(5-98)

For light aircraft the calculated or measured landing distance is unfactored, i.e. $f_{land} = 1.0$.

According to eq. 5-98 the wing loading limit is independent of the engine thrust, provided reverse thrust is not used. However, the usable value of C_L-max can be limited by climb requirements in the approach or landing configuration. Provided the airplane polar is known, these limitations can be worked out in the same way as example 3 of Section 5.4.2.

5.5. AIRCRAFT SYNTHESIS AND OPTIMIZATION

From the previous chapters it will have become clear that the performance aims may be realized in various ways, by choosing suitable combinations of wing loading, aspect ratio and thrust loading. A frequently used systematic approach to this problem is the parametric investigation. A parametric investigation is a design study based on generalized calculation methods and sizing procedures, making it possible to vary configuration parameters and to quantify their effect on the design. The most relevant aircraft parameters to be studied are mentioned in the introduction to Section 5.4. Engine cycle parameters like design Turbine Entry Temperature, Overall Pressure Ratio and bypass ratio, may be subjected to parametric investigations in order to gain an insight into the most suitable type of engine to be installed. As there are many types of parametric design studies, depending on the accuracy required and the phase of the design, some simplified examples to demonstrate the principle will be given below. The list of references relating to this chapter contains a number of more complete studies.

5.5.1. Purpose of parametric studies

Parametric studies can be useful in many types of design problems, provided the design criteria can be quantified in terms of minimum weight, cost and/or noise figures. However, in airplane design they require a considerable amount of computational work, as all variations will have far-reaching consequences. For this reason, computerized design studies are increasingly becoming a prerequisite for advanced and complex aircraft. However, the system must be devised, monitored and utilized by experienced designers in order to define the design problems and the interfaces between various technical disciplines in the design an essential for avoiding unrealistic results.

Although the optimum choice of aircraft design parameters has been a major design problem since the early days of aviation, the generalized approaches described in the published literature are scanty and have a very limited validity due to the necessary simplifications and lack of flexibility. The following design goals can be achieved with a computerized system:

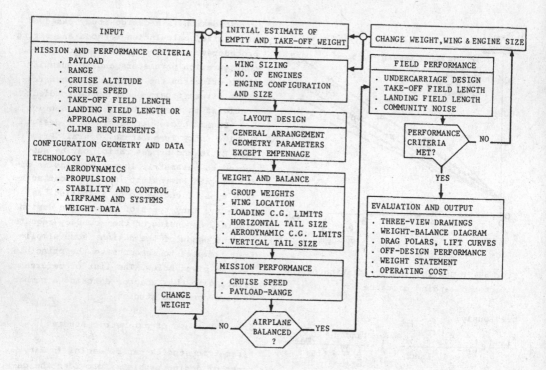

Fig. 5-18. Example of a generalized design procedure (Ref. 5-45, modified)

a. Determination of combinations of parameters, characterizing designs that satisfy specified operational requirements.
b. Calculation of values for the configuration parameters already mentioned, resulting in the most favorable objective function, e.g. takeoff weight or operating costs.
c. Sensitivity studies to assess the effects of minor changes in the shape or geometry, material properties, drag coefficients, etc.
d. Mission/performance analysis and trade-off studies to investigate the effect of variations in the performance requirements.
e. The effect of certain technological constraints in terms of weight and cost penalties. It should be realized that the validity of computerized studies is determined entirely by the accuracy of the input data and design methods available. In that respect they are no better than manual computations. However, improved data can/

easily be incorporated in a computerized system. On the other hand, the design of the program and evaluation of the output may be time-consuming and costly.

5.5.2. Basic rules

Variation of design parameters in effect means designing many aircraft layouts. It is therefore necessary to lay down basic rules governing the design process for each variant. A generalized procedure as depicted in Fig. 5-18 will be explained. Input data are performance criteria and technological constraints. Performance data are generally mission data (payload and range) and performance constraints (field length, approach speed). Cruise performance may be considered as a constraint or as a fixed mission requirement, depending on the application.
The first design step will be to generate initial configuration characteristics such

as takeoff weight, wing loading and thrust loading, on the basis of semi-statistical formulas of the type presented in this chapter. Configuration and layout design activities generally refer to the work on the drawing board; for computerized design this must be translated into a mathematical procedure for defining the external geometry of the major aircraft components. This is a difficult part of the program and in many cases it will, for example, be acceptable to use the geometry of a fuselage design generated outside the numerical program.

The next step is to calculate group weights and the empty weight, desirable loading center of gravity limits and empennage size*. Sufficient data are then available to calculate the maximum range for the design payload and a conclusion can be drawn about whether the fuel weight is sufficient or not. In the first attempt, this will not be the case and the airplane is not balanced with respect to the all-up weight. The fuel and takeoff weights must be changed and the procedure repeated, until the design characteristics have converged sufficiently. Field performance and noise characteristics may now be calculated and if there is some deficiency, the wing size must be changed. Except in the case of a fixed engine, some scaling of the engine size may be feasible as well. The design procedure is repeated again, until all performance requirements are met. The output is presented in the form of drawings, diagrams and characteristic data. Excellent examples of such a program are described in Refs. 5-40, 5-43 and 5-45.

Two comments must be made on the type of studies discussed.

1. The prediction methods for weight and aerodynamic characteristics must not only be accurate in absolute values, but must also predict the effect of design changes accurately. More accurate data than those presented in this chapter are generally

*These subjects will be dealt with in Chapter 8

required.

2. The evaluation of computer output is a time-consuming problem. The computer graphics facility is a useful tool here (cf. Ref. 5-41).

5.5.3. Sizing the wing of a long-range passenger transport

This section is concerned with investigating the effect of wing area variation on the takeoff weight (Fig. 5-19) of a large, long-range passenger transport, equipped with turbofan engines with a total thrust of 176,000 lb (80,000 kg) at SLS (ISA). The airplane will be designed to fly a payload of 126,000 lb (57,000 kg) over a range of 4,000 n.m. (7,400 km); M = .85 at 35,000 ft (10,700 m). Fuel reserves must be available for two hours holding at 95% of $(L/D)_{max}$, plus 5% tolerance on the total fuel weight. The field length is limited to 3,000 m (10,000 ft) and 2,000 m (6,500 ft) for takeoff and landing respectively. For simplicity, climb performance requirements are summarized in the condition that at 85% of the ISA-SLS thrust the second segment climb gradient (FAR 25.121 b) must be achieved.

The all-up weight at takeoff is initially estimated at 705,000 lb (320,000 kg), with a wing area of 5,900 sq.ft (550 m^2). A subdivision is made as follows:

a. A fixed weight of 390,000 lb (177,000 kg) for the payload and payload service items, the fuselage group, the propulsion group, the undercarriage, fixed equipment and systems.

b. A variable weight affected by the wing and tailplane size and fuel quantity required.

The wing plus tailplane structure weight is estimated after several iterations by a simple formula given in Fig. 5-19. Reserve fuel for 2 hours holding is estimated as follows:

$$W_f = \frac{2}{.95} C_T \left(C_D/C_L\right)_{min} W_{land} \qquad (5-99)$$

while

173

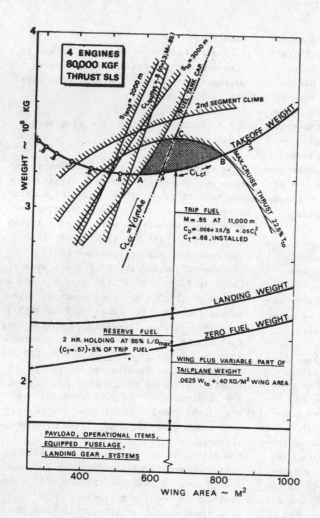

Fig. 5-19. Effect of wing area on the weight breakdown of a large passenger transport (project)

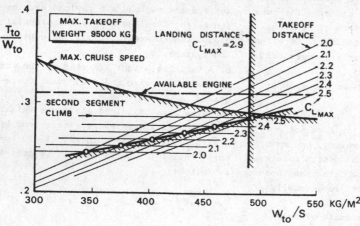

Fig. 5-20. Wing loading vs. thrust loading diagram of the three-engine airliner in Fig. 11-1

174

$$C_D/C_{L_{min}} = 2\sqrt{C_{D_o}/\pi Ae}\qquad\qquad (5-100)$$

It is noted that C_D is a function of the wing area (cf. Section 5.3.2.). Cruise fuel is derived from the Bréguet equation:

$$W_f/W_{to} = 1-\exp\left|-\frac{M}{R}\frac{C_L/C_D}{C_T/\sqrt{\theta}}\right|\qquad (5-101)$$

For an initial cruise altitude of 35,000 ft (10,700 m) the dynamic pressure at M = .85 is 267 lb/ft^2 (1,300 kg/m^2). The effects of climb and descent, flying horizontally at constant engine rating, intake losses, power offtakes and bleed air, etc., are summarized in an equivalent C_T = .68, which is 10% above the uninstalled C_T.

The wing loading for minimum thrust in cruising flight according to eq. 5-35 is 108 lb/ft^2 (525 kg/m^2). According to Fig. 5-19 this value is quite close to the wing loading for minimum all-up weight.

The limitations to the takeoff weight corresponding to each performance requirement can be found from the equations derived in the previous sections. For example, the condition that M = .85 must be achieved with max. cruise thrust can be converted into a limitation of the all-up weight, derived from eq. 5-32:

$$W \leqslant \tfrac{1}{2}\gamma p M^2\left|\pi Ae\ S\left(\frac{T}{\tfrac{1}{2}\gamma p M^2}-C_{D_o}S\right)\right|^{\frac{1}{2}}\qquad (5-102)$$

Substitution of C_D S according to the expression in Fig. 5-19 results in the weight limit indicated in this figure. Other boundaries are derived in the same way from the second segment climb requirement and the field length limitation.

In addition, the wing must have sufficient volume to contain the fuel. It can be shown (Appendix B, Section B-3) that the maximum wing tank volume is proportional to the thickness/chord ratio, the wing area squared and inversely proportional to the span. For a given aspect ratio and section shape it follows that $S^{3/2}$ is the factor of proportionality. The available tank volume restricts the fuel load and consequently the all-up weight. In the case of t/c = 10% (mean value) this requirement appears to set a lower limit to the wing area. If t/c is increased to 12%, to be achieved for the same M with advanced wing sections or more sweep, the tank volume is no longer the limiting factor. Another operational limit originates from the requirement that it must be possible to maneuver the airplane at cruise altitude with n = 1.3 without entering the buffet condition* at an assumed conservative C_L = .6 at M = .85. This leads to a maximum cruise lift coefficient of .462, or a wing loading limit of 600 kg/m^2 (123 lb/sq.ft). As in the previous case this value can be improved by suitable wing design.

The resulting diagram indicates a zone of acceptable combinations for S and W_{to} (ABC). In point A the takeoff weight is a minimum and for a long-range aircraft this is also an approximation for the condition of minimum direct operating costs (d.o.c.). Point B results in the most favorable field performance, but decreased ride comfort and increased costs. Point C is the maximum range condition, provided the extra weight is converted into fuel.

Additional factors in the final choice of the wing area will be discussed in Sect. 7.2.

5.5.4. Wing loading and thrust (power) loading diagrams

A convenient way of illustrating how the powerplant and wing size may be chosen is the wing loading vs thrust (power) loading diagram. An example is shown in Fig. 5-21, pertaining to the short haul passenger aircraft project (185 pax.), referred to in Fig. 1-5. Instead of the thrust loading we have used the T/W-ratio.

Each point in the diagram in effect represents a different airplane design, with different weight distribution. In the case of a constant design payload and range,

*See Section 7.2.2. of Chapter 7

175

the fuel weight and takeoff weight are different for each combination. Alternatively, for a constant all-up weight and payload, the fuel weight and range vary with the empty weight. Lines of constant takeoff weight, constant range or constant d.o.c. can be plotted, using the design procedure described in Section 5.5.2. The diagram is then suitable for simplifying the final choice of wing size and engine thrust (power). In many cases, the trend is to take the highest practicable value of the wing loading and the minimum-size engine within· the region of possible combinations.

Inspection of the performance equations derived in the previous sections reveals that in all formulas except C_D the wing area and engine thrust (power) are combined with the weight*. The exception is eq. 5-13 for the zero-lift drag coefficient. In a first-order analysis Fig. 5-20 is therefore valid for different values of the takeoff weight, except the boundary derived from the high-speed performance requirement.

The following comments can be made on the limitations in the diagram:

a. The max. cruise speed requirement (given M, given cruise altitude) limits the thrust loading to a value which appears to be relatively insensitive to W/S. The wing loading for minimum T/W is 670 kg/m^2 (137.4 lb/sq.ft) but the gain in T/W for wing loadings above 500 kg/m^2 (102.5 lb/sq.ft) is small.

b. The second segment climb requirement is derived from FAR Part 25.121, using eqs. 5-51 and 5-29. The lift/drag ratio with flaps in the takeoff position (see Fig. 5-9) and the thrust lapse rate with speed are assumed to be independent of the wing loading, but for large W/S variations this is no longer true. The required T/W is calculated for different values of C_L-max,

*Although in a more detailed analysis this is no longer true, the conclusion that one diagram can be used for different takeoff weights remains acceptable.

i.e. different flap deflections.

c. The takeoff balanced field length of 1,800 m (5,900 ft) is calculated according to eq. 5-91, to find a wing loading limitation for each C_L-max. The intersections of these lines with those of the climb requirement define the limiting combinations of wing and thrust loading for variable (critical) flap settings. Calculation of the all-engine takeoff with eq. 5-80 indicates that this case is not critical for the case considered.

d. The W/S limit for landing was derived from eq. 5-98, assuming an average deceleration of .38 g during the landing run. Although for the landing and approach conditions certain climb requirements must be met, it was found that these were not critical and they have therefor been omitted from the diagram to improve its clarity.

The actual choice of wing area and engine thrust is subject to several considerations outside the field of performance (cf. Chapters 6 and 7). In the case of Fig. 5-20 the available engine type resulted in a takeoff T/W ratio of .31 for the initially estimated takeoff weight of 95,000 kg (209,440 lb). A weight increase of 5% to 100,000 kg (220,460 lb) is feasible for a constant wing loading of 490 kg/m^2 (100 lb/sq.ft) without running into a cruise speed limitation. The extra weight might be used in the form of fuel or payload, provided there is no space limitation.

As already stated, the fuel weight and range for the different airplane designs, represented by Fig. 5-20, will vary as a consequence of fluctuations in empty weight. This makes an evaluation of the design parameters more difficult, in spite of the simplicity of the representation. The result of a more detailed design procedure for a very large passenger transport is presented in Fig. 5-21. In this example the design payload and range are constant, and the weight distribution is calculated for each combination of wing loading and thrust loading, resulting in the indicated

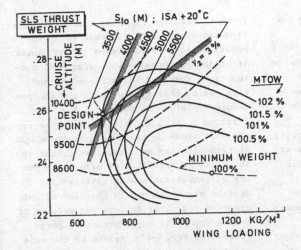

Design data: cruise M=.9; payload: 145,000 kg (320,000 lb); range: 7600 km (4100 n.m.); 4 engines of bypass ratio 12; wing aspect ratio 7

Fig. 5-21. Wing loading vs. thrust/weight ratio for a very large long-haul passenger transport

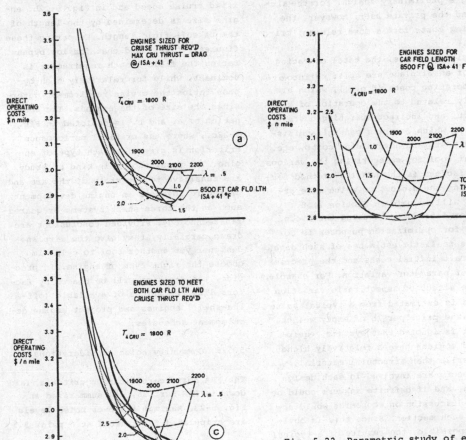

Fig. 5-22. Parametric study of engine cycles for a cargo airplane (Ref. 5-36)

177

lines of constant takeoff weight. The minimum all-up weight is achieved at the clearly impracticable wing loading of 1125 kg/m^2 (231 lb/sq.ft). The indicated design point is based on only two performance limitations, again for the sake of simplicity. The figure illustrates that decreasing the wing loading to 700 kg/m^2 (144 lb/sq.ft) results in a weight penalty of 2%, provided the thrust is chosen near the value for minimum takeoff weight for specified wing loading.

5.5.5. Optimization for low operating costs

In the previous examples the takeoff weight was considered as an important factor in judging a preliminary design. For the airline and the private user, however, the operating costs form a more relevant criterion.

For design purposes, the total operating costs of an airplane are split up into direct operating costs (d.o.c.), which are directly related to the operation of the aircraft, and indirect operating costs (i.o.c.), which comprise general administrative costs and advertising costs, etc. A useful tool for categorizing the various d.o.c. factors is the ATA 1967 method for estimating the d.o.c., to which more attention will be paid in Section 11.8.

The major difficulty in using the d.o.c. formula for optimization purposes is to generate realistic estimates of such data as aircraft initial costs and the precise effect of parameter variation. For example, when the effect of aspect ratio variation on costs is estimated from a typical value of airplane price per lb of empty weight, the fact is ignored that systems, equipment and engines have a relatively higher price per lb than structural material. Cost aspects are involved in each design decision, and if definite answers could be given a discussion on economics would appear in each section. Since this is obviously impossible, the results of a typical d.o.c. optimization study are presented here instead.

In Sect. 4.4. the importance of Turbine Entry Temperature, Overall Pressure Ratio and bypass ratio were discussed in the context of engine performance (specific fuel consumption; specific thrust), weight and size. Using this type of data and the procedure explained in Section 5.5.2. as starting points, the effect of engine cycle variations on airplane drag and weight distribution can be investigated, taking into account installation effects. A number of propulsion system optimization studies are mentioned in the list of references. Fig. 5-22 depicts the results of such a study for a jet-propelled cargo transport. Fig. 5-22a. refers to the case where the engines are sized for the specified cruise speed and in Fig. 5-22b. engine size is determined by the length of the takeoff field length. Comparing these figures, it is found that for low bypass ratios the field length requirement is dominant, while for relatively high bypass ratios the cruise requirement determines the size of the engines. The combination of a. and b. is depicted in Fig. 5-22c., where the critical performance criterion is given for each type of engine. The results of this kind of study are sensitive to the state of the art and disregard advances in engine development such as the three-shaft systems or geared-fan technology. Provided conclusions are drawn carefully, they give the airplane designer yet another tool to help him choose the right type of engine. In practice, the decision will be biased by factors like the prices of available (off-the-shelf) engines and project engine development schedules.

5.5.6. Community noise considerations*

The FAA noise certification criteria, laid down in FAR Part 36, are summarized in Fig. 5-23. Maximum fly-over noise levels are stipulated for takeoff at a point 3.5

*Definitions of noise terminology is found in Ref. 5-50

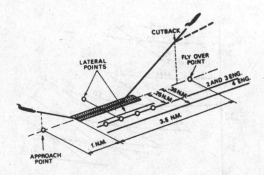

a. Reference points for noise measurements

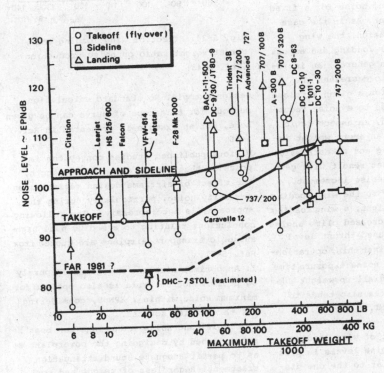

b. Noise levels: regulations and measurements (Ref. NASA SP-265, Ref. 5-55, Flight, Nov. 1972, Flight Manuals F-28, DC-10/30, B747/200B, A-300B brochure)

Fig. 5-23. Noise certification according to the FAR 36 airworthiness regulations

n.m. from the start of the takeoff roll and for landing at a point 1 n.m. from the threshold. A limit is also imposed on the sideline noise, which is measured on a line to the side of, and parallel to, the runway. The noise limits are compared with figures for some current airplane types. The basic methods of obtaining desirable noise levels were dealt with in Section 4.4.3.: engine design, installation and noise attenuation (acoustic lining), gen-

eral arrangement of the airplane, performance and special flight procedures. The performance aspects will be considered from two points of view.

a. Aircraft with Short Take-Off and Landing (STOL) field lengths making steep gradient approaches may exhibit a considerable community noise reduction in comparison to current conventional subsonic transports. However, the effects of reducing the field length on airplane configuration are far-

179

reaching and generally a considerable direct operating cost penalty must be expected.

The present line of thought (cf. Ref. 5-55) tends to favour increased climb and descent angles, without necessarily aiming at very short field lengths. The result may be a relatively conventional airplane with Reduced Take-Off and Landing field lengths (RTOL) of the order of 3,000 to 5,000 ft (900 to 1500 m).

b. Aircraft performance may be optimized so as to achieve minimum noise for a fixed field length requirement. As in the case of performance optimization, the wing loading, thrust (power) loading and aspect ratio can be varied. In general, an increment in the aspect ratio decreases the drag and thrust level for a given airplane lift and therefore leads to a noise reduction. On the other hand, an aspect ratio increment will lead to an empty weight increment and both landing and takeoff weight will be affected. The net result will generally be a thrust and noise increment. Similar arguments apply to the wing loading. For a given aircraft weight, a wing loading reduction results in increased climb angles after takeoff and decreased thrust level during approach. As against this, operational speeds are lower and noise exposure time increases. The increased all-up weight due to the increased wing area also tends to increase the thrust level.

An example of the effect of wing loading and thrust loading on noise levels is presented in a diagram similar to the one discussed in Section 5.5.4.: Fig. 5-24. The relative reduction in EPNdB relative to the FAR 36 requirement is shown for a specific

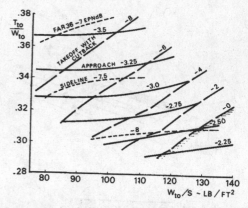

Fig. 5-24. Effect of wing loading and thrust/weight ratio on noise production (Ref. 5-43)

design example. No standard calculation methods for this type of curve can be given here, but an example is presented in Ref. 5-56.

Various published studies convey the impression that, for a given field length, the effect of airframe design variables on community noise, particularly during the approach, is not impressive. The following conclusions relating to a medium-haul high-subsonic transport airplane are quoted from Ref. 5-51:

1. An optimum airplane design based purely on performance criteria is also optimum for minimum noise within 1 EPNdB, considering all three noise sources.

2. Significant noise reductions can best be accomplished by designing the powerplant so as to permit adequate sound attenuation treatment. Regardless of weight and cost, the noise attenuation available by this method is limited to approximately 10 to 12 EPNdB.

Chapter 6. Choice of the engine and propeller and installation of the powerplant

SUMMARY

In this chapter it will be assumed that the total thrust (power) to be supplied by the engines is approximately known. The various considerations which govern the choice of the number and type of engines are discussed. A short survey is given of propeller coefficients, diagrams and methods for controlling the propeller blade angle. Methods are presented for choosing the diameter, shape and number of blades, and some attention is paid to the location and installation of propeller and jet engines, thrust reversers and Auxiliary Power Units.

NOMENCLATURE

AF	- Activity Factor of a propeller blade	rpm	- propeller speed (revolutions per minute)
a	- speed of sound	S	- area
B	- number of propeller blades	S_c	- cross-sectional area of the body in the slipstream of a propeller
b	- chord of a propeller blade element		
c_D	- drag coefficient	T	- thrust
C_{L_i}	- integrated design lift coefficient of a propeller blade	T_{eff}	- effective propeller thrust
		TAF	- Total Activity Factor of a propeller
c_{l_i}	- design lift coefficient of a propeller blade element	V	- flight speed
c_P	- Power coefficient	V_{cr}	- cruising speed
c_S	- Speed/Power coefficient	V_{tip}	- helical tip speed of the propeller blade
c_T	- Thrust coefficient		
D	- propeller diameter; inlet diameter; nacelle diameter	W	- aircraft weight
		x	- fore-and-aft location of nacelle leading edge relative to the wing leading edge
F	- slipstream factor		
f_s	- drag area $\sum(C_D S)$ of all parts exposed to the slipstream of one propeller	z	- vertical distance of thrust line to the chord
		α	- angle of attack of a propeller blade element
h	- factor allowing for retardation of the propeller flow	$\beta, \beta_{.70}$	- geometric pitch or blade angle (at 70 or 75% R)
HSC	- high-speed cruise	$\beta_{.75}$	
LRC	- long-range cruise	ε	- tilt-down angle of engine thrust line
L	- nacelle length		
M	- Mach number	λ	- engine scaling factor
M_{tip}	- Mach number corresponding to V_{tip}	η	- propeller efficiency
n	- propeller speed (revolutions per second)	θ	- relative ambient temperature (temperature/ambient temperature at sea level ISA)
P	- probability of engine failure; engine shaft power transmitted to the propeller	φ	- effective pitch angle; toe-in angle of nacelle
		ρ	- air density
P_{to}	- takeoff power of one engine	σ	- relative air density (air density/air density at sea level ISA)
R	- propeller radius; Reynolds number		
r	- radial distance from the propeller axis of rotation	ω	- rotational speed of propeller

6.1. INTRODUCTION

In order to be able to choose the appropriate type of engine, the total thrust (power) required must be known first, the deciding factor here being the specified performance. As explained in the previous chapter, the engine thrust (power) will also be affected up to a certain extent by the geometry and aerodynamic characteristics of the wing, which may possibly not yet have been finalized at the stage of the design when the engine type is chosen. If the number of engines has in fact been chosen, the approximate thrust (power) to be supplied per engine will already be known. Most designers will prefer to choose an engine which has already been developed and tested, but in some cases the airplane design will be based on an engine project

for which certain characteristics are
still subject to variation. The choice is
generally limited to only a few types and
in some cases there may even be only one
engine which is regarded as suitable. Aircraft design studies are sometimes intended to investigate the possibilities of a
new type of engine and that implies that
there is no choice at all*. When two or
more engine types are considered, an assessment may be made on the basis of a
rough comparison of technological and e-
conomical factors. If this comparison does
not provide a clear basis for choice, various designs should be developed to the extent that all the important consequences
which the choice of engine has on the overall design will be clearly shown.
In the case of a propeller aircraft, the
design of the propeller should fit both
the engine characteristics and the performance of the aircraft. The optimum propeller must do full justice to the performance of the engine and for this reason
a standard propeller is not used on high-
performance aircraft. The geometry of the
propeller is also important in view of the
clearance between the propeller and the
airframe or the ground. Jet aircraft offer
a relatively high degree of freedom in the
location of the engines. Although the final answer regarding the details of the
location will have to come from aerodynamic
research, even the preliminary design
should already be aimed at achieving the
right basic configuration. Although the
thrust reversal system will generally not
be designed by the aircraft manufacturer,
its presence should be taken into account
at an early stage. This also applies to
the Auxiliary Power Unit (APU), which is
generally used in modern transport aircraft.

6.2. CHOICE OF THE NUMBER OF ENGINES AND THE ENGINE TYPE

*cf. also Section 6.2.3.: rubberizing of
turbo engines

6.2.1. Engine installation factors

Several general aspects of engine location
in relation to the number of engines were
mentioned in Section 2.3. We shall now fo-
cus our attention on several considerations on which the final choice of the
number of engines will have to be based.
Fig. 6-1 shows the values generally used

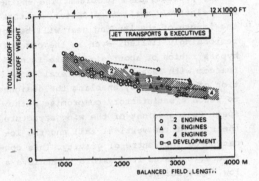

Fig. 6-1. Thrust loading, number of engines
and takeoff field length required

for the takeoff thrust/weight ratio in the
case of civil jet aircraft. The length of
the takeoff runway has proved to be the
most useful parameter for comparison purposes. Takeoff performance will also be
influenced by the wing loading, flap design and, to a lesser degree, the bypass
ratio. The shaded regions show the combinations of thrust loading and the number
of engines usually employed for different
design requirements as regards takeoff
field length. It is also generally found
that two engines are used for short-haul
aircraft and three or four for medium and
long ranges. However, it should be pointed
out that Fig. 6-1 should not be regarded
as in any way mandatory with regard to the
choice of the number of engines. For propeller aircraft a simple correlation of the
type shown in Fig. 6-1 is not readily obtainable.
The limited availability of suitable engine types
may sometimes be a decisive factor, forcing the designer to choose an aircraft configuration which is

possibly not ideal. In such a case it will be worthwhile to consider a variant based on a hypothetical engine, delivering a thrust in the desired order of magnitude. A comparative study of both versions provides a good indication of the price that will have to be paid for the choice one is "forced" to make.

Mounting four jet engines on the wing generally presents minor problems. The engines are relatively small in diameter, so that the length of the landing gear will be correspondingly limited, even in the case of bypass ratios of about six. There is some freedom with regard to the lateral location of the engines, enabling the designer to reach a satisfactory compromise with respect to the weight of the wing structure, the size of the vertical tail and the location of the centre of gravity. This configuration makes it possible to achieve a low empty weight.

When three engines are used, there is always the problem of installing the third engine, for this will have to be placed in the plane of symmetry (cf. Section 2.3.2. and Fig. 2-17). Whichever location is chosen, it is almost certain that the weight penalty caused by installing the central engine will be increased relative to that of the other two engines.

If only two engines are installed the fan diameter will be relatively large, and in order to avoid a high landing gear, location on the rear of the fuselage may be worth investigating. If this is not a favorable location, a high wing may solve the problem of engine-to-ground clearance, though this will result in increased structural weight.

6.2.2. Engine failure

a. Probability of engine failure.
Although the modern turbine engine is very reliable, the possibility of an engine malfunction must never be ignored. In unfavorable circumstances the engine may have to be stopped or may be unable to operate. This will not only lead to a considerable decrement in thrust, but also to yawing and rolling moments, as well as extra drag of the nacelle of the dead engine and drag resulting from the asymmetrical flight condition. The airworthiness authorities have therefore laid down a number of airworthiness criteria relating to takeoff speeds takeoff distances, climb performance, etc., which must be met when engine failure has occurred. In addition, regulations also require that the aircraft should be capable of being stopped safely if failure occurs below the decision speed V_1. Although only 50 percent of the thrust will be available after failure of one engine out of two, the airplane must be designed to provide an acceptable level of safety. If the probability of malfunction per flying hour is taken as P, then the probability that no malfunction will occur will be 1-P. Since P is extremely small in relation to 1 - the order of magnitude comes to $.5 \times 10^{-3}$ to 10^{-4} - an approximation may be introduced for aircraft with two, three and four engines, resulting in the figures in Table 6-1.

Failure of	Probability of engine failure (per flying hour)		
	1 engine	2 engines	3 engines
twin-engine aircraft	2P	P^2	-
three-engine aircraft	3P	$3P^2$	P^3
four-engine aircraft	4P	$6P^2$	$4P^3$

Table 6-1. Probability of engine failure related to the number of engines.

Table 6-1 shows that during a given period of time, the probability of malfunction in one engine is twice as great in the case of a four-engined aircraft as in the case of a twin-engined aircraft, and the likelihood that two engines may fail is even six times as great. Although engine malfunction in itself need not have fatal con

sequences, failure of both engines in a
twin-engined aircraft will obviously be
extremely critical. However, the reliability of today's engines is such that a
probability of P^2 may be practically ignored for a short time period such as the
takeoff phase and the safety level of a
twin-engined passenger aircraft is now
generally considered to be acceptable for
passenger transports which are not intended for extended overwater flights. On the
other hand, $6P^2$ is not a negligible quantity and this is why a four-engined aircraft must meet certain requirements regarding performance and flying characteristics with two engines inoperative in the
en route configuration. The three-engined
aircraft is an intermediate case. If two
engines should fail, say during a transatlantic flight, the DC-10 will be able to
continue on one engine by flying low and
dumping fuel. Failure of two engines out
of three or four during takeoff is extremely unlikely and need not be considered, provided the failure of any one engine
does not entail a situation in which a
second engine is likely to fail.

b. Engine failure during or shortly after
takeoff.
Under these circumstances, a transport aircraft should always have a sufficiently
large reserve of thrust to enable it to
continue the takeoff safely, and for this
reason it must be equipped with at least
two engines.
The requirements for small aircraft (FAR
Part 23) are less severe. Although the aircraft should remain controllable in case
of failure, no requirements are laid down
for its performance during takeoff. Safety
is ensured by allowing for a generous margin in the takeoff safety speed. There are
requirements with regard to climb performance, but aircraft with a marginal performance do not have to comply with these,
provided the stalling speed does not exceed 61 knots (113 km/h); see FAR Part
23.67 and Section 5.4.3. of this textbook.

c. Engine failure during cruising flight.
This case may lead to a forced descent to
a lower altitude, which will depend on the
thrust still available. In the case of
three- and four-engined aircraft the service ceiling after engine failure will be
adequate, so that mountains can be crossed
with a wide margin. This is not always the
case with twin-engined aircraft and the
route to be flown will be the main factor
in determining whether a service ceiling
of, say, 12,000 feet (3600 m) would still
be acceptable. During development of the
Lockheed L-1011 and the Douglas DC-10 Trijets, this consideration contributed to
the choice of the three-engined layout.
Fig. 6-2 shows the influence of the alti-

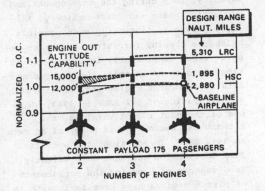

Fig. 6-2. Direct operating costs and number of engines (Ref. SAE Paper No. 710423)

tude requirement on the direct operating
costs of an aircraft project. When a cruising level of 15,000 ft (4575 m) after engine failure is stipulated, the thrust will
have to be adapted accordingly. In that
case the difference in direct operating
costs (d.o.c.) for all versions is very
small. However, when the acceptable limit
is put at 12,000 ft (3660 m), the choice
of the engines will be based on other criteria and the initial engine costs will be
reduced, resulting in lower depreciation.
Twin-engined aircraft may not be operated
on routes where more than one hour's flying from an airport would be involved at

any point. This implies that a twin-engined aircraft cannot be considered for extended overwater flights.

d. Flying qualities after engine failure. In the case of failure of a wing-mounted engine, a yawing moment of such magnitude will be created that the area of the fin and rudder will have to be largely designed to cope with it. This will be discussed more fully in Chapter 9.

6.2.3. Engine performance and weight variations

a. Engine thrust (power) growth.
Transport aircraft shows a steady rise in capacity, both during the development and production phase and after delivery. Aircraft growth is frequently made possible by engine growth, for which the thrust or power rises considerably in the course of time. To cite an example: the Turbomeca Astazou turboprop engine started life with a power rating of 562 ESHP in 1963, while one of the most recent versions, the Astazou XX, has a takeoff power of as much as 1442 ESHP. In addition, the specific fuel consumption has gone down slightly. Some aspects of boosting the performance of piston engines have been dealt with in Section 4.2.2. For gas turbines the growth is obtained by raising the Turbine Entry Temperature, improving cycle efficiencies, increasing the diameter of the air intake and sometimes by increasing the Overall Pressure Ratio. None of these leads to any appreciable increase in engine weight, if anything, they result in a considerable decrease in specific weight. In the case of propeller aircraft, it is recommended that this growth element be anticipated by making allowances for it in propeller design and location so as to ensure efficient absorption of the increased power.

b. Rubberizing of turbo engines.
Engine development is usually a long-term process, but as against this the airplane designer may sometimes have to consider engines which are still in the project phase. This, in turn, may imply that the size has not yet been frozen. In order to adjust the thrust level desired to the specific airplane project application, the engine geometry may be scaled up or down identically, without changing the major thermodynamic characteristics like design Turbine Entry Temperature and Overall Pressure Ratio. This technique is referred to as "rubberizing" the engine for a constant technology.

When the size of engine components is increased, various cycle efficiencies will improve slightly as a result of the increased Reynolds number. Provided the thrust (power) variation is of the order of, say, 10 to 20 percent, the influence of rubberizing on specific thrust and specific fuel consumption is hardly noticeable and is frequently ignored. The growth of engine weight with size is much more difficult to assess. If the well-known square-cube law held good for the turbojet, we would find that an inlet diameter growth with a factor λ would entail growth factors λ^2 for the airflow and thrust and λ^3 for engine mass. As a consequence, the engine weight would be proportional to $T^{1.5}$ and the specific engine weight would vary in proportion to $T^{.5}$. However, a more detailed

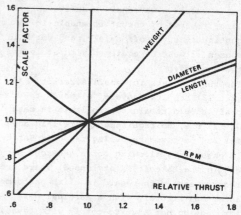

a. Rubberizing effects for a turbojet aircraft (example)

Fig. 6-3. Engine scaling effects

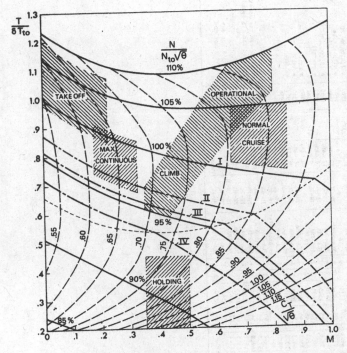

NOTE

1. The following ratings apply to Sea Level ISA conditions:

  I: Maximum takeoff

  II: Maximum continuous

  III: Maximum recommended climb

  IV: Maximum recommended cruise

2. The subscript "to" refers to Sea Level static takeoff conditions

b. Non-dimensional presentation of engine performance of a low-bypass jet engine

Fig. 6-3. (continued)

breakdown of the engine weight into component weights shows that in actual practice the growth of weight with size is considerably overestimated by the square-cube law (e.g. Refs. 6-4 and 6-5). A more realistic relationship is therefore:

$$\text{engine weight} \doteq (\text{thrust})^n \qquad (6-1)$$

in which n has a typical value of 1.07 to 1.14. The example in Fig. 6-3a shows that one should also be aware of the change in engine rotational speed. Reasoning on the basis of the theoretical result of square-cube law, or the more moderate sensitivity given by (6-1), it may be concluded that from the point of view of minimum engine weight it would be advantageous to generate a specified total thrust with a large number of small engines instead of a small number of larger engines, but the advantage in installed engine weight is then much less or may even disappear. Moreover, if real engines of different thrust level are compared (Table 6-2), a systematic trend

cannot be observed in the specific weight for various sizes. It is therefore justifiable to conclude that in practice the number of engines to be chosen cannot be based on simplified theoretical considerations. There is no clear evidence to show that the cost of engines on a basis of dollars per pound of thrust is significantly affected by the number of engines required to develop that thrust. Although for a given number of engines to be produced the engine costs per lb of thrust (or per hp) will decrease with increasing engine size, this effect is compensated for by the fact that the smaller engine will probably be produced in larger numbers, resulting in lower development and production costs per engine.

6.2.4. Choice of the engine type

After preliminary performance calculations have been carried out and the number of engines has been decided upon, the thrust (power) per engine will be roughly known. The choice between engines satisfying the requirements as regards required thrust

Engine Manufacturer	Engine Type	Net Thrust kg Takeoff SL/ISA static	Cruise 11,000 m M=.80	Cruise SFC kg/kg.h	Bypass ratio SL/ISA static	Air flow SL/ISA static T.O. kg/sec	Pressure ratio total	Pressure ratio fan	Turbine Entry Temp. SL/ISA T.O. K	Internal intake diam. cm	Length m	Dry weight kg	No. Spools	Fan stages	Compressor	Turbine	Comb. Chamber	Turbine Cooling	Date first run	Remarks
Turboméca	Arbizon III	400	133	1.14	0	6	5.5	-			1.51	115	1	0	1LA-1HC	1	ann.	no	1970	* 6000m/600 km/h.geared
Turboméca	Astafan II	710 Dry	230*	.63*	6.5	31	9	1.32			1.90	195	1	1	2LA-1HC	3	ann.rev.	air	1969	geared aft fan
Pratt & Whitney Can.	JT-15D-1	1000	223	.88	3.2	35.4	10.2	1.52		56	1.51	229	2	1	1HC	1H-2L	8 can.	no	1967	
Pratt & Whitney	JT-12A-6	1360	363	1.06*	0	22	6.5	-		55.3	1.60	203	1	0	4LA-4HA	1H-2L	ann.	no	1958	
Ivchenko	AI-25	1500	443*	.84*	8	42	9	1.7	1235	45.7	1.99	290	2	1	3LA-8HA	1H-3L	ann.	no	1966	
Rolls-Royce (BS)	Viper 20-F.20	1545	467	1.26	0	20	5.6	-		56	2.16	270	1	0	7		ann.		1965	
Garret-Airesearch	TFE 731-1	1585	408	.83	2.7	51.3	19	1.39	1135	47	1.26	272	2	1	4LA-1HC	2	ann.	air	1969	geared
General Electric	CF-700 2B	1950	428	.93	1.9	38	6.8		1285	71.6	1.91	334	1	0	8A	2	can.	air	1962	aft fan
Avco Lycoming	S02 D	2800	609	.78	6.1	108	10.3			91.3	1.62	526	2	1	7LA-1HC	1H-3L	ann.		1970	geared
Rolls-Royce (BS)	M-45H-01	3520	1245*	.73*	2.8	104	18		1285	102.4	2.96	673	2	1	5LA-7HA	1H-1H-1L	ann.	air	1968	* 6,100m/Mv.65
Rolls-Royce	Spey Mk555-15	4470	1680*	.79*	1	91	16		1330	90.9	2.79	990	2	1	4LA-12HA	8		air	1965	* 7,620m/Mv.70
Rolls-Royce	Trent RB 203.08	4530	1315*	.72*	3	136	16			82.4	2.09	805	3	1	4MA-5HA	2H-2L	can.	air	1967	* 7,620m/Mv.70
Rolls-Royce	Avon Mk524B	4760			0	83	8.7		990	98.3	3.20*	1516	1	0	16	1H-2L	8 can.	no	1957	* with nozzle
Rolls-Royce	Spey Mk511-5	5170	1975*	.79*	.64	92	19		1100	99.1	2.91	1470	2	1	5LA-12HA	2H-2L	10 can.	air	1960	
Soloviev	D-20P	5400	1100	.78*		113	14	2.6	1350	82.4	3.30	1585	2	0	3LA-8HA	1H-2L	12 can.	no	1957	
Pratt & Whitney	JT-3C-7	5440	1540	.90	0	91	13			93	4.25*	1431	2	0	9LA-7HA	1H-2L	8 can.	no	1957	
Pratt & Whitney	JT-8D-7	6350	1540	.81	1.1	141	15.8			98.8	3.75	1680	2	1	4LA-7HA	1H-3L	9 can.	air	1960	* 7, 20 m/Mv.70
General Electric	CJ-805-21	6800	≈1300*	.76	0	188	12*	1.6		108	3.93	1520	2	0	17	1H-2L	8 can.		1961	* 10,000 m/Mv.82
Soloviev	D-30P	6800	≈1300*	.77	1.5	125	18.6		1300	98.5	3.34	2060	2	1	4LA-10HA	2H-2L	12 can.	air	1960	
Rolls-Royce	Conway RCo. 12	7940	2075	.89	.3	127	15.1		1315	95.5	3.66	2310	2	1	7LA-9HA	2H-2L	10 can.	no	1965	* Excl. aft fan
Pratt & Whitney	JT-4A-11	7940	2130	.92	0	115	12			103.5	3.71	1890	2	0	8LA-7HA	1H-3L	8 can.	air	1957	
Pratt & Whitney	JT-3D-3	8170			1.4	204	13			135		2073	2	1	as above+fgt.	1H-2L	8 can.	no	1958	Mk 509
Pratt & Whitney	JT-3D-5A	9520					12			135		2310	2	1			8 can.			
Rolls-Royce	Conway RCo.42-3	9900	2420	.82	0.6	165	15.1			114.2	3.90	1940	2	0	6LA-6HA	1H-3L	10 can.	air	1960	
S.N.E.C.M.A.	M-56-20	10000	2320	.67	4.0	323	18	1.55	1480	157.5	2.90	2120	3	1	6MA-8HA	2H-1H-4L	ann.	air	proj.	400m/sec fan tip speed
S.N.E.C.M.A.	M-56-40	10000	2205	.64	4.8	334	18	1.55	1520	162.8	3.36	2200	3	1	2LA-6HA	2H-1H-4L	ann.	air	proj.	310m/sec fan tip speed
Kuznetsov	NK-8-4	10500	2750	.78	1	232	23.2	2.15	1145	144.2	3.36	2150	2	1	6HA	1H-2L	8 can.	air		
Soloviev	D-30K	11500	2750	.67	2.3	272	20		1420*	145.5	5.10	2150	2	1	3LA-11HA	2H-4L	12 can.	air		
Rolls-Royce	RB.211-22	19050	4020	.61	4.8	602	25	1.5	1570	217.2	4.61	3267	3	1	7MA-6HA	1H-1H-3L	ann.	air	1968	* 1485 K hot day
General Electric	CF-6-6G	19500		.64	5.6	655	29			219	4.90	3540	3	1	1LA-16HA	2H-5L	ann.	air	1968	* 1595 K hot day
Pratt & Whitney	JT-9-D-15	20500	4350	.62	4.4	687	22		1425	237	3.26	3970	2	1	3LA-11HA	2H-4L	ann.	air	1966	

Table 6-2. Principal characteristics of turbine engines

TURBOPROP ENGINES

Engine Manufacturer	Engine Type	Performance, ISA Power Takeoff SL/static	Power Cruise 20,000ft 245kts	ESFC cruise 245kts kgf/pk/h	Engine cycle Air-flow kg/s	OPR	TET (K)	Max. RPM	Propeller Gear ratio	RPM	Dimensions length m	width/diam. m	Dry weight kg	No. spools	Engine-assembly Compressor	Turbine	Comb. Chamb.	Date first run	Remarks
Rover	Model TP-90	110	49	.609	.85	2.8	-	47,000	.0540	2538A	.90	.52	91	1	1C	1	1 can	1960	NS
Allison	Model 250(T-63)	250	149	.295	1.35	6	1170	48,950	-	- C	.95	.40	62	2	6LA-1HC	2H-2LF	ann.	1959	turboshaft
Turboméca	Orédon III	E 350	190	.242	1.35	7.5	-	59,100	.1024	-	1.09	.37	82	1	2LA-1HC	3	ann.	1964	turboshaft
Turboméca	Astazou II	E 555	307	.361	2.50	5.7	-	43,500	.0558	2425C	1.91*	.46	122	1	1LA-1HC	3	ann.	1960	*incl. prop.
Pratt & Whitney	PT 6A-6	E 578	332	.279	2.25	5.7	-	33,000	.0663	2200C	1.58	.48	122	2	3LA-1HC	1H-1LF	ann.	1963	turboshaft
Turboméca	Astazou XIV A	E 600	321	.251	2.50	7.5	-	43,000	.0415	1783C	1.43	.68	160	1	2LA-1HC	3	ann.	1966	R
Turboméca	Bastan IV	E 935	552	.251	4.50	5.8	1143	53,500			1.55	.75	222	1	1LA-1HC	3	ann.	1959	R
AVCO Lycoming	T-53-L-7	E 1150	730*	.261*	5.00	6.0		25,240	.0479	1700	1.49	.58	245	1	5LA-1HC	2H+2LF	ann.	1959	*15,000 ft aircooled
AiResearch	TPE 331-20	600	355	.255	2.50	7.82	1240	41,730		2000C	1.18	.55	129	1	2C	3	ann.	1963	β-control
General Electric	T-58-GE-8	1250	780	.247	5.62	8.30	1150	19,500		2000C	1.39	.41	130	1	10A	2H+1LF	can	1959	turboshaft
AVCO Lycoming	T-5313A	1400	800	.240	5.50	7.40	1210	25,240		1680C	1.48	.58	274	2	5LA-1HC	2H+2LF	ann.	1967	R
AVCO Lycoming	T-5313B	1400	800	.240	5.50		1210			1680C	1.48	.58	312	2	5LA-1HC	2H+2LF	ann.	1967	R
Rolls-Royce	Dart Mk511-7E	E 1740	930	.308	9.30	5.50	1130	14,500	.086		2.43	.96	494	1	1LC-1HC	2	7can	1953	
AVCO Lycoming	T-5321A	1800	920	.229	5.80		1325	21,300		C			306	2	5LA-1HC	2H+2LF	ann.	1969	aircooled, R
Rolls-Royce	Dart Mk528	E 2105	1170	.266	10.70	5.75	1160	15,000	.0929		2.50	.96	560	1	1LC-1HC	3	7can	1956	
			Cruise 30,000ft	ESFC 300 kts															
General Electric	T-64-GE-6	2650	1350	.200	11.10	12.60	1200	13,640			2.80	.74	490	2	14A	2H+2LF	ann.	1959	
Napier	Eland N.El. 1	3000	1230	.229	14.10	7.00		12,500	.0714	893A	3.10	.92	715	1	10A	3	6can	1952	
Rolls-Royce	Dart Mk542-10	E 3025	1450	.213	12.25	6.35	1225	15,000	.0775		2.53	.96	625	1	1LC-1HC	3	7can	1960	
Allison	501-D-13	E 3750	1450	.215	14.75	9.25	1249	13,820		A	3.69	.69	797	1	14A	4	6can	1954	
Allison	T-56-A-10W	E 4050	1450	.224	14.65	9.25	1249			1021C	3.71	.69	838	1	14A	4	6can	1953	
Bristol Siddeley	Proteus Mk765	E 4400	2155	.185	20.90	7.10	1150		.0863		3.13	1.10	1315	2	12LA-1HC	2H-2L	8can	1958	
Rolls-Royce	Tyne R.Ty.12	E 5505	2400	.185	21.20	13.50	1242	15,250	.064	976A	2.77	1.03	1000	2	6LA-9HA	1H-3L	10can	1959	
Rolls-Royce	Tyne R.Ty.20	E 6100	2500	.193	21.20	13.50	1273	15,250	.064	976A	2.76	1.40	1085	2	6LA-9HA	1H-3L	10can	1962	aircooled

Table 6-2. (Continued)

189

(power), fuel consumption and noise production, etc., is usually very limited. Table 6-2 presents a survey of the most important characteristics of some types of engines.

Engines, like aircraft, are subjected to extensive airworthiness tests, cf. for instance ICAO Circular 51-AN/4612. The designer must consider whether the engine chosen has been designed for civil or military use, when testing and certification are planned, which regulations will apply, and so on. To ensure smooth progress of the aircraft project, it is desirable to use an engine which is sufficiently far developed; there is generally a preference for a type which has already been built and tested (off-the-shelve engines).

Even a preliminary design requires a complete engine type specification or brochure, which should preferably contain at least the following data:
- limits regarding the engine output (ratings) and the operational conditions (temperatures, altitudes, speeds),
- thrust and fuel consumption for various engine ratings, altitude and airspeed,
- influence of airbleed and power off-take on engine performance,
- installation data: weight, dimensions, location of centre of gravity,
- noise levels, particularly for civil jet aircraft.

Engine performance is sometimes presented in the form of non-dimensional curves; an example is given in Fig. 6-3b. These curves not only represent the thrust lapse with altitude, airspeed and engine rating, but may also be used for the purpose of rubberizing the engine.

The choice of the engine and the design of the aircraft are so closely interrelated that it is difficult to make an appropriate choice. The designer may obtain a fair picture by an overall comparison of specific fuel consumption, thrust variation with altitude, specific weight and dimensions, engine configuration, initial cost and average time between major overhauls, etc. Sometimes an important part is played by contacts with the engine manufacturer, or economic and political factors affecting aviation. The survey of engine technology in Chapter 4 may possibly help the designer to make up his mind.

Summing up the arguments in the previous sections we may conclude that:
a. the number of engines may be decided entirely by the limited choice from competitive types of power plant;
b. the minimum number of engines with which the specified performance can be obtained will, as a rule, also be the most favorable number;
c. three-engined aircraft will be at a slight disadvantage technically as compared with four-engined aircraft;
d. a design study can only be started when sufficiently detailed engine performance and installation data are known.

The best engine for a particular application can only be arrived at by long and close collaboration between aircraft and engine designers.

6.3. CHARACTERISTICS, CHOICE AND INSTALLATION OF PROPELLERS

6.3.1. General aspects

The design and production of modern propellers is so specialized in nature that there are only a few manufacturers in this field. The aircraft designer draws up the propeller specification and has a number of possibilities to choose from. He should be aware of:
- the airplane's performance
- the engine characteristics and engine control
- propeller noise and vibrations
- the installed weight of the propellers
- the influence of the propellers on flying qualities
- the structural limits imposed by the layout of the aircraft.

In the preliminary design stage the choice of the propeller is mainly associated with the following characteristics:
a. The blade angle control system: fixed or

constant speed propeller, β-control, etc.
b. The blade shape: the planform of the
blades, aerofoil shape and, incidentally,
the twist along the blade
c. The number of blades per propeller
d. The propeller diameter and rotational
speed (gearing).

The losses incurred by transmitting (crank)
shaft power into useful thrust horsepower,
is expressed in the propeller efficiency,

$$\eta = \frac{\text{useful power output}}{\text{shaft power input}} = \frac{TV}{P} \qquad (6-2)$$

Propeller efficiency is zero under static
conditions; its maximum value is obtained
in high-speed flight and amounts to between
approximately 85 and 92% depending upon the
blade shape.
Since a high propeller efficiency is desired in all the important phases of flight,
the efficiency constitutes the most important design criterion for the propeller. It
is essential to realize that cruising flight
is not the only decisive factor, for the
takeoff and climb performance may be equally important, particularly in case of engine failure.
This section will not deal with propeller
theory, and the mechanical design of propellers will only be referred to very
briefly; the list of references may be
consulted for literature on these subjects.
The application of propeller theory to
various airplane design aspects is covered
very adequately, for example, in Ref. 6-20.

6.3.2. Propeller coefficients and diagrams

The characteristic angles relating to the
propeller blade element are shown in Fig.
6-4. The propeller rotates with an angular
velocity ω, while the airspeed is V. When
the plane of the propeller is normal to
the airspeed, the resulting speed at the
blade element at a distance r from the
axis of the propeller will have an effective pitch angle φ with the circumferential
velocity ωr

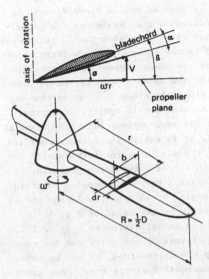

Fig. 6-4. Propeller blade element and local
direction of flow

$$\phi = \arctan \frac{V}{\omega r} \qquad (6-3)$$

In terms of the rotational speed and the
propeller diameter,

$$\phi = \arctan \frac{V}{nD} \times \frac{1}{\pi r/R} \qquad (6-4)$$

For $r = R$, ϕ represents the pitch angle of
the helical path described by the propeller
tip. The decisive factor here is the advance ratio,

$$J = \frac{V}{nD} \qquad (6-5)$$

The angle between the blade chord and the
propeller plane is the geometric pitch or
blade angle β. The angle of incidence of
the blade element relative to the airstream
is

$$\alpha = \beta - \phi \qquad (6-6)$$

or, in combination with (6-4) and (6-5):

$$\alpha = \beta - \arctan \frac{J}{\pi r/R} \qquad (6-7)$$

The angle α is not equal to the local effective angle of attack, since an extra

191

speed is induced at the blade element by the trailing vortex system of the propeller, but (6-7) does show that J is the deciding factor as far as the working condition of the blade element is concerned. Assuming the induced velocity to be constant along the blade and noting that each blade element should operate at approximately the same optimum effective angle of attack, it is easily shown that according to (6-7) β will have to decrease with increasing r. For this reason propeller blades always have a pronounced built-in twist angle. It is usual to define the blade pitch by the value of β at 70% or 75% of the radial distance: $\beta_{.70}$ or $\beta_{.75}$. If the effects of viscosity and compressibility are disregarded, it can be shown by dimensional analysis that for a given blade geometry the following propeller coefficients are determined only by J and

$\beta_{.75}$:

the Power Coefficient: $C_P = \dfrac{P}{\rho n^3 D^5}$ (6-8)

the Thrust Coefficient: $C_T = \dfrac{T}{\rho n^2 D^4}$ (6-9)

or in British units:

$$C_P = 50 \; \frac{Bhp}{\sigma (rpm/1000)^3 (D,ft)^5} \quad (6-8a)$$

$$C_T = 1.515 \; \frac{T,lb}{\sigma (rpm/1000)^2 (D,ft)^4} \quad (6-9a)$$

Provided the tip speed is sufficiently far below the speed of sound and the blades are not stalled, the influence of the Mach and Reynolds number will be negligible. From this it follows that the power and thrust coefficients of geometrically similar propellers, within a certain region

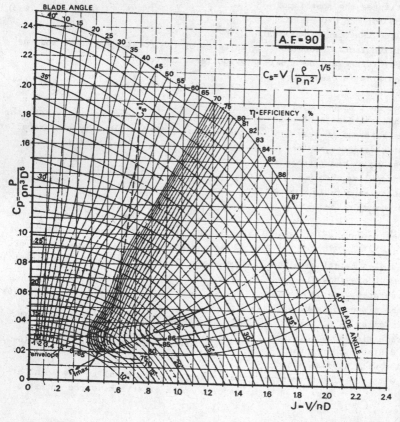

a. Two-blade propeller
(Ref.: NACA WR 286)

Fig. 6-5. Performance diagrams of isolated propellers

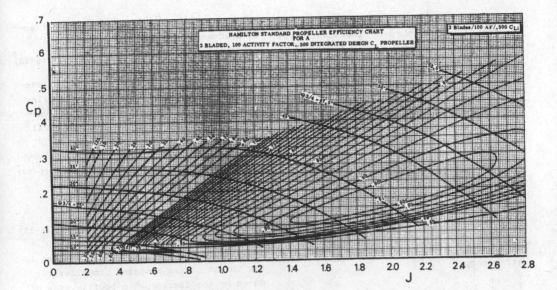

b. Three-blade propeller (Ref. Hamilton Standard PDB 6101)

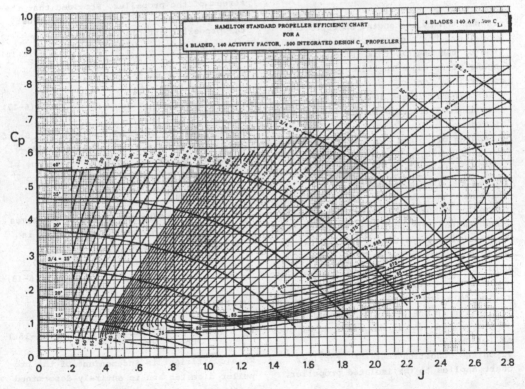

c. Four-blade propeller (Ref. Hamilton Standard PDB 6101)

Fig. 6-5 (concluded)

of M and R (per unit length) can be represented in a single non-dimensional propeller diagram. Examples of such diagrams are depicted in Fig. 6-5. In some diagrams the thrust coefficient is specified instead of η. The propeller efficiency is derived from the power and thrust coefficients and the advance ratio as follows:

$$\eta = \frac{C_T}{C_P} J \qquad (6-10)$$

The propeller thrust may be obtained directly from C_T:

$$T = C_T \, \rho n^2 D^4 \qquad (6-11)$$

or from the power input and the propeller efficiency

$$T = \frac{\eta P}{V} \qquad (6-12)$$

The propeller power P is the net power transmitted by the engine shaft to the propeller. For a given operational condition of the engine, this is found from the engine performance specification, making corrections for:
- aerodynamic losses in the intake
- mechanical friction losses in the reduction gear*
- tapping off bleed air from the compressor for anti-icing, airconditioning, etc.
- power extraction to drive engine accessories and airplane systems.

The thrust must be clearly defined (by the propeller manufacturer) and is generally taken as the tension force in the shaft of the isolated propeller (η_{is}) that is the efficiency without an engine nacelle placed behind it. In the case of performance calculations this efficiency should be corrected to allow for the effects of the installation in the aircraft. The principle of the method used in Ref. 6-16 works out as follows.

1. On the condition that the propeller chart applies to the isolated propeller,

*Frequently included in the specified engine performance

the efficiency is determined for an effective advance ratio, which is given by:

$$J_{eff} = (1-h) \, J \qquad (6-13)$$

The factor h takes account of the retardation of the airflow through the propeller disk, caused by the presence of the fuselage or nacelle behind it. Its numerical value is:

$$h = .329 \, \frac{S_c}{D^2} \qquad (6-14)$$

2. A slipstream factor F is defined,

$$F = \frac{\text{effective propeller thrust}}{\text{isolated propeller thrust}} = \frac{T_{eff}}{T_{is}} \qquad (6-15)$$

The effective propeller thrust is equal to the thrust of the isolated propeller reduced by the increment in profile drag of those parts that are exposed to the slipstream of the propeller. Provided that the airplane drag is defined for zero thrust, the propeller efficiency is equal to:

$$\eta_{eff} = F \, \eta_{is} \qquad (6-16)$$

A usual approximation for F is:

$$F = 1 - 1.558 \, \frac{\sigma f_s}{D^2} \qquad (6-17)$$

where f_s is the profile drag area $\sum(C_D S)$ of the aircraft parts immersed in the slipstream of one propeller. In the absence of better data, C_D may be assumed equal to .004, based on the wetted area.

A propeller coefficient which is sometimes used to choose the propeller diameter is the Speed/Power Coefficient:

$$C_S = V \sqrt[5]{\frac{\rho}{Pn^2}} \qquad (6-18)$$

or in British units:

$$C_S = .638 \, \frac{\text{mph} \, \sigma^{1/5}}{\text{Bhp}^{1/5} \, (\text{rpm})^{2/5}} \qquad (6-18a)$$

This coefficient is independent of the propeller diameter and is entirely determined by the operational conditions: the flying speed, altitude, engine power and rpm. In some outdated propeller diagrams C_S will be found as a parameter and these can be used directly for the choice of J and hence D. Manipulating (6-5), (6-8) and (6-18), we find:

$$C_p = \left(\frac{J}{C_S}\right)^5 = \frac{P \; n^2}{\rho \; V^5} J^5 \qquad (6-19)$$

and for given flight conditions we may draw the corresponding line of C_S = constant in the propeller diagram as in Fig. 6-5a; in the example this has been done for C_S = 1. From a comparison of this line with those for constant η it may be concluded that in this case the maximum efficiency (77.5 percent) is obtained at J = .51, β = 15° and C_p = .033. The diameter which will give the maximum efficiency can be deduced from this condition. The usefulness of the coefficient C_S is limited, since various operational conditions have to be considered when choosing a propeller. In such a case it will be desirable to have the propeller work under the best possible conditions, both in cruising flight and during takeoff and climb. A procedure which may be followed for choosing the propeller geometry is given in Section 6.3.4.

Propeller performance may be determined in various ways:

a. By using propeller diagrams supplied by the manufacturer, for example, Ref. 6-21. Propeller diagrams may also be found in various NACA Reports and handbooks. This method is generally satisfactory for a preliminary choice.

b. By calculation, using generalized methods, such as:

- the approximations given in Ref. 6-15 (light aircraft) or Ref. 6-17 (transport aircraft)
- the SBAC standard method (Ref. 6-16), which also allows for the installation of the propeller
- the method given in Ref. 6-26, which also makes it possible to calculate propeller noise characteristics.

An advantage of generalized methods is that systematic variations in blade shape may be accounted for, thus making opti-

mization possible.

The way in which the propeller data - and more particularly the propeller diagrams - should be used depends on the particular application and the blade angle control, as will be explained in the next section.

6.3.3. Blade angle control

The following distinctions are made with regard to propeller control:

a. Propellers with constant pitch during flight:
- fixed pitch propellers, with the pitch built in during manufacture
- adjustable propellers, with the blade angle adjustable on the ground but fixed in flight.

b. Propellers with variable blade angle during flight:
- adjustable propellers, for which the blade angle may, within given limits, be set at some pre-determined values, e.g. fine and coarse pitch
- controllable pitch propellers, for which the blade angle, within given limits, may be set by the pilot at any desired value
- constant speed propellers, for which the rate of revolution is kept constant at a value set by the pilot. The blade angle is controlled automatically, so that there will be equilibrium between the power supplied by the engine and the power absorbed by the propeller
- constant speed propellers with blade angle control (β-control), for which the blade angle may be set and controlled directly by the pilot, while the rate of revolution during flight remains constant. Engine power is governed by an automatic fuel control system.

Adjustable and controllable pitch propellers are hardly ever used nowadays. Fixed pitch propellers are still fitted to small aircraft used for flying instruction, touring and business flights. Controllable pitch propellers are in use on some powered sailplanes, the pilot being able to feather the blades in order to reduce the drag when the engine is switched off. Propeller-

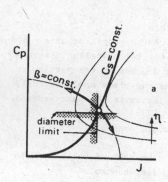

a. FIXED PITCH PROPELLER: Choice of propeller diameter and blade angle.

1. Speed-power coefficient: $C_S = V\left(\dfrac{\rho}{Pn^2}\right)^{1/5}$ e.g. for maximum cruising speed and appropriate engine rating.

2. $C_p = (J/C_S)^5$, to be computed as $f(J)$, as in Fig. 6-6a.

3. Diameter limit: $D = \dfrac{108.4\sqrt{\theta}}{n}\sqrt{M_{tip}^2 - M^2}$; $M_{tip} < .80$ to $.85$; calculate corresponding C_p and J-limits.

4. Choose β and D in diagram.

5. For performance calculations only the line for β = constant is used.

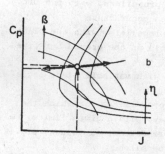

b. CONSTANT SPEED PROPELLER - POWER CONTROL:

Calculation of efficiency for specified shaft power, propeller diameter and rpm.

1. Power coefficient: $C_p = \dfrac{P}{\rho n^3 D^5}$

2. Advance ratio : $J = V/nD$

3. Read η from the diagram.

4. For variation of flight speed C_p remains approximately constant (piston engines) or increases slightly with speed (turboprop engines).

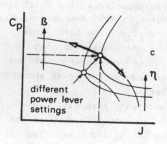

c. CONSTANT SPEED PROPELLER - β CONTROL:

Effect of flight speed and engine control on power and propeller efficiency.

1. For specified V the power required is determined by using the airplane polar (horizontal flight).

2. Assuming η, the engine power for equilibrium is found and J and C_p are calculated.

3. η is read from diagram and if necessary the procedure is repeated.

4. For variable speed β remains constant. The engine power is controlled automatically to keep n constant until the engine and the propeller power are matched.

UNITS:

V in m/s (1 m/s = 1.942 kts) ; D in m (1 m = 3.28 ft)

P in kgm/s (1 kgm/s = .01316 hp) ; ρ in kg sec^2/m^4

(1 kg sec^2/m^4 = .019 slug/ft^3)

Fig. 6-6. Applications of the propeller diagram

powered transport aircraft are always fitted with constant speed propellers; these are often equipped with blade angle control. Turbopropeller engines have to be fitted with constant speed propellers.

a. Fixed pitch propellers.

The advantage of fixed pitch propellers is

obviously that they are simple to produce, hence cheap, while they are also light and require no maintenance. For these reasons they are used on small aircraft with engines of up to about 200 hp and even more in the case of agricultural aircraft.

An aircraft with a fixed pitch propeller, however, is very much at a disadvantage where flexibility in performance is concerned. During the takeoff and subsequent climb the rpm of the engine is limited by the power which the propeller is able to absorb, and consequently the power cannot be fully utilized. Decreasing the blade angle will improve the situation during takeoff and low-speed flight, but in high-speed flight the engine will be overspeeded and must be throttled back.

When a preliminary choice of the blade angle and diameter has been made, it may be assumed that the aircraft will fly at maximum cruising speed with the rpm and power values recommended by the engine manufacturer. The equations required to estimate the cruising speed are presented in Sections 5.3.4. and 5.4.1. If at this stage the drag polar is still unknown, the cruising speed may be estimated by comparison with similar aircraft or it can be deduced from the design specification. The Speed/Power Coefficient C_S can then be determined by (6-18) and the efficiency will be dependent only on the diameter. When this is chosen as large as possible, the propeller blade angle will be small and at low speeds it will be possible to obtain a high rpm, and consequently high engine power. The propeller diameter is limited by the permissible Mach number at the tip ($M_{tip} < .80$ to $.85$), or by practical limits, such as the clearance from the ground. Fig. 6-6a illustrates the choice by means of a propeller diagram.

For a more definite propeller design, the aerodynamic characteristics of the aircraft should be known, but these will only be determined at an advanced stage. It will be necessary to carry out performance calculations under varying flight conditions in order to find the most favorable compromise. Once the angle of incidence of the blade has been chosen, the only remaining important factor in the propeller diagram will be the line related to the fixed propeller pitch angle β.

b. Constant-speed propellers.

As far back as the 19th century the French pioneer Alphonse Pénaud proposed a propeller with adjustable blades, but it was not until about 1925 that the constant-speed propeller proved to be a practical proposition and it came into service around 1935. The performance of aircraft improved to such an extent that since that time the constant-speed propeller has come into general use. The pilot sets the rpm of the engine to the value desired and for each flight speed the propeller blade angle is automatically adjusted in such a way that the propeller is able to absorb the power of the engine. The main difference between this and the fixed pitch propeller is that the engine power available can be fully utilized at each airspeed. There are various methods for obtaining an automatic system of adjustment. For example, oil pressure can be used to twist the blade in one direction, while a counterweight at the hub of the blade permits movement in the opposite direction. Fig. 6-7a shows a simple system but many variants have been derived from this. The sketch in Fig. 6-7b gives an example of the mechanism of a pitch-control system. Electrical systems

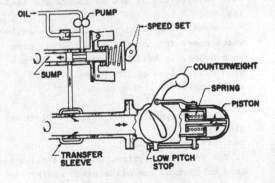

a. Non-reversing propeller and governor

Fig. 6-7. Mechanical principle of constant-speed propellers

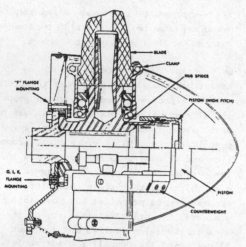

Hartzell Model HC-82X-1 Constant Speed

b. Blade angle control

Fig. 6-7 (concluded)

are used as well. Constant-speed propellers
are often provided with:
- a feathering device, which allows the
blade to take up an angle of attack of 80
to 90 degrees when the engine is inoperative, thus considerably reducing the drag
of the propeller, while engine damage due
to windmilling is avoided;
- a facility for turning the blade to a
negative pitch angle, which is used to obtain reverse thrust and so reduce the landing run;
- automatic restriction of the blade angle
which comes into action when the oil pressure drops, in order to safeguard against
high revolutions of the propeller with
small angles of attack.

c. Constant-speed propellers with blade angle control.
Two versions are known:
1. The β-control system, where the pilot is
able to control the blade pitch angle both
in flight and on the ground
2. A system in which control of the propeller blade is only possible during maneuvering on the ground.

When the β-control is in operation during
the flight phase and the pilot wants to increase the engine power, the direct result
of moving the power level will be an increase in the blade pitch. Since the propeller torque will then increase, the rpm
will tend to drop. A "speed set" will now
transmit a signal to the fuel control system and more fuel will be fed to the engine. Engine power will increase until the
rpm again reaches the pre-set value.
Fig. 6-8 illustrates how this method of en-

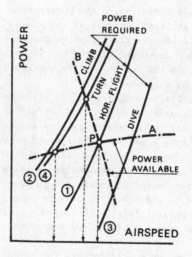

A: Power control mode
B: Blade angle control mode (β constant)

Fig. 6-8. Comparison of normal power control and β control (Ref.: SAE Paper No.
670244)

gine control works in practice. It should
be noted that it is only applicable to
turboprop engines where the propeller rpm
and power can be varied independently.
Curve 1 shows the power required for steady
horizontal flight. Curves 2, 3 and 4 represent the total power required during
stationary climb and descent and in a stationary turn. The speed in horizontal flight
is determined by the equilibrium between the
power available and the power required
(point B). The difference between the lines
of power available with the conventional

method of adjustment (line A) and with β-control (line B) results from the automatic adaptation of the power to the flying speed in the case of β-control. To take an example of the β-control system: when the flying speed in a stationary condition is reduced, the propeller blade angle of attack will increase for constant blade setting and so will the propeller-torque (equation 6-7). The engine control system will react to this change and will cause the power to increase to the extent that the rpm remains constant. On the other hand, when the speed is increased, engine power will automatically drop. Fig. 6-8 shows that with the use of β-control the loss in speed resulting from a transition to a climb or turn is small as compared with that which occurs with power control by means of a conventional constant speed propeller. Consequently, with a pre-set flight condition the margins relative to the stalling speed will be greater and the pilot will have to make fewer speed corrections. This will increase flying comfort and safety. In descents there will be a lower speed build-up, which reduces the danger of overloading the structure.

The β-control mode is only operational when the engine is running within the predetermined limits. When, during some phase of the flight, the permissible Turbine Entry Temperature is reached, another control system is activated and the engine-propeller combination will function on the lines of the conventional control system. The β-control facility will therefore not affect the ultimate performance in conditions where the engine is operating at its limits, e.g. during takeoff or maximum continuous operation.

The differences which exist between the conventional control and β-control system are illustrated once more in the propeller diagrams in Figs. 6-6b and 6-6c.

6.3.4. Propeller geometry

Apart from the considerations which will follow, there may be certain overriding factors which govern the choice of a propeller, such as the low costs of spare parts for existing type of propeller, or the possibility of varying the gear ratio between the engine and the propeller, etc. For a final propeller design specification it will always be necessary to consult the propeller manufacturer.

a. Propeller diameter and tip speed.
The diameter is undoubtedly the most important design parameter of the propeller and in a sense comparable to the span of the wing, albeit with definite limitations. It can be shown theoretically that propeller efficiency grows with increasing propeller diameter, provided it is possible to vary propeller rpm and blade shape freely (Ref. 6-14). In actual practice, however, the propeller speed is determined by the operating regime of the engine and the propeller reduction which cannot, as a rule, be chosen arbitrarily. The most important factors governing the choice of the propeller diameter are:
- the propeller performance (efficiency) under varying conditions,
- the permissible tip speed in connection with propeller noise and performance,
- practical limits such as ground clearance and the clearance between the tips and the aircraft structure (Section 6.4.1.), and
- the weight of the propeller installation.
The propeller tip speed is the resulting velocity of the propeller tip relative to the airflow. If the induced velocity is ignored, Fig. 6-4 shows that:

$$V_{tip} = \sqrt{v^2 + (\pi n D)^2} \qquad (6-20)$$

Using (6-5), this expression may be rewritten as:

$$M_{tip} = M \sqrt{1 + \left(\frac{\pi}{J}\right)^2} \qquad (6-21)$$

Although in the case of propellers with thin blades (thickness/chord ratios about 6 per cent) and small camber, the efficiency only shows an appreciable decline at

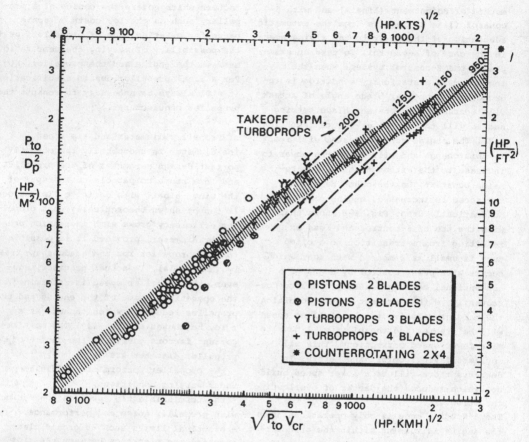

NOTE: P_{to} = takeoff bhp per engine

Fig. 6-9. Diagram to facilitate the choice of the disk loading

M_{tip} > .90 to .92, a practical limit of .85-.90 must be observed in order to keep the propeller noise down to an acceptable level.

b. Static tip speeds (V_{tip} = πnD) of some 800 to 1000 ft/s (250 to 300 m/s) have long been regarded as normal values. Now that considerable attention is being paid to reducing the airplane noise level, tip speeds of 500 to 700 ft/s (150 to 200 m/s) are being aimed at, while 800 ft/s (250 m/s) is regarded as the upper limit. Tip speed is the most important factor in the suppression of noise. A low tip speed, however, will only give acceptable performance when

the diameter is sufficiently large. Accordingly, in the design of propeller aircraft for short runways, the accent lies on the development of slowly revolving propellers with a large diameter and special shape of blade (cf. References 6-24 and 6-25).

c. A preliminary choice of propeller diameter may be made on the basis of one or more of the following methods:
Method A: the propeller disk loading P/D^2 at takeoff is used as a parameter. In Fig. 6-9 this quantity is plotted vs. the parameter $\sqrt{P_{to} V_{cr}}$ for a large number of aircraft. This correlation is based on the ob-

servation that during cruising flight J/C_p does not vary appreciably, and for given altitude and engine speed the disk loading should be proportional to the parameter $\sqrt{PV}$.

Method B: The diameter is deduced from a specified M_{tip}, which is considered to be acceptable. For this purpose we can derive from (6-21):

$$D = \frac{a}{\pi n} \sqrt{M_{tip}^2 - M^2} \qquad (6-22)$$

Assume $M_{tip} \leqslant .80$ for low-speed propellers with relatively thick aerofoils with thickness/chord ratios of about 10 per cent, and $M_{tip} \leqslant .85$ to .90 for high-speed propellers with relatively thin blades (thickness/chord ratios about 6 per cent). The speed of sound is taken at the appropriate design (cruising) altitude.

For low-speed aircraft with piston engines the rpm during takeoff will be considerably higher than in cruising flight, so the former will be the determining factor. In the case of high-speed propeller aircraft the cruising flight will generally be decisive.

Method C: The influence of the diameter on efficiency under various flight conditions is calculated on the basis of the propeller diagram for one or more blade shapes and number of blades per propeller, as appropriate. It will be necessary to estimate the flight speed of the aircraft in the flight phases chosen. Although these calculations are more involved than those used in the approximate methods mentioned above, a better picture of the performance trade-off is obtained.

Fig. 6-10 gives an example of a small commercial aircraft with two turboprop engines of 650 shp each. The influence of the propeller diameter on efficiency was investigated and it was found that D = 7 ft 7 in. (2.33 m) is the optimum value for the cruise and D = 8 ft 5 in. (2.57 m) for the climb. Admittedly, the static thrust improves rapidly with increasing diameter, but at 88 knots (45 m/s) there will be little gain with D > 8 ft (2.44 m). With

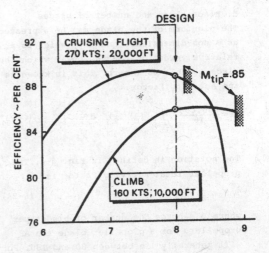

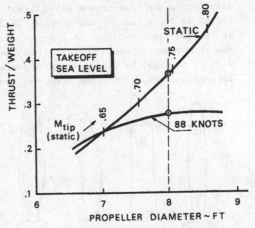

Fig. 6-10. Propeller diameter optimization (3 blades, AF = 140, C_{L_i} = .5)

this diameter the tip speed at takeoff will be 830 ft/s (254 m/s), which is considered an acceptable value for this type of aircraft. As indicated in the figure, D was chosen as 8 ft in this case.

It should be noted that the slipstream correction, approximated by (6-13) through (6-17), is dependent on the diameter. The result of this will be that, from the performance point of view, the optimum diameter of the propeller as installed on the aircraft will be greater than for the isolated propeller.

d. Blade shape and number of blades

The planform of the blade may be expressed
as a non-dimensional quantity which is a
relative yardstick for the power which the
blade is able to absorb. This is known as
the Activity Factor:

$$AF = \frac{100,000}{16} \int_{.15}^{1.0} (\tfrac{b}{D})(\tfrac{r}{R})^3 \, d(\tfrac{r}{R}) \qquad (6-23)$$

The notation is defined in Fig. 6-4. The
propeller Total Activity Factor is:

$$TAF = B \times AF \qquad (6-24)$$

where B denotes the number of blades per
propeller. For a propeller blade the AF
will generally lie between 80 and 180. For
example, in the case of the Helio Courier,
the AF is 90, while for the Lockheed C-130
it is 162. From (6-23) it is apparent that
the AF may be increased by making the blade
wide at the tip. In this connection there
are various shapes of blade (Fig. 6-11) and

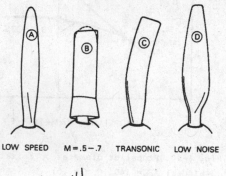

LOW SPEED   M = .5 − .7   TRANSONIC   LOW NOISE

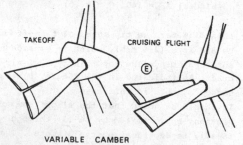

TAKEOFF   CRUISING FLIGHT

VARIABLE CAMBER

Fig. 6-11. Some propeller blade shapes

their application depends on the operational conditions and the engine power.

Type A is used on low-speed aircraft with
engines in the lower power bracket. For an
AF of up to about 90 a tapered blade form
is chosen; a blunt tip is generally adopted for an AF between 90 and 115. The shank
of the blade has a section which is nearly
circular or elliptical, evolving to an
aerofoil with a thickness/chord ratio of 8
to 12 per cent at the tip. The resulting
Mach number at the tip is limited to about
.80 and the flying speed to about M = .4
Type B is a practical shape for speeds between M = .4 and .6. Examples of aircraft
using this type are the Lockheed Electra
and C-130. For an AF of 115 to 140 almost
prismatic blades with rectangular tips are
used, while for AF values greater than 140,
blades are designed with inverted taper,
the tip chord being larger than the root
chord. As a result of the relatively large
local chord the lift coefficient at the tip
will be low and the critical flight Mach
number high. This is also achieved by the
use of a thin blade section (about 6 per
cent) and the relatively low rpm. Streamline fairings (cuffs), which improve the
inlet conditions of the turboprop engine
particularly with reverse thrust, may be
fitted at the shank of the blades.
Type C is an example of the types of blade
evolved around 1950-1960, intended for
high-subsonic flight speeds of up to M =
.95. The tip speeds are transonic and the
blades are very thin: about 6 per cent at
the root down to 2 per cent at the tip.
Sections of this type are now of historical interest only, since the turbofan engine is a more efficient prime mover for
speeds at the high-subsonic level.
Type D is a special design for a quiet,
slowly revolving propeller (Ref. 6-25).
Here the chord is changed in such a way
that in the static condition the region
of the blade which is stalled is kept as
small as possible.
Type E is a propeller with variable blade
camber which has been the subject of research by Hamilton Standard* and may be of

*cf. Av.Week & Space Techn. 11/24/69, pp.56-65

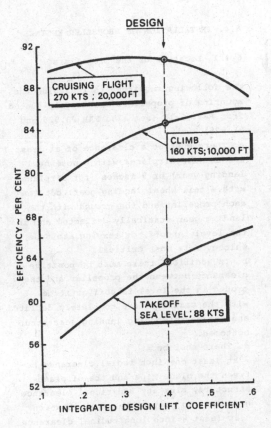

$$C_{L_i} = 4 \int_{.15}^{1.0} c_{\ell_i} \left(\frac{r}{R}\right)^3 d\left(\frac{r}{R}\right) \qquad (6\text{-}25)$$

Fig. 6-12. Effect of C_{L_i} on propeller performance (TAF = 350; B = 3; D = 8 ft, 2.44 m)

interest for propeller-powered V/STOL aircraft designs. The variable angle of incidence of the pairs of blades relative to each other is optimum both for the takeoff and cruising flight. The slot action between the blades may be compared to that of wing trailing edge flaps.

e. The camber of a blade element is represented by the design-lift coefficient c_{ℓ_i}. This is the value of c_ℓ for the airfoil which gives the lowest profile drag. The integrated design-lift coefficient is used for the propeller blade. This is defined as follows:

For the notations, see Fig. 6-4.

The blade camber C_{L_i} is generally chosen between .4 and .6. For optimum cruise a fairly low value of C_{L_i} has been found to be advantageous, but a higher value of C_{L_i} leads to higher thrust at low speeds. Fig. 6-12 shows the effect of blade camber on efficiency for the airplane design mentioned previously.

Although the TAF and the C_{L_i} do not entirely define the shape of the blade profile - thickness, taper and twist are also important - these parameters will suffice for analyzing the airplane performance at the preliminary design stage. Even when the aerodynamic characteristics of the aircraft are not yet accurately known, it is still possible to check the influence of the TAF and C_{L_i} on the performance of the propeller for some of the flight conditions chosen or laid down in the specification. Such a comparison will only be of any significance when data on propellers with geometrically comparable blade shapes are available. Useful sources in this respect are mentioned in References 6-21 and 6-26. The propeller diagrams shown in Fig. 6-5b and c have been taken from Ref. 6-21. This publication presents diagrams for three AF and four C_{L_i} values for both three- and four-bladed propellers.

Fig. 6-13 shows how propeller efficiency is influenced by the number of blades and the blade AF in the case of the aircraft mentioned. As can be seen, the four-bladed propeller is at a slight advantage under all conditions so far as performance is concerned. If the TAF is chosen slightly larger in the case of the three-bladed propeller, however, the disadvantage in the takeoff will be eliminated, while the propeller will be both simpler and lighter. The four-bladed version shows 1 to 1½ percent improvement in efficiency in cruising flight and climb. From Fig. 6-9 it will be observed that the transition from three to four blades occurs at about 1500 hp engine power. A four-bladed propeller will also be

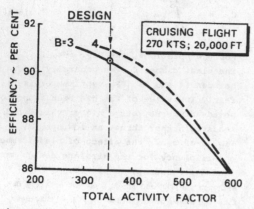

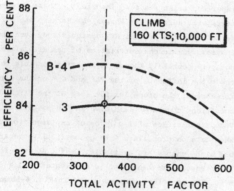

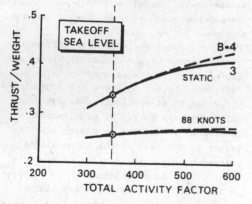

Fig. 6-13. Effects of AF and number of blades on propeller performance (C_{L_i} = .4, D = 8 ft, 2.44 m)

chosen when the diameter is restricted for structural reasons, because in this case the efficiency of the three-bladed propeller would be too low.

6.4. INSTALLATION OF PROPELLER ENGINES

6.4.1. Location of the propellers

The following requirements regarding the mounting of propellers are largely quoted from Ref. 6-18 (see also FAR 23.925 and 25.925).

a. There must be a clearance of at least 7 inches (for airplanes with a nosewheel landing gear) or 9 inches (for airplanes with a tail wheel landing gear) between each propeller and the ground with the landing gear statically deflected and in the level takeoff, or taxying attitude, whichever is most critical.

b. In addition, there must be positive clearance between the propeller and the ground in the level takeoff attitude with the critical tire completely deflated and the corresponding landing gear strut bottomed.

c. There must be
- at least one inch radial clearance between the blade tips and the airplane structure, plus any additional clearance necessary to prevent harmful vibration
- at least ½ inch longitudinal clearance between the propeller blades or cuffs and stationary parts of the airplane
- positive clearance between other rotating parts of the propeller or spinner and stationary parts of the airplane.

The values stated above are absolute minima; in order to limit cabin noise level it is advisable to provide a propeller tip / fuselage clearance of at least 4 inches (10 cm) plus .65 inch (1.65 cm) per 100 hp of one engine. It should also be kept in mind that in the case of engine power growth the optimum diameter of the propeller is likely to increase.

d. In the case of seaplanes and flying boats there must be a clearance of at least 18 inches (46 cm) between each propeller and the water, but it is recommended that this clearance should be at least 40 per cent of the propeller diameter.

e. When two propeller disks are adjacent to each other, the relative distance between the two, as seen in front view, should be at least 9 inches (23 cm). Over-

lapping of propeller disks is discouraged, although this will probably not apply to STOL aircraft with deflected slipstream.
f. Another important ruling (FAR 23.771 and 25.771) stipulates that the primary controls, excluding cables and control rods, must be located with respect to the propellers, so that no member of the minimum flight crew, or part of the flight deck controls, lies in the region between the plane of rotation of any inboard propeller and the area generated by lines passing through the center of the propeller hub making an angle of 5 degrees forward and aft of the plane of rotation of the propeller.
This latter requirement is depicted in Fig. 6-14 and is in practice more decisive

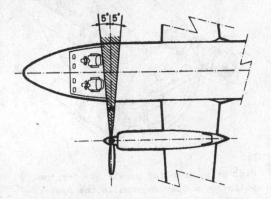

Fig. 6-14. Region in which flight crew members, flying controls and instruments may not be located

for the propeller location than for the layout of the flight crew compartment. It is also advisable to avoid siting passenger seats in this region, although this will not always be possible. In addition, it is preferable to locate the propellers in such a way that cargo holds, toilets and suchlike will not be in the plane of the propellers. It will be obvious that they should not be placed too close to cabin doors. The fuselage will have to be reinforced locally in view of the possibility that lumps of ice may be thrown from the propellers.

6.4.2. Tractor engines in the nose of the fuselage

In the vertical direction the position of the engine will be dictated by the downward view of the pilot (Section 3.4.1.), as well as by the required clearance between the propeller blade tips and the ground. It may sometimes be desirable to adapt the length of the nosewheel strut in order to give the aircraft a slight tail-down attitude.
When piston engines are used there should be a proper supply and exit of cooling air to and from the engine and oil cooler, as well as an adequate supply of engine air to the carburetor.
The exhaust gases of turboprop engines should not be blown against any part of the aircraft, nor penetrate into the cockpit or cabin compartments. Tilting down the engine thrust line may have a favorable effect on the longitudinal stability (Fig. 6-15), but this should be limited to a few degrees in order to avoid excessive variation in the angle of attack between the airflow and the propeller axis.

6.4.3. Wing-mounted tractor engines

Examples of some engine arrangements are shown in Fig. 2-10.
High-wing monoplanes will allow a certain measure of freedom with regard to the vertical location of the engine before a limit is imposed by the ground clearance of the propeller. If the landing gear has to be retracted into the engine nacelle this will largely determine the position of the en-

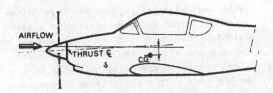

Fig. 6-15. Propeller thrust line tilt-down for improved power-on longitudinal stability

gine and its jet-exhaust pipe. Where this is not the case, it is preferable to place the thrustline at about the level of the wing-chord or a little below it. The exhaust gases must be directed in such a way that they do not impinge on the wing flaps in the extended position.

Low-wing monoplanes will often present a problem when it comes to obtaining sufficient ground clearance for the propeller. The engine nacelle, on the other hand, should not be placed too high on the wing since that would create excessive drag. An acceptable compromise may be created by placing the thrustline in such a position that its extension would be at a tangent to the top of the local aerofoil. In the case of turboprop engines the exhaust gases will be directed over the wing through extended exhaust pipes.

When the direction of the airflow is not perpendicular to the propeller disks the blades will be subject to alternating loads with a period equal to the time of revolution of the propellers (1P). This periodic loading may be reduced by a correct choice of the angle of incidence of the propeller axis in relation to the wing. It will be advisable to contact the propeller manufacturer at an early stage.

6.5. INSTALLATION OF TURBOJET ENGINES

6.5.1. General requirements

A number of aspects relating to the location of the engines has already been discussed in Section 2.3.

Engines buried within the fuselage can only be used when the payload has a relatively small volume and enough space is available for the engine and its inlet and exhaust ducts and for the wing center section. Generally speaking, this will only be the case with small private aircraft and trainers, as far as civil aviation is concerned. The relatively large diameter of modern high bypass engines virtually prohibits the installation of engines buried

in the wing roots, as in the Hawker Siddeley Comet and Tupolev 104. These engines are generally installed in pods, which are provided with detachable or hinged panels for ease of maintenance. The engines and their pods are connected to the airframe by means of pylons. The pylons of fuselage-mounted engine pods are more heavily loaded as a consequence of bending and - unlike wing-mounted podded installations - contain heavy forgings (Fig. 6-16).

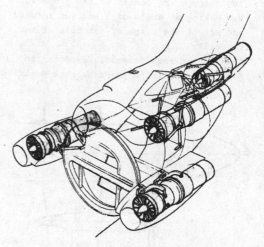

Fig. 6-16. Mounting provisions for the Rolls-Royce Spey engines in the Hawker Siddeley Trident 3E rear fuselage (Ref. Aircraft Engineering, April 1969)

In order to ensure favorable working conditions for the engine the air should reach the compressor with a minimum of pressure losses and fluctuations. The velocity of the incoming air will be about M = .4 to .5 in the plane of the compressor and it follows that at high-subsonic speeds the inlet air should be decelerated in a diffusor. When the shape of the inlet is optimized for this condition, there will be a possibility that during takeoff the air may separate inside the inlet duct, resulting in inlet pressure fluctuations and thrust loss. Possible solutions are a relatively blunt and long inlet lip or the

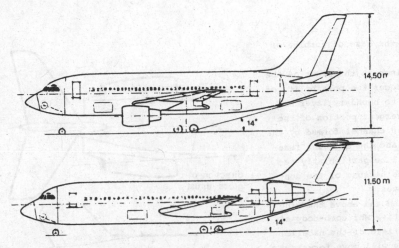

14,50 m

11.50 m

Fig. 6-17. Installa-
lation of high-by-
pass-ratio engines
on the wing and the
rear fuselage

provision of auxiliary inlet doors which
admit air only at low speeds.

The absence of large propellers, the relatively good accessibility and the compactness of podded jet engines allow the designer some measure of freedom in locating
the nacelles in a favorable position, the
bypass ratio being an important parameter
here. Fig. 6-17 shows two design sketches
of a short-range airliner powered by two
engines with a bypass ratio of 6.5. The
figure shows that when the engines are
mounted below the wing this will lead to
a fairly high undercarriage and a large
vertical tailplane. In this case it would
be preferable to place the engines at the
tail of the fuselage, provided this configuration did not result in an unacceptably large c.g. travel (cf. Section 8.5.3.).
An alternative solution was sought on the
Fokker-VFW 614 short-haul aircraft, where
the engines are located above the wing.
Some problems associated with this layout
are discussed in Ref. 6-41.

With respect to the location of the engine
inlet we quote the following requirement
from FAR 25.1091:

"The airplane must be designed to prevent
water or slush on the runway, taxiway, or
other airport operating surfaces from being directed into the engine air inlet
ducts in hazardous quantities, and the air

inlet ducts must be located or protected
so as to minimize the ingestion of foreign
matter during takeoff, landing, and taxying".

6.5.2. Fuselage-mounted podded engines

The following points will have to be considered when choosing the location of the
engines:

a. A good transfer of the loads acting on
the engine - approximately 10 times the installed weight, forwards as well as downwards - necessitates the adoption of the
type of construction shown in Fig. 6-16.
Another solution would be to use an almost
solid bulkhead at the location of the
fittings which pick up the outriggers
carrying the engine. An opening the size
of a door might still be acceptable. Such
a frame would generally serve as the rearmost limit of the passenger cabin, thereby
fixing the maximum forward location of the
engines.

b. When there is only a thin-walled cabin
structure between the engine and the cabin
interior, it will be undesirable to have
passengers seated in the plane of either
the turbines or the compressor, although
this may be acceptable in the case of engine with a very high bypass ratio, provided there are sufficient safeguards a-
gainst any engine parts penetrating the

207

passenger cabin in the case of turbine or
compressor failure.

c. Placing the engines at the sides of the
aft fuselage introduces the possibility of
a drag problem due to boundary layer separation in the divergent portion of the
convergent-divergent channel formed by the
nacelle, the pylon and the adjacent fuselage wall. This will be particularly serious at high speeds because of the generation of shock waves.

d. The nacelle behind and above the wing
may create an effective Whitcomb-body effect. The pressure field of the nacelle
at high Mach numbers will move forward the
local shock wave on the inboard part of
the wing in front of the nacelle. This
will postpone the breakaway of the wing
boundary layer induced by the shock waves.
For an optimized design this effect may
increase the drag-critical Mach number of
the aircraft (Ref. 6-32).

e. Since the nacelles are located behind
the aerodynamic center of the wing, they
will cause the aerodynamic center to move
backwards. This stabilizing effect is
partly compensated by an increment in the
downwash at the horizontal tailplane due
to the low effective aspect ratio of the
nacelle/pylon combination.

f. At large angles of attack, particularly
when the airflow over the wing has separated, the wake created by the nacelles
and the pylons may greatly reduce the
effectiveness of the horizontal tailplane.
For this reason the location of the engines will be an important factor in connection with the deep stall problem (cf.
Section 2.4.2.).

The aerodynamic design group's task is to
take maximum advantage of the favorable
effects and suppress the unfavorable aspects as far as possible. Fig. 6-18 shows
that this may result in a complexly shaped
nacelle. In this example, the nacelle forms
an angle of 3 degrees with the fuselage
datum line, while in plan view the shape
has been adapted to the local contour of
the fuselage. If just any nacelle is taken

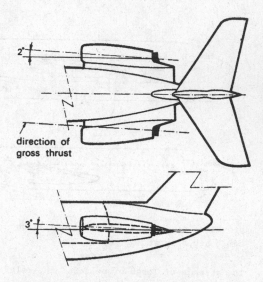

direction of
gross thrust

Fig. 6-18. Example of the shape of engine
nacelles mounted to the rear sides of the
fuselage

at random, the interference drag may a-
mount to 40-50 per cent of that of the
nacelle plus pylon, but this value can be
reduced to 10-20 per cent by appropriate
aerodynamic development. As to the distance between the centerline of the nacelle and the local fuselage contour,
something like 75 to 80 percent of the
maximum diameter of the nacelle will be an
adequate value. In the example given the
jet pipe is toed slightly outwards in order to reduce the moment arm of the thrust
about the center of gravity. This reduces
the yawing moment created in the event of
engine failure.

For speeds in excess of M = .90, it will
be necessary to adapt the shape of the
fuselage in accordance with Whitcomb's
"area rule" (example in Ref. 6-39). In the
case of aircraft with a long fuselage and
a relatively small nacelle diameter, the
requirement that the fuselage boundary
layer should not enter the duct is decisive in determining the location of the
inlet. At large angles of attack or yaw
the wake of the fuselage, wing or other
parts of the aircraft may enter the inlet

duct. The designer should be aware that
even in this case the fluctuations in inlet pressure must not exceed the limit
laid down by the engine manufacturer.

6.5.3. Wing-mounted podded engines

The following considerations will influence the location of the engines:

a. The reduction in the bending moment at
the wing root increases proportionally as
the engines are placed further outwards
until the taxi loads become predominant
and the trend is reversed. In addition,
when the spanwise coordinate exceeds a
certain value, the area of the vertical
tail surfaces must be increased in order
to compensate for the adverse yawing moment in the case of engine failure.

b. With swept-back wings it is possible
to effect corrections in the location of
the c.g. by changing the position of the
engines in a spanwise direction, since
for given chordwise location that will
also entail a shift in the longitudinal
direction.

c. The shape of the flow channel between
the nacelle and the wing will be the decisive factor as far as the interference
drag between the two is concerned. Fig.
6-19 shows that when an unfavorable position is chosen, the drag penalty may be
considerable. As a general rule, the front
and back of the nacelle should not coincide with or lie near the leading and
trailing edge of the wing. Incidentally,
by designing a suitable shape for the nacelle and the pylon the result shown in
Fig. 6-19 can be considerably improved.

d. The hot jet blast is generally not
allowed to impinge on any part of the aircraft. This necessitates a local interruption of the flaps, resulting in a typical C_L-max penalty of about .07 relative
to the uninterrupted flap. However, the
efflux of high bypass engines is fairly
cool and may not be harmful to the structure.

e. When the horizontal tail surfaces are
mounted to the fuselage, the jet efflux

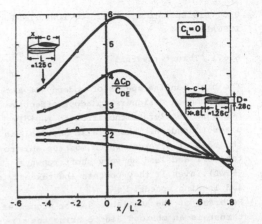

C_{DE} = external drag of isolated nacelle
ΔC_D = installed drag of nacelle, including interference
wing sweep angle = 27°
inlet velocity = 50% of airspeed

Fig. 6-19. Location effects on interference
drag of a wing-mounted nacelle (Ref. 6-28)

will be close to the stabilizer at large
angles of attack. The pumping effect of the
jet will induce a downwash which will decrease the effectiveness of the stabilizer.
This effect can be reduced by locating the
engine or engines further outboard.

Reference 6-35 contains a survey of the results obtained by Lockheed during investigation of the most favorable location of
the engine on the wing, using various configurations. Fig. 6-20 shows a location
which may be regarded as one of the most
suitable solutions, although it should be
noted that this may not apply when a different type of engine is chosen. The structural arrangement of a typical podded installation of a high bypass engine is depicted in Fig. 6-21. A comprehensive discussion on engine installation considerations and the aerodynamic development of
configurations with rear fuselage- and
wing-mounted engines can be found in Refs.
6-32 and 6-33.

6.6. MISCELLANEOUS ASPECTS OF POWERPLANT INSTALLATION

6.6.1. Thrust reversal

The high landing speeds of modern jet aircraft compel designers to adopt effective means to decelerate the aircraft rapidly after touchdown. This is also desirable in the case of aircraft which must be able to take off and land on very short runways (STOL), even if they possess low takeoff and landing speeds.

Reversal of the available jet or propeller thrust is an obvious aid to bring the aircraft to a stop. The main advantages are:

a. When the aircraft is landed on wet or snow-covered runways it is only possible to obtain a considerably reduced deceleration by the use of wheelbrakes and aerodynamic drag. Fig. 6-22 shows the ratio of the landing run length with thrust reversers to the distance required in the case that no reversing is used. As can be seen, the use of thrust reversers leads to a considerable reduction in the landing run. In such cases they contribute to the safety of operation.*

b. Thrust reversers are also used during day-to-day landings on dry runways although, as shown in Fig. 6-22, the reduction in the landing run is less pronounced. One advantage is claimed to be that there will be less wear on the wheelbrakes as well as on the tires.

c. On some types of aircraft it is possible to operate the thrust reversers in flight as well as on the ground, thus providing the pilot with an effective means for controlling the glide-path.

Two types of thrust reversers for jet engines are shown in Fig. 6-24. In the case of engines with a high bypass ratio and separate hot and cold nozzles, weight can be saved by merely reversing the fan stream.

*The use of thrust reversers is generally not allowed in defining the certificated field performance, cf. Section 11.7.

The thrust of the hot gases will thus be eliminated, which makes for a simple system (Fig. 6-25).

Fig. 6-23 shows the maximum percentage of engine thrust which may be reversed when using different systems. A typical value would be 45 per cent. Thrust reversers are designed by the engine manufacturer or by specialist firms. The following points are of importance to the airplane designer:

a. The weight of the reverser system installation is considerable and increases with the bypass ratio. It is generally of the order of 15 to 20 per cent of the bare engine weight.

b. There should be the necessary space to fit the system and this will often entail a lengthening of the nacelle.

c. The loss in thrust and increase in fuel consumption resulting from the use of the system may amount to 1 or 2 per cent in cruising flight.

d. Precautions should be taken against overheating and excitation of parts of the aircraft by the diverted hot gases.

e. At low speeds the gases which are diverted in a forward direction may enter the inlet duct, resulting in overheating of the engine. The likelihood that this will take place increases with the effectiveness of the reverser system. When four engines are mounted on the wings there is also a possibility that the gases of one engine will enter the inlet duct of the adjacent engine. For this reason, thrust reversers are only activated above a certain critical speed of about 40 to 70 knots (cancellation or cut-off speed). The cancellation speed will generally be lower for engines at the sides of the rear fuselage.

With the very high bypass ratios which may be used in the future, the conventional thrust reverser system will be inadequate and too heavy. Several firms have designed fans with variable blade adjustment (Fig. 6-26). Apart from other advantages (cf. Section 4.4.3.a.), it is possible to reverse about 60 per cent of the thrust by turning the rotor blades to a negative

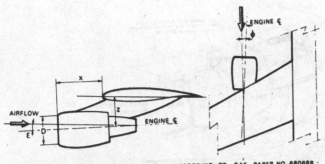

OPTIMUM FOR THE LOCKHEED 1011, ACCORDING TO SAE PAPER NO. 680688 :
$\varepsilon = 4°$, $\phi = 2°$, $x = 1.85\,D$, $z = .95\,D$

Fig. 6-20. Typical installation of a wing-mounted, high-bypass-ratio engine pod

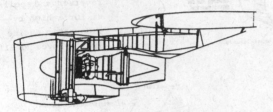

Fig. 6-21. General Electric CF-6 engine installation on the McDonnell Douglas DC-10 wing

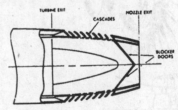

Fan reverser effectiveness: 60%

Fig. 6-23. Reverse thrust of bypass engines (Ref. 6-45)

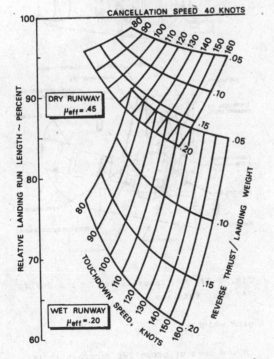

Fig. 6-22. Reduction of the landing run due to thrust reverser action

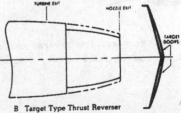

A. Cascade Type Thrust Reverser

B. Target Type Thrust Reverser

Fig. 6-24. Thrust reverser configurations for single-flow jet engines (Ref. 6-43)

211

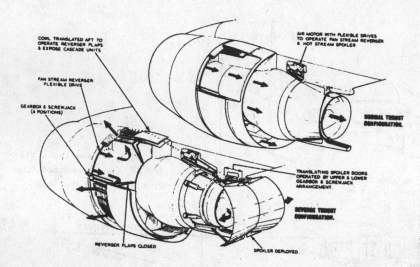

COWL TRANSLATED AFT TO OPERATE REVERSER FLAPS & EXPOSE CASCADE UNITS

AIR MOTOR WITH FLEXIBLE DRIVES TO OPERATE FAN STREAM REVERSER & HOT STREAM SPOILER

FAN STREAM REVERSER FLEXIBLE DRIVE

GEARBOX & SCREWJACK (6 POSITIONS)

NORMAL THRUST CONFIGURATION.

TRANSLATING SPOILER DOORS OPERATED BY UPPER & LOWER GEARBOX & SCREWJACK ARRANGEMENT.

REVERSE THRUST CONFIGURATION.

REVERSER FLAPS CLOSED

SPOILER DEPLOYED.

Fig. 6-25. Thrust reverser and spoiler for a high bypass ratio engine (Rolls-Royce RB 211)

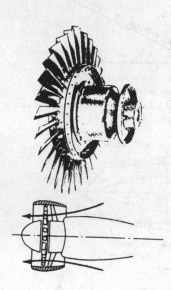

Fig. 6-26. Proposal for reverse thrust on a variable - pitch fan by Dowty - Rotol

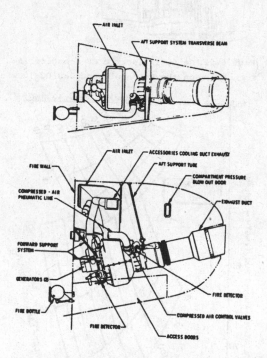

AIR INLET

AFT SUPPORT SYSTEM TRANSVERSE BEAM

AIR INLET

ACCESSORIES COOLING DUCT EXHAUST

FIRE WALL

AFT SUPPORT TUBE

COMPARTMENT PRESSURE BLOW OUT DOOR

COMPRESSED - AIR PNEUMATIC LINE

EXHAUST DUCT

FORWARD SUPPORT SYSTEM

GENERATORS (2)

FIRE DETECTOR

FIRE BOTTLE

COMPRESSED AIR CONTROL VALVES

FIRE DETECTOR

ACCESS DOORS

Fig. 6-27. APU installation in the Boeing 747 rear fuselage

pitch angle. According to the manufacturer it will also be possible to reverse the thrust at very low speeds. This system accounts for only 8 per cent additional engine weight.

In the case of propeller aircraft it is possible to obtain negative thrust by

212

changing the angle of pitch of the propeller blades from positive to negative. This requires a hydromechanical system which is operated by means of the throttle and ensures that the fuel supply is adapted to the blade angle. Negative thrust may also be used for reverse travel on the ground. The braking thrust of propellers (on piston engines) can be calculated by using the data given in Ref. 6-47.

6.6.2. Auxiliary Power Units (APU)

Almost every new transport aircraft is equipped with a relatively small gas turbine, which supplements the main engines for the following purposes.
a. Supplying pressurized air and power for operating the air conditioning system while the aircraft is on the ground.
b. Supplying power for the electrical system.
c. Starting the main engines.
d. Supplying for maintenance work outside the hangar.
The APU may, in principle, also be used for the pressurization and air conditioning systems during the takeoff, or as an emergency power supply for the electrical system during flight. This requires separate certification.
The principle advantage of the APU is that it makes the aircraft independent of power supplies at airports and so increases its flexibility. Since no auxiliary ground equipment is required, the aircraft is more easily accessible for embarking and disembarking passengers, as well as for servicing purposes, and comfort during stops is also improved. The APU will inevitably increase the empty weight, complexity and initial price, but in spite of this even small business jets are also optionally equipped with an APU.

Up to a certain point the choice of the APU is subject to the same factors as those which apply to the main engines. An assessment can be made from the few standard types which are offered, but in many cases

the requirements are such that a new unit will have to be designed by a specialized (engine) manufacturer to fit the project. Table 6-3 presents a survey of the main

AIRPLANE TYPE	BOEING 727	BOEING 747
APU MODEL	GTCP 85-98	GTCP 660-4
MANUFACTURER	AiResearch	AiResearch
WEIGHT - DRY	290 lb	543 lb
INSTALLED	(132 kg)	(246 kg)
	600 lb	1110 lb
	(272 kg)	(503 kg)
POWER	450 hp	2500 hp
AIR FLOW CAPACITY	110 lb/min	527 lb/min
	(50 kg/min)	(248 kg/min)
MTBUR*	1955 h	1980 h

*Mean Time Between Unscheduled Removals, time period Jan.-Sept. 1970

Table 6-3. Main characteristics of two APU installations (Ref. 6-54)

characteristics of two units. From this we may deduce that the installed weight of the APU is fairly high. This is mainly due to the sound-proofing required and to the inlet and outlet ducts. Although the weight only amounts to roughly .5 per cent of the empty weight, its effect on the location of the center of gravity should not be underestimated, because the tail of the fuselage is generally the best location for the APU for the following reasons:
a. As a rule, there will be sufficient space in the tail of the fuselage and a drastic adaptation of the rear fuselage can be avoided.
b. The APU should be isolated from the aircraft by means of a firewall; in the tail of the fuselage this can be done at little expense in weight.
c. The engine gases can easily be exhausted into the open air.
d. The location of the installation in the rear of the aircraft gives the designer some freedom in the choice on inlet and exhaust, which leads to a reduction of the noise level of the APU.
e. Accessibility is good, particularly when

the APU is mounted inside the rear extremity of the fuselage.

This location may have certain disadvantages:

a. With the engines mounted on the wing a separate fuel feed to the rear fuselage will be required.

b. Noise levels are high near passenger doors located in the rear fuselage.

Fig. 6-27 shows an example of an APU installation.

Possible alternative locations are as follows:

a. In the case of a high-wing aircraft with the landing gear attached to the fuselage, the streamline fairings housing the retracted gear.

b. Any appropriate space near the centre-section of the wing and the main landing gear wheelbay.

In the case of large aircraft it may become feasible to install two APUs in order to increase the regularity of the service. In this case it would appear suitable to place them in streamline bodies at the wing trailing edge, since these would act as "Whitcomb bodies" and so increase the drag-rise Mach number, as on the General Dynamics CV-990.

Chapter 7. An introduction to wing design

SUMMARY

The basic requirements for wing design are associated with performance and operational aspects, flying characteristics and handling, structural design and considerations of general layout design.

Conditions are derived for optimizing the wing loading of long-range aircraft and compared with constraints on the wing loading imposed by low-speed performance requirements, available tank volume and buffet margins for high-speed aircraft.

The information on stall handling requirements, stall characteristics of airfoil sections and stall progression on wings is applicable to all conventional wing designs.

Radical differences are shown between low-speed and high-subsonic aircraft with respect to planform shape and airfoil section design.

Definitions of critical Mach numbers are discussed and an approximate method is presented to find combinations of wing sweep and thickness ratio to attain a specified high-speed Mach number. Low-speed problems of swept-wing aircraft are dealt with qualitatively.

An assessment of high-lift technology is followed by recommendations regarding the arrangement of ailerons and spoilers and the choice of the dihedral and the wing/fuselage incidence, together with some considerations relating to structural design.

NOMENCLATURE

A — aspect ratio of wing

b — span (no subscript: wing span); width

C_D — drag coefficient of aircraft

C_{D_p} — profile drag coefficient

$\bar{C}_{D_p}$ — mean value of C_{D_p} for variable air-craft components

C_{D_v} — vortex-induced drag coefficient

C_{D_o} — zero-lift drag coefficient

C_L — aircraft lift coefficient

C_L^* — C_L in flight condition with horizontal fuselage reference line

C_{L_α} — $dC_L/d\alpha$, lift-curve slope

C_p — pressure coefficient

C_{p_i} — C_p in incompressible flow

C_T — thrust specific fuel consumption

c — chord (no subscript: wing chord)

c_{d_p} — two-dimensional profile drag coefficient

c_ℓ — two-dimensional lift coefficient

c_m — two-dimensional pitching moment coefficient

D — drag

f_{fix} — drag area of fixed aircraft components

g — acceleration due to gravity

h_{to} — takeoff height

i_w — angle of incidence of wing relative to the fuselage reference line

K_g — gust alleviation factor

k_{to} — factor of proportionality for the takeoff distance

L — lift

L.E. — leading edge

M — Mach number

M_{cr} — critical Mach number

M_{cr_D} — drag-divergence Mach number

M_{cr_L} — lift-divergence Mach number

M_D — Design Diving Mach number

M_{MO} — Maximum Operating Mach number

M_n — component of M normal to the leading edge

M_∞ — M referred to undisturbed flow conditions

M^* — equivalent Mach number characterizing the extent of supercritical flow in the design condition

N_e — number of engines

n — load factor = L/W

P_{to} — total static (equivalent) engine power of all engines at sea level

p — ambient pressure

P_o — p at sea level

P_∞ — p of undisturbed flow

q — dynamic pressure = $\frac{1}{2}\rho V^2$

q_∞ — q of undisturbed flow

S — area (no subscript: wing area); distance

S_{to} — takeoff distance

T_{to} — static takeoff thrust of all engines at sea level

T.E. — trailing edge

t — (maximum) thickness of an airfoil section

V — flight speed

V_D — Design Diving Speed

V_E — Equivalent Air Speed (EAS)

V_{MO} — Maximum Operating Airspeed

V_{NE} — Never-Exceed Speed

V_S — minimum speed in a stall

V_2 — takeoff safety speed

W — aircraft weight

w_E — design gust speed (EAS)

α — angle of attack

α_{ℓ_o} — zero-lift angle of attack of a section

α_{o_1} — change in wing zero-lift angle of attack per degree of positive twist

Γ — wing dihedral

γ — ratio of specific heats of air

Δ — increment

δ — relative ambient pressure = p/p_o

ε — wing twist

θ — relative ambient temperature = (static) temperature divided by temperature at sea level

Λ — angle of sweep

λ — wing taper ratio

ρ — density of air

ρ_o — ρ at sea level

σ — relative density = ρ/ρ_o

φ — vortex-induced drag factor; scaling factor

Subscripts

BO — buffet onset

BP — buffet penetration

cr - cruising flight; critical
f   - fuselage; flap
fix - fixed aircraft components
h   - horizontal tailplane
n   - nacelle(s); normal to leading edge
r   - wing root

t   - wing tip
to - takeoff
v   - vertical tailplane
var - variable aircraft components
w   - wing

7.1. INTRODUCTION AND GENERAL DESIGN RE-QUIREMENTS

The following basic requirements form the point of departure for wing design.
a. The aircraft must satisfy the perform-ance figures laid down in the design spec-ification and within these limits it must achieve the best economic yield and oper-ational flexibility.
b. Flight characteristics must be satis-factory both at high and low flying speeds, at high and low altitudes and in the var-ious configurations (flap angles, power settings).
c. It must be possible to design a struc-ture within the external lines and the gen-eral arrangement which satisfies demands regarding strength, rigidity, weight, serv-ice life, accessibility, development and manufacturing costs.
d. Sufficient space must be provided for fuel and to permit the attachment and re-traction of the main undercarriage.

Whether or not all requirements can be satisfied also depends on various other factors, such as the engine thrust (power) and fuel consumption, the design of the empennage, the weight distribution, etc.
Each project will differ in the degree of freedom available to the designer to make a choice from the opportunities available to him. It is, for example, common prac-tice to choose the type of powerplant at a fairly early stage of the design, but it may be necessary to compromise on the wing design as a result. As against this, it is very likely that the design of the tail-plane will not be started until the wing design has been established.

a. Performance requirements.
These may be subdivided as follows.
1. Minimum requirements and rules for es-tablishing the performance relating to the safe operational use of the aircraft. Most of these are laid down in the airworthiness regulations and any concessions in this field are generally not permitted.
2. Design requirements laid down in the de-sign specification, relating to the trans-port capacity and such aspects of economic operation as cruising speeds and a range of cruising altitudes, maximum range with full payload and airfield performance (run-way length and elevation). When it appears to be impossible to meet all requirements simultaneously, it may be necessary to re-consider the choice of the powerplant or revise certain demands.
Chapter 5 presents a methodology for find-ing combinations of wing loading, aspect ratio and maximum lift coefficients with flaps deflected, which permit the perform-ance goals to be achieved. The results still have to be evaluated and a choice must be made from a large number of pos-sibilities.
This chapter is intended to form a link between the initial performance feasibility study and the final stage where a fairly complete performance assessment is carried out for one or more configurations. The choice will be narrowed down by considera-tions of optimization, available high-lift technology and structural weight, etc. The use of statistics (see for example the data in Table 7-1) is probably not essential if enough time is available to investigate many possibilities. However, designing un-der the constraint of insufficient time a-vailable for a thorough job may greatly

simplify the designer's task in the more usual case where he is expected to be finished even before he has had an opportunity to recognize the important problems.

b. Flying qualities.
The following flight characteristics are particularly affected by wing design:
1. Stalling speeds and handling of the aircraft prior to and during the stall. The stalling speed is determined by the wing loading and the maximum lift coefficient, the stalling behavior by the planform, airfoil section(s) and twist.
2. The phenomenon of buffet on high-speed civil aircraft, which should be experienced only occasionally, i.e. during maneuvers or in gusty weather.
3. High-subsonic aircraft may inherently suffer from several types of longitudinal instability (tuck under, speed instability), lateral-directional stability problems (poor Dutch roll damping, wing drop or wing rocking), and lateral control deficiencies (aeroelastic deformation at high EAS, aircraft dynamics at high lift).
In the case of low-speed aircraft the proposals drawn up by the preliminary design engineer may be quite adequate to obtain inherently good flight characteristics. However, opportunities to provide high-speed aircraft with good low-speed flight characteristics are often conflicting, while in addition theoretical methods to ensure accurate prediction of flight characteristics are lacking. In this case the final wing design will of necessity be made at a stage where the wind tunnel can be used, while the completion may even take place during the period of flight testing. Artificial devices such as stick pushers, Mach trimmers and yaw dampers are usually indispensable on high-speed aircraft.

c. The wing structure.
The prerequisites for the good structural design of a wing for low-speed aircraft are usually present if the main members which introduce high loads (engine mountings, undercarriage supports, wing tanks) are suitably arranged in relation to the primary structure, so as to avoid complex members for the transmission of these loads. On high-speed aircraft the structural design may be complicated by aeroelastic effects - for example, various forms of flutter or aileron reversal may occur, wing twist induced by bending of the swept-back wing may cause reduced longitudinal stability. These snags can only be prevented by careful analysis and by such measures as shifting the aeroelastic axis, repositioning the powerplant and using high-speed ailerons and spoilers. The objective of preliminary structural wing design is to provide a good point of departure for the detailed design.

Small variations in the wing shape may sometimes have far-reaching effects in all areas. Design requirements differ from project to project and they are frequently conflicting. Since most aspects of wing design are closely related, a good syntnesis is only arrived at after consultation with various specialists. No attempt has therefore been made to present this chapter as a kind of "universal design procedure". Wing design is a highly iterative process, particularly in the preliminary stage; the following comments may help to speed it up.
1. It is often convenient to make a distinction between:
- wing size (area),
- basic wing shape (planform, sections, twist), and
- high-lift devices.
Wing size and high-lift performance are closely related to performance, while shape parameters primarily affect the stalling properties. The aspect ratio spoils the simplification: it is a shape parameter affecting performance.
2. In the case of low-speed aircraft it is probably best to determine the aspect ratio first. The wing loading and type of high-lift devices are dealt with next, and the basic shape is finally evaluated mainly on the basis of the stalling characteristics. Small variations in the wing size have only a minor effect on the stalling characteristics.
3. In the case of high-speed jet aircraft the span loading and wing loading may be tackled first, using Sections 7.2. and 7.5.4. as background information; the aspect ratio will be found from this. Short-haul transports usually have a wing loading

based on field performance and the type of wing flaps must be decided at an early stage. The emphasis lies on cruise performance in the case of long-range aircraft and the criterion in Section 7.2.1. is fairly decisive in this respect.

4. The wing sweep and mean thickness/chord ratio of high-subsonic aircraft are based primarily on the Mach number in high-speed flight. Various combinations are possible, and the combination of wing span, root thickness ratio, sweep and taper should be checked against statistical data on cantilever ratios.

5. On high-subsonic, long-range aircraft the high-lift configuration is likely to be decided after a satisfactory wing shape for high-speed flight has been obtained.

6. A final check on low-speed performance, fuel tank volume and buffet margins may lead to corrections of the wing area which have only minor effects on high-speed performance.

The data in Table 7-1 may be useful at every stage of the design.

7.2. WING AREA

The choice of the wing area is mainly based on performance requirements, although structural (weight) aspects are by no means unimportant. In performance considerations the wing area usually appears in combination with the All-Up Weight (wing loading, W/S).

The choice of the wing area is important for laying down the cruise conditions on which the choice of the wing shape will primarily be based. The wing loading is subject to optimization from the point of view of minimum fuel consumption and to constraints imposed by other criteria. The final choice of the wing area will be decided by the aerodynamic performance of the high-lift system (Section 7.6.).

7.2.1. Wing loading for optimum cruising conditions

The primary variables affecting cruise performance for a given aircraft geometry are the cruising speed and altitude. Conditions

have been derived in Section 5.4.2. for which the parameter ML/D* is maximum, corresponding to an optimum cruising speed and altitude (see Fig. 5-10). In the case of a specified cruising speed, the designer will try to find the best combination of wing loading and cruise altitude so that minimum operating costs can be achieved. A simplified example of a representative wing sizing study has been discussed in Section 5.5.3., where it was concluded that for the long-range aircraft project considered there the condition for minimum drag is representative of the minimum takeoff weight (see Fig. 5-19). As the direct operating costs are approximately proportional to the MTOW, it may be argued that the lift/drag ratio at a given cruising speed is a fair indication of the optimum design for long-range aircraft.

In order to get some insight into the major factors involved, simple generalized results may be obtained by assuming the aircraft's wetted components to be composed of two major groups:

a. "Fixed" components, which are assumed to be unaffected by variations in the wing size. The fuselage is an example of this group and, in the case of a fixed engine type, this also applies to the engine installation and the nacelles. The total profile drag area of these items will be denoted as f_{fix}:

$$f_{fix} = \sum (C_{D_p} S)_{fix} \qquad (7-1)$$

b. Variable components, the size of which is affected directly by variations in wing size, e.g. the wing itself and the horizontal tailplane. Their profile drag coefficient will be assumed constant:

$$\bar{C}_{D_p} = \frac{\sum (C_{D_p} S)_{var}}{S} \qquad (7-2)$$

*This parameter is a fair indication of the cruise fuel required, provided the corrected specific fuel consumption $C_T/\sqrt{\theta}$ is approximately independent of the altitude and Mach number.

AIRCRAFT TYPE	1st flight prototype	Aspect ratio A	Taper ratio λ	Sweep angle $\Lambda_{.25}$ deg.	Geom. twist ε_g deg.	Dihedral Γ deg.	Profile type and streamwise thickness root %	tip %	$(\overline{t/c})$ %	V_{mo}*** km/h EAS	M_{mo}	V_D km/h EAS	M_D	Flap type* T.E./L.E.	$(\overline{c_f/c})$ stream-wise	b_f/b %	Flap angle takeoff deg.	landing deg.	$C_{L_{max}}$ (flight test) takeoff	landing
JET TRANSPORT AIRCRAFT																				
Yakovlev YAK 40	1966	9	.396	0	–	6°30'	632−015	0	12,5	528	.70	–	–	P	31,5	67	20	40	–	2.10
VFW-Fokker 614	1971	7.22	.402	15	3°45'	3°	a*.4 mod.	651A−012	13,5	680	–	615	.740	F1	31	69	20	40	2.12	2.37
Grumman Gulfstream II	1966	5.97	.370	25	–	3	13 Naca-6 series	a*.5 mod.	13	611	.860	–	–	F1-2	30	73	25	40	2.16	1.807
Fokker-VFW F 28	1967	7.47	.355	16	–	2°30'	12.5	10	11,8	652*	.830	760	.860	P1	32	69.5	25	42	2.16	2.53
BAC One-Eleven Srs 200/400	1963	8.00	.321	20	–	2	13.65	11	11,65	630	.750	–	.890	P1	32	67	18	45	1.86	2.40
McDonnell-Douglas DC-9 srs 10	1965	8.56	.246	24	-7.4	3	13.65	9.6	11,6	630	.780	–	.890	S2	36	75	15	50	–	2.40
McDonnell-Douglas DC-9 srs 30	1966	8.72	.226	24	-7.4	3					.840	–	.890	S3/I	30	67	15	50	2.45	2.98
Tupolev Tu 134/134A	1964	7.42	.287	35	–	-1°30'	14	11.5	12	648	.817	722	–	S2	29	60.5	10	30	1.51	1.67
Boeing 737 srs 100/200	1967	8.83	.251	25	-2	6	65−212	65−212	9,8	556	.840	–	.890	F3/I,II	27	66	10	45	1.20	3.05
Aérospatiale Caravelle	1955	8.02	.354	20	-5	3	11.5	9	11	722	.810	732	.870	S2	28	69	15	43	1.74	2.10
Hawker Siddeley Trident 2E	1967	6.57	.240	35	-2	5	13	9	10	575	.880	–	.950	F3/I,II	30	68	25	40	1.93	2.40
Boeing 727 srs 100/200	1963	7.67	.323	32	-3	3	12.5		11	722	.900	787	.950	F3/I	31	75	15	–	2.05	2.75
Tupolev Tu 154	1968		.250	35	–	7	12	10.16	10	710*	.900	751	.950	F1/I,II	30	67	15	50	1.76	2.70
Boeing 707/720	1957/60	8.14	.250	35	–	7			11,1	630	.880	770	.950	S2/II	30	73	25	50	1.87	1.94/2.06
McDonnell-Douglas DC-8 srs 10,50,61	1958/60/66	7.11	.244	30	-3°40'	6°30'	12.5	9.75	10	562	.860	704	.940	S3/II	30	73	23	50	1.87	2.03
McDonnell-Douglas DC-8 srs 62/63	1966/67	7.30	.194	30	–	3	0013 mod.	0010 mod.	11,5			760*	.890	S2/II	29	62		45	–	2.05
BAC VC-10 srs 1100/Super VC-10	1962/64	7.49	.273	32°30'	-5°90'	-1°26'	12.5	10						S2/I	29	66			–	2.27
Lockheed L-300 Starlifter	1963	7.90	.350	25	–	3	15	10	14,2	656*	.840		.950	S2	24	72	–	30	–	2.32
Ilyushin IL 62	1963	6.075	.262	35	–	–	12.4	9	13,5	668	.840			F2/II	32,5	82	22	45	–	2.36
BAC Three-Eleven	1969/70	8.00	.255	25	–	-7°	12.5	10	10,7		.880	787	.950	S2/II	26	77	22	35	2.70	3.19
300B	1972	8.60	.364	28	-3°40'	7°31'/5°30'	12.2	8.4	11		.880			S2/I	25†	77	22	42	2.46	3.00
Lockheed 1011	1972	7.16	.296	35	-3°40'	-7°	13.44	8	9,4	695	.920	824	.970	S2/II	30	77.5	15	55	2.15	2.73
McDonnell-Douglas DC-10 srs 10	1970	6.90	.250	35	-3°40'	-5° 6'	12.5	9.7	11,5	648	.825	745/726*	.875	S3/II	30	72	30	40	1.89	2.98
McDonnell-Douglas DC-10 srs 30	1970	7.21	.230	35										F1/II	24				2.21	2.63
Boeing 747/747A	1969/70	6.96	.309	37°30'																
Lockheed L500 Galaxy	1968	6.75	.256	25			63A 112	63A 309												
JET EXECUTIVE AIRCRAFT																				
Potez-Air Fouga CM170 Magister	1952	7.42	.400	9°90'	–	0	64−219	64−212	15,3	706	–	740	.820	S1	25	59.5	15	40	–	1.70
Aermacchi MB 326	1957	7.45	.600	8°92'	-1°01'	2°55'	63A-213.7	63A-212	12,9	806	–	–	.800	S1	25	56	20	–	–	1.72
Cessna Model 500 Citation	1969	7.45	.490	10°10'	-2°38'	3°6'	23014 mod.	23012	13	531*	.700	–	–	S1	23	73	15	40	1.55	2.40
SNIAS 600 Corvette	1970	6.20	.447	20°6'	–	13	13.65	11.50	16	816**	.71	695	.82	S2	31	53	20	40	–	1.93
Gates Learjet 25 (A/T31)	1954	5.02	.510	13°	1°	3	2418 mod.	2412 mod.		705**	.705**	843	.850	S1	28	61	20	40	1.37	1.37
Aerocommander Jet Commander	1969	6.19	.333	1°9'	-7	2°30'	64A 109	64A 109	12	617	.765	833	.890	P	25	60	20	60	–	2.00
Dassault Minifalcon 10	1963	6.00	.263	27°45'		6	64₁−212	64₁−212		667	.800	–	.990	S2/I	31,5	60	20	55	2.04	2.35
MBB 320 Hansa Jet	1970	6.25	.410	-15		5	63A−1−7.1 mod.	63A−1−7.1 mod.		825	.87	700	.830	S2	22,5	67	20	55	1.61	1.91
Piaggio Douglas PD 808	1964	5.77	.313	1°90'	-2°30'	3	DES0009−11	DES0009−11	9	852**	–	788	.850	S1/I	30	61	15	–	–	1.71
North Am. Rockwell Sabre Liner	1958	6.25	.300	20		2	40°/11° mod.	40°/11° mod.	12,5	617	.763	–	.850					50	1.84	2.35
Hawker Siddeley HS.125 srs 400B	1962	6.40	.312	30		2	14	11	9,3	685**	.755	685	.825	S1/II	29	65	15	50	–	2.19
Dassault Fan Jet Falcon 20.F	1963	5.47	.335	30	-2	2	10.5	9.7	10,5	710**	.870	787	.90	S2/II	25	61	10	40	–	2.32
Lockheed Jetstar	1957						63A 112								21	69.5				1.67

Legend:

*CAS or IAS	*p - plain or split flap
**maximum level flight speed	S - slotted flap; 1, 2 or 3: number of slots
***V_{NE} is taken equal to V_{MO}	F - Fowler flap
	I - slat
	II - droop nose, leading edge flap or slot

Table 7-1. Wing design data

PROPELLER AIRCRAFT

AIRCRAFT TYPE	1st flight prototype	Aspect Ratio A	Taper Ratio λ	Sweep angle Λ.25 deg.	Geom. twist εg deg.	Dihedral Γ deg.	Profile type and streamwise thickness root %	tip %	(t/c) %	VMO (EAS) km/h	MMO	VD (EAS) km/h	Hp	flap type T.E./L.E.	(cf/c) streamwise %	bf/b %	Flap angle takeoff deg.	Flap angle landing deg.	CLmax flight test takeoff	CLmax flight test landing
TURBOPROP TRANSPORT AIRCRAFT																				
LET L-410 Turbolet	1969	9.30	.500	0	-2°30'	1°.5'	63A 418	63A 412	15	385	-	518	-	S-2	29	66	18/8	45-70	-	2.45
Short Skyvan Srs 3	1963	11.30	1.0	0	-	2° 2'	63A srs	63A srs	14	402	-	445	-	S-1	30	69	-	20/45	2.07	2.72
IAI Arava	1969	10.00	1.0	0	-	1°30'	63 (215) A 417 (mod.)		17	350	-	466	-	S-2	-	79	-	-	-	2.45
Beriev Be 30	1967	9.04	.475	0	-	3°			16	480	-	-	-	S-2	38.4	62.5	20	-	-	2.55
DHC 6 Twin Otter Srs 300	1965	10.05	.333	0°53'	-2°	3°	6A srs meanline:0016 (mod.) thickness distribution		15	398	-	518	.7	S-2	20	96.3	20	60	2.2	2.35
Handley Page Jetstream	1967	10.01	.400	0°54'	-2°	5°	NACA		15	422	-	496	-	S-2	20	61	-	-	1.93	-
Swearingen Metro	1969	7.71	.400	0	-	7°	65₂A 215	64₂A 415	14	386	-	-	-	S-1	24.8	62	15	25/35	1.55	3.15
SNIAS (NORD) 262 Fregat	1962	8.72	.380	0	-	4°30'	63A 418	23012 mod.		444	-	494	-	S-1	36	66	20	23/60	2.70	3.80
DHC-7 STOL	1975	10.00	.521	0°23'	-1°	4°	23016.5	4412	14.3	443	.5	533	-	P-1	32	66	16.5	26°20'/40	2.435	2.943
HP Dart Herald srs 200	1958	10.20	.400	0°13'	-2°	-4°30'	64₂-421 mod.	64₂-415 mod.	18	421	-	528	-	F-1	31.3	64.7	15	-	2.11	2.88
Fokker VFW F 27 Friendship	1955	12.00	.400	0°23'	-2°	7°	64₂-421 mod.	4412	18	417	-	-	-	F-1	31.5	64.7	-	-	-	2.747
Hawker Siddeley 748 Srs 2A	1960	11.97	.386	2°54'	-	-	23018		15	450	-	546	.601	F-1	18	65	45°/30°	98°/65°	-	2.70
Antonov AN 24V srs 11	1960	11.77	-	9°11'	-	6°19'	64A 218 mod.	64A 416 mod.	16	245	.475	-	-	S-2	30	64	15°	40	-	2.46
NAMC YS-11A	1961	10.81	.340	0	-	4°30'	63-X15	63-X13	16	645	-	717	.711	F-1	25	65	30°	40	-	2.54
Breguet 941C STOL	1961	6.56	.530	=0	-	6°	0014-1.10	0012-1.10	13	675	-	756	-	S-2	31	63				
Vickers Vanguard	1959	9.10	.380		-										32					
Lockheed L188 Electra	1957	7.50	.40		-															
Antonov AN 10	1957	12.03	-		-		64A 318	64A 412	14	550	.660	611	.710	S-2	30	70	15	36	2.17	2.28
Ilyushin IL-18	1957	10.00	.333		-3°	2°30'	25017	4413	15	533	.650	593	.700	S-2	32	67	15	45	-	2.51
Lockheed L100 Hercules	1954	10.07	.513	9°3'										F-1						
Bristol Britannia	1951	9.53	.300	35	-2°51'															
Tupolev Tu-114	1957	8.39	.350																	
Antonov AN 72	1965	12.02	.360		-2°30'					740	-	-	-	S-2	22	62.7				
PISTON ENGINED GENERAL AVIATION AIRCRAFT																				
Beagle Pup B.121	1967	8.04	.550	2°56'	-2°	6°30'	63₂-615	63₂-615	15	259	-	314	-	S-1	21	59	10	40	1.60	1.87
Beagle B.206	1961	10.00	.400	≈0	-1°12'	5°	23015	4412 mod.	13.5	372	-	483	-	S-2	28	50.5	65%	100%	1.22	2.10
Beechcraft Queen Air Model 65	1958	7.51	-.42		-6°48'	6°30'	23018	23012	15	384	-	433	-	S-1	-	-	20	30	-	1.75
Beechcraft Musketeer	1961	7.50	.679	0	-	6°	63₂A 415	63₂A 415	14	235	-	-	-	S-1	22	54.4	20	-	-	1.80
Beechcraft Bonanza 35A-TC	1945	6.10	.457	0	-5°	6°	23012 mod.	23010.5	14.3	402	-	283	-	S-1	23	48.7	-	-	-	1.85
Beechcraft Baron D 55	1960	7.16	.410	0	-	6°	23016.5	23009 mod.	14	390	-	320	-	P-1	22	65.5	20	40	1.99	1.76
Beechcraft B 206C Junior	1962	6.90	.650	-3°	-1°	2°30'	64-215	64-212	9	230	-	261	-	P-1	23.5	69.5	20	40	1.73	-
Messerschmitt-Bölkow Bo 209 Monsun	1969	6.90	.687	1°24'	-	1°44'	64-215	2412	12	274	-	280	-	P-1	27.5	61.5	20	40	2.10	-
Cessna Model 150	1957	7.00	.672	0	-3°	1°30'	2412	2412 - symm.	12	196	-	-	-	P-1	27.5	46.1	20	40	1.87	1.87
Cessna Model 172	1955	7.52	.726	0	-1°	3	2412	2412 - symm.	12	224	-	-	-	S-1	32.9	65.0	20	40	1.82	1.82
Cessna T 210 Centurion	1965	7.66	.650	0	-2°30'	0/5°	64₂A215(αr-.5)	64₂A215(αr-.5)	13.5	322	-	-	-	S-1	27.5	70.5	20	42	2.00	2.00
Cessna Mod. 337 Super Skymaster	1961	7.18	.650	0	-	1°	2412	2409	14	322	-	428	-	S-1	33	71	26	-	-	2.36
Cessna Model 401/402	1965	7.2	.679	0	-	1°	23018	23009	18	269	-	-	-	S-2/1	33	71	26	-	-	6.42
Dornier DO27	1955	8.50	1.0	0	-3°	1°	23018	23018	18	-	-	304	-	S-2/1	27	74	-	50	2.12	2.12
Dornier DO28 Skyservant	1959	6.58	1.0	0	-0°12'	5°	23012	23018	18	348	-	446	-	S-1/1	25	50	-	50	-	-
Helio H-295 Super Courier STOL	1965	6.80	-.460	-2°30'	-	5°	USA 35B mod.	23012	14	314	-	365	-	P-1	18	60	10-25	50	1.71	1.74
Piper PA 23 Aztec D	1961	7.28	64₂A 215		-1°	7°	65₂-415	USA 35B mod.	14	245	-	-	-	P-1	57	57				
Piper PA 24-260 Comanche C	1956	5.63	1.0	0°30'	-2°30'	7°	65₂-415	65₂-415	13.5	-	-	-	-	P-1	65.6	65.6			1.74	-
Piper PA 28-140C Cherokee	1964	7.25	.372		-2°30'	7°	63₂-415	63₂-212	13.5	-	-	-	-	S-1	-	-				2.15
Piper PA 31 Navajo	1964	7.25	.372	0°30'	-	7°	63-517	65₂-212	15.3	365	-	-	-	S-1	-	-				
SNIAS GY-80 Horizon	1962	7.20	.570		-2°30'	7°	63-517	64₁3.6						F-1						

Table 7-1 (continued)

In what follows we will arbitrarily assume the fuselage, vertical tailplane and nacelles as fixed items and the wing and horizontal tailplane as variable items. The horizontal tailplane area is assumed proportional to the wing area, and allowance is made for roughness and interference drag by using suitable multiplication factors. Thus we have:

$$f_{fix} = (C_{D_p} S)_f + (C_{D_p} S)_v + (C_{D_p} S)_n \qquad (7-3)$$

$$\bar{C}_{D_p} = (C_{D_p})_w + (C_{D_p})_h \frac{S_h}{S} \qquad (7-4)$$

while it is assumed that both f_{fix} and $\bar{C}_{D_p}$ are independent of S. These assumptions can be interpreted as a linearization of a curve of wetted area vs. wing area (Fig. 7-1), which is valid in a limited range of wing area variations.

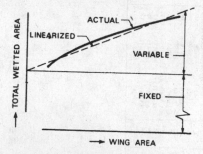

Fig. 7-1. Linearization of the wetted area with wing area variation

It should be noted that the present considerations are strictly valid only for the conventional general arrangement of aircraft, where the payload is contained by the fuselage, the wing contains (most of) the fuel, and the engines are installed in pods. The example in Section 2.3.2. (Fig. 2-13) shows that aircraft of entirely different conceptions may nevertheless have essentially similar range performance, provided their wetted area (or internal volume) and span loading are equal.
The drag coefficient at the design cruising C_L may now be written as follows:

$$C_D = \bar{C}_{D_p} + \frac{f_{fix}}{S} + \frac{C_L^2}{\pi A \varphi} \qquad (7-5)$$

where the last term represents the vortex-induced drag. In this relation C_L-variations are effected by variations in the wing size and the factor φ is not identical to the Oswald factor. It is assumed that all profile drag contributions can be minimized for the cruising condition to an approximately constant value by suitable design, notably camber and twist.
In horizontal cruising flight the drag/lift ratio may be obtained from (7-5):

$$\frac{C_D}{C_L} = \left(\bar{C}_{D_p} \frac{p_o S}{W} + \frac{f_{fix} p_o}{W}\right) \tfrac{1}{2}\gamma\delta M^2 + \frac{1}{\tfrac{1}{2}\gamma\delta M^2 \, \pi A \varphi} \frac{W}{p_o S} \qquad (7-6)$$

Fig. 7-2 gives an example of the effect of wing loading and cruise altitude on the L/D ratio. It can readily be shown that for a given altitude the L/D ratio is maximum if the wing loading is defined as follows:

$$\left(\frac{W}{p_o S}\right)_{opt} = \tfrac{1}{2}\gamma\delta M^2 C_{L_{opt}} = \tfrac{1}{2}\gamma\delta M^2 \sqrt{\bar{C}_{D_p} \, \pi A \varphi} \qquad (7-7)$$

and the minimum drag is therefore:

$$\left(\frac{C_D}{C_L}\right)_{min} = \tfrac{1}{2}\gamma\delta M^2 \frac{f_{fix} p_o}{W} + 2\sqrt{\bar{C}_{D_p}/(\pi A \varphi)} \qquad (7-8)$$

Equations 7-7 and 7-8 show that the maximum L/D ratio increases with increasing aspect ratio, provided the wing loading and cruise altitude are allowed to increase accordingly. For a given value of A, flight at higher altitude requires the wing loading to be reduced.
Substitution of typical values $\bar{C}_{D_p} = .0095$ and $\varphi = .95$ into (7-7) yields a simple guideline for the optimum lift coefficient for long-range flight:

$$C_{L_{opt}} = .17 \sqrt{A} \qquad (7-9)$$

resulting in an optimum wing loading, which is related to the span loading W/b^2 as follows:

$$\left(\frac{W}{p_o S}\right)_{opt} = .014 \frac{(\delta M^2)^2}{W/(p_o b^2)} \qquad (7-10)$$

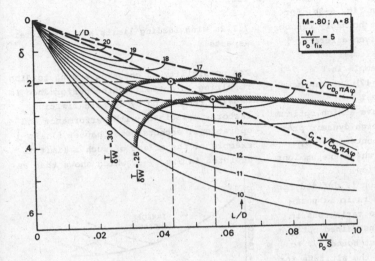

M=.80; A=8

$$\frac{W}{p_o f_{fix}} = 5$$

$C_L = \sqrt{C_{D_o} \pi A \varphi}$

$C_L = \sqrt{C_{D_p} \pi A \varphi}$

Fig. 7-2. Example of the effect of wing loading and cruise altitude on the lift/drag ratio for a high-subsonic long-range jet transport aircraft

Note that this condition applies to a representative mean All-Up Weight.

The well-known condition for minimum drag (constant wing loading):

$$C_L = \sqrt{C_{D_o} \pi A \varphi} \qquad (7-11)$$

resulting in

$$\left(\frac{C_D}{C_L}\right)_{min} = 2\sqrt{\frac{\overline{C_{D_o}}}{\pi A \varphi}} = 2\sqrt{\frac{\overline{C_{D_p}} + f_{fix}/S}{\pi A \varphi}} \qquad (7-12)$$

is obtained at a cruising altitude defined by:

$$\delta = \frac{W/S}{\frac{1}{2}\gamma p_o M^2 \sqrt{C_{D_o} \pi A \varphi}} \qquad (7-13)$$

The curves for constant L/D in Fig. 7-2 have not been drawn for higher altitudes as in this region C_L is too high and the same performance can be obtained for lower C_L at a lower altitude.

An absolute optimum for the wing loading and altitude cannot be obtained from these considerations as conditions (7-7) and (7-11) are incompatible and L/D continues to increase with altitude. However, in the case of a fixed powerplant, the cruising altitude is limited by the available thrust:

$$\delta > \left(\frac{L}{D} \cdot \frac{T}{\delta W}\right)^{-1} \qquad (7-14)$$

Since we know from gas turbine theory that in an isothermal atmosphere* T/δ is constant for a given engine rating and Mach number, we obtain a thrust boundary in the W/S-δ diagram as indicated in Fig. 7-2. The following "optimum" wing loading and cruise altitude are now obtained from (7-7), (7-8) and (7-14):

$$\left(\frac{W}{P_o S}\right)_{opt} = \frac{2 \overline{C_{D_p}}}{\frac{T}{\delta T_{to}} \frac{T_{to}}{W} \frac{2}{\gamma M^2} - \frac{f_{fix} P_o}{W}} \qquad (7-15)$$

$$\delta_{opt} = \frac{2\sqrt{\overline{C_{D_p}}/(\pi A \varphi)}}{\frac{T}{\delta T_{to}} \frac{T_{to}}{W} - \frac{1}{2}\gamma M^2 \frac{f_{fix} P_o}{W}} \qquad (7-16)$$

and the highest obtainable L/D ratio for a given powerplant is:

$$(L/D)_{max} = \frac{1}{2}\sqrt{\frac{\pi A \varphi}{\overline{C_{D_p}}}} \left\{1 - \frac{1}{2}\gamma M^2 \frac{f_{fix} P_o}{W} \middle/ \left(\frac{T}{\delta T_{to}} \frac{T_{to}}{W}\right)\right\} \qquad (7-17)$$

*This condition simplifies the analysis, but the results are essentially similar for other atmospheric models.

223

where $T/(\delta T_{to})$ is defined by the engine
rating and M, as shown in Fig. 6-3b, for
example. A similar thrust boundary can be
derived for propeller-driven aircraft.

Referring back to the example in Fig. 7-2
and to (7-15) through (7-17), we can make
the following observations:
1. High-speed aircraft have a high optimum
wing loading due to the high dynamic pressure. This tendency is counteracted by
flying at high altitudes where the ambient
pressure and density are low.
2. If the optimum condition is defined for
a given altitude, the L/D ratio is unaffected by relatively large variations in
the wing loading. If a constrained optimum
is defined along the thrust boundary, to
be obtained by optimizing the altitude for
each wing loading, the L/D ratio is much
more sensitive to wing loading variations.
3. For a given cruise altitude the optimum
wing loading increases with the aspect ratio, whereas according to (7-15) the "absolute" optimum is not affected by this.
Both the best cruising altitude and the
maximum L/D ratio increase sensitively with
increasing aspect ratio.
4. For a given cruise altitude the size of
the aircraft, represented by W/f_{fix}, does
not affect the optimum wing loading, contrary to the "absolute" optimum, which is
definitely affected. For example, if the
aircraft is stretched in such a way that
W/f_{fix} decreases, the optimum wing loading
increases and the altitude decreases. The
opposite is true if the fuselage is
shortened, as demonstrated by the Boeing
747 SP variant.
5. Aircraft with high bypass engines have
relatively low values of $T/(\delta T_{to})$ due to
the thrust decay with speed. Their optimum
cruise altitude is low and the wing loading high, although this effect is partly
cancelled by the relatively low thrust
loading required for adequate low-speed
performance.
Obviously, all conclusions stated are valid on the condition that the simplifying
assumptions mentioned previously are realistic.

7.2.2. Wing loading limits and structural aspects

The possibility of choosing a wing loading
resulting in optimum cruise performance is
usually restricted by certain limiting
factors, associated with performance and
operational aspects (see Chapter 5). An
example of the effect of such a limit is
depicted in Fig. 7-3, which shows that an

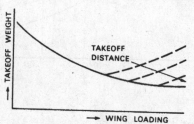

Fig. 7-3. Effect of a performance constraint on the takeoff weight

increase in the wing loading results in a
decreasing takeoff weight up to a point
where the takeoff field length requirement
becomes critical. A larger engine must be
chosen or a more complicated flap system
designed if the wing loading continues to
increase, resulting in a progressive takeoff weight increment. In addition, considerations associated with structural weight
reduction may bias the designer's decision.

a. Takeoff field length.
The takeoff distance required is approximately proportional to the sum of the kinetic and the potential energy at the takeoff height and to the thrust loading. For
jet aircraft:

$$S_{to} = k_{to} \frac{W_{to}}{T_{to}} \left(\frac{V_2^2}{2g} + h_{to} \right) \qquad (7-18)$$

The constant of proportionality k_{to} is
primarily dependent on the type of operation - reflected in the airworthiness
rules - the number of engines, the bypass
ratio and the size of the aircraft. The

designer may derive k_{to} from statistical data on existing aircraft or from calculations of the type discussed in Section 5.4.5. A typical result is k_{to} = 2.2, 2.0 and 1.8 for twin-, three- and four-engine transport aircraft, respectively. For propeller aircraft the power loading must be used instead of the thrust loading.

The limitation in the takeoff safety speed V_2 can be translated into a wing loading limit as follows:

$$W_{to}/S = \tfrac{1}{2}\rho V_2^2 \, C_{L_{max}} \, (V_S/V_2)^2 \qquad (7-19)$$

where minima for V_2/V_S are specified in the regulations.

It is noted that the performance of the high-lift system* has an appreciable effect on the wing loading limit (see Section 7.6.).

b. Landing field length.

A semi-empirical method for estimating the wing loading limit to cope with a specified landing distance is derived in Section 5.4.6. Suitable design of the high-lift system is again a powerful means for attaining the optimum wing loading. For long-range aircraft this requirement is usually not critical in view of the relatively low landing weight (see Section 8.2.4.).

c. Fuel tank volume.

The internal volume of a wing of given shape is proportional to $S^{3/2}$ and the available fuel tank volume thus decreases rapidly with increasing wing loading. In the example in Fig. 5-19 it was shown that for the long-range aircraft considered this condition determines the wing area.

An approximate wing area limit for a given fuel quantity may be obtained with Appendix B, Section B-3. Preliminary estimation of the fuel weight has been discussed in Section 5.4.2., while data on the specific gravity of fuel are given in Table 8-14. It should be noted that the internal wing volume

*$C_{L_{max}}$ in (7-19) refers to the flap deflection angle for takeoff

can be increased by increasing the sweepback angle, resulting in a thicker wing for a given high-speed design condition (see Section 7.5.4a.).

d. High-speed buffet boundaries: transport aircraft.

Limitations to the wing loading can be derived from the requirements in Section 7.5.2b., provided the lift coefficients for buffet onset $C_{L_{BO}}$ and for maximum buffet penetration $C_{L_{BP}}$ are known. The maneuver requirements will limit the wing loading or the cruise altitude as follows:

$$\frac{W/S}{P_o} \leqslant .538 \, \delta M^2 \, C_{L_{BO}} \qquad (n = 1.3) \qquad (7-20)$$

and:

$$\frac{W/S}{P_o} \leqslant .438 \, \delta M^2 \, C_{L_{BP}} \qquad (n = 1.6) \qquad (7-21)$$

Assuming a gust alleviation factor of .8 and using (7-23), it can readily be shown that a specified gust speed of 41 ft/s (12.5 m/s) can be coped with if the wing loading is limited as follows:

$$\frac{W/S}{P_o} \leqslant .7 \, \delta M^2 \left(C_{L_{BP}} - .03 \, \frac{C_{L_\alpha}}{M\sqrt{\delta}} \right) \qquad (7-22)$$

For a given wing shape, these buffet boundaries are a function of the Mach number only. The lift-curve slope is a function of the Mach number, the aspect ratio and the sweepback angle (see Appendix E, Section E-4.1.). It is obvious that the various limits can be determined only after design of the wing shape and prediction of the buffet boundaries.

e. Some structural aspects.

Variation of the wing area will obviously entail a variation of its structural weight. It is obvious that the wing loading resulting in a minimum All-Up Weight will therefore be higher than that for the best range performance, provided other design considerations do not impose constraints on the optimum. This influence of the wing structure weight is particularly pronounced in the case of short-range aircraft, where the emphasis lies on reducing the empty weight and fuel economy is of secondary importance.

The size and takeoff weight of the largest transport aircraft have increased steadily since the dawn of aviation. Whereas the aircraft with which the Wright brothers made their first powered flight had a wing loading of only 1.5 lb/sq.ft (7.3 kg/m^2), the loading amounts to a hundred times this figure for modern transports like the McDonnell Douglas DC-10 (see also Fig. 1-6). The effect of aircraft growth on the empty weight has been the subject of many theoretical considerations, based on the square-cube law. Applying this law to the aircraft structure, it is concluded that when its linear dimensions are enlarged by a scale factor φ, the wing area will be increased proportional to φ^2, but the volume - and hence the structure weight - will increase by φ^3, provided the specific density and stress level of the structural material are constant. If the wing loading is assumed constant, the All-Up Weight will increase proportional to φ^2 and consequently the structure weight grows more rapidly than the All-Up Weight. A point will be reached where the aircraft is unable to carry more than the structure and no margin is left for useful load, engines or equipment. F.W. Lanchester (1868-1946) drew attention to this difficulty in the early 1900s and several others have made (widely divergent) predictions of a practical limit to the size of aircraft.

The square-cube law is based on many simplifying assumptions and has been defeated by the ingenuity of designers. Aircraft will not be scaled up according to a geometric similarity and stress levels have increased considerably. Nevertheless, the actual wing structural weight fraction will tend to increase with the size of the aircraft, unless the wing loading is increased (Section 8.4.1.). A statistical plot (Fig. 7-4) shows that for propeller aircraft the wing loading trend increases in proportion to $W^{1/3}$, while for jet aircraft it is approximately proportional to $W^{1/5}$. From other statistical material it can also be observed that the cruising speed of propeller aircraft increases considerably with size, as opposed to jet transports which usually operate at high-subsonic speeds where the flight Mach number is limited by compressibility effects.

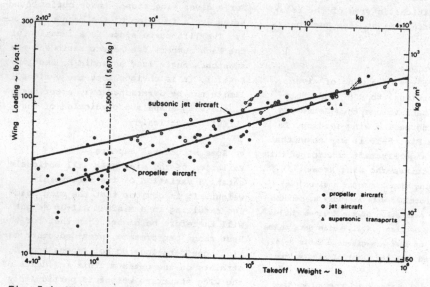

Fig. 7-4. Wing loading trends.

226

The wing size not only has a direct influence on the structural weight, it also affects the gust loads. This is observed from the normal load factor resulting from a sharp-edged gust velocity:

$$n = 1 + K_g C_{L_\alpha} \tfrac{1}{2}\rho_0 w_E v_E \frac{S}{W} \qquad (7-23)$$

where K_g is a gust alleviation factor which is mainly dependent on the altitude and the wing loading. Since the weight of the wing is approximately proportional to $\sqrt{n}$, a high wing loading is favorable for achieving a low structural weight, provided the gust load is the deciding factor for a critical load. Apart from the weight aspect a reduction of gust loads offers the occupants a smooth ride when flying through turbulent air.

7.3. SOME CONSIDERATIONS ON LOW-SPEED STALLING

Although the stall is outside the normal pattern of transport flight, stall incidents occur from time to time. It appears that the probability of the occurrence of stall is one in 10^5, but the probability of reaching the stall warning is between one in 10^2 to 10^3 (Ref. 7-103). Acceptable handling characteristics at the stall must therefore be shown during certification tests, while the outcome of flight tests also forms the basis for establishing the minimum flight speed(s). The regulatory requirements for stalling behavior are stated in FAR 23.201 through 23.207 (BCAR Section K Ch. K 2-11) for light aircraft, and FAR 25.201 through 25.207 (BCAR Section D Ch. D 2-11) for transport aircraft. These regulations stipulate that acceptable stalling behavior must be demonstrated in straight flight and in a coordinated turn, for the operational flap settings, center of gravity positions, undercarriage up and down and specified power. In addition, satisfactory stall behavior must be shown during powered flight with one engine inoperative.

The aim of this section is to provide some generalized guidelines for achieving a wing design with inherently acceptable stalling characteristics. It is emphasized that in the preliminary design stage the available methods to achieve this are rather limited, particularly for sweptwing aircraft. The discussion is confined to the "conventional" type of stalling at angles of attack between about 10 and 20 degrees, the wing shape being the major factor affecting this stall. Some remarks on post-stall behavior and deep-stall, relevant to a restricted category of aircraft, are given in Section 2.4.2c. In that context the shape of the wing is of limited interest only.

7.3.1. Stall handling requirements and stall warning

The purpose of specifying certain acceptable stall characteristics is to minimize the chance of an inadvertent entry into a stall and to ensure recovery from it if the pilot stalls the aircraft intentionally. During flight testing the aircraft will also be deliberately stalled with fully throttled engines in order to establish the stalling speed, which constitutes the most important basis for the low-speed performance.

a. Stalling behavior.
During the standard stalling maneuver the engines are fully throttled and the elevator control is pulled back so that the aircraft is decelerating at approximately 1 knot/sec (.5 m/sec^2). With the usual three control system it must be possible to produce and correct angles of roll and yaw by the normal (unreversed) use of the controls up to the moment when the stall becomes apparent. The stall should preferably be characterized by a distinct and initially uncontrollable nose-down pitching motion, and during recovery to level flight it must be possible to prevent angles of roll and yaw of more than approximately 15 degrees (light aircraft) or 20

227

degrees (transport aircraft). In addition, dynamic stall tests must be executed at higher rates of deceleration - up to 4 knots/sec, 2 m/sec^2 - with engines operating at specified power levels and with an inoperative engine.

It is generally conceded that a "good stall" is non-existent and many attempts have been made to prevent stalling altogether by limiting the elevator control power in such a way that the angle of attack for maximum lift can never be attained ("stall proofing"). This approach is usually rejected as the limited control power conflicts with other maneuverability requirements in unstalled flight, and is difficult to achieve in practice in view of the differences in aircraft configuration. In addition, it results in degraded performance, and the conventional approach is therefore to minimize the consequences of a stall rather than prevent it altogether.

b. Stall warning.
The occurrence of a stall must be preceded by an appropriate stall warning in a range of speeds of about 10% EAS above the stalling speed, i.e. in a range of angles of attack between about 80% and 100% of the stalling angle. Large increases in elevator control forces or control stick movements are generally unmistakable warnings. A gradually increasing, small amplitude oscillation in roll and pitch may also be acceptable, provided it is initially controllable by the pilot. Vibrations or buffeting of the aircraft and the control stick on approaching the stall are also considered adequate stall warnings, provided they cause no danger of structural damage.

Mechanical and electronic devices (stick shakers, warning lights, audible alarms) have been developed to provide stall warning when adequate characteristics cannot be made inherent to the design. A survey of the most common devices is presented in Ref. 7-75.

7.3.2. Design for adequate stall characteristics

The stalling characteristics of an aircraft may be markedly different in varying conditions of flap setting and engine power, depending on the engine location relative to the wing. Small variations in the external shape may have a great effect on stalling, as illustrated by Ref. 7-100, for example. Stalling characteristics cannot be predicted accurately by theoretical methods, while wind tunnel experiments have limited value in view of the effects of differences in the Reynolds number, surface roughness and structural details. An additional complication is that a design criterion cannot be easily defined, as the requirements give a qualitative description (except for the limits on roll and yaw angles) and are to some extent open to differences in interpretation.

Nevertheless, a few general principles can be given which permit the achievement of a basically good design by suitably shaping the wing, which is the primary element affecting the stalling characteristics, and carefully locating the horizontal tailplane. The wing loading sets the stalling speed, while its planform, section shape and twist determine the initial occurrence and progression of the separation and the violence of the motions associated with wing stalling. The horizontal tail load will contribute to the desirable pitch-down behavior in a way which is determined primarily by its location relative to the wing (cf. Section 2.4.2.). Finally, the type of flow around the control surfaces in the condition of a stalled wing determines the effectiveness of control deflections while the pilot is trying to keep the aircraft on course.

The progression of the stall can usually be predicted for straight wings of moderate to high aspect ratio and the wing shape can be chosen so that its behavior will be satisfactory. These aims are much more difficult to achieve on swept wings, not only because of the marked three-dimensional character of the flow, which is difficult to predict, but also because the shape of swept wings is usually dictated primarily by considerations of high-speed flight, with the result that the stalling properties of the basic wing may tend to be compromised. Where inherently acceptable characteristics cannot be achieved, some form of stall protection system may be adopted. These systems limit the maximum usable angle of incidence by a large nose-down control input of short duration, counteracting the action taken by the pilot to produce higher angles of incidence ("stick-pusher"). These devices are generally incorporated on the basis of investigations outside the field of preliminary aerodynamic design.

7.3.3. Stalling properties of airfoil sections

The airfoil section is fundamental to wing design and much research has been done - and will continue to be done - to correlate the geometric properties of wing sections to the stalling properties and the shape of the lift and pitching moment curves. Following the classical exposition of McCullough and Gault (Ref. 7-74), three representative types of airfoil stall are considered: trailing edge stall, leading edge stall and thin airfoil stall. The course of events which determines the type of flow and the resulting lift and pitching moment curves is illustrated in Fig. 7-5. Although every airfoil cannot be classified uniquely in one of these stalling categories, the following description of the aerodynamic phenomena is generally valid for most commonly used airfoils.

TYPE I: TRAILING EDGE STALL. This type of stall is characteristic of most airfoil sections with thickness/chord ratios* of approximately 15% and above. The flow at large angles of attack is characterized by a progressive thickening of the turbulent boundary layer on the upper surface. As the angle of attack is increased to about 10 degrees (B) flow separation starts at the trailing edge and moves gradually forward. This is associated with a decreasing lift curve slope, although initially the increasing lift near the leading edge is predominant and the lift gradient remains positive. A maximum lift coefficient of approximately 1.5 is obtained on a symmetrical airfoil of 18% thickness when the separation region reaches the mid-chord point (C). Beyond maximum lift (D) the forward progression of separation continues at about the same rate as prior to the stall, and the peak of the lift curve is rounded. The variation of the pitching moment with lift is smooth; there is no sudden break at the stall.

TYPE II: LEADING EDGE STALL. Airfoils with thickness/chord ratios of about 9 to 12 percent experience an abrupt separation of the flow near the leading edge. On these sections separation of the laminar boundary layer may occur well before the attainment of maximum lift and prior to transition to a turbulent boundary layer. Transition occurs in the shear layer thus formed, and the expansion of the turbulent motion spreads at such an angle that reattachment of the flow quickly occurs, enclosing a "short bubble" and subsequently forming a turbulent boundary layer (B). The pressure distribution is affected by the short bubble only locally and the peak suction is not greatly altered by the short bubble, which is very small in comparison with the wing chord. An increase of the angle of attack (C) moves the separation point in a region of sharp airfoil curvature, so that the turbulent shear layer fails to reattach. At this critical incidence a complete disruption of the

*Defined in Appendix A, Section A-2.1.

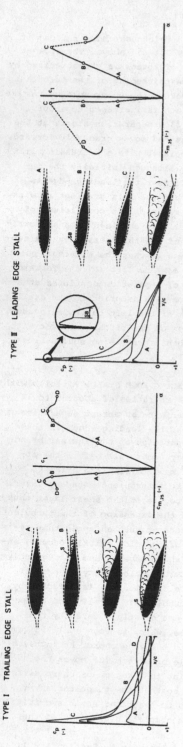

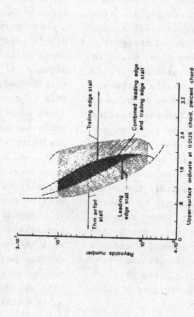

TYPE I TRAILING EDGE STALL

TYPE II LEADING EDGE STALL

TYPE III THIN AIRFOIL STALL

b. Stalling characteristics correlated with Reynolds number and airfoil geometry (Ref. 7-81)

a. Upper surface pressure distributions, growth of the boundary layer and separation regions and lift and pitching moment curves.

NOTES: 1. Drawings are not exactly to scale

2. S = separation, SB = short bubble, LB = long bubble

Fig. 7-5. Representative types of low-speed airfoil stall

230

flow occurs over the entire upper surface; the short bubble is said to burst (D). The leading edge pressure peak collapses and the pressure is subsequently redistributed along the chord into a more or less flattened form which is characteristic of complete separation.

The lift and pitching moment curves exhibit abrupt changes when the angle of attack for maximum lift is exceeded. There is little or no rounding of the lift curve and a sudden negative shift of the pitching moment resulting from the rearward shift of the centre of pressure is observed.

TYPE III: THIN AIRFOIL STALL. On very thin sections of thickness/chord ratios of less than about 6 percent and on round noses a small separation bubble occurs at very small angles of attack (S). At a certain critical angle of attack the short bubble breaks down, but the flow subsequently reattaches downstream, forming a "long bubble" which causes a slight reduction in the lift-curve slope (B). With increasing angle of attack the point of flow reattachment progressively moves backward until it coincides with the trailing edge and maximum lift is reached at this condition (C). The lift curve is characterized by a rounded peak, while the pitching moment curve shows a pronounced negative trend near maximum lift.

The pressure distribution associated with a long bubble has a reduced level suction peak which extends over the length of the bubble.

In addition to the three types of stall, various forms of combinations of trailing and leading edge stall can be observed. These are characterized by a semirounded or relatively sharp peak of the lift curve. The various types of stall should not be too closely allied to the thickness ratio. It has been shown by Gault (Ref. 7-81) that there is a relationship between stalling characteristics, airfoil nose geometry and Reynolds number. In Fig. 7-5b the upper surface ordinate at 1.25 percent chord

has been used to characterize the leading edge shape. The correlation shows that distinct areas are present where any of the three types of stalling dominates, but in addition certain airfoils show a combination of trailing edge and leading edge types of stall. The stalling characteristics are difficult to predict and very sensitive to minor variations in airfoil geometry.

The flow field around an airfoil with leading-edge and/or trailing-edge high-lift devices ("multi-element" airfoils) is considerably more complex than that of a basic section. Boundary layer separation may occur on any of the airfoil components, there is a strong interaction of the flow around these components and the resulting flow fields are very difficult to analyze. A theoretical prediction of the stalling characteristics of wings with flaps deflected has a very limited value in the preliminary design stage. Some comments on the effect of trailing-edge flap deflection on the type of stall are made in Appendix G, Section G-2.3.

7.3.4. Spanwise progression of the stall

If flow separation starts at the wing tip and progresses inboard, the stall is likely to be characterized by a violent roll without warning, as the stalled region exerts a large rolling moment. Adequate aileron control power may also be lost because of the separation of the flow on the aileron. Nevertheless, a stall starting at the tip may be acceptable, provided the tip sections have a flat-top lift curve while the inboard sections have a sharp lift curve. In this case the wing tip maintains enough lift beyond the stall and large rolling moments are not generated.

An initial stalling region near the wing root may result in a large wake hitting the horizontal tailplane - depending upon its location relative to the wing - thus producing tailplane buffet. This may be considered as a desirable feature - if limited - on small aircraft, but it will be unacceptable if excessively violent buffeting is likely to cause structural damage. A stall confined to a limited inboard area will also be accompanied by large regions of unstalled flow, and this will generally result in an appreciable loss in maximum lift. In view of the extra wing area required, the early root stall may not be an efficient solution for transport aircraft. In addition, the wake of the stalled inner wing may blanket the vertical tail and as a result rudder control may be lost, resulting in a directionally uncontrollable aircraft. A point in favor of root stalling is the resulting nose-down pitching moment which is caused by a decrease in downwash at the tail and the loss in lift over the forward part of a sweptback wing.

A wing stall starting at about 40 percent semi-span is probably a desirable feature for transport aircraft. An increment of several degrees in the angle of attack should be required for the stall to progress from the root to the tip in order to prevent the whole wing from stalling simultaneously and aileron control being abruptly lost.

The effects of wing/fuselage interference, the presence of outboard nacelles and the effects of slipstream may have a large, unpredictable influence on the stalling properties, as shown by several examples mentioned in the references.

7.4. WING DESIGN FOR LOW-SUBSONIC AIRCRAFT

The class of aircraft referred to as "low-subsonic" is interpreted as aircraft with an operational flight envelope which does not allow critical compressibility effects to occur up to the Design Diving Speed or Mach number. These aircraft have maximum level flight Mach numbers of less than about $M = .6$ and dive Mach numbers of up to about $M = .7$, depending primarily on the wing thickness ratio. Straight wings can be used and the airfoil thickness/chord ratio may be varied within certain

limits in order to obtain a favorable interplay between aerodynamic and structural requirements.

The term "straight wing" may be interpreted as a wing with zero sweepback of a spanwise line interconnecting corresponding points at the tip and root sections, and must not be confused with the term "untapered wing". The aerodynamicist will favor a wing with zero sweepback of the quarter-chord line, as this considerably simplifies the aerodynamic analysis. However, the location of the wing root cannot always be freely chosen in view of considerations pertaining to the general layout or structural design. In order to get the airplane balanced it is necessary to find an optimum location of the center of gravity relative to the aerodynamic center. To this end, the aerodynamic center may be shifted backwards or forwards by applying sweepback or sweep forward, respectively, while the center of gravity is shifted over a smaller distance.

7.4.1. Planform

The planform of a wing is defined as the shape of the wing when viewed from directly above, as shown in Fig. 7-6. Planform is directly related to aspect ratio and

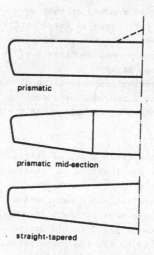

prismatic

prismatic mid-section

straight-tapered

STRAIGHT WINGS

Fig. 7-6. Basic planform shapes for straight wings

taper * and the main aerodynamic characteristics influenced by planform are the induced drag coefficient and the stalling characteristics.

Great variations in the planform can sometimes be observed in the final design compromise adopted by different airplane designers, even though the design specifications may be almost the same. The choice of the basic wing shape, however, to be made by the individual designer or design team, does not offer quite so much scope as might be expected after observing all the shapes which have been actually adopted on aircraft. Prior to a preliminary design effort, various design offices have gained experience with certain shapes, or may have carried out test programs to investigate aspects of aerodynamic performance, stability and control and structural design. One or more acceptable concepts will emerge from such a program, to be further evaluated during preliminary design.

It is observed that nowadays there are basically three forms of straight wings: the tapered wing, the untapered (rectangular) wing and the wing with a prismatic inner portion and a tapered outboard portion (Fig. 7-6). Tapered wings have been adopted for the majority of aircraft since they offer an efficient solution on account of their low induced drag, high maximum lift, low structural weight and good stowing provisions for the undercarriage. Acceptable stalling characteristics can be obtained, provided the wing is not too sharply tapered.

The untapered wing is attractive from the point of view of manufacture, since only one airfoil contour is involved; this simplifies jigging as there are no compound curvatures. It is aerodynamically inferior to the tapered wing, but may nevertheless be the logical choice for inexpensive private aircraft, where the utilization factor is low and initial cost and cheapness

*The terminology used in this section is explained in Appendix A.

of components are important. Untapered wings are well suited for the application of efficient full-span flaps, where the structural complication is outweighed by the relative simplicity of constant-chord flap segments. Untapered cantilever wings are generally of relatively low aspect ratio to save weight, but braced wings may have a high aspect ratio in spite of the absence of any taper (e.g. Short Skyvan). Wings with a prismatic inboard section have good aerodynamic characteristics and offer some advantages for the structural design and manufacturing of the mid-section, particularly in the case of twin-engine aircraft with wing-mounted nacelles.

7.4.2. Aspect ratio

The aspect ratio denotes the ratio of the wing span to the mean geometric chord. For a given wing area, it provides a direct measure for the wing span:

$$b = \sqrt{AS} \qquad (7-24)$$

Instead of the aspect ratio, which determines the vortex-induced drag coefficient, use is sometimes made of the span loading, which is related to A and the wing loading:

$$\frac{W}{b^2} = \frac{W/S}{A} \qquad (7-25)$$

It can readily be shown that the span loading is a direct measure for the vortex-induced drag as a fraction of the weight, if the dynamic pressure is fixed. The span loading will therefore be a good criterion in design studies where the restricted field length imposes a limit on the stalling speed and the aim is to lift as high a takeoff weight as possible under the adverse condition of engine failure. This criterion will be used in Section 7.5.4. to make an initial choice of A and the span for jet-propelled transports.

The climb requirements for propeller transports work out slightly differently. Using the flight mechanics of Section 5.4.3., it can be shown that the parameter

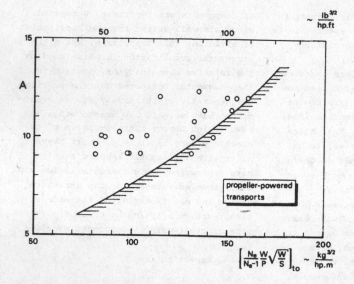

Fig. 7-7. Recommended lower limit for the aspect ratio of propeller transports.

$$\frac{N_e - 1}{N_e} \frac{P_{to}}{W_{to}} \bigg/ \sqrt{\frac{W_{to}}{C_{L_{max}} S}}$$

is of primary importance here. The reciprocal value of this factor is shown in Fig. 7-7 - where variations in C_L-max are ignored - which may be used as an indication for the minimum acceptable value of A.

It is recognized that for wing loadings satisfying (7-7) the maximum L/D ratio and the range for given cruising speed are quite sensitive to A. Good range performance may thus be obtained for a highly loaded, high aspect ratio wing, but an efficient flap system will be required to ensure acceptable stalling speeds in this case. The concept of a high aspect ratio wing is therefore a logical one for transport aircraft where the emphasis lies on high cruising efficiency; however, sophisticated high-lift devices are inherent to this design concept.

For light aircraft, the wing loading is usually fairly low, the complication of a sophisticated high-lift system generally being considered as undesirable by most manufacturers. Consequently, the optimum cruising speed for long-range flight may be too slow to make it an attractive speed, particularly when the altitude is limited to some 10,000 ft (3,000 m). These aircraft are usually flown at the maximum cruise rating and it can readily be shown that a large increase in A results in a relatively modest gain in speed and range. For example, increasing the aspect ratio of an aircraft with W = 3,300 lb (1,500 kg), S = 160 sq.ft (15 m^2), P = 180 hp and C_{D_O} = .025 from 6 to 10 results in a speed increment at sea level from 128 knots (238 km/h) to 134 knots (248 km/h), a gain of only 4%.

A high aspect ratio may result in a low drag in the landing configuration, which tends to flatten the approach glide, makes judgment of the landing point more difficult to the pilot and gives the aircraft a tendency to "float" after the landing flare. In addition, a high aspect ratio wing does not favor good maneuverability in roll due to its large damping and the reduced effectiveness of the small-chord ailerons (Ref. 7-96).

In conclusion, moderate aspect ratios between 7 and 9 are usually applied for twin-engine general aviation aircraft; for single-engine aircraft these values are usually somewhat lower, e.g. between 5.5

and 8.

7.4.3. Thickness ratio

The desirable high aspect ratio for low-
speed transport aircraft can be achieved
only if sufficient structure height is a-
vailable at the wing root, where the bending moment during flight is maximum. In
this connection use is often made of the
cantilever or overhang ratio, which is
defined as the structural wing semi-span,
divided by the maximum root thickness. The
expression for the wing weight in Section
8.4.1. shows that the structural wing
weight fraction increases linearly with
increasing cantilever ratio, provided all
other parameters remain constant.
The cantilever ratio is plotted in Fig.
7-8 for aircraft of various weight categories. Transport aircraft usually have
values of between 18 and 22; ratios in
excess of 25 are rare, even in the case of
supersonic transport aircraft. A rather
lower value is found for trainers, probably because of the high maneuver load fac-

tors for which their wings must be designed.

The favorable trend of low induced drag of
a high aspect ratio wing is partly cancelled out by a profile drag increment if
the thickness ratio is allowed to increase
in proportion to A. Maximum lift is also
affected by the thickness ratio, as shown
by Fig. 7-9. The trend for basic wing sections is readily explained by recalling
that for thin wings the leading edge type
of stall dominates, while for thick sections the trailing edge stall is predominant (see Fig. 7-5). The highest maximum
lift of conventional, standard NACA airfoil sections is achieved for thickness
ratios of 12 to 15 percent chord, where a
combined stall will be observed. Recent
developments in sections for low-speed
aircraft show that higher maximum lift coefficients can be obtained with special
sections having a thickness ratio of about
17 percent (Ref. 7-55). The maximum lift
coefficient with trailing-edge flaps deflected has a relatively flat maximum for

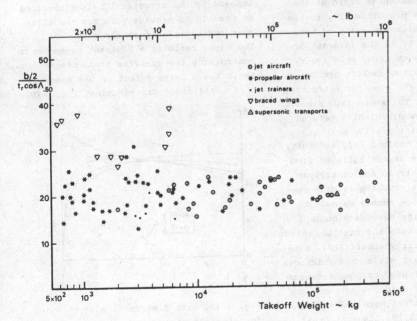

Fig. 7-8. Cantilever ratio vs. aircraft size

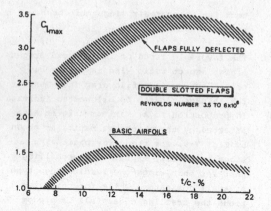

Fig. 7-9. Trends of best maximum lift values of NACA sections vs. thickness ratio.

thickness ratios between 15 and 20 percent. The reasons for this are explained in Appendix G, Section G-2.3.

It is concluded that the root section thickness of transport aircraft should be chosen such that a good cantilever ratio is obtained. For a given taper ratio this implies that the thickness ratio at the root will increase proportionally to the aspect ratio. A thickness ratio of between 15 and 20 percent is in the interest of good performance when using relatively simple trailing-edge high-lift devices and provides adequate room for retracting the undercarriage. Thickness ratios above 20 percent may show diminishing returns due to the increasing profile drag and the relatively low maximum lift and this, in turn, limits the aspect ratio to a maximum of approximately 13 for cantilever wings. For aircraft cruising at Mach numbers above .5 a check should be made to ensure that up to the dive Mach number, which is about .1 above the cruising speed, the flow is essentially subcritical. This condition imposes a limit on the thickness ratio, which is affected to some extent by the airfoil section shape. A method for estimating the critical Mach number is given in Section 7.5.1d. Light aircraft wings have lower aspect ratios and the best

root sections are approximately 15 percent thick.

Tip sections (without flaps) should be between 10 and 15 percent in order to attain a high maximum lift. This reduction relative to the root is also in favor of low structural weight. A minimum practicable thickness should be present on light aircraft in order to provide adequate room for control system elements.

7.4.4. Wing taper

The taper ratio λ has a great effect on the spanwise lift distribution. The spanwise position of the center of pressure of a half wing moves in the direction of the wing root as λ decreases and the root bending moment due to lift decreases accordingly. Since the structural height of the wing root also increases - for a given wing area, span and section shape - a highly tapered wing can be built lighter and with much more torsional rigidity than a rectangular wing. In the case of small aircraft, a practical lower limit to λ is imposed by the structural height required at the tip to provide room for the ailerons and their control elements.

The taper ratio is a dominant parameter in controlling the spanwise stall progression, as it has a large effect on the spanwise lift coefficient distribution (Fig. 7-10).

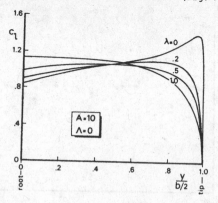

Fig. 7-10. Lift distribution at $C_L = 1$ for straight wings with various taper ratios

A first-order approximation of the location where the c_ℓ-distribution is at its peak is:

$$\eta = 1 - \lambda \qquad (7-26)$$

for untwisted, straight-tapered wings. Hence, a wing with constant section properties and a taper ratio of .4 will tend to stall first at 60 percent semi-span, fairly close to the inner part of the aileron. If a wing is tapered sharply, there will also be a notable reduction in the maximum lift coefficient near the tip due to the locally reduced Reynolds number, thus aggravating the tendency towards early tip stall. Although precautions can be taken to shift the initial separation point inboard by means of section shape variation and twist, the amount of taper has a definite limit. Realizing that the vortex-induced drag of a tapered wing is minimum for a taper ratio of about .4 and is insensitive to relatively large deviations from this value, it is concluded that for straight wings taper ratios appreciably below .4 are of little use (Fig. 7-11).

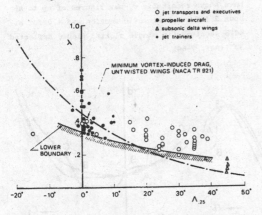

Fig. 7-11. Taper ratios of straight, swept and delta wings.

Once the taper is chosen, the wing geometry is known, provided the area and aspect ratio have also been selected. For straight-tapered wings the tip and root chords are given by:

$$c_r = \frac{2}{1 + \lambda} \frac{S}{b} \qquad (7-27)$$

$$c_t = \lambda \, c_r \qquad (7-28)$$

and the plan view of the wing may be provisionally drawn.

It will be noted that wings with taper both in planform and in the sectional thickness ratio, airfoil sections between fairing stations will be sligly distorted when linear lofting is used for structural simplicity. A tapered wing with respectively 18 percent and 12 percent thickness ratios at the root and tip does not have a 15 percent thickness ratio at the section midway between root and tip, but at a station closer to the tip.

7.4.5. Airfoil selection

In selecting the airfoil sections the designer must give consideration to several general requirements.

1. The basic airfoil must have a low profile drag coefficient for the range of lift coefficients used in cruising flight.
2. For the inboard sections with flaps extended, the drag must be low in high lift conditions, particularly during the takeoff climb.
3. The tip section should have a fairly high maximum lift coefficient and gradual stalling characteristics.
4. The inboard wing section should have high maximum lift with flaps extended.
5. The critical Mach number should be sufficiently high to ensure that critical compressibility effects are avoided in the case of aircraft reaching dive Mach numbers of approximately .65.
6. The pitching moment coefficient should be of low to moderate magnitude to prevent a high trim drag and torsional moments at maximum dynamic pressure.
7. The aerodynamic characteristics should not be extremely sensitive to manufacturing variations in the wing shape, contaminations and dirt, etc.
8. The wing sections should have the largest possible thickness ratio in the interest of low structural weight. Sufficient internal space must be provided for

237

fuel, main gear, mechanical controls and possibly other components.

All these requirements cannot be satisfied by one single airfoil. Spanwise variation of the sectional shape and some measure of compromise will therefore generally be accepted.

For low-subsonic aircraft the selection is usually made from NACA standard sections, to which modifications may be made if necessary, usually during the stage of detailed aerodynamic design. The effect of systematic variations in the profile shape has been the subject of thorough investigations by the NACA, resulting in many series of satisfactory airfoil sections. The relevant findings have been presented in the form of very complete information. Refs. 7-5 and 7-51 in particular are very useful tools for the designer, and the use of this data has greatly simplified the choice of a suitable airfoil for conventional aircraft. The observations made in Ref. 7-5 (Chapter 7) are fairly complete as far as wing sections are concerned, while Ref. 7-96 gives a systematic treatment of the effect of airfoil variation on stalling characteristics, using the successful method or R.F. Andersen (Ref. 7-70). A survey will follow of the most commonly used NACA sections*, examples of

which are shown in Fig. 7-12.

a. The NACA four-digit wing sections basically constitute a synthesis of early Göttingen and Clark Y sections, empirically developed on a basis of pre-war experimental data. Both the thickness distributions and the mean lines are defined in the form of polynomials; the sections have a near-elliptic shape. The maximum camber is at approximately the mid-chord position.

Although the sections in the 4-digit series are by no means low-drag profiles, the drag increase with lift is fairly gradual. The cambered sections have relatively high maximum lift and the stalling is fairly docile. These properties have marked the 2412 and 4412 sections, for example, as being suitable tip sections for the wings of light aircraft and tailplanes. In view of the gradual changes in drag and pitching moment with lift, the 4-digit sections are frequently used for light trainers, which often fly in different conditions.

Recent experiments with 16 percent thick sections have shown remarkable c_ℓ-max figures of up to about 2.1 for a basic 6716 section and 4.0 for a 4416 section with single slotted flaps, deflected 30° (Ref. 7-64).

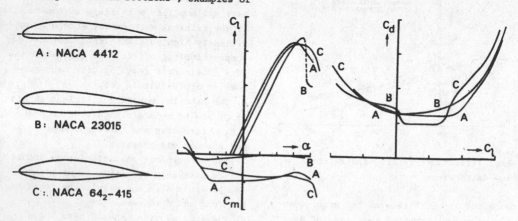

Fig. 7-12. Characteristics of NACA standard series airfoils at R = 6 x 10^6 (Ref. 7-5).

*Terminology explained in Appendix A, Section A-2.3.

238

b. The NACA 5-digit wing sections have the same thickness distribution as the 4-digit series, but the mean lines are different, having their maximum ordinate further forward. The well-known 230-series airfoils have the maximum camber at the .15 chord point. These sections have the highest maximum lift of the standard NACA sections, but the stalling behavior is not particularly favorable and rather sensitive to scale effects. For wings where high lift performance is a prerequisite, the 230-series sections have been frequently used, sometimes combined with a 4-digit section at the tip.

c. The NACA 6-series ("laminar flow") of wing sections is the outcome of a succession of attempts to design airfoils by (approximate) theoretical methods, aimed at achieving low profile drag in a limited range of lift coefficients: the "sag" or "bucket" in the low-drag range. The laminar boundary layer over the forward part of the section is stabilized by avoiding pressure peaks, keeping the local velocities low and applying a favorable pressure gradient over the forward part of the upper surface. The extent of laminar layer is limited by the separation of the turbulent boundary layer over the rear part. The low supervelocities on these airfoils also favor the attainment of a high critical Mach number.

Due to the relatively sharp nose of thin laminar flow sections, their maximum lift is notably below that of the 4- and 5-digit series, although the difference for the thicker cambered sections is negligible; these sections also exhibit a docile stall. The profile drag, although very low under ideal conditions, is sensitive to surface roughness, excrescences and contaminations. Special structures are therefore needed to maintain the laminar flow, and on practical wing constructions of transport aircraft the potentially large extent of laminar layer will not normally be realized.

A great advantage of the 6-series is that sectional properties have been tested extensively and reported systematically. The designer is thus provided with a tool for establishing the best sectional shape by systematically varying the shape parameters. A modification to the standard series is the A-series (Ref. 7-52) in which the sharp trailing edge angle is replaced by a larger one, resulting from straight contours which run from 80 percent chord backwards.

Having decided on the series of sections to be used, the designer will have to choose the various shape parameters.
THICKNESS/CHORD RATIO: see Section 7.4.3.
LOCATION OF MAXIMUM THICKNESS: the further back this point is chosen, the lower the minimum profile drag (of smooth profiles) and the higher the critical Mach number at the design c_ℓ. This works out at the expense of c_ℓ-max and profile drag at high lift. For these reasons, the 63- and 64-sections are the most popular amongst the 6-series airfoils.
(MAXIMUM) CAMBER: determines the angle of attack for zero lift, the pitching moment coefficient, the lift coefficient for minimum profile drag and c_ℓ-max. A large camber is in the interest of high c_ℓ-max, but the tail load required to trim the aerodynamic pitching moment may cause too much extra drag. The camber is usually chosen so that in normal cruising flight the section operates at its design c_ℓ. Little camber is used on trainers, where a requirement exists for acceptable characteristics in inverted flight.
SHAPE OF THE MEAN LINE: a forward location of the point of maximum camber results in a high c_ℓ-max with a leading-edge type of stall at normal thickness ratios. A lower c_ℓ-max and a more gradual stall are obtained when the maximum camber is further back.

7.4.6. Stalling characteristics and wing twist

The following observations, originally

made in Ref. 7-71, are considered as a
good starting point for achieving the desirable stalling characteristics of
straight wings, specified in Section 7.3.:
1. The point of initial stalling should be
sufficiently far inboard, the best location being at about 40 percent semi-span
from the root.
2. Stall progression should be more rapid
to the inboard than to the outboard sections.
3. The margin in c_ℓ at 70 percent of the
semi-span from the root - approximately at
the inboard end of the aileron - should be
at least .1 in the condition where separation occurs first.

Considering the practical measures available to produce the desirable characteristics, it is possible on straight wings of
moderate to high aspect ratios to choose
a suitable spanwise variation of the local
and maximum lift coefficient. An example
is shown in Fig. 7-13. The designer is also

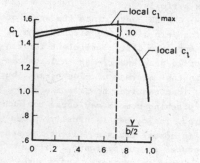

Fig. 7-13. Calculated spanwise lift distribution at high incidence (flaps ups)

concerned with the abruptness of the stall
progression at slightly higher angles of
attack and he may try to gain an impression of the chordwise stall progression by
using Fig. 7-5 and the measured sectional
characteristics.
The spanwise lift distribution* is influenced primarily by the wing planform and

*A survey of prediction methods is given
in Appendix E, Section E-4.2.

wing twist (washout). The type of airfoil
section has little effect in the linear
range of incidences in view of the generally small spanwise variations in the lift-
curve slope. The curve of local c_ℓ-max is
determined exclusively by the local sectional shape. The effect of variations in
the taper ratio, aspect ratio and washout
on stalling characteristics, C_L-max and
induced drag may therefore be investigated
without making any decision on the section
to be used, which saves a lot of work.
The large number of wing shapes to be studied may be reduced by making several restrictions, for example:
1. Although the aspect ratio is important
from aircraft performance considerations,
its effect on the stalling characteristics
is generally small.
2. An aerodynamic washout of more than a-
bout five degrees results in unacceptably
large induced drag increments of the order
of 5 to 10 counts.
3. Low taper ratios may be used only on
wings with thick root sections of the 4-
and 5-digit series and tip thickness ratios of about 12 percent.
4. The effect of wing taper on C_L-max may
be much larger for 5-digit series airfoils
than for 4- and 6-series, particularly if
the Reynolds number at the tip is below 2
million.

There is no chain of logic to show how the designer
will arrive at a (provisional) solution. The basic
wing shape will finally be developed after many
hours of wind tunnel work. Modifications to the
basic standard section may appear desirable and
during flight tests stall control devices may prove
unavoidable, even though theoretical predictions
and wind tunnel tests have not shown any deficiencies.
It should also be noted that most theoretical methods do not take account of any wing/fuselage interference effects. For high-wing aircraft, where the
flow interaction is confined mainly to the less
critical lower surface, these effects are small.
The low-wing position introduces the largest interference effects, but these can usually be controlled by local fuselage contour modifications

and/or adequate root fairings. Established theoretical methods for predicting the characteristics of faired wing/fuselage combinations are not available and the final geometry of these fairings is determined by wind tunnel or flight experiments. Propeller slipstream effects near maximum lift generally result in unstalling the wing in the areas directly influenced by the slipstream. In flight tests aimed at establishing stalling speeds these effects are small as a power-off condition is then required. Power-on stalls may, however, cause a completely different type of behavior, depending on the configuration.

7.5. WING DESIGN FOR HIGH-SUBSONIC AIR-CRAFT

In the aerodynamic design of high-performance aircraft, emphasis is placed on speed as a major factor contributing to the economy and operational suitability of the conceptual design. If we assume for the moment that the type of powerplant to be installed and its rating are fixed, the main factor in determining the speed is the drag area in the case of relatively slow transports and light aircraft, as shown in Fig. 7-14. For a given flying al-

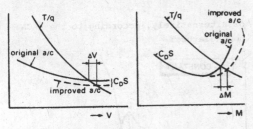

a. LOW-SUBSONIC AIRCRAFT   b. HIGH-SUBSONIC AIRCRAFT

Fig. 7-14. The basic high-speed design problems of improving low-subsonic and high-subsonic aircraft.

titude any improvement in the speed ΔV may be obtained by a reduction of the drag area $C_D S$ - for example by minimizing the wetted area, streamlining, reducing interference, optimizing the wing loading and aspect ratio, etc.

For high-subsonic aircraft these aims are still present but, contrary to the situation with low-speed aircraft, the region of compressible flow is penetrated intentionally in order to attain as high a cruising speed as possible. The problems of aerodynamic design in these aircraft mainly relate to attaining a high critical Mach number, avoiding undesirable flight characteristics at off-design conditions (maneuvering, gustiness, speed overshoots) and providing good low-speed characteristics of the sweptback wing.
The basic opportunities for attaining a high cruise Mach number are the adoption of sweepback (or sweep forward), reduction of the thickness/chord ratio, design of improved airfoil sections and optimum distribution of spanwise camber and twist.
The use of a moderate wing loading and aspect ratio is of some help, but may conflict with other objectives of performance optimization. The application of wing/fuselage blending, fairings and anti-shock bodies may be considered, provided they do not conflict with the structural and layout design.
The aerodynamic design problems are by no means different in the case of a new aircraft which is not required to fly faster than the aircraft to be replaced. The advanced technology may then be used to increase the thickness and/or aspect ratio, or to reduce the angle of sweepback.
This section will be confined to the contribution of the wing only; compressibility effects in other areas, such as the nacelle/airframe junction, may have an appreciable effect (see Section 6.5.).

7.5.1. Wing sections at high-subsonic speeds

a. Subcritical speeds.
In the subcritical region the flow around an airfoil in two-dimensional flow is subsonic throughout. The effects of compressibility on the pressure distribution are well described by potential flow methods, known as the Prandtl-Glauert correction,

$$c_p = \frac{c_{p_i}}{\sqrt{1 - M_\infty^2}} \qquad (7\text{-}29)$$

or more accurate approximations, e.g. the von Karman-Tsien relation. In (7-29) c_p is the pressure coefficient,

$$c_p = \frac{p - p_\infty}{q_\infty} \qquad (7\text{-}30)$$

and the subscript i refers to the incompressible situation ($M_\infty \ll 1$).

Fig. 7-15 shows both the effect of sub-

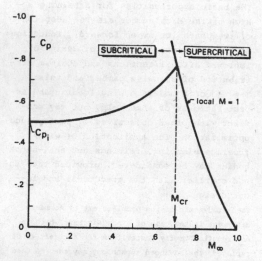

Fig. 7-15. Variation of the pressure coefficient with Mach number and determination of the critical Mach number

critical compressibility on a local pressure coefficient and the boundary of the subcritical region, which is defined by

$$c_{p_{cr}} = \frac{2}{\gamma M_\infty^2} \left[\left\{ \frac{2 + (\gamma-1) \, M_\infty^2}{\gamma + 1} \right\}^{\gamma/(\gamma-1)} - 1 \right] \quad (7\text{-}31)$$

This equation can be derived by using the Bernoulli equation for compressible flow and substituting M = 1 for the local velocity (see, for example, Ref. 7-7).

The critical Mach number of an airfoil section is defined as the free stream Mach number for which sonic flow is reached at the point of minimum pressure. Provided

the value of c_p in Fig. 7-15 refers to this pressure, the critical Mach number is defined by the intersection of both curves, assuming a constant angle of attack. The critical Mach number is thus easily determined from the low-speed pressure distribution, and for NACA standard airfoils this method is used to compute M_{cr} (Ref. 7-51).

Fig. 7-16 shows that the effects of subcritical compressibility on the profile drag and pitching moment coefficients at constant angle of attack are small, while the lift coefficient - and hence the lift-curve slope - are affected in a similar fashion to c_p.

b. Supercritical speeds.

In the case of positive α, regions of supersonic flow will appear on the upper surface when M_{cr} is exceeded. As soon as this region terminates in a shock wave of appreciable strength, which thickens the boundary layer, the drag increases noticeably. The drag-divergence (or drag-critical) Mach number is defined as M_∞ for which:

$$\Delta c_{d_p} = .002 \qquad (7\text{-}32)$$

or alternatively, according to the NACA nomenclature:

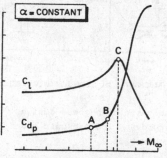

A: critical pressure - M_{cr}
B: drag-divergence - M_{cr_D}
C: lift-divergence - M_{cr_L}

Fig. 7-16. Section lift and drag coefficients vs. Mach number

$$dc_{d_p}/dM_\infty = .10 \qquad (7-33)$$

An increase in M_∞ above this speed (point B in Fig. 7-16) results in a progressive rise in drag. The lift continues to increase until - at point C - a shock appears on the lower surface of the section. At this point - the lift-critical or lift-divergence Mach number - the lift coefficient diverges from its previous trend. As from this point onwards, dc_ℓ/dM_∞ is negative, a flight regime of longitudinal instability ("tuck under") becomes manifest. This must be neutralized by an artificial stabilization system ("Mach trimmer") or by aerodynamic means.

The presence of strong shock waves may lead to separation of the flow and large pressure fluctuations, experienced as an excitation of the wing. This phenomenon, referred to as buffet, constitutes a limitation to the operational flight regime.

c. Trends in high-speed section design. Much work is being devoted to the development of airfoil sections, the objective being to increase the thickness ratio for given design conditions (drag-divergence Mach number and lift coefficient) and to improve the off-design characteristics. Early attempts in this direction were based on designing for low supervelocities, in order to postpone supercritical flow to high Mach numbers. They were followed by designs where regions of local supersonic flow were admitted on the forward part of the airfoil, terminating in near-isentropic and shock free compression. These designs allow greater thickness ratios to be used for the same design Mach number and c_ℓ.

The aerodynamic analysis of mixed (subsonic and supersonic) flows is only possible with sophisticated methods, and a certain amount of empiricism is still common practice here. Several aspects of existing aerodynamic design concepts will be further explained, using the examples in Fig. 7-17. It is not suggested here that in modern section design a choice must be

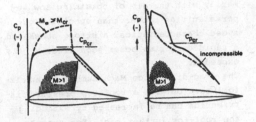

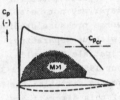 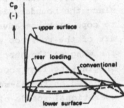

a. Conventional section with roof-top pressure distribution

b. Peaky upper surface pressure distribution

c. Supercritical upper surface pressure distribution

d. Rear loading airfoil compared with conventional airfoil (lower surface)

Fig. 7-17. Aerodynamic design concepts for high-subsonic airfoil sections in the design condition

made from the various concepts; a mixture of these will be present in a practical wing.

ROOF-TOP PRESSURE DISTRIBUTIONS have a gradually changing or approximately constant upper surface pressure over the forward part of the section, which delays the critical Mach number by virtue of a uniform velocity at the design condition (Fig. 7-17a). Slightly above this speed large regions of supersonic flow will appear and the associated suction forces then occur near the crest* of the airfoil or behind it. The NACA 6-series have this type of pressure distribution at subcritical speed for a limited region of c_ℓ values. Since the pressure distribution is designed pri-

*The highest point of the airfoil relative to the free flow direction.

243

marily with the aim of obtaining low supervelocities rather than special high-
speed characteristics, a strong shock and
a rapid drag rise occur soon after M_{cr} is
exceeded.

The drag-divergence Mach number of an airfoil section with a roof-top pressure distribution can be increased by extending
the roof-top further back. For example, a
crest at about 60 percent chord results
in a ΔM_{cr_D} value of about .04 relative to
a section with the crest at 30 percent.
This effect cannot be exploited too far
since the boundary layer may not be able
to cope with the adverse pressure gradient
aft of the roof-top without separating.
The wing of the Aerospatiale Caravelle,
originating from about 1950 (see Fig. 2-3)
is based on the 65-series airfoils; the
operational speeds of this aircraft are
relatively low, and allow a reasonable
thickness ratio of 12 percent.

A PEAKY PRESSURE DISTRIBUTION (Fig. 7-17b),
pioneered by Piercy at the NPL, and by
others, intentionally creates supersonic
velocities and suction forces close to
the leading edge. The airfoil nose is
carefully designed so that near-isentropic
compression and a weak shock are obtained.
The suction forces have a large forward
component and the drag rise is postponed
to high speeds. As compared with conventional sections of the same thickness ratio, the value of M_{cr_D} is approximately
.03 to .05 higher and the off-design behavior is improved. This type of airfoil
has been used on the BAC 1-11, VC-10 and
DC-9 aircraft. The technique employed in
designing peaky airfoils was highly empirical.

Recent advances in high-speed airfoil development have resulted in SUPERCRITICAL
SECTIONS - first proposed by R.T. Whitcomb -
which have a relatively flat upper surface
contour (Fig. 7-17c). With these sections
a much greater extent of shock-free supersonic flow can be created than in the
peaky design. The amount of flattening of

the upper surface is limited by the pressure rise which the boundary layer can
accept without separation. An extensive
NASA program has been conducted to investigate the potentials of this wing technology (Ref. 7-39). Large gains in drag-
divergence Mach number and off-design
performance are claimed if rear loading
is also used in designing supercritical
airfoils.

REAR LOADING (Fig. 7-17a) is a method for
improving high-speed performance by generating lift at the rear part of the airfoil, mainly by pronounced camber of the
lower surface.

The effect of rear loading may be explained in different ways. For a given
thickness ratio and c_ℓ the supervelocities
at the upper surface can be reduced, and
M_{cr_D} can be increased, by generating higher pressures at the lower surface. Alternatively, if an airfoil is considered for
which the upper surface has a critical
flow condition, the rear part is contoured
in such a way that a higher lift is generated, maintaining the same thickness ratio and M_{cr_D}. Finally, for a given M_{cr_D}
and c_ℓ the front part of the airfoil may
be beefed up until near-sonic flow is
created at the lower surface, while the
associated suction forces are cancelled
out by high pressures near the trailing
edge. Sections of high thickness ratio can
thus be obtained for given design conditions.

The extent to which the rear loading can
be accommodated is limited by the nose-
down pitching moment and trim drag associated with the rear location of the
center of pressure. It may also prove difficult to install an effective flap system in the sharp, cambered rear part of
such a section. The European Airbus A-300
wing is an example of a wing design with
a limited amount of rear loading (Ref.
7-28).

d. Criteria for section characteristics
in design and off-design conditions.

The conditions to be considered in selecting airfoil sections are illustrated in
Fig. 7-18. The sections should be selected

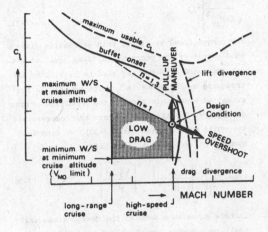

Fig. 7-18. Criteria for the selection of
high-speed sections

primarily in order to achieve low drag at
the high-speed cruise Mach number and the
highest lift coefficient relevant to this
speed. This point is labeled Design Condition in the figure. The lift coefficient
in this condition corresponds with the
maximum cruise altitude specified for the
aircraft and the highest wing loading anticipated for cruising flight at that altitude.
The drag characteristics of an airfoil
section selected on this basis will generally also be satisfactory at lower lift
coefficients and reduced Mach numbers
(shaded area in the figure) which correspond to lower cruising altitudes and
speeds, down to the long-range cruise
Mach number. This region of cruising conditions must be specified for any high-
speed aircraft design in order to check
the drag of the profile sections. Off-
design conditions outside this region result from aircraft maneuvering and gustiness and may be associated with alleviated
requirements relating to drag. A distinction may be made here between two types of
maneuvers: overshoots in speed to increased

Mach numbers, without significant change
in c_ℓ, and pull-ups or turning flights
with increased c_ℓ at constant Mach number.
The overshoot in speed has to be demonstrated in certification flight testing to
show compliance with requirements relating
to stability and maneuvering characteristics up to the Dive Mach number M_D. In
this Mach number region the interaction
between shocks of increasing strength and
the boundary layer will result in flow
separations. A rapid rise in drag beyond
the design Mach number caused by these effects may be conducive to smaller overshoots in speed, but a rapid lift divergence may be unacceptable as it can adversely affect stability. Shock-induced
separations and shock waves hitting the
tailplane cause buffeting vibrations,
which should remain sufficiently mild in
the dive and the ensuing pull-up maneuver
to restore the normal flight condition.
Buffet will also occur if c_ℓ is increased
for constant M during pull-up or turning
maneuvers and in gusty weather. Buffet onset will be experienced as a light vibration when the boundary layer starts to
separate. The violence of the excitation
will subsequently increase with the angle
of attack through conditions of light and
moderate to heavy buffet - a condition
which defines the upper limit to the range
of useful lift coefficients at high speeds.
It is generally acknowledged, though not
considered as a formal requirement, that
for transport aircraft during cruising
flight a maneuver load factor of at least
1.3 g must be available without any
buffeting vibrations occurring. It is
sometimes additionally laid down that load
factors of up to 1.6 or normal gust velocities up to 12.5 m/s (41 ft/s) and a wave-
length of 33 m (110 ft) must be covered
with only a moderate amount of buffet,
this condition being taken as the maximum
penetration of the buffet regime acceptable for civil operations.
An important additional requirement for
the airfoil section is therefore that it
should be capable of producing lift with-

out flow separation up to 1.3 times the design c_ℓ. Cruising performance may be seriously impaired if the buffet boundaries do not permit this reserve in c_ℓ in the design condition. A reduced cruising altitude is generally unfavorable and since the optimum cruise wing loading is high for high-speed transports (Section 7.2.1.), much attention is devoted in the design stage to the development of section and wing shapes which are suitable both at design and off-design lift coefficients providing generous buffet margins.

It should be noted that the criteria presented here are basically applicable to the three-dimensional wing, or rather the complete aircraft. It will be shown, however, that for high aspect ratio wings these requirements can be transformed into airfoil selection criteria.

e. Thickness ratio and drag-divergence Mach number.

An objective of high-speed airfoil design is to obtain a section with the highest possible thickness for a specified combination of M_∞ and c_ℓ. In view of the very complex character of mixed flow, concise methods for making predictions of aerodynamic characteristics, such as drag- and lift-divergence Mach numbers, are not available. Faced with this problem, the preliminary design engineer will have to consult an aeronautical department or establishment equipped with facilities to tackle the problems of supercritical flow.

An estimation of the permissible thickness ratio of a section for a given value of M_{cr_D} may be obtained from available design charts or from the following suggested approach. Sectional data at low speeds ($M_\infty \ll 1$) indicate that the lowest pressure coefficient of symmetrical sections at zero lift is approximately:

$$\left(c_{p_i}\right)_{min} = \text{constant} \cdot \left(\frac{t}{c}\right)^{1.5} \tag{7-34}$$

where the constant is determined by the thickness distribution. Using (7-29) and (7-31) it is easily found that for a given critical Mach number the permissible thickness is represented by:

$$t/c = \text{constant} \left\{ \frac{2}{\gamma M_{cr}^2} \left[1 - \left| \frac{2 + (\gamma-1) M_{cr}^2}{\gamma+1} \right|^{\gamma/(\gamma-1)} \right] \right. \\ \left. \sqrt{1 - M_{cr}^2} \right\}^{2/3} \tag{7-35}$$

For conventional sections, this expression gives a good result if the constant is taken equal to .24. More advanced sections may also be tackled by introducing a Mach number M^*, which represents the extent of supersonic flow at the condition of drag divergence. Substituting M^* into the compressible Bernoulli equation and into (7-30), we find for symmetrical sections at zero lift:

$$t/c = .30 \left\{ \left[1 - \left| \frac{5 + M^2}{5 + (M^*)^2} \right|^{3.5} \sqrt{\frac{1 - M^2}{M^2}} \right]^{2/3} \right\} \tag{7-36}$$

In this equation M denotes the design (drag-critical) Mach number for which the airfoil is to be designed.

The factor M^* in (7-36) has no physical meaning and is merely a figure defining the aerodynamic sophistication employed to obtain supercritical flow at the design condition. Good results are obtained by taking:

$M^* = 1.0$, conventional airfoils; maximum
    t/c at about .30c

$M^* = 1.05$, high-speed (peaky) airfoils,
    1960-1970 technology   (7-37)

$M^* = 1.12$ to 1.15, supercritical airfoils

It is difficult to make adequate allowance for the effects of airfoil camber and lift. Provided the airfoil operates at the design c_ℓ, it is possible to use an approximation by reducing M^* in (7-36) by .25 times the design c_ℓ for c_ℓ up to .7.

7.5.2. Wing design for high speeds

a. Simple wing sweep theory and its limitations.

Sweeping back the wing postpones the effects of critical compressibility to a certain extent, an effect which can be explained by the "simple sweep concept". This assumes an infinitely long sheared wing (Fig. 7-19a), for which the supervelocities and the pressure distribution are determined solely by the velocity com-

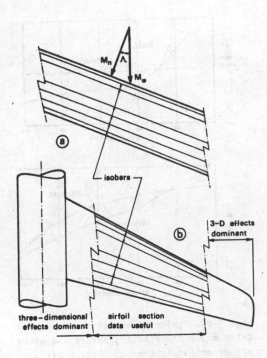

Fig. 7-19. The simple wing sweep theory
for infinitely long and high aspect ratio
wings

ponent normal to the leading edge,
$M_n = M_\infty \cdot \cos\Lambda$. The section normal to the
leading edge is thus the relevant shape
to be considered. The simple sweep concept yields the following relationships.
1. The thickness ratio and effective angle of attack of the normal section are
greater than those of the streamwise section by an amount equal to $1/\cos\Lambda$.
2. The wing lift coefficient C_L is equal
to $c_\ell (\cos\Lambda)^2$, where c_ℓ is the normal section lift coefficient.
3. The critical Mach number of the wing is
$M_{cr}/\cos\Lambda$, where M_{cr} is the normal section's critical Mach number at the normal
section's c_ℓ.
These relationships show that potentially
the critical compressibility effects may
be postponed to a free stream Mach number
which is increased relative to that for a
straight wing by a factor $(\cos\Lambda)^{-1}$. The
normal section shape, however, has to be

designed to cope with a design value of
c_ℓ which is greater than the wing lift coefficient by a factor $(\cos\Lambda)^{-2}$. For example, a wing sweptback by 35 degrees, and
operating at M = .85 and C_L = .4, will
have normal sections designed for operation at M = .7 and c_ℓ = .6.

In early applications of sweptback wings
it became apparent that the gain actually
obtained in M_{cr} was smaller than predictions based on the simple sweep theory had
indicated; in effect a factor $(\cos\Lambda)^{-1/2}$
was achieved, rather than $(\cos\Lambda)^{-1}$. The
reasons for this observation stem from the
finite span and the detrimental effects of
fuselage and nacelles. For an untwisted
sweptback wing with constant section
shapes the lift is concentrated towards
the tip and the outboard sections work at
relatively high c_ℓ values, resulting in a
local reduction in M_{cr}. In addition, root
and tip effects decline the isobars in a
direction normal to the flow, so that the
sweep effect is, in fact, reduced. The
points of lowest pressure will be shifted
backwards near the root and forwards near
the tip. The natural curved path of the
flow over a swept wing is hampered by the
fuselage and nacelles, if any, and this
results in a system of expansion waves,
terminating in a spanwise shock. The various effects combine to form a complex
pattern of shocks and expansion waves,
appearing first on the outboard wing and
next on the rear and front part of the
inner wing. The complete picture is sometimes referred to as a λ-shock. The techniques developed to restore the full effect of wing sweep are therefore based on
eliminating these detrimental effects by
reshaping the wing and bodies in the areas
affected (Fig. 7-19b). The mid-section of
a high aspect ratio wing, however, is only slightly affected by three-dimensional
effects, provided suitable measures have
been taken to counteract the root and tip
effects. Two-dimensional section shapes
and data may therefore be used to define
this part of the wing.

b. Aerodynamic optimization of tne wing
and drag-divergence Mach number.
The objective of high-speed wing design is
to obtain a pattern of approximately
straight isobars swept back at an angle at
least equal to the wing sweepback angle,
the upper surface generally being critical
for the drag divergence*. If this aim is
achieved, the flow will be approximately
two-dimensional and the drag-divergence
will occur at the same Mach number every-
where along the span. A detailed examina-
tion of the very complex optimization procedures of this type is outside the scope
of this book, but it is considered appropriate to mention some of the measures
which may be taken, although not all of
them are required for each design.
1. The sweep angle and thickness ratio between approximately 30 and 80 percent
semi-span from the root are based on a
pressure distribution obtained from the
simple sweep concept.
2. The points of maximum thickness at the
root and tip are shifted forwards and
backwards, respectively. Streamwise tips
are used on the BAC VC-10.
3. The lift on the inboard wing is increased by a negative twist (washout). The
example in Fig. 7-20 shows a linear lofted
twist compared with the more complex twist
distribution required to increase the
critical Mach number.
4. For low-wing designs the pressures tend
to be increased at the lower surface. This
may be turned to advantage by locally
thickening the lower part of the section
and bending the nose of the root section
slightly upwards. This results in a root
section with negative camber which is a
few percent thicker than the outboard wing
(Fig. 7-20).
5. The sweep angle near the root section
may be increased by introducing a kink in
the leading edge (Fig. 7-21). Incidentally, the kink in the trailing edge, observed

*Wings with rear loading sections may
form an exception in view of the near-
critical conditions at the lower surface

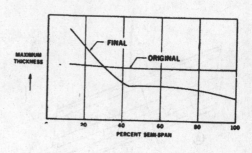

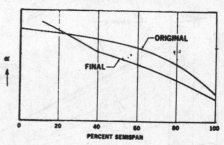

Fig. 7-20. Typical spanwise thickness and
twist angle distribution before and after
aerodynamic optimization (Ref. 7-24).

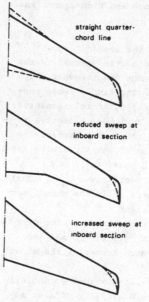

SWEPTBACK WINGS

Fig. 7-21. Sweptback wing planforms.

248

in many high-subsonic transport, is used
to provide internal space for retracting
the undercarriage (see Section 7.8.2.).
6. Adequate wing/fuselage fairings must be
incorporated as a means of obtaining a
good area distribution (see Fig. 2-9c).
The cylindrical fuselage is generally
maintained in the interest of interior
layout but if the speed is increased a-
bove, say, $M_\infty = .90$, wing/fuselage blending is generally regarded as unavoidable
(Fig. 3-5b).
Some examples of swept wing planforms are
shown in Fig. 7-21. These wings are usually composed of at least two sections,
with linear lofted contours between the
intermediate profiles. This sophistication in aerodynamic design results in compound curvatures, giving rise to complications in the structural design and the
manufacturing process. An early example
of an aerodynamically efficient design,
the crescent wing (Handley Page Victor,
see Fig. 2-12 and Ref. 7-137), has never
been widely adopted, probably due to its
structural complexity.

For the purpose of preliminary wing optimization
it may be desirable to derive the combinations of
$\overline{t/c}$ and $\Lambda_{.25}$ required to achieve a specified drag-
divergence Mach number. Equation 7-36 may be
modified by substitution of $M\cos\Lambda_{.25}$ and $\overline{t/c}$ x
$(\cos\Lambda)^{-1}$ where M and t/c are used. For wings with
symmetrical sections at zero lift we then find:

$$\frac{\overline{t}}{c} = \frac{.30}{M} \left\{ (M\cos\Lambda_{.25})^{-1} - M\cos\Lambda_{.25} \right\}^{1/3} x$$

$$\left[1 - \left\{ \frac{5 + (M\cos\Lambda_{.25})^2}{5 + (M^*)^2} \right\}^{3.5} \right]^{2/3} \qquad (7-38)$$

Here M refers to the drag-divergence Mach number,
defined by $\Delta C_{D_p} = .002$, and $\overline{t/c}$ is the thickness
ratio at about 50 percent semi-span from the root.
The values for M^* are given by (7-37) for zero
lift and should be reduced by $.25 C_L (\cos\Lambda_{.25})^{-2}$ to
account for lift and camber at the design condi-
tion.
The drag-critical Mach number to be selected depends on the amount of compressibility drag the

designer is prepared to accept during cruising
flight. It is fair to assume that in high-speed
flight over relatively short distances a penalty
of up to some 20 to 30 counts will be acceptable.
In this case the drag-divergence Mach number may
be taken as being equal to or slightly below the
high-speed Mach number.
For example, if a high-speed cruising flight Mach
number of .8 is specified at $C_L = .3$, we may assume for a straight wing with a peaky-type airfoil
$M^* = 1.05 - .25(.3) = .975$. Substitution of this
value, together with M = .8 and $\Lambda = 0$ into (7-38)
results in a permissible thickness ratio of only
9.7 percent. For a sweep angle of 30° we obtain
$M^* = .95$, and a permissible thickness ratio of 12
percent. This would be increased to 15.3 percent
if an advanced airfoil could be designed with
$M^* = 1.20$ at zero lift. This may be compared with
the allowable 11 percent thickness for a conventional airfoil with the maximum thickness at about
30 percent chord.

The permissible thickness of a wing may well be
determined by the requirement of freedom from buffet. As the prediction of buffet boundaries is as
yet based on wind tunnel measurements; theoretical
design methods for matching a wing to a specified
buffet boundary cannot be presented in this book.
The designer should probably follow an intuitive
approach, based on the available information relating to buffet boundaries of existing aircraft
types or (new) wing sections.

7.5.3. Low-speed problems of high-speed wings

In the same way as straight wings, the
flow over swept wings may separate first
at the leading edge or at the trailing
edge. Leading edge separation results in
a leading edge vortex, which cleans up the
flow over the inner wing, but stalls the
outer wing. At the lift coefficient where
this stalling occurs, the forces and
pitching moment on the wing are characterized by a break (Fig. 7-22), indicating
that the tip stall causes an undesirable
pitch-up, for which the remote position of
the tip is mainly responsible. The primary
causes of this stalling behavior are:

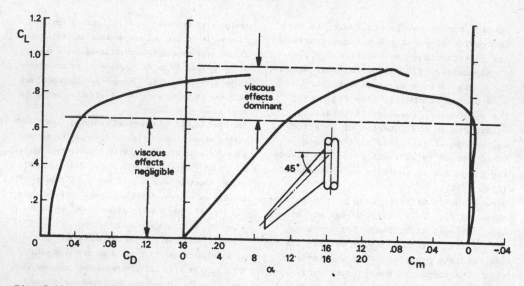

Fig. 7-22. Aerodynamic characteristics at high lift of a typical sweptback wing configuration (Ref. 7-89)

1. the highly loaded tip of swept wings,
2. changes in the chordwise pressure distribution due to sweep, and
3. a spanwise flow of the boundary towards the tip, cleaning up the flow over the inboard wing and making it difficult to stall.

Several investigators have made attempts to correlate the type of separation with the angle of sweepback, the nose sharpness and the Reynolds number (Refs. 7-82 and 7-92). Others have derived conditions for a stable pitching moment curve at the stall (Refs. 7-82 and 7-84), a typical outcome of which is shown in Fig. 2-25b. The highest aspect ratio for a longitudinally stable wing at the stall is found to increase with increasing taper (i.e. decreasing λ) and to decrease with the angle of sweepback. The location of the horizontal tailplane is another factor, determining whether the longitudinal stability improves or deteriorates if the tail contribution is taken into account (see Section 2.4.2b.).

The objective of low-speed design for transport aircraft is to avoid separated flow and to increase the range of useful lift coefficients. The obstacles to achieving this goal are the three-dimensional nature of the viscous flow and the incompatibility of the measures to be taken with high-speed design requirements. As theoretical methods are not available, an experimental program is required, where a choice will have to be made from various methods of triggering the flow. For example:

1. Washout and increasing camber towards the tip.
2. Leading edge modifications, such as nose droop or an increased nose radius on the outer wing.
3. Leading edge devices postponing the stall. The inboard end of a slat almost invariably fixes the separation.
4. Fences, acting as barriers to the cross flow.
5. Pylons for wing-mounted engines, shedding a vortex at high incidences. The "vortilons" on the DC-9 wing have the same effect (Ref. 7-13).
6. Discontinuities in the leading edge: saw tooth, chord extensions.
7. Vortex generators on the outboard wing, in front of the aileron.

It will be clear that these opportunities for design render a simple pitch-up criterion almost useless.

In spite of the potentially unfavorable stalling characteristics of swept wings, reasonably high aspect ratios can be accepted if suitable measures are taken. In view of the small thickness at the root, the span is in any case limited by considerations of structural design, and the swept wings of transport aircraft have a cantilever ratio of the same order as that used for propeller aircraft (Fig. 7-8). The maximum usable lift (flaps retracted) is reduced relative to straight wing designs when trailing edge separation occurs. A theoretical reduction factor of $\cos^2\Lambda$ might be expected according to the simple sweep theory, but in reality the trend is more nearly approximated by a $\cos\Lambda$ effect due to the postponed stalling of the inner wing.

The stalling behavior of swept wings with high-lift devices depends to a great extent on the sophistication of these devices. For simple trailing-edge devices (plain and split flaps) the maximum lift increment is greatly reduced, but the performance of optimized, multi-element airfoils on swept wings is quite good. The stalling behavior is generally not very different from that of the basic wing, or slightly better. Leading-edge devices have a powerful effect on the longitudinal stability at the stall and can be used to advantage.

7.5.4. Planform selection

a. Angle of sweepback.
The previous section has made it clear that the wing sweepback should not be more than the minimum required. For cruising speeds of up to about M = .65 to .70 compressibility effects can be catered for with a straight wing of acceptable thickness ratio. Increasing the Mach number makes it desirable to use sweepback (or sweep forward) in order to avoid severe compressibility problems in a dive. At cruising speeds of M = .75 to .80 a straight wing may only be acceptable if it is very thin and this requires a low aspect ratio (e.g. Learjet 24). The angle of sweepback increases progressively for

cruising speeds in excess of M = .80 (see Table 7-1).
It should be noted that different combinations of $\Lambda_{.25}$ and t/c are possible for a given drag-divergence Mach number. Increasing the angle of sweep allows a thicker wing for the same design condition, permitting more fuel to be carried if required. In addition, sweepback reduces the lift-curve slope and causes the wing to be twisted if it is bent upwards by a lift load. As a result the gust loads are reduced and sweptback wings take the bumps in gusty weather more smoothly. The best choice is the result of a trade-off between structural weight, maximum usable lift and high-speed performance at design and off-design conditions. It may also be affected by considerations of layout design - for example, if the location of the retractable main undercarriage conflicts with that of the rear spar (see Section 7.8.2.).

b. Taper ratio.
The considerations relating to straight wings in Section 7.4.4. are equally valid for swept wings. In addition we may note that for untwisted wings the taper ratio for minimum induced drag decreases with the sweepback angle (see Fig. 7-11). The trend shows that actual designs have higher taper ratios, probably to avoid excessive washout and the associated drag penalty.
A highly tapered wing is favorable for reducing the pitch-up tendency and has a high torsional rigidity. This is a desirable feature, particularly for high-speed wings as it minimizes aeroelastic problems.

c. Aspect ratio and span.
As in the case of straight wings, an optimum aspect ratio can be determined which results in a good balance between fuel weight (fuel cost) and empty weight (initial cost). The aspect ratio is structurally limited by the range of acceptable cantilever ratios (Fig. 7-8), while a performance limit stems from the climb gradi-

ent requirement with one engine inoperative.

Realizing that high-subsonic aircraft are
jet-propelled, an approximate range for
the span loading for this category is derived from statistical data, as follows:

$$\frac{W/b^2}{\frac{1}{2}\rho V_2^2} = .18 \text{ to } .20 \qquad (7-39)$$

It can be shown* that this parameter is
proportional to the ratio of induced drag
to weight at the takeoff safety speed V_2,
which is related to the takeoff field
length by (7-19). Hence, even without considering the flap effectiveness, a reasonable choice of the span loading can be
made with (7-39).

For twin-engine transport aircraft the
thrust available after engine failure imposes a rather sharp limit on the span
loading. The following recommended limit
is derived from statistical data:

$$W/b^2 < \frac{1}{2}\rho V_2^2 \left(1.45 \frac{T_{to}}{W_{to}} - .215\right) \qquad (7-40)$$

where T_{to} refers to standard sea level
conditions. Aircraft with a higher span
loading are likely to be penalized in
takeoffs from hot and high airfields.

7.6. HIGH-LIFT AND FLIGHT CONTROL DEVICES

7.6.1. General considerations

In Section 7.2. it was argued that in most
designs an attempt will be made to achieve
a favorable wing loading, based on considerations of fuel usage, structural weight
and acceptable comfort in turbulent
weather. High-lift devices are required to
prevent the flight speeds from reaching
unacceptable values during takeoff, approach and landing. These devices have
proved to be of vital importance to an
aircraft's operational characteristics
and economic yield and much work has been

*See Chapter 11, equation 11-2.

done to foster their development.
Although the improvement in C_L-max since
the DC-6, for example, may not appear excessive at first sight, the application
of swept wings with thin sections has ob-
scured the actual progress made. The multi-element systems at present in use permit a sectional c_ℓ-max of about 5.5 to be
obtained during the landing, and this appears to be about the limit of passive
systems*. The designer may choose from a
large collection of feasible high-lift
systems, although in the case of a specific project this freedom will be limited, since incremental drag, mechanical
complexity, development and maintenance
costs and structural weight are all factors to be considered.

The additional wing lift contributed by
high-lift devices is basically obtained
through:
- increased airfoil camber,
- boundary layer control resulting from
improved pressure distributions, reener-
gizing or removing low energy boundary
layers,
- an increment of the effective wing area
in the case of flaps extending the chord
when deflected.
Not all existing configurations combine
these actions simultaneously. Trailing-
edge flaps increase the camber and improve
the flow at the trailing edge, but tend to
promote leading edge stall on thin sections and may cause a reduction in the
stalling angle of attack. Leading-edge
high-lift devices postpone or eliminate
leading edge stall, but they have little
effect on the airfoil camber as a whole,
although locally the camber is increased.
Fig. 7-23 shows that these differences
result in large variations in the range
of operational angles of incidence at low
speeds.

Most flap systems are offered with a

*"passive" means that no boundary layer
control by suction or blowing is used.

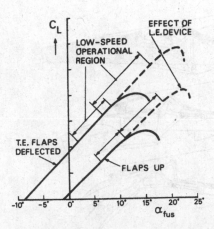

Fig. 7-23. Lift curves with and without high-lift devices.

$$C_{L_{max}} = \frac{W/S}{\tfrac{1}{2}\rho V_S^2} \qquad (7\text{-}41)$$

Generally speaking, C_L-max is less for small airplanes than for large ones, as a result of a less sophisticated wing design and lower Reynolds numbers. There are obviously many other details which may lead to departures from the quoted figures, such as differences in the relative flap chord, flap span, airfoil type and interference with engine nacelles, etc It should also be noted that the values stated apply to the trimmed aircraft, and hence they include the tailplane load required for longitudinal equilibrium. A detailed aerodynamic prediction method, taking account of the most relevant geometric design features, is presented in Appendix G-2.

7.6.2. Trailing-edge flaps

Typical lift and profile drag increments caused by flaps on airfoil sections are compared in Fig. 7-24.

choice of several discrete deflection angles. Deflections of up to about 25 degrees are used on takeoff, while higher values are used during the approach and landing. For each item of performance there is an optimum flap angle which constitutes a tradeoff between high lift and low drag - see Section 11.7.1. Though the operational flexibility of the aircraft is improved by increasing the freedom of choice, the cost and time involved with the certification tests and data handling may become prohibitive.

Typical deflection angles are shown in Table 7-2 for several commonly used high-lift configurations, together with a range of C_L-max values, defined as follows:

HIGH-LIFT DEVICE		TYPICAL FLAP ANGLE		$c_{l_{max}}/\cos\Lambda_{.25}$	
TRAILING EDGE	LEADING EDGE	TAKEOFF	LANDING	TAKEOFF	LANDING
PLAIN	-	20°	60°	1.40-1.60	1.70-2.00
SINGLE SLOTTED	-	20°	40°	1.50-1.70	1.80-2.20
FOWLER*	-	15°	40°	2.00-2.20	2.50-2.90
DOUBLE SLOTTED**	-	20°	50°	1.70-1.95	2.30-2.70
	SLAT			2.30-2.60	2.80-3.20
TRIPLE SLOTTED**	SLAT	20°	40°	2.40-2.70	3.20-3.50

* SINGLE SLOTTED
** WITH VARYING AMOUNTS OF CHORD EXTENSION (FOWLER MOTION)

Table 7-2. Typical maximum lift coefficients for wings with high lift devices.

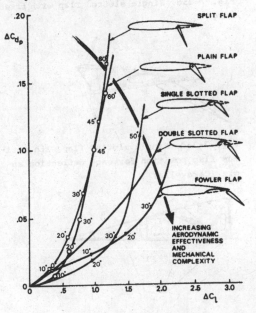

Fig. 7-24. Trends in performance of trailing-edge flaps.

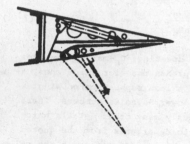

Fig. 25a. Split flap

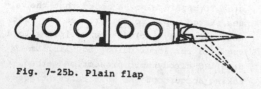

Fig. 7-25b. Plain flap

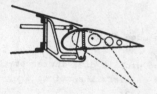

Fig. 7-25c. Single slotted flap with fixed hinge

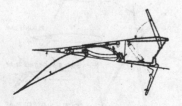

Fig. 7-25d. Single slotted flap with optimum flap position for each deflection angle (Caravelle)

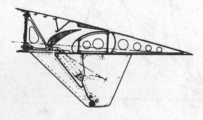

Fig. 7-25e. Double slotted flap with fixed hinge and fixed vane (Douglas DC-9)

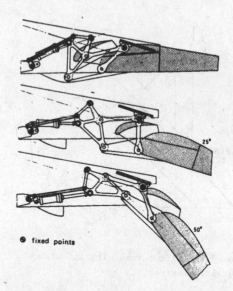

⊕ fixed points

Fig. 7-25f. Double slotted flap with four-bar motion (Douglas DC-8)

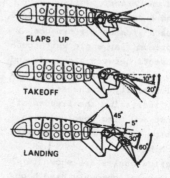

FLAPS UP

TAKEOFF

LANDING

Fig. 7-25g. Double slotted flaps with individual adjustment of flap segments and drooped aileron (GAF N-22 Nomad)

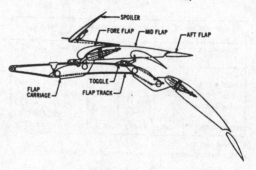

Fig. 7-25h. Triple slotted flap (Boeing 727)

254

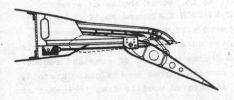

Fig. 7-25j. Single slotted Fowler flap

SPLIT FLAPS consist of a stiffened plate on the lower surface, hinged just aft of the rear spar, by means of a piano hinge (Fig. 7-25a). The drag due to flap deflection is large, particularly in the case of small deflections, thus making the split flap less suitable for takeoff. Although its structural simplicity and low weight are attractive, this type of flap must now be considered as obsolete.

PLAIN FLAPS are hardly more than a hinged part of the trailing edge (Fig. 7-25b). The best performance is obtained with a sealed gap. If the deflection exceeds 10 to 15 degrees, the flow separates immediately after the knuckle, the lift effectiveness drops progressively and the drag increases to values comparable to those of the split flap.

SINGLE SLOTTED FLAPS derive their favorable action from a specially contoured slot through which air is admitted from below the wing. The upper surface boundary layer is stabilized by the suction on the flap's leading edge and diverted at the trailing edge of the basic wing. A new boundary layer is formed over the flap which permits an effective deflection of up to 40 degrees. The performance is sensitive to the shape of the slot, which is determined by the kinematics of deflection. A simple fixed hinge (Fig. 7-25c) is not very effective, but the achievement of an optimum slot shape requires a system which can only be realized by means of a track and flap carriage assembly (Fig. 7-25d). Single slotted flaps with fixed hinge are commonly used on light aircraft.

DOUBLE SLOTTED FLAPS are markedly superior to the previous type at large deflections, because separation of the flow over the flap is postponed by the more favorable pressure distribution. Various degrees of mechanical sophistication are possible:
a. Flaps with a fixed hinge and a fixed vane of relatively small dimensions (Fig. 7-25e) are structurally simple, but may have high profile drag during takeoff. If the vane is made retractable for small flap deflections, the flap is effectively single-slotted during the takeoff, which improves the drag and hence the climb performance. The external flap supports cause a noticeable parasite drag, which may be acceptable only on short ranges.
b. Double slotted flaps may be supported on a four-bar mechanism (Fig. 7-24f). During extension the slot shape closely approximates to the aerodynamic optimum, but the flap supports cause a disturbance in the flow through the slot. The parasite drag penalty in the en route configuration being negligible, this configuration is probably most suitable for application on three- or four-engine long-range transports.
c. If the two flap elements are independently adjustable, the maximum deflection may be increased up to 70 degrees. This rather complex system (Fig. 7-25g) is occasionally used on STOL aircraft, such as the DHC-7. The figure shows the system used on the GAF N-22 Nomad full span flaps with aileron control at high speeds, augmented by spoiler ailerons at large flap deflections.

TRIPLE SLOTTED FLAPS are used on several transport aircraft with very high wing loadings. In combination with leading-edge devices, this system represents almost the ultimate achievement in passive high-lift technology, but Fig. 7-25h shows that complicated flap supports and controls are required.

THE FOWLER FLAP is a slotted flap travelling aft on tracks over almost its entire chord and subsequently deflecting to its maximum angle (Fig. 7-25j). The upper wing skin continues to approximately 90 to 95 percent of the chord. The favorable performance is derived from the effective wing area extension, yielding a gain in lift for very little extra drag which makes the Fowler flap particularly suitable for twin-engine aircraft. Configurations have been devised with one, two and even three slots, depending upon the magnitude of the required lift. These systems not only have considerable chord extension due to the basic flap motion, but the flap sections move relative to each other and the extended length of the chord is therefore in excess of the nested length. The structural complications of these multi-track supports and their weight penalty are usually the limiting factor here. The shape of the slots must be very carefully optimized to obtain good performance.

7.6.3. Leading-edge high-lift devices

The increased circulation associated with the deflection of an effective trailing-edge device induces an upwash at the nose. The local suction peak increases and on airfoils which are liable to leading edge stall the flow will separate at an angle of attack which is below that of the basic wing. Leading-edge high-lift devices (Fig. 7-26) are intended primarily to delay the

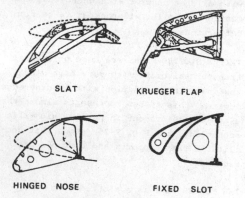

SLAT   KRUEGER FLAP

HINGED NOSE   FIXED SLOT

Fig. 7-26. Leading-edge high-lift devices.

stalling to higher angles of attack. LEADING-EDGE SLATS are small, highly cambered airfoils forward of the wing leading edge, which experience large suction forces per unit of area and reduce the suction forces on the basic airfoil. The associated profile drag increment and pitching moment changes are small and for optimized configurations a C_L-max increment due to a full-span slat between .5 and .9 can be obtained without any appreciable increase in tail load and trim drag. In view of the large stalling angle of attack, the angles of pitch during takeoff and landing are large and proper attention must therefore be paid to the visibility from the cockpit, particularly if the angle of sweep is large and the aspect ratio low.

Slats may be used to advantage by varying their effectiveness in spanwise direction in order to delay the onset of the tip stall and the associated pitch-up of swept-back wings. This may be achieved by varying the gap width, the slat deflection angle or the relative slat chord in spanwise direction.

KRUEGER FLAPS perform in the same way as slats, but they are thinner and more suitable for installation on thin wings. Krueger flaps are often used on the inboard part of wings, in combination with outboard slats, to obtain positive longitudinal stability in the stall.

(PLAIN) LEADING-EDGE FLAPS (hinged noses) are less effective than slats. They are mechanically simple and rigid and particularly suitable for thin airfoil sections. FIXED SLOTS are the simplest devices for postponing leading edge stall, but their profile drag penalty is generally prohibitive for effective cruising, except on some low-speed STOL aircraft.

Leading-edge devices are beneficial if there is any possibility of leading edge stall occurring, but with the comparatively thick sections used on light aircraft and small propeller transports, slats or leading-edge flaps are not normally considered necessary. Stall prevention on the outboard wing should be obtained by a proper choice of the sectional shape, moderate taper ratio and washout. If necessary, an increased leading-edge radius, drooped nose or a part-span slot may be considered.

7.6.4. Flight control devices

A typical arrangement of high lift and flight control devices on a swept wing is shown in Fig. 7-27, to which the following comments can be added.

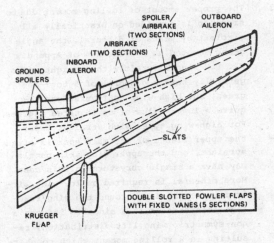

Fig. 7-27. Arrangement of high-lift devices and flight control surfaces in plan view.

a. Ailerons.
The aileron span is chosen as small as possible in order to obtain the largest possible flap span, within the limits imposed by roll control requirements. Slotted ailerons, which are drooped with the flaps, are sometimes used on straight wings, where the object is to attain the highest possible lift with the additional advantage of reducing the induced drag. Their maximum symmetrical deflection is generally limited to some 20 degrees, unless roll control is augmented by means of spoilers. Drooped ailerons on highly swept wings are likely to be fairly ineffective, because they introduce a large pitching moment and

the gain in lift is largely reduced by the associated trim load.
If the aileron of a thin wing is deflected at a high EAS, the load on it causes the wing to twist in the opposite direction. This could, in some cases, offset the load produced by the aileron, resulting in a roll in the wrong direction (aileron reversal). For this reason some high-speed aircraft employ ailerons located closer to the wing root, where adequate torsional rigidity is present. The outboard ailerons operate only at low speeds, while the small-span "high-speed" ailerons are operative both at low and high speeds. A suitable location for the high-speed aileron is behind the (inboard) engines, since the flap will have to be interrupted there to keep it free from the jet efflux. In addition, to reduce the size of the ailerons on high-speed aircraft, these are frequently assisted by spoiler-ailerons.

The area of conventional ailerons may be estimated in the preliminary design stage from statistical data of the parameter $S_a \ell_a/S_b$ (Fig. 7-28), which is a measure of the rolling moment, for a given aileron deflection.

b. Spoilers may be fitted for various reasons; they often combine more than one function and usually occupy a substantial part of the flap span, just behind the rear spar. As shown in Fig. 7-27, the upper skin consists of stiffened panels which can be deflected in almost upright position. Immediately after landing, or in the event of an aborted takeoff, they may be activated either by the pilot or auto-

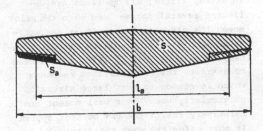

Fig. 7-28. Dimensions related to the aileron geometry.

matically (pre-armed). As a result of the interrupted airflow over the flaps, the wing loses a large part of its lift, which increases the normal force on the tires and makes braking more effective. In addition, they create considerable drag and these combined effects increase the deceleration by some 20 percent.

Outboard spoilers may be used in flight when an appreciable increment in drag is required to obtain a high rate of descent or improved speed stability with a constant angle of descent. Inboard spoilers are not deflected in this case to avoid disturbing the flow over the empennage and prevent buffeting. For this reason the inboard spoilers are only installed to decrease the lift on the ground and are referred to as ground spoilers (liftdumpers), while the others are known as (in-)flight spoilers. When acting as drag-producing devices, these spoilers are referred to as airbrakes (speedbrakes) which may be installed both on the upper and lower surfaces of the wing (Fig. 7-25d).

The effectiveness of roll control may be increased by operating the outboard spoilers simultaneously with the upward-moving aileron. Their motion is proportional to the control wheel movement when it is turned beyond a predetermined angle of rotation. These devices, referred to as lateral spoilers, may be used as airbrakes by deflecting them on both sides.

A recent development is the use of spoilers as elements of a system known as Direct Lift Control (DLC), which has been adopted on the Lockheed 1011. During the approach, in steady flight, all spoilers are deflected several degrees and when the pilot moves the elevator the spoilers are activated accordingly, resulting in an instantaneous gain or loss in lift. This system counteracts the slow elevator response which is common to most large aircraft, particularly when their tail moment arm is relatively small. The DLC control system is most effective when the flaps are deflected and it is therefore used only in the low-speed configuration.

7.7. DIHEDRAL, ANHEDRAL AND WING SETTING

Once the wing shape and the high-lift system have been selected, the position of the wing relative to the fuselage can be laid down. Its vertical position has been dealt with in Section 2.2., while the longitudinal position will depend on considerations of center of gravity position (Chapter 8) and horizontal tail size (Chapter 9).

7.7.1. The angle of dihedral (anhedral)

The proper amount of rolling moment due to sideslip is provided on practically all aircraft by means of dihedral, the angle being denoted as Γ (Fig. A-4 of Appendix A). Low-wing aircraft with straight wings generally have a dihedral of 5 to 7 degrees, but high-wing configurations require a lower value of up to three degrees. For high-wing types with straight taper the upper flange of the main spar is often straight, and the upper wing skin panels may have a single curvature in this case. More dihedral is required on low-wing aircraft, for the flow around the wing/fuselage intersection in a sideslip induces a non-symmetric wing lift distribution, resulting in a rolling moment which counteracts the effect of the dihedral. Similarly, the wing/fuselage interaction contributes to the desired rolling moment on high-wing configurations. The air forces on a swept wing are affected by a sideslip to the extent that the same effect is obtained as that exerted by the dihedral on a straight wing. This results in an effective dihedral which increases with the lift coefficient and may become too large at high lift coefficients. The yawing motion, following a side gust, is accompanied by a rolling motion (Dutch roll) which may be poorly damped or even unstable and consequently objectionable to pilots. This is one of the reasons that for most swept wing aircraft it is necessary to resort to the use of a device, known as a yaw damper, which essentially consists of a rate gyro transmitting a signal to a servomechanism. The rudder is then operated in a direction so as to oppose the yawing motion. The yaw damper is incorporated in the autopilot, and on aircraft with wing-mounted engines it has the additional function of suppressing the yawing motion following engine failure.

An alternative method to combating the lateral-directional stability problems with swept wings is to reduce the dihedral, or even to use anhedral. This can be realized without any layout problem on high-wing aircraft (Fig. 2-4); the application of anhedral on some Russian low-wing designs by Tupolev necessitated a special undercarriage configuration to ensure a minimum wing tip clearance with respect to the runway in takeoff and during landing in crosswind conditions (see also Section 10.3.1.).

7.7.2. Wing/body incidence

The wing setting relative to the fuselage is usually chosen such that during cruising flight with a representative All-Up Weight at a representative altitude the fuselage axis or cabin floor will be horizontal. An expression for the wing incidence is derived in Appendix E (Section E-9.3.), taking into account the effects of wing/fuselage interference, body lift and trim load. If these effects are ignored, the following expression is found:

$$i_w = \frac{C_L^*}{C_{L_\alpha}} + \alpha_{o_1} \varepsilon_t + \left(\alpha_{\ell_o}\right)_r \qquad (7-42)$$

where C_L^* is the lift coefficient for which the fuselage reference line is horizontal, α_{o_1} the change in zero-lift angle of the wing per degree of positive twist at the tip and $\left(\alpha_{\ell_o}\right)_r$ the zero-lift angle of the root section. The lift-curve slope can be obtained from Section E-4.1., while α_{o_1} is approximately -.4 for straight-tapered wings with linear twist. The wing/body incidence may be affected to some extent by considerations of tail clearance

during takeoff and landing (Section 10.3.1.). Deviations of more than about 2 degrees from a level cabin in cruising flight should, however, be avoided.

7.8. THE WING STRUCTURE

The outline of the wing, both in planform and in the cross-sectional shape, must be suitable for housing a structure which is capable of doing its job. As soon as the basic wing shape has been decided, a preliminary layout of the wing structure must be indicated, which is expected to lead, after further refinement and detail design, to a sufficiently strong, stiff and light solution, with a minimum of manufacturing problems.

7.8.1. Types of wing structure

Three basic types of primary wing structure can be distinguished (Fig. 7-29):

MASS BOOM types of wing structure, where the flanges of one or two spars take the normal forces resulting from bending, while the torsional load is carried by either the spar webs (differential bending) or the combination of the spars and skin covers (shear). The single spar wing uses the spar web and the nose section or the section aft of the spar to combine into a box which provides torsional stiffness. Mass boom structures are mainly used on slow aircraft with thick wings (t/c above 15 percent) and low wing loadings (up to 30 lb/sq.ft, 150 kg/m^2). Advantages of this arrangement are as follows.
a. Tapered booms are easy to produce and may be adapted to the local stress level desired. If the booms are stabilized against buckling by means of closely spaced ribs, high stress levels are attainable.
b. The skin is subject only to shear forces and with the use of closely spaced ribs stringers may be dispensed with, which simplifies the manufacturing of the ribs.
c. Attachment to the fuselage is a simple

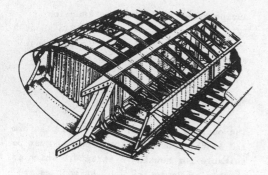

a. Typical single spar mass boom structure

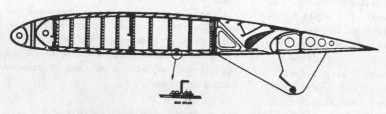

b. Typical wing section of a box beam structure

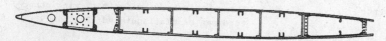

c. Example of a multispar structure

Fig. 7-29. Examples of basic types of wing structure

matter: only two main frames are required.
d. Weight penalties for openings and in-spection doors will be modest.
The disadvantages, however, may be of de-cisive importance:
a. Failure of a spar boom is catastrophic. Due to the absence of fail-safe character-istics, the mass boom wing structure is no longer used in new transport aircraft de-signs.
b. Due to the high stresses in the spar booms the deflections under bending loads are large.
c. The skin plays no part in absorbing the bending moment so that it is not used very efficiently.
d. If a two-spar configuration is used, the spar height is less than the airfoil thickness. The forces in the spar booms due to bending are thus increased and more

material will be required.
e. Many ribs are required to stabilize the spar booms.
f. The skin will buckle when loaded if no stringers are used; this will adversely affect the aerodynamic cleanness.

BOX BEAM structures incorporate skin pan-els, which are stressed not only to take shear forces, but also the end load due to bending. The example in Fig. 7-29b shows a three-spar box beam, but various alternatives are known, such as the two-spar box beam with or without a stiffened nose section.
From the point of view of fail-safe design the stressed skin structure is much better than the mass boom type. The skin can be divided into several planks (multi-load path) coupled with spanwise splice members.

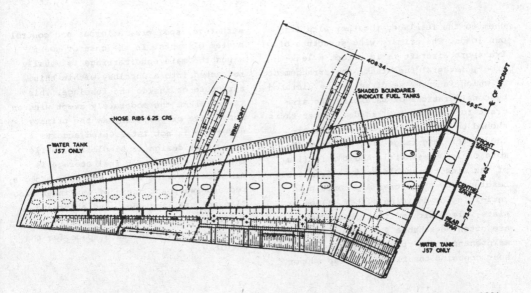

Fig. 7-30. DC-8 wing structure in plan (Ref.: The Aeroplane, 15 Aug. 1958, page 230)

The web can be designed to accommodate an in-service crack, as indicated in the figure.

By and large, the pros and cons of the box beam structure contrast with those of the mass boom type. The advantages of the box beam will be evident when considerable skin thickness is required to obtain sufficient torsional rigidity on wings designed for high speeds and/or thin, high aspect ratio wings. In lightly loaded wings, however, the stress levels in the upper skin will be kept fairly low to avoid buckling and the difference in weight will be small as compared with the mass boom type. Integral construction is often employed on highly loaded wings.

MULTI-SPAR structures are used on very thin wings, sometimes in conjunction with closely spaced ribs. This type of structure is most unsuitable for the provision of cutouts, such as would be required for stowing the landing gear or for inspection facilities.

7.8.2. Structural arrangement in plan

Fig. 7-30 shows the plan view of a typical wing of a high-subsonic transport aircraft (DC-8: wing loading approximately 100 lb/sq.ft, 500 kg/m^2; thickness ratio 12 percent).

The rear spar must be located at a suitable chordwise station, leaving sufficient space for the flaps and for housing the controls to operate the flaps, ailerons and spoilers. A rearward shift of this spar increases the cross-sectional area of the torsion box - and incidentally the fuel storage space - but the reduction in the sectional height will make it less efficient in bending. Similar criteria apply to the front spar when it is moved forward. It is noted that the best flap chord for simple plain and split flaps is about 25 percent wing chord, but highly efficient slotted and Fowler flap systems are more effective with flap chords of up to 35 or even 40 percent of the wing chord. The front spar is located at about 15 percent chord, the rear spar at 55 to 60 percent. About 5 to 10 percent chord should be available between the nested flap and the rear spar for control system elements. The central part of the wing, bounded by the front and rear spars, takes the loads from the nose and rear sections and carries them to the fuselage, together with its own loads. The primary wing structure of transport aircraft is in effect a leakproof, integral fuel tank, the arrangement of which in spanwise direction is dictated by considerations of balancing the aircraft for various fuel loads. Center tanks should be avoided from the outset, although for long-range aircraft they are more or less essential. About 3 ft (1 m) of the wing tip should be free from fuel. Leading-edge structures may contain an anti-icing or de-icing system, flaps or slats, bleed air and fuel pipes, and they are often detachable for inspection and maintenance. The structure behind the rear spar contains the flaps and flap support structure, spoilers, ailerons and control system elements. In the case of swept wings the main undercarriage is usually retracted into a wheelbay within this structure or inside the fuselage. This layout makes the moderately swept wing an attractive proposition, as the primary structure is not interrupted and its structural design is simplified, while the loss in space for fuel storage is minimized. A kink in the trailing edge, a reduced sweepback angle of the inner wing flap and trailing edge, and a forward shift of the rear spar are usually required to create adequate space for the wheelbay.

Chapter 8. Airplane weight and balance

SUMMARY

The sensitivity of airplane performance and operating economy to the empty weight is discussed and the value of weight-saving is demonstrated.

An accurate weight prediction in the preliminary design stage is a most effective way to control the weight; it begins with a consistent scheme for weight subdivision and limitations. Considerations are presented for making a sound choice of the operational weight limitations.

Some general remarks on weight prediction methods are followed by a comprehensive collection of available and consistent methods, useful for most categories of modern civil aircraft. Attention is paid both to simple approximate methods and to more detailed procedures, for which detailed design information must be available.

The load and balance diagram is introduced to illustrate the flexibility of loading an airplane. The effect of the general arrangement and layout of the aircraft on the problem of obtaining adequate balance in all likely flight conditions is discussed and a procedure suggested for establishing a suitable longitudinal wing location and center of gravity range.

Many references to literature are given, as well as a large collection of data on weights and center of gravity ranges of airplane types in present service.

NOMENCLATURE

AC	alternating current
A_i	capture area of inlet
APS	Aircraft Prepared for Service
APU	Auxiliary Power Unit
AUW	All-Up Weight
a	constant factor in statistical weight equation
BOW	Basic Operating Weight
B_p	number of propeller blades per propeller
b	span (no index: wing span); factor of proportionality in statistical correlation
b_{ref}	reference span
b_s	structural span ($b_s = b/\cos \Lambda_{\frac{1}{2}}$)
$\bar{c}$	length of mean aerodynamic chord
c.g.	center of gravity
D	selling price of payload
DC	Direct Current
D_p	propeller diameter
ESHP	Equivalent Shaft Horse Power (takeoff, standard atmosphere)
h	height; depth
h_h	height of horizontal tailplane above fin root
k	factor of proportionality
k_w	factor of proportionality for the weight of a group of items
l	length; moment arm; distance between end faces of a prismoid
l_h	horizontal tail length (cf. Chapter 9)
l_t	distance between 1/4-chord points of wing and horizontal tailplane root (see Fig. D-2)
MAC	Mean Aerodynamic Chord
MLW (MRLW)	Maximum (Regular) Landing Weight
MTOW (MRTOW)	Maximum (Regular) Takeoff Weight
MZFW	Maximum Zero Fuel Weight
m_i	ratio of actual to estimated weight for a sample point
N	number of an item present in the airplane
$n_1, n_2, \ldots, n_m$	exponent of a weight parameter
n_{ult}	ultimate load factor
OEW	Operational Empty Weight
P_{el}	total electrical generator power (kVA)
P_{to}	takeoff horsepower per engine (sea level, static)
R_B	maximum range with maximum payload (Fig. 8-3)
R_D	maximum range with maximum fuel (Fig. 8-3)
R_{ref}	reference range
S	(projected) area of a surface (no index: wing area); standard error of prediction
S_1, S_2	areas of parallel end faces of a prismoid
S_G	gross shell area of the fuselage
S_{he}	exposed horizontal tailplane area
SIV	Standard Item Variations
SMC	Standard Mean Chord
S_{wet}	wetted area
T_{to}	takeoff thrust per engine (sea level, static)
t_r	(absolute) maximum thickness of root chord
U	annual airplane utilisation
u.c.	undercarriage
V	speed; volume
V_b	blockspeed
V_D	Design Dive speed
V_{max}	maximum horizontal flight speed
W	weight
$\dot{W}_{ba}$	rated bleed airflow of APU
$\dot{W}_{f_{to}}$	fuel flow per engine, corresponding to P_{to} or T_{to}
W_{DE}	Delivery Empty Weight
W_{OE}	Operating Empty Weight
W_G	Gross Weight
W_{ZF}	Maximum Zero Fuel Weight
X	X-axis; parameter for wing weight estimation example
X_{LEMAC}	coordinate of MAC leading edge
x	coordinate of weight contribution; sample value of X
Δx	range of x-coordinates for the c.g.
x_{OE}	airplane c.g. position for the OEW
Y	weight of an airplane part
Y_i	actual (measured) value of Y for a sample
δ	maximum deflection angle; incidence variation

$\Lambda_{\frac{1}{2}}$	sweepback angle at 50% chord (no index: wing)	geo	geometric shape
ϕ	average load factor	h	horizontal tailplane
$\phi_1, \phi_2, \ldots, \phi_m$	parameters for general weight estimation formula	hc	horizontal tail controls
		i	inlet; installation; sample
		ieg	instruments and electronics group

Subscripts

		ld	lift dumper
		LEMAC	Leading Edge of MAC
APU	Auxiliary Power Unit	n	nacelle (group)
APUG	APU Group	p	propeller
ba	(APU) bleed airflow	pax	passengers
cc	cabin crew	pc	passenger cabin
cf	cabin floor	pg	propulsion group
cg	center of gravity	s	structure; slat
ch	cargo hold	sb	speed brake
d	intake duct	sc	surface controls group
e	engine(s)	tail	horizontal plus vertical tail
el	electrical system	thr	thrust reverser
er	bending moment relief due to engine(s)	to	takeoff
		uc	undercarriage
f	fuselage; flaps; fuel	v	vertical tailplane
fc	flight crew	w	wing; weight
fg	fuselage group	wc	toilet/watercloset compartment
ft	fuel tank	wg	wing group
		wt	water tank for injection fluid

8.1. INTRODUCTION; THE IMPORTANCE OF LOW WEIGHT

Weight minimization of an airplane design is a subject of the utmost importance. Although reduction of weight is generally obtained only at some initial cost penalty, the effects on total operating costs are paramount for most high-performance designs, particularly for large and complex airplanes. In many cases the increased weight of one component means added weight elsewhere, leading to the well-known snowball effect of weight growth. The opportunities to achieve a weight reduction and the associated costs depend upon the phase of the design process.

a. During the initial conceptual design the choice of the airplane layout, geometry and detailed configuration affects weight. The design layout should be carefully optimized and high accuracy of the initial weight prediction is a prerequisite. Weight pre-

diction is necessary not only to make an assessment of the design qualities, but also to set a goal for the structural and systems design offices. The initial weight prediction must be a realistic challenge to both. This type of work, being the normal task of the preliminary design office, involves virtually no extra costs.

Weight reductions or increments are generally evaluated for constant design performance, unless limited engine performance does not permit this. Any component weight increase is therefore associated with a takeoff weight increase. If, however, the component weight growth is caused by a design change to improve performance - e.g. improved high-lift devices, increased wing span - the final result may well be a takeoff weight reduction.

Sensitivities to structure weight increments are shown in Table 8-1 for some typical missions. In the case considered, a 10% structure weight growth was followed by a re-

AIRPLANE CATEGORY	DESIGN RANGE	EFFECT ON MTOW
subsonic transport	250 nm	6.5 %
	1000 nm	6.9 %
	3000 nm	7.0 %
supersonic transport	3000 nm	9.4 %
transport VTOL	250 nm	6.9 %
military VTOL	250 nm	9.5 %

Table 8-1. Effect of a 10% increase in structure weight on maximum takeoff weight for a constant mission (Reference: ICAS Paper No. 66-1)

sizing process in order to maintain the specified payload-range and other performance. The data presented are illustrative only, the sensitivity also depending on where the weight growth occurs.

An even more sensitive case is that of a fixed takeoff weight. A typical weight breakdown* of a conventional design of a medium subsonic propeller turbine transport is as follows:

OEW   : 61% MTOW
Payload: 22% MTOW
Fuel   : 17% MTOW (5% reserve, 12% trip
                   fuel)

Increasing the OEW by 5% MTOW to 66% MTOW initially reduces the payload by 22% for the same range or the range by 42% for the same payload. Obviously, the final result may be quite different if the airplane is redesigned to accommodate the reduced payload.

b. During detail design it is essential to save every small item of weight that can possibly be saved, in order to ensure a high standard of weight prediction accuracy and to continuously monitor the weight, using an effective weight control system. In most airplane development programs weight reduction programs must be started occasionally in order to redress unfavorable weight creep which may have become apparent. Some cost may be involved in the form of additional manpower, but this is often a small

*Definitions in Section 8.2.2.

portion of the penalties incurred by an overweight design.

In order to save weight, the use of advanced materials and sophisticated manufacturing techniques may be considered, resulting in a reduction of the amount of material required. The weight saving may be used to reduce the takeoff weight or to increase the payload or fuel load. However, the cost involved may lead to a noticeable increase in the price of the airplane and an assessment of the value of the weight saving should be made.

c. During negotiations with potential customers the question will arise whether or not the design is subject to special requirements, resulting in a weight penalty. For a civil transport aircraft the weight/cost tradeoff can be based on the value of payload to the operator, which in turn depends upon the productivity of the aircraft and the frequency with which the payload availability can be sold. The revenue on a pound of capacity payload on a critical

AIRPLANE CATEGORY	PERCENTAGE OF MTOW			
	airframe structure	propulsion group	fixed eq. and serv.	empty weight
PASSENGER TRANSPORTS				
short-haul jets	31.5	8.0	13.5	53.0
turboprops	32.0	12.5	13.5	58.0
pistons	29.5	20.5	15.5	65.5
long-haul jets	24.5	8.5	9.0	42.0
turboprops	27.0	12.0	12.0	51.0
pistons	25.5	17.5	11.0	54.0
FREIGHTERS				
short-haul turboprops	35.0	13.0	8.0	56.0
long-haul turboprops	26.5	10.0	7.0	43.5
EXECUTIVE JETS	27.5	8.0	15.5	51.0

Table 8-2. Typical average empty weight fractions for several categories of transport aircraft

GROUP WEIGHT BREAKDOWN			
AIRPLANE TYPE:		DATE:	
ENGINE TYPE:		NAME:	
GROUP INDICATION	WEIGHT ( )	MOMENT ARM	
		x( )	z( )
AIRFRAME STRUCTURE	( )	( )	( )
WING GROUP			
TAIL GROUP			
BODY GROUP			
ALIGHTING GEAR GROUP			
SURFACE CONTROLS GROUP			
ENGINE SECTION OR NACELLE GROUP			
PROPULSION GROUP	( )	( )	( )
ENGINE INSTALLATION AND AFTERBURNERS			
ACCESSORY GEAR BOXES AND DRIVES			
SUPERCHARGERS (FOR TURBO TYPES)			
AIR INDUCTION SYSTEM			
EXHAUST SYSTEM			
OIL SYSTEM AND COOLER			
LUBRICATING SYSTEM			
FUEL SYSTEM			
WATER INJECTION SYSTEM			
ENGINE CONTROLS			
STARTING SYSTEM			
PROPELLER INSTALLATION			
THRUST REVERSERS			
AIRFRAME SERVICES AND EQUIPMENT	( )	( )	( )
AUXILIARY POWER PLANT GROUP			
INSTRUMENTS AND NAV. EQPT. GROUP			
HYDRAULIC AND PNEUMATIC GROUP			
ELECTRICAL GROUP			
ELECTRONICS GROUP			
FURNISHING AND EQUIPMENT GROUP			
AIRCONDITIONING AND ANTI-ICING GROUP			
MISCELLANUOUS			
BASIC (EMPTY) WEIGHT	( )	( )	( )
OPERATIONAL ITEMS	( )	( )	( )
CREW PROVISIONS			
PASSENGER CABIN SUPPLIES			
POTABLE WATER AND TOILET CHEMICALS			
SAFETY EQUIPMENT			
OIL, RESIDUAL FUEL, WATER/METHANOL			
CARGO HANDLING EQUIPMENT, MISC.			
OPERATIONAL EMPTY WEIGHT			

Table 8-3. Weight breakdown for civil conventional airplanes (AN 9103 D, modified)

sector* is given by:

revenue = constant V_b U D ϕ $/annum   (8-1)

For a utilization U = 3000 hours per annum, a load factor ϕ = .55 and a passenger revenue D = .45$ per sh.ton nm, the revenue is $167 per lb per annum for an average block speed V_b = 450 kts (835 kmh). During a 12-year service period this amounts to $2000

*i.e. a sector where the MTOW is restricted by field length or other operational limitations.

per lb. In practice, the money lost by the operator will be a mixture of lost passenger revenue, cargo revenue or extra fuel. Moreover, not all sectors are critical all the time and therefore the value of a pound of empty weight will be a fraction of the stated value, e.g. 10-15%. However, the lesson to be learnt from these figures is that the empty weight must not increase to such an extent that the capacity payload cannot be achieved.

Basic Empty Weight is composed of airframe

structure weight, propulsion group weight and the weight of airframe equipment and services (see Table 8-3). Many contributions to these weight groups are affected to some extent by the airplane layout; this applies to the structure weight in particular. A survey of empty weight fractions for several airplane categories is given in Table 8-2. In comparing the structure weight fraction of modern airplane types with oldtimers, it is sometimes noted that this fraction has not improved, probably for the following reasons:

a. The relative fuel and propulsion group weights must be taken into account. Due to the improved engines of recent years, these weight fractions have decreased and, for the same payload, the structure weight fraction tends to increase.

b. The wings of modern high-subsonic airplanes are swept and relatively thin. More emphasis has to be laid on stiffness requirements, leading to weight penalties.

c. Improved high-lift devices have resulted in higher unit structure weights for the wings.

d. Fatigue life requirements are critical for modern transport airplanes, leading to limitations in the stress levels, e.g. in pressurized cabins. An example of the empty weight penalty to improve the airframe lifetime is shown in Fig. 8-1.

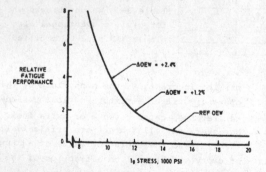

RELATIVE
FATIGUE
PERFORMANCE

ΔOEW = +2.4%

ΔOEW = +1.2%

REF OEW

1₀ STRESS, 1000 PSI

Fig. 8-1. Effect of level flight stress level on OEW and fatigue performance (Reference: AIAA Paper No. 66-882)

e. More stringent airworthiness requirements concerned with the safety and comfort level have resulted in more complex and heavier structure and airframe systems.

Weight prediction in the preliminary design is necessary to performance prediction, center of gravity determination and design of the undercarriage, and also to provide the various design departments with realistic design weights and weight limits. The following sections demonstrate how a weight prediction may be obtained. The publications mentioned in the list of references contain a large amount of information, although in many cases the methods presented are based on a particular category of aircraft and state of the art and do not take changes in technology into account; for these reasons their value is very limited. A good weight estimate starts, however, with clear definitions and effective subdivision of the items to be considered and that is the subject dealt with in the following section.

8.2. WEIGHT SUBDIVISION AND LIMITATIONS

The airplane is composed of a large number of parts which can be combined into groups according to several schemes. The weight of these groups and several combinations of groups are of importance in the design, certification and operation of the airplane. Unfortunately, there is little international agreement in the matter of weight terminology, leading to some confusion with regard to the actual meaning of terms. In this section the most commonly used weight terminology is explained and various relationships demonstrated. Not only the weight subdivision is dealt with, but also various limitations on several characteristic weights. The proposed scheme is representative of a wide range of commercial airplane designs. Some considerations are offered to enable a reasonable first choice of the various limits to be made.

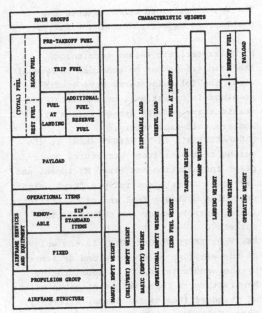

* STANDARD ITEM VARIATIONS

Fig. 8-2. Weight groups and characteristic weight terminology (not to scale)

8.2.1. Weight subdivision

Figure 8-2 gives a subdivision into main groups and combinations of these in characteristic weight terms.

a. Main groups.

Various terms are defined below and in Table 8-3. More detailed weight breakdowns will be presented in Section 3.4.

AIRFRAME STRUCTURE: the wing group, the tail group, the body group, the alighting gear group and the engine nacelle group. The surface controls group may be classified as airframe structure or as a part of the airframe services.

PROPULSION GROUP: the engine(s), items associated with engine installation and operation, the fuel system, and thrust reversing provisions.

AIRFRAME EQUIPMENT AND SERVICES: APU(s) instruments, the hydraulic, electric and electronic systems, furnishings and equipment, air-conditioning, anti-icing systems

and other equipment. A further subdivision into fixed and removable equipment is useful for obtaining an accurate and repeatable empty weight definition.

FIXED EQUIPMENT AND SERVICES are considered an integral part of a particular aircraft configuration. These include the weight of fixed ballast (if present) and the fluids which are contained in a closed system (such as hydraulic fluid).

REMOVABLE EQUIPMENT AND SERVICES are those items of equipment or system fluids that are not considered an integral part of a particular aircraft configuration. Removable separating walls, passenger seats*, floor covering, basic emergency equipment and the like are typical examples.

Aircraft of a given type are usually ordered by different operators, each of whom has particular requirements with regard to equipment and services. Various items of equipment are not delivered by the airplane manufacturer (government- or buyer-furnished equipment). Consequently, a standard or basic airplane configuration is usually defined with items of equipment or systems fluid which do not vary for aircraft of the same type: Standard Items. Those items which operators choose to add to, omit from or change in the Removable Equipment are referred to as Standard Item Variations (SIV). These items are the Operator's responsibility.

OPERATIONAL ITEMS are those items of personnel, equipment and supplies that are necessary on a particular operation, unless already included in the Basic Empty Weight**, i.e. crew, manuals, crew bags, catering supplies, water supplies, oil, unusable fuel***, additional emergency equipment, tool kits, etc. These items may vary for a particular airplane configuration according to the operator's allowances for the service intended. However, a minimum crew is defined for each airplane by government regulations. Alternative terms: Operators Items, A(ircraft) P(repared for) S(ervice) Items.

PAYLOAD: all commercial load which is

*Passenger seats are occasionally considered as Operational Items
**See Para. b of this Section and Fig. 8-3
***Unusable fuel is occasionally included in the Removable Equipment

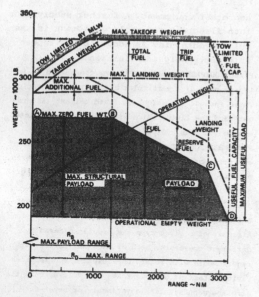

Fig. 8-3. Derivation of the payload-range diagram

carried: passengers and their baggage, cargo and mail. Alternative terms: traffic load, transport load. The payload must never exceed the maximum volumetric payload capacity or the maximum structural payload (Section 8.2.2.). The Estimated Normal Payload is a payload selected by an operator according to the seating pattern and for the passenger and cargo unit weights allowable, and is used for statistical and other related purposes.

TOTAL FUEL: all usable fuel, engine injection fluid and other consumable propulsion agents. A subdivision is made into:
- fuel consumed during engine runup and taxying prior to takeoff,
- trip fuel, i.e. the fuel consumed during flight up to the moment of touchdown in the landing,
- reserve fuel*, determined by the Operator in accordance with the relevant Operational Rules, and
- additional fuel (available for further flights, for example).

*Reserve fuel may be burnt during taxying after landing

The Total Fuel load cannot exceed the Usable Fuel Capacity (Section 8.2.2.).

b. Characteristic weight terminology. The main groups can be combined in several ways; the following terms have specific meanings in operational use. Several definitions are deduced directly from Fig. 8-2 and are self-explanatory.

MANUFACTURER'S EMPTY WEIGHT (MEW) is the weight of the airframe structure, powerplant installation and Fixed Equipment and Services. It is essentially a "dry" weight, excluding unusable fuel and oil, anti-icing fluid, potable water and chemicals in toilets.

DELIVERY EMPTY WEIGHT (DEW) is the weight of the airplane as produced and delivered by the manufacturer. It is equal to the Manufacturer's Empty Weight plus the weight of Standard (Removable) Items. Alternative term: Empty Weight Dry.

The condition at the time of determining the MEW must be one that is well defined and easily repeatable. The Empty Weight is established experimentally by weighing the delivered airplane and is laid down in the certification documents.

The Tare Weight is the nominal (calculated) Empty Weight; it differs from the experimentally determined Empty Weight due to the Manufacturing Variation.

BASIC EMPTY WEIGHT (BEW) is the Delivery Empty Weight plus or minus the net weight of the SIV. It is also equal to the Manufacturer's Empty Weight plus the weight of the Removable Items. Alternative term: Basic Weight.

The Fleet Empty Weight is an average value of the Basic Empty Weight of a fleet or group of airplanes of the same model and configuration, with similar Services and Equipment, operated by the same airliner.

OPERATIONAL EMPTY WEIGHT (OEW): the weight of the airplane without Payload and Fuel. Alternative terms: APS Weight, Basic Operational Weight (BOW), Weight Empty Equipped.

ZERO FUEL WEIGHT: the OEW plus payload. It must not exceed the Maximum Zero Fuel Weight (Section 8.2.2.). Alternative term: Empty Tank Weight. Occasionally the term

Zero Wing Fuel Weight is used to denote
the airplane weight less fuel in the wing.
This concept is important in the case of
airplanes with fuel tanks both in the wing
and in the fuselage or elsewhere.
TAKEOFF WEIGHT: the total weight of a dis-
patch-loaded aircraft at the moment of
brake release or start of the takeoff run.
It depends upon the loading condition and
must not exceed the Maximum Takeoff Weight
and the operational weight limitations in
relation to the available performance (Sections 8.2.2. and 8.2.3.).
RAMP WEIGHT: the Takeoff Weight plus the
weight of fuel required for engine run-up
and taxiing prior to takeoff. It must not
exceed the Maximum Ramp Weight (Section
8.2.2.). Alternative term: Taxiing Weight.
LANDING WEIGHT: the airplane weight at the
moment of touchdown in the landing. The
Landing Weight must not exceed the Maximum
Landing Weight or the operational weight
limitations referred to in Sections 8.2.2.
and 8.2.3.
GROSS WEIGHT: the total airplane weight at
any moment during the flight. It decreases
during the flight due to fuel and oil consumption and may also vary due to payload
dropping or refuelling during flight. Alternative term: All-Up Weight (AUW).
At the moment of brake release, the Gross
Weight is equal to the Takeoff Weight;
during flight it is referred to as En-
Route Weight or In-Flight Weight. It is
limited by several weight restrictions
mentioned in Section 8.2.2.
OPERATING WEIGHT (the author recommends
the term Zero Payload Weight): the weight
of the laden airplane, but without payload. Consequently, the Operating Weight
is equal to the OEW plus the Total Fuel.
It cannot exceed the Usable Fuel Capacity
(Section 8.2.2.).
DISPOSABLE LOAD AND USEFUL LOAD: the variable load (Payload and Fuel) that can be
taken on a particular flight operation.
Both items are determining factors for the
payload-range capabilities and the economic yield of the airplane, as explained in
Section 8.2.3. These loads are limited by

the Permissible Load, i.e. the difference
between the Maximum Takeoff Weight and the
Basic Empty Weight.

8.2.2. Weight limitations and capacities

Restrictions must be imposed on several
variable characteristic weights in order
to avoid overloading of the structure or
unacceptable performance or handling qualities during any phase of day-to-day operation. Weight limits are therefore established during the process of design and
certification for airworthiness and laid
down in the Flight Manual (FM) and other
documents associated with the Certificate
of Airworthiness (C of A).
Although definite weight limits are not
established until certification tests,
reasonable values must be anticipated during preliminary design in order to ensure
satisfactory operational flexibility.

a. Maximum and Minimum Weights.
Maximum weights must be established for
each airplane operating condition (ramp
or taxiing, takeoff, en-route flight, approach and landing) and loading condition
(zero fuel condition, center of gravity
position, weight distribution). They should
not exceed the least of the following:
1. The greatest weight for which compliance
with the relevant structural and engineering requirements has been shown. This
weight limit is usually designated as a
Design Weight, e.g. the Design Take-Off
Weight.
2. The greatest weight for which compliance
with the relevant handling requirements in
flight has been demonstrated, a weight restriction which is particularly important
for aerobatic airplanes, where certain aerobatic maneuvers can only be executed with a
limited Gross Weight.
3. The greatest weight for which performance
data have been established.
4. A weight selected by the Applicant.*
These maximum weights are determined by the

*generally the airplane manufacturer

design and are invariable during service. As a rule, the four criteria mentioned above will not differ greatly, except in the case of an airplane type for which certification is requested under several different regulations or airworthiness categories. The following maximum weights must be established on most commercial airplane types.

MAXIMUM RAMP WEIGHT: the maximum weight authorized* for ground maneuvering prior to brake release in the takeoff. Alternative term: Maximum Design Taxying Weight.

MAXIMUM TAKEOFF WEIGHT (MTOW): the maximum weight authorized* at takeoff brake release. It is frequently fixed by structural requirements in the takeoff and occasionally by the Maximum En Route Weight. The Useful Load corresponding to the MTOW is the Permissible Load or Maximum Useful Load.

MAXIMUM LANDING WEIGHT (MLW): the maximum weight authorized* at the landing. It generally depends on the landing gear strength or the landing impact loads on certain parts of the wing structure. The MLW must not exceed the MTOW.

MAXIMUM (IN-) FLIGHT WEIGHT (MFW): the maximum weight at which flight other than takeoff and landing is permitted. The Maximum Flight Weight and the MTOW are generally equal, except when provisions for refueling in flight exist. The "flaps-up" condition is assumed, unless otherwise stated. Alternative term: Maximum En Route Weight.

MAXIMUM WEIGHT: the greater of the MTOW and the MFW. The Maximum Design Weight is the maximum of the corresponding Design Weights.

MAXIMUM ZERO FUEL WEIGHT (MZFW): the maximum weight of an aircraft less the weight of the Total Fuel Load and other consumable propulsive agents in particular sections of the aircraft that are structurally limited by this condition. At this weight the subsequent addition of fuel will not result in the aircraft design strength being exceeded. The weight difference between the MTOW or MLW and the MZFW may be utilized only for the addition of fuel.

*by relevant government regulations

The empty tank case can be a critical loading case in certain critical areas of the structure at positive load factors, as there is no relieving load due to the fuel mass. For some aircraft there may be a limit to the weight for initiating fuel transfer between tanks (Maximum Design Fuel Transfer Weight, MFTW).

b. Loading restrictions and minimum weight. Mass distribution: for each airplane weight the mass distribution must be such that:
1. the center of gravity remains within the appropriate limits for ensuring satisfactory handling qualities (Section 8.5.2. and Chapter 9);
2. compliance with structural requirements such as allowable floor loads, external loads, braking loads, deceleration loads in emergency landings, etc. can be shown.

Minimum Weight: flight is prohibited if not clearly impracticable at weights below the Minimum Weight, which is not less than the greatest of the following weights:
1. the Design Minimum Weight, i.e. the lowest Gross Weight at which compliance with the structural loading conditions can be demonstrated;
2. the lowest Gross Weight for which compliance with the relevant handling requirements can be demonstrated;
3. a weight selected by the Applicant.

c. Capacities.
SPACE LIMITED PAYLOAD (SLP): the maximum payload available for passengers, passengers' baggage and/or cargo when restricted by seating limitations and/or volumetric limitations in the cabin, cargo and/or baggage compartments. It must not exceed the Maximum Payload.
Alternative terms: Volumetric Payload Capacity, Maximum Volumetric Capacity Payload.

MAXIMUM (STRUCTURAL) PAYLOAD: the maximum weight of passengers and their baggage, cargo and/or mail that can be loaded in the Aircraft Prepared for Service without exceeding the MFZW.

MAXIMUM SEATING CAPACITY: the maximum number of passengers anticipated for certifi-

cation. This limit may, for example, correspond to the number of available emergency exits.
MAXIMUM CARGO VOLUME: the maximum space available in the cargo compartment(s).
USABLE FUEL CAPACITY: the maximum volume of fuel that can be carried for a particular operation, less drainable and trapped unusable fuel remaining after a fuel run-out test has been effected.
Alternative term: Standard Fuel Capacity. The OEW plus the (weight corresponding to the) Usable Fuel Capacity is the Maximum Operating Weight. The author recommends the term: Maximum Zero Payload Weight.

d. Weight restrictions based on available performance.
For each flight the highest operational weights must be determined for which all performance requirements can be met, taking off from the field of departure and landing at the scheduled field of destination or alternate. Appropriate atmospheric conditions (pressure altitude, temperature, wind speed and direction) and field characteristics (runway length, inclination, clearways, stopways) must be accounted for. The wing flap angle(s) may be chosen such that the most favorable condition is obtained.
The weight permitted for takeoff and landing is the lowest of the following:
1. the weight restricted by the available takeoff and landing field length;
2. the weight restricted by available climb performance as laid down in the W(eight) A(ltitude) T(emperature) diagrams[*];
3. the weight limited by obstacle clearance requirements;
4. the weight limited by airport runway or platform loading restrictions;
5. the weight limited by local noise regulations (if any).

8.2.3. Operational weights and the payload-

[*]cf. Section 11.6.2.

range diagram

For each regular flight operation, the operator must guarantee that all limitations applying to the weights and capacities mentioned previously have been respected. The interaction of the various weight criteria results in limitations to the operational weights determined in accordance with the Operational Rules (Operational Weights, Regular Weights). The actual Take-off Weight, Landing Weight and Payload for a particular flight must never exceed the limiting weights defined below.
OPERATIONAL LANDING WEIGHT (OLW) is the maximum weight authorized for landing and is the lowest of:
1. the Maximum Landing Weight;
2. the permissible Landing Weight based on available performance;
3. the Maximum Zero Fuel Weight plus the fuel load on landing.
Alternative (British) term: Maximum Regular Landing Weight (MRLW).
OPERATIONAL TAKEOFF WEIGHT (OTOW) is the maximum weight authorized for takeoff and is the lowest of:
1. the Maximum Take-Off Weight (MTOW);
2. the permissible Takeoff Weight based on available performance;
3. the Operational Landing Weight plus trip fuel;
4. the MZFW plus the fuel on takeoff;
5. the Takeoff Weight restricted by the Maximum Operating Weight (Useful Fuel Capacity).
Alternative (British) term: Maximum Regular Take-Off Weight (MRTOW).
WEIGHT-LIMITED PAYLOAD is the weight remaining after the Operating Weight is deducted from the OTOW.
The payload that can be carried is shown in Fig. 8-3 in relation to the flight range and can be computed from the inter-relationships between the previous operational weights and other limitations.
For short stages (line AB) the Payload must not exceed the Space Limited Payload or Maximum Structural Payload or is limited by the allowable floor loading. In the

example given the Maximum Structural Payload is assumed to be the limiting criterion.

For flight ranges below R_B, additional fuel can be carried up to the OTOW determined by the MLW, as indicated. This is possible only if the MZFW plus fuel reserve is less than the MLW, as is usually the case. For intermediate and long stages the payload is weight-limited and thus equal to the difference between the OTOW and the Operating Weight. Between points B and C, the MTOW is assumed to limit the takeoff weight in the case considered. Point B corresponds to the maximum flight range (R_B) with maximum payload appropriate to the relevant cruising conditions and reserve fuel policy. Increasing the range beyond R_B is possible only by an exchange of payload against fuel.

At point C (range R_C) the usable fuel load is limited by the fuel tank capacity and the Operating Weight reaches its limit. Further increase of the range is possible only by reducing the takeoff weight - and thus the hourly fuel consumption - resulting in a progressive payload reduction with increasing range. At point D there is no payload left and R_D is the maximum flight range (without payload), usually specified for a long-range cruise technique.

For normal commercial use the region CD is of minor importance and R_C is frequently referred to as the maximum range. If sufficient space is available, both R_C and R_D may be increased by additional internal fuel tank capacity for relatively low structural weight and drag penalties. A range increment may also be achieved with external fuel tanks.

8.2.4. The choice of weight limits

A compromise must be made between various conflicting requirements in order to design an airplane with adequate operational flexibility, satisfying the requirements of several operators on a variety of routes, under different atmospheric and field conditions. Discussions with operators may lead to incompatible requirements, in which case several versions of basically the same airplane may be considered, with differences in MZFW and Useful Fuel Capacity, for example. Fig. 8-4 illustrates the flexibility of a particular airplane design. A discussion on this subject can also be found in Ref. 8-11.

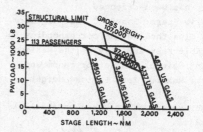

Fig. 8-4. Payload-range diagrams for several versions of the same airplane type

The Maximum Take-Off Weight is generally determined by structural design, and occasionally by handling criteria. In the pre-design stage the MTOW is calculated by adding together the OEW, the Estimated Normal Payload and the Total Fuel Load required at the design range*. Empty weight predictions (Section 8.4.) and performance calculations (Section 11.5.) are required to accomplish this. The airplane layout, geometry and engine thrust must be balanced in such a way that for the airport elevations and atmospheric temperatures most frequently encountered the takeoff weight is not unduly limited by restrictions based on available performance. An increment in MTOW might be envisaged for operators whose low-speed performance requirements are not so stringent, in order to provide greater ranges and/or payload capability. In such a case increments in the MZFW, the MLW and the DEW will be desirable to cater for the increased loads on certain critical regions of the structure.

*The word "design" here refers to the values laid down in the design specification

274

Certain light airplane categories are defined by limitations on the MTOW, e.g. airplanes with MTOW up to 6,000 lb or 12,500 lb (2,722 kg or 5,670 kg). The associated differences in airworthiness requirements have far-reaching effects on the design and operation of such airplanes (cf. Section 5.4.).

The Maximum Zero Fuel Weight must be sufficiently high to ensure that this structural limit will not create a payload restriction for the usual payload specific densities and loading conditions. An increment in MZFW increases the loading flexibility at the cost of a structure weight penalty, mainly in the wing and wing-support structure. Usually this also entails an increase in the MLW. For preliminary design the Space Limited Payload can be estimated with the data in Sections 3.2.5. and 8.4.4. The MZFW is thus assumed equal to the OEW plus the payload corresponding to a high-density version of the passenger seating arrangement.

The Maximum Landing Weight should always be higher than the MZFW plus the regular fuel reserve, otherwise the payload will frequently be limited by the MLW. For certain categories of light airplanes the MLW and the MTOW are equal (BCAR group K); for others the MLW may be slightly lower, namely down to 95% MTOW (cf. FAR 23.473). The difference between the OTOW and the actual takeoff weight required for a specified flight may be used for additional fuel (Fig. 8-3). This is of particular concern to operators flying short route sectors without refueling. For short-haul aircraft this type of flight execution is common practice and a fairly high MLW is desirable, namely of the order of 90 to 95% of the MTOW. Sufficient fuel tank capacity should accordingly be provided. A fuel jettisoning system is required on most multi-engine aircraft unless the airplane meets certain landing and approach climb requirements at a weight equal to the MTOW minus the fuel consumed during a

specified 15-minute flight (FAR 25.1001). Choosing the MLW above this characteristic weight makes a fuel dump system superfluous. In earlier regulations this characteristic (landing) weight was equal to the MTOW divided by 1.05.

The position may be summarized by saying that for commercial use the MLW is usually between the MZFW plus a normal fuel reserve and approximately 95% of the MTOW. The following statistical relationship can be used to make a reasonable choice of the MLW:

$$\frac{MLW - MZFW}{MTOW - MZFW} = .2 + .9 \exp (-R_B/R_{ref}) \qquad (8-2)$$

$$\text{for } R_B > 172 \text{ nm (320 km)}$$

where R_B is the design range for max. payload (cf. Fig. 8-3) and R_{ref} is 1000 nm if R_B is in nm, or 1854 km if R_B is in km.

8.3 METHODOLOGY OF EMPTY WEIGHT PREDICTION

Most airplane manufacturers develop their own methods of empty weight prediction, usually basing them on extensive experience with a limited and well-defined category of airplanes. The more generalized and simpler methods presented in this textbook have a predictive accuracy of the order of 5 to 10% standard error for the major groups. In the terminology of Section 5.2.1., they may be classified as methods for configuration selection as far as the structure weight is concerned. If the method is applied judiciously, the result will be quite realistic.

Part of the discrepancy between the predicted weight of a particular aircraft part and the actual weight can be ascribed to different definitions of such weight items. Typical sources of accounting differences can be observed in the wing-fuselage interconnection and carry-through structure and fillets, in retractable undercarriage structures such as wheel doors, and in several items of furnishing and equipment. If a consistent weight prediction method is used, the final addition of all items to the empty weight will be not greatly in error. Although the accuracy of prediction does not necessarily improve when more details or parameters are introduced, the designer generally aims at a fairly detailed knowledge of weight penalties and contributions for the purpose of center of gravity determination, optimization and weight control. It is shown in several reports of the Society of Allied Weight Engineers (SAWE) that there is at this stage of the design an optimum number of parameters which yields a minimum prediction error (cf. Ref. 8-38).

The subdivision of the empty weight as presented in Table 8-3 is based on the AN* 9103-D Group Weight Statement, slightly adapted for civil use. For center of gravity determination (Section 8.5.) the moment arm and static moment of each item can be substituted. More detailed weight breakdowns are frequently required; for several weight groups these will be given in Section 8.4.

Although a wealth of detailed published information on weight engineering can be found in various Technical Papers of the SAWE, consistent and up-to-date methods for calculating the empty weight of the various categories of modern civil airplanes are very scanty indeed. Several examples in the list of references (Ref.8-47/ 8-63) may be used, but in most cases the designer will have to develop ad hoc methods. The information contained in Table 8-5 may be helpful for checking the accuracy of these methods by comparing the results with data on several existing airplanes.

In preliminary aircraft design weight prediction is always a mixture of rational analysis and statistical methods, the reason being that many design details are still not known at that moment. Statistical weight equations for many components are usually written in the exponential form,

*American Normalisation

as follows:

$$\text{Weight} = \text{constant} \cdot \phi_1^{n_1} \, \phi_2^{n_2} \, \phi_3^{n_3} \, \cdots \, \phi_m^{n_m}$$

(8-3)

The constant of proportionality and the exponents $n_1, n_2, \ldots n_m$ of the design parameters $\phi_1, \phi_2, \ldots \phi_m$ are determined by standard regression analysis techniques, using weight and geometric data of actual components, and subject to the condition of minimum standard deviation. The problem may arise that the magnitude and/or algebraic sign of the exponents is unacceptable for use in design analysis, the results being clearly irrational. Staton explains in Ref. 8-41 how constrained regression analysis can be used to solve this problem. A much better approach, though not always feasible owing to lack of verificatory data, is to break the item considered down into the sum of several other contributions. For each contribution relevant parameters must be chosen on a rational basis. If a weight contribution Y is expressed as a linear function of a parameter X, assumed to be known without error, linear regression analysis can then be used to obtain the best curve fit.

For example, on the assumption of a constant average stress level, the material required to resist the wing bending moment is proportional to the root bending moment times the structural span divided by the root chord thickness. For a large proportion of the wing structural weight the following parameter is therefore relevant:

$$X = \frac{n_{ult} \, W_G \, b^2}{t_r \cos^2 \Lambda_{\frac{1}{2}}}$$

(8-4)

The parameter X is calculated for N sample airplanes (values x_i), for which the wing weight has known values y_i. The regression line of Y upon X is fitted by the method of least squares, according to the equation:

$$.Y = a + bX$$

(8-5)

where

$$a= \frac{\sum x_i \sum x_i y_i - \sum y_i \sum x_i^2}{\left(\sum x_i\right)^2 - N \sum x_i^2} \qquad (8\text{-}6)$$

and

$$b= \frac{\sum x_i \sum y_i - N \sum x_i y_i}{\left(\sum x_i\right)^2 - N \sum x_i^2} \qquad (8\text{-}7)$$

For limited ranges of X and Y a linear function may be satisfactory, but if a considerable variation in the actual size of the item exists, a better result is usually obtained with:

$$Y = kx^n \qquad (8\text{-}8)$$

On a log-log scale this relation is linear:

$$\log Y = \log k + n \log X \qquad (8\text{-}9)$$

and again linear regression analysis can be used.
The standard error of a prediction method is:

$$S= \sqrt{\frac{1}{N-1}\left[\sum m_i^2 - \frac{\left(\sum m_i\right)^2}{N}\right]} \qquad (8\text{-}10)$$

where m_i is the ratio of actual to estimated weight of the sample.
References 8-41 and 8-44 give more information on the use of statistics and various types of regression analysis. A certain amount of care should always be taken when using statistical methods. A check must be made to see if the airplane being analyzed falls within the range of data points that were used to develop the method. The choice of parameters to be used is always somewhat arbitrary and due attention must be paid to data points that are far from the regression line. They may indicate that alternative correlations should be investigated. Finally, all parameters used in weight prediction must be well defined and not give rise to misinterpretation or vagueness.
Finally, it should be realized that many weight prediction methods apply to a limited category of airplanes. Occasionally they may be adapted to other categories simply by modifying the factor of proportionality, provided that the basic expression has a rational background and derivation.

8.4. WEIGHT PREDICTION DATA AND METHODS

8.4.1. Airframe structure

a. Structure weight prediction based on the aircraft specific density. An intriguing approach to structure weight estimation, applicable to conventional configurations, is made by Caddell in Ref. 8-39, who uses the aircraft density, i.e. the design gross weight divided by the total airplane volume. If his line of thought is adopted, the structural weight fraction of transport-type turbine-powered airplanes can be expressed in terms of the ultimate load factor, the fuselage dimensions and the MTOW:

$$\frac{W_s}{W_{to}} = k_s \sqrt{n_{ult}} \left(\frac{b_f h_f l_f}{W_{to}}\right)^{.24} \qquad (8\text{-}11)$$

where
k_s = .230 for b_f, h_f and l_f in ft and W_s and W_{to} in lb,
k_s = .447 for b_f, h_f and l_f in m and W_s and W_{to} in kg, and
n_{ult} corresponds to the MTOW.
Although this simple expression yields a reasonably accurate prediction, it is useless for design optimization, as the effects of the airplane layout are not accounted for. The only alternative is a detailed assessment of the contributions of all structural components or groups. The subdivision in Table 8-4 can be used to collect structural weight data. A compilation of structure weight data for existing aircraft is presented in Table 8-5. According to (8-11) the ultimate load factor affects structural weight to a considerable extent. The rules for establishing the ultimate load factor (1.5 times the limit load factor) are laid down in the various airworthiness regulations. It should be

WING GROUP						
CENTER SECTION-BASIC STRUCTURE						
INTERMEDIATE PANEL-BASIC STRUCTURE						
OUTER PANEL-BASIC STRUCTURE (INCL. TIPS ....)						
SECONDARY STRUCTURE (INCL. WING FOLD MECH. ....)						
AILERONS (INCL. BALANCE WEIGHT ....)						
FLAPS - TRAILING EDGE						
- LEADING EDGE						
SLATS						
SPOILERS, SPEED BRAKES, LIFT DUMPERS						
FENCES AND VORTEX GENERATORS						
STRUTS						
TAIL GROUP						
STABILIZER-BASIC STRUCTURE						
FINS-BASIC STRUCTURE (INCL. DORSAL ....)						
SECONDARY STRUCTURE (STAB. AND FINS)						
ELEVATOR (INCL. BALANCE WEIGHT ....)						
RUDDERS (INCL. BALANCE WEIGHT ....)						
BODY GROUP						
FUSELAGE OR HULL-BASIC STRUCTURE						
BOOMS-BASIC STRUCTURE						
SECONDARY STRUCTURE - FUSELAGE OR HULL						
- BOOMS						
- SPEED BRAKES						
- DOORS, PANELS AND MISC.						

ALIGHTING GEAR GROUP - LAND (TYPE ....)

LOCATION	WHEELS, BRAKES, TYRES, TUBES, AIR	STRUCTURE	CONTROLS	TOTAL	
MAIN					
NOSE					
TAIL (BUMPER)					

ALIGHTING GEAR GROUP - WATER

LOCATION	FLOATS	STRUTS	CONTROLS		

SURFACE CONTROLS GROUP		
COCKPIT CONTROLS		
AUTOMATIC PILOT		
SYSTEM CONTROLS (INCL. POWER AND FEEL CONTR.)		
ENGINE SECTION OR NACELLE GROUP		
INBOARD		
CENTER		
OUTBOARD		
DOORS, PANELS AND MISC.		
TOTAL, AIRFRAME STRUCTURE		

Table 8-4. Airframe Structure Group weight breakdown according to AN-9103-D (modified)

AIRPLANE CATEGORY AND TYPE		MTOW	WING GROUP		TAIL GROUP		FUSELAGE GROUP		LANDING GEAR		SURFACE CONTROLS		NACELLE GROUP	
		10^3 lb	10^3 lb	%	10^3 lb	%	10^3 lb	%	10^3 lb	%	10^3 lb	%	10^3 lb	%
LIGHT SINGLES RECIPROCATING	Cessna - 150A	1.50	0.213	14.2	0.041	2.73	0.166	11.1	0.106	7.07	0.031	2.07	0.024	1.60
	- 172B	2.20	0.236	10.7	0.061	2.77	0.253	11.5	0.122	5.55	0.031	1.41	0.031	1.41
	- 180D	2.65	0.254	9.58	0.059	2.23	0.270	10.2	0.119	4.49	0.036	1.36	0.037	1.40
	- 182D	2.65	0.254	9.58	0.061	2.30	0.273	10.3	0.136	5.13	0.036	1.36	0.036	1.36
	- 185	3.20	0.266	8.31	0.071	2.22	0.290	9.06	0.132	4.13	0.036	1.13	0.041	1.28
	- 210	2.90	0.261	9.0	0.071	2.45	0.316	10.90	0.207	7.14	0.044	1.52	0.031	1.07
	Beechcraft J-35	2.90	0.379	13.1	0.058	2.00	0.200	6.90	0.205	7.07	0.056	1.93	0.062	2.14
	Saab Safir	2.66	0.276	10.4	0.060	2.26	0.386	14.5	0.119	4.47	**	-	**	-
LIGHT TWINS RECIPROCATING	Cessna C-310	4.83	0.454	9.40	0.118	2.44	0.319	6.60	0.263	5.45	0.066	1.37	0.129	2.67
	Beechcraft G-50	7.15	0.656	9.17	0.156	2.18	0.495	6.92	0.447	6.25	0.120	1.68	0.261	3.65
	-65	7.37	0.670	9.09	0.153	2.08	0.601	8.15	0.444	6.02	0.132	1.79	0.285	3.87
	-95	4.00	0.458	11.5	0.079	1.98	0.276	6.90	0.218	5.45	0.073*	1.83	0.180	4.50
	D-18S	8.75	0.858	9.81	0.177	2.02	0.733	8.38	0.560	6.40	0.115*	1.31	0.311	3.55
	E-18S	9.70	0.874	9.01	0.180	1.86	0.768	7.92	0.585	6.03	0.115*	1.19	0.331	3.41
	De Havilland Dove	8.80	0.930	10.6	0.196	2.23	0.745	8.47	0.391	4.44	**	-	0.220*	2.50
JET TRAINERS	Cessna T-37	6.44	0.531	8.24	0.128	1.99	0.839	13.0	0.330	5.12	0.154	2.39	-	-
	Fouga Magister	6.28	1.089	17.3	0.165	2.63	0.743	11.8	0.459	7.31	0.260	4.14	-	-
	Canadair CL-41	6.50	0.892	13.7	0.201	3.09	0.955	14.7	0.318	4.89	0.172	2.65	0.040	0.62
JET EXECUTIVES	H. Siddeley - 125	21.200	1.968	9.28	0.608	2.87	1.628	7.68	0.659	3.11	0.217	1.02	**	-
	Jet Commander 1121	16.000	1.322	8.26	0.425	2.66	1.622	10.1	0.443	2.76	0.223	1.39	0.35	2.19
	N.Am.Sabreliner	16.700	1.753	10.5	0.297	1.78	2.014	12.1	0.728	4.36	0.344	2.06	0.315	1.89
	Lockheed Jetstar	30.680	2.827	9.21	0.879	2.87	3.491	11.4	1.061	3.46	0.768	2.50	0.792	2.58

* estimated   ** included in other items

Table 8-5. Weight breakdown of the structure group weight

PROPELLER TRANSPORTS

AIRPLANE CATEGORY AND TYPE		MTOW	WING GROUP		TAIL GROUP		FUSELAGE GROUP		LANDING GEAR		SURFACE CONTROLS		NACELLE GROUP	
		10^3 lb	10^3 lb	%	10^3 lb	%	10^3 lb	%	10^3 lb	%	10^3 lb	%	10^3 lb	%
RECIPROCATING — 2 ENGINES	De Havilland DHC-4	24.000	2.925	12.2	0.790	3.29	2.849	11.9	1.23	5.13	0.326	1.36	0.781	3.25
	Saab Scandia	35.273	4.195	11.9	0.584	1.66	2.773	7.86	1.841	5.22	0.369	1.05	1.479	4.19
	H. Page Herald	37.500	4.365	11.6	0.987	2.63	2.986	7.96	1.625	4.33	0.364	0.97	0.830	2.21
	S.A. Twin Pioneer	14.600	2.121	14.5	0.576	3.95	1.381	9.46	0.703	4.82	0.300	2.05	0.230	1.58
	Canadair CL-21	32.500	3.99	12.3	1.055	3.25	3.260	10.0	1.609	4.95	0.371	1.14	1.29	3.97
RECIPROCATING — 4 ENGINES	Douglas DC-6B	81.500	7.506	9.21	1.406	1.73	5.471	6.71	4.165	5.11	1.052	1.29	2.871	3.52
	DC-7C	143.000	11.100	7.76	1.900	1.33	8.450	5.91	5.130	3.59	1.215	0.85	4.130	2.89
	Lockheed L-749	102.072	11.102	10.9	2.059	2.02	7.407	7.26	4.782	4.68	1.488	1.46	3.869	3.79
	L-1049	137.500	11.542	8.39	2.604	1.89	12.839	9.34	5.422	3.94	1.685	1.23	4.420	3.21
TURBOPROPELLER — 2 ENGINES	Nord 262	23.050	2.698	11.7	0.805	3.49	3.675	15.9	1.085	4.71	0.408	1.77	0.236	1.02
	Fokker F-27/100	39.000	4.408	11.3	0.977	2.51	4.122	10.6	1.940	4.97	0.613	1.57	0.628	1.61
	F-27/200	43.500	4.505	10.4	1.501	2.42	4.303	9.89	1.825	4.20	0.620	1.43	0.667	1.53
	F-27/500	45.000	4.510	10.0	1.060	2.35	5.142	11.4	1.865	4.14	0.626	1.39	0.668	1.48
	Grumman Gulfstream	33.600	3.735	11.2	0.874	2.60	3.718	11.1	1.207	3.59	0.461	1.37	1.136	3.38
	Short Skyvan	12.500	1.220	9.76	0.374	2.99	2.154	17.2	0.466	3.73	0.265	2.12	0.254	2.03
TURBOPROPELLER — 4 ENGINES	Breguet 941	58.421	4.096	7.01	1.387	2.37	6.481*	11.1	2.626	4.94	1.056	1.81	1.200	1.46
	H.S. Argosy	82.000	10.800	13.2	1.300	1.59	11.100	13.5	3.180	3.88	**	-	1.810	2.62
	Vickers Viscount 810	69.000	6.25	9.06	1.245	1.80	6.900	10.0	2.469	3.58	0.824	1.19	4.930	3.18
	Bristol Brit. 300	155.000	13.433	8.67	3.202	2.07	11.100	7.16	5.785	3.73	1.221	0.79	7.350	3.98
	Brit. 320	184.523	14.199	7.69	3.221	1.75	11.750	6.38	6.500	3.52	2.048	1.11	6.834	3.33
	Canadair CL-44C	205.000	15.710	7.66	3.749	1.83	20.524	10.0	7.083	3.46	2.146	1.05	6.043	2.95
	CL-44D	205.000	15.588	7.60	3.540	1.73	16.047	7.83	7.300	3.56	1.830	0.89	4.417	4.14
	Lockheed Electra	106.700	7.670	7.19	1.924	1.80	9.954	9.33	3.817	3.58	***	-	2.675	1.77
	C-130E	151.522	11.697	7.72	3.425	2.26	14.340	9.46	5.341	3.53	1.702	1.12	2.675	1.77
	C-133A	275.000	27.403	9.96	6.011	2.19	30.940	11.3	10.635	3.87	1.804	0.66	3.512	1.28

* tail booms (2,360 lb) included   ** included in other items   *** no data available

Table 8-5. (Continued)

AIRPLANE CATEGORY AND TYPE	MTOW	WING GROUP		TAIL GROUP		FUSELAGE GROUP		LANDING GEAR		SURFACE CONTROLS		NACELLE GROUP	
	10^3 lb	10^3 lb	%	10^3 lb	%	10^3 lb	%	10^3 lb	%	10^3 lb	%	10^3 lb	%
2 ENGINES													
VFW-Fokker 614	40.981	5.767	14.1	1.121	2.74	5.233	12.8	1.620	3.45	0.745	1.82	0.971	2.37
Fokker-VFW F-28/1000	65.000	7.330	11.3	1.632	2.46	7.043	10.8	2.759	4.24	1.387	2.13	0.834	1.28
F-28/2000	65.000	7.347	11.3	1.632	2.46	7.649	11.8	2.759	4.24	1.400	2.15	0.834	1.28
F-28/5000	70.800	8.223	11.6	1.632	2.31	7.043	9.95	2.759	3.90	1.665	2.35	0.849	1.20
F-28/6000	70.800	8.244	11.6	1.632	2.31	7.649	10.8	2.789	3.94	1.674	2.36	0.849	1.20
BAC 1-11/300	87.000	9.643	11.1	2.369	2.72	9.713	11.2	2.865	3.29	1.481	1.76	**	-
1-11/400	87.000	9.670	11.1	2.419	2.78	9.743	11.3	2.899	3.33	1.207	1.39	**	-
Mc D. Douglas DC-9/10	91.500	9.470	10.3	2.630	2.87	11.206	12.2	3.660	4.00	1.264	1.38	1.417	1.55
Boeing 737-100M	97.800	9.968	10.2	2.700	2.76	12.380	12.7	3.687	3.77	1.589	1.62	***	-
737-200	100.000	10.613	10.6	2.718	2.72	12.108	12.1	4.354	4.35	2.348	2.35	1.392	1.39
Aerospat. Caravelle VIR	110.230	14.735	13.4	1.957	1.77	11.570	10.5	5.110	4.63	2.063	1.87	1.581	1.43
Airbus A300B/2	304.000	44.131	14.5	5.941	1.95	35.820	11.8	13.611	4.47	5.808	1.94	7.039	2.32
3 ENGINES													
H. Siddeley 121-1C	115.000	12.600	11.0	3.225	2.80	12.469	10.8	4.413	3.84	1.792	1.56	**	-
121-1E	134.000	13.462	10.0	3.341	2.49	13.328	9.95	5.073	3.79	1.689	1.26	**	-
Boeing 727-100	161.000	17.764	11.0	4.133	2.57	17.681	10.9	7.211	4.48	2.996	1.86	3.864*	2.40
727-100C	160.000	17.492	10.9	4.142	2.59	20.044	12.5	6.860	4.29	2.957	1.85	3.839	2.40
4 ENGINES													
Boeing KC-135	297.000	25.251	8.50	5.074	1.71	18.867	6.35	10.180	3.43	2.044	0.69	2.575	0.87
707-121	246.000	24.024	9.76	5.151	2.09	20.061	8.15	9.763	3.97	2.044	0.83	4.639	1.89
707-320	311.000	29.762	9.57	5.511	1.77	21.650	6.96	12.700	4.08	2.400*	0.77	4.497	1.45
707-320C	330.000	32.255	9.77	6.165	1.87	26.937	8.16	12.737	3.86	3.052	0.92	4.183	1.27
707-321	301.000	28.647	9.52	6.004	1.99	22.129	7.35	11.122	3.70	2.408	0.80	5.119	1.70
720-022	203.000	22.850	11.3	5.230	2.58	19.035	9.38	8.110	4.00	2.430	1.21	4.510	2.22
747-100	710.000	86.402	12.2	11.850	1.67	71.845	10.1	31.427	4.43	6.982	0.98	10.031	1.41
747-200B	775.000	92.542	11.9	11.842	1.53	72.053	9.30	32.693	4.22	7.073	0.91	10.136	1.31
Mc D. Douglas DC-8-10	273.000	26.235	9.61	4.740	1.74	21.495	7.87	10.185	3.73	2.000*	0.73	3.505	1.28
DC-8-55	328.000	34.759	10.6	4.889	1.49	22.248	6.78	11.255	3.43	2.253	0.69	4.685	1.43
BAC VC-10-1101	312.000*	34.672	11.1	6.958	2.23	25.113	8.05	10.489	3.36	***	-	**	-
G. Dynamics 880	184.500	17.669	9.58	4.247	2.30	13.699	7.42	6.203	3.36	***	-	3.685	2.00
990	253.000	26.871	10.6	5.326	2.11	16.673	6.59	8.718	3.44	***	-	6.772	2.68

* estimated   ** included in other items   *** no data available

Table 8-5. (Continued)

taken as the larger of the maximum positive gust or the maneuver load factor for the applicable weight at the most critical flight altitude (approximately 20,000 ft for pressurized transports). For further details see Appendix C.

b. Wing group.

A reasonably accurate wing weight estimate can be made in preliminary design as the loads on the wing are fairly well known at the design stage. Usually the bending moment in flight is assumed to be decisive for most of the primary structure. For a certain category of high-speed aircraft, however, torsional stiffness requirements may become dominant and the extra structure weight required to safeguard against flutter may amount to as much as 20% of the wing weight. The location of the inertia axis of the wing plus wing-mounted engines is of importance. A fairly large portion is also made up of secondary structure and non-optimum penalties, such as joints, non-tapered skin,

undercarriage attachments, etc.

The derivation of a typical wing weight prediction method is explained in Ref. 8-101, the results of which are summarized in Appendix C. If sufficient data are not available to apply this method, the following simplified approximation can be used for civil airplanes with Al-alloy cantilever wings. The following basic expression is valid for the case of a wing-mounted retractable undercarriage, but not for wing-mounted engines:

$$\frac{W_w}{W_G} = k_w b_s^{.75} \left[1 + \sqrt{\frac{b_{ref}}{b_s}} \right] n_{ult}^{.55} \left(\frac{b_s/t_r}{W_G/S} \right)^{.30} \quad (8-12)$$

where b_{ref} = 6.25 ft or 1.905 m for b_s in ft or m, respectively, while b_s = $b/\cos\Lambda_{\frac{1}{2}}$, the structural wing span. The factor of proportionality is as follows:

Light aircraft, $W_{to} \leqslant 12,500$ lb (5670 kg):
$k_w = 1.25 \times 10^{-3}$; $W_G \equiv$ MTOW in lb, b_s in ft, S in ft^2, W_w in lb.

$k_w = 4.90 \times 10^{-3}$; $W_G \equiv$ MTOW in kg, b_s in m, S in m^2, W_w in kg.

Transport category aircraft $W_{to} > 12,500$ lb (5670 kg):

$k_w = 1.70 \times 10^{-3}$; $W_G \equiv$ MZFW in lb, b_s in ft, S in ft^2, W_w in lb.

$k_w = 6.67 \times 10^{-3}$; $W_G \equiv$ MZFW in kg, b_s in m, S in m^2, W_w in kg.

The weight given by (8-12) includes high-lift devices and ailerons. For spoilers and speed brakes, if incorporated, 2% should be added. Reduce W_w by 5% or 10% for 2 or 4 wing-mounted engines, respectively and by 5% if the main undercarriage is not mounted to the wing. For braced wings a reduction of approximately 30% relative to (8-12) can be assumed. This figure includes the strut, contributing about 10% of the total wing group weight.

Wing optimization studies must be sensitive to variations in the external geometry, configuration and operational characteristics. It is generally recognized that for modern wing designs the weight of high-lift devices should be determined separately. The method in Appendix C meets these requirements and predicts the wing weights with a standard prediction error of 9.64%.

On inspection of (8-12), the observation can be made that the structural weight fraction, for a given cantilever ratio b_s/t_r and wing loading W_G/S, increases with the wing span. This unfavorable scale effect, associated with the square-cube law (cf. Section 7.2.2.), can be counteracted by increasing the wing loading. This is one of the reasons why large aircraft usually have high wing loadings. Decreasing the cantilever ratio is unfavorable as it results in a drag increment; its value is usually between 35 and 45 (see Fig. 7-8.).

c. Tail group.

This weight is only a small part - about 2 to 3% - of the MTOW but on account of its remote location it has an appreciable effect on the position of the airplane's center of gravity. Accurate weight prediction is difficult due to the wide variety of tailplane configurations and the limited knowledge of strength, stiffness and other conditions which will govern the design.

For relatively low-speed, light aircraft (V_D up to 250 kts EAS), the maneuvering loads are most important and the specific tailplane weight is affected by the load factor as follows:

$$W_{tail} = k_{wt} \left| n_{ult} \, S_{tail}^{2} \right|^{.75} \qquad (8-13)$$

where $k_{wt} = .04$; W_{tail} in lb and S_{tail} in ft^2
$k_{wt} = .64$; W_{tail} in kg and S_{tail} in m^2

It is interesting to note that for this category the specific tailplane weight obeys the square-cube law, the weight being proportional to the cube while the area is proportional to the square of the linear dimension. If the tailplane area is not (yet) known, the total tailplane weight may be assumed between 3½ and 4% of the empty weight. For transport category aircraft and executive jets the Design Dive speed appears to have a dominant effect (Fig. 8-5):

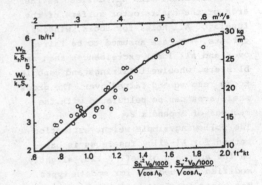

Fig. 8-5. Normalized specific horizontal tailplane weight

$$\frac{W_h}{S_h} = k_h \cdot f \left(\frac{S_h^{.2} \, V_D}{\sqrt{\cos \Lambda_h}} \right) \qquad (8-14)$$

$$\frac{W_v}{S_v} = k_v \cdot f \left(\frac{S_v^{.2} \, V_D}{\sqrt{\cos \Lambda_v}} \right) \qquad (8-15)$$

where V_D is expressed in terms of EAS and k_h and k_v are correction factors for the tailplane configuration:

$k_h = 1.0$ - fixed stabilizer, 1.1 for variable-incidence tails; for a bullet of appreciable size 8% should be added.

$k_v = 1.0$ - fuselage-mounted horizontal tailplanes

$k_v = 1 + .15 \dfrac{S_h h_h}{S_v b_v}$ - fin-mounted stabilizers (e.g. T-tails) b_v defined in Fig. 9-20.

Fig. 8-5 demonstrates that the scale effect on specific tailplane weight $(S^{.2})$ applies to medium-sized airplanes, but disappears for very large aircraft.

d. Body group.

The fuselage makes a large contribution to the structural weight, but it is much more difficult to predict by a generalized method than the wing weight. The reason is the large number of local weight penalties in the form of floors, cutouts, attachment and support structure, bulkheads, doors, windows and other special structural features.

Fuselage weight is affected primarily by the gross shell area S_G, defined as the area of the entire outer surface of the fuselage. All holes for doors, windows, cutouts, etc. are assumed to be faired over and all local excrescences such as blisters, wheelwell fairings and canopies to be removed and faired over. The gross shell area can be calculated with the methods of Appendix B.

The following simple weight estimation method for Al-alloy fuselages is based on the approach of Ref. 8-113, slightly modified and updated for modern types. The basic fuselage weight is:

$$W_f = k_{wf} \sqrt{V_D \frac{l_t}{b_f + h_f}} \, S_G^{1.2} \qquad (8\text{-}16)$$

The Design Dive speed V_D is expressed in terms of EAS. For definitions of l_t, b_f and h_f see Appendix D (Fig. D-2). The constant of proportionality is:

$k_{wf} = .021$ - W_f in lb, V_D in kts and S_G in ft^2

$k_{wf} = .23$ - W_f in kg, V_D in m/s and S_G in m^2

To the basic weight given by (8-16), 8% should be added for pressurized cabins, 4% for rear fuselage-mounted engines, 7% if the main landing gear is attached to the fuselage, and an extra 10% for freighter

aircraft. If there is no attachment structure for the landing gear nor a wheelbay, 4% may be subtracted from the basic weight. Most of the more detailed prediction methods are based on the approach in Ref. 8-115 applicable to semi-monocoque structures. The calculation of the shell weight according to this method, supplemented with some recent data to estimate various weight penalties, is given in Appendix D.

For tail booms (8-16) can be used for each boom separately. In this case l_t is defined as the distance between the quarter-chord points of the local wing chord and the horizontal tailplane. Add 7% for a main landing gear wheelbay and undercarriage attachment.

e. Alighting gear group*.

The undercarriage has a well-defined set of loading conditions and weight prediction can therefore be dealt with on a analytical basis. To this end the weight of each gear must be subdivided into:
- wheels, brakes, tires, tubes and air
- main structure, i.e. legs and struts
- items such as the retraction mechanism, bogies, dampers, controls, etc.

The first part of the weight prediction process is to decide upon tire and wheel size, inflation pressure, location of the gears, length of the legs, etc. This subject will be treated in Chapter 10, an example of a weight prediction method is given in Ref. 8-125.

The weight of conventional undercarriages may be found by summation of the main gear and the nose gear, each predicted separately with the following expression:

$$W_{uc} = k_{uc} \left\{ A + B.W_{to}^{3/4} + C.W_{to} + D.W_{to}^{3/2} \right\} \qquad (8\text{-}17)$$

where $k_{uc} = 1.0$ for low-wing airplanes and
$k_{uc} = 1.08$ for high-wing airplanes

Table 8-6 gives suggested values of the factors A, B, C and D, based on a statistical evaluation of data on undercar-

*Only conventional undercarriages will be dealt with

A/C CATEGORY	U.C. CONFIGURATION		A	B	C	D
JET PROPELLED TRAINERS AND EXECUTIVES	RETRACTABLE	MAIN	33 (15.0)	.04 (.033)	.021	-
		NOSE	12 (5.4)	.06 (.049)	-	-
ALL OTHER CIVIL TYPES	FIXED	MAIN	20 (9.1)	.10 (.082)	.019	-
		NOSE	25 (11.3)	-	.0024	-
		TAIL	9 (4.1)	-	.0024	-
	RETRACTABLE	MAIN	40 (18.1)	.16 (.131)	.019	$1.5\ (2.23).10^{-5}$
		NOSE	20 (9.1)	.10 (.082)	-	$2\ (2.97).10^{-6}$
		TAIL	5 (2.3)	-	.0031	-

COEFFICIENTS CORRESPOND TO WEIGHTS IN LB(KG)

Table 8-6. Coefficients for the calculation of the landing gear weight

riage weights of existing airplanes. Fig. 8-6 compares the result of (8-17) with data for existing airplanes. Up to 100,000 lb (45,000 kg) takeoff weight the weight fraction decreases with increasing airplane size. The main reasons are that for large airplanes a larger part of the gear structure can be highly stressed, while the use of higher inflation pressures on large aircraft saves some weight as well. For main landing gears the weight fraction does not appreciably decrease at takeoff weights above 100,000 lb (45,000 kg), but for nose gears there is still a reduction of the weight fraction up to very large airplane sizes like the B-747 and C-5A.

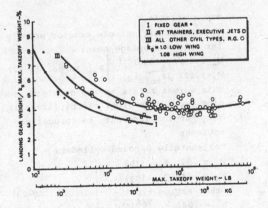

Fig. 8-6. Landing gear weight fraction

It can be argued that in many aircraft the critical load is formed by the landing impact load and that the MLW should therefore be used to predict the undercarriage weight. A reasonable approximation

for the weight of retractable undercarriages is 4.7% of the MLW.

f. Surface controls group.
The weight of surface controls is generally of the order of .8 to 2% of the takeoff weight. An approximation is:

$$W_{sc} = k_{sc}\ W_{to}^{2/3} \qquad (8-18)$$

The factor k_{sc} can be determined from known weights of airplanes in the same category with a similar flight control system. Alternatively, for W_{sc} and W_{to} in lb, we have:
k_{sc} = .23 - light airplanes without duplicated system controls,
k_{sc} = .44 - transport airplanes and trainers, manually controlled, and
k_{sc} = .64 - transport airplanes, with powered controls and trailing-edge high-lift devices only.
Multiply these values by .768 when W_{sc} and W_{to} are in kg. Add 20% for leading-edge flap or slat controls and 15% for lift dumper controls, if used.
If sufficient data are available, a more detailed analysis can be made. To this end the surface controls group weight is subdivided into:
1. cockpit controls:
.056 $W_{to}^{3/4}$ lb (.046 $W_{to}^{3/4}$ kg)
for $W_{to} \leq 25,000$ lb (11,340 kg)
110 lb (50 kg)   (8-19)
for $W_{to} > 25,000$ lb (11,340 kg)
2. automatic pilot:
17 $W_{to}^{1/5}$ lb   (9 $W_{to}^{1/5}$ kg)   (8-20)
for transport and executive aircraft
3. system controls:
.008 W_{to} for light airplanes with   (8-21)
single flight control systems
For transport aircraft a prediction of the system controls weight can be made with the aid of Table 8-7. In the absence of better data these formulas may be also used for trainers and executive airplanes.

g. Engine section or nacelle group.
The following statistical data may be used if no details of the engine installation

SYSTEM COMPONENT	METHOD		NOMENCLATURE
MANEUVERING CONTROL SYSTEM (elevator, rudder, ailerons, spoilers)	manually operated duplicated controls	$\frac{.2}{(.154)} \times W_{to}^{.67}$	W_{to} = Max. Takeoff Weight $\sim$ lb (kg)
	duplicated powered controls, single hydr. power system	$\frac{.42}{(.318)} \times W_{to}^{.65}$	
	duplicated powered controls, dual hydr. power system	$\frac{1.06}{(.773)} \times W_{to}^{.60}$	
TRAILING EDGE FLAP CONTROL SYSTEM	rotating flaps (cylinder actuation)	$\frac{1.38}{(5.569)} \times (S_f \sin \delta_f)^{.92}$	S_f = total projected flap area $\sim$ ft^2 (m^2)
	translating (Fowler) flaps (screwjack actuation)	$\frac{2.73}{(11.02)} \times (S_f \sin \delta_f)^{.92}$	δ_f = maximum flap deflection angle
LEADING EDGE FLAP OR SLAT CONTROL SYSTEM		$\frac{3.53}{(11.23)} \times S_s^{.82}$	S_s = total projected slat area $\sim$ ft^2 (m^2)
VARIABLE INCIDENCE STABILIZER CONTROLS	$k_{hc} (S_{he} V_{max}^{.5} \sin \delta_h)^{.88}$ single powered: k_{hc} = .31 (1.52) dual powered: k_{hc} = .44 (2.16) δ_h = total range of hor. tailplane incidence variation		S_{he} = exposed hor. tail area $\sim$ ft^2 (m^2) V_{max} = max. hor. flight speed $\sim$ kts (m/s) TAS
SPEED BRAKE CONTROLS		$\frac{10}{(40.4)} \times S_{sb}^{.92}$	S_{sb} = speed brake wetted area $\sim$ ft^2 (m^2)
LIFT DUMPER CONTROLS		$\frac{5}{(20.2)} \times (S_{\ell d} \sin \delta_{\ell d})^{.92}$	$S_{\ell d}$ = total area of lift dumpers $\sim$ ft^2 (m^2)
DIRECT LIFT CONTROL SYSTEM: no data available			$\delta_{\ell d}$ = maximum lift dumper deflection angle

ALL COMPONENT WEIGHTS IN LB (KG)

NOTES:

1. Most formulas are approximations of the curves in SAWE Technical Paper No 812
2. Coefficients in brackets refer to the metric system

Table 8-7. Weight of system controls (transport aircraft)

WEIGHT CONTRIBUTION	METHOD
ENGINE MOUNTS AND VIBRATION ABSORBERS	5% of engine plus propeller installation weight
NACELLE STRUCTURE, PYLONS AND STRUTS, ENGINE COWLINGS, FLAPS AND BAFFLES	$.03 \sqrt{V_D} S_{wet}^{1.3}$ (lb); $V_D \sim$ kts EAS; $S_{wet} \sim$ sq.ft $.405 \sqrt{V_D} S_{wet}^{1.3}$ (kg); $V_D \sim$ m/s EAS; $S_{wet} \sim$ m^2 $S_{wet} \sim$ total area per nacelle wetted by the cold airflow, both internally and externally*
GAS GENERATOR COWLING AND PLUG	3 lb/sq.ft (14.6 kg/m^2) of wetted area
NOISE SUPPRESSION MATERIAL (EXTRA WEIGHT)	.35 lb/sq.ft (1.71 kg/m^2) $\sim$ nacelle walls 1.75 lb/sq.ft (8.53 kg/m^2) $\sim$ splitter plates
FIREWALLS AND SHROUDS FOR FIRE PROTECTION	1.13 lb/sq.ft (5.51 kg/m^2)

*for straight jet engines the external nacelle area plus the inlet duct area

Table 8-8. Data for estimating the nacelle group weight

are available.

Light aircraft, single tractor propeller in the fuselage nose:

$$W_n = 2.5 \sqrt{P_{to}} \text{ (lb)}$$
$$W_n = 1.134 \sqrt{P_{to}} \text{ (kg)} \qquad P_{to} \text{ in hp} \quad (8\text{-}22)$$

This weight refers to the complete engine section in front of the firewall.

Multi-engine aircraft, reciprocating engines:

horizontally opposed cylinders -
$$W_n = .32 \ P_{to} \text{ (lb)}$$
$$W_n = .145 \ P_{to} \text{ (kg)}$$

Other engine types -
$$W_n = .045 \ P_{to}^{5/4} \text{ (lb)}$$
$$W_n = .0204 \ P_{to}^{5/4} \text{ (kg)} \qquad (8\text{-}23)$$

All weights per nacelle

P_{to}: takeoff bhp per engine

Aircraft with turboprop engines:
$$W_n = .14 \text{ lb } (.0635 \text{ kg}) \text{ per takeoff ESHP} \qquad (8\text{-}24)$$

Add .04 lb (.018 kg) per ESHP if the main landing gear is retractable into the nacelle and .11 lb (.05 kg) per ESHP for overwing exhausts (cf. Fig. 2-10, Lockheed Electra). Aircraft with pod-mounted turbojet or turbofan engines:

$$W_n = .055 \; T_{to}$$
high bypass turbofans with
short fan duct -
$$W_n = .065 \; T_{to}$$
(8-25)

This value includes the pylon weight and extended nacelle structure for a thrust reverser installation. In the absence of thrust reversing a reduction of 10% may be assumed.

If a more detailed weight analysis taking into account the configuration and geometry of the nacelle and engine mounting is desirable, some degree of structural design must be attempted first. The subdivision and weight data in Table 8-8 may then be used to calculate the weight. The weight penalty due to noise suppression material obviously depends upon the amount of suppression desired; the engine manufacturer should be consulted for detailed data. For a typical "quiet" turbofan pod, acoustic lining may be required over 50% of the nacelle area. A typical weight penalty is 20% of the nacelle weight, apart from the extra weight of the engine itself.

8.4.2. The propulsion group

Project designs are normally based on existing engine types or paper studies of engines in an advanced state of development. Thus a specification of the definitive engine weight W_e is usually available comprising:
1. engine weight, bare and dry,
2. standard engine accessories and
3. additional weight contributions such as gas generator cowling and/or noise suppression material.

During parametric investigations it may be convenient to employ more general information and the engine weight data in Chapter 4 may be used:
- reciprocating engines: Section 4.2.2. and

Fig. 4-12,
- turbojet and turbofan engines: Section 4.4.3. and equation 4-36,
- turboprop engines: Section 4.5.2. and equation 4-40.

Detailed methods for the computation of turbojet engine weights will be found in References 8-129 through 8-136.

If sufficient details of the powerplant installation are not available, a first approximation for the propulsion group weight is obtained by assuming that part of this weight contribution is proportional to the engine weight, while propeller weight is proportional to the power to be absorbed:

propeller aircraft -
$$W_{pg}=k_{pg}N_e(W_e+.24P_{to}) \; (lb)$$
$$W_{pg}=k_{pg}N_e(W_e+.109P_{to}) \; (kg)$$
(8-26)

P_{to}: takeoff hp per engine
jet aircraft -
$$W_{pg}=k_{pg}k_{thr}N_eW_e$$
(8-27)

where

k_{pg} = 1.16 for single tractor propeller in fuselage
= 1.35 for multi-engine propeller airplanes
= 1.15 for jet transports, podded engines
= 1.40 for light jet airplanes, buried engines

k_{thr} = 1.00 with no thrust reversers
= 1.18 with thrust reversers installed

Add 1.5% for jets and 3% for propeller aircraft with a water injection system. The term .24 (.109) P_{to} in (8-26) for propeller aircraft represents the propeller installation weight in lb (kg).

Instead of the simple approximation given above, Table 8-9 can be used to analyze the powerplant weight in more detail. Weight data for some present-day aircraft are presented in Table 8-10.

A large contribution to the powerplant group included in Table 8-9 is made by the fuel system, comprising:
1. fuel tanks and sealing,
2. pumps, collector tanks and plumbing,
3. distribution and filling system, and
4. fuel dump system (if used).

WEIGHT CONTRIBUTION	METHOD			REMARKS AND NOMENCLATURE
	TURBOJET/TURBOFAN	TURBOPROP	RECIPROCATING	
ENGINE INSTALLATION	$N_e W_e$			consult engine manufacturer's brochure
ACCESSORY GEAR BOXES AND DRIVES, POWER PLANT CONTROLS, STARTING AND IGNITION SYSTEM	$36 \, N_e \, W_{f_{to}}$ pneumatic or cartridge starting system	$\dfrac{.4}{(.181)} \times N_e \, P_{to}^{.8}$ add 30% for beta control		W_e = definitive weight $\sim$ lb(kg) per engine N_e = number of engines $\dot{W}_{f_{to}}$ = fuel flow/engine during takeoff $\sim$ lb/sec (kg/sec) P_{to} = takeoff BHP per engine
AIR INDUCTION SYSTEM	podded engines: included in nacelle group buried engines : $\dfrac{11.45}{(29.62)} \times (\ell_d N_i A_i^{.5} k_{geo})^{.7331}$	included in nacelle group	$\dfrac{1.03}{(.467)} \times N_e \, P_{to}^{.7}$	ℓ_d = duct length $\sim$ ft (m) N_i = number of inlets A_i = capture area per inlet $\sim$ sq.ft(m^2) k_{geo} = 1.0: round or one flat side = 1.33: two or more flat sides
EXHAUST SYSTEM	tailpipes: 3 lb/sq.ft (14.63 kg/m^2) silencers: .01 $N_e \, T_{to}$			T_{to} = takeoff SLS thrust/ engine assumed inlet Mach number: .4
SUPERCHARGERS	–	–	$\dfrac{.455}{(.435)} \times (N_e W_e)^{.943}$	for separate superchargers
OIL SYSTEM AND COOLER	$(.01 \text{ to } .03) \, N_e \, W_e$**	$.07 \, N_e W_e$	radial: .08 $N_e W_e$ hor. opposed: .03 $N_e W_e$	** additional system; basic system supplied by engine manufacturer
FUEL SYSTEM	integral tanks: $\dfrac{80}{(36.3)} \times (N_e + N_{ft} - 1) + \dfrac{15}{(4.366)} \times N_{ft}^{.5} \, V_{ft}^{.333}$ bladder tanks: $\dfrac{3.2}{(.551)} \times V_{ft}^{.727}$		single engine: $\dfrac{2}{(.3735)} \times V_{ft}^{.667}$ multi engine: $\dfrac{4.5}{(.9184)} \times V_{ft}^{.60}$	N_{ft} = total number of fuel tanks ($N_{ft} \geqslant N_e$ for airworthiness) V_{ft} = total fuel tank volume, U.S. gal. (liters)
WATER INJECTION SYSTEM	$\dfrac{8.586}{(1.561)} \times V_{wt}^{.687}$	(optional)		V_{wt} = total water tank capacity $\sim$ U.S. gal. (liters)
PROPELLER INSTALLATION*	–	$k_p \, N_p \, (D_p \, P_{to} \sqrt{B_p})^{.78174}$ k_p = .108 (.124) | k_p = .144 (.165)		N_p = number of propellers D_p = propeller diameter $\sim$ ft (m) B_p = number of blades / propeller
THRUST REVERSERS	$.18 \, N_e W_e$		–	optional

ALL WEIGHTS IN LB (KG)

*From SAWE Technical Paper No. 790

Note: coefficients in brackets refer to the metric system

Table 8-9. Weight analysis of the propulsion group

It will be observed that, for a given integral fuel tank capacity, the number of fuel tanks and the number of engines are primary parameters for determining the fuel system weight.

8.4.3. Airframe services and equipment

In the pre-design phase, with few details of the design of the airframe services and equipment,* their weight is very difficult to predict. The initial prediction error may be very large, as demonstrated by the examples quoted in Ref. 8-151. As soon as preliminary discussions with system (component) manufacturers have been held, the initial weight prediction must be revised.

*A subdivision is shown in Table 8-3.

	ENGINE INSTALL. 10^3 LB	FUEL SYSTEM 10^3 LB %*	EXHAUST + THRUST REV. 10^3 LB %*	OTHER ITEMS 10^3 LB %*	PROPULSION GROUP 10^3 LB %*
JET AIRCRAFT					
Atlas Airbus A-300 B2	16.825	1.257 7.47	4.001 23.8	.814 4.84	22.897 136
Boeing 707/320 C	17.368	2.418 13.9	3.492 20.1	.798 4.59	24.247 140
727/100	9.325	1.143 12.2	1.744 18.7	.250 2.68	12.759 137
737/200	6.217	.575 9.25	1.007 16.2	.378 6.08	8.177 132
747/100	34.120	2.322 6.81	6.452 18.9	.802 2.35	43.696 128
Fokker VFW F-28 Mk 1000	4.495	.545 12.1	.127 2.82	.215 4.78	5.227 116
Lockheed Jetstar	1.750	.360 20.6	** -	.365 20.9	2.475 141
McDonnell Douglas DC-8/55	16.856	3.107 18.4	4.964 29.4	1.580 9.37	26.507 157
DC-9/10RC	6.160	.510 8.28	.658 10.7	.409 6.64	7.737 126
North Am. T-39A Sabreliner	.959	.190 19.8	** -	.152 15.8	1.301 136
Aerospatiale Caravelle VI R	7.055	.518 7.34	.975 13.8	.179 2.54	8.727 124
VFW Fokker 614	3.413	.162 4.75	.119 3.49	.690 20.2	3.763 110
Cessna T-37	.751	.224 29.8	** -	.221 29.4	1.196 159
Northrop T-38A Talon	1.038	.285 27.4	** -	.307 29.6	1.630 157
PROPELLER AIRCRAFT			PROPELLER(S)		
Bristol Britannia 300A	11.192	1.329 11.9	3.557 31.8	3.820 34.1	19.898 178
Canadair CL-44C	12.800	1.755 13.7	5.006 39.1	3.134 24.5	22.695 177
Fokker VFW F-27 Mk 100	2.427	.390 16.1	.918 37.8	.612 25.2	4.454 184
Grumman Gulfstream I	2.688	.133 4.95	1.002 37.3	.698 26.0	4.521 168
Lockheed C-130 E	7.076	1.695 24.0	4.573 64.6	1.874 26.5	15.268 216
L-1049 E	14.256	.893 6.26	2.980 20.9	2.547 17.9	20.682 145
Beechcraft 95 Travel Air	.519	.083 16.0	.162 31.2	.109 21.0	.873 168
G-50 Twin Bonanza	1.008	.137 13.6	.258 25.6	.207 20.5	1.610 160
E-18S	1.352	.274 20.3	.334 24.7	.321 23.7	2.281 169
Cessna 310-C	.852	.076 8.92	.162 19.0	.160 18.8	1.250 147
Beechcraft Bonanza J-35	.432	.030 6.94	.073 16.9	.045 10.4	.580 134
Cessna 150A	.194	.020 10.3	.025 12.9	.034 17.5	.273 141
175B	.312	.030 9.61	.038 12.2	.047 15.1	.427 137
18S	.428	.024 5.61	.072 16.8	.056 13.1	.580 135

* percent of engine installation weight   ** not specified; included in other items

Table 8-10. Propulsion group weight breakdown for existing aircraft types

The data and methods in this section are based primarily on statistical correlations. There is, however, not always a functional relationship between the parameter on which the correlation is based and the actual weight contribution. Consequently, if some weight item is related to the takeoff weight or the empty weight and the first and second estimation of these characteristic weights are different, it may be unnecessary to reiterate the complete weight estimation, provided the estimates do not differ greatly.

It should be noted that for several individual weight contributions a marked discrepancy between the calculated value according to the present methods and the actual value for existing aircraft may be observed. This will be caused to a large extent by differences in de definitions of these items. However, the total estimated systems and equipment weight will be reasonably representative of the actual weight of the operational airplane. In some cases, particularly for wide-body aircraft, the weight estimate may be somewhat conservative due to recent improvements in systems design technology. Typical averages for the total airframe services and equipment weight are:

light single-engine private airplane: 8% W_{to}
light twin-engined airplanes :11% W_{to}
jet trainers :13% W_{to}
short-range transports :14% W_{to}
medium-range transports :11% W_{to}
long-range transports : 8% W_{to}

A collection of weight data is presented in

	AIRPLANE TYPE	MTOW	A.P.U.-GROUP	INSTR. NAV.EQPT.	HYDR. PNEUM.	ELEC-TRICAL	ELEC-TRONICS	FURNISH. EQPT.	AIRCOND. ANTI-ICE	MISC.	TOTAL
JET TRANSPORTS	Atlas Airbus A-300 B2	302,000	983	377	3,701	4,923	1,726	13,161	3,642	732	29,245
	BAC 1-11 Srs 300	87,000	457	182	997	2,317	1,005	4,933	1,579	-	11,465
	Boeing 707/320 C	330,000	151	515	1,086	4,179	2,338	9,527	3,608	-389	21,015
	707/321	301,000	-	561	498	3,959	1,716	14,854	3,290	-	24,878
	720/022	203,000	-	555	505	4,070	1,200	13,055	2,890	-	22,275
	727/100	160,000	60	756	1,418	2,142	1,591	10,257	1,976	85	18,285
	727/100C	160,000	52	802	843	3,617	1,559	6,729	2,401	75	16,078
	737/200	100,400	836	625	873	1,066	956	6,643	1,416	75	13,539
	747/100	710,000	1,130	1,909	4,471	3,348	4,429	37,245	3,969	-421	54,380
	Fokker VFW F-28 Mk 1000	65,000	346	302	364	1,023	869	4,030	1,074	-	8,008
	Mk 2000	65,000	353	309	366	1,045	869	4,614	1,111	-	8,667
	Lockheed Jetstar	30,680	-	153	262	973	318	1,521	510	560	4,297
	McDonnell Douglas DC-8/55	328,000	-	1,271	2,196	2,398	1,551	14,335	3,144	57	24,952
	DC-9/10 RC	91,500	818	719	714	1,663	914	7,408	1,476	24	13,736
	North Am. T-39A Sabreliner	16,700	-	122	116	720	407	857	333	-	2,555
	Aerospatiale Caravelle VI R	114,640	-	236	1,376	2,846	1,187	6,481	1,752	-	13,878
	VFW Fokker - 614	40,981	305	215	403	1,054	436	2,655	719	49	5,836
PROPELLER TRANSPORTS	Bristol Britannia 300A	155,000	-	505	650	1,800	1,040	6,866	3,000	-	13,861
	Canadair CL-44C	205,000	-	858	630	3,040	1,229	12,349	2,536	-	20,662
	CL-44D	205,000	-	783	640	2,875	1,046	3,155	4,090	-	12,589
	Fokker VFW F-27 Mk 100	39,000	-	81	242	835	386	2,291	1,225	-	5,060
	Mk 500	45,000	-	126	256	840	329	3,035	1,257	-	5,843
	Grumman Gulfstream I	33,600	355	97	235	966	99	415	755	6	2,929
	Lockheed C-130 E	151,522	466	665	671	2,300	2,432	4,765	2,126	62	13,487
	L-1049 E	133,000	-	503	654	1,505	1,371	7,405	3,298	-	14,736
	Nord 262	23,050	-	133	765		238	1,324	527	33	3,020
	Vickers Viscount 702	50,044	-	154	331	2,048	447	2,519	1,516	-	7,015
JET TRAINERS	Beechcraft MS 760	7,650	-	70	-	284	158	169	48	30	759
	Cessna T-37	6,436	-	132	56	194	86	256	69	3	796
	Northrop T-38A Talon	11,651	-	211	154	296	246	460	142	24	1,539
LIGHT TWINS	Beechcraft 95 Travel Air	2,900	-	49	-	96	26	194	48	25	438
	G-50	7,150	-	80	-	184	9	333	81	27	834
	E-18 S	9,700	-	100	-	295	63	524	144	58	1,184
	Cessna 310 C	4,830	-	46	-	121	-	154	46	65	498
SINGLE ENGINE A/C	Beechcraft Bonanza J-35	2,900	-	16	-	72	-	174	12	7	281
	Cessna 150A	1,500	-	7	2	41	-	42	4	-	96
	172B	2,200	-	7	3	41	-	99	4	-	154
	180D	2,650	-	8	3	59	-	105	6	-	181
	210A	2,900	-	16	4	60	-	116	12	20	228

ALL WEIGHTS IN LB

Table 8-11. Airframe services and equipment group weight breakdown

Table 8-11. Several items will be discussed in greater detail in the paragraphs below.

a. APU group.
An APU is installed in most modern transport aircraft and also in some jet executives. The installed weight may be based on the dry weight of the APU:

$$\text{Weight} = k_{APU} W_{APU} \qquad (8\text{-}28)$$

The installation factor accounts for the inlet and exhaust ducting mounting frames, silencers, fire protection and accessories, and is generally of the order of 2.0 to 2.5. The APU engine weight is mainly a function of the airflow capacity and power delivery. The bleed airflow requirement is approximately .025 lb/min per cu. ft (.4 kg/min per m^3) of passenger cabin volume or 1.1 lb/min (.5 kg/min) per passenger in the high-density layout.

The APU engine weight can be obtained from the APU specification once the engine has been chosen. The following relationship may be used instead:

$$\left. \begin{array}{l} W_{APU} = 16 \ \dot{W}_{ba}^{3/5} \quad (W_{APU} \text{ in lb, } \dot{W}_{ba} \text{ in} \\ \qquad\qquad\qquad\qquad \text{lb/min}) \\ \\ W_{APU} = 11.7 \ \dot{W}_{ba}^{3/5} \quad (W_{APU} \text{ in kg, } \dot{W}_{ba} \text{ in} \\ \qquad\qquad\qquad\qquad \text{kg/min}) \end{array} \right\} \quad (8\text{-}29)$$

Recent APU engines used on wide-body transports have a specific weight of only 65% of this value, due to improved materials and cycle efficiencies and increased cycle pressures and turbine temperatures.

b. Instruments, navigational equipment and electronics groups.

Requirements for the instruments and NAV/COM equipment (avionics) are usually listed in the design specification. The minimum equipment required for safe operation is supplemented by a choice of optional equipment to improve the operational flexibility. The effects of airplane size are found mainly in the weight of wiring and the flight control system, which increases in size and complexity when the aircraft is scaled up. NAV/COM equipment is partly or fully duplicated on modern transports and even triplicated on recent large transports. A weight estimate may be based on the unit weight of each item of equipment, as obtained from manufacturers, as well as on data for airplanes designed for similar operational capabilities. If these data are not available, the following statistical correlations may be used for the combined weight of instruments and avionics.

Single-engine propeller aircraft: 8 lb (3.6

kg) per pilot, for instruments and 20-30 lb (9-13.6 kg) for radio, which is optional on private aircraft but compulsory on trainers, commuters and taxi aircraft. Propeller-powered utility airplanes up to 12,500 lb (5,670 kg) takeoff weight, VFR operations:

$$\left. \begin{array}{l} 40 \quad + .008 \ W_{to} \qquad \text{(lb)} \\ \\ 18.1 + .008 \ W_{to} \qquad \text{(kg)} \end{array} \right\} \quad (8\text{-}30)$$

Low-subsonic transports with manual flight control system, intended for IFR operations and equipped with single NAV/COM equipment:

$$\left. \begin{array}{l} 120 \ + 20 \quad N_e + .006 \ W_{to} \qquad \text{(lb)} \\ \\ 54.4 + 9.1 \ N_e + .006 \ W_{to} \qquad \text{(kg)} \end{array} \right\} \quad (8\text{-}31)$$

where N_e is the number of engines per aircraft. This equation also gives reasonable results for low-subsonic jet trainers.

For high-subsonic jet transports with predominantly duplicated NAV/COM equipment, jet executives and high-subsonic trainers, the weight of the instruments and electronics group is:

$$W_{ieg} = k_{ieg} \ W_{DE}^{5/9} \ R_D^{1/4} \qquad (8\text{-}32)$$

where W_{DE} is the Delivery Empty Weight and R_D the maximum range (Fig. 8-3)
$k_{ieg} = .575$ for W_{DE} and W_{ieg} in lb, R_D in nm
$k_{ieg} = .347$ for W_{DE} and W_{ieg} in kg, R_D in km
These data do not include the autopilot weight, which is considered part of the surface control system weight in the present subdivision.

c. Hydraulic, pneumatic and electrical groups.

On light aircraft (MTOW up to 12,500 lb or 5,670 kg) the hydraulic system is generally restricted to a brake system and flap and undercarriage operation. For some categories a good correlation was found for the combined weight of hydraulic and electrical systems:
utility aircraft -

$$\left. \begin{array}{l} \text{weight} = .00780 \ W_E^{6/5} \text{ lb } (W_E \text{ in lb)} \\ \\ \text{weight} = .00914 \ W_E^{6/5} \text{ kg } (W_E \text{ in kg)} \end{array} \right\} \quad (8\text{-}33)$$

jet trainers -
$$\text{weight} = .064 \ W_E \qquad\qquad (8-34)$$
propeller transports -
$$\text{weight} = .325 \ W_e^{4/5} \ \text{lb} \ (W_E \text{ in lb}) \left.\vphantom{\begin{matrix}a\\b\end{matrix}}\right\}$$
$$\text{weight} = .277 \ W_e^{4/5} \ \text{kg} \ (W_E \text{ in kg}) \qquad (8-35)$$

A subdivision for jet transports and jet executives appears desirable. The hydraulic and pneumatic power system weight is mainly affected by:

1. the number of functions to be powered, i.e. powered or non-powered controls, operation of spoilers, etc.,
2. the extent of duplication or even triplication,
3. the operating hydraulic or pneumatic pressure, as well as other details of the system design,
4. the airplane size and geometry as related to the length of the plumbing,
5. the relative quantity of pneumatic functions, if any, and
6. the state of the art.

The combined weight of the hydraulic plus pneumatic system may be assumed to be $1\frac{1}{2}\%$ of the DEW or, alternatively:
no powered controls -
$$\text{weight} = .004 \ W_{DE} + 100 \quad (\text{lb})\left.\vphantom{\begin{matrix}a\\b\end{matrix}}\right\}$$
$$\text{weight} = .004 \ W_{DE} + \ 45 \quad (\text{kg}) \qquad (8-36)$$
boosted controls, only some essential functions duplicated -
$$\text{weight} = .007 \ W_{DE} + 200 \quad (\text{lb})\left.\vphantom{\begin{matrix}a\\b\end{matrix}}\right\}$$
$$\text{weight} = .007 \ W_{DE} + \ 91 \quad (\text{kg}) \qquad (8-37)$$
powered controls, fully duplicated system -
$$\text{weight} = .011 \ W_{DE} + 400 \quad (\text{lb})\left.\vphantom{\begin{matrix}a\\b\end{matrix}}\right\}$$
$$\text{weight} = .011 \ W_{DE} + 181 \quad (\text{kg}) \qquad (8-38)$$
powered controls, triplex system -
$$\text{weight} = .015 \ W_{DE} + 600 \quad (\text{lb})\left.\vphantom{\begin{matrix}a\\b\end{matrix}}\right\}$$
$$\text{weight} = .015 \ W_{DE} + 272 \quad (\text{kg}) \qquad (8-39)$$
For jet freighters these figures are roughly 30% higher, due partly to the somewhat lower empty weight and partly to the extra services required for loading and unloading. Some weight reduction is possible for an increasing number of pneumatic system functions.

The electrical system weight is affected mainly by:

1. the total electrical power required, which is primaliry determined by the galley power, electronic equipment and fuel system power,
2. whether or not the primary system is an A.C. or D.C. system[*],
3. the size of the airplane, in view of the length of wiring,
4. the amount of system duplication and the standby systems,
5. whether or not electrical power is generated by the A.P.U., and
6. the state of the art.

the following statistical relationships are suggested:
primary system D.C. -
$$W_{el} = .02 \ W_{to} + 400 \qquad (\text{lb})\left.\vphantom{\begin{matrix}a\\b\end{matrix}}\right\}$$
$$W_{el} = .02 \ W_{to} + 181 \qquad (\text{kg}) \qquad (8-40)$$
primary system A.C., total electrical power generated up to 400 kVA -
$$W_{el} = 36 \quad P_{el} \ (1-.033 \sqrt{P_{el}}) \ (\text{lb})\left.\vphantom{\begin{matrix}a\\b\end{matrix}}\right\}$$
$$W_{el} = 16.3 \ P_{el} \ (1-.033 \sqrt{P_{el}}) \ (\text{kg}) \qquad (8-41)$$
In the absence of better information the electrical power generation may be obtained from statistical data in publications like Jane's All the World's Aircraft or from correlations with the passenger cabin volume V_{pc}:
if no electrical power is generated by the APU, V_{pc} up to 8,000 cu.ft (227 m^3) -
$$P_{el} = .016 \ V_{pc} \ (V_{pc} \text{ in cu.ft})\left.\vphantom{\begin{matrix}a\\b\end{matrix}}\right\}$$
$$P_{el} = .565 \ V_{pc} \ (V_{pc} \text{ in } m^3) \qquad (8-42)$$
if electrical power generation by the APU is included -
$$P_{el} = .3 \quad V_{pc}^{.7} \quad (V_{pc} \text{ in cu.ft})\left.\vphantom{\begin{matrix}a\\b\end{matrix}}\right\}$$
$$P_{el} = 3.64 \ V_{pc}^{.7} \quad (V_{pc} \text{ in } m^3) \qquad (8-43)$$
These figures on electrical systems are based on 1950-1965 technology. Recent developments have indicated that considerable improvements in system weights are possible by applying advanced techniques - like multiplexing[**] and high-speed generators.

---

[*]Most present-day transport aircraft feature A.C. primary systems
[**]Aviation Week of October 28, 1968, pp. 157-161: a weight reduction of 400 lb (181 kg) was achieved on the Boeing 747

GROUP	DESCRIPTION	METHOD	REMARKS
FLIGHT DECK ACCOMMODATIONS	flight crew seats, instrument panels, control stands, sound proofing, insulation, trim, floor covering, lighting and wiring, miscellaneous equipment	jet a/c : $29 (16.5) \times W_{DE}^{.285}$ propeller a/c: $16 (9.1) \times W_{DE}^{.285}$	W_{DE} = Delivery Empty Weight~ℓb(kg)
PASSENGER CABIN ACCOMMODATIONS	passenger and attendants' seats	Table 3-2	
	galley (pantry) structure and provisions	main meal galley: $250\ell b(113.4kg)$ each snack pantry : $100\ell b(45.3kg)$ each coffee bar : $65\ell b(29.5kg)$ each	galley inserts, potable water and toilet chemicals not included
	lavatory and toilet provisions, water system (dry)	medium/long-haul: $300\ell b(136.0kg)$/toilet short-haul : $165\ell b(75.0kg)$/toilet commuters : $85\ell b(38.5kg)$/toilet	
	floor covering	jet aircraft: $.18 (1.25) \times S_{cf}^{1.15}$ propeller aircraft: $.135 (.94) \times S_{cf}^{1.15}$	S_{cf} = cabin floor area, galleys and toilets included~sq.ft (m^2)
	soundproofing and insulation, wall covering, curtains, screens, window shades, ceiling, lighting panels, hatracks, partitions and doors; wardrobe and stowage provisions, freight hold linings and partitions	$.30 (6.17) \times (V_{pc}+V_{ch})^{1.07}$ $.14 (3.69) \times (V_{pc}+V_{ch})^{1.14}$	V_{pc} = passenger cabin volume, galleys and toilets included~cu.ft (m^3) V_{ch} = total cargo hold volume ~ cu.ft (m^3)
CARGO ACCOMMODATIONS	cargo restraints and handling provisions	$.08$ ℓb/cu.ft $(1.28kg/m^3)$ of V_{ch}	
	container or pallet cargo handling provisions	2.8 ℓb/sq.ft $(13.67kg/m^2)$ of freight floor area for convertible passenger/cargo versions	
(STANDARD) EMERGENCY EQUIPMENT	fixed oxygen system, portable oxygen sets	short or no overwater flights, cruise altitude up to 25,000ft(7620m): $20+.5N_{pax}$ ~ ℓb($9.1+.227N_{pax}$ ~ kg) above 25,000ft(7620m): $30+1.2N_{pax}$ ~ ℓb($13.6+.544N_{pax}$ ~ kg) extended overwater flights: $40+2.4N_{pax}$ ~ ℓb($18.1+1.09$ N_{pax} ~ kg)	N_{pax} =max.no. of passengers for certification (pressure cabins)
	fire detection and extinguishing system, portable extinguishers	jet a/c : $.0012$ W_{to} turboprop a/c: $.0030$ W_{to} reciproc. a/c: $.0060$ W_{to}	
	escape provisions (evacuation slides and ropes)	1 ℓb(.453 kg) per occupant	Other provisions in Operational Items

ALL WEIGHTS IN LB (KG)

NOTE: coefficients in brackets refer to the metric system

Table 8-12. Furnishing and equipment group weight for transport and executive aircraft

d. Furnishing and equipment group.
Light single-engine aircraft: this weight group consists mainly of the weight of seats, wall and floor covering, and some miscellaneous contributions. The weight is approximately 13 lb (5.9 kg) per seat, plus 25 lb (11.3 kg) per row of two seats, plus an additional 5 lb (2.3 kg).

Light twin-engine aircraft: 15 lb (6.3 kg) per seat, plus 1 lb per cu.ft (16 kg per m^3) of cabin plus cargo compartment volume. Jet trainers, equipped with two ejection seats: 6.5% of the Delivery Empty Weight. Civil freighters: 3 lb per sq.ft (14.7 kg/

ITEM	SUBDIVISION	METHOD	REMARKS, SYMBOLS
CREW PROVISIONS	flight and cabin crew with baggage, flight equipment	$\frac{205}{(93)} \times N_{fc} + \frac{150}{(68)} \times N_{cc}$	N_{fc}, N_{cc} = number of flight/ cabin crew members respectively
PASSENGER CABIN SUPPLIES	removable galley bar equipment, meal service, consumable food, drinks, beverages pillows, papers and magazines, entertainment	commuters: 1 ℓb (.453 kg) x N_{pax} transport aircraft, snacks only : 5ℓb (2.27kg).N_{pax} main meal, short-range:14ℓb (6.35kg).N_{pax} , long -range:19ℓb (8.62kg).N_{pax}	N_{pax} = number of passengers, all-tourist. First class: all data 5ℓb (2.27kg) per passenger higher
POTABLE WATER AND TOILET CHEMICALS		short range : 80N_{wc} or 1.5N_{pax}∿ℓb (36.3 N_{wc} or .68N_{pax}.∿kg) short/medium-range: 120N_{wc} or 3.0N_{pax}∿ℓb (54.4 N_{wc} or 1.36N_{pax}∿kg) long-range : 200N_{wc} or 6.5N_{pax}∿ℓb (90.7 N_{wc} or 2.95N_{pax}∿kg)	
SAFETY EQUIPMENT	life jackets, fire axes, emergency navigational equipment	short or no overwater sectors: 2N_{pax}∿ℓb (.907N_{pax}∿kg) extended overwater flights: 7.5N_{pax}∿ℓb (3.4N_{pax}∿kg)	N_{wc} = number of toilets/water closets; data based on all- tourist layout
OIL RESIDUAL FUEL WATER/ METHANOL	residual fuel	gas turbine engines: reciprocating: $\frac{.81}{(.151)} \times V_{ft}^{2/3}$ .008 W_{to}	V_{ft} = total fuel tank capacity ∿ U.S. gal. (liters) W_{to} = Max. Takeoff Weight ∿ ℓb (kg)
	residual oil	turboprop engines: $\frac{.81}{(.151)} \times V_{ft}^{2/3}$	
	engine oil consumed	.045 W_{f}	W_f = fuel weight ∿ ℓb (kg)
	water/methanol	optional	
CARGO HANDLING EQUIPMENT	pallets, containers, cargo tiedown eqpt.	Fig. 3-20	ALL WEIGHTS IN LB (KG)

NOTE: coefficients in brackets refer to the metric system

Table 8-13. Data for estimating the weight of Operational Items (transport aircraft)

m^2) of main freightfloor area.
Passenger transports and jet executives: a rough approximation is obtained with the statistical expression:

weight = .211 $W_{ZF}^{.91}$ (1b)
weight = .196 $W_{ZF}^{.91}$ (kg)   (8-44)

where W_{ZF} is the Maximum Zero Fuel Weight. The furnishing and equipment weight forms a very substantial contribution, of the order of half the fuselage structure weight. Instead of using (8-44) the designer may prefer to use a more detailed

estimation by breaking down the weight into several individual contributions. A proposed subdivision and calculation methods are presented in Tables 8-12 and 3-2. It should be noted that several items such as the weight of seats depend on the required standard of comfort and the type of interior; these may be subject to customer requirements (Standard Items Variation).

e. Air-conditioning and anti-icing group. The weight of the air-conditioning and pressurization system depends on many factors:

1. the type of system used: air cycle or vapor cycle, use of ram air or engine bleed air, etc.,

2. design requirements, in terms of air-conditioning airflow per unit of time, air temperature, humidity and cabin pressure differential, cargo compartment air-conditioning,

3. the amount of system duplication,

4. the airplane size, or more specifically the cabin volume and length, and the subdivision into zones,

5. the state of the art.

Factors affecting the anti-icing and de-icing system weight are:

1. type of system (electrical, hot-air, rubber boots),

2. dimensions, mainly the length or span of the airplane parts concerned, and

3. the type of operation, viz. IFR or VFR flights.

For the combined system, the following data can be used:

Light single-engine aircraft - 2.5 lb (1.1 kg) per seat. Multi-engine unpressurized aircraft and jet trainers - 1.8% of the Delivery Empty Weight

pressurized transports and executive aircraft -

$$\left. \begin{array}{l} \text{weight} = 6.75 \ l_{pc}^{1.28} \ (lb)-l_{pc} \ \text{in ft} \\ \text{weight} = 14.0 \ l_{pc}^{1.28} \ (kg)-l_{pc} \ \text{in m} \end{array} \right| (8-45)$$

f. Miscellaneous.

This item refers to auxiliary gears, photographic equipment, external paint, manufacturing variation, unaccounted items, unexpected weight growth, etc. No systematic data are available, but in general a figure of up to 1% of the Delivery Empty Weight is typical for existing aircraft.

8.4.4. Useful Load and the All-Up Weight

a. Operational Items.

Due to the large variation in operational conditions and requirements applying to passenger service, considerable variations in the weight of operational items can be observed. The data in Table 8-13 are re-presentative of but by no means mandatory for the transport aircraft category. For private aircraft and jet trainers the only item of interest is the residual fuel and oil.

It should be noted that the data of Table 8-13 are generally applicable to modern, gas turbine powered aircraft. Considerably higher weight values are applicable to older piston-engine powered transport aircraft.

b. Payload and fuel.

Some data on specific gravity of fuels and civil payload will be found in Table 8-14.

	LB	KG	
PASSENGERS :	165	75	
PASS. BAGGAGE :	40	18	- TOURIST CLASS
	60	27	- FIRST CLASS
BAGGAGE SPEC. DENSITY :	12 LB/FT^3 (192 KG/M^3)		

FUEL	SPECIFIC HEAT		SPECIFIC WEIGHT*	
	BTU/LB	KCAL/KG	LB/U.S. GAL	KG/LITER
GASOLINE :	18,700	10,389	5.85	.701
JP - 3 :	18,000	10,000	6.32	.767
JP - 4 :	18,550	10,305	6.50	.779
JP - 5 :	18,400	10,222	6.84	.820
LUBRICATING OIL SPECIFIC WEIGHT: 7.5 LB/U.S. GAL (.9 KG/LTR)				

* AT 59°F (15°C)

Table 8-14. Standard weights of payload, fuel and oil

The data presented in this Section 8.4 and Appendices C and D are sufficiently complete to enable the designer to make a fairly accurate prediction of the OEW of a civil airplane. Of necessity, the procedure is based on an initial estimate of the various characteristic weights, as obtained, for example, with Sections 5.2 and 8.2.

The more detailed weight prediction will result in a value for the OEW that is essentially different from the first "guesstimate". The designer must therefore decide whether he should modify the Useful Load (i.e. fuel and/or payload) or the MTOW. Fresh calculations of the weight distribution will then be necessary until the designer is satisfied with the convergence.

	COMPONENT	C.G. LOCATION
STRUCTURE	WING (HALF)	straight wing: 38-42% chord from LE at 40% semi-span from centerline swept wing: 70% local distance between front and rear spar, measured from front spar, at 35% semi-span from centerline
STRUCTURE	FUSELAGE	distance from fuselage nose, in % of fuselage length (excl. spinner) single tractor engine : 32 - 35 wing-mounted propeller engines : 38 - 40 wing-mounted jet engines : 42 - 45 rear fuselage mounted pods : 47 jet engine buried in fuselage : 45
STRUCTURE	TAILPLANE (HALF)	42% chord from LE at 38% semi-span from root chord. Fin, T-tail configuration: 42% chord from LE at 55% of height from root chord
STRUCTURE	NACELLES	40% of nacelle length from nose, spinner excluded
STRUCTURE	SURFACE CONTROL SYSTEM	100% MAC from LEMAC, autopilot excluded
STRUCTURE	ALIGHTING GEAR	at airplane c.g., or determined from location and weight of main and nose undercarriage
ENGINES AND ACCESSORIES		from engine manufacturer's data
AIRFRAME SERVICES AND EQUIPMENT		from educated guess, taking into account location of main elements and functions to be powered
FURNISHING		from subdivision of Table 8-11 and cabin layout
FILLED FUEL TANK		for prismoid with height ℓ and parallel end faces with area S_1 and S_2 (see Fig. 8-4), at distance $\frac{\ell}{4}\frac{S_1 + 3S_2 + 2\sqrt{S_1 S_2}}{S_1 + S_2 + \sqrt{S_1 S_2}}$ from plane S_1

NOTE: more accurate estimates can be made by further breakdown of each item into several contributions

Table 8-15. Approximate location of the center of gravity for several groups

If the designer decides to alter the Useful Load, the specified transport performance (payload-range) may no longer be achieved, while in the second eventuality the takeoff weight may become too high for acceptable takeoff, landing and/or climb performance. Sufficient aerodynamic data must be available to evaluate the design performance (Chapter 11).
In this textbook we will proceed with the layout design, assuming that the previous weight prediction has not entailed major difficulties and that only minor changes in the weight distribution are required. However, there may be occasions when weight evaluation and changes result in a considerable increase of the takeoff weight.

8.5. CENTER OF GRAVITY

Each airplane must be designed in such a way that good stability and control properties and adequate flexibility in loading conditions are obtained. By suitable arrangement of the design layout and acceptable tailplane size, acceptable fore-and-aft limits of the center of gravity must be established, taking into account the following aspects:
1. fore-and-aft position of the wing relative to the fuselage,
2. provision of suitable locations for payload and fuel,
3.. design of the horizontal tailplane, the elevator and the longitudinal flight con-

Table 8-16. Center of gravity limits for several types of aircraft.

AIRPLANE TYPE	FORWARD takeoff landing	FORWARD flight	REAR takeoff landing	REAR flight	RANGE takeoff landing	RANGE flight	PAY-LOAD % OEW	$\frac{S_h \ell_h}{S c}$	hor. tail type **	$C_{L_{max}}$ ***
2 JET ENGINES										
Aerospatiale Corvette SN601	–	20.0	–	36.0	–	16.0	28.3	.64	V	2.40
A.C. Jet Commander 1121	20.0	20.0	36.0	36.0	16.0	16.0	20.6	.64	F	1.66
Lear Jet 25	9.0	9.0	30.0	30.0	21.0	21.0	35.6	.64	F	1.39
H. Siddeley HS-125 1A/1B	18.0 *	18.0 *	37.5 *	37.5 *	19.5 *	19.5 *	14.0	.69	F	2.44
Dassault Mystère 20F	14.0	16.0	28.5	28.5	14.5	12.5	23.1	.66	V	2.30
H.F.B. Hansa	13.0	11.7	23.0	21.7	10.0	10.0	30.9	.71	F	2.00
Fokker VFW F-28 Mk1000	18.0	17.0	35.0	37.0	17.0	20.0	42.0	.97	V	2.53
BAC 1-11 Srs. 400	15.0 *	14.0 *	39.0 *	41.0 *	24.0 *	27.0 *	35.3	.85	V	2.38
Sud. Av. Caravelle 10R	25.0	25.0	41.5	41.5	16.5	16.5	32.3	.56	F	2.10
McD. Douglas DC-9/10	16.3	15.0	39.0	40.0	22.7	25.0	42.4	1.15	V	2.40
DC-9/33F	5.9	3.1	34.7	34.7	28.8	31.6	70.8	1.18	V	2.98
Boeing 737/100	15.0	15.0	35.0	35.0	20.0	20.0	49.4	1.14	V	3.10
Airbus A-300 B2	11.0	11.0	31.0	31.0	20.0	20.0	37.4	1.07	V	2.65
3 OR 4 JET ENGINES										
Lockheed 1011 Tristar	–	12.0	–	32.0	–	20.0	36.1	.93	A	2.57
Boeing 707/120	16.0	16.0	34.0	34.0	18.0	18.0	38.2	.61	V	1.86
720/022	15.0	15.0	31.0	31.0	16.0	16.0	31.9	.59	V	2.26
747/200B	–	12.5	–	32.0	–	19.5	45.8	1.00	V	2.55
McD. Douglas DC-8/21	16.5	16.5	32.0	32.0	15.5	15.5	27.0	.58	V	2.10
Lockheed C-141A	19.0	19.0	32.0	32.0	13.0	13.0	50.3	.51	V	2.32
Lockheed C-5A	19.0	19.0	41.0	41.0	22.0	22.0	67.9	.64	V	2.60
1 PROPELLER ENGINE										
Fokker S-11 Instructor	21.5	21.5	27.0	27.0	5.5	5.5	22.2	.43	F	1.25
Cessna 172, Normal Cat.	15.6	15.6	36.5	36.5	20.9	20.9	64.3	.59	F	2.14
177, Normal Cat.	5.0	5.0	28.0	28.0	23.0	23.0	58.6	.60	A	1.86
177, Utility Cat.	5.0	5.0	18.5	18.5	13.5	13.5	58.6	.60	A	1.86
206 Skywagon	12.2	12.2	39.4	39.4	27.2	27.2	67.3	.77	F	2.16
Beechcraft B-45 Mentor	20.1	19.0	28.0	28.0	7.9	9.0	17.4	.54	F	2.01
Piaggio P-148 (3 seater)	22.3	22.3	30.7	30.7	8.4	8.4	26.3	.43	F	1.90
Pilatus PC-6-M2 Porter	11.0	11.0	34.0	34.0	23.0	23.0	79.9	.67	A	2.28
Saab 91-B Safir	17.9	17.9	27.1	27.1	9.2	9.2	23.9	.64	F	–
De Havilland DHC-2 Beaver	17.4	17.4	40.3	40.3	22.9	22.9	49.3	.76	F	–
2 PROPELLER ENGINES										
Cessna Model 337	17.3	17.3	30.9	30.9	13.6	13.6	37.1	.51	F	1.78
Piper PA 30C Twin Comanche	12.0	12.0	27.8	27.8	15.8	15.8	40.7	.44	A	1.66
Beechcraft Queen Air M. 80	16.0	16.0	29.9	29.9	13.9	13.9	44.9	.73	F	1.88
Dornier Do 28-D-1	10.7	10.7	30.8	30.8	20.1	20.1	34.7	.67	A	2.36
DHC-6 Twin Otter	20.0	20.0	36.0	36.0	16.0	16.0	74.0	.93	F	2.37
Nord 262	16.0	16.0	30.0	30.0	14.0	14.0	48.0	.96	F	2.23
Fokker VFW F-27 Mk 200	20.0	18.7	38.0	40.7	18.0	22.0	55.8	.96	F	2.94
Hurel Dubois HD 32	23.5	23.5	46.5	46.5	23.0	23.0	36.5	1.32	F	2.70
Convair 240	15.0	8.5	31.0	33.0	16.0	24.5	33.9	1.07	F	2.33
340	13.0	8.5	34.0	35.0	21.0	26.5	56.0	1.03	F	2.61
H. Siddeley Andover C.Mk 1	13.3 *	13.3 *	36.0 *	36.0 *	22.7 *	22.7 *	53.6	1.09	F	2.88
4 PROPELLER ENGINES										
Bréguet 941	23.0	23.0	32.0	32.0	9.0	9.0	70.0	1.05	V	7.19
Douglas DC-6	16.0	12.0	33.0	35.0	17.0	23.0	29.4	1.04	F	2.77
Lockheed 188C Electra	15.0	13.0	32.0	33.0	17.0	20.0	53.6	.80	F	2.54
Bristol 175 Britannia	13.0 *	12.0 *	34.5 *	35.5 *	21.5 *	23.5 *	37.3	.97	F	2.56
Lockheed L-1049 H	18.0	15.0	32.0	34.0	14.0	19.0	44.9	1.15	F	2.60
L-1649 A	15.0	12.0	32.0	34.0	17.0	22.0	17.5	1.12	F	2.53
C-130 E	15.0	15.0	30.0	30.0	15.0	15.0	51.4	1.00	F	2.28
Canadair CL-44 C	12.2	12.2	30.5	31.4	18.3	19.2	22.8	1.14	F	2.56

* per cent SMC

** F = fixed stabilizer, V = variable incidence stabilizer, A = all-movable tail

*** flap angle for landing

Table 8-16. Center of gravity limits for several types of aircraft

295

trol system, and

4. location of the undercarriage legs.
The designer's freedom of choice is greatly limited by the conditions imposed upon
the c.g. location. This applies in particular to the location of fixed items like
engines and cargo holds, both having a
considerable bearing on the balance of
the airplane.

The c.g. must be established in both the
longitudinal and the vertical direction.
Having obtained the weight distribution,
the designer can produce a sideview with
a suitable system of coordinate axes and
the centers of gravity plotted on it for
each individual item (Table 8-15). Tables
similar to Table 8-3 are filled out and
the final result of this tabulation yields
the horizontal and vertical location of
the c.g. for the Operating Empty Weight
condition:

$$x_{cg} = \frac{\sum x_i W_i}{\sum W_i} \qquad (8-46)$$

$$z_{cg} = \frac{\sum z_i W_i}{\sum W_i} \qquad (8-47)$$

Fore-and-aft shift of limits for the c.g.
due to different loading conditions must
be estimated and indicated in the drawing.
The designer must demonstrate that in all

likely loading conditions the actual c.g.
will remain inside the fore-and-aft limitations, without undue penalties in the
form of loading restrictions. Some procedures used to obtain a good wing location
and c.g. range will be discussed in the
following sections.

8.5.1. The load and balance diagram

It is not sufficient to determine the c.g.
for only one condition, e.g. the fully
loaded condition. What is important for an
analysis of the stability and control properties are the most critical fore-and-aft
locations.
Assuming that the general (airplane) arrangement and layout design have been decided upon, the loading flexibility can be
illustrated in a weight and balance diagram (Fig. 8-7). Both the likely operational c.g. shift and the "aerodynamic" limits
are indicated in this diagram as a function
of the weight. The Mean Aerodynamic Chord
(MAC)[*] or the Standard Mean Chord (SMC)[*]
are used to define the c.g. position.
The Aircraft Prepared for Service, but
without fuel and payload (OEW), is represented by point A in Fig. 8-7. A margin
of a few per cent MAC is usually assumed

[*]Definitions are given in Appendix A.

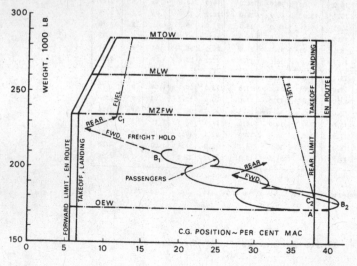

Fig. 8-7. Load and balance
diagram for a short-haul
wide-body airliner

on both sides of the standard location to allow for variation in the weights and locations of the airframe services, equipment and Operational Items (cf. Section 8.2.1.), undercarriage retraction, etc. The loading of passengers in the cabin and baggage or cargo in the compartments increases the weight and causes a shift of the c.g. which will depend upon the location of the variable load relative to point A.

For passenger transports it is not usual to assume that the passenger seating is completely arbitrary. The procedure to be described in Section 8.5.2. is a statistical prediction of the most practicable envelope of passenger loading distributions, assuming a free choice of the seats. In view of its shape the passenger loading envelope is sometimes referred to as "loading potato" or "passengers loop". The effects of loading the cargo compartments is indicated by shifting the most extreme points of the loading envelope, points B_1 and B_2. In the example B_2 is located outside the allowable c.g. limits and hence for low payloads there must always be some load in the forward cargo hold. At a later stage of the design loading restrictions of this type are translated into load sheets used by the operator's personnel to control the c.g. Assuming that points C_1 and C_2 represent the most extreme locations, the fuel load can be inserted by further envelopes, originating at C_1 and C_2. In the example shown it is assumed that a fuel management procedure has been developed to minimize the c.g. shift, resulting in approximately straight lines in the diagram. On a swept wing this is not always entirely feasible; the use of reserve tanks to minimize the c.g. shift can then be considered. As opposed to this, the BAC-Sud Concorde features a system for transferring fuel from the wing tanks to a rear fuselage tank, in order to cope with the a.c. shift of the airplane caused by the transition from subsonic to supersonic flight.

Variations in the c.g. have an effect not only on stability and control characteristics, but also on the tail maneuver loads and the ground loads acting on the nose undercarriage. A loading envelope must therefore be established, defining possible combinations of c.g. and Gross Weight, for which certification is requested. The principal aim during preliminary design is to ensure that the most likely operational loadings are not unduly penalized by complicated and/or stringent loading restrictions.

8.5.2. Loading flexibility and restrictions

The type of operation envisaged for the airplane determines to what extent special loading restrictions are acceptable. Though a generous allowable c.g. travel improves the operational flexibility, there are some pitfalls as well:
1. A large c.g. shift entails large variations in the stability and maneuver margins and generally makes the design of the tailplane and control system more difficult. Added complexity, weight and skin friction drag will be the inevitable result.
2. A large stability margin entails appreciable trim drag.
3. A c.g. location that is well forward of the airplane's neutral point must be balanced by a download on the tail. The maximum lift coefficient (flaps down) may be considerably reduced by this.
A collection of data on c.g. locations is presented in Table 8-16. The following comments will make them easier to interpret. For passenger transports a generous c.g. travel of the order of 20-25% MAC is desirable. Different seating arrangements and cabin layouts, payload growth, alternate conversions into Quick Change (QC) or combined passenger/freight transport, and increased fuel tank capacity on later versions are all aspects to be considered. Airlines object strenuously to assigning passengers definite seats, but balancing with freight is acceptable up to a certain point. Fuel management procedures are ac-

ceptable, but fixed ballast should be
avoided altogether.

For light aircraft it is desirable to analyze all likely loading conditions. For
passenger transports the following "window
seating rule" is observed (example in Fig.
8-8).

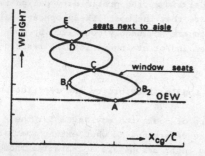

Fig. 8-8. The window seating procedure

In the empty aircraft the passengers are
assumed to occupy the seats nearest to the
window first. When these are filled (A → C),
the rows next to the window seats are occupied (C → D), followed finally by those
nearest to the aisle (D → E). The window
seats are occupied in two different ways:
- starting from the front (A → B_1 → C)
- starting from the rear (A → B_2 → C)
In this way a loop is formed. Here the most
forward and rearward points (B_1 and B_2)
correspond to the situation where all window seats in front of or behind A respectively, are occupied. The loops C → D and
D → E are computed in a similar manner.
The example applies to a cabin layout with
5-abreast seating, resulting in a small
upper loop for a single third row of seats.
Other aircraft arrangements may give quite
different results (Section 8.5.3.).

General aviation aircraft (commuters, utility aircraft, air taxis) usually need a
fairly large c.g. travel for versatility
in operation. For example, the Pilatus
Porter has a travel of 23% MAC, the Scottish
Aviation Jetstream as much as 30% SMC. A
large travel is not required on light private aircraft as the payload is relatively
low and located close to the c.g. of the

empty aircraft. Only simple loading procedures will be acceptable. Light aircraft
for 2 persons need a c.g. travel of only
5 to 10% MAC. Statistical variation of the
occupant weights and variations in equipment and furnishing should be accounted
for.

In the case of freighter aircraft the large
variety in payload characteristics means
that there are too many conceivable loading
conditions for all of them to be considered
in the preliminary design stage. It is
suggested that reasonable fore-and-aft c.g.
limits be chosen on the basis of statistical evidence or stability and acceptable
control characteristics, assuming the horizontal tailplane design to be fixed. A
balance diagram can be drawn to limit the
cargo c.g. as a function of its weight
(Fig. 8-9). The relevant equation is:

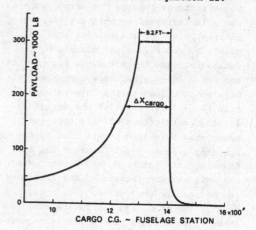

Fig. 8-9. Typical loading limits for a
large freighter

$$x_{cargo} = \frac{W_{OE}}{W_{cargo}} \left(x_{cg_{limit}} - x_{OE} \right) + x_{cg_{limit}} \quad (8-48)$$

where W_{OE} and x_{OE} are the OEW and the associated c.g. position, while $x_{cg_{limit}}$ represents the fore-and-aft c.g. limits,
corresponding to the weight $W_{OE} + W_{cargo}$.
The cargo loading flexibility,

$$\Delta x_{cargo} = \Delta x_{cg} \left(1 + \frac{W_{OE}}{W_{cargo}} \right) \quad (8-49)$$

298

is therefore proportional to the allowable c.g. travel. The asymmetric shape of the diagram is caused by the c.g. location of the empty airplane, which is very near the aft limit in this particular example. On a fully loaded airplane the freight must always be unloaded from the aft section of the freighthold. Whether or not this is acceptable depends upon the location of the loading doors and the type of operation.

A noticeable c.g. shift will also occur during ground loading and unloading of transport aircraft. To avoid an unexpected turnover, the c.g. must always be inside the triangle interconnecting the wheel contact points. Passenger access and freighthold doors must be located not too far to the rear end of the fuselage if a nosewheel undercarriage is envisaged.

8.5.3. Effects of the general arrangement and layout

The effect of engine location on the balance is illustrated in Fig. 8-10 for three typical general arrangements. To give the comparison point, the designs have approximately equal weights, wing area, payload and fuel. The differences in engine location for Configurations 2 and 3 have resulted in a shift of the wing backwards or forwards relative to Configuration 1.

Configuration 1 is generally the easiest to balance as the cabin volumetric center is close to the c.g. of the empty aircraft. The c.g. travel is large for a mixed class layout if one of the cabin compartments is empty. Cargo can be used to balance the aircraft by suitable distribution between the forward and rear cargo holds, which are of roughly equal volume. The c.g. corresponding to the OEW is usually located at 25-30% MAC.

Configuration 2 is generally the most difficult layout for which to obtain a satisfactory balance, particularly on short-haul aircraft, as both the engines and the payload masses are relatively large.

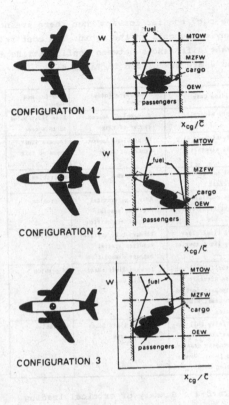

Fig. 8-10. Effect of the general arrangement on load and balance

The c.g. of the empty aircraft is very near the rear limit (approximately 35-40% MAC)[*]. Most of the payload is ahead of the c.g. and the most forward location corresponds to the MZFW.

The c.g. shift of Configuration 2 is 3 to 5% MAC larger than for an otherwise comparable Configuration 1, but an acceptable loading flexibility is nevertheless feasible in most cases, in particular when a T-tail is adopted, being very effective under normal flight conditions. For combined passenger/cargo transport, however, this configuration is generally considered undesirable, because the cargo, when located in front of the passenger cabin, will bring

[*]It is known that the Tupolev 154 has its c.g. at 50% MAC in the empty condition (cf. The Aeroplane, Oct. 1966, page 18)

299

the c.g. too far forward when there are not
many passengers in the cabin. The most relevant differences between configurations 1
and 2 are summarized in Table 8-17.

Loading Case	Underwing Podded Layout	Rear-Engined Layout
Empty aircraft	No problem. Approx. center of range	Rear limit based on this case
Very high-density full aircraft	No problem. Approx. center of range	Forward limit based on this case
Full one-class aircraft	No problem.	No problem
Full tourist. Very light first class	Rear c.g. critical Requires special baggage disposition	No problem
Full first class Very light tourist	Forward c.g. critical Requires special baggage disposition	No problem
Partially full one-class aircraft; passengers seating from rear. Window seating only	Forward limit based on this case	No problem
Partially full one-class aircraft; passengers seating from front. Window seating only	Forward limit based on this case	No problem

Table 8-17. Summary of critical loading
conditions for two airplane configurations
(Reference: Aerospace Engineering, October
1960, page 74)

Configuration 3. Although the example
assumes jet engines, this layout is chosen
more frequently in propeller aircraft, particularly if the propellers are in front of
the flight deck. The situation is to some
extent the reverse of Configuration 2, but
the balancing problems are generally less
difficult. The empty aircraft has its c.g.
at about 20-25% MAC, the critical loading
case being the high-density layout, resulting in the rear c.g. position. The
relatively large tail arm aids in solving
stability and control problems and the
horizontal tailplane can be kept relatively small, particularly for T-tail designs.
The provision of a cargo compartment in
the fuselage nose in front of the cockpit
will improve the balance.

8.5.4. Design procedure to obtain a balanced aircraft

There may be various ways in which the airplane can be balanced, depending upon the
degree to which the design has been frozen:
1. The c.g. of the empty aircraft can be
optimized by locating the wing appropriately in the longitudinal direction.
2. The c.g. range can be reduced by a favorable cabin layout and suitable location
of the engines, cargo compartments, fuel
tanks and certain items of airframe services and equipment.
3. Suitable tailplane and control system
design and undercarriage location should
provide acceptable fore-and-aft c.g. limits.

The first step in c.g. calculation is usually made
in the stage when the wing location has to be determined. Wing location may be fixed in some cases
by consideration of wing attachment provisions.
For example, in the case of the HFB Hansa business
jet, a mid-wing layout, the condition was imposed
that the wing box had to be located aft of the passenger cabin. This could be solved only by using a
swept-forward wing. For low-subsonic aircraft, however, wing sweep should be limited to some 5 or 10
degrees to avoid undesirable aerodynamic effects.

A proposed design procedure to balance the
wing location, tail size and c.g. travel
will be discussed in Section 9.5.2. In a
first approach it is generally sufficient
to aim at a reasonable c.g. travel and to
choose the wing location accordingly. This
simple procedure is illustrated by Fig.
8-11.

Step 1.
The airplane is subdivided into at least
the following groups:
1. The fuselage group* - containing parts
whose location is fixed relative to the
fuselage, e.g. the fully furnished and

*Note that the terms "fuselage group" and
"wing group" as used here have different
meanings from those in Section 8.4.2.

300

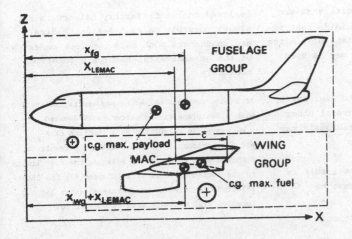

Fig. 8-11. Determination of
the wing location

equipped fuselage, several airframe services, fuselage-mounted engines, the horizontal and vertical tailplane, the nosewheel u.c., but not the payload nor the
main u.c., as the latter is more or less
fixed relative to the MAC·(assume 10-15%
behind the aft c.g. limit or 45-50% MAC
approximately for a nosewheel u.c.).
2. The wing group* - parts whose location
is fixed relative to the wing: the wing
structure, fuel system, main u.c. legs
(even if attached to the fuselage) and
wing-mounted jet engines. Occasionally,
wing-mounted engines should be considered
as part of the fuselage group if, for example, the propeller plane has a fixed
location next to a cargo compartment (cf.
Section 6.4.5.).
3. The (variable) payload.
4. The (variable) fuel load.
Step 2
The fuselage group is drawn and the X-axis
usually assumed parallel to the cabin floor
or the propeller axis. Using the data of
Section 8.4. and Table 8-15, the c.g. of
the complete group is computed in both the
X- and the Z-direction.
Step 3
The empty wing group is drawn on a separate

(transparent) sheet and the root chord, the
tip chord and the MAC* or SMC* are indicated. The c.g. is computed and expressed
as a fraction of the MAC relative to the
MAC Leading Edge (LEMAC).
Step 4
Assume a reasonable location for the c.g.
of the empty aircraft (x_{OE}) relative to
the MAC, using statistical data and the
considerations presented in Section 8.5.3.
Step 5
Calculate the coordinate of the wing leading edge (X_{LEMAC}) relative to the fuselage
coordinate system:

$$X_{LEMAC} = x_{fg} - x_{OE} + \frac{W_{wg}}{W_{fg}} (x_{wg} - x_{OE}) \qquad (8-50)$$

where "fg" and "wg" respectively denote
the fuselage and wing groups.
Step 6
Compute a load and balance diagram of the
type illustrated in Fig. 8-7 by considering
various possible combinations of payload
and fuel loading. The window seating rule
(Section 8.5.2.) is applied to passenger
transports.
Step 7
Allowing for weight growth, various cabin
layouts, fuel capacity growth, Standard
Items Variations and design tolerances,

*Note that the terms "fuselage group" and
"wing group" as used here have different
meanings from those in Section 8.4.2.

*Defined in Appendix A.

estimate the fore-and-aft limits that are
likely to yield acceptable loading restrictions and a horizontal tailplane of limited
size. The data in Table 8-16 may be used
for comparison.

Step 8

In the case of an unacceptable c.g. shift,
a revised choice of x_{OE} or several other
design revisions discussed previously may
improve the situation.

Step 9

Repeat the procedure until the result is
considered satisfactory. It must be
realized that satisfactory balance is not
always possible on each design. Radical
design changes such as a complete redesign
of the general arrangement are sometimes
necessary.

If a more complete balancing process in combination
with a tailplane sizing procedure is desired, the
c.g. travel is calculated as a function of x_{LEMAC}
and the horizontal tailplane area. The result is
combined with an "X-plot", relating the c.g. limits
to the horizontal tail size (Chapter 9). The interrelationship between undercarriage design and c.g.
location will be dealt with in Chapter 10.

Chapter 9. Preliminary tailplane design

SUMMARY

The aerodynamic design of the tailplane is based on many specific requirements regarding its functions, which are to provide equilibrium in steady flight (trim), to ensure that this condition is stable and that disturbances are well damped, and to generate aerodynamic forces for maneuvering the aircraft. The control forces involved must be acceptable to pilots, with the airplane both in trimmed and out-of-trim conditions.

Design requirements for longitudinal stability and control characteristics - basically those specified in the airworthiness regulations - form the starting point for the derivation of limits to the location of the center of gravity in connection with the size of the horizontal tailplane. Attention is paid to the reduction of control forces by aerodynamic balancing and to some aspects of dynamic behavior, the latter affecting the tailplane design of large aircraft with irreversible control systems. A detailed procedure is presented, together with design data, for obtaining a balance between the required and available c.g. range, and the wing location.

Design criteria, considerations and methods are presented for estimating the minimum size of the vertical tailplane and the rudder control capacity. Control after failure of an engine on multi-engine transports, directional stability and landings in crosswind are considered as the most pertinent aspects. Design recommendations relevant to light aircraft are made in order to ensure recovery from spins.

NOMENCLATURE

A	-	aspect ratio (no subscript: wing aspect ratio)
a.c.	-	aerodynamic center
B_p	-	number of blades per propeller
b	-	span (no index: wing span); width
C_h	-	hinge moment coefficient of a control surface = hinge moment/ (local dynamic pressure x control surface area x mean control surface chord)
C_{h_α}; C_{h_δ}	-	$\partial C_h/\partial\alpha$ and $\partial C_h/\partial\delta$, respectively
C_L	-	lift coefficient, based on local dynamic pressure and gross area of airfoil
C_{L_α}	-	lift-curve slope = $dC_L/d\alpha$
$C_{L_{h_\alpha}}$; $C_{L_{h_\delta}}$	-	$\partial C_{L_h}/\partial\alpha_h$ and $\partial C_{L_h}/\partial\delta_e$, respectively
C_m	-	pitching moment coefficient, $C_m = M/(\tfrac{1}{2}\rho V^2 S\bar{c})$
C_{m_α}; $C_{m_{\dot\alpha}}$	-	$\partial C_m/\partial\alpha$ and $\partial C_m/\partial\left(\frac{\dot\alpha\bar{c}}{2V}\right)$
$C_{m_{ac}}$	-	pitching moment coefficient about the aerodynamic center
C_{m_q}	-	$\partial C_m/\partial\left(\frac{q\bar{c}}{2V}\right)$
C_N	-	normal force coefficient = normal force/(local dynamic pressure x reference area)
C_n	-	yawing moment coefficient = $N/(\tfrac{1}{2}\rho V^2 Sb)$
C_{n_β}	-	coefficient of directional stability = $\partial C_n/\partial\beta$
C_y, C_{y_v}	-	side force coefficient = $Y/(\tfrac{1}{2}\rho V^2 S)$ and $Y_v/(\tfrac{1}{2}\rho V_v^2 S_v)$, respectively
$C_{y_{v_\alpha}}$	-	lift-curve slope of vertical tailplane = $\partial C_{y_v}/\partial\alpha_v$
$C_{\frac{1}{2}}$	-	cyclic damping of an oscillation = $T_{\frac{1}{2}}/P$
$\bar{c}$	-	length of the MAC
c.g.	-	center of gravity
D_p	-	propeller diameter
F_e	-	elevator control force, positive forwards (push)
f	-	frequency (sec^{-1})
G	-	elevator control gear ratio
g	-	acceleration due to gravity
h	-	height of fuselage; fin span
I_y	-	polar moment of inertia about the lateral axis

i_h	-	horizontal tailplane incidence
K_y	-	non-dimensional radius of inertia in pitch, $K_y^2 = I_y g/(W\bar{c}^2)$
k	-	correction factor
L	-	lift
L_α	-	factor characterizing the longitudinal maneuverability
ℓ	-	length (of a moment arm); distance
M	-	pitching moment; Mach number
$M_{.25}$	-	pitching moment about the mean quarter-chord point of the wing
MAC	-	Mean Aerodynamic Chord
m	-	non-dimensional factor defining the vertical position of the horizontal tailplane
N	-	yawing moment
N_β	-	$\partial N/\partial\beta$
N_e	-	yawing moment caused by asymmetrical engine thrust
n	-	normal load factor = L/W
n_L	-	limit load factor
n_{z_α}	-	normal load factor gradient = $\partial n/\partial\alpha$
P	-	engine power; period of an oscillation
P_{br}	-	brake horsepower
q	-	angular pitching velocity; dynamic pressure
r	-	factor defining the longitudinal position of the horizontal tailplane = $2\ell_h/b$
S	-	area (no subscript: gross wing area)
SMC	-	Standard Mean Chord
T	-	thrust
$T_{\frac{1}{2}}$	-	time to half amplitude of an oscillation
ΔT_e	-	asymmetry in the thrust caused by engine failure
V	-	(true) airspeed
V_a	-	approach speed
V_{cr}	-	cruising speed
V_{FE}	-	design speed with flaps extended (EAS)
V_{max}	-	maximum level flight speed
V_{MC_A}	-	minimum control speed in the air (EAS)
V_R	-	rotation speed
V_S	-	minimum speed in a stall
V_{tr}	-	trim speed

W — (airplane) weight

W_e — Operational Weight Empty

W_{land} — Maximum Landing Weight

W_{Pmax} — Maximum Payload Weight (structural limit)

W_{to} — Maximum Takeoff Weight

x — longitudinal coordinate in airplane axis system

$x_{cg};\Delta x_{cg}$ — longitudinal coordinate and travel of airplane c.g., respectively

Y — side force (normal to the aircraft's plane of symmetry)

Y_β — $\partial Y/\partial\beta$

Y_e — yawing moment arm of critical engine

z_T — distance of thrust vector below c.g.

$\alpha;\dot\alpha$ — angle of attack; $d\alpha/dt$

β — angle of sideslip

Δ — shift; additional contribution (e.g. $\Delta_T...$= increment due to thrust)

δ — control surface deflection angle

ε — downwash angle

ζ — damping ratio of an oscillation

η — tailplane effectiveness (reduction) factor

$\theta;\dot\theta$ — angle of pitch; $d\theta/dt$

Λ — sweepback angle (no subscript: wing sweepback angle)

λ — taper ratio (no subscript: wing taper)

μ — tire-to-runway friction coefficient

μ_c — airplane mass ratio = $W/(\rho g S\bar c)$

$\rho;\rho_0$ — air density and ρ at sea level, respectively

σ — sidewash angle

τ — change in zero-lift angle per degree control deflection

ϕ — angle defining the location of the horizontal tailplane relative to the wing root chord

φ — angle of bank

ω — angular frequency of an oscillation

ω_n — undamped natural frequency of an oscillation

Subscripts

A-h — aircraft without horizontal tailplane

A-v — aircraft without vertical tailplane

ac — aerodynamic center

cg — center of gravity

e — elevator; (critical) engine

f — fuselage; flap at the trailing edge

fn — nose section of fuselage

fs — side view of fuselage

g — (main) landing gear

h — horizontal tailplane

i — interference

r — maneuver point

n — neutral point; nacelle(s)

p — propeller(s)

R — takeoff rotation

r — rudder

s — slipstream of propeller(s); leading-edge high-lift device (slat)

T — thrust

to — takeoff

tr — trimmed condition

v — vertical tailplane

0 — sea level; landing configuration

1 — takeoff configuration

.25 — quarter-chord line or point

A prime denotes the stick-free condition or a special definition of the vertical tailplane

9.1. INTRODUCTION TO TAILPLANE DESIGN, CONTROL SYSTEMS AND STABILIZATION

The functions allotted to the fixed and movable tail surfaces are as follows:
a. To ensure equilibrium of moments in steady flight by exercising a force at a given distance from the center of gravity.
b. To ensure that this equilibrium is stable, which implies that arter a disturbance the equilibrium is restored and that there is adequate damping for the

rapid suppression of oscillations.

c. To generate forces for maneuvering the aircraft: rotation during takeoff, control of the flight path, flareout during landing and taxying.

These useful attributes are counteracted to some extent by a large item on the debit side. Although the total tail surface area of small aircraft will not exceed some 25 to 30 percent of the wing area, ratios of 40 to 50 percent are no exception in the case of some high-speed and STOL aircraft. In cases like these the structural weight and drag reach such high values that it will unquestionably prove worthwhile to investigate how the tail surfaces can be reduced to a minimum during the preliminary design stage.

The tailless aircraft shows that the functions of the horizontal tail may be taken over by other elements, nor will it always be necessary to locate the tail surfaces behind the wing. The canard type of configuration is sometimes adopted to obtain particular characteristics. Section 2.6.3. shows why this rather unconventional configuration may sometimes have certain advantages. It will henceforth be assumed that the arrangement of the empennage has (provisionally) been decided upon, for example on the basis of the considerations dealt with in Section 2.4. Using the methods given it will nevertheless be possible to obtain a general idea of the influence which the location of the horizontal tail surface will have on its dimensions. A final decision cannot be taken until the structure has been fully developed.

A survey of some methods which enable the designer to size the tail surfaces and determine their principal characteristics will be presented here, taking a number of important design criteria as point of departure. The survey makes no claim to completeness, and the approach recommended is not necessarily representative of that chosen in most design departments. In many cases the knowledge gained through experience, as well as the designer's own intui-

tion, may play a much more important role than may be evident from the present text. Apart from some rather detailed procedures fast, simple methods are also given for a first estimation of the size and shape of the tailplane.

During the preliminary design stage the tail surfaces may present one of the most difficult problems in the dimensioning of the main parts of the aircraft, and this, in turn, may lead to many iterations.

a. The aerodynamic characteristics are very sensitive to design details and can only be calculated very approximately during the stage when "paper designs" are being made. For example, the asymmetrical flow pattern during flight with one engine inoperative and the aircraft in a sideslip, does not lend itself to calculations. In addition, nonlinear behavior is a frequent occurrence and compressibility effects may have a much more radical and unpredictable influence on the flying qualities than on performance.

b. Most requirements have to be satisfied for a wide range of operational conditions. Not only is it difficult to anticipate which conditions will be critical, but in addition there may be conflicting solutions, resulting in an unsatisfactory compromise.

c. Contrary to the dimensioning of the wing of conventional aircraft, where reliable calculations may be carried out on the basis of relatively simple relationships from flight mechanics, the analysis of the flying characteristics calls for determination of the dynamic behavior. In the case of large, high-speed aircraft attention should also be given to aeroelasticity.

d. The design criteria relating to the flying qualities of civil aircraft are mostly outlined in fairly general terms and may be subject to different interpretations. The pilot's verdict plays an important part and for this reason a flight simulator is sometimes introduced during the design stage.

e. The design of the empennage is closely tied up with that of the surface control system. The distribution of masses and its variations - e.g. center of gravity locations, moments of inertia - for various loading conditions will also have to be known. Proposals for further modifications may nevertheless follow from the dimensioning of the tail surface.

Several attempts have been made to combine existing knowledge on aerodynamic characteristics relating to tailplane design into a consistent set of handbook-type methods. The most important of these are the USAF Stability and Control Datcom (Ref. 9-39) and the sheets prepared by the Engineering Sciences Data Unit (ESDU), formerly by the Royal Aeronautical Society (Ref. 9-41). The Royal Aircraft Establishment (RAE) has recently drawn up a computer program based on these procedures (Ref. 9-46). A more concise collection of prediction methods can be found in Appendices E and G of the present textbook.

The detailed design of the surface control system is generally not regarded as part of the work of the preliminary design engineer, with the usual exception of light aircraft which have relatively simple mechanical systems. Nevertheless, it is essential to decide how the aerodynamic loads exerted on the movable portion of the tailplane will be felt by the pilot.
a. When manual control is adopted there will be a direct mechanical transmission of forces from the control surfaces to the flight deck controls and back by means of rods and/or cables (reversible system). The aerodynamic forces acting on the control surface will be felt by the pilot directly or indirectly (via tabs), although friction in the system will obscure his perception.
For a given type of control, the stick forces increase with the size, equivalent airspeed and load factor. Methods are required to reduce these forces, e.g. overhanging balance, balancing tabs and spring tabs.
b. Power-assisted controls may be used when the desired control forces and system linearity cannot be satisfactorily obtained by aerodynamic means. The rudder or elevator is operated by means of a pneumatic or hydraulic ram which exerts a multiple of the force applied by the pilot (boost ratio). Control feel will still be provided without excessive control forces.
c. Power(-operated) controls are used on many large and high-speed transports and even on relatively small high-subsonic executive aircraft. In these systems the control surfaces are moved by electrical, hydraulic or pneumatic means without direct physical effort by the pilot. Control feel is obtained by artificial means. A survey of the advantages and disadvantages of the various systems can be found in Ref. 9-6.

The majority of present-day high-speed aircraft are controlled by power-assisted or power-operated control systems, because aerodynamic balancing of the control surfaces demands considerable time in development, while the result is often sensitive to manufacturing tolerances, and modifications are sometimes required during flight evaluation. In the case of very large, fast aircraft with highly sweptback wings and effective flaps, it will not always be possible to achieve satisfactory characteristics for all operational conditions and aircraft configurations. Electronic systems have been developed - and are in operational use - which improve the stability (Stability Augmentation Systems, SAS) and provide more effective response to control wheel movements as compared with the natural behavior of the aircraft. Development work is also being carried out to find out whether a worthwile gain can be obtained from a fully automatic stabilization system. An appreciable reduction in tail area may be feasible if the natural "aerodynamic" stability is dispensed with and if the whole aircraft is designed from the outset with this approach in mind (Control Configured Vehicle). The actual gain from such a system will be determined by the tail design requirements which stem from conditions other than stabilization (maneuvering, takeoff rotation and landing flareout) and may thus vary widely for different types of aircraft.
Although these aspects are closely tied up with the design of the tail surfaces, they

will only be referred to as occasion demands. Ref. 9-11 investigates this subject more fully on the basis of a practical approach.

9.2. STATIC LONGITUDINAL STABILITY AND ELEVATOR CONTROL FORCES

A condensed account will be given of the major requirements relating to static longitudinal stability and control and to the influence of tail surface design on these aspects. For more detailed information the reader should consult the relevant literature. For the sake of clarity the derivation of most of the formulas has been omitted since they can be found in practically every textbook on the present subject.

9.2.1. Stick-fixed static stability and neutral points

In flight conditions which may persist for any length of time, as well as during the takeoff, approach and landing phases, the aircraft will have to be statically stable with fixed as well as free elevator controls. Under these circumstances a state of equilibrium is achieved by adjusting the elevator or the stabilizer, if the latter is adjustable. The aircraft will possess static longitudinal stability if, from this trimmed state, a disturbance in the angle of attack results in a pitching moment which tends to restore the aircraft to the state of equilibrium. In Fig. 9-1 the trimmed condition is defined by the equilibrium of forces*. Ignoring the contribution of drag forces, we have for the total lift:

*Departing from convention, we have taken lift forces instead of normal forces, since this is acceptable for small angles of attack. The reference line is taken through the leading edge of the MAC (LEMAC), parallel to the fuselage reference line (e.g. the cabin floor)

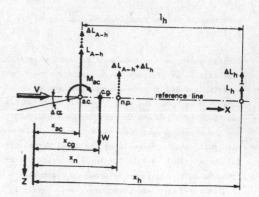

Fig. 9-1. Geometric definitions, forces and moments

$$L = L_{A-h} + L_h = W \qquad (9-1)$$

and the moment about an axis through the aerodynamic center, normal to the X-Z plane:

$$M = M_{ac} + W (x_{cg} - x_{ac}) - L_h \ell_h \qquad (9-2)$$

or, in nondimensional form:

$$C_m = C_{m_{ac}} + C_L \frac{x_{cg} - x_{ac}}{\bar{c}} - C_{L_h} \frac{S_h \ell_h}{S \bar{c}} \left(\frac{V_h}{V}\right)^2 \qquad (9-3)$$

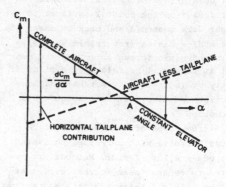

Fig. 9-2. Pitching moment curve and trimmed condition

Fig. 9-2 shows the pitching moment coefficient of the aircraft without the horizontal tail, the tailplane contribution and the aircraft pitching moment coefficient. The trimmed condition is represented by point A, where $C_m = 0$. A disturbance by an instantaneous change in the angle of attack $\Delta\alpha$ causes incremental lift forces on the aircraft less the horizontal tail, and the horizontal tailplane, whose resultant acts through the neutral point. By definition the stick-fixed neutral point is the position of the center of gravity for which $dC_m/d\alpha = 0$, for constant elevator angle. The degree of static stability is normally defined as the distance of the center of gravity ahead of the neutral point, expressed as a fraction of the Mean Aerodynamic Chord (MAC) or the Standard Mean Chord (SMC), referred to as the static margin:

$$\frac{x_n - x_{cg}}{\bar{c}} = -\frac{dC_m}{dC_N} = -\frac{dC_m/d\alpha}{dC_N/d\alpha} \approx -\frac{dC_m/d\alpha}{dC_L/d\alpha} \qquad (9-4)$$

The condition for static stability is $dC_m/d\alpha < 0$ and the center of gravity must therefore be in front of the neutral point to achieve this.

It can be shown that the neutral point is related to the aerodynamic center of the aircraft, less tail, as follows:

$$\frac{x_n - x_{ac}}{\bar{c}} = \frac{C_{L_{h_\alpha}}}{C_{L_\alpha}} \left(1 - \frac{d\epsilon}{d\alpha}\right) \frac{S_h \ell_h}{S \bar{c}} \left(\frac{V_h}{V}\right)^2 \qquad (9-5)$$

where

$$C_{L_\alpha} = \left(C_{L_\alpha}\right)_{A-h} + C_{L_{h_\alpha}} \frac{S_h}{S} \left(1 - \frac{d\epsilon}{d\alpha}\right)\left(\frac{V_h}{V}\right)^2 \qquad (9-6)$$

This expression is valid only in the absence of compressibility effects and effects due to powerplant operation. Some additional details will be given in Section 9.2.5.

A simple criterion for the size of the horizontal tail may be obtained by assuming that the stick-fixed static margin should at least have a certain specified minimum value, from which a condition for the aft c.g. location is found:

$$\frac{x_{cg}}{\bar{c}} = \frac{x_{ac}}{\bar{c}} + \frac{x_n - x_{ac}}{\bar{c}} - \frac{x_n - x_{cg}}{\bar{c}} \qquad (9-7)$$

The aft c.g. location is depicted in Fig. 9-3 as a function of the horizontal tail

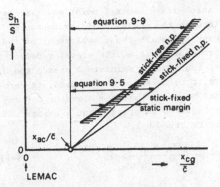

Fig. 9-3. Stick-fixed and stick-free neutral points and aft c.g. limits vs. tail volume size.

The aerodynamic center (aircraft less tail) is mainly determined by the wing shape, but important contributions due to the fuselage and nacelles must be taken into account. Appendix E provides methods and data for estimating its location. The term $x_n - x_{ac}$ in (9-7) is obtained from (9-5) and (9-6), for which the required coefficients can be obtained from Appendix E.

A minimum required value of 5 percent of $\bar{c}$ is frequently used for $x_n - x_{cg}$, but it should be realized that for most aircraft the effects of compressibility and power result in large variations. In addition, an analysis of static margins at the aft c.g. of existing aircraft shows a considerable spread, which must be attributed to the following points.

a. The stick-fixed static margin is not really a significant criterion, but rather an outcome of the tail sizing process according to other criteria (stick forces, dynamic stability, etc.).

b. The use of the MAC in static stability considerations is arbitrary. It has been found that aircraft with a relatively small MAC - or a large value of $\ell_h/\bar{c}$ - have large static margins. A better result would be obtained by relating the static margin to the tail moment arm ℓ_h, which is more significant for c.g. shifts due to different loading conditions. A static margin of 1.5 to 2 percent of ℓ_h at the aft c.g. appears to be a good assumption, if, in spite of the previous objections, the tail size is based on static stability.

The stick-fixed neutral point may also be defined as the position of the center of gravity for which the displacement of the control column (i.e. the elevator deflection) required to maintain a steady speed, above or below the trimmed speed, is zero. It can be shown that the elevator deflection required to bring about a change in speed is given by:

$$\frac{d\delta_e}{dV} = \frac{2C_L}{C_{L_{h_\delta}}} \frac{(x_n - x_{cg})S}{V_{tr}} \frac{S}{S_h \ell_h} \qquad (9-8)$$

where V_{tr} is the speed in trimmed flight. The elevator and stick movements required to change the trimmed speed by a small amount are thus proportional to the stick-fixed static margin. Since the pilot regards the stick forces as more important than the control displacements, no limits are imposed on $d\delta_e/dV$.

9.2.2. Stick-free static stability and neutral point; the stick force gradient

In the case of a manual (reversible) control system the free elevator will assume a position where the aerodynamic hinge moment will be zero, provided any friction in the system is neglected. This will result in a different position for each angle of attack, and consequently the stabilizing action of the tail surface will be changed. There is an analogy with the case where the controls are fixed, so for the stick-free neutral point we have:

$$\frac{x_n' - x_{ac}}{\bar{c}} = \frac{C_{L_{h_\alpha}}'}{C_{L_\alpha}} \left(1 - \frac{d\epsilon}{d\alpha}\right) \frac{S_h \ell_h}{S\bar{c}} \left(\frac{V_h}{V}\right)^2 \qquad (9-9)$$

where

$$C_{L_{h_\alpha}}' = C_{L_{h_\alpha}} - C_{L_{h_\delta}} \frac{C_{h_\alpha}}{C_{h_\delta}} \qquad (9-10)$$

and

$$C_{L_\alpha}' = \left(C_{L_\alpha}\right)_{A-h} + C_{L_{h_\alpha}}' \frac{S_h}{S} \left(1 - \frac{d\epsilon}{d\alpha}\right) \qquad (9-11)$$

Hence the stick-free stability is identical to the stick-fixed stability, except that the tailplane lift gradient is effectively reduced by the factor $C_{L_{h_\delta}} C_{h_\alpha}/C_{h_\delta}$. Since for a normally balanced elevator $C_{h_\delta} < 0$ and $C_{L_{h_\delta}}$ is always positive, the sign of C_{h_α} will decide whether or not the stability will be increased when the control is left free. If $C_{h_\alpha} > 0$ the control will trail against the wind, stability will be increased and the stick-free neutral point will be further back than it would be with the stick fixed. If $C_{h_\alpha} < 0$ the control trails with the wind and the stick-free stability is less than in the stick-fixed situation.

It can be shown that the stick-free stability is related to the stick force gradient, i.e. the change in stick force required for a small change in forward speed is given by the following relationship:

$$\frac{dF_e}{dV} = -2 \frac{W}{V_{tr}} \frac{S_e \bar{c}_e}{S_h \ell_h} \frac{\bar{c} \, G \, C_{h_\delta}}{C_{L_{h_\delta}}} \frac{x_n' - x_{cg}}{\bar{c}} \qquad (9-12)$$

where the stick gearing G is defined as the ratio of the elevator rotation to the control column displacement. Normal values of G will be .6 to .9 rads/ft (2 to 3 rads/m).

It can be seen that the stick force gradient is positive if the pilot has to push the control dolumn forward ($dF_e > 0$) to increase the flight speed from a trimmed condition ($dV > 0$).

For transport aircraft the FAR 25.173 regulation stipulates that for all weights between the MLW and the MTOW the stick force gradient must amount to at least one lb per six knots (.15 kg per m/s) of speed increment from the trimmed speed. Assuming that the elevator span is equal to that of the horizontal stabilizer, we may assume:

$$\frac{C_{L_{h_\delta}}}{C_{L_{h_\alpha}}} \approx \sqrt{\frac{S_e}{S_h}} \qquad (9-13)$$

Substituting this in (9-12), we obtain the following condition for the stick-free static margin:

$$\frac{x_n'-x_{cg}}{\bar{c}} > - \frac{(dF_e/dV)_{min}\, V_{max}\, \sqrt{A_h}\, \ell_h\, C_{L_{h_\alpha}}}{2\, W_{land}\, \sqrt{S_h}\, (S_e/S_h)^{3/2}\, \bar{c}\, G\, C_{h_\delta}}$$

$$(9-14)$$

where the maximum level flight speed and the MLW are assumed to form the most demanding situation. The aft c.g. limit corresponding to (9-14) has been plotted in Fig. 9-4.

Another requirement is formulated as a limitation of the control force (FAR 25.175) within a limited region of approach speeds $(1.1\, V_{S_o} < V < 1.8\, V_{S_o})$. This should not exceed 80 lb (36 kg) at a trimmed speed of $1.4\, V_{S_o}$. Then, by approximation, we have:

$$\frac{dF_e}{d(V/V_{S_o})} \leqslant 200 \text{ lb (91 kg)} \qquad (9-15)$$

This requirement is equivalent to a forward limit of the c.g. position:

$$\frac{x_n'-x_{cg}}{\bar{c}} < -.7 \left[\frac{dF_e}{d(V/V_{S_o})}\right]_{max} \times$$

$$\frac{\sqrt{A_h}\, \ell_h\, C_{L_{h_\alpha}}}{W_{land}\, \sqrt{S_h}\, (S_e/S_h)^{3/2}\, \bar{c}\, G\, C_{h_\delta}}$$

$$(9-16)$$

This requirement is generally less critical than the limits imposed by the maximum permissible stick force per g, which will be dealt with in the following section.

9.2.3. Stick-fixed and stick-free maneuver points and maneuver control forces

The maneuver stability of a conventional aircraft must satisfy the condition:

$$\frac{d\delta_e}{dn} < 0 \qquad (9-17)$$

which implies that, in order to pull the aircraft out of a dive at a steady speed,

the pilot must pull the control column towards him. The maneuver stability is made up of two components, one resulting from the change in incidence involved and one from the angular rotation in pitch as the aircraft follows a curved flight path. The stick-fixed maneuver point is defined as the position of the center of gravity when the stick displacement per g is zero. This point is related to the stick-fixed neutral point by the following relationship:

$$\frac{x_m-x_n}{\bar{c}} = - \frac{C_{m_q}\, \rho g S \bar{c}}{4W} \qquad (9-18)$$

where the stability derivative C_{m_q},

$$C_{m_q} = \frac{\partial C_m}{\partial \left(\frac{q\bar{c}}{2V}\right)} \qquad (9-19)$$

represents the aerodynamic damping in pitch, primarily due to the change in horizontal tailplane angle of attack. The following equation can be used to estimate the stick-fixed maneuver point:

$$\frac{x_m-x_n}{\bar{c}} = .55\, \rho g\, \frac{S_h\, \ell_h^2\, C_{L_{h_\alpha}}}{W\bar{c}} \qquad (9-20)$$

This equation shows that the stick-fixed maneuver point lies behind the stick-fixed neutral point (Fig. 9-4) and that the dis-

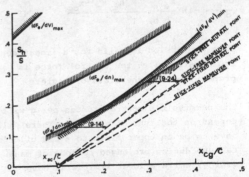

Fig. 9-4. Limitations for the horizontal tail area and c.g. position for aircraft with manual control, based on stick force requirements

311

tance between the two points decreases
with altitude.

There are no civil requirements which stipulate limits regarding $d\delta_e/dn$. However, as
will be seen from Section 9.3.2., the maneuver stability is important in relation
to the longitudinal short period oscillatory response of an aircraft to control
movements.

The stick-free maneuver point is defined
as the position of the center of gravity –
with free elevator control – for which the
stick force per g is zero in the case of a
steady pull-out of an airplane, trimmed
for level flight at the same speed. Again
using (9-13), the location of this point
can be related to the stick-free neutral
point:

$$\frac{x_m' - x_n'}{\bar{c}} = .55 \; \rho g \; \frac{S_h \ell_h^2 \; C_{L_{h_\alpha}}}{W\bar{c}} \qquad (9\text{-}21)$$

where

$$C_{L_{h_\alpha}}' \approx C_{L_{h_\alpha}} \left\{ 1 - \frac{\bar{S}_e}{S_h} \frac{C_{h_\alpha}}{C_{h_\delta}} \right\} \qquad (9\text{-}22)$$

The maneuver margin is the distance by
which the c.g. lies ahead of the maneuver
point, and it represents a direct measure
of the stick force per g:

$$\frac{dF_e}{dn} = - \; W \; \frac{S_e \bar{c}_e}{S_h \ell_h} \; \frac{\bar{c} \, G \, C_{h_\delta}}{C_{L_{h_\delta}}} \; \frac{x_m' - x_{cg}}{\bar{c}} \qquad (9\text{-}23)$$

This expression is valid for a linear control system, i.e. the control force is
proportional to the elevator angle.

The permissible control forces per g will
depend on the type of aircraft. There will
have to be an upper limit to avoid pilot's
fatigue during prolonged periods of maneuvering, while a lower limit will have to
be set in order to prevent excessive
stresses in the structure and over-sensi-
tivity in rough air. The limits taken from
the military requirement MIL-F-8785 (Ref.
9-18) may also be usefully applied to

	Unit	Control Wheel	Control Stick
Maximum dF_e/dn	lb/g	$\frac{120}{n_L - 1}$ but ≤ 120	$\frac{56}{n_L - 1}$ but ≤ 28
	kg/g	$\frac{54.4}{n_L - 1}$ but ≤ 54.4	$\frac{25.4}{n_L - 1}$ but ≤ 12.7
Minimum dF_e/dn	lb/g	$\frac{.45}{n_L - 1}$ but ≥ 6	$\frac{21}{n_L - 1}$ but ≥ 3
	kg/g	$\frac{20.4}{n_L - 1}$ but ≥ 2.7	$\frac{9.5}{n_L - 1}$ but ≥ 1.4

Table 9-1. Longitudinal stick force limits
according to MIL-F-8785 B(ASG), Ref. 9-18

transport aircraft; see Table 9-1. The factor n_L in this table is the limit load
factor, specified in the airworthiness regulations for maneuver loads. For transport
aircraft weighing more than 50,000 lb
(22,680 kg) this load factor is equal to
2.5, while for lighter aircraft it is a
function of the weight.

The control force limits may be used to
supply corresponding limits to the stick-
free maneuver margin. Substituting (9-13)
into (9-22), and rearranging, we obtain:

$$\frac{x_m' - x_{cg}}{\bar{c}} = - \; \frac{dF_e/dn \; \sqrt{A_h} \; \ell_h \; C_{L_{h_\alpha}}}{W \; \sqrt{S_h} \; (S_e/S_h)^{3/2} \; \bar{c} \, G \, C_{h_\delta}} \qquad (9\text{-}24)$$

Flight at sea level will be the deciding
case for the forward limit, while the aft
limit will be determined by flight at high
altitude. An example of the permissible
range of c.g. positions is shown in Fig.
9-4 as a function of the relative tailplane area.

The range of allowable c.g. locations for transport
aircraft with manual control systems should not be
unduly limited by the permissible stick forces per
g. In particular, the forward limit is determined
by considerations of controllability in the low-
speed configuration (Section 9.4.). If, therefore,
we assume that for this c.g. limit the stick force
per g reaches its maximum value, we may derive a
condition for the aft c.g. limit from the value of
minimum stick force per g by subtracting the maximum and minimum values obtained for (9-23) within
the flight envelope, with the following result for

the aft c.g. limit:

$$\frac{x_{cg}}{\bar{c}} = \frac{x_m'}{\bar{c}} - .6\,\frac{\Delta x_{cg}}{\bar{c}} - .88\,\frac{(\rho_0-\rho_{min})\,g\,S_h\,\ell_h^2\,C_{L_{h_\alpha}}}{w\,\bar{c}}$$

$$(9-25)$$

where x_m' is the stick-free maneuver point at sea
level. The forward limit is located at a distance
Δx_{cg} ahead of the aft limit.
Equation 9-25 shows clearly that, in order to obtain acceptable maneuvering stick forces, the aft
c.g. limit should be forward of the stick-free
maneuver point by an amount which is directly related to the c.g. travel. For aircraft with a large
c.g. travel we conclude that the maneuver margin at
the aft c.g. will also be relatively large, and
this is one of the reasons why a generalized minimum for the static margin is not relevant, as already stated in Section 9.2.1.

9.2.4. Reduction of control forces

The control forces for a manual control system are dependent on the hinge mo-
ment coefficients C_{h_α} and C_{h_δ} and the a-
mount of stick-free stability provided by
the aircraft. For given values of C_{h_α} and
C_{h_δ} the stick force per g increases proportionally to the All-Up Weight of the
aircraft and the linear dimensions, as
shown by (9-23), although even in the case
of light aircraft it will be necessary to
reduce the control forces. Incidentally,
aerodynamic balancing is applied not only
to manually controlled surfaces, but also
to power-assisted or power-operated surfaces, in order to reduce the power requirements.

Various methods of balancing are available,
but the reduction of C_{h_δ}, for example, must
not be carried too far, because if C_{h_δ} becomes positive the control surface is overbalanced. The degree of control force reduction is limited by non-linearities in
the control force/deflection relationship,
by the variation in characteristics between aircraft of the same production
batch due to production imperfections, and
by small changes in the shape of the control surface due to the formation of ice.

In addition, certain tab systems exhibit a
slow response to the pilot's commands.

The hinge moment coefficient C_{h_δ} - the control "heaviness" parameter - is of primary
importance in reducing the control forces.
In order to get some idea of the reduction
in C_{h_δ} required to meet the stick force requirements for various airplane sizes, it
may be useful to impose the condition that
the maximum and minimum permitted values
of the stick force per g should be obtained at the forward and aft c.g. limit,
respectively. The corresponding value of
C_{h_δ} is obtained from (9-23) by ignoring
the effect of altitude variation on the
stick-free maneuver point:

$$C_{h_\delta} = \frac{\left\{(dF_e/dn)_{max} - (dF_e/dn)_{min}\right\}\sqrt{W/S}\,\sqrt{A_h}}{w^{3/2}\,(S_e/S_h)^{3/2}\,\sqrt{S_h/S}\;G\;\Delta x_{cg}/\bar{c}}\;\times$$

$$C_{L_{h_\alpha}}\,\ell_h/\bar{c} \qquad\qquad (9-26)$$

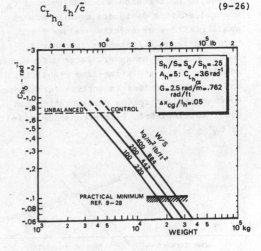

Fig. 9-5. Typical hinge moment coefficient
required to obtain maximum and minimum
stick force per g at forward and aft c.g.
limit for transport aircraft with control
wheels

This result has been plotted in Fig. 9-5
for a typical combination of parameters vs.
the All-Up Weight of transport aircraft,
for various wing loadings. The figure suggests that for weights in excess of about

313

50,000 lb (23,000 kg) the reduction of C_{h_δ} becomes problematic in view of the minimum value recommended in Ref. 9-28. However, various servo-tab and spring-tab systems are available, which may extend the range of airplane sizes for which manual control can be applied up to the Douglas DC-8 class of high-speed transports, though this will entail an increasing number of control design problems.

The hinge coefficient C_{h_α} - the control "floating" parameter - is decisive for the change in stick force resulting from the response of the aircraft to the control movement. When the aircraft responds to the elevator deflection, its angle of attack is changed and the control force required to maintain a steady maneuver is either greater or less than the control force required to initiate the maneuver, depending on the sign and magnitude of C_{h_α}. Control surface designers usually aim at small positive or negative values for C_{h_α} to avoid large differences in control forces during a maneuver, large differences in stick-fixed and stick-free stability, and large variations in control forces due to changes in airplane configuration and power setting. Ref. 9-28 recommends a limit of $C_{h_\alpha} < -\frac{1}{3} C_{h_\delta}$.

Various methods for obtaining satisfactory values of C_{h_δ} and C_{h_α} will now be reviewed briefly (see Figs. 9-6 and 9-7). Methods for estimating these coefficients can be found in Refs. 9-24 and 9-41, but high prediction accuracies must not be expected. THE PLAIN (UNBALANCED) CONTROL SURFACE has an increase in linear lift with deflections of up to 10 to 15 degrees. It has values of C_{h_δ} and C_{h_α} that are too large to give acceptable forces on most aircraft. OVERHANGING (OR SET-BACK HINGE) BALANCE - A blunt-nose balance has a greater effect on hinge moments than a sharp-nose balance, but it may produce non-linear characteristics at relatively small deflections. Unsealing the gap between the control and the airfoil tends to increase the balance.
- Small values of C_{h_α} cannot always be obtained without overbalancing.
- Set-back hinge balance is commonly used on aircraft with manual controls, sometimes in combination with other methods. SEALED INTERNAL BALANCE
- It is structurally complicated to install a flexible, airtight seal and to provide enough deflection without the balance interfering with the inside part of the fixed structure.
- Balancing effectiveness is comparable to

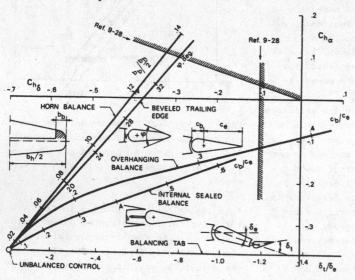

Fig. 9-6. Effect of aerodynamic balance on hinge moment coefficient (example)

314

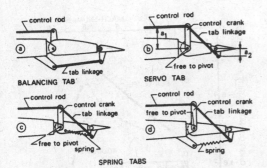

BALANCING TAB   SERVO TAB

SPRING TABS

Fig. 9-7. Mechanical principle of tab control systems

the overhang balance, except that non-
linearities do not occur.

HORN BALANCE

- Balance weight and aerodynamic loads are
not evenly distributed in spanwise direction, and this may give rise to torsion
loads in the control surface.
- For an unshielded horn balance the effect
of balancing on C_{h_α} is large. This may give
rise to an unfavorable relation between C_{h_δ}
and C_{h_α}. This effect is less for a shielded
horn balance.
- The formation of ice on an unshielded
horn may give rise to control problems.

BEVELED TRAILING EDGE BALANCE

- The characteristics are nonlinear, unless the control gap is sealed.
- It may be difficult to reproduce the
specified trailing-edge angle in the production stage.

BALANCING (OR GEARED) TAB (Fig. 9-7a)

- On deflection of the control surface the
tab is geared to move in the opposite direction, thus producing an aerodynamic moment about the hinge line, which assists
the pilot in moving the control.
- A balancing tab has the effect of reducing the value of C_{h_δ} without appreciably changing C_{h_α}, since the airfoil shape
is not varied, except when the control surface is deflected. This type of balance
may thus be used in conjunction with other
methods to adjust C_{h_δ} while leaving C_{h_α}
unchanged.
- A tab chord of 20 percent of the control

chord gives the least reduction in control
effectiveness due to the action of the tab.
- The control forces may, if so desired,
be increased by using an anti-balance tab,
i.e. a tab which is geared to move in the
same direction as the control surface.
- For a given tab chord/control chord ratio the amount of C_{h_δ} reduction is determined by the fixed ratio of the tab and
control deflection angles. This is usually
made adjustable in the design stage.
- The balancing tab is frequently used as
a trim tab.

THE SERVO TAB (Fig. 9-7b)

- The pilot's control is connected directly to the tab, while there is no rigid
connection with the control surface. The
control force depends on the hinge moments
of both the control surface and the tab to
an extent determined by the ratio a_1/a_2.
For small a_1/a_2 very low stick forces are
obtained.
- For low a_1/a_2 ratios the control is easily overbalanced as the tab is in the
boundary layer of the stabilizer and the
control.
- Control effectiveness at low speeds may
be inadequate and the servo tab is easily
overbalanced at the stall. It is mainly
for these reasons that the servo tab is
not used on many modern aircraft.

THE SPRING TAB (Fig. 9-7c and d)

- The principle is basically the same as
that of the servo tab, but in addition
there is a spring connecting the tab with
the fixed airfoil or the main control surface.
- At high speeds the tab carries a large
control force in relation to the spring
force. The action of the spring tab is
then comparable to that of the servo tab.
- At low speeds the spring force is large
in relation to the lift on the tab and
system c. behaves like a plain control,
while system d. acts more or less like a
geared tab.
- The effective C_{h_δ} decreases with EAS in
a way which is mainly determined by a_1/a_2
and the spring stiffness. This causes the
control forces to vary only moderately with

315

speed.
- The control forces are relatively independent of the C_{h_α} and C_{h_δ} values of both surfaces.
- The spring tab system cannot be overbalanced even at the stall, since the tab deflects only when there is a load on the control surface.
- Spring tabs may be pre-loaded to prevent them from coming into operation for small stick forces, in order to keep the spring tab out of action at low speeds.
- Practical spring tab systems are structurally much more complicated than Fig. 9-7 might suggest.
- The response to control commands is slow in view of the floating action of the main control surface.

9.2.5. Effects of compressibility and powerplant operation

When a condition of equilibrium is disturbed, resulting in variations in the airspeed, attention must be paid to the fact that compressibility and engine operation will make the aerodynamic coefficients dependent on the flying speed. Both effects are of a highly complex nature and will be discussed only very briefly. The effects of aeroelasticity on stability will not be dealt with here, but these are quite important in the case of large, high-speed aircraft.

a. Compressibility.
In making an assessment of the influence of the Mach number on the various factors contributing to stability, it will be noted that at subcritical speeds - i.e. where there are no shock waves - the neutral point will generally move forward with increasing M. Fig. 9-8 shows that $C_{m_\alpha} = C_m/\partial\alpha$ becomes less negative, which implies that the subcritical compressibility has a destabilizing effect. The available methods of analysis give only a generalized influence of Mach effects on C_{m_α} and, in addition, a term $\partial C_m/\partial M$ must be added to the stability equation.

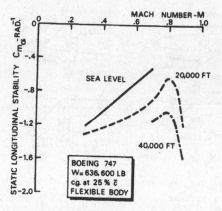

Fig. 9-8. Effects of compressibility and airplane flexibility on static stability

At Mach numbers above the lift-divergence condition, C_{L_α} decreases with M and $\partial C_{L_\alpha}/\partial M$ becomes negative. A shock wave which first appears on the inboard wing causes a rearward shift of the neutral point and the lift loss causes a reduction in the downwash at the tailplane. The result of these phenomena may be a large nose-down pitching moment which will force the aircraft into a high-speed dive from which recovery is very difficult ("tuck under"). Most high-subsonic aircraft have a Mach trimmer incorporated in the autopilot to cure this misbehavior.

b. Engine operation: propeller aircraft.
The direct effects are:
1. A force normal to the flow, a component of which is acting in the plane of the propeller if it is at an angle of attack to the flow. For a propeller ahead of the c.g. the effect will be destabilizing. In Appendix E, Section E-8 this contribution is referred to as $\Delta_p x_{ac}$.
2. If the c.g. is above the thrustline of the propellers, the thrust will supply a moment which will vary with the airspeed, resulting in a destabilizing effect which will become greater with increasing C_L. A stabilizing effect will be obtained when the c.g. lies below the thrust-line; an example is shown in Fig. 6-15.

Indirect effects are:

1. A shift in the a.c. of the wing.

2. An increase in downwash resulting from an increase in the circulation over the wing, which has a destabilizing effect.

3. Increased dynamic pressure behind the propeller. Its influence on stability will depend on the tail load and the location of the tailplane in relation to the propeller slipstream, and may have either a stabilizing or a destabilizing influence. On the whole, engine operation will generally have a destabilizing effect and will be greatest for relatively high power outputs and low flying speeds during takeoff and aborted landings.

Ref. 9-34 gives a relatively simple method for taking account of slipstream effects on the location of the neutral point. This method may be approximated by increasing the downwash term in (9-5) and (9-6) relative to the power-off condition by an amount:

$$\Delta_s \frac{d\epsilon}{d\alpha} = 6.5 \left| \frac{\rho \, P_{br}^{\,2} \, S^3 \, C_L^{\,3}}{\ell_h^{\,4} \, w^3} \right|^{\frac{1}{4}} \left| \sin(6\phi) \right|^{2.5}$$

(9-27)

where P_{br} is the shaft horsepower of one engine and ϕ defines the position of the horizontal tailplane relative to the wing (see Fig. E-13 and Fig. 2-24):

$$\phi = \sin^{-1} \frac{m}{r} \qquad (9-28)$$

Equation 9-27 is valid for $0° < \phi < 30°$, while $\Delta_s \, d\epsilon/d\alpha$ may be assumed zero outside this range.

c. Engine operation: jet aircraft.
Direct effects:

1. Normal force on the engine inlet which will have a destabilizing effect when it is ahead of the c.g. and a stabilizing effect when it is aft of the c.g.

2. As in the case of propeller aircraft there will be a destabilizing factor when the c.g. lies at a distance z_T above the line of thrust. This effect may be regarded as a forward shift of the neutral point in the case of variations in the airspeed with constant thrust:

$$\frac{\Delta_T x_{ac}}{\bar{c}} = - \sum \frac{T}{w} \frac{z_T}{\bar{c}} \qquad (9-29)$$

where the summation is made for all operating engines.

An indirect effect of engine operation on jet aircraft is an increase in downwash when the exhaust efflux passes a relatively short distance below the tail. Engine operation will have a destabilizing effect on aircraft with engines mounted below the wing.

9.3. SOME ASPECTS OF DYNAMIC BEHAVIOR

When dealing with large high-speed aircraft, cruising at high altitudes and equipped with effective flaps for takeoff and landing, it will be desirable to know some of the dynamic characteristics at an early design stage, since these will generally be the deciding factors for the size of the tailplane. Only a condensed account of the dynamic stability with fixed controls will be given here.

The overall dynamic longitudinal motion of an aircraft can generally be broken down into the short period (SP) oscillation and the phugoid.

9.3.1. Characteristics of the SP oscillation

The SP oscillation is a periodic motion which takes place at nearly constant speed and is subject to both angular and normal accelerations. This motion is generally highly damped and stable when the c.g. lies ahead of the stick-fixed neutral point. It is not so much the stability which matters, but rather the character of the oscillation which is defined by the following parameters.

The PERIOD P is generally of the order of a few seconds, but for large aircraft it may amount to some ten seconds. The period increases with a rearward shift of the c.g.

317

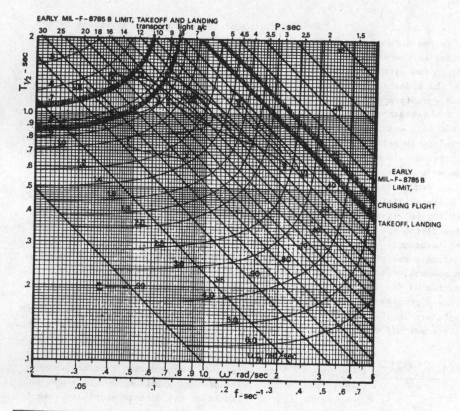

$$C_{\frac{1}{2}} = \frac{T_{\frac{1}{2}}}{P} = \frac{\ln 2}{2\pi} \frac{\sqrt{1-\zeta^2}}{\zeta}$$

$$\omega = \frac{2\pi}{P} = \frac{\ln 2}{T_{\frac{1}{2}}} \frac{\sqrt{1-\zeta^2}}{\zeta}$$

$$\omega_n = \frac{\omega}{\sqrt{1-\zeta^2}} = \frac{\ln 2}{\zeta \, T_{\frac{1}{2}}}$$

I	DC-9	landing,	$x_{cg} = 40\%$	$\bar{c}$,	W =	74,000 lb
II		idem ,	$x_{cg} = 15\%$	$\bar{c}$,	idem	
III	DC-8-62,	landing,	$x_{cg} = 32.3\%$	$\bar{c}$,	W =	240,000 lb
IV		idem ,	$x_{cg} = 18.2\%$	$\bar{c}$,	idem	
V	idem ,	25,000ft,	$x_{cg} = 34.6\%$	$\bar{c}$,	W =	250,000lb
VI	idem ,	idem ,	$x_{cg} = 16.8\%$	$\bar{c}$,	idem	
VII	DC-8-63,	landing,	$x_{cg} = 34.6\%$	$\bar{c}$,	W =	250,000 lb
VIII	idem ,	idem ,	$x_{cg} = 13\%$	$\bar{c}$,	idem	
IX	idem ,	25,000ft,	$x_{cg} = 34.6\%$	$\bar{c}$,	idem	
X	idem ,	idem ,	$x_{cg} = 13\%$	$\bar{c}$,	idem	

Fig. 9-9. Relationships between the parameters characterizing the short period longitudinal oscillation

and with CRITICAL DAMPING the motion becomes aperiodic ("dead-beat motion").
The TIME TO HALF AMPLITUDE $T_{\frac{1}{2}}$ is the time which has elapsed when the amplitude of the oscillation has decreased to one half its initial value.
CYCLIC DAMPING is defined as:

$$C_{\frac{1}{2}} = T_{\frac{1}{2}}/P \qquad\qquad (9-30)$$

RELATIVE DAMPING is the actual damping as a fraction of the critical damping.
The FREQUENCY ω is the angular speed in radians per second: $\omega = 2\pi/P$. Another definition takes the frequency as being

$f = 1/P$.

The UNDAMPED NATURAL FREQUENCY ω_n is the frequency of a short period oscillation in the absence of aerodynamic damping. Fig. 9-9 shows a pictorial representation of the relationships between the parameters defined above, including data on some high-subsonic aircraft.

The frequency of the short period oscillation should be in the right proportion to the maneuverability, which is often expressed in the following parameters:

$$L_\alpha = \frac{\text{increase in lift}}{\text{increase in } \alpha \times \text{mass} \times \text{speed}} =$$

$$= \frac{\rho g S V \, C_{L_\alpha}}{2W} \qquad (9-31)$$

and

$$n_{z_\alpha} = \frac{\text{increase in normal load factor}}{\text{increase in angle of attack}} \approx$$

$$\approx \frac{C_{L_\alpha}}{C_L} \qquad (9-32)$$

These parameters appear to have some significance in considerations of the piloting techniques involved in precise flight path control.

9.3.2. Criteria for acceptable SP characteristics*

Several attempts have been made in the past to derive criteria which may be useful to designers for ensuring good SP oscillation characteristics. Civil requirements are qualitative in character, but the military authorities have been more specific in setting limits to the undamped natural fre-

---

*Very recently, at the moment of final production of this chapter, a similar approach has been published for military aircraft in J. of Aircraft, June 1975, entitled: "Use of short period frequency requirements in horizontal tail sizing", by D.J. Moorhouse and W.M. Jenkins

---

quency and the damping ratio in early MIL-F-8785B requirements (Fig. 9-9). Their present requirements, summarized in Fig. 9-10, may be used for civil aircraft as well.

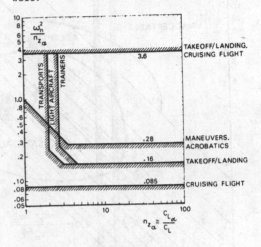

Fig. 9-10. Longitudinal short period requirements, adapted from MIL-F-8785B (Ref. 9-18)

Alternatively, the criteria proposed by Shomber and Gertsen (Ref. 9-15), which are shown in Fig. 9-11, must be considered as an even better starting point for the preliminary design of civil aircraft. These are based on the opinions of pilots who had to perform specific tasks in flight simulators and expressed their views in the form of Cooper Ratings (CR). The characteristics are regarded as satisfactory within a limited region of combinations of L_α/ω_n and ζ identified by CR < 3.5. The linearized equations of motion for the longitudinal short period oscillation may be used to find an approximate expression for L_α/ω_n in terms of the damping ratio:

$$\frac{L_\alpha}{\omega_n} = \frac{4\zeta}{2 - \dfrac{C_{m_q} + C_{m_{\dot\alpha}}}{C_{L_\alpha} \, K_y^2}} \qquad (9-33)$$

The coefficients C_{m_q} and $C_{m_{\dot\alpha}}$ represent the aerodynamic damping in pitch due to the rate of pitch and change of incidence, and

319

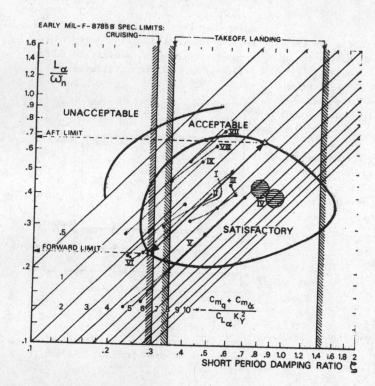

$\frac{L_\alpha}{\omega_n}$

UNACCEPTABLE

ACCEPTABLE

AFT LIMIT

SATISFACTORY

FORWARD LIMIT

$-\dfrac{C_{m_q} + C_{m_{\dot{\alpha}}}}{C_{L_\alpha} K_Y^2}$

SHORT PERIOD DAMPING RATIO ζ

Fig. 9-11. Longitudinal short period oscillation criteria for transport aircraft as proposed by Shomber and Gertsen (Ref. 9-15)

K_y is the inertia radius about the lateral axis as a fraction of the MAC. For a given type of aircraft ζ is a function of the stick-fixed maneuver margin, while K_y is dependent upon the mass distribution. Lines have been drawn in Fig. 9-11 according to (9-33), which show that aircraft with a given mass distribution at a given altitude approximately follow this relationship when their c.g. location is varied. The figure thus defines acceptable fore-and-aft c.g. limits, which may be expressed as limits to the stick-fixed maneuver margin:

$$\frac{x_m - x_{cg}}{\bar{c}} = \frac{1}{2} \frac{K_y^2}{\mu_{c}} \left(\frac{\omega_n}{L_\alpha}\right)^2 C_{L_\alpha} \qquad (9-34)$$

where

$$K_y^2 = \frac{I_y}{W} \frac{g}{\bar{c}^2}$$

and

$$\mu_c = \frac{W}{\rho g S \bar{c}} \qquad (9-35)$$

Using the approximation

$$\frac{C_{m_q} + C_{m_{\dot{\alpha}}}}{C_{L_\alpha}} = -2.2 \frac{C_{L_{h_\alpha}}}{C_{L_\alpha}} \frac{S_h}{S} \left(\frac{\ell_h}{\bar{c}}\right)^2 \left(1 + \frac{d\epsilon}{d\alpha}\right) \qquad (9-36)$$

Fig. 9-11 may be used to find maximum and minimum values of L_α/ω_n, corresponding to the forward and aft c.g. limit for satisfactory dynamic characteristics. Generally speaking, flight at high altitude and high wing loading will determine the forward c.g. limit, since in this condition damping is poor, while low-speed flight at low weight will determine the aft c.g. limit. An example of acceptable combinations of S_h/S and $x_{cg}/\bar{c}$ is shown in Fig. 9-12.

9.3.3. A simple criterion for the tail-plane size

The time to half amplitude of the SP oscillation can be approximated by:

320

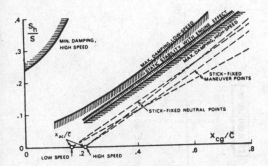

Fig. 9-12. Example of acceptable combinations satisfying SP requirements

$$T_{\frac{1}{2}} = \frac{4 \ln 2}{L_\alpha \left(2 - \dfrac{C_{m_q} + C_{m_{\dot{\alpha}}}}{C_{L_\alpha} K_Y^2} \right)} \qquad (9-37)$$

This relationship has been plotted in Fig. 9-13, together with values of $T_{\frac{1}{2}}$ for some airplanes of different classes in the landing configuration. The trend which these points seem to indicate suggests

that $T_{\frac{1}{2}}$ may be a useful criterion for sizing the horizontal tailplane. Inspection of (9-36) and (9-37) reveals that $T_{\frac{1}{2}}$ is primarily determined by the term

$$L_\alpha \frac{C_{m_q} + C_{m_{\dot{\alpha}}}}{C_{L_\alpha} K_Y^2} = -1.1 \frac{\rho V}{I_y} S_h \ell_h^2 C_{L_{h_\alpha}} \left(1 + \frac{d\epsilon}{d\alpha} \right)$$

$$(9-38)$$

which, in fact, represents the aerodynamic damping moment in relation to the polar moment of inertia in pitch. Fig. 9-14 is based on this relation and supplies a simple method for estimating the horizontal tailplane size.* This approach also shows the importance of the tail moment arm.

Analysis of the SP oscillation shows that an important part is played by the parameter $I_y/(\ell_h W)$. From Fig. 9-15 it can be seen that the length of the fuselage and the engine location are deter-

*Some data for estimating the horizontal tailplane lift gradient can be obtained from Section E-10.1. of Appendix E

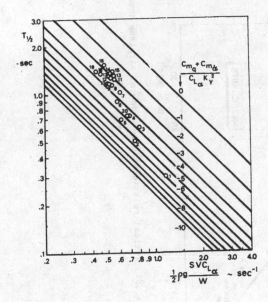

1. Queen Air
2. Beaver
3. Lear Jet 24
4. F-27
5. HS-125
6. DC-9/10
7. DC-8/21
8. VC-10
9. CV-880
10. C-141
11. Jetstar
12. B707-120
13. DC-8/62
14. B727-100
15. DC-8/63
16. B727-200
17. B747
18. C-5A
19. B737-200
20. MS Paris

Note: the points indicated refer to the approach speed at sea level, at MLW

Fig. 9-13. Time required to damp the short period motion

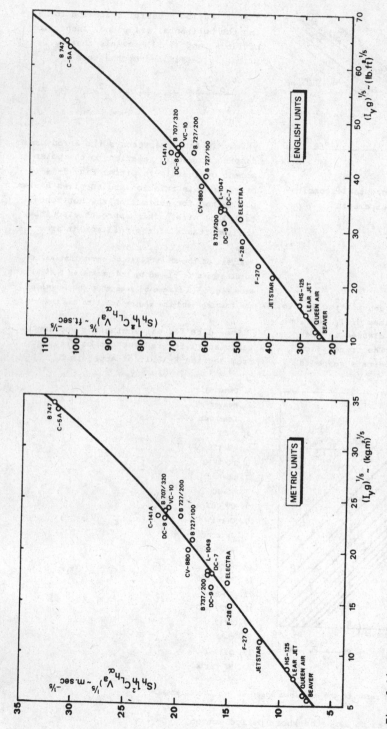

Fig. 9-14. Statistical relationship between the airplane inertia and aerodynamic damping: approach at sea level

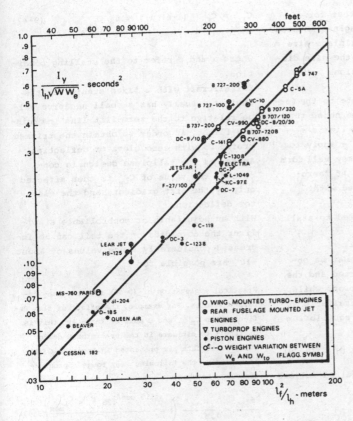

$$\frac{I_y}{l_h\sqrt{W W_e}} - seconds^2$$

B 747
C-5A
B 727-200
VC-10
B 707/320
B 727-100
B 707/120
CV-990
DC-8/20/30
B 737-200
B 707-720B
DC-9/10
C-141
CV-880
C-130B
JETSTAR
DC-7
ELECTRA
L-1049
F-27/100
KC-97E
DC-7
C-119
DC-3
LEAR JET
C-123B
HS-125
MS-760 PARIS
SI-204
D-18 S
QUEEN AIR
BEAVER
CESSNA 182

O WING MOUNTED TURBO-ENGINES
● REAR FUSELAGE MOUNTED JET ENGINES
▽ TURBOPROP ENGINES
● PISTON ENGINES
O---O WEIGHT VARIATION BETWEEN W_e AND W_{to} (FLAGG. SYMB.)

$$l_f/l_h - meters$$

Fig. 9-15. Polar moment of inertia about the lateral axis as affected by airplane size for various categories

mining factors here and problems may therefore be encountered with the SP oscillation in a particular class of high-subsonic jet transports. It is not always possible to solve such problems in the entire range of speeds, weights, and altitudes by good aerodynamic design. It may then become desirable or even necessary to use equipment which supplies artificial stabilization (SAS, Stability Augmentation System).

9.3.4. The phugoid

The phugoid is a motion during which speed variations occur which generate tangential and normal forces at an almost constant angle of attack. The aircraft has a constant energy level and oscillates in such a way that an interchange takes place between potential and kinetic energy. Although the period of the phugoid will generally be of such magnitude that the

pilot can still correct any instability by elevator deflections, a stable phugoid is regarded as desirable under normal conditions and particularly during IFR flights. In a preliminary design this may basically be achieved by:
- taking account of power effects as far as possible (see Section 9.2.5.) and ensuring that the static stability remains positive,
- paying due regard to the influence of compressibility in the case of high-subsonic aircraft.

9.4. LONGITUDINAL CONTROL AT LOW SPEEDS

For some aircraft the aft c.g. limit may be imposed by the condition that recovery must be possible from a deep stall (see Section 2.4.2c.). From the available literature it can be concluded that for aircraft with T-tails and jet engines mounted on the rear fuselage, the tailplane size may be about 15% larger than the size dictated by stability considerations at a given aft c.g. location.

The criteria given below relate to the forward c.g. location. It will be noted that on aircraft with CCV, similar limits - based on control requirements - apply to the aft c.g. location, which may well form the critical design condition if "aerodynamic" stability is dispensed with.

9.4.1. Control capacity required to stall the aircraft

The maximum lift coefficient must be obtained with flaps fully deflected and the c.g. in its most forward position, while some reserve will have to be available for maneuvering. The condition of equilibrium in Fig. 9-16 results in:

$$L_h \ell_h'' = M_{.25} + L_{A-h}(x_{cg} - .25\bar{c}) \quad (9-39)$$

The horizontal tail volume required follows from:

$$\frac{S_h \ell_h}{S\bar{c}} = \frac{C_{L_{max}}}{\eta_h C_{L_h}}\left(\frac{C_{m.25}}{C_{L_{max}}} + \frac{x_{cg}}{\bar{c}} - .25\right) \quad (9-40)$$

where

$$\eta_h = \frac{\ell_h''}{\ell_h}\left(\frac{V_h}{V}\right)^2 \quad (9-41)$$

The tailplane lift coefficient, which is positive in the upward direction, amounts to:

$$C_{L_h} = C_{L_{h_\alpha}}(\alpha - \varepsilon + i_h) + \delta_e C_{L_{h_\delta}} \quad (9-42)$$

where α and ε refer to the stalling condition.

For aircraft with a fixed stabilizer the angle i_h usually has a small negative value relative to the zero-lift line in cruising flight, in order to obtain the trimmed cruise C_L with zero elevator deflection. As far as the tailplane design is concerned, the value of C_{L_h} is then affected only by the lift gradients and the elevator deflection.

With an adjustable or controllable stabilizer the downloads on the tail can be increased considerably since values of about $-10°$ are possible for i_h.

Practical application of (9-40) requires many detailed studies. For example, the effect of flap deflection on $C_{m.25}$ and ε must be known and this is difficult to estimate in the preliminary design stage. Some data are presented in Appendix G. Alternatively, the following very rough approximations may be used:

$$\frac{C_{m.25}}{C_{L_{max}}} = \frac{C_{m.25}}{\Delta_f C_{L_{max}}}\left(1 - \frac{1.5\cos\Lambda_{.25} + \Delta_S C_{L_{max}}}{C_{L_{max}}}\right) \quad (9-43)$$

where Δ_f and Δ_S denote the effects of flap and slat deflection on the coefficients to which they are attached. Typical values for $C_{m.25}/\Delta_f C_{L_{max}}$ are:

- .18 , split flap, plain flap
- .26 , single slotted flap
- .385, double slotted flap, fixed hinge
- .415, single slotted Fowler flap
- .445, double slotted Fowler flap
- .475, triple slotted Fowler flap

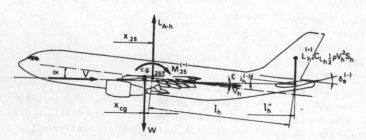

Fig. 9-16. Longitudinal equilibrium at the stall: landing configuration

The following figures are proposed for C_{L_h}

$-.35 \cdot A_h^{1/3}$, fixed stabilizer

$-.8$, adjustable stabilizer

-1.0 , controllable stabilizer.

For η_h we may assume:

$.95$, fin-mounted stabilizer

$.85$, fuselage-mounted stabilizer.

The condition of adequate control power required to reach $C_{L_{max}}$ in the landing configuration may be depicted in an S_h/S vs. $x_{cg}/\bar{c}$ diagram; see Fig. 9-18.

9.4.2. Control capacity required for take-off rotation and landing flareout

It must be possible to generate a sufficiently large tail load to ensure that at the rotation speed the aircraft will have an angular velocity of $\dot{\theta}_R$ within a limited period of time, e.g. one second. From Fig. 9-17 we may derive the condition:

$$\frac{S_h \ell_h}{S\bar{c}} = \frac{C_{L_{max}}}{\eta_h \, \eta_q \, C_{L_h}} \left| \frac{C_{m.25}}{C_{L_{max}}} - \left(\frac{V_{S_1}}{V_R}\right)^2 \times \right.$$

$$\left. \frac{x_g - z_T \sum T/W - x_{cg}}{\bar{c}} \right| + \frac{C_{L_R}}{C_{L_h}} \left(\frac{x_g}{\bar{c}} - .25\right) \quad (9-44)$$

where V_{S_1}, $C_{L_{max}}$ and $C_{m.25}$ refer to the takeoff flap setting and

$$\eta_h = \frac{x_h - x_g}{\ell_h} \left(\frac{V_h}{V_R}\right)^2 \quad (9-45)$$

and

$$\eta_q = 1 + \frac{C_{L_{h_\alpha}}}{C_{L_h}} \frac{\dot{\theta}_R \, (x_h - x_g)}{V_R} \quad (9-46)$$

As a minimum for the rotation speed we may take: $V_R = 1.05 \, V_{S_1}$ for transport aircraft; for light aircraft an explicit requirement will be found in FAR 23.51 regarding the minimum speed at which rotation can take place.

The value for C_{L_h} follows from (9-42), where α and ε relate to the initiation of the rotation and the aircraft is trimmed for $1.3 \, V_{S_1}$. When the stabilizer is adjustable it may be possible to obtain a reasonably high value, such as $C_{L_h} = -1.0$. In the case of propeller aircraft, C_{L_h} and η_h are more difficult to determine because of powerplant operation, ground effect, etc. Various data needed for further evaluation of (9-44) often only become available after wind tunnel tests. An example of the resulting S_h/S-limit is shown in Fig. 9-18.

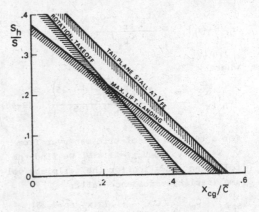

Fig. 9-18. Tailplane size and c.g. limitations based on controllability at low speeds

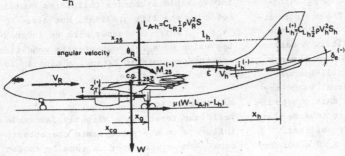

Fig. 9-17. Takeoff rotation

9.4.3. Out-of-trim conditions

When excursions are made from the trimmed conditions, variations in the local flow conditions at the tailplane will occur, but these must not lead to tailplane stall. The critical condition here will generally be experienced in the landing configuration with the most forward c.g. location. Typical maneuvers to be considered are a level speed increase from a trimmed condition at $1.2\ V_S$ or a push-over maneuver at $n = .5g$ from a trimmed condition at $1.3\ V_S$, up to the flap-extended design speed V_{FE} (Ref. 9-32).

Taking, for example, an aircraft with a variable-incidence tailplane, trimmed at $1.2\ V_S$, which is accelerated at $n = 1$ to V_{FE}, the tailplane angle of attack is obtained from:

$$\alpha_h = (\alpha_h)_{tr} - \frac{(C_L)_{tr} - C_L}{C_{L_\alpha}}\left(1 - \frac{d\varepsilon}{d\alpha}\right) \quad (9\text{-}47)$$

where

$$(\alpha_h)_{tr} = \frac{C_{m.25} + (C_L)_{tr}\left(\frac{x_{cg}}{\bar{c}} - .25\right)}{\eta_h\, C_{L_{h_\alpha}}\, S_h\, \ell_h/(S\bar{c})} \quad (9\text{-}48)$$

For a typical out-of-trim maneuver from $V_{tr} = 1.2\ V_S$ to $V_{FE} = 1.8\ V_S$ we find the following condition for the tailplane volume (variable incidence tail):

$$\frac{S_h \ell_h}{S\bar{c}} = \frac{C_{m.25}/C_{L_{max}} + .7\,(x_{cg}/\bar{c} - .25)}{\eta_h\, C_{L_{h_\alpha}}\left[\dfrac{(\alpha_h)_{min}}{C_{L_{max}}} + .39\,\dfrac{1 - d\varepsilon/d\alpha}{C_{L_\alpha}}\right]} \quad (9\text{-}49)$$

where $C_{m.25}$ is the aerodynamic pitching moment, tail-off, at $(C_L)_{tr}$. The minimum value of α_h to be accepted depends upon the detail design of the stabilizer. A margin should be available to allow for the reduction in the tailplane stalling angle due to ice formation. A typical value may be $(\alpha_h)_{min} = -.25$ rad., but this can be increased by negative camber or inverse slats on the leading edge of the stabilizer. An example of limitation of S_h/S according

to (9-49) is plotted in Fig. 9-18. Similar limitations may also be derived for other out-of-trim maneuvers. It will be obvious that a detailed study of flow conditions will be required at a later stage of the design.

9.5. PRELIMINARY DESIGN OF THE HORIZONTAL TAILPLANE

The design of the tailplane is always an iterative process. It is usual to make an initial choice of certain shape parameters such as aspect ratio and thickness ratio, etc. The choice of the type of aerodynamic balance, whether the stabilizer will be fixed or adjustable, and the type of control system is much more difficult, and more data will generally be required than are available to the preliminary design engineer. Once these decisions have been taken, it may well be that previous assumptions regarding the tailplane shape or even the wing location will have to be revised.

9.5.1. Tailplane shape and configuration

a. Aspect ratio A_h.
This factor is of direct influence because of its effect on the lift-curve slope. For manual control systems the c.g. range satisfying the stick-force requirements will be widened, or the required tailplane size may be reduced. For aircraft with a fixed stabilizer the forward c.g. limitation required to cope with the stall is favorably affected with increasing A_h. Equation 9-49 for variable incidence tailplanes contains not only the lift gradient, but also $(\alpha_h)_{min}$, which decreases with an increasing value of A_h. If out-of-trim conditions are the predominant factor, a high A_h is not always desirable.

b. Taper ratio λ_h.
Tailplane taper has a slightly favorable influence on the aerodynamic characteristics. A moderate taper is usually chosen

to save structural weight.

c. Angle of sweep Λ_h.
In the case of high-speed aircraft the tailplane angle of sweep, in combination with its thickness ratio, is chosen so that at the design diving Mach number strong shocks are not jet formed. The same procedure as applied to the wing (Section 7.5.2b) will then result in a thinner section and/or larger Λ_h as compared with the wing.
Positive sweepback is occasionally used on low-speed aircraft to increase the tailplane moment arm and the stalling angle of attack, although the result is a decrease in the lift-curve slope. Up to about 25 degrees of sweepback there is still an advantage. The sweepback angle may be determined by the condition of a straight elevator hingeline, which is sometimes imposed in the interest of structural simplicity.

d. Airfoil shape.
The basic requirements are that the airfoil section should have a high c_{l_α} and a large range of usable angles of attack. Frequent use is made of approximately symmetrical airfoils with a thickness ratio of 9 to 12 percent and a large nose radius, e.g. NACA 0012. Tailplane stalling - at the lower surface - can be postponed by adopting negative camber (e.g. NACA 23012 upside down), upward nose droop, an increase in the nose radius or by means of a fixed slot.

e. Dihedral.
The position of the tailplane relative to the propeller slipstream or jet efflux may make it desirable to shift it slightly in an upward direction. This may be achieved by using a certain degree of dihedral.

f. Control and trim system.
A variable-incidence (adjustable) tailplane has the advantages that at high-subsonic speeds adjustment of the tailplane is more effective than trimming by means

of the elevator - which may cause shock waves - and that the c.g. travel can be extended forward. An all-flying (controllable) tail has the additional advantage that it improves both maneuverability and control in out-of-trim conditions. A geared elevator, deflecting in the same direction, will extend the c.g. range in the forward direction through an increased download. All-flying tails on transport aircraft are usually power-operated. A controllable tailplane is sometimes used on small aircraft and frequently on gliders. An anti-balance tab is recommended in order to obtain acceptable control forces and control force stability.
Trim tabs, balance tabs or variable-incidence stabilizers can be used to reduce the control forces to zero. Large trim changes may be induced by the deflection of effective flaps, and in some aircraft there is a coupling between the flap controls and the (trim) tab controls or the stabilizer incidence controls.

g. Elevator area and deflections.
A large elevator area - as a fraction of the tailplane area - promotes good controllability at forward c.g. locations, but for manual control systems the stick forces will increase. Similar arguments apply to the maximum elevator deflection. The danger of tailplane stall and elevator lock-over grows greater with increasing elevator chord and deflection.

9.5.2. Design procedures

The objective of preliminary tailplane design is not only to arrive at an acceptable guess with regard to shape and size but, in addition, to ensure a good balance between the wing location in longitudinal direction, the c.g. range required, the c.g. range available and the tailplane design. In view of the many parameters involved, various solutions may be proposed and the design finally selected will be the one which gives a small tailplane and adequate flexibility to suit a variety of cus-

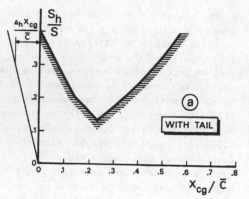

Aerodynamic c.g. limits

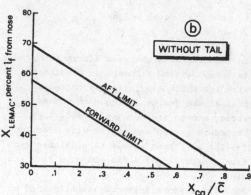

Limits based on loading requirements: aircraft without horizontal tailplane

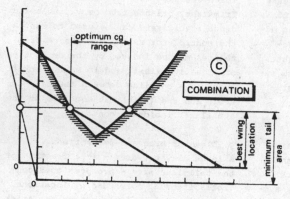

Balancing

Fig. 9-19. Balancing the airplane

tomer requirements and future aircraft growth.

A design procedure will be suggested below. The basic relationships required can be found in this chapter, while methods for predicting aerodynamic characteristics are presented in Appendices E and G, and in the references. As a considerable amount of work is involved in the detailed procedure, a simple approach will be presented first.

a. Quick method.

It can be argued that longitudinal stability is affected primarily by S_h/S, whereas the volume $S_h \ell_h/(S\bar{c})$ is more relevant to controllability. By inspecting the data for comparable aircraft (Table 9-2) the area and aspect ratio may be chosen so that $A_h^{1/3} S_h/S$ and $A_h^{1/3} S_h \ell_h/S\bar{c}$ both have acceptable values.

A check may be made with Fig. 9-14, which relates particularly to large transport aircraft. The tailplane lift-curve slope for low aspect ratios is approximately:

$$C_{L_{h_\alpha}} = \frac{2\pi}{1 + \frac{3}{A_h \cos\Lambda_h}} \qquad (9-50)$$

and the moment of inertia can be obtained from Fig. 9-15.

Table 9-2. Horizontal tailplane design data

Aircraft type	Design Dive Speed V_D kts EAS	M_D	$\frac{S_h}{S}$	A_h	λ_h	Λ_h deg.	Airfoil Section root/tip	Average t/c %	Tail Type *	i_{max} + deg.	i_{max} - deg.	$\frac{S_h l_h}{S c}$	Γ_h deg.	$\frac{S_e}{S_h}$	Hinge Pos. $z\,c_h$	Balance Ratio $z\,c_e$	$\delta_{e\,max}$ + deg.	$\delta_{e\,max}$ - deg.	Tabs **
Wright Flyer (1903)	–	–	.163	5.72	1	0	curved "plates"	few	M	–	–	−.357	0	0	0	–	–	–	–
Scottish Aviation Pup	–	–	.229	4.40	1	0		12	F	–	–	.730	0	.377	60	–	25	27	T
Cessna 177	–	–	.202	4.00	1	0	NACA 0012/0009	10.5	M	–	–	.600	0	0	25	–	–	–	S
Aérospatiale N-262	260	–	.285	3.84	.568	8.2	23015 mod. inv.	15	F	–	–	1.015	0	.242	75	31.1/36.2	22	25	T/S
Fokker-VFW F-27 Mk200	288	.51	.229	6.00	.400	0	NACA 63A-014 inv.	14	F	–	–	.962	6	.198	78	42.7	15	40	T
Lockh. Hercules C-130B	325	.64	.313	5.20	.358	7.5	NACA 23012 inv.	12	F	–	–	1.000	0	.332	65	–	15	30	T/S
Lockh. Electra 188A	–	.711	.246	5.25	.335	8.6			F	–	–	.800	8.5	.247	65	–	15	30	T
Bristol Britannia 310	301	–	.282	5.14	.500	6	RAF 30 mod.	13	F	–	–	1.040	0	.300	66	30	15	35	T/S
Cessna Citation 500	–	–	.260	5.20	.500	–	NACA 0010/0008	9	F	–	–	.755	9	.310	67	–	9	24	–
Hawker Sidd. HS-125/400B	370	.825	.260	4.00	.580	19.6		10	V	2.5	7.5	.677	0	.408	58.7	25	15	25	S
Aérospat. SN-600 Corvette	–	–	.283	4.79	.523	28.6	neg. cambered	9	V	3.0	6.0	.640	0	.320	64	–	25	25	–
Yakovlev YAK40	330	–	.237	4.33	.430	11.5		10	V	1.5	12.0	.621	0	.377	70	–	20	25	–
VFW-Fokker 614	390	.74	.207	4.50	.400	20	NACA mod.	10.45	V	2.67	8.33	.834	10.5	.273	67	45	15	25	T
Fokker-VFW F-28 Mk1000	375	.83	.281	3.83	.480	27.5	NACA 65-011	11	F	–	–	.972	0	.197	78	–	12	30	–
Aérospat. Caravelle 10-R	410	.87	.255	4.03	.326	30	DSMA:9.5%/8%	10	V	–	10	.995	0	.235	75	41	–	25	S
BAC-111/200-400	–	.86	.204	3.38	.600	25		8.75	V	1	9	.908	0	.274	65	35	15	25	S
McDonnell Douglas DC-9 project	–	.89	.256	4.21	.310	32	12%/9%	8.7	V	2	10	.941	10	.280	70	35	15	25	S
McDonnell Douglas DC-9/10	398	.89	.242	4.93	.352	31.6		10.5	V	1	9	1.147	7	.280	70	35	20	20	S
Boeing 737/100	–	.89	.295	4.16	.380	30		8.5	V	4	12.5	1.140	7	.224	75	–	16	26	S
Boeing 727/100	–	.95	.319	3.40	.400	35		8.5	V	0	12	.902	−3	.230	75	–	–	–	S
Boeing 727/200	–	.95	.221	3.40	.400	35		8.5	V	–	–	1.062	−3	.230	75	–	16	26	S
Dassault Mercure	420	.90	.221	3.79	.360	32.5		8?	V	3	12	1.050	0	.251	75	–	–	–	–
Airbus A-300B	–	.90	.275	4.13	.500	32.5		11.6	M	0.5	14	1.068	6	.295	70	–	15	25	–
Boeing 707/320	–	.95	.267	3.37	.421	35	BAC-317	8.75	V	2	10	.630	7	.251	75	35	10	25	T/S
McDonnell Douglas DC-8 proj	435	.95	.216	4.04	.329	35	DSMA-89/-90	8.75	V	2	10	.626	10	.225	75	35.2	16.5	27	S
McDonnell Douglas DC-8/10	–	.95	.215	4.04	.329	35	DSMA-89/-90	8	V	0	14	.590	10	.225	77	–	0	25	S
Lockheed L-1011 Tristar	–	.95	.203	4.00	.333	35		9	V	5	15	.928	3	.215	77	30	8.5	11	–
McDonnell Douglas DC-10/10	445	.95	.371	3.78	.375	35		9	V	3	12	.855	10	.235	75	31	17	23	–
Boeing 747/100-200	–	.97	.346	3.60	.264	37		9	V	4	12	1.000	8.5	.185	77	–	10	20	T
Lockheed C-5A	410	.875	.156	4.89	.364	24.5		10	V	–	–	.700	−4.5	.268	65	30	10	20	–

* F = fixed; M = all-flying; V = variable incidence

** S = servo tab; T = trim tab

Table 9-2. Horizontal tailplane design data

b. Detailed procedure.

The results of stability and control considerations are collected in S_h/S vs. $x_{cg}/\bar{c}$ plots (Fig. 9-19a), indicating a region where S_h/S may be chosen freely. It is advisable to choose the most critical mass (distribution) for each limit.

For aircraft with REVERSIBLE CONTROLS the flying speeds are generally not very high and compressibility effects need not always be taken into account.

1. Calculate the lift gradient $(C_{L_\alpha})_{A-h}$ and the aerodynamic center x_{ac} for the aircraft without horizontal tailplane.

2. Determine $d\epsilon/d\alpha$ and the tailplane lift-curve slope and calculate C_{L_α} and x_n as a function of S_h/S, using (9-5) and (9-6).

3. Choose a value for C_{h_δ} - using, for example, Figs. 9-5 and 9-6 - and a (low) value for C_{h_α}.

4. Calculate the stick-free lift-curve slopes of the tailplane and the aircraft, and the stick-free neutral point with (9-9) through (9-11).

5. Choose a value for G and calculate the forward and aft c.g. limits determined by control force stability requirements, using (9-14) through (9-16).

6. Determine the stick-free maneuver points at the maximum cruising altitude and at sea level with (9-21) and (9-22).

7. Calculate the forward and aft c.g. limits imposed by the maximum and minimum stick force per g requirements, respectively, using (9-24) and Table 9-1.

8. Use (9-40) through (9-43) to estimate S_h/S vs. the c.g. position which permits C_L-max to be obtained with flaps in the landing position.

9. Limits for S_h/S vs. $x_{cg}/\bar{c}$ may also be derived from (9-44) through (9-46) and from (9-49), provided sufficient data are available.

10. If it is found that any of the conditions in relation to the control forces unduly restricts the forward c.g. limit, the value of C_{h_δ} should be revised.

Aircraft flying at high-subsonic speeds or large, high-speed propeller aircraft often have IRREVERSIBLE CONTROLS. Sub-critical compressibility effects on the aerodynamic properties should be taken into account as far as possible.

1. and 2. These steps in the procedure are the same as those above, but all coefficients are calculated for high-speed, high-altitude cruising as well as for low speeds. Typical angles of attack are about 10 degrees without slats and 15 degrees with slats extended.

3. In the case of jet aircraft determine the shift of the neutral point resulting from engine action, using (9-29) for engines placed below the c.g. For propeller aircraft a similar forward shift is obtained from (9-27) and (9-5).

4. Locate the aft c.g. limit about .015 ℓ_h ahead of the stick-fixed neutral point with engines operating at low speeds and high power.

5. Calculate the stick-fixed maneuver points at the maximum cruising altitude and at sea level.

6. Determine I_y at the maximum takeoff weight, using Fig. 9-15. Provided the angle of sweep is not excessive, the influence of the fuel can be ignored.

7. Calculate $(C_{m_q} + C_{m_{\dot{\alpha}}})/(C_{L_\alpha} K_y^2)$ with (9-36) and enter this value in Fig. 9-11 (in the example 3.0). The maximum and minimum values of L_α/ω_n for satisfactory SP characteristics are then read off.

8. The minimum and maximum stick-fixed maneuver margins are calculated with (9-34) and subtracted from the x_m values to obtain aft and forward c.g. limits for each condition.

9. and 10. These steps are identical to steps 8 and 9 for aircraft with reversible controls.
In the case of aircraft with a T-tail and engines pod-mounted on the rear fuselage, the tailplane area for the aft c.g. limit is increased by, say, 15 percent to allow for the deep stall case.

The most critical requirements for S_h/S are now selected and plotted in a diagram (Fig. 9-19a) which defines the "aerodynamic" limits to the c.g. range. This diagram must be balanced with the c.g. range required to obtain good loading flexibility (Section 8.5.2.). The following graphical solution is suggested.

1. In a diagram (Fig. 9-19b) on transparent paper the c.g. range of the aircraft without horizontal tailplane is represented as a function of the position of the wing in the longitudinal direction.

2. The magnitude of the c.g. shift as a result of the tailplane weight variation with S_h/S is indicated in the S_h/S vs. $x_{cg}/\bar{c}$ diagram (Fig. 9-19b):

$$\frac{\Delta_h}{\bar{c}} \frac{x_{cg}}{\bar{c}} = \frac{S_h \ell_h}{S\bar{c}} \frac{\text{specific tailplane weight}}{\text{wing loading}} \qquad (9-51)$$

3. Fig. 9-19b is now placed over Fig. 9-19a in such a way that the X axes are parallel. The aircraft will be balanced (Fig. 9-19c) when the following points of intersection occur at an identical value of S_h/S:
- the point of intersection between the acceptable and desired aft positions of the c.g.,
- the point of intersection between the acceptable and desired forward positions of the c.g., and
- the point of intersection of the vertical axis in Fig. 9-19b and the line given by (9-51) in Fig. 9-19a.

4. The optimum values for S_h/S and the balanced location of the wing may now be read off (Fig. 9-19c).

If it proves difficult or impossible to obtain a satisfactory balance, the following modifications may be considered:
1. altering the design of the tailplane;
2. changing the position of the tailplane, for example by increasing ℓ_h, if necessary by extending the length of the fuselage;
3. changing the mass distribution, particularly that of the fixed equipment-fixed ballast is to be avoided - or altering the arrangement of the fuselage interior (see also Section 8.5.2.);
4. limiting the c.g. range by imposing loading restrictions.
If none of these measures yields a satisfactory solution, a complete revision of the general arrangement may be required. In this case it may be found that an alternative disposition of the powerplant is unavoidable.

9.6. DESIGN OF THE VERTICAL TAILPLANE

The design of the vertical tailplane is more complicated than that of the horizontal tailplane. It is generally quite difficult to calculate the lateral-directional aerodynamic characteristics, since they are closely connected with a complicated asymmetrical flow field behind the wing/fuselage combination, which meets the oncoming air at an angle of sideslip.

The following broad design criteria are relevant in this respect:
a. The vertical tailplane must not stall as a result of an oscillation after deflection of the rudder or a sudden engine failure.
b. After failure of the critical engine, multi-engined aircraft must remain controllable to ensure steady flight.
c. It should be possible to land transport aircraft in crosswinds of up to 30 knots (55 km/h).
d. The aircraft must possess positive directional and lateral static stability and the short-period lateral/directional oscillation (Dutch roll) must be well damped. Some degree of spiral instability will be acceptable for aircraft fitted with an automatic pilot.

The effectiveness of the vertical tailplane is difficult to calculate as a result of the large variety in shapes. The NACA has recommended certain definitions which lead to reasonable solutions (Fig. 9-20). For

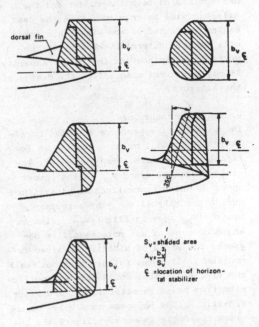

Fig. 9-20. Aerodynamic definitions of vertical tailplane area and aspect ratio

comparison purposes it should be pointed out that the R.Ae.S. Data Sheets use another definition, while some manufacturers base the area on structural considerations. Attention is drawn to the special case of a T-tail (Fig. 2-26), where the aerodynamic aspect ratio is considerably larger than the geometric.

9.6.1. Control after engine failure: multi-engine aircraft

Official requirements can be found in FAR 23.149 and 25.149, BCAR Sections K and D, Ch. 02-8 Paragraph 4.
Engine failure must be investigated for multi-engine aircraft in all configurations in a specified range of speeds, while, in addition, failure of two engines in the en route configuration must be catered for in the case of a four-engine aircraft.
Since engine failure generally causes a disturbing yawing moment and - although to a lesser degree - also a rolling moment, deflections of both the rudder and the ailerons will be required. When the dead engine is fitted to the wing, this case will largely determine the size of the vertical tailplane and the rudder capacity. It may sometimes also affect the design of the ailerons.

a. Transient behavior.
The pilot will experience a yawing oscillation which will be more violent at low flying speeds (Refs. 9-49 and 9-50). The minimum control speed V_{MC} is the lowest speed at which control can be established during the takeoff run - by aerodynamic means - V_{MC_G}, or in flight V_{MC_A}, with an angle of bank of not more than five degrees. The angle of sideslip involved immediately after the failure must not stall the fin, since that would prevent equilibrium from being re-established. Transient sideslip angles may peak to a value of 1.6 times the angle in steady flight, and angles of up to 20 to 25 degrees are no exception with wing-mounted engines. This

"overshoot" may be restricted by such measures as:
- The use of a yaw damper (see Section 7.7.1.) which superimposes rudder deflections on those initiated by the pilot.
- The use of a pneumatic system which derives a control signal from the pressure differential in the compressors of corresponding engines (Hawker Siddeley 125).
The following measures will reduce the likelihood of vertical tailplane stall:
1. The choice of a moderate aspect ratio - such as $A_v < 1.8$ - and a fairly large fin.
2. The use of a swept-back leading edge.
3. The addition of a dorsal fin (Fig. 9-20), which does not contribute materially to the directional stability at small angles of sideslip, but will postpone the stall on the lower part of the fin at large angles.

b. Steady flight after engine failure.
A good starting point for sizing the vertical tailplane and rudder for aircraft with an outboard engine is formed by the condition of equilibrium which must be established some time from the engine loss. We have to consider the following forces and moments in sideslipping flight with one engine inoperative (Fig. 9-21):
1. Contribution of the asymmetry in the thrust (ΔT_e) at a distance y_e from the plane of symmetry. Provided the aircraft is fitted with an automatic feathering system, the drag of the dead engine (and propeller) can be ignored and the disturbing yawing moment will be:

$$N_e = \Delta T_e \, y_e \qquad (9-52)$$

2. Side force and yawing moment on the aircraft without the vertical tailplane (subscript A-v):

$$Y_{A-v} = \frac{\partial Y_{A-v}}{\partial \beta} \, \beta \qquad (9-53)$$

$$N_{A-v} = \frac{\partial N_{A-v}}{\partial \beta} \, \beta \qquad (9-54)$$

3. Contribution of the aircraft weight:

$$W \sin\varphi = L \tan\varphi \approx L\varphi \text{ for } \varphi \leqslant 5^\circ \qquad (9-55)$$

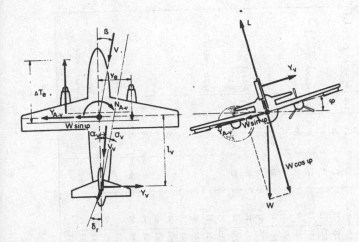

Fig. 9-21. Flight on asymmetric power

where the difference between the angles of bank and roll has been ignored.

4. Side force and yawing moment due to the vertical tailplane:

$$Y_v = \frac{\partial Y_v}{\partial \alpha_v} \alpha_v + \frac{\partial Y_v}{\partial \delta_r} \delta_r \qquad (9-56)$$

$$N_v = -Y_v \ell_v \qquad (9-57)$$

where:

$$\alpha_v = -(\beta - \sigma_v) \qquad (9-58)$$

The yawing moment due to aileron deflection has been neglected. Since the position of the airplane c.g. has little influence during flight with engine failure, ℓ_v is taken relative to the mean quarter chord point. In the case of a propeller aircraft, as depicted in Fig. 9-21, the sidewash angle σ_v will have a negative sign, with the result that the effect of the yawing moment due to engine failure will in fact be increased, particularly in the case of a high-wing aircraft (Ref. 9-54).

The simplest expression for the equilibrium condition is obtained by eliminating φ. After introduction of the usual nondimensional coefficients (see the Nomenclature) and some algebraic manipulations, we find the following condition for the vertical tailplane area required:

$$\eta_v C_{Y_{v_\alpha}} \frac{S_v}{S} = \frac{C_L \frac{Y_e}{\ell_v} \frac{\Delta T_e}{W} + \beta \left(C_{n_\beta}\right)_{A-h} \frac{b}{\ell_v}}{\tau_v \delta_r - (\beta - \sigma_v)} \qquad (9-59)$$

where

$$\eta_v = \left(\frac{V_v}{V}\right)^2 \qquad (9-60)$$

The numerator represents - in nondimensional form - the yawing moments due to engine failure and sideslip on the aircraft, less vertical tailplane. The first term of the denominator is the change in zero-lift angle of the tailplane caused by rudder deflection, and the bracketed term defines the direction of the local airflow, which is dependent on the direction of rotation of the revolving propeller(s). If the operating propeller(s) revolve(s) clockwise - viewed from behind the aircraft - the critical engine will be the outboard port engine. This asymmetry is not present on jet aircraft.

Fig. 9-22 shows that a negative sideslip angle makes it possible to use a smaller tailplane as compared with $\beta = 0$. However, the condition imposed by the angle of bank of five degrees limits the usable sideslip angle, and for $\Delta T_e y_e C_L/(W\ell_v) > .15$ (approximately) this limit will considerably increase the control capacity required. If

Table 9-3. Vertical tailplane design data

Aircraft type	V_D (kts EAS)	M_D	V_{MC} (kts EAS)	Max. cross-wind (kts)	$\dfrac{S_v}{S}$	A_v	Λ_v (deg.)	Airfoil Section root/tip	Average t/c (%)	$\dfrac{S_v l_v}{Sb}$	$\dfrac{S_r}{S_v}$	Hinge Position root/tip $z c_v$ (%)	Balance Ratio $z c_r$ (%)	$\pm\delta_r$ max. (deg.)	Tabs *	Remarks
Wright Flyer (1903)					.045	2.91	0	"flat plates"	–	.0133	1	30	42.8			biplane in front
Scottish Aviation Pup				30	.182	.89	33		12	.0750	.350	64		25	T	Horn balance
Cessna 177		.60		20	.107	1.41	35	NACA 0009/0006	7.5	.0411	.368	60				Horn balance
Scottish Av. Jetstream			85		.212	1.44	43			.0820	.350	65				
Aérospatiale N262	260		90	30	.184	1.60	9.5	NACA 0012 mod.	12	.0763	.270	68				
Fokker-VFW F 27 Mk 200	288	.51	78	30	.203	1.55	3.3	NACA 63A-015 mod.	15	.0765	.218	76	31.2		T/C	
Lockheed Hercules C-130E	325	.64	93.5		.180	1.84	18.8	NACA 64A-015	15	.0575	.239	75	44.6		T/C	
Lockheed Electra 188A		.711	110		.145	1.93	14.0	NACA 0012	12	.0707	.250	69		20	T/C	
Bristol Britannia 310	300				.197	1.65	10	RAF30 mod.	13	.0774	.225	60/67	25.8	35 +23.5 −30	T	$\delta_r \pm 16°$ for CL-44
Aérospat. Corvette SN-600					.177	.88	50.3		9	.0720	.250	72	30	25	T/S	
Cessna Citation 500	370		90		.191	1.58	33.0	NACA 0012/0008	10	.0806	.220	75	25	30	T	
Hawker Siddeley HS-125/400	370	.825	90	25	.161	1.19	52		11.5	.0548	.197	72	25	28.5	T	Approx. T-tail
Yakovlev YAK-40	330	.74			.162	.81	47.5		10	.0442	.222	78		30	T	T-tail
VFW-Fokker 614	390	.83			.174	1.28	32.4		10.15	.0682	.321	78		35	T/S	T-tail
Fokker-VFW F 28 Mk1000	375	.87	71	30	.203	1.00	40	NACA mod.	11	.0910	.187	76/40		33		T-tail
Aérospat. Caravelle 10.R	410	.86			.106	1.24	37.9	NACA 65-011	12.5	.0379	.258	70	30	24		T-tail
BAC-111/200,400		.89			.132	.91	41.0		11	.0482	.254	68				T-tail
McDonnell Douglas DC-9/10	400				.192	.95	43.5	DSMA	12	.0810	.270	75	30			T-tail
Boeing 737/100	400	.89			.268	1.88	35		9	.1117	.250	71.5	35	30	T/C	
Dassault Mercure					.232	1.96	35		9	.1025	.221	80				
Boeing 727/100	420	.95			.238	.78	55		12.5	.0905	.168	70			T	Split rudder
Airbus A-300B	420	.90	103		.204	1.62	40		10	.1020	.248	65				T-tail
Boeing KC-135					.143	1.49	31		10	.0628	.250	65				
Boeing 707/120			107.5		.148	1.62	31		10	.0656	.282	65		20	S	
Boeing 707/320B		.95	122		.144	1.81	31			.0626	.242	65		25	S	
McDonnell Douglas DC-8/10,50	405	.95	$<V_{LOF}$	34	.122	1.91	35	DSMA-111/–112	9.85	.0494	.269	65		25	S	
BAC VC-10/1101	380	.94			.142	1.10	38.5			.0453	.251		37.2	32.5	C	
Lockh. Tristar L-1011/1	435	.95		30	.231	1.83	35		10	.0830	.161	70				T-tail
McDonnell Douglas DC-10/10		.95		30	.221	1.92	40	∿122/∿102	11	.0811	.145	62				Split rudder
Boeing 747/100,200	445	.97	103/138	30	.196	1.38	44			.0990	.173	77	42	23/46.5	T	Tandem rudder
Lockheed C-5A	410	.875		43	.191	.84	34.9			.0951	.191			25		T-tail / Split rudder

* C = control tab; S = servo tab; T = trim tab

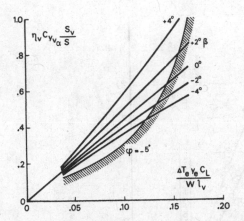

Fig. 9-22. Effect of sideslip on vertical tailplane area required to cope with engine failure

this criterion is exceeded to any great extent, a method of reducing the yawing moment will have to be found. The possibility of interconnecting the propeller shafts (Brêguet 941) and/or adopting an unconventional method of directional control - such as cyclic pitch of the propellers, control by means of air jets or a propeller at the tail (as in helicopters) - will have to be investigated in this case.

Practical application of (9-59) is hampered by the difficulty that σ_v cannot be predicted with any accuracy for propeller aircraft. In addition, a combination of W and C_L has to be substituted, which means, in effect, that a choice of V_{MC} must be made. The statistics in Fig. 9-23 and Table 9-3 are intended to replace a detailed study by a rapid estimation of the vertical tail area required, or rather the tailplane area on which the rudder is fitted. Large tailplanes can be avoided in the following ways:

1. Ensuring that the critical engine is not located further outboard than is strictly necessary. Another way of reducing y_e is to adjust the angle of the jet efflux (see Fig. 6-18).
2. Choosing a favorable location for the vertical tailplane, or possibly by using twin tails.
3. Employing a special rudder design, e.g. one consisting of two segments (double-hinged or varicam rudder, see the De Havilland Canada DHC-7 and the

DC-10) or a rudder provided with guide vanes (Fokker P 301 STOL project).
4. Choosing an aspect ratio which is not too low. A T-tail will increase the effective aspect ratio, but for structural reasons the geometric aspect ratio will generally not exceed 1.2.
5. Increasing the tail moment arm (longer fuselage, swept-back fin).

An approximate limit would be a rudder chord of 30 to 35 percent of the tailplane chord with a maximum rudder deflection of 25 to 30 degrees. In the case of manual control the rudder deflection may be limited by the requirement that the pressure on the rudder pedals should not exceed 180·lb (81.6 kg). In connection with Fig. 9-23 it should be noted that the choice of S_v may have been decided by other criteria in the case of several aircraft. In the absence of relevant data, the rudder deflection of some aircraft has been assumed at 25 degrees. These factors may explain the fairly wide disparities in the lower limit values for a number of types.

c. Drag caused by an inoperative engine.

A detailed account of this drag increment is given in Appendix G, Section G-8. Only one contribution will be mentioned here: the vortex-induced drag resulting from the side force on the vertical tailplane, which may be expressed as a fraction of the total vortex-induced drag:

$$\frac{\Delta C_D}{C_L^2/(\pi A)} = \left(\frac{\Delta T_e}{W}\frac{y_e}{\ell_v}\frac{b}{b_v}\right)^2 \qquad (9-61)$$

This expression shows that ΔC_D is proportional to $(y_e/\ell_v\, b_v)^2$, which proves that the recommendations set out above deserve special attention.

9.6.2. Lateral stability

a. Directional stability.
Every aircraft must possess positive stick-fixed and stick-free directional stability in normal operational conditions, for all aircraft configurations. The following applies to the stick-fixed directional stability, which may be used as a preliminary design tool:

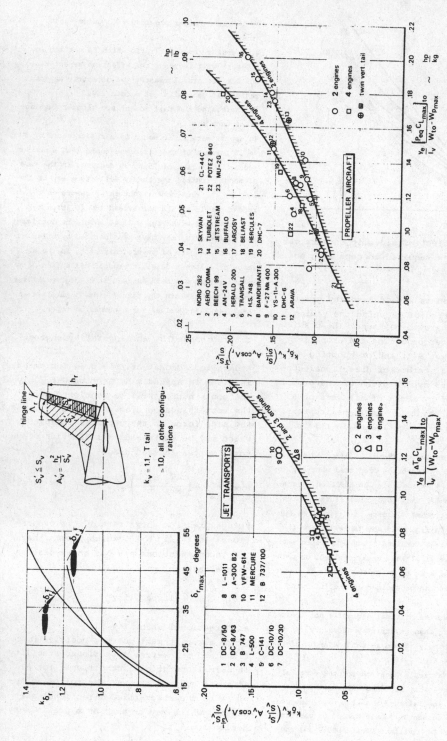

Fig. 9-23. Rapid estimation of vertical tailplane size for multi-engined aircraft with wing-mounted engines

$$C_{n_\beta} = \left(C_{n_\beta}\right)_{A-h} + \eta_v \, c_{y_{v_\alpha}} \frac{S_v \ell_v}{Sb} \left(1 - \frac{d\sigma_v}{d\beta}\right)\left(\frac{V_v}{V}\right)^2 > 0 \tag{9-62}$$

where η_v is defined by (9-60). The contribution of engine nacelles to the directional stability is generally small and may be ignored. The same applies to the wing contribution, provided the sweep angle is not too large. Wing/fuselage interference is manifest both in $\left(C_{n_\beta}\right)_{A-h}$ and σ_v; these effects will be combined in a single coefficient $\Delta_i C_{n_\beta}$, which is mainly dependent on the location of the wing in relation to the fuselage in Z-direction. The vertical tail volume required to obtain.a specified C_{n_β} is obtained from:

$$\frac{S_v \ell_v}{Sb} = \frac{C_{n_\beta} - \left(C_{n_{\beta_f}} + C_{n_{\beta_p}} + \Delta_i C_{n_\beta}\right)}{C_{y_{v_\alpha}}(V_v/V)^2} \tag{9-63}$$

The degree of directional stability cannot be deduced in a straightforward manner from existing requirements. A minimum value of $C_{n_\beta} = .03$ is sometimes recommended, but this is generally inadequate for ensuring a well-damped Dutch roll. In the case of single-engine subsonic aircraft C_{n_β} is often found to lie between .04 and .10, while for transport aircraft values generally range from .10 to .25. For aircraft with swept wings C_{n_β} is a function of C_L, while a part is also played by deflection of the flaps.

An estimate of the tail volume required may be made with the help of Fig. 9-24, which applies to aircraft with engines either mounted close to or buried in the fuselage. Aircraft with wing-mounted engines are not mentioned here, since flight with engine failure will mostly be the deciding case.

The following aerodynamic data[*] have been used to derive the contributions to C_{n_β} in Fig. 9-24:

[*]derived from Ref. 9-1.

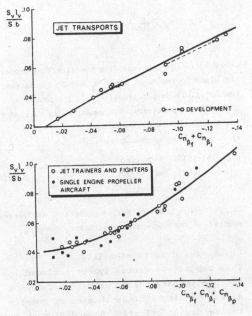

Fig. 9-24. Rapid estimation of vertical tail volume for aircraft with fuselage-mounted engines

$$C_{n_{\beta_f}} = -K_\beta \frac{S_{fs} \ell_f}{Sb} \left(\frac{h_{f_1}}{h_{f_2}}\right)^{\!\frac{1}{2}} \left(\frac{b_{f_2}}{b_{f_1}}\right)^{1/3} \tag{9-64}$$

where, for $\ell_f/h_{f_{max}} \geqslant 3.5$,

$$K_\beta = .3 \frac{\ell_{cg}}{\ell_f} + .75 \frac{h_{f_{max}}}{\ell_f} - .105 \tag{9-65}$$

The various dimensions are defined in Fig. 9-25

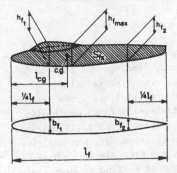

Fig. 9-25. Fuselage geometry in relation to the yawing moment due to sideslip.

$$c_{n_{\beta_p}} = -.053 \, B_p \sum \frac{\ell_p \, D_p^2}{Sb} \qquad (9\text{-}66)$$

where ℓ_p is the distance from the (tractor) propeller plane to the c.g., while B_p and D_p denote the number of blades per propeller and the propeller diameter, respectively. The summation is carried out for all operating propellers.

Finally we have approximately:

$$\Delta_i c_{n_\beta} = \begin{array}{ll} -.017, & \text{high wing} \\ +.012, & \text{mid wing} \\ +.024, & \text{low wing} \end{array} \qquad (9\text{-}67)$$

If the design provides for twin vertical tailplanes mounted on the horizontal stabilizer, use $\Delta_i c_{n_\beta}$

b. Lateral dynamic stability.

This subject is generally not included in the preliminary design process because of a lack of data and the absence of quantitative design criteria. Some design objectives are laid down in Ref. 9-13 with respect to the damping of the Dutch roll oscillation and the amount of spiral stability required. It should be noted that the dihedral of the wing has a great influence on lateral stability, and since this parameter can be varied without greatly affecting the general arrangement, some measure of freedom is available to the designer, even at a later stage of the project. As explained in Section 7.7.1. most high-speed aircraft with swept wings are e-quipped with a yaw damper to obtain an adequately damped Dutch roll mode.

9.6.3. Crosswind landings

In order to ensure regularity in air transport operations, landings may have to be carried out at crosswind velocities of up to, say, 30 knots (15.5 m/s). The associated angle of sideslip to be dealt with will be of the order of 12 to 15 degrees for conventional transport, but may rise to 20 degrees for STOL aircraft. Equation (9-59) may be used to find a lower limit to the vertical tail size by taking $\Delta T_e = 0$:

$$\eta_v \, c_{y_{v_\alpha}} \frac{S_v}{S} = \frac{\beta \, (c_{n_\beta})_{A\text{-}v} \frac{b}{\ell_v}}{\tau_v \, \delta_r - (\beta - \sigma_v)} \qquad (9\text{-}68)$$

where τ_v takes account of the nonlinearity in the c_{y_v} vs. δ_r curve when the rudder deflection is large. Although this criterion looks quite simple, a detailed knowledge of the various factors is required to make it usable, since small variations in the terms in the denominator have a great effect on the result. Accurate wind-tunnel data are therefore required. Ensuring control over the airplane on a wet runway in a crosswind is another important design problem. It should be noted that the operation of thrust reversers may have an adverse effect.

9.6.4. The spin

Aircraft to be certified in the Aerobatic Category and some types in the Utility Category must have safe spinning characteristics.

The loading of the airplane, which determines the inertial moments about the three axes, dictates what controls have to be used for recovery from a spin. Particularly in the case of light aircraft, the rudder is usually the most effective control since it provides the yawing moment required to oppose the spin rotation. Additional elevator deflection may be required to apply an antispin yawing moment on aircraft where the mass is heavily concentrated in the wing (engines, fuel, tip tanks).

The shielding effect of the horizontal tail is responsible for a dead air region over much of the vertical tail (see Fig. 2-27). As a rule, it can be said that at least one-third of the rudder area should remain outside this wake, assuming that for an angle of attack of 45 degrees the forward and aft boundaries of the wake form an angle of 60 and 30 degrees respectively with the horizontal tail. In addition, there should be a substantial amount of fixed area beneath the horizontal tail to provide damping of the spinning motion. More accurate design criteria have been developed by the NACA; these involve the polar moments of inertia about the longitudinal and lateral axes. A detailed account of this is given in the references on airplane spinning. Spin recovery characteristics can also be affected to a large extent by antispin fillets and dorsal fins (Ref. 9-56).

9.6.5. Preliminary design of the vertical tailplane

a. Aspect ratio A_v.

The aspect ratio, which is defined in Fig. 9-20, has a direct effect on the tailplane contribution to C_{n_β}, which is approximately proportional to $A_v^{1/3}$. A high aspect ratio tailplane is effective at small angles of sideslip, but it has a small stalling angle of attack. A low A_v is required with a high-mounted horizontal tailplane to provide adequate rigidity of the fin without an excessive weight penalty.

b. Taper ratio λ_v.

The tailplane contribution to lateral stability is only slightly affected by taper, which is applied mainly to reduce weight and/or increase the rigidity of the fin. Little taper is possible on T-tails.

c. Angle of sweep Λ_v.

Design considerations are similar to those for the horizontal tailplane (Section 9.5.1c) but in addition it should be noted that in the case of a fin-mounted horizontal tailplane the moment arm of both tailplanes is increased. Most designers of private aircraft seem to prefer using a swept vertical tailplane in order to please their customers.

d. Airfoil section.

The usual sections for a vertical tailplane are symmetrical, have thickness ratios of about 12 percent and a relatively large nose radius to permit a large range of angles of attack.

e. Design procedures.

A simple approach is to choose the tailplane volume $S_v \ell_v / (Sb)$ by comparing design data of aircraft with a similar general layout in the same airworthiness category (Table 9-3). The location of the engines (wing- or fuselage-mounted), the wing (low, high or mid-wing), and the vertical position of the horizontal stabilizer are decisive factors. The process is iterative, in view of the fact that the moment arm ℓ_v cannot be determined accurately before the layout of the tailplane has been chosen. A more detailed procedure may be started if the designer is prepared to make assumptions regarding the (relative) rudder area and maximum deflection.

For multi-engine aircraft with wing-mounted engines a lower limit for $S_v{}'$ can be derived from Fig. 9-23. A subsequent check should be made to ascertain that the volume is not less than the value dictated by Fig. 9-24.[*] If sufficient data are available, the crosswind criterion given by (9-68) should also be checked.

For aircraft with engines mounted on or buried inside the fuselage Fig. 9-24[*] can be used to estimate the volume parameter, and the crosswind criterion given by (9-68) may be checked as soon as the required aerodynamic data are available.

Light aircraft, which must demonstrate good spinning characteristics, should satisfy the criteria laid down in Section 9.6.4. It is advisable to apply the NACA criteria summarized in Refs. 9-57 and 9-58.

[*] The various contributions to C_{n_β} are given by (9-64) through (9-67).

Chapter 10. The undercarriage layout

SUMMARY

The basic requirements for the design of an undercarriage are that it must be capable of
absorbing a certain amount of energy, both vertically and horizontally, and that during
taxying, liftoff and touchdown no other part of the aircraft will touch the ground. No
instabilities must occur, particularly during maximum braking effort, crosswind landings
and high-speed taxying. In addition, the undercarriage characteristics must be adapted
to the load-carrying capacity of the airfields from which the aircraft is intended to
operate. This chapter indicates how these requirements can be translated into an acceptable initial choice of the undercarriage layout, without going into the details of its
structural design.

NOMENCLATURE

A	- aspect ratio of wing		UCI	- Unit Construction Index
a_x	- deceleration during braking		u.c.	- undercarriage (assembly)
b	- wing span		V	- velocity
CBR	- California Bearing Ratio		W	- weight
C_L	- lift coefficient		w	- velocity of descent
C_{L_α}	- lift-curve gradient ($C_{L_\alpha}=dC_L/d\alpha_f$)		x,y,z	- dimensions of an u.c. leg
$\bar{c}$	- mean aerodynamic chord		x_{cg}	- coordinate of c.g., measured in percent of $\bar{c}$

A   - aspect ratio of wing
a_x - deceleration during braking
b   - wing span
CBR   - California Bearing Ratio
C_L - lift coefficient
C_{L_α} - lift-curve gradient ($C_{L_\alpha}=dC_L/d\alpha_f$)
$\bar{c}$ - mean aerodynamic chord
c.g.   - center of gravity
D   - diameter; distance between tire imprints; drag
E   - kinetic energy
ESWL   - Equivalent Single Wheel Load
e   - deflection
g   - acceleration due to gravity
h   - height above ground plane
k_{sg} - gear stiffness ratio
L   - radius of relative stiffness; lift
LCN   - Load Classification Number
L_w - static load on a wheel
l   - length
N_s - number of shock struts per main u.c.
n_x,n_y - load factors in longitudinal and lateral direction, respectively
P   - load on an undercarriage leg
p   - inflation pressure; fraction of $C_{L_{max}}$
r   - radius of loci of force intersection with ground plane
S_B - wheel base of a (dual) tandem u.c.
S_D - center distance between two most widely spaced contact areas
S_T - spacing of two dual wheels
s   - stroke of shock absorber
s_t - maximum tire deflection
T   - thrust
t   - track of main u.c.; time

UCI   - Unit Construction Index
u.c. - undercarriage (assembly)
V   - velocity
W   - weight
w   - velocity of descent
x,y,z - dimensions of an u.c. leg
x_{cg} - coordinate of c.g., measured in percent of $\bar{c}$
α_f - angle of attack relative to fuselage reference line
β - angle defining main wheel location of a tail wheel u.c.
Γ - angle defining position of wing tip in front view
δ - thickness of runway construction
δ_f - deflection angle of wing flap
η - efficiency of energy absorption
θ - angle of pitch
Λ - angle of sweep
λ - reaction factor
ψ - overturning angle
ϕ - angle of roll
μ - friction coefficient

Subscripts

cg   - center of gravity
crit - critical (stalling)
cr   - cruising
g   - (landing) gear
LOF   - liftoff
m   - main u.c.
n   - nose u.c.
s   - shock absorber; static
TD   - touchdown
t   - tire
to   - takeoff
uc   - undercarriage
w   - wheel
.25   - quarter-chord line

10.1. INTRODUCTION

Although the dimensions of the undercarriage are modest as compared to those of the wing or fuselage, it must not be regarded merely as an accessory, particularly since its function is much more that of an integral part of the structure. Its

weight constitutes some 3 to 5% of the Maximum Takeoff Weight, which is equivalent to about one-third to half the structural weight of the wing.
The maintenance costs associated with the undercarriage, such as the inspection and replacement of tires and brakes, represent a considerable item in the total mainte-

nance bill. This is particularly hard to accept because the undercarriage contributes virtually nothing to the flying and economic capabilities of the aircraft, so it is not surprising that many - unsuccessful - attempts have been made to eliminate it, such as takeoff carts, skid landing gear, aircushions and the like. At present it is unlikely that there will be a breakthrough in this direction. In the basic design costs may be reduced by aiming at simplicity and compactness, such as a simple retraction system, devoid of complicated kinematics and duplication of retracting jacks. It is not the job of the preliminary aircraft design engineer to investigate the details of the undercarriage, nor will he design the details of the hydraulic equipment and air conditioning system. Unlike this equipment, which can be stowed in irregularly shaped and widely dispersed volumes, the relationship between the wheels, the points of attachment of the gear to the airframe and the principle of the kinematics of retraction must be carefully tailored to the general arrangement and layout of the aircraft.

The following functional requirements have a bearing on undercarriage layout in the preliminary design phase:

1. During the phases of takeoff rotation and liftoff and landing flare-out and touchdown, only the wheels should be in contact with the ground. There should be adequate clearance between the runway and all other parts of the aircraft, such as the rear fuselage, the wingtips and the tips of propellers or engine pods.

2. The inflation pressure of the tires and the configuration of the landing gear should be chosen in accordance with the bearing capacity of the airfields from which the aircraft is designed to operate.

3. The landing gear should be able to absorb the normal landing impact loads and possess good damping characteristics. When taxying over rough ground no excessive shocks should be transmitted by the landing gear.

4. Braking should be efficient, the maximum braking force allowed by the condition of the runway being the limiting factor. During crosswind landings and high-speed taxying there should be no tendency to instabilities such as canting of the aircraft or groundlooping.

5. Suitable structural elements should be provided in the aircraft to serve as attachment points for the landing gear, and there should be sufficient internal space for retraction.

In Section 2.5., which deals with the general arrangement of the undercarriage, the choice between nosewheel (tricycle), tailwheel or tandem (bicycle) landing gear has been discussed. The general conclusion is that the tricycle gear has superseded the tailwheel type almost completely, mainly for reasons of improved stability on the ground, braking and steering. On specialized designs, however, there may be indications that a tailwheel or a bicycle undercarriage will be the best choice.

The information presented in this chapter is mainly a summary of data from References 10-1 through 10-4. For more detailed considerations with regard to the structural design of landing gear, the reader should consult Ref. 10-1, which is a standard textbook on landing gear.

10.2. TAILORING THE UNDERCARRIAGE TO THE BEARING CAPACITY OF AIRFIELDS

10.2.1. Runway classification

For light aircraft operating from grass or unsurfaced runways, the tire inflation pressure is generally limited to some 60 lb/sq.in. (4 kg/cm^2) for grass, as indicated in Table 10-1. When the load per tire exceeds approximately 10,000 lb (4,500 kg), and the tire pressure is more than about 60 lb/sq.in., consideration must be given to the limited bearing capacity of runways. The stresses in the runway or the possibility of damage caused by the undercarriage, are dependent upon the undercarriage con-

343

TYPE OF LANDING SURFACE	Maximum tire pressure kg/cm^2	lb/sq.in.
Large, properly maintained airports (concrete runway)	8.5 - 14	120 - 200
Small tarmac runway, good foundation	5 - 6.3	70 - 90
Small tarmac runway, poor foundation	3.5 - 5	50 - 70
Hard grass, depending on soil	3.2 - 4.2	45 - 60
Wet, boggy grass	2.1 - 3.2	30 - 45
Hard desert sand	2.8 - 4.2	40 - 60
Soft, loose desert sand	1.8 - 2.5	25 - 35

Table 10-1. Tire pressure recommendations (Ref. 10-1)

figuration and the type of runway structure. Two classes of runway are generally distinguished:

a. Runways with rigid pavement, consisting of discrete slabs of concrete, laid on a relatively soft subsoil. Failure occurs by fracture across the corner, especially when the subsoil is bad, e.g. clay. A sandy subsoil provides a good foundation.

b. Runways with a flexible pavement, consisting of a relatively thick layer of asphalt or tarmacadam on a base of gravel or sand, often with an intermediate layer of crushed stone. The total thickness is about twice that of a rigid pavement. Failure occurs because of local indentation.
Apart from these airfields, which are most commonly used in civil aviation, airfields with a simply prepared surface of, for example, fine gravel are becoming increasingly important for aircraft designed for operation in less accessible regions (Ref. 10-2).

In order to avoid damage to runways airfields have been classified according to various characteristic parameters:
CBR: California Bearing Ratio
UCI: Unit Construction Index
RLI: Runway Loading Index
LCN: Load Classification Number
On some airfields the bearing capacity of the runways is simply given as the maximum allowable gross weight of the aircraft.
The CBR index gives us the bearing capacity

of the runway, expressed as a percentage of the bearing capacity of a surface consisting of a hard type of stone. This index, which is fully explained in Ref. 10-2, is used to characterize the substructure of runways with a flexible pavement. Airfields without a hard upper layer are also classified according to the CBR index.
The UCI is an index used in the United States and shows affinity to the LCN, which will be discussed later. A method of determining the UCI graphically is given in Ref. 10-2. The RLI also shows some affinity to the LCN, but will not be discussed here. The LCN rating method has been introduced by the ICAO on the basis of much theoretical and experimental work, and is now widely accepted in many countries. Permissible values of the LCN have been assigned to all major runways, and aircraft have to be designed in such a way that the undercarriage will not exceed the lowest LCN value of the airfields from which the aircraft is likely to operate. Typical values of LCN for various aircraft categories are given in Table 10-2.

AIRCRAFT TYPE	MTOW (LB)	TIRE PRESS. (PSI)	LCN
Fokker F-27 Mk 500	45,000	80	19
Fokker F-28 Mk 2000	65,000	100	27
McDonnell D. DC-9/10	90,700	129	39
Boeing 707/320	300,000	135	58
McDonnell D. DC-10/10	410,000	175	88

Table 10-2. Load Classification Number of several airliners - main landing gear

A large number of tests on both rigid and flexible pavements have shown that the bearing capacity of a runway is dependent upon the wheel load and the inflation pressure of the tire(s). The relationship is reproduced in Fig. 10-1, which can be used directly for single wheels to find the limiting values of the wheel load for given values of the inflation pressure and LCN. For multiwheel undercarriages the concept

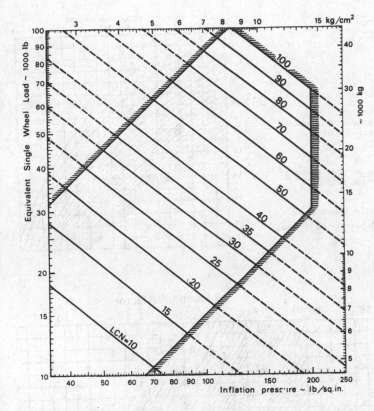

NOTE

The data given in the ICAO Aerodrome Manual (Ref. 10-3) are slightly different

Fig. 10-1. Load Classification Number for various combinations of tire pressure and wheel load (Ref. 10-6)

of the ESWL, as explained in the following section, can be employed in using this diagram.

10.2.2. The Equivalent Single Wheel Load (ESWL)

The ESWL of a group of two or more wheels which are relatively close together, is equal to the load on an isolated wheel, having the same inflation pressure, and causing the same stresses in the runway material as those due to the group of wheels. This equivalent wheel load accounts for the fact that a given loading, spread over a number of contact areas, causes lower stresses in the runway material than would be the case when the same load is concentrated on a single wheel. Typical reduction factors - i.e. the actual static load on the leg divided by the ESWL - are 4/3 for dual wheel layouts and 2 for twin

tandem (bogie) layouts (Ref. 10-4). Various methods have been developed for calculating the ESWL, depending on the mechanical characteristics of the runway and the undercarriage layout. The methods described below for rigid and flexible pavements and a method for graval runways can be found in Refs. 10-3 and 10-4.

a. Rigid pavements.
Runway characteristics are generally represented by the radius of relative stiffness L of the concrete. In the absence of more adequate information the following approximation may be used:

$$L = \text{constant} \times \delta^{3/4} \qquad (10-1)$$

where δ denotes the thickness of the runway construction. The constant takes account of the stiffness of the substructure; typical values are:
constant = 8.0 when L and δ are in inches,

345

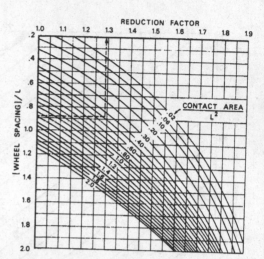

CONTACT AREA = TOTAL CONTACT AREA OF ALL WHEELS OF ONE UNDERCARRIAGE ASSEMBLY

$$\text{EQUIVALENT SINGLE WHEEL LOAD} = \frac{\text{TOTAL LOAD ON ONE U.C. ASSEMBLY}}{\text{REDUCTION FACTOR}}$$

Fig. 10-2. Equivalent Single Wheel Load assessment curves - rigid pavements - dual wheel undercarriages (Ref. 10-3)

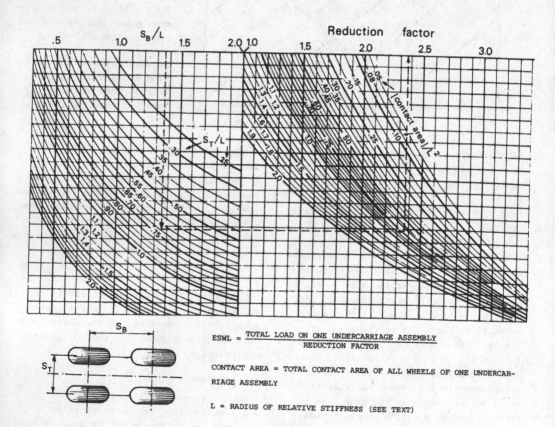

$$\text{ESWL} = \frac{\text{TOTAL LOAD ON ONE UNDERCARRIAGE ASSEMBLY}}{\text{REDUCTION FACTOR}}$$

CONTACT AREA = TOTAL CONTACT AREA OF ALL WHEELS OF ONE UNDERCARRIAGE ASSEMBLY

L = RADIUS OF RELATIVE STIFFNESS (SEE TEXT)

Fig. 10-3. Equivalent Single Wheel Load assessment curves - rigid pavements - dual and tandem undercarriages (Ref. 10-3)

or 10.1 when L and δ are in cm, for a bad, soft substructure;

constant = 6.1 when L and δ are in inches, or 7.7 when L and δ are in cm, for a good, hard substructure.

When the thickness of the rigid pavement is not specified, L is usually assumed equal to 45 in. (115 cm), which is equivalent to δ = 10 in. (25 cm) on a bad substructure. These assumptions will keep the designer on the conservative side. In the case of large aircraft the method may lead to excessive requirements where the undercarriage assembly is concerned, and more precise data should be collected in this case.

Using the calculated or chosen value for L and the geometrical data of the landing gear, the ESWL may now be calculated on the basis of Fig. 10-2 or 10-3. Strictly speaking the ESWL of the nosewheel leg should also be determined, but this is seldom a critical case. For dual tandem assemblies with a pair of wheels at each end of the axle, each pair of wheels is replaced by one equivalent wheel with the same ESWL as the pair of wheels, using Fig. 10-2. The ESWL of the four replacing wheels is then computed on the basis of Fig. 10-3.

b. Flexible pavements.

The graphical method of ESWL determination for flexible pavements is shown in Fig. 10-4.

The ESWL is considered to be equal to the wheel load from one tire when the pavement thickness is equal to or less than half the distance between the closest contact area edges, D/2. The contact area for each tire is equal to the wheel load divided by the tire inflation pressure. Assuming the tire imprint to be an ellipse with the major axis equal to 1.4 times the minor axis, we can derive:

$$\frac{D}{2} = \frac{S_T}{2} - \sqrt{\frac{L_w}{1.4 \, \pi p}} \tag{10-2}$$

where S_T is the distance between the imprint centers, L_w the load per tire and p the inflation pressure. When the pavement thickness equals or exceeds twice the center distance of the two most widely spaced contact areas (i.e. S_T for dual wheel assemblies or S_D for twin tandem wheel assemblies), the ESWL is considered to be equal to the load from one undercarriage assembly.

Between the two limits for δ it is assumed that log ESWL varies linearly with log δ.

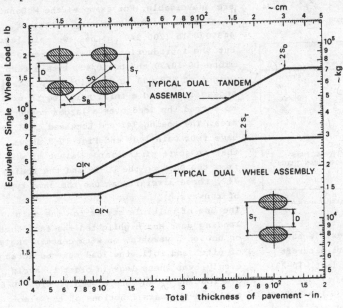

Fig. 10-4. Multiple wheeled undercarriages on flexible pavement-Equivalent Single Wheel Loads at varying depths

Instead of the graphical construction, one of the following expressions may be used:

dual wheel assemblies:

$$\log \text{ESWL} = \log L_w + .3 \frac{\log \delta - \log D/2}{\log 2S_T - \log D/2} \quad (10\text{-}3)$$

twin tandem wheel assemblies:

$$\log \text{ESWL} = \log L_w + .6 \frac{\log \delta - \log D/2}{\log 2S_D - \log D/2} \quad (10\text{-}4)$$

An example of the determination of LCN values will be given for the McDonnell Douglas DC-10/10. For an all-up weight of 388,000 lb (176,000 kg) it is assumed that the load on one main gear is equal to 46% of the weight, or 178,500 lb (81,000 kg). Tire pressure is 170 lb/sq.in. (12 kg/cm^2). For a rigid pavement of 12 in. (30.4 cm) thickness, the radius of relative stiffness for a good subsoil is L = 6.1 (12)$^{3/4}$ = 39.6 in. (100 cm). For S_B = 64 in. (163 cm) and S_T = 54 in. (137 cm) we find S_B/L = 1.63 and S_T/L = 1.37. The total contact area per undercarriage assembly is 178,500/170 = 1050 sq.in. (6774 cm^2), hence (contact area)/L^2 = .677. Using Fig. 10-3, we obtain a reduction factor of 3.40, resulting in an ESWL of 178,500/3.4 = 52,500 lb (23,800 kg). Fig. 10-1 is then used to read LCN = 76.

For a flexible pavement of δ = 43.3 in. (110 cm) thickness, the distance $S_D = (64^2 + 54^2)^{\frac{1}{2}}$ = 83.74 in. (213 cm) - see Fig. 10-4. Hence, we find D/2 =

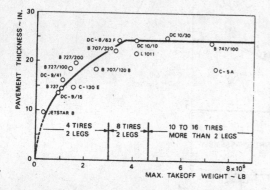

Fig. 10-5. Thickness of a flexible pavement required for several transport aircraft types - CBR of sublayer = 15 (Ref. 10-23)

$54/2 - \{178,500 / (4 \times 1.4\pi \times 170)\}^{\frac{1}{2}}$ = 19.3 in. (49 cm),

for a load per tire of 44,625 lb (20,250 kg). Using the logarithmic relationship of ESWL vs. δ, we find with Fig. 10-4: ESWL = 74,800 lb (33,930 kg) and using Fig. 10-1, we find LCN = 100.
These results are in fair agreement with the data quoted by the airplane manufacturer.

10.2.3. Multiple wheel undercarriage configurations

The ability to utilize runways of moderate quality and the ever-increasing all-up weight of transport aircraft have resulted in the use of twin, tandem, dual tandem and dual twin tandem wheel patterns. The application of multiple wheel undercarriages also results in a gain in safety, a flat tire being of little or no consequence. In addition, tandem or twin tandem gears are superior in taxying over obstacles in that the fore and aft wheels meet these obstacles at different times, thus only raising the airplane about half the total obstacle height.
Fig. 10-5 shows that for takeoff weights of up to about 450,000 lb (200,000 kg) the use of two main undercarriage legs appears to be adequate; for higher weights radical changes in the undercarriage configuration are unavoidable. For example, the McDonnell Douglas DC-10/10 with a ramp weight of 443,000 lb (200,940 kg) has two main legs, but the introduction of the long-range versions DC-10/30 and DC-10/40, with a ramp weight of 558,000 lb (253,105 kg), made it necessary to add a third main leg in order to spread the load over a larger contact area. The Boeing 747 and Lockheed C-5A even have four main legs and Fig. 10-5 shows that in spite of the large weight growth from the DC-8 to the DC-10 and the Boeing 747, these aircraft can use the same type of runways.
The use of multiple struts for the main landing gear has highlighted the fact that on uneven pavements, or with certain angles of pitch and roll, the load must be spread evenly over these legs in order to avoid overloading of any one assembly. Fig. 10-6 explains the main functions of the oleo-

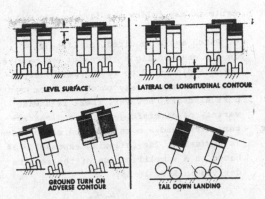

Fig. 10-6. Load equalizing system of the Boeing 747 main undercarriage system

pneumatic load equalizing system of the Boeing 747. An oil pipe connects the shock struts of the pair on each side, so that uneven loads are balanced out.

The Lockheed C-5A, having four main gear units (Ref. 10-29), each comprising a six-wheel bogie, illustrates an extreme in flotation capability. At reduced takeoff weights this aircraft may even be operable from unprepared runways.

The multiple gear develops higher ground turning torques, particularly if the aircraft is pivoted about one main gear. Steerable main landing gear systems may be employed to improve the turn radius, to avoid excessive side loads and to reduce tire wear from scrubbing. Detailed design considerations of these specialized designs, however, are considered to be outside the scope of this book.

10.3. DISPOSITION OF THE WHEELS

The following sections will deal with the choice of the wheel location. Boundaries will be derived for the intersection of the wheel load with the ground, based on conditions of stability during taxying, liftoff and touchdown. Assuming that for a given undercarriage assembly the center of area of the tire imprints coincides with these intersections, conditions will thus be obtained for the location of the tires. In

most cases it is fair to approximate the resulting ground contact points by the wheel axis or the pivot point of a bogie assembly.

10.3.1. Angles of pitch and roll during takeoff and landing

The landing gear legs should be sufficiently long to allow for any combination of the pitch angle (θ) and roll angle (ϕ) which may occur in normal operational use, without the risk that parts of the aircraft will come into contact with the ground. It is generally acknowledged that the available angle of pitch on liftoff and touchdown should at least be equal to, or preferably exceed, the limits imposed by performance or flight characteristics. A geometric limitation to the pitch angle will be detrimental to the liftoff speed and hence to the takeoff distance. A geometrical roll angle limitation may result in an undesirable operational limit in the case of crosswind landings.

The geometric limits may be reproduced in a ϕ-θ diagram (Fig. 10-7). The various

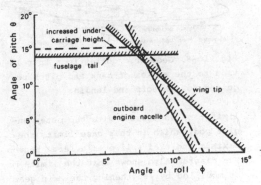

Fig. 10-7. Limitations on the angles of pitch and bank determined by the airplane geometry

boundaries define the point where the rear fuselage tail (bumper, if present), the wingtip, engines suspended below the wing, trailing-edge flaps, or any other part of the aircraft, just touches the ground plane. For a given aircraft geometry and length of the main undercarriage, the limit for θ will follow directly from the side elevation. The condition that the tip of the wing just touches the ground (Fig. 10-8 *) will be:

$$\tan\phi = \tan\Gamma + \frac{2h_g}{b-t} - \tan\theta \tan\Lambda \qquad (10-5)$$

Similar conditions may be deduced for other parts of the aircraft. Fig. 10-7 shows a

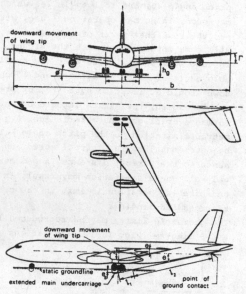

Fig. 10-8. Geometric definitions in relation to the angles of bank and pitch required for takeoff and landing

limitation associated with the outer engine pod, which in this case limits the maximum angle of roll over a large region. The diagram also shows that the limits may be modified by lengthening the main gear leg.

Accurate determination of the desired values for ϕ and θ can only be carried out when the performance calculations of the design have been completed. A provisional

*The difference between the angle of roll and angle of bank has been ignored here - which is acceptable for small angles of pitch.

estimate may follow the lines given below.

a. Pitch angle required for liftoff.

In the case of light aircraft, the required pitch angle of 12 to 14 degrees will generally be fairly easy to obtain. For transport aircraft, however, there is a wide variety of aircraft geometry and characteristics, and a more detailed calculation is advisable. The following expression is based on a simplification of the results derived in Ref. 10-25:

$$\theta_{LOF} = \alpha_{LOF} + \frac{d\theta}{dt}\left(\frac{2\ell_1}{V_{LOF}} + \sqrt{\frac{\ell_2}{g}\frac{C_{L_{LOF}}}{dC_L/d\alpha}}\right) \qquad (10-6)$$

where ℓ_1 and ℓ_2 are dimensions defined in Fig. 10-8, and θ_{LOF} refers to the condition where the undercarriage is fully extended.

In (10-6) the value to be used for α_{LOF} is the highest angle of attack to be anticipated for normal operational use. The second term is a correction to allow for the climb angle of the center of gravity associated with the extending undercarriage. The last term represents an increment to the pitch angle which will be needed, since immediately after rotation the rear of the fuselage will still be moving in the direction of the runway (Ref. 10-25), an effect which is particularly noticeable on large aircraft. Typical values to be used for the rate of rotation are:
$d\theta/dt = 3$ to 4 degrees/sec for large transports (DC-8, B 707, 747 class)
$d\theta/dt = 4$ to 5 degrees/sec for small transports (F-28, BAC 1-11 class).

Prediction methods for the aerodynamic data required to use (10-6) can be found in Appendices E, G and K. The amount of work involved in applying these methods is not always justified at the undercarriage design stage and the simplified approximation given below may be acceptable instead.
In the case of propeller aircraft, the effect of the propeller slipstream on lift will be quite appreciable. Assuming that slats are not present, a reasonable guess is obtained from:

350

$$\theta_{LOF} = 7\left(1 + \frac{3}{A}\right) \qquad \text{(deg.)} \qquad (10\text{-}7)$$

For jet transports the liftoff condition can be derived from the stalling angle of attack (see Fig. 10-9). Ignoring tailplane trim and ground

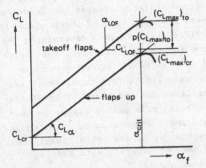

Fig. 10-9. Determination of the liftoff angle of attack from the lift-curves for cruising flight and takeoff

effect, we may obtain:

$$\alpha_{LOF} = \frac{1}{C_{L_\alpha}} \left[\left(C_{L_{max}}\right)_{\delta_f=o} - C_{L_{cr}} - p\left(C_{L_{max}}\right)_{to} \right] \qquad (10\text{-}8)$$

where it has been assumed that in the takeoff configuration flap deflection has no appreciable effect on the critical angle of attack, and that the fuselage is horizontal during cruising flight. Introducing an approximation for the lift-curve slope of high aspect ratio wings,

$$C_{L_\alpha} = \frac{2\pi \cos \Lambda_{.25}}{1 + 2/A} \qquad (10\text{-}9)$$

and allowing for ground effect, we find:

$$\alpha_{LOF} = 9.12 \left\{ \left(1 + \frac{2}{A}\right) \left[\frac{\left(C_{L_{max}}\right)_{\delta_f=o}}{\cos\Lambda_{.25}} - p\frac{\left(C_{L_{max}}\right)_{to}}{\cos\Lambda_{.25}} - \frac{C_{L_{cr}}}{\cos\Lambda_{.25}} \right] - \frac{1}{A} \right\} \qquad \text{(deg.)} \qquad (10\text{-}10)$$

The factor p allows for the margin relative to $C_{L_{max}}$ which should be adhered to in any case during liftoff. It is dependent on the aerodynamic characteristics, and may be assumed equal to .15 to .20. In the absence of better information, the designer may use:

$$\frac{\left(C_{L_{max}}\right)_{\delta_f=o}}{\cos\Lambda_{.25}} = 1.50 - \text{no leading-edge devices} \qquad (10\text{-}11)$$

$$= 2.10 - \text{with leading-edge devices}$$

b. Pitch and roll angles during landing. In the case of jet aircraft, the largest angle during landing is generally less than that during takeoff, because for fully deflected flaps the critical angle of attack of the wing will be smaller by some degrees than in takeoff. For propeller aircraft, we may take almost the same values in landing and takeoff.
For the desirable angle of roll we may assume:

8° - transport aircraft
15° - light aircraft

10.3.2. Stability at touchdown and during taxying: tricycle undercarriages

Although attention should be given to various types of static and dynamic instability during the design stage, only the measures which have to be taken to prevent the aircraft from canting over will be mentioned here. We shall use the plan view of the aircraft and assume a statically compressed landing gear (Fig. 10-11). The aircraft will cant over about any of the lines connecting the tire contact areas, if the resultant of air and mass forces intersects the ground at a point which lies outside the triangle formed by these connecting lines (Fig. 10-10). In that case the ground is unable to exert a reaction force which opposes the tendency to cant over.

a. Condition at touchdown.
The most unfavorable condition will be a landing with the centre of gravity in its most rearward and highest location. When there are no retarding forces (spin-up load), only a vertical force will be present which intersects the ground plane at

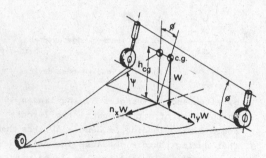

Fig. 10-10. Condition for stability at touchdown and during taxying

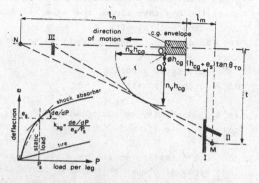

M : chosen position of main u.c. leg

N : chosen position of nose u.c. leg

I : limit of main leg position determined by touchdown angle of pitch

II : limit for the main leg position for given N, to attain stability against turnover

III: limit for the nosegear location for given M, to attain stability against turnover

Fig. 10-11. Limits for the undercarriage disposition based on stability considerations

a certain distance behind the projection of the center of gravity on the ground. The main gear must be placed at least this distance behind the aft c.g. position, nence:

$$\ell_m \geqslant (h_{cg} + e_s) \tan \theta_{TD} \qquad (10-12)$$

where e_s is the static deflection of the tire plus shock absorber and h_{cg} the c.g. height during taxying.

b. Avoiding sideways turnover of the air-craft.

Forces acting sideways on the airplane may be the result of a crosswind, an angle of yaw relative to the runway or a high-speed turn during taxying. In addition, taxying over uneven surfaces creates the danger of turnover. A simple rule of thumb states that the angle ψ (Fig. 10-10), determining the tendency to overturn sideways, must never exceed 60°, 55° to 57° probably being a maximum safe limit. This condition fixes a lower limit for the track. This statistical working rule cannot always be used, however, and a somewhat more detailed investigation may be desirable, particularly in the case of aircraft with fuselage-mounted main gear.

Fig. 10-11 shows a case with an unfavorable c.g. location, that is, the extreme forward and lateral position with asymmetrical loading (point 0). As a result of the force $n_y W$, directed sideways, the aircraft will assume an angle of roll which is mainly dependent on the stiffness of the landing gear and the track. The angle of roll may be deduced from the balance of forces, with the following result:

$$\tan\phi = 4 k_{sg} \frac{e_s}{t} \frac{h_{cg}}{t} n_y \qquad (10-13)$$

where k_{sg} is an undercarriage stiffness parameter defined in Fig. 10-11, while

e_s = static deflection (tire plus shock strut)

s = track

h_{cg} = height of c.g. above the ground during taxying and

n_y = lateral load/weight

As a result of the angle of bank the c.g. will be displaced sideways over a distance $h_{cg} \tan\phi$ (point Q). If the weight is represented by a vector having a length h_{cg} from the c.g., then with a sideways loading $n_y W$ the intersection of the resultant force with the ground will lie at a distance $n_y h_{cg}$ towards the side from point Q. When there is only an inertia force in the longitudinal direction $n_x W$ during braking, the point of intersection will lie at a distance $n_x h_{cg}$ ahead of 0.

In view of the danger of turning over the most unfavorable case will be on a dry hard surface when the adhesion forces between the tire and the ground

will be maximum. In such a case, values of .5 to
.6 may be found for n_x, while n_y = .5. For a dry
grass surface we may take $n_x = n_y$ = .35 as the
limit. Since the loads in all directions will be
of the same order of magnitude, a circle may be
taken as the extreme limit of the points of intersection mentioned above. This circle will have 0
as its centre, while the radius is given by:

$$r = n_y h_{cg} (1 + 4 k_{sg} \frac{e_s}{t} \frac{h_{cg}}{t}) \qquad (10-14)$$

When the line interconnecting the mainwheel with
the nosewheel contact areas touches this circle,
a condition is found that defines the lower limit
of the track of the landing gear, provided the
nosewheel location is fixed.

Equation 10-13 demonstrates that under a
given lateral force the angular roll is inversely proportional to the square of the
track. In a condition where a narrow track
is dictated by the general arrangement, the
aircraft becomes liable to lateral "wal-
lowing" and a landing gear with high stiffness (i.e. small value of k_{sg}) and a short
stroke (i.e. small value of e_s) is desirable.

When using (10-13) and (10-14) it may be assumed
that the static deflection of the tire is about 1/3
of the maximum, while that of a conventional shock
absorber will be about 3/4. The factor k_{sg} will be
1 for a landing gear where the compression has a
linear relationship with the wheel load (spring
without pre-tensioning, pneumatic tire), a typical
range of values for a strut with hydraulic shock
absorber will be k_{sg} = 1/3 to 1/2.

10.3.3. Gear length, wheelbase and track: tricycle undercarriages

The most favorable location of the wheels
can be determined by indicating limits in
the three-view drawing of the aircraft
(Fig. 10-12).

a. Disposition in elevation.
1. During takeoff and landing the rear
of the fuselage should remain clear of the
ground by an amount equal to at least the
maximum tire deflection, or about 2% of
$\ell_1 + \ell_2$ (see Fig. 10-8). The attitudes of
the aircraft are known from performance
and aerodynamic data or may be estimated
with the approximations given in Section
10.3.1. In this case the landing gear is
completely extended.
2. During taxying (landing gear statically
compressed) there should be a propeller
clearance of at least 7 in. (18 cm). In the
case of an inflated tire and a fully deflected shock absorber, the propeller(s) or
any other part of the aircraft must remain
clear of the ground*.
3. To avoid a tail slam on touchdown, the
center of the tire contact areas must be
located just behind the intersection of the
normal from the rear center of gravity to
the ground. In the absence of adequate information, the pitch angle on touchdown
θ_{TD} may be assumed equal to θ_{LOF}.

b. Disposition in plan.
1. When the load on the nosewheel is less
than about 8% of the MTOW, controllability
on the ground and stability during taxying
will suffer, particularly in crosswind conditions. When the static load on the nosewheel exceeds about 15% of the MTOW, the
load during heavy braking may become excessive, braking may be less efficient, and
too much effort may be required for steering. Although the margin between these limits seems to be generous, it should be remembered that variations in the location
of the center of gravity have a great effect on the nosewheel load. Fig. 10-12
relates the limits for the nosewheel disposition to the airplane load and balance
diagram.

The figures stated above should be considered as
recommendations and not as requirements. Structural
considerations may be conclusive in deciding where
to place the nose gear.

2. Once the location of the nosewheel has
been chosen, the maximum overturning angle
of 57° - or alternatively the graphical
method given in Section 10.3.2. - will
provide a lower limit for the track of the
main gear. If the location of the main

*see also Section 6.4.1.

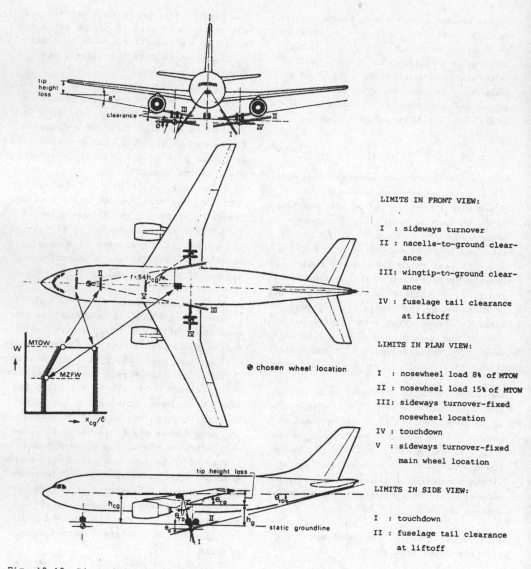

LIMITS IN FRONT VIEW:

I   : sideways turnover
II : nacelle-to-ground clear-
      ance
III : wingtip-to-ground clear-
      ance
IV : fuselage tail clearance
      at liftoff

LIMITS IN PLAN VIEW:

I   : nosewheel load 8% of MTOW
II : nosewheel load 15% of MTOW
III : sideways turnover-fixed
      nosewheel location
IV : touchdown
V   : sideways turnover-fixed
      main wheel location

LIMITS IN SIDE VIEW:

I   : touchdown
II : fuselage tail clearance
      at liftoff

Fig. 10-12. Disposition of the wheels in the three-view drawing

gear, and thus the track, have been fixed, a lower limit for the wheelbase will follow, although this is less sharply defined.
3. Occasionally there may be arguments in favor of shifting the main wheels further back than required for stability during landing. Examples are:
- freighters with the loading door in the rear of the fuselage,
- aircraft with engines at the rear which have a backward c.g. when empty.
The object in these two cases is to ensure safe loading without having to place ballast in the nose or fit a support at the tail of the fuselage. However, there are serious objections to such a backward shift:
- Rotation in the takeoff will be more dif-

ficult, with the result that the aircraft will unstick later, while accurate longitudinal control will be more difficult for the pilot, unless the elevator power is considerably increased (Ref. 10-14).
- The load on the nosewheel will be increased and it may not be possible to stop efficiently unless brakes are also fitted to the nosewheel.

c. Disposition in front view.
The main factors to be considered in determining the landing gear height follow from the considerations in Section 10.3.1. with regard to the available angles of roll and pitch. The minimum track of the main wheels follows from the required stability against overturning sideways.

Once all limits have been established, the designer may choose the shortest landing gear which satisfies every requirement. There may, however, be good reasons for departing from this rule:
- With the track chosen a main leg of minimum length may be too short to ensure that the wheels will lie in the desired location when retracted.
- The leg will be too short to accommodate the shock absorber (see Section 10.5.2.).
- There may be reason to anticipate fuselage stretch, requiring a higher landing gear for the same rotation angle (see Fig. 2-8).
If the main gear is unacceptably high, possibilities of modifying the design of the aircraft must be investigated. Depending on the critical limitations, this may involve:
- changing the contour of the rear fuselage,
- increasing the angle of incidence of the wing relative to the fuselage,
- increasing the dihedral of the wings and/ or
- re-locating an engine.

The length of the nosewheel leg will sometimes follow directly from the required clearance of a propeller in the nose of the aircraft (see Section 6.4.1.). In other cases the length is generally based on the requirement that the fuselage should be horizontal or slightly tilted nosedown when the aircraft is on the ground.

10.3.4. Disposition of a tailwheel undercarriage

The governing factor here will be the landing, since the aircraft touches down near the condition of stalling. Assuming the fuselage reference line to be horizontal during cruising, an approximation for the critical angle of attack is:

$$\alpha_{crit} = \left(12 - 10\ C_{L_{cr}}\right)\left(1 + \frac{3}{A}\right) \quad (deg.) \quad (10\text{-}15)$$

With the landing gear fully extended, at least this angle should be available with he aircraft tail down.
The deciding factor for the location of the main wheels will be the stability attainable when full braking is applied during landing. The resultant force of the load acting on the wheels perpendicular to the ground and the friction with the ground must therefore create a tail-down moment. The condition for the angle β in Fig. 10-13

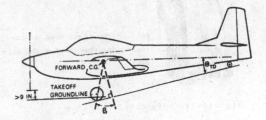

Fig. 10-13. Main wheel disposition for tailwheel undercarriages

will then be:
$$\tan\beta > \mu \quad (10\text{-}16)$$

Assuming μ to be limited to .25 to .35 as a result of the runway condition (e.g. grass) or the limited braking torque, β should be at least 14° to 17°, the latter value generally being taken as the lower limit. When the main gear is placed too far for-

ward, the aircraft cannot be brought into
a favorable attitude for taking off. The
clearance of the propeller to the ground
in the takeoff attitude must be at least
9 in. (23 cm.) - see also Section 6.4.1.

10.4. TYPE, SIZE AND INFLATION PRESSURE OF
THE TIRES

Fig. 10-14 shows some examples of tire

Normal, grooved tire, wear-
resistant and having good characteristics on wet runways

Chined tires, used on aircraft
with rear-mounted engines

Anti-shimmy tire, used on castoring nosewheels of light aircraft

Fig. 10-14. Tire tread patterns

treads, the choice of which will depend on
the application envisaged. The grooved
pattern normally used is provided with
ribs. This has been found necessary in
order to obtain good adhesion on wet runways and to minimise the effects of the
cutting action of stones or flints in the
runway surface. Aircraft with engines
located in the wing roots or at the sides
of the rear fuselage use tires provided
with a chine, which serves to direct the

water on a wet runway sideways.
Treads with two contact areas are intended
to counteract the tendency to shimmy; they
are mainly used on light aircraft with a
single, castoring nosewheel.
Shimmy is a violent, self-excited oscillation about
the swivel axis, caused by positive trail in the
case of a castoring wheel. The problem can be overcome by incorporating friction or hydraulic damping,
or heavy self-centering. Alternatively, twin wheels
may be employed to achieve the same effect as a
single wheel with two contact areas.
Manufacturers of tires issue catalogs in
American, British and French standard
sizes. Definitions of tire sizes and a
survey of applications in aircraft can be
found in Ref. 10-30.
The choice of the size of tire is simplified by using Fig. 10-15 which gives representative values for the static loading.
Since the characteristics of tires are continually being improved, the data in Fig.
10-15 should be checked against the latest
tire manufacturer's data to make sure that
a certain type is still in production and
its characteristics have not been changed.

10.4.1. Main wheel tires

The choice of the main wheel tires is generally made on the basis of the static
loading case. The load is determined by
the weight of the aircraft, the number of
main wheel legs, the number of wheels per
leg and the location of the leg in relation to the center of gravity. The aircraft is considered to be taxying without
braking, at low speed, and hence the wheel
load follows from the static equilibrium.
From Fig. 10-15 we may derive the following
expression for the total main gear load:

$$P_m = \frac{\ell_n}{\ell_m + \ell_n} W \qquad (10-17)$$

The critical condition is the aft location
of the center of gravity at the Maximum
Takeoff Weight*. If the data required to
carry out a calculation are not available,

*More precisely: the Maximum Ramp Weight

356

it may be assumed that when two main legs are used, each of them will have to carry 46% of the MTOW.

When the characteristics of the shock absorber in the landing gear leg are known, it will be possible to determine the energy absorption required for the impact during landing. The maximum dynamic load can then be calculated. Alternatively, for a preliminary design it is justifiable to assume that the design of the shock absorber will be adapted to the energy which the tire is able to absorb at maximum deflection. In that case the choice of the tire may be based on the static load it can carry.

For undercarriage legs with a single axis the total load on the leg is divided equally over the tires. In bogie assemblies the load per wheel depends on the position of the bogie pivot point. If the pivot point is midway between the front and rear wheel axles the wheel loads are equal in the static case, but the front wheels will

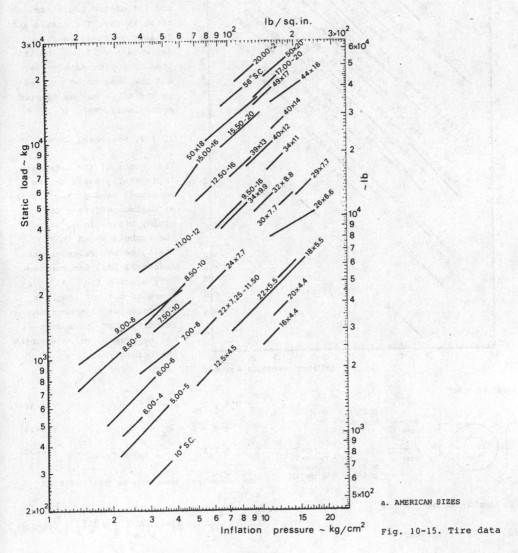

a. AMERICAN SIZES

Fig. 10-15. Tire data

357

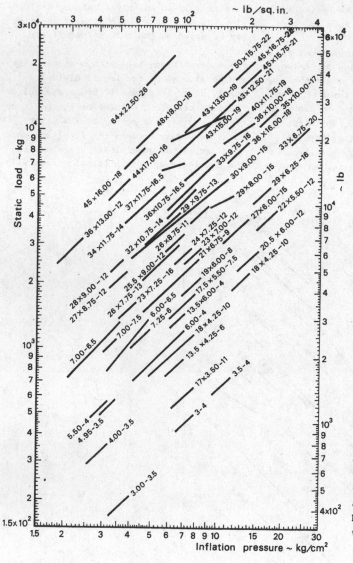

~ lb./sq.in.

Inflation pressure ~ kg/cm²

Static load ~ kg

be overloaded during braking. This overloading can be reduced by
- a low pivot point position
- a rearward shift of the pivot point, and/or
- the use of compensating linkages.

The bogie pivot is usually placed so that the distance between it and the front and rear wheel axles is about 55 and 45% of the bogie arm, respectively. The static and dynamic overload on any of the wheels will then be limited to approximately 10%, and the load to be carried by the tires may be adapted accordingly (Ref. 10-1).

10.4.2. Nosewheel tires

The choice of the tire size is generally based on the nosewheel load during braking at maximum effort - the steady braked load. Using the symbols shown in Fig. 10-16, we may calculate the nosewheel load for constant deceleration from the equations of motion. Ignoring the aerodynamic moment and assuming that the nosewheel has no brakes, we have during the braked roll:

b. BRITISH SIZES

	D_t × b_t - d
British sizes	36 × 10.00 - 18
	6.00 - 4
American sizes	27 type I
	15.50 - 20 type III
	26 × 6.6 type VII
	32 × 11.50 - 15 type VIII

c. EXPLANATION OF THE TIRE CODE

Fig. 10-15 (continued)

$$\frac{a_x}{g} W = \mu P_m + D - T \qquad (10\text{-}18)$$

$$0 = W - L - P_m - P_n \qquad (10\text{-}19)$$

$$0 = P_m \ell_m + \mu P_m h_{cg} - P_n \ell_n \qquad (10\text{-}20)$$

The nosewheel load can be derived from this:

$$\frac{P_n}{W} = \frac{\ell_m}{\ell_m + \ell_n}\left(1 - \frac{L}{W}\right) + \frac{h_{cg}}{\ell_m + \ell_n}\left(\frac{a_x}{g} - \frac{D-T}{W}\right) \qquad (10\text{-}21)$$

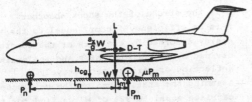

Fig. 10-16. Forces acting on the aircraft during a braked roll

As both D and L are positive, the maximum nosewheel load occurs at low speed. Reverse thrust decreases the nosewheel load and hence the condition T = 0 results in the maximum load:

$$\frac{P_n}{W} = \frac{\ell_m + \frac{a_x}{g} h_{cg}}{\ell_m + \ell_n} \qquad (10-22)$$

or:

$$\frac{P_n}{W} = \frac{\ell_m + \mu h_{cg}}{\ell_m + \ell_n + \mu h_{cg}} \qquad (10-23)$$

Provided the necessary data are available to calculate the friction coefficient (see Appendix K), equation 10-23 can be used, whereas (10-22) is more suitable when these data are not available. Typical values are
a_x/g = .35 - dry concrete, simple brake system
a_x/g = .45 - dry concrete, automatic brake pressure control.
The design condition is the forward c.g. limit at a high takeoff weight.

10.4.3. Inflation pressure

When the size of the tire is to be determined there will be a limited choice regarding the inflation pressure. Assuming that the wheel load and configuration of the landing gear are constant, the weight and volume of the tires will decrease with an increase in pressure. The wheels will take up less space when retracted while the drag of the extended undercarriage will be reduced as a result of the smaller frontal area. There will, however, be secondary factors which the designer should

note:
a. With increasing inflation pressure the contact area with the runway will decrease. As explained in Section 10.2. the bearing capacity of the runway will therefore impose a limit on the inflation pressure.
b. There must be sufficient space for fitting internal brakes inside the wheel rims and this will set a lower limit to the diameter of the hub as well as the wheel flange width. The total brake energy to be absorbed must therefore be determined, and for transport aircraft this may be assumed at approximately 600 lb.ft per lb (183 kgm per kg) of Maximum Landing Weight for jet aircraft and 450 lb.ft per lb (137 kgm per kg) for propeller aircraft*. A brake manufacturer should be consulted to insure that this amount can be accommodated within the tire and wheel size. The choice of a lower inflation pressure will result in a larger tire size and hence more space will be available for the brakes. If an acceptable size for the tires and wheels still cannot be arrived at after this, a non-symmetrical wheel hub arrangement may be required.

c. The contact area of the tires with the ground is inversely proportional to the inflation pressure, and braking will be less effective with high inflation pressures. For example, Ref. 10-27 quotes for the tire/runway friction coefficient at zero rolling velocity, on dry concrete:

$$\mu_{static} = .93 - constant. p \qquad (10-24)$$

where the constant is equal to .0011 sq.in. per lb or .0155 cm^2 per kg.
d. The aircraft will frequently have to be operated from wet runways. Assuming a tire with a suitable tread pattern, the aquaplaning velocity is found to depend almost entirely on the inflation pressure alone (Ref. 10-27):

*These data are reasonably accurate, provided the landing speed of the design has a normal value as compared with existing aircraft in the category considered.

$$V_{aqua} = \text{constant} \sqrt{p} \qquad (10\text{-}25)$$

where the constant is equal to 9 if p is in lb/sq.in. and V in knots, or 17.5 if p is in kg/cm^2 and V in m/s.

The tire pressure also has an influence on the drag caused by water or slush on the runway. Although the performance of jet transport aircraft with their relatively high power loadings and rolling speeds is sensitive to this effect, it is no design practice to adapt the tire pressure accordingly.

10.5. GEAR GEOMETRY AND RETRACTION

Having chosen the disposition of the undercarriage and the tire dimensions, the airplane project design engineer must pay attention to the shock strut configuration and dimensions and, in the case of retractable gear, to the basic solution for the kinematics of the retraction and the space required for wheelbays.

10.5.1. Energy absorption on touchdown

The maximum kinetic energy of the aircraft normal to the runway to be absorbed when touching down is:

$$E = \frac{W}{2g} w^2 \qquad (10\text{-}26)$$

where w is the ultimate velocity of descent. Assuming conservatively that this energy will have to be absorbed completely by the main undercarriage, thus ignoring the energy transmitted to the atmosphere, the required stroke of each shock absorber is derived from:

$$E = N_s P_s \lambda (n_t S_t + n_s S) \qquad (10\text{-}27)$$

where

N_s is the number of main gear shock absorbers

P_s is the static load per leg

λ is a reaction factor, or ratio of maximum load to static load per leg

S_t is the maximum tire deflection

S is the stroke of the shock absorber

η is an efficiency factor, equal to the energy absorbed by the tire or the absorber divided by the product of P and the maximum deflection or stroke, respectively.

Assuming the static load on the main undercarriage equal to 92% of W, we find for the required shock absorber stroke*:

$$S = \frac{1}{n_s} \left(\frac{w^2}{1.84g \lambda} - n_t S_t \right) \qquad (10\text{-}28)$$

It is recommended to add one inch to this value to allow for inaccuracies in the method and miscellaneous factors. The ultimate descent velocity w is specified in the airworthiness regulations: FAR 23.473 and 25.473, BCAR Section D Chapter D3-5 Par. 4 and Section K Chapter K3-5 Par. 2. For light aircraft:

$$w = \text{constant} \times (W/S)^{.25} \qquad (10\text{-}29)$$

where the constant is 4.4 if the wing loading is in lb/sq.ft and w in ft/s, or .9 if the wing loading is in kg/m^2 and w in m/s. For transport aircraft w is generally 12 ft/s (3.66 m/s), although in the BCAR requirements it is dependent on the stalling speed.

The reaction factor λ may be assumed equal to 2 to 2.5 for transport aircraft and 3.0 for light aircraft.

The efficiency data to be used in (10-28) may be assumed as follows:

tires: $n_t = .47$

air springs: $n_s = .60$ to .65

metal springs with oil damping: $n_s = .70$

liquid springs: $n_s = .75 - .85$

oleo-pneumatic absorbers: $n_s = .80$.

The maximum tire deflection can be obtained from the tire handbook, or alternatively from the approximation:

$$S_t = \text{constant} \frac{\lambda L_w}{p \sqrt{D_t b_t}} \qquad (10\text{-}30)$$

where L_w is the static load per wheel and D_t and b_t are the tire diameter and maximum width. The

*more precisely: the projection of the stroke normal to the ground.

360

constant of proportionality in (10-30) is equal to
.5. **A simpler assumption is to take** s_t
equal to three times the static deflection of the
tire.

10.5.2. Dimensions of the gear

Useful statistical data can be found in
Refs. 10-4 and 10-30, which are summarized
as follows.

a. The diameter of the cylinder of a telescopic main gear unit may be taken as

$$D = .5 + .03 \sqrt{P_s} \text{ in., for } P_s \text{ in lb}$$
$$D = 1.3 + .11 \sqrt{P_s} \text{ cm, for } P_s \text{ in kg}$$
$$(10-31)$$

where P_s is the maximum unfactored vertical
load per leg.

b. Clearance allowances between the tire
and the adjacent parts of the aircraft are
based on
- the maximum dimensions of the inflated
tire
- a growth allowance due to service
- the effect of centrifugal forces at high-
speed rolling, which increase the diameter.
Although the growth of tires depends to
some extent on the type used, a 4% growth
in maximum width and 10% in diameter during
use will be good average values. The clearance around the tire required in connection
with centrifugal forces may be taken from
Fig. 10-17.

c. The distance between the tire center
lines of twin tires S_T shall be at least
1.18 times the maximum grown width of the
tire. The minimum distance between the axle centers of tires in tandem shall be e-
qual to the maximum grown tire diameter
plus twice the radial clearance allowance
according to Fig. 10-17. As the wheelbase
of the bogie should be as short as possible
in order to minimize tire scrubbing and
bending moments, a value of 1.2 times the
tire grown diameter can be taken as a
workable ratio.

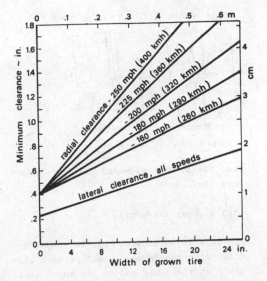

Fig. 10-17. Minimum clearance between tires
and parts of the aircraft

d. The wheel spacing on a landing gear u-
nit, with the wheels located on each side
of the shock absorber strut, may be derived
from (10-31) and Fig. 10-17 by adding twice
the lateral clearance to the cylinder diameter. This dimension determines the total
width of a dual wheel assembly.

e. In the case of a dual tandem bogie assembly, there should be a lateral spacing
between the wheel center lines S_T of 1.8
times the maximum tire width.

f. The length of the leg is determined
mainly by the shock absorber stroke (see
Section 10.5.1.) and the amount of over-
lap of the sliding assembly. Minimum values for this length are indicated in Fig.
10-18 for two types of simple telescopic
gears.
Ref. 10-4 presents statistical data for
determining the dimensions X, Y and Z. It
is generally found that the undercarriage
leg length required is about 3 times the
shock absorber stroke for dual and multi-
wheel assemblies, plus a tire radius for
the single wheel layout.

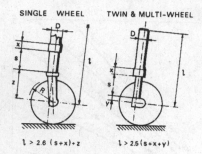

SINGLE WHEEL   TWIN & MULTI-WHEEL

$l > 2.6 (s+x)+z$ $l > 2.5 (s+x+y)$

Fig. 10-18. Dimensions of undercarriage legs (Ref. 10-4)

10.5.3. Gear retraction

Nowadays fixed landing gear is almost exclusively used on many small private aircraft at the lower end of the speed scale, i.e. cruising speeds of less than 135 to 160 knots (250 to 300 km/h). It is relatively easy to design a robust fixed landing gear and this may be the right choice for aircraft which have to use rough airfields. It may sometimes be attractive to fit a semi-retractable landing gear.

Table 10-3 shows that even at speeds around

	Cardinal RG (Retractable Gear)	Cardinal 177 (Fixed Gear)
Gross weight	2,800 lb	2,500 lb
Empty weight	1,630 lb	1,480 lb
Baggage	120 lb	120 lb
Fuel capacity	51 gal	50 gal
Top speed at sea level	176 mph	151 mph
Cruise speed (75% power at 7,000 ft*)	166 mph	142 mph
Range at 75% power at 7,000 ft*, no reserve)	765 sm	690 sm
Rate of climb at sea level	860 fpm	840 fpm
Service ceiling	16,900 ft	14,600 ft
Takeoff over 50-ft obstacle	1,585 ft	1,400 ft
Landing over 50-ft obstacle	1,350 ft	1,220 ft
Stall speed, flaps up, power off	66 mph	63 mph
Stall speed, flaps down, power off	57 mph	53 mph

Power:

✿ Cardinal RG: 200 hp Lycoming IO-360A1B6

✿ Cardinal 177 180 hp Lycoming O-360A2F

* Cardinal 177 cruise speed and range are calculated at an altitude of 8,000 ft.

Table 10-3. Performance comparison of Cessna Cardinals

160 knots (300 km/h) there is a considerable improvement in cruising performance with the use of a retractable landing gear. Al-

though the two versions of the Cessna Cardinal shown are also slightly different in other respects, the gain in cruising speed is caused primarily by gear retraction. Incidentally, the manufacturer appears to stress the improved appeal of the RG version more than the actual gain in performance.

There are almost as many gear retraction schemes as there are different aircraft designs. A review of known retraction solutions should be made before the actual geometric design is started; Refs. 10-1 and 10-13 are valuable contributions in this field.

Most retraction mechanisms are derived from the four-bar linkage and the designer must have very good reasons to deviate from this. A suitable pivot point must be chosen for the leg which, at the same time, gives the required wheel positions and allows adequate length of the leg. A retraction mechanism, generally consisting of a folding stay member and a retraction jack, is then required. Provided a reasonable looking mechanism has been found, a retraction curve should be constructed, indicating the retraction load on the jack and a number of checks will have to be made. For example:

a. Make sure that the jack has adequate dead length.

b. The efficiency of the retraction geometry - i.e. the work done by the jack divided by the maximum jack force times the total jack travel - should be at least about 50%. Excessive variations in the jack force during retraction must be avoided.

c. The points where the undercarriage is to be suspended must be arranged as close as possible to the wing spars, major frames or ribs or any other strong major structural member.

d. There must be adequate clearance between the members of the retraction mechanism.

e. The leg must be adequately supported against forces in a plane normal to the plane of retraction.

A solution for various retraction problems

may sometimes be found by arranging the
swivel axis so that the wheel goes up a
spiral path, with the result that it is
rotated during retraction. Shock absorber
contraction during retraction can be used
to facilitate stowage, particularly when
an articulated landing gear is used. The
bogie undercarriage may have an extra degree of freedom available in that the bogie unit can swivel with respect to the
main member, thus requiring a minimum of

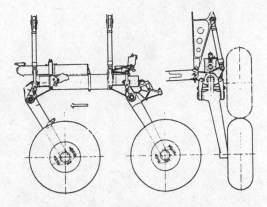

Fig. 10-20. The jockey undercarriage of
the Bréguet 941

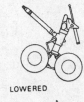

LOWERED

PARTIALLY RETRACTED

RETRACTED

Fig. 10-19. Retrac-
tion sequence of a
typical bogie under-
carriage

space when retracted (Fig. 10-19).
Jockey undercarriages (Fig. 10-20) are
sometimes used on rough-field landing gear;
they require only a small streamline fairing when retracted and may therefore be
particularly suitable for fuselage-mounted
gears.
Omission of a main wheel door may be considered on aircraft where the drag penalty
thus incurred is of less importance than
the gain in weight and retraction space
(e.g. the Boeing 737, Yak 40).

Chapter 11. Analysis of aerodynamic and operational characteristics

SUMMARY

The object of the design synthesis process dealt with in the previous chapters is to a-
chieve the goals laid down in the design specification. The first cycle of the iterative
design process will be concluded with an analysis of the operational characteristics for
the purpose of investigating to what extent the design requirements have been met.

Some general comments on the prediction of aerodynamic characteristics are made in this
chapter. Definitions and subdivisions of the drag according to several schemes are discussed. The choice of operational limit speeds and the determination of n-V diagrams are
then briefly reviewed. A procedure to analyze the flight profile, reserve fuel quantity
and payload-range characteristics is given, followed by some general aspects of climb and
field performance.

The chapter concludes with certain aspects of operating economy and some critical notes
on the use of standard formulas for estimating direct operating costs.

NOMENCLATURE

A   - wing aspect ratio
A_j - factor in drag coefficient contribution
a   - speed of sound
B_j - factor in drag coefficient contribution
b   - wing span
C   - rate of climb
CAS - Calibrated Air Speed
C_D - drag coefficient
C_{D_o} - zero-lift drag coefficient
$\bar{C}_F$ - mean skin friction drag coefficient
C_L - lift coefficient
C_m - pitching moment coefficient
C_T - Thrust Specific Fuel Consumption (TSFC)
D   - drag; distance
D_j - factor in drag coefficient contribution
DOC - Direct Operating Costs
EAS - Equivalent Air Speed
e   - Oswald factor
F   - (hourly) fuel consumption
g   - acceleration due to gravity
h   - (pressure) altitude
IOC - Indirect Operating Costs
K_g - gust alleviation factor
L   - lift
M   - Mach number
M_{cr_D} - drag-critical Mach number
MTOW - Maximum Take-Off Weight
N   - engine rpm
N_e - number of engines per aircraft
n   - load factor; coefficient in mathematical representation of the compressibility drag
P   - power
p   - static pressure
R   - range; Reynolds number
ROI - Return On Investment
S   - wing area; field length; distance traveled during takeoff

T   - thrust; temperature
TAS - True Air Speed
t   - time
V   - airspeed
V_E - equivalent airspeed
W   - weight
w   - gust velocity (EAS)
α - angle of attack
β - factor of proportionality in the drag coefficient
γ - ratio of specific heats; climb angle
δ - relative atmospheric pressure
δ_e - elevator deflection
Δ - increment
θ - relative atmospheric temperature
ρ - atmospheric density
μ - coefficient of frictional retardation

Subscripts

am - air maneuver
b   - beginning of cruising flight; block-to-block (time)
C   - design cruising condition
cr - cruising flight
cl - climb
D   - design diving condition
des - descent
e   - end of cruising flight
gm - ground maneuver
j   - drag contribution
MO - Maximum Operating (limit speed)
R   - rotation
S   - stalling
to - takeoff
uc - undercarriage
uu - moment of retraction of uc
v   - vortex-induced
wet - wetted (area)
o   - sea level
2   - takeoff safety (speed)

11.1. INTRODUCTION

The previous chapters were intended to provide the designer with background in-

formation and basic (semi-)statistical methods and procedures for sizing the major parts of the airplane and allocating and disposing them in relation to each other. These considerations are intended to be of some help in obtaining a reasonable insight into the design process and in finalizing the first design iteration in a reasonable space of time.

Referring back to Section 1.5 of the first chapter and in particular to Figs. 1-7 and 1-9, we find that the designer's next step is to analyze the design by estimating some of the aerodynamic characteristics and determining major performances and certain flying qualities. The primary aim of this step is to verify whether the design meets the initial specification. Generally speaking, this comparison will reveal that design improvements are required but experience gained during the design stage may well lead to the conclusion that the design specification should be changed instead of or in combination with the aircraft. In any case, the design analysis and evaluation will finally result in an initial baseline design which takes into account the most recent views on the design requirements.

Although many aspects can only be considered after detailed study and testing, estimates are usually made of several performance aspects, such as cruising performance, airfield performance, design speeds and Direct Operating Costs (DOC). A report on the background information for a design study might contain the following elements.

a. A prediction of the lift curve, the drag polar, and the pitching moment curve for representative cruising conditions.

b. Calculations of the cruising speed and/or maximum flight speed and the specific range (i.e. distance traveled per unit weight of fuel) for several flight speeds and/or cruising altitudes.

c. Considerations for choosing the most relevant structural design speeds (V_C and V_D) and determining the limitations of operational speeds vs. pressure altitude, i.e. the flight envelope.

d. Calculations of V-n diagrams for maneuvering and gust loads for several representative operational conditions.

e. Computations of the maximum rate of climb and ceilings with all engines operating and with a failed engine (as appropriate).

f. A prediction of the C_L-α curves and the airplane drag polar for several positions of the high-lift devices, with undercarriage up and down.

g. Limits for the takeoff and landing weight, based on the airworthiness requirements in relation to climb performance, i.e. WAT curves.

h. Calculations of the takeoff and landing field length required in accordance with the airworthiness rules appropriate to the particular airplane category.

j. Payload-range diagrams for several cruising conditions, due allowances being made for reserve fuel.

k. An estimate of the operating costs as a function of the stage length, using a suitable standard method.

Many other questions can be raised and the designer must make a decision on the basis of what he considers necessary and appropriate in the preliminary design stage. For example, a detailed study of some principal stability and control characteristics can be carried out, particularly if there are sound reasons for doubting the adequacy of the empennage and control surface design. However, the designer should also realize that many assumptions have to be made which must be verified later on and decide whether much time should be spent in studies the conclusions of which are subject to considerable inaccuracy. Several topics in this chapter will be touched upon in a rather superficial manner since a thorough treatment might necessitate as many chapters as there are sections in this chapter. More detailed information can be obtained from the publications listed in the references and also in Appendices E, F, G and K.

11.2. TERMINOLOGY RELATING TO THE DETERMINATION OF DRAG

One difficulty that arises when drag prediction methods are compared or drag measurements interpreted is that the terminology often creates confusion. Several schemes (Fig. 11-1) are possible for sub-

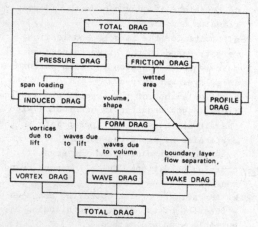

Fig. 11-1. Drag breakdown of a body without internal flow

dividing the drag of an airplane, as explained in Refs. 11-1 and 11-2 and the following sections.

11.2.1. Pressure drag and skin friction drag

The total action that a fluid exerts on a body can be considered as the resultant of the elementary forces exerted on all points of the surface, both normally and tangentially. Considering the drag on a closed body (i.e. one with no internal flow due to powerplant installation or internal systems), the SKIN FRICTION or SURFACE DRAG is generated by tangential forces, while PRESSURE DRAG is caused by the normal forces. The lift can be subdivided in a similar manner, but it is usually considered a component of only pressure forces. The pressure distribution is also affected by the boundary layer and regions of separated flow. In wind tunnel experiments

the pressure forces can be obtained directly by measuring the distribution of the pressure at several small holes, situated at the surface of the body, while friction drag is obtained by subtracting the pressure drag from the total drag. Assuming the mean friction drag coefficient based on the wetted areas of the various aircraft parts to be denoted $\bar{C}_F$, the skin friction drag can be written:

$$D_F = \bar{C}_F \ \tfrac{1}{2}\rho V^2 \ S_{wet} \qquad (11-1)$$

where $\bar{C}_F$ is typically of the order of .003 to .005 for most subsonic aircraft. The wetted area of a body or wing is therefore the main geometric parameter determining the skin friction drag.

11.2.2. Wake drag, vortex-induced drag, and wave drag

According to the law of conservation of energy, the work produced in overcoming the aerodynamic drag of a body moving at constant speed is equal to the energy increase in the surrounding fluid. The forms of energy transmitted to the fluid are vortex-induced energy, wake energy, and shock wave energy. The aerodynamic drag can be subdivided accordingly (Fig. 11-1). VORTEX-INDUCED DRAG is part of the pressure drag, corresponding to the kinetic energy - distributed throughout the fluid - associated with the trailing vortices shed by a lifting wing of finite aspect ratio. For high aspect ratio wings with relatively thick airfoil sections, the vortex-induced drag can be computed accurately with lifting-line or lifting-surface theories. However, when section leading edges are sharp, angles of sweepback large and/or aspect ratios low, the leading-edge suction force begins to break down at some critical lift coefficient and the vortex-induced drag increases considerably (see Fig. 7-22 and Ref. 11-12).
For elliptical wings of high aspect ratio, the vortex-induced drag is given by:

$$\frac{D_v}{W} = \frac{C_L^2 \; \tfrac{1}{2}\rho V^2 S}{\pi A \; W} = \frac{1}{\pi} \frac{W/b^2}{\tfrac{1}{2}\rho V^2} \qquad (11-2)$$

This equation illustrates that, for given flight conditions, the span loading W/b^2 determines the induced drag.

WAKE DRAG is caused by the boundary layer and regions of separated flow. The main source of drag generated by the boundary layer is shear action, resulting in skin friction drag; the pressure drag is generally an order of magnitude less. Separation drag, however, is predominantly caused by forces normal to the surface. For a well-streamlined body at small angles of attack, the skin friction drag is the dominant part of the wake drag.

SEPARATION DRAG increases sharply when the stall is approached and is also manifest in areas near ill-shaped wing body junctions, blunt bases, sharp corners, etc. Flow separation can be predicted and analyzed theoretically in only a few simple cases and drag prediction is therefore of an empirical nature. It cannot be estimated with good accuracy for project design.

WAVE DRAG is another part of the pressure drag, associated with the work produced by compression of the fluid at high (local) flow velocities, which manifests itself in the form of shock waves. A complicating factor is that strong shock waves may induce flow separation, resulting in an increase in both the wake drag and the vortex-induced drag. At supersonic speeds, assuming linear theory, wave drag can be subdivided into wave drag due to the body volume and wave drag due to lift, but at transonic speeds, when mixed flow is present, this subdivision is less evident.

It will be clear from the foregoing that the subdivision of drag into vortex-induced, wake and wave drag components is not very well defined in cases where appreciable interactions between the various flow fields occur. These interference effects result in drag increments, frequently referred to as INTERFERENCE DRAG.

11.2.3. Form drag, profile drag and induced drag

The pressure drag component of the wake drag and the wave drag due to volume (if present) may be combined into the FORM DRAG. For a specified angle of attack this contribution depends upon the shape or form of the body. For a nonlifting body, shedding no vortices, the form drag is equal to the pressure drag on the body. When a two-dimensional wing is placed in a wind tunnel, it experiences skin friction plus form drag, a combination called the PROFILE DRAG or section drag. This term is also used to define the skin friction plus form drag of three-dimensional bodies, with or without lift.

INDUCED DRAG is the resultant of vortex-induced drag and wave drag due to lift. In the absence of shock waves, profile drag is equal to wake drag and for a wind tunnel model it can be derived from static and total pressure distribution measurements in the wake of the body. The induced drag is found by subtracting the wake drag from the total drag.

11.2.4. Zero-lift drag and lift-dependent drag

The subdivision shown in Fig. 11-1 is useful for the purpose of theoretical and semi-empirical drag computation and for the interpretation of wind tunnel measurements. When it comes to the analysis of flight test results, only the total lift and drag are available, as these are derived directly from measurements of the aircraft motion (cf. Ref. 11-13). In view of the fact that the vortex-induced drag coefficient is essentially proportional to C_L^2, it is usual to plot measured C_D values versus C_L^2 (Fig. 5-4b). In the range of normal operational values of C_L a straight-line approximation is found to be acceptable. The subdivision of drag into zero-lift drag and lift-dependent drag,

$$C_D = C_{D_o} + \frac{C_L^2}{\pi Ae} \qquad (5-9)$$

originates from this concept. The meaning of the Oswald factor e is explained in Section 5.3.1.

It is regrettable that the term $C_L^2/\pi Ae$ is frequently referred to as "induced drag coefficient", which makes confusion with the previous scheme unavoidable. Lift-dependent drag comprises not only vortex-induced drag and lift-induced wave drag but also the variation of the profile drag with the angle of attack.

According to (5-9) the condition of minimum drag is $C_L = 0$, while this should in fact be some positive value of C_L, referred to as $C_{L_{ref}}$ in Fig. 11-2. The fol-

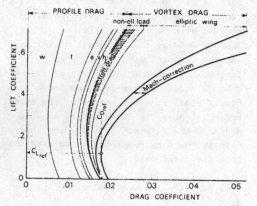

Fig. 11-2. Drag buildup by analysis (w = wing; f = fuselage; e = engine installation; v = vertical tailplane; h = horizontal tailplane)

lowing representation of the drag polar yields a more accurate approximation, particularly for low values of C_L:

$$C_D = C_{D_{ref}} + \beta\left(C_L - C_{L_{ref}}\right)^2 \qquad (11-3)$$

The drag for $C_L = C_{L_{ref}}$ is referred to as the reference drag or basic drag.

11.2.5. Breakdown for drag analysis

In the preliminary design stage, when no wind tunnel data are available, traditional drag prediction methods are used, based on the following procedure:

1. Calculate the drag of each major part separately, assuming that they are isolated from each other.
2. Add all contributions.
3. Make corrections for the interaction of the flow fields.

The last term is sometimes called "interference drag", but the term "interference corrections" is more justifiable. A classical subdivision of the total drag is as follows:
- (minimum) wing profile drag,
- (minimum) parasite drag*, i.e. the drag of all other airplane items plus corrections for interference,
- lift-dependent drag.

Most individual drag contributions can be approximated as follows:

$$C_{D_j} = A_j + B_j C_L + D_j C_L^2 \qquad (11-4)$$

Fig. 11-2 shows the variation of several contributions with C_L. The minimum value of the wing profile drag - the basic profile drag - corresponds to the design lift coefficient. Similarly, a flow condition can be defined for minimum profile drag for other airplane parts; this is usually at a positive value of C_L which is representative of cruising conditions.

TRIM DRAG is defined as the drag due to the horizontal tailplane load required to ensure longitudinal equilibrium at a specified center of gravity position. This drag component is composed of vortex-induced drag due to tail load and a profile drag increment due to elevator deflection. It takes into account the wing lift increment or reduction needed to obtain a given C_L, resulting from the horizontal tail download or upload.

Compressibility correction is not a pure drag contribution, as compressibility affects both the vortex-induced and profile drag. Most drag predictions are based on a low speed drag polar (M ≤ .5) and a sep-

*The term "parasite drag" is not uniquely defined in the literature and is too arbitrary to serve any useful purpose nowadays

arate correction is made for compressibility effects at M > .5.

A detailed subdivision, useful for the estimation of a drag polar, is presented in Appendix F, Table F-1.

11.2.6. Bodies with internal flow

All subdivisions treated previously are strictly applicable only to closed bodies without internal flow. The effects of engine power on the aerodynamic properties can be fairly complicated, as explained in some detail in Appendix F, Section F-5.6. If the effects of powerplant operation on lift and drag are appreciable, lift curves and drag polars must be established for various engine operational conditions. In view of the limited scope of this chapter no details will be given here. The interested reader is referred to Ref. 11-6 and the references mentioned in Appendix F.

11.3. DETERMINATION OF AERODYNAMIC CHARACTERISTICS

In order to analyze the performance and flight characteristics of an aircraft in the preliminary design stage, several aerodynamic characteristics must be determined:

C_L vs. α : the lift curve
C_D vs. C_L : the drag polar
C_m vs. C_L or C_m vs. α (for δ_e = constant): pitching moment curves.

These relationships must be determined for the cruise configuration (en route) and for various flap deflection angles. In the case of high-subsonic speeds the Mach number must also be treated as a separate variable.

The abovementioned relationships are adequate for performance calculation and determination of the neutral point, with the stick fixed. Prediction methods and data can be found in Appendices E, F and G. For the calculation of the stick-free longitudinal stability the elevator hinge moments must be known and additional stability derivates are required for an assessment of dynamic longitudinal stability, lateral stability, and control properties. In view of the limited space available in this book no attempt has been made to present calculation methods for these properties. Valuable standard methods are available in the form of Data Sheets issued by the Engineering Sciences Data Unit (ESDU) and the USAF Stability and Control DATCOM (Ref. 11-4).

The designer may be faced with a number of difficulties when he tries to predict the aerodynamic properties of his project:
a. The external shape is not yet fully defined in the preliminary design stage; hence pressure distributions cannot be calculated and semi-empirical prediction methods must be developed. The effects of surface imperfections, excrescences, etc., must be estimated on a statistical basis.
b. Wind tunnel measurements are sometimes available, but their interpretation may be difficult and their applicability limited due to the appreciable differences in Reynolds number compared with free flight conditions.
c. Theoretical prediction methods generally yield good results for conventional shapes, but drag due to flow separations and shock waves is very difficult to predict and much remains to be done in this field for aerodynamicists.

11.3.1. Reynolds number effects

For each airplane configuration the operational variation in airspeed and altitude results in Reynolds number variations. In the preliminary design stage only one polar is usually generated for each configuration. A representative value for R is therefore chosen for each polar curve, but a simple refinement on this particular point is also possible, as discussed in Appendix F, Section F-3.2.

11.3.2. Mach number effects

A more fundamental difficulty is how compressibility effects on drag have to be taken into account. The critical Mach number M_{cr_D}, at which a sharp increase in the drag occurs (drag rise), as defined in Section 7.5.1b may be used to characterize the drag due to compressibility. For Mach numbers up to this value the drag increment is referred to as the drag creep. An arbitrary mathematical representation may take the following form:

$$\Delta C_D = .002 \left\{ 1 + n \frac{M_{cr_D} - M}{\Delta M} \right\}^{-1}$$

$$\text{for } M \leqslant M_{cr_D} \qquad (11-5)$$

while the drag rise can be represented as follows:

$$\Delta C_D = .002 \left\{ 1 + \frac{M - M_{cr_D}}{\Delta M} \right\}^{n}$$

$$\text{for } M \geqslant M_{cr_D} \qquad (11-6)$$

These expressions, like the symbols n and ΔM, have no physical significance, but they can be used to approximate the actual shape of the drag curve (Fig. 11-3). A suitable combination of ΔM and n can be derived from experimental data, if available. Assuming, for example, $\Delta M = .05$ and $n = 2.5$, it is found that $dC_D/dM = .10$ for $M = M_{cr_D}$. In this case the two definitions

of M_{cr_D} given in Section 7.5 are identical.

11.3.3. Low speed polars

The aerodynamic characteristics for the configuration with high-lift devices in various positions and undercarriage up or down must be available before we can analyze the low-speed performance. Estimation of these characteristics on a basis of theoretical analysis is usually not possible in the conceptual design stage and the designer must rely on semi-empirical procedures. A fairly complete survey of this subject is presented in Ref. 11-15, while some prediction methods have been summarized in Appendix G. The designer must be prepared to accept considerable inaccuracy in the results of these handbook-type methods and it is very desirable that wind tunnel experiments are made available as early as possible in the development stage.

Calculations of low-speed polars (undercarriage up) for a hypothetical design presented in Fig. 12-1 have resulted in Fig. 11-4. Drag due to engine failure is not included in this figure but it is important for takeoff and approach climb performance in the case of an inoperative outboard engine. Some information on this subject is presented in Section G-8 of Appendix G. It has been found convenient

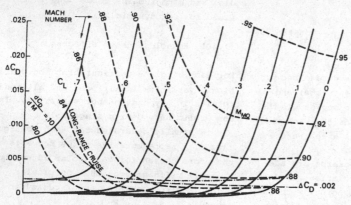

Fig. 11-3. Drag rise of a long-range aircraft

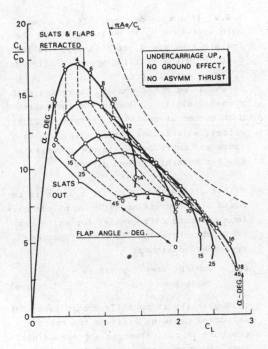

Fig. 11-4. Lift-to-drag ratio for the design in Fig. 12-1.

to consider the induced drag of the vertical tail due to the associated side force to be equivalent to a loss of thrust, by multiplying the thrust available by a factor:

$$\frac{\text{thrust} - \Delta\text{drag}}{\text{thrust}} = 1 - \text{constant} \cdot \frac{T}{W} C_L \quad (11-7)$$

For the project shown in Fig. 12-1 a typical value is .92 at V_2. Hence, the asymmetric drag is equal to 8% of the total thrust in this condition.

11.4. THE FLIGHT ENVELOPE

The main background to this subject is found in:
FAR 23.335, 23.1505, BCAR Section K Ch. K3-2 and K7-2 for light aircraft;
FAR 25.335, 25.1505, BCAR Section D Ch. D3-2 and D7-2 for transport aircraft.
The determination of performance capabil-

ities is preceded by a proper choice of the operational variation of flight speeds and altitude - an example is given in Fig. 11-5 - referred to as the flight envelope. The following definitions refer to jet transports.

a. Maximum Operating Limit Speed (V_{MO}) or Mach Number (M_{MO}) is the EAS, CAS or Mach number (whichever is applicable to the altitude) which should not be exceeded in any flight regime. It is so selected that the aircraft remains free from buffeting or undesirable flying qualities associated with compressibility up to this speed. V_{MO} must not exceed V_C.

b. The Design Cruising Speed (V_C) is the maximum EAS in level flight at which the structure is designed to withstand particular loads specified in the airworthiness regulations, e.g. a gust load at 50 ft/s (15.3 m/s) vertical gust velocity. V_C must be high enough to permit economic climb and cruise performance, but low enough to avoid excessive structure weight penalties in order to cater for gust loads. At high altitude V_C becomes related to the

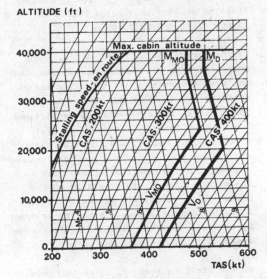

Fig. 11-5. Example of a flight envelope.

373

design cruising Mach number for optimum
long-range cruise capability. A calculation should be made of the maximum speed
in cruising flight, using a performance
diagram (Fig. 11-6), for various atmos-

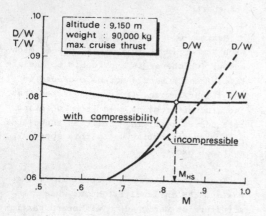

Fig. 11-6. Determination of the cruise
speed

pheric temperatures and altitudes - see
also Fig. 5-10. The choice of V_{MO}/M_{MO} at
cruising altitude must be such that high-
speed cruise conditions are not unduly
limited by this placard speed. For high-
subsonic jet aircraft V_{MO} usually becomes
limiting at altitudes below 20,000 to
25,000 ft (6,000 to 7,500 m).

c. The Design Diving Speed (V_D) or Mach
number (M_D) is the maximum speed (EAS, CAS)
in level flight at which the structure is
designed to withstand particular loads
specified in the airworthiness regulations,
e.g. a gust load at 25 ft/s (7.55 m/s)
vertical gust velocity. It must be sufficiently greater than V_C to provide for safe
recovery from inadvertent upsets at V_C.
The margin $M_D - M_{MO}$ is generally of the
order of .05 to .08 for aircraft cruising
at high-subsonic speeds. It has to be shown
by analysis that the aircraft is free from
undamped flutter vibrations up to 1.2 V_D.

d. The stalling speed V_S is the minimum
speed with power off, encountered in the
stalling maneuver referred to in Section

5.4.4. (flaps up or down). It is acceptable to assume V_S to be about 94% of the
1-g stalling speed, if a significant reduction in the stalling speed relative to
V_S-1g is anticipated.
It should be noted that V_S increases with
pressure altitude. Due to the increasing
Mach number at the stall, compressibility
effects will reduce C_L-max at high altitudes and therefore V_S is not a constant
EAS for most high-speed aircraft.

e. For pressurized aircraft a limit is imposed on the cruising altitude by the maximum pressure differential for which the
cabin structure and the pressurization
system are designed:

$$\Delta p_c = \text{ambient static pressure} - \text{cabin pressure} \qquad (11-8)$$

This quantity is normally chosen such that
at normal cruising altitude the cabin
pressure is equivalent to a pressure altitude of approximately 6,000 ft (1,830 m),
i.e. δ = .80 in the ISA. Hence,

$$\Delta p_c = (.80 - \delta_{cr}) \, p_0 \qquad (11-9)$$

The maximum allowable cabin pressure altitude is 8,000 ft (2,440 m), i.e. δ = .67 in
the ISA. Thus the maximum altitude limited
by the cabin design is defined by

$$\delta = .67 - \frac{\Delta p_c}{p_0} = \delta_{cr} - .13 \qquad (11-10)$$

f. Diagrams for maneuver and gust loads vs.
EAS, referred to as q-V or n-V diagrams,
can be drawn once the design airspeeds
have been determined. The conditions for
defining the gust speeds, design speeds,
and normal load factors are defined in the
airworthiness regulations, FAR 23.321-341,
25.321-341, BCAR Ch. D3-2 and Ch. K3-2.
The maneuver load factor is a simple
straightforward function of the aircraft
All-Up Weight for the appropriate airworthiness category. The gust load is determined by:

$$n = 1 \pm K_g \frac{dC_L}{d\alpha} \tfrac{1}{2}\rho_0 w \, V_E \frac{S}{W} \qquad (7-23)$$

The lift-curve slope $dC_L/d\alpha$ can be obtained from Appendix E, Section E-4.1. It is affected by the compressibility of the air. The gust alleviation factor K_g can be taken from the rules if better methods are not available to the designer. Critical conditions for the gust loads are frequently met at an altitude of about 20,000 ft (6,000 m) with approximately 50% of the fuel load burnt off.

11.5. FLIGHT PROFILE ANALYSIS AND PAYLOAD-RANGE DIAGRAMS

Range performance has a direct effect on the transportation costs through the aerodynamic quality of the aircraft and its engine characteristics. The payload-range diagram - explained in Section 8.2.3. - must be drawn to check whether the specified performance aims can be achieved with the MTOW assumed previously.

Fig. 11-7 shows a typical breakdown of the

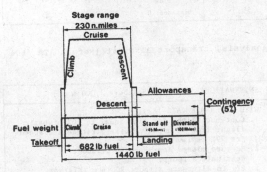

Fig. 11-7. Typical fuel breakdown for an executive aircraft (Ref. SAWE Paper No. 996)

total fuel load for an executive aircraft. For civil operations the basic mission profile analysis for a specified stage length can be carried out with the data of Table 11-1. Minimum fuel reserve loads are specified in FAR 91.23 (IFR operations) and 91.207 (VFR operations), but in airline practice the quantity of reserve fuel is usually larger. Typical policies are given in Table 11-2.

In the next part of this section it is assumed that the principles of flight mechanics and performance calculation are known or at least available to the reader. A selected list of publications on this subject can be found in the references.

11.5.1. Operational climb

In view of the limited equivalent cabin rate of climb and the general desirability of obtaining a high block speed, the operational climb speed is usually taken greater than the speed for maximum rate of climb. A climb at constant EAS results in a steadily increasing TAS and M with increasing altitude. Above a certain altitude it will be necessary to continue the climb with constant M in order to avoid undesirable compressibility effects. For small climb angles the rate of climb is given by:

$$C = \frac{(T - D) \, V}{W(1 + \frac{V}{g}\frac{dV}{dh})} \qquad (11-11)$$

The correction term for accelerated flight depends on the flight procedure. For flight in the ISA, it can be shown that:

$$\frac{V}{g}\frac{dV}{dh} = \begin{array}{llll} .5668 \ M^2 & - & \text{EAS constant} \\ = -.1332 \ M^2 & - & \text{M} \quad \text{constant} \\ = .7 \quad \ M^2 & - & \text{EAS constant} \\ = 0 & - & \text{M} \quad \text{constant} \end{array} \begin{array}{l} \Big\}\ \text{tropo-} \\ \text{sphere} \\ \Big\}\ \text{strato-} \\ \text{sphere} \end{array}$$

Similar corrections for flight at constant CAS can be obtained from Ref. 11-23. The operational climb is computed by means of a step-by-step calculation procedure, for an appropriate engine rating which is usually specified by the engine manufacturer.

11.5.2. Cruise performance

In view of the large number of variables associated with cruising in various operational conditions (weight, altitude, speed, engine rating) it is desirable to carry out the analysis using nondimensional parameters. This will be shown here for jet aircraft, but the reader may work it out for propeller aircraft, if desired.

PROFILE SEGMENT	PERFORMANCE CONFIGURATION	ENGINE POWER/THRUST	ALTITUDE	SPEED	REQUIRED DATA	REMARKS
GROUND MANEUVER (gm)	MTOW/MLW flaps down	as required $\mu = .02$	0 ft	-	Time: T_{gm} [1]) Fuel: F_{gm} [1])	1) ATA '67: 14 min. ground idle + 1 min. takeoff
TAKEOFF	MTOW t.o. flaps	takeoff, all engines	0-35 ft	0 to $V_2 + 10$ kts	Distance: 0	
INITIAL CLIMB AND ACCELERATE	t.o. flaps, u.c. up	takeoff, all engines	35-400 ft	$V_2 + 10$ kts	Time: T_{to} Fuel: F_{to} Distance: D_{to}	2) below 10,000 ft V_{cl} may not be more than 250 kts (IAS)
	flaps retracting		400-2,000 ft	accelerate to $1.2V_S$(flaps up)		
EN ROUTE CLIMB TO CR. ALTITUDE, ACCELERATE (cl)	en route	max. operational climb	2,000 ft	accelerate to V_{cl} 2) 3)	Time: T_{cl} Fuel: F_{cl} Distance: D_{cl}	3) - max. cabin rate of climb: 300 f p m - at high altitude climb at M_{cl},
			2,000 ft to cruise altitude	V_{cl} 2) 3) and acc. to V_{cr}		4) cruise altitude such that $D_{cl} + D_{des} \leq D_{cr}$
CRUISE (cr)	en route	as required, up to max. cruise	cruise alt. 4) long-range: max. of two step climbs	V_{cr} (M_{cr}) long-range, cost-econ. or high-speed	Time: $T_{cr} = D_{cr}/V_{cr}$ Fuel: F_{cr} Distance: D_{cr}	5) $D_{cr} = 1.02$ x actual cruise distance + 20 n m
DESCENT AND DECELERATE (des)	en route	flight idle	cruise alt. to 15,000 ft	V_{des} 2)6)	Time: T_{des} Fuel: F_{des} Distance: D_{des}	6) max. cabin rate of descent 300 fpm $\theta_{cr} - \theta_{des} \leq 4^o$ (θ=fuselage inclination angle)
	appro. flap below 10,000ft	as required for 3°	15,000 ft to 3,500 ft	initial appro. speed		
LANDING	landing flap, u.c. down	glide slope	3,500 ft to 0 ft	final appro. speed to 0		
AIR MANEUVERS (am)	en route	as required	cruise	V_{cr}	Time: T_{am} 7) Fuel: F_{am} 8) Distance: 0	7) $T_{am} = 6$ min 8) $F_{am} = (T_{am}/T_{cr}) F_{cr}$

Table 11-1. Typical basic flight profile analysis, transport aircraft (Refs.: ATA '67 method, SAWE Technical Report 619)

DOMESTIC OPERATIONS (up to 3,000 n m)	INTERNATIONAL OPERATIONS (above 3,000 n m)
ATA '67: 1. Fly for 1 hour at normal cruise altitude at fuel flow for end of cruise weight, speed for 99% of maximum range speed. 2. Exercise a missed approach and climbout at destination. 3. Fly to and land at alternate airport 200 n m distant. OTHER PROCEDURE 1. Execute missed approach at destination airport. 2. Cruise to alternate airport 200 n m distant at cruise speed used in basic mission (altitude optional). 3. Hold at alternate airport for 45 minutes, altitude 1,500 ft. 4. Descend and land at alternate airport. 5. Contingency fuel equal to 50% of holding fuel.	WITH ALTERNATE (ATA '67): 1. Continue flight for time equal to 10% of basic flight time at normal cruise altitude and speed for 99% maximum range. 2. Execute missed approach and climbout at destinate airport. 3. Fly to alternate airport 200 n m distant. 4. Hold at alternate airport for 30 minutes at 1,500 ft above the ground. 5. Descend and land at alternate airport. WITHOUT ALTERNATE: Continue basic flight profile for two hours.
	Conditions for flight to alternate airport and holding: 1. Cruise thrust or power setting may be equal to 99% of max. subsonic range, cruise speed used in basic mission (Table 11-1). 2. Holding thrust or power setting shall be for maximum endurance or 110% of min. drag or power speed, whichever is greater. 3. Cruise altitude shall be optimum for best range, except that it shall not exceed the altitude where cruise ditance equals the climb plus descent distance.

Table 11-2. Typical reserve fuel policies

Assuming steady horizontal flight, the lift coefficient is

$$C_L = \frac{W/S}{\frac{1}{2}\gamma p M^2} = \frac{W}{\delta W_{to}} \frac{1}{M^2} \frac{W_{to}/S}{\frac{1}{2}\gamma P_o} = \frac{W_{to}/S}{\frac{1}{2}\gamma P_o} f_1(\frac{W}{\delta W_{to}}, M) \tag{11-12}$$

A general expression for the drag polar (Fig. 5-6) is:

$$C_D = f(C_L, M) = f_2(\frac{W}{\delta W_{to}}, M) \tag{11-13}$$

The thrust required is obtained from the condition T = D, hence

$$\frac{T}{\delta T_{to}} = \frac{C_D}{C_L} \frac{W}{\delta W_{to}} \frac{W_{to}}{T_{to}} = \frac{W_{to}}{T_{to}} f_3(\frac{W}{\delta W_{to}}, M) \tag{11-14}$$

For a given engine the nondimensional performance can be specified as follows:

$$\frac{T}{\delta T_{to}} = f_T(\frac{N}{\sqrt{\theta}}, M) \tag{11-15}$$

$$\frac{C_T}{\sqrt{\theta}} = f_C(\frac{N}{\sqrt{\theta}}, M) \tag{11-16}$$

and the specific fuel consumption is obtained from these equations by eliminating $N/\sqrt{\theta}$:

$$\frac{C_T}{\sqrt{\theta}} = f(\frac{T}{\delta T_{to}}, M) = f_4(\frac{W}{\delta W_{to}}, M) \tag{11-17}$$

The effects of intake losses, bleed air takeoff, and power extraction are ignored in this elementary analysis, which is probably not acceptable in practice. Finally, the range parameter V/F is found by combining (11-12) through (11-17):

$$W \frac{V}{F} = \frac{WV}{C_T T} = a_o \frac{M}{C_T/\sqrt{\theta}} \frac{C_L}{C_D} = a_o f_5(\frac{W}{\delta W_{to}}, M) \tag{11-18}$$

It is thus possible to represent the specific range for all cruise conditions in the form of a set of curves, as exemplified by Fig. 11-8.
The distance flown in cruising flight is:

$$R = \int_{W_e}^{W_b} \frac{V}{F} dW = \int_{W_b/W_{to}}^{W_e/W_{to}} W \frac{V}{F} d(-\ln W/W_{to}) \tag{11-19}$$

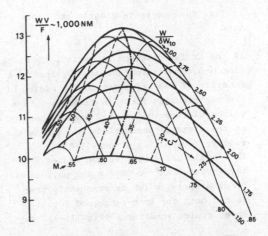

Fig. 11-8. Generalized range performance of a long-range jet transport

where WV/F is given by (11-18). Different cruise procedures can be assumed, e.g. flight at constant Mach number, constant engine rating, or constant lift coefficient, resulting in slightly different ranges for a given fuel weight. With the initial conditions known, the range can be computed numerically for several values of W_e/W_{to} or, alternatively, the cruise fuel required for various distances can be determined.

11.5.3. Descent

The procedure for analyzing the descent (Table 11-1) is based on several assumptions regarding the descent speed, the use of airbrakes, and engine ratings. It may be argued that in the pre-design stage several assumptions cannot be substantiated and a simpler approach may be acceptable. A good approximation is usually found by assuming that the amount of fuel used in a descent over a given distance is equal to the fuel used during an extended cruising flight over the same distance. The time to descend is derived directly from the assumed cabin rate of descent, taking into account the permitted slope of the cabin floor.

11.5.4. Payload-range diagram and block time

If the cruise distance is varied, the block time and total trip fuel are obtained as a function of the trip distance by adding of the contributing items in Table 11-1. The weight of reserve fuel may now be computed on the basis óf assumptions concerning the reserve fuel policy (Table 11-2).
The total fuel weight is subtracted from the Useful Load* to obtain the payload vs. range diagram, which is frequently presented both for long-range and for high-speed cruise conditions (Fig. 11-9). Pay-

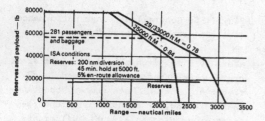

Fig. 11-9. Payload-range diagram for the Airbus A-300B

load-range diagrams (example in Fig. 11-9) must be compared with the design requirements and a decision made whether to alter the MTOW or not. Any such corrective action will have farreaching consequences òn all calculations carried out previously, particularly the structural weight prediction and the loading and balance of the complete airplane.

11.6. CLIMB PERFORMANCE

11.6.1. Maximum rate of climb, time to climb, and ceilings

For most conventional subsonic aircraft the angle of climb is limited to roughly 10°-15° and (11-11) can be used to compute the rate of climb as a function of the flight speed. The usual assumption is made

*See Section 8.2.

that climb performance is defined for constant CAS; for low-subsonic speeds this is approximately equivalent to constant EAS. The maximum rate of climb must be determined by a numerical process as the simplifying analytical process, assuming thrust or power available to be invariable with speed, usually leads to considerable errors.
The minimum time to climb to a specified altitude can be approximated by

$$ t_{cl} = \int_{o}^{h} \frac{dh}{C_{max}} \qquad (11-20) $$

where C_{max} is the maximum rate of climb for the particular altitude. However, C_{max} is frequently defined for a constant CAS (or EAS) and this does not necessarily result in the optimum climb procedure.
The energy method in Ref. 11-23 is the most practical procedure for obtaining accurate results.
Absolute and service ceilings are defined as those altitude where C_{max} = 0 and 100 ft/min (.5 m/s), respectively. The ceiling is found by computing C_{max} for several altitudes and plotting this result vs. the altitude. For multi-engine aircraft the ceilings are determined with all engines operating and with one engine inoperative. The results must be compared with airworthiness and operational requirements.

11.6.2. Takeoff and landing climb

As stated in Section 5.4.3., airworthiness requirements for each airplane category specify the minimum permissible climb performance, with and without engine failure. A summary for turbine-powered transport aircraft is presented in Table 11-3. Although these data pertain to the FAR 25 regulations, the equivalent BCAR Section D requirements are very similar.
The takeoff climb is divided into a number of nominally distinct segments, all with engine failed (Fig. K-1 of Appendix K):
a. A first segment, virtually defining a climb potential immediately after liftoff,

PHASE OF FLIGHT		AIRPLANE CONFIGURATION						MINIMUM CLIMB GRADIENT %		
		flap setting	u.c.	engine thrust (power)		speed	altitude	$N_e=2$	$N_e=3$	$N_e=4$
TAKEOFF CLIMB POTENTIAL ("first segment")		t.o.	$\downarrow$	one engine out	t.o.	V_{LOF}	$0 \rightarrow h_{uu}$ 1)	0	.3	.5
TAKEOFF FLIGHT PATH	"second segment"	t.o.	$\uparrow$	one engine out	t.o.	V_2 2)	$h_{uu} \rightarrow 400$ ft	2.4	2.7	3.0
TAKEOFF FLIGHT PATH	final takeoff ("third segment")	en route	$\uparrow$	one engine out	max. cont.	$V \geq 1.25 V_S$	400→1,500ft	1,2	1.5	1.7
APPROACH CLIMB POTENTIAL		approach 3)	$\uparrow$	one engine out	t.o.	$V \leq 1.5 V_S$	0 1)	2.1	2.4	2.7
LANDING CLIMB POTENTIAL		landing	$\downarrow$	all engines takeoff 4)		$V \leq 1.3 V_S$	0 1)	3.2	3.2	3.2

Nomenclature:

V_{LOF} - liftoff speed

V_2 - takeoff safety speed

V_R - rotation speed

V_S - stalling speed

u.c. - undercarriage position

h_{uu} - height at which u.c. retraction is completed

N_e - number of engines per a/c

1) out of ground effect
2) defined in Section 2 of Appendix K
3) flap setting such that $V_S \leq 1.10\ V_S$ for landing
4) more precisely: the engine power (thrust) available 8 seconds after throttle opening to takeoff rating
5) takeoff requirements are at actual weight, other requirements at landing (touchdown) weight

Table 11-3. Summary of climb requirements for turbine-powered transport category aircraft (FAR 25)

while the undercarriage is still extended and high-lift devices are in the takeoff position. The generally favorable effects of ground proximity on climb performance are disregarded.

b. A second segment, extending from the point where the undercarriage is retracted up to 400 ft (120 m). During this phase the high-lift devices are in the takeoff position, the engine(s) operate(s) at takeoff rating.

c. A third segment, extending from 400 ft (120 m) to a height of at least 1500 ft (450 m), during which the airplane is accelerated, flaps retracted and engine power (thrust) reduced to the maximum continuous rating.

Minimum permissible climb performance is also specified for the approach and landing configuration (Table 11-3):

a. The approach climb potential, applying to the approach configuration with one engine inoperative. For a specified landing weight this requirement may limit the flap deflection angle in the approach, and as a result of this the landing performance will be affected since there is a relation between the stalling speeds in the approach and landing configurations.

b. The landing climb potential is intended to ensure safe wave-off and climbout after a baulked landing. In view of the rapid response of modern turbo engines, this requirement is usually not critical.

The available angle of climb is determined from the relationship:

$$\tan\gamma = \frac{T}{W} - \frac{C_D}{C_L} \tag{11-21}$$

assuming that the effect of acceleration due to flight with constant CAS is negligible. For a specified position of the high-lift devices and ambient conditions, the available γ may be computed for various values of the takeoff weight, taking into account the variation of stalling speed with weight. For each gradient requirement a weight limit is thus obtained, defining the condition in which the available climb gradient equals the minimum required value. The results are usually presented in the form of Weight limits for variable Altitude and Temperature, usually referred to as WAT curves (Fig. 11-10).

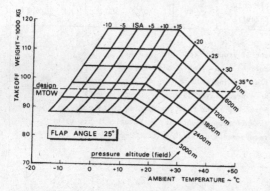

Fig. 11-10. WAT curves for the aircraft shown in Fig. 12-1

The following additional points may be mentioned.
a. Appropriate drag penalties must be accounted for in the engine-failure case (Appendix G, Section G-8).
b. In the case of a takeoff climb gradient limit the flight speed may be increased above V_2-min (Appendix K); this will usually improve the climb gradient, particularly for jet aircraft. However, owing to the connection between V_2 and the takeoff performance, the required field length will progressively increase.
c. In cases where the takeoff weight is unduly limited by the WAT requirements - on high and hot airfields - it will be appropriate to investigate the possibility of several intermediate positions of the high-lift devices.

11.7. AIRFIELD PERFORMANCE

11.7.1. Takeoff field length

Analysis of the takeoff characteristics is the most complicated item of performance; this applies particularly to transport aircraft, where the possibility of engine failure must be considered. Detailed consideration is given to this subject in Appendix K. For small aircraft the matter is much simpler and it is left to the reader to translate the requirements into a practical process for computating the field length required. The main differences in relation to transport category airplanes are as follows:
- the takeoff height is 50 ft (15,24 m),
- the takeoff safety speed is 1.3 V_S (FAR 23),
- emergency braking is not considered (FAR 23) or only to a limited extent (SFAR 23). In order to examine the operational flexibility and to optimize the performance, the effect of several variables must be considered.

a. Position of high-lift devices.
Increasing the flap deflection angle results in increased values of C_L for stalling and liftoff; V_S and V_2 will consequently decrease and the takeoff run will be shortened. However, the L/D ratio and the climb gradient deteriorate, and the airborne distance increases. For a given T/W ratio, an optimum flap angle can be defined, resulting in a minimum field length. The flap deflection is limited by the climb requirements with one engine failed (Section 11.6.2.) and for low T/W the optimum flap angle is not feasible (Fig. 11-11a).

b. Takeoff weight.
The takeoff reference speed variation in relation to the stalling speed for varying T/W-ratio is discussed in Section K-2 of Appendix K. For a specified flap angle three regimes can be distinguished in the takeoff distance vs. weight relationship (Fig. 11-11b):

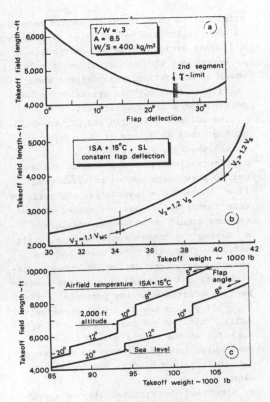

Fig. 11-11. Effect of operational variations on takeoff performance (a, b and c apply to different aircraft)

- For moderate P/W or T/W ratios $V_2 = 1.2 V_S$ and the field length is proportional to V_S^2 - and hence to W/S - and to the thrust or power loading. Hence S_{to} is approximately proportional to W^2.
- For high T/W (low takeoff weight) the margin relative to V_{MC} forms a lower limit to V_R and V_2. In this region it is mainly the P/W or T/W ratio that governs the field length and S_{to} is approximately a linear function of the weight.
- For low P/W or T/W ratios the second-segment climb gradient requirement cannot be satisfied; V_R and V_2 must be increased relative to V_S and the field length increases rapidly with W. For a given flap angle the takeoff weight cannot be increased beyond a value where the maximum climb gradients only just satisfy the min-

imum required values, e.g. for V_2 equal to the speed for maximum climb gradient. In the case of a variable flap setting, low T/W values are possible by gradually decreasing the flap angle.

c. Temperature and altitude variation. Increasing ambient temperatures reduce the air density and thrust (power)*; the field length is thus increased. For similar reasons, the field length increases with altitude.
The complete picture of field length vs. weight is shown in Fig. 11-11c, illustrating the flexibility of the design to variations in operational conditions.

11.7.2. Landing field length

Some basic principles of landing performance estimation have been explained in Section 5.4.6. The designer should consult the airworthiness requirements to define the operational conditions for which the analysis must be made. For transport aircraft the "arbitrary" landing distance is generally used, i.e. the shortest possible landing distance, to be demonstrated in ideal conditions on a dry runway, multiplied by a factor 5/3. British requirements additionally specify a "reference" landing distance, both for dry and for wet runways, in conditions that are representative of day-to-day operation. The following refinements on the simple analysis in Section 5.4.6. may be introduced.
a. The landing weight limitation must be established, using WAT curves, or the maximum flap deflection determined for a given landing weight.
b. The flare-out may be analyzed by a step-by-step type of analysis (cf. Ref. 11-60), assuming any suitable control law, e.g. a linear increase of the pitch angle with time from the steady approach condition to the touchdown. Analytical solu-

*depending upon the possibility of flat rating or water injection (cf. Section 4.4.6).

tions may be acceptable if the accuracy required is not very high (Refs. 11-58, 11-59).

c. The flight path angle at touchdown is of the order of ½ degree, corresponding to a rate of descent of 1 to 2 ft/s (.3 to .6 m/s).

d. The braked run after touchdown can be analyzed in much the same way as in the case of the aborted takeoff (Section 4 of Appendix K). According to Ref. 11-52, typical time delays are:

2 seconds from touchdown to brake operation,

3½ seconds from touchdown to spoilers effective,

7 seconds from touchdown to reverse thrust effective, with a cancellation speed of 45 - 50 kts.

The use of reverse thrust is not permitted in the case of FAR 25 requirements. The BCAR allow the use of reverse thrust, but the reference factor is higher in this case.

11.8. SOME ASPECTS OF OPERATING ECONOMY

11.8.1. Economy criteria

The operating costs of airlines are generally divided into Direct Operating Costs (DOC) and Indirect Operating Costs (IOC). The DOC are broadly defined as the costs that are associated with flying operations, and the maintenance and depreciation of the flying material while the IOC include the operator's other costs associated with maintenance and depreciation of ground properties and equipment, servicing, administration, and sales.

The revenues are usually referred to the passenger-mile or ton-mile production. For a given type of operation the ratio of the revenues to the operating costs is affected primarily by the load factor, the actual number of seats sold as a percentage of the seats available. For most scheduled airlines the average load factor is of the order of 50-60%. The question

arises as to what criterion should be used in the preliminary design stage to compare different designs on an economic basis.

a. We can use the DOC per aircraft mile, which is of particular importance when traffic is scarce. This criterion is important for a private aircraft.

b. The DOC per seat-mile and per ton-mile, on the other hand, are of particular interest when traffic is dense.

c. Alternatively, we may look at the number of passengers needed to break even (i.e. DOC + IOC = Revenue),as the ratio of this factor to the maximum number of passengers should be as low as possible.

d. Taking the comparison between aircraft still further, the total costs and the revenues per annum define the annual profitability and relating this to the investment associated with the purchase of the aircraft yields the Return on Investment (ROI). This type of cost analysis is typical for an operator's evaluation of a particular type of aircraft on his network, but it is seldom used in preliminary design.

e. Sometimes the profitability is approximated by the "profit-potential", defined as the area in the payload range diagram in which the operation yields a profit (Fig. 11-12).

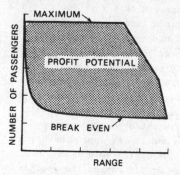

Fig. 11-12. Definition of profit potential

Although the DOC constitute only one aspect of the economic profitability of an airliner or private aircraft, most attention will have to be paid to this aspect, for the reason that several factors contributing to the DOC are directly related to the technical conception and operational characteristics of the airplane and as such are under the direct control of the design team. Nevertheless the economic aspects of aircraft operation will not be dealt with fully here. We will confine ourselves to the methods available to the designer to determine the position of his design relative to other designs or types of aircraft.

A breakdown of DOC into several cost items can be found on Table 11-4, while Fig. 11-13 shows the relative magnitude of the DOC contributions.

It is surprising that very little generalized information has been published on the subject of Indirect Operating Costs. There may be various reasons for this situation:
a. The IOC are not under the control of the airplane designer, although he may incidentally wish to estimate IOC in order to assess the profitability of an airplane design.
b. IOC will be affected by the operator's type of organisation and managerial policy, while considerable differences in passen-

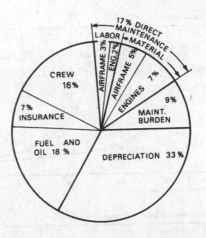

Fig. 11-13. Typical DOC breakdown (1967 ATA method, at design range = 2880 nm, from Ref. 11-92)

ger servicing can also be observed between operators.
c. Published data are very scarce and available essentially only for aviation in the United States.
A subdivision of IOC is presented in Table 11-4; a statistical estimation method is published in Ref. 11-82.

DIRECT OPERATING COSTS [1]	INDIRECT OPERATING COSTS [2]
FLYING OPERATIONS	MAINTENANCE - GROUND PROPERTY AND EQUIPMENT
flight crew costs	direct maintenance
fuel and oil	maintenance burden
hull insurance costs	SERVICING - FLIGHT OPERATIONS
DIRECT MAINTENANCE - FLIGHT EQUIPMENT	passenger service
labor - airplane	aircraft servicing [3]
material - airplane	traffic servicing
labor - engine	ADMINISTRATION AND SALES
material - engine	servicing administration
DEPRECIATION - FLIGHT EQUIPMENT	reservation and sales
	advertising and publicity
	general administration
	DEPRECIATION - GROUND PROPERTY AND EQUIPMENT

NOTES

1) Subdivision according to ATA-method

2) Subdivision according to the Aeroplane, Dec. 22, 1966, page 14

3) Including landing fees; in other subdivisions this item is considered as part of the DOC

Table 11-4. Subdivision of operating costs

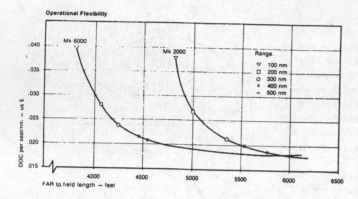

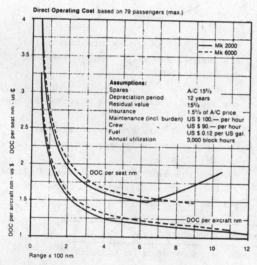

Fig. 11-14. Direct Operating Costs of two versions of the Fokker F-28 (1973 data)

structors (Ref. 11-68) and the Air Transport Association of America (Ref. 11-83). The 1967 ATA Direct Operating Cost prediction method is frequently employed as the basis for the presentation of cost data. Such design and economic criteria are useful principally for establishing first-order trends and will be of little help in designing the aircraft. An example of an economic comparison between two versions of the Fokker F-28 Fellowship, based on the ATA formula, is presented in Fig. 11-14. The DOC is usually expressed as a function of the stage length, primarily because the block speed is a function of range. For short ranges the block speed and the productivity decrease, while the increased ground maneuver time has an unfavorable effect on utilization. The cost per aircraft mile and the number of passengers rise sharply at stage distances below approximately 200 nm.

11.8.2   Estimation of DOC for preliminary design

The objectives of a standardized method for the estimation of aircraft operating costs are:
a. to provide a ready means for comparing the operating economics of competitive aircraft and/or aircraft designs under a standard set of conditions, and
b. to assist airlines and the aircraft manufacturer in assessing the economic suitability of an airplane for operation on a given route.
Standardized methods have been developed by the Society of British Aircraft Con-

CREW COSTS form a substantial part of the DOC, but are essentially outside the control of the designer.

FUEL AND OIL are directly affected by the aircraft and engine performance, but also by the fuel price, which varies significantly geographically and with time. It is not necessary to dwell here on the recent (1974) fuel cost explosion, which has entailed considerable difficulties for many operators.

384

MAINTENANCE COSTS have been based tradi-
tionally on such parameters as airplane
empty weight, engine thrust and airframe
or engine initial cost. It should be noted
that the correlations used in the ATA
method are purely statistical. There is no
direct relationship, for example, between
the aircraft weight and ease and cost of
maintenance. On the contrary, provisions
intended to simplify maintenance by means
of improved accessibility will tend to increase the empty weight, but this will result, wrongly, in increased maintenance
costs if the ATA method is used incorrect-
ly.

Ideally, the maintenance cost item should
be broken down into various components, as
systems vary widely in their maintenance
costs, partly due to their difference in
complexity and sensitivity to hourly and
cyclical effects. A typical example is engine maintenance, which appears to be very
sensitive to the complexity of the engine
and the TET (Ref. 11-95), while the introduction of a new category of engine
will also entail higher maintenance.

DEPRECIATION, like insurance, is in reality an annual cost. In the ATA-67 method
the appreciation period is taken as 12
years. The lifetime of the aircraft, in
terms of both flying hours and flight cycles, must obviously exceed this period.

UTILIZATION is principally affected by the
elapsed time during which the airplane is
on the ground due to traffic requirements,
loading, unloading and refueling, and regular maintenance, and due to delays caused
by the weather and unscheduled maintenance.
Utilization is also affected by the airline's network and flight planning. The
general trend given in the ATA method is
a curve of annual utilization vs. block
time, showing a variation between 3,000 hr/
year for a block time of one hour, to
4,500 hr/year for a block time of 8 hours.
If the number of flights per day is fixed,
however, the utilization is proportional
to the block time. It should be noted that

measures to improve the utilization, e.g.
by reducing the air maneuver time, transition time between two flights, etc., are
not reflected in the statistical correlation. To appreciate potential improvements
in utilization, a detailed assessment must
be made of the effects involved.

AIRPLANE COSTS (engines included) are the
other primary factor governing the depreciation costs, which must be estimated in
the preliminary design stage. As in the
case of weight estimation, various degrees
of detail may be aimed at, varying from a
first-order statistical relationship of
the type given in Fig. 11-15, to a detailed initial cost breakdown drawn up by
the factory's cost accounting department.
A decision must be made on the number of

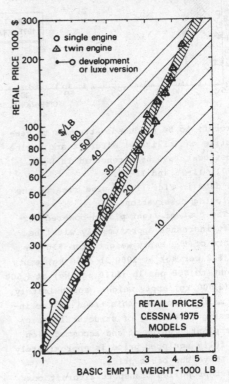

a. Retail prices of light aircraft with
reciprocating engines

Fig. 11-15. Aircraft prices

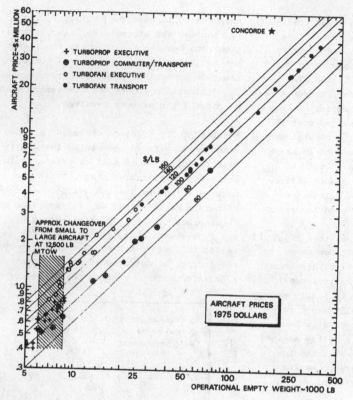

b. Prices of transport category aircraft

Fig. 11-15. (continued)

aircraft to be produced. Attempts to generalize on (military) aircraft and engine first cost have been reported in Refs. 11-71, 11-76 and 11-79.

The data in Fig. 11-15[*] give rise to the following observations.

a. The cost of light piston-powered aircraft increases approximately with the square of the empty weight from $10 per lb ($22 per kg) at 1000 lb (450 kg) empty weight to $50 per lb ($110 per kg) at 4,500 lb (2,000 kg) empty weight, approximately. The main reason for this trend is the increasing complexity of structural design, powerplant, equipment and accommodation. The fully equipped price is approximately 20% above the basic price.

b. Jet-propelled transport aircraft cost between approximately $120 per lb ($265 per kg) of empty weight for the smaller

*Price levels are for the 1973-1975 period.

types to $100 per lb ($220 per kg) for large transports. This level is fairly constant due to the comparable degree of complication and sophistication of these aircraft, irrespective of their size.

c. Jet-propelled business aircraft form the costliest category; their price depends particularly on the type of interior and completeness of the equipment and may go as high as $160 per lb ($350 per kg).

d. Turboprop transport aircraft cost approximately $80 per lb ($176 per kg) of empty weight.

A word of warning is appropriate when Fig. 11-15 is used. There is no explicit functional relationship between aircraft weight and first cost. Structure weight, for example, may be reduced by the use of special manufacturing methods or materials, but this will generally raise the costs instead of reducing them, contrary to what might be concluded from the figure. On the other hand, putting more sophisticated equipment into an aircraft will raise both its empty weight and its cost, but the cost will increase more rapidly than the weight.

For transport-type aircraft profitability is a major economic factor and it can be argued that the price level will be related not only to the costs of the technical development, but also to the productivity. Hence, instead of plotting the aircraft first costs vs. the empty weight, attempts can be made to correlate first costs with the product of payload and cruising speed, as this factor affects revenue to a considerable extent. Other refinements may also be justified, such as breaking down the aircraft first cost into airframe costs and engine costs. In any case the cost level expressed in $ per unit of empty weight must be used with great care.

The designer should be reluctant to compare the results of the ATA formula with actual airline costs as reported to the CAB. Standardized methods do not take account of factors such as fleet size, fleet mix, route structure, actual winds encountered, variations in labor rate and fuel costs with time*, etc., all of which can have a significant effect on costs and will vary from one airline to another. In view of the limited applicability of standard methods in general a number of restrictions must be placed on their use:

*At the time of writing this chapter (1975) the general situation is that the ATA-67 formula underestimates crew costs by about 40%, and fuel prices have more than doubled since 1967, while maintenance costs are overestimated by 30%, depreciation by 25% and insurance by 100% (approximate figures).

a. If absolute levels of DOC are required for comparing study airplanes with operational airplanes, or for distinguishing between airplanes with different numbers of engines, generalized methods are generally inadequate.
b. Misleading information may be produced as to the relative value of improvements in specific fuel consumption and engine specific weight when the ATA-67 method is used.

Where incremental DOC are of main importance, standard methods may have some utility by providing a simplified tool to evaluate the relative importance of aircraft and engine first costs, fuel costs and maintenance costs. Thus the DOC analysis yields weighting factors for several types of studies:
a. Certain categories of parametric investigations where a decision is sought concerning the relative desirability of airplane design and sizing options - an example is given in Section 5.5.5. These cases are restricted to airplane comparisons made on a consistent basis of technology level and airframe and engine price level.
b. Optimization studies of the cruising speed. Cruise fuel costs are minimum for the airspeed for maximum distance traveled per lb of fuel (Fig. 11-8); increasing the airspeed results in increased cruise fuel cost. However, due to a reduced block time, the depreciation per flight can be reduced, provided the yearly utilization remains at least constant. At some cruise speed intermediate between the long-range and high-speed cruise conditions a minimum DOC is reached. For long-range aircraft the cost-economical cruise speed is close to the long-range cruise condition, for short-haul aircraft it is close to the high-speed cruise condition.

Chapter 12. Evaluation and presentation of a preliminary design

SUMMARY

The presentation of a preliminary design comprises layout drawings, a summary of the principal characteristics and geometry, and the results of a preliminary estimate of performance, flight characteristics and operating costs.

This concluding chapter gives a framework for an initial qualitative assessment of the structural integrity and a design philosophy in the interest of low structural weight. A checklist is presented for completing the layout drawings and mention is made of points to be considered when initial conclusions have to be drawn regarding the feasibility and further development of a project.

12.1. PRESENTATION OF THE DESIGN

The initial baseline design was characterized in Section 1.2. as a feasible configuration which the designer anticipates will satisfy the design requirements. The major part of this textbook is devoted to providing the design engineer with directives for obtaining such a feasible preliminary design. As already indicated in Fig. 1-3, the subsequent elaboration of the design may require parametric design studies and the investigation of alternative design solutions. The design activities will subsequently be intensified in the configuration development phase, when a large engineering effort is required to arrive at a final detailed project definition.

The initial baseline design thus being the starting point for further studies, it is essential that all results of the design activities, together with the relevant background information, are summarized and incorporated in a status report. The following documents are usually produced by a preliminary design department and incorporated in the report in which a new design is presented:

a. Layout drawings (example in Fig. 12-1) intended to give an impression of the detailed arrangement of both the interior and the external shape, as well as some of the major elements of systems and installations. The location of major structural elements will have to be indicated to make it unlikely that major problems will arise during the subsequent stage of structural design.

b. A summary of the principal characteristics and geometry, including a three-view drawing. This summary is comparable to the short descriptions given in, for example, Jane's All the World's Aircraft.

c. Results of a preliminary estimate of performance, flight characteristics and operating costs.

The initial baseline design is frequently used during a period of discussions with interested potential users of the aircraft.

For this reason and in order to compare a transport aircraft design with existing and competitive new designs, the designer may add to the presentation a survey of the major characteristics of comparable designs. Finally, he will arrive at conclusions as regards the suitability and feasibility of the design and state his view on how to proceed with the project. The presentation of a project is a stage in the project when the designer is inclined to reflect on the suitability of the general arrangement and the overall integrity of the aircraft. The final sections of this book are therefore intended to give a framework for an overall structural assessment.

12.2. EXTERNAL GEOMETRY AND STRUCTURAL ARRANGEMENT

The preliminary design engineer or design team is not directly responsible for the detailed structural features of the project. However, there is a definite interface between the external geometry, i.e. the outline or outside contours of the airplane, and the structural arrangement. Both have a significant effect on weight and weight distribution. In the early stage of preliminary design the basic principles of the structural layout must be laid down in order to ensure that a light structure can be fitted inside the external lines. Only the major structural elements need be considered; Fig. 12-2 gives an example of the type of drawing defining an initial structural arrangement. The design is still sufficiently flexible to allow certain modifications in the interest of a good general design philosophy, to be formulated as follows.

a. Structural integrity of the airframe must be insured in the event of failure of a single primary structural element or the occurrence of partial damage in extensive structures, such as skin panels. Adequate residual strength and stiffness and a slow rate of crack propagation must be ensured.

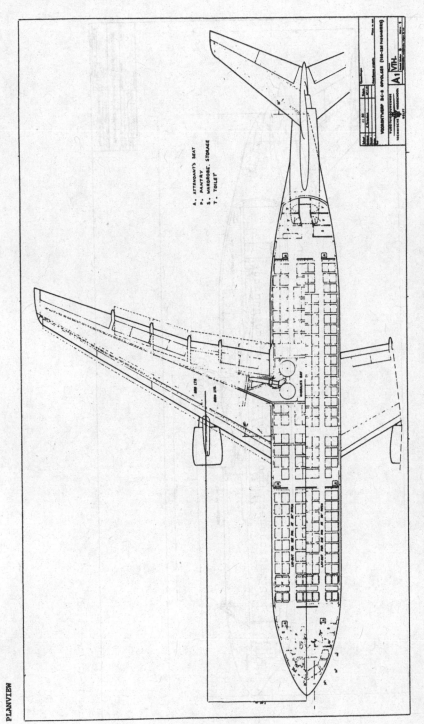

A. ATTENDANT'S SEAT
P. PANTRY
S. WARDROBE. STORAGE
T. TOILET

Fig. 12-1. Layout drawings of a preliminary design of a hypothetical short-haul airliner

SIDEVIEW

Fig. 12-1. (continued)

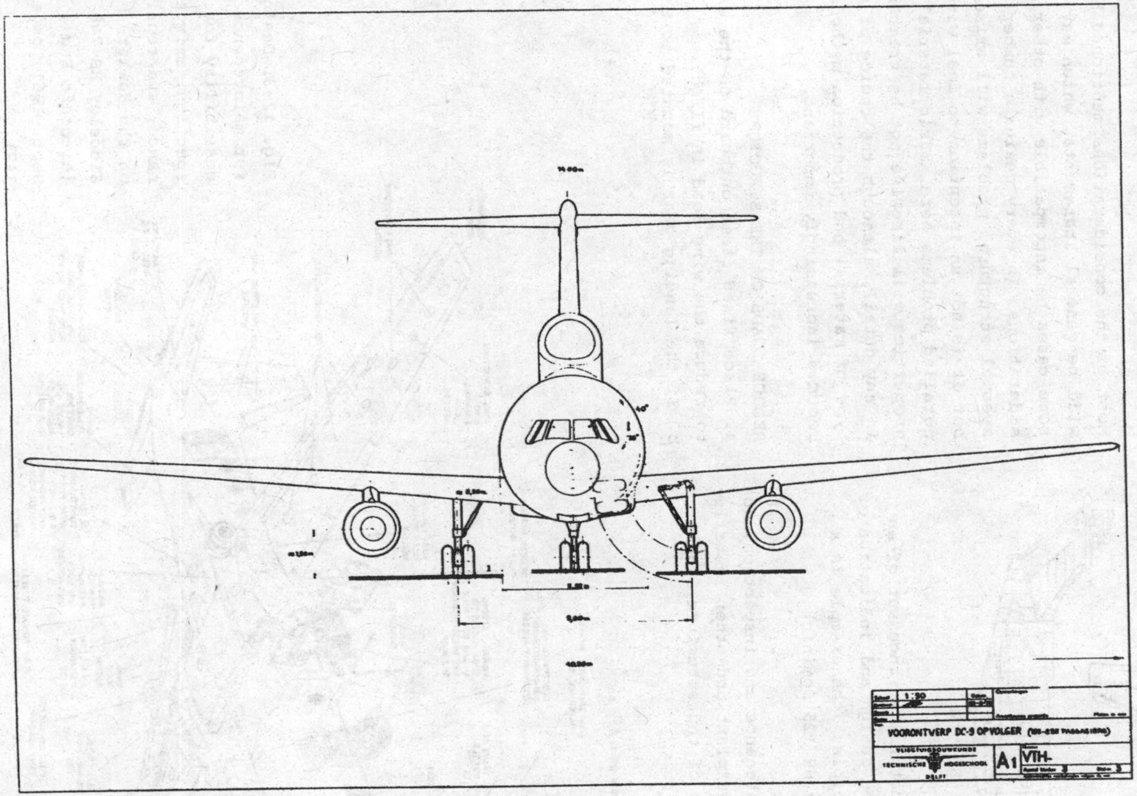

Fig. 12-1. (continued)

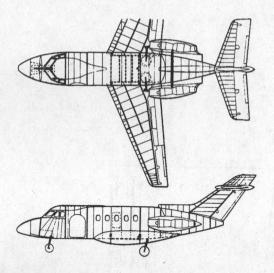

Fig. 12-2. Structural arrangement of an initial baseline design and indication of the basic principles (the example is a Hawker Siddeley HS 125 Dominie)

b. Ease of maintenance and inspection, not only of the powerplant and other installations, but also of the structure itself (Fig. 12-3).

c. Minimum structural weight, involving an accurate assessment of loading and strength requirements, a functional and simple structural arrangement and appropriate choice of materials.
d. Design for a specified life of the structure, in terms of operational hours, (e.g. 30,000 - 60,000), numbers of flight cycles or takeoffs and landings (e.g. 30,000).

Most of the considerations mentioned below will be general statements, which may in some cases be incompatible with other considerations. In every design a number of special structural problems will arise, but it is not an intention to deal with detailed problems here. Basic material properties will likewise not be treated in any detail, although the choice of the type of material and production methods becomes increasingly important.

REDUCE LOADS ON THE STRUCTURE
a. Place fixed items outboard on the wing to reduce the wing load in flight.
b. If the landing gear is mounted to the

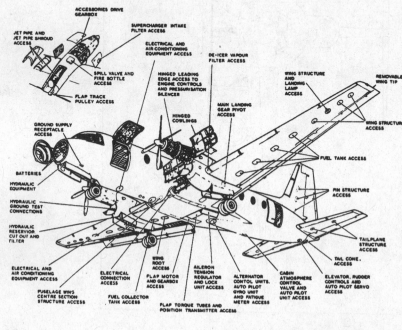

Fig. 12-3. Design for maintenance: accessibility for inspection, maintenance, and repair on the Hawker Siddeley HS 748 (Aircraft Eng., Sept. 1961, page 249)

wing, keep it in a position such that landing-impact bending and shear loads are not more critical than flight loads for most of the wing structure.

c. Avoid mounting heavy masses to the wing behind the elastic axis as these may lead to aeroelastic problems.

d. Place wing tanks outboard to reduce the bending load in level flight, but keep about 3 ft (.9 m) of wing tip structure free from fuel, in view of the fire risk caused by electrostatic loads.

e. Locate heavy masses in the fuselage near the c.g. to reduce the inertia loads.

f. Tailplane loads can be reduced by keeping the tail arm sufficiently long.

g. Use undercarriage legs of minimum length.

h. Avoid large nosewheel loads by properly locating the nosewheel leg.

KEEP LOAD PATHS AS SHORT AS POSSIBLE

a. Limit the sweep angle of the wing to the minimum value required. Use straight wing center sections and avoid any reduction in wing thickness in the plane of symmetry, unless simplification and weight reduction can be achieved elsewhere (Fig. 12-4).

b. Keep the support structure for landing

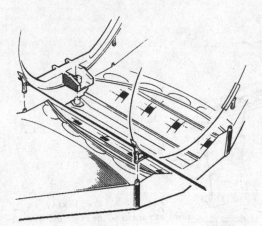

Fig. 12-4. Reduction in wing thickness in the plane of symmetry makes it possible to avoid interruption of the fuselage structure of the HS 125 (see also Fig. 12-2)

gear and engine nacelles close to the primary structure.

c. Provide a continuous wing structure through the fuselage and avoid ring bulkheads if other solutions are possible.

d. To maintain structural continuity where the (low) wing is connected to the fuselage, a central keel member, extending over the length of the center section, may be used.

e. Locate the wing root so that the loads from the wing shear webs can be easily distributed over the fuselage shell. Make sure that major frames do not interfere with openings such as emergency exits, windows, etc.

f. Heavy masses or gear attachment structures should not be cantilevered from bulkheads.

CUTOUTS AND DISCONTINUITIES

a. Large cutouts should be kept out of highly stressed regions, e.g. the wing root or the fuselage structure immediately behind the wing fuselage connection.

b. Sharp corners in cutouts must be avoided altogether, especially in pressurized fuselages.

c. The height of windows in pressure cabins should preferably be larger than the width.

d. Locate parts of the flight control system (cables, push rods), fuel lines, air ducts, etc. outside the primary wing structure to simplify the provision of inspection facilities.

e. Avoid cutouts in the primary wing structure for engine intake ducts and exhaust pipes.

f. Special attention must be paid to buried engines in view of the need to inspect and remove them.

g. Avoid abrupt changes in the load path of wing, tail and landing gear attachment to the fuselage.

h. Try to locate wheelbays outside the primary wing box, particularly in the case of a stressed wing skin.

j. Reduce the number of joints to a bare minimum, compatible with manufacturing requirements and maintenance considerations.

MAKE THE STRUCTURE FUNCTIONAL AND AVOID COMPLEXITY

a. Avoid complicated mechanisms employing tracks, rollers, etc. This applies particularly to flap operating and undercarriage retraction mechanisms. Fig. 7-25e is a good example of simplicity in design of a highly effective flap system. An example in which structural complexity cannot be avoided is shown in Fig. 12-5.

b. Reduce regions of compound curvature to a minimum and avoid doors with appreciable double curvature.

c. Use configurations with low noise exposure to primary structural members. High acoustical loads on skin panels will result from noise levels above 145 dB. Bonded honeycomb structure is suitable in regions of high acoustic loads due to the absence of rivet holes and favorable damping properties.

MAKE MEMBERS PERFORM MULTIPLE FUNCTIONS

a. Combine wing ribs supporting the engine(s), landing gear, and flaps.

b. Ensure that bending material also provides torsional stiffness.

c. Use common fuselage bulkheads to support the horizontal and vertical tailplane carry-through.

d. Use existing bulkheads to divide fuel tank bays and for attachment of items of equipment, undercarriage legs, etc.

e. Avoid an extreme number of keels, floors, frames, etc.

SAFETY

a. Multipath and safe-life structure must be provided throughout in the primary structure of transport category aircraft.

b. Provide adequate safeguard against shedding propeller or turbine blades or fragments of blades. Avoid primary structural elements in regions that are likely to be struck in the event of engine or propeller disintegration.

c. Main gear or support structure must not damage fuel tanks in the event of a crash landing.

d. Sufficient clearance must be provided between the loaded and deformed fixed structure and moving parts of, for example, surface controls.

e. Flaps and elevators on both sides of the plane of symmetry must be interconnected.

f. The engine suspension system must be capable of absorbing a certain percentage of the limit loads even after failure of any one link.

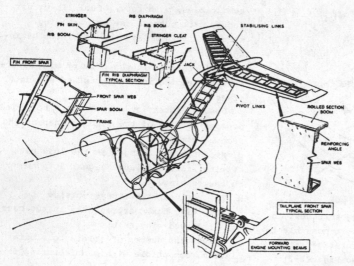

Fig. 12-5. Complexity in structural design cannot always be avoided altogether rear fuselage and empennage structure of the BAC 1-11 (Aircraft Eng. May 1962)

12.3. LAYOUT DRAWINGS

Detailed drawings have to be produced on a sufficiently large scale and to embody both the internal arrangement and the most relevant contour lines (Fig. 12-1). Several cross-sections are usually also drawn (Fig. 12-6). Attention must be paid to the following items.

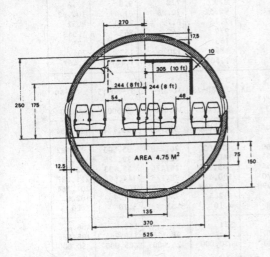

All dimensions in centimeters, except where indicated.

Fig. 12-6. Fuselage cross-section for tourist (left) and economy (right) layout for the design shown in Fig. 12-1

FUSELAGE
Passenger cabin: arrangement of seats, layout of cross-section, windows, entrance doors, emergency exits, lavatories, pantries, cloakrooms, baggage holds, overhead racks, cabin separations, and pressure bulkheads.
Freight and cargo holds: dimensions, hatches and freight doors.
Flight deck: location of the pilot's seat and instrument panel, location and arrangement of the windscreen, primary flight controls (schematic), accessibility.
Structural layout: location of major frames, longerons, major bulkheads, and keels.

POWERPLANT AND NACELLES
Engines and nacelles: external lines of the nacelles, propeller(s), intake and exhaust ducts, engine mounting frames, pylons, panels and hinged doors for engine maintenance, firewalls, thrust reversers, cooling system. Fuel tanks: location, dimensions, volume, accessibility.

WING
Root and tip section, leading and trailing edges, mean aerodynamic chord, high-lift devices in retracted and extended positions, ailerons, spoilers, speedbrakes. Structural layout: provisions for engine and undercarriage mounting, flap and aileron controls, wing spars and heavily loaded ribs, bracing elements (if used), flap tracks, hinges and supports.

TAILPLANE AND FLIGHT CONTROL SYSTEM
Root and tip section, leading and trailing edges, mean chord, elevator and rudder deflection angles, hinge location, incidence (range for adjustable tails). Major elements of the structure and the flight control system.

UNDERCARRIAGE
Location of the gear, wheel positions fully extended, fully deflected and statically loaded by the MTOW. Principle of the retraction system; retracted position, wheelbays.

SYSTEMS
Although systems design takes place in a fairly advanced stage of the design, space must be provided for some primary elements of these systems, notably those of the environmental control system and the APU, while bays must be provided for the hydraulic and/or pneumatic system and NAV/COM equipment.

Layout drawings for a hypothetical short-haul airliner are presented in Fig. 12-1 to illustrate the result of a preliminary

	dimen-sion	Airbus A-300B2	Boeing Advanced 727-200	Mc Donnell Douglas DC-9/30	Dassault Mercure project	Project Fig.12-1
Wing span	ft	147.02	107.93	93.34	100.16	132.13
Wing area	ft^2	2,795	1,697	1,000	1,247	2,064
Wing aspect ratio	–	7.71	7.67	8.73	8.00	8.46
Length overall	ft	167.08	153.08	114.25	111.48	159.18
Height overall	ft	54.30	33.47	27.48	37.25	45.41
Length fuselage	ft	161.90	136.90	106.89	110.16	139.89
Outside fuselage width	ft	18.49	12.33	10.95	12.79	17.21
Approx. height of cabin floor above apron	ft	15.4	8.9	7.2	9.8	12.8
Cabin length	ft	119.67	92.59	70.98	81.47	104.92
Cabin heigth	ft	8.33	7.15	6.75	7.20	8.52
Floor area	ft^2	979	645	863	1.548	
Passenger accommodation { all-tourist	–	259	163	105	134	185
{ high density	–	286	189	120	155	235
Total cargo hold volume	ft^3	4,229	1,503	892	1,181	2,467
Powerplant: number of engines	–	2	3	2	2	3
type	–	GE-CF-6-50A	JT-8D-15	JT-8D-7	JT-8D-15	M-56-40
SL/ISA static takeoff thrust	lb	48,953	15,500	14,000	15,500	22,050
Operational Empty Weight	lb	176,406	101,000	55,788	63,727	123,484
Maximum Weight Limited Payload	lb	69,900	39,030	26,240	35,500	52,922
Maximum Zero Fuel Weight	lb	240,353	140,000	82,030	99,226	176,406
Total Fuel Capacity	lb	75,854	65,270	24,476	22,933	52,922
Maximum Takeoff Weight	lb	291,070	190,518	98,126	144,664	211,687
Maximum Landing Weight	lb	264,609	159,868	93,495	108,049	201,764
Maximum Wing Loading	lb/ft^2	104.0	112.4	98.0	91.9	102.6
Maximum Thrust Loading	–	2.97	4.10	3.51	3.70	3.20
Maximum cruising speed	kts	506	521	491	500	486
at weight of	lb	291,070		98,126	110,254	198,456
and at cruise altitude of	ft	25,000	30,000	25,000	25,000	30,000
Cruise Mach number at these conditions	–	.84	.84	.816	.83	.825
Maximum Operating Mach number (M_{MO})	–	.86	.90	.89	.85	.85
Maximum Operating Limit Speed (V_{MO}, EAS)	kts	360	390	340	380	380
Takeoff safety speed at MTOW	kts	141	159	147		132
Approach speed at MLW	kts	132	137	129	124	132
FAR Takeoff Field Length at SL/ISA, MTOW	ft	6,030	8,740	6,810	6,510	5,250
FAR Landing Field Length at SL, MLW	ft	5,840	5,330	4,920	4,740	5,110
Maximum Payload Range (high-speed cruise)	nm	865	1,365	910	180	738
Maximum Fuel Range (long-range cruise)	nm	2,240	2,415	1,465	825	1,920
Range with maximum number of passengers in all-tourist class	nm	1,185	1,510	1,105	657	1,205

including normal reserves

Table 12-1. Principal characteristics of the design in Fig. 12-1 compared with present-day aircraft types

design exercise. The design specification of this aircraft is summarized in Fig. 1-5, while a three-view drawing can be found in Fig. 2-4.

12.4. CONCLUSION

Any assessment of the qualities and characteristics of a preliminary design should be accompanied by an appreciation of other design projects and/or design specifications of aircraft types, considered as competitors or types to be replaced by the new design. These comparisons must be made on the basis of a well-defined set of ground rules and similar methods of analysis. The results of such an assessment constitute an essential part of the basis on which further decisions have to be made. Table 12-1 summarizes the major items to be considered.

In reporting his activities, the designer will not only describe the most relevant characteristics of the design, but he will also pay attention to those aspects which he thinks are of vital importance to the successful further development of the project, particularly if there are reasons for doubts about the major predictions and assumptions on which the design characteristics are based. The designer may also anticipate certain design modifications, worth considering in the subsequent "parametric design phase" (cf. Fig. 1-3), for example:

- Is the general arrangement suitable, so that no unsatisfactory characteristics have to be anticipated? The various aspects pointed out in Chapter 2 may be considered here once more.
- Is the engine type to be installed matched to the airplane? In the case of performance deficiencies in certain areas, investigate whether the engine manufacturer is prepared to accept modifications.
- Is it necessary to modify the main wing proportions, such as area, aspect ratio or taper ratio? Critical performance items may be improved in this way, without unduly penalizing other characteristics.

- Is the choice of the high-lift system satisfactory? Only after performance analysis it is possible to be conclude whether the flap system should be more sophisticated or whether alternatively, a simplification may be acceptable.
- Do opportunities exist to reduce the drag by improving the aerodynamic design or the wetted area?
- How satisfactory is the weight breakdown? A final weight analysis, drawn up after the performance estimation has been made, may indicate significant differences compared with other aircraft. If so, an explanation must be sought.
- Can any measures be indicated to reduce the noise production? This point will be of particular concern to the designer of an airplane type, the airworthiness standards for which can be expected to be revised in the near future (cf. Section 5.5.6.).

Practically all decisions to be made in the preliminary design stage have implications with respect to the operating costs and benefits and although a quantitative evaluation is not always possible, the designer should be attentive to improving the economic potential of the design. It is only fairly recently that much emphasis has been placed on the environmental aspects. Every new civil airplane type must now satisfy the broad requirement that it will be acceptable not only to the passenger or the owner, but to society as a whole. A creative approach to any aspect tending to reduce noise, pollution and energy, space and land requirements will be most welcome.

References

CHAPTER 1.

Textbooks and general literature on aircraft design

1-1 F.K. Teichmann: "Airplane design manual". Pitman Publishing Corp., Third Edition, 1950.

1-2 G. Corning: "Supersonic and subsonic airplane design". Published by the author, 1953.

1-3 J. Hay Stevens: "The shape of the aeroplane". Hutchinson & Co. Ltd., London, 1953.

1-4 G.S. Schairer: "Economic considerations of a jet transport airplane". SAE Paper No. 426, Jan. 1955.

1-5 R.B. Morrison, and M.J. Ingle: "Design data for aeronautics and astronautics". John Wiley and Sons, Inc., New York and London, 1961.

1-6 E.E. Sechler, and L.G. Dunn: "Airplane structural analysis and design". Dover Publications Inc., New York, 1963.

1-7 A.C. Kermode: "The aeroplane structure". Second Edition, Pitman & Sons Ltd., London, 1964.

1-8 K.D. Wood: "Aerospace vehicle design". Volume I: Aircraft design. Johnson Publishing Company, Boulder, Colorado, Second Edition, 1966.

1-9 R. Miller and D. Sawers: "The technical development of modern aviation". New York, Prager, 1968.

1-10 F. Maccabee (ed.): "Light aircraft design handbook". Loughborough University of Technology, Dept. of Transport Technology, TT 6801, Second Edition, Feb. 1969.

1-11 J.M. Swihart (ed.): "Jet transport design". AIAA Selected Reprint Series, Vol. 8, 1969.

1-12 M.O. Wilmer: "Some aspects of optimisation in the design of civil aircraft". Two-day convention on economic factors in aviation. The R.Ae.S., 13-14th May, 1970.

1-13 E.V. Krick: "An introduction to engineering and engineering design". J. Wiley & Sons, Inc., New York, Second Edition 1969.

1-14 D.D. Hufford, J.A. Ross and K.W. Hoefs: "The economics of subsonic transport airplane design, evaluation and operation". SAE Paper No. 710423.

1-15 L.M. Nicolai: "Design of airlift vehicles". USAF Academy, Dept. of Aeronautics, Aero 464, July 1972.

Trends in aircraft design and technological forecast

1-16 J.C. Floyd: "Some current problems facing the aircraft designer". J.R.Ae.S. 1961, pp. 613-631.

1-17 G. Schairer: "Aircraft design, present and future". ICAS Paper No. 64-533.

1-18 Various contributors: "Cheap short-range air transport". A Symposium, J.R.Ae.S., Nov. 1965, pp. 737-758.

1-19 M.B. Morgan: "Some aspects of aircraft evolution". 5th ICAS Congress, Paper 66-1, 1966. (See also Aerospace Proceedings 1966 Vol. 1, MacMillan London, 1967).

1-20 P.G. Masefield: "Light aviation problems, prospects and performance". Journal of the Royal Aero. Soc.,

Nov. 1966.

1-21 F.W. Kolle and D.R. Blundell: "Evolution and revolution with the jumbo trijets". Astronautics and Aeronautics, Oct. 1968, pp. 64-69.

1-22 J.C. Brizendine and C.R. Strang: "The future transport world". J. of Aircraft, Nov.-Dec. 1967, pp. 481-486.

1-23 Handel Davies: "Some thoughts about the future of European aeronautics". The Aeronautical Journal, May 1968, pp. 385-395.

1-24 D. Küchemann and J. Weber: "Analysis of some performance aspects of various types of aircraft, designed to fly at different ranges and speeds". Progress in Aeronautical Sciences, Vol. 9, 1968, pp. 329-456.

1-25 E. Ufer: "Das Reiseflugzeug von Morgen". Deutscher Aerokurier, Jan. 1970, pp. 8-10.

1-26 M.B. Morgan: "The impact of research and development programmes in various technical areas on the economics of future aircraft design". Two-day convention on economic factors in aviation, R.Ae.S., 13th - 14th May 1970.

1-27 F.A. Cleveland: "Size effects in conventional aircraft design". AIAA Paper No. 70-940.

1-28 J.E. Steiner: "Aircraft development and world aviation". The Aeronautical Journal, June 1970, pp. 433-443.

1-29 Various contributors: "Vehicle technology for civil aviation, the seventies and beyond". NASA SP-292, Nov. 1971.

Airplane design and development

1-30 G.R. Edwards: "Problems in the development of a new aeroplane". Journal of the Royal Aero. Soc. Vol. 53, pp. 197-252, 1949.

1-31 E. Weining: "Design factors in the development of light aircraft". Aero. Eng. Review, July 1950, pp. 18-19.

1-32 Anon.: "Design of aircraft for commercial use". SAE Preprint, 1959.

1-33 Anon.: "Design for the future". An outline of the research, design and development organisation of Bristol Aircraft Limited, undated.

1-34 H. Mansfield: "Billion Dollar Battle". David McKay Comp. Inc., New York.

1-35 J.E. Steiner: "Planning a new commercial aircraft". Astronautics and Aeronautics, Sept. 1967, pp. 52-60.

1-36 W.M. Magruder: "Development of requirement, configuration and design for the Lockheed 1011". SAE Paper No. 680688.

1-37 J. Spintzyk: "Ueber Methoden der Systemtechnik beim Flugzeugentwurf". Zeitschrift für Flugwissenschaften, June 1968, pp. 206-212.

1-38 D.M. Ryle: "Parametric design studies". AGARD Lecture Series No. 16, April 1969.

1-39 J.T. Stamper: "Management in design". The Aeronautical Journal, Vol. 73, March 1969, pp. 174-185.

1-40 E.S. Bradley, W.M. Johnston and G.H. von Keszycki: "Passenger transport at low supersonic speeds". AIAA Paper No. 69-776.

1-41 R.H. Lange: "Parametric analysis of ATT configurations". AIAA Paper No. 72-757.

1-42 D.L.I. Kirkpatrick and D.H. Peckham: "Multivariate analysis applied to aircraft optimisation - some effects of research advances on the design of future subsonic transport aircraft". R.A.E. Techn. Memo Aero 1448, Sept. 1972.

1-43 A.W. Bishop: "Optimisation in aircraft design - the whole aircraft". Symposium of the Royal Aeron. Soc. on "Optimisation in aircraft design". 15th November 1972.

1-44 W.M. Eldridge et al: "Conceptual design studies of candidate V/STOL lift fan commercial short haul transport for 1980-85 V/STOL lift fan study". NASA CR-2183, Feb. 1973.

1-45 R.G. Knight, W.V. Powell Jr. and J.A. Prizlow: "Conceptual design study of a V/STOL lift fan commercial short haul transport". NASA CR-2185, April 1973.

1-46 R.R. Heldenfels: "Integrated, computer-aided design of aircraft". AGARD CP-147, Vol. 1, Oct. 1973.

Broad design requirements and market analysis

1-47 B.S. Shenstone: "Why airlines are hard to please". J. of the Royal Aero. Soc., May 1958, p. 319-336.

1-48 W.P. Kennedy: "Short-haul air transportation". Aerospace Engineering, Dec. 1961, pp. 24, 25, 72-74.

1-49 R. Nivet: "Airlines' approach to aircraft selection". J. of the Royal Aeron. Soc., Vol. 66, Dec. 1962, pp. 751-759.

1-50 B.P.G. de Bray: "Aerodynamic effects of top dressing operations". J. of the Royal Aeron. Soc., Oct. 1962, pp. 631-636.

1-51 J.M. McMahon: "Agricultural aircraft for the future - the fixed wing aircraft". J. of the Royal Aeron. Soc., Dec. 1962, pp. 776-777.

1-52 R.D. Fitzsimmons: "Design and economic suitability of present and future conventional aircraft in Boston - to - Washington service". AIAA Paper No. 66-945.

1-53 K.M. Trützscher: "Marktforschung für das Europäische Airbus Projekt". Jahrbuch der WGLR 1967, pp. 59-67.

1-54 Anon.: "The short-haul market". The Aeroplane, July 31, 1968, pp. 4-11.

1-55 D.J. Lloyd-Jones: "Airline equipment planning". J. of Aircraft, Jan./Feb. 1968, pp. 60-63.

1-56 H.M. Drake, G.C. Kenyon and T.L. Galloway: "Mission analysis for general aviation in the 1970's". AIAA Paper No. 69-818.

1-57 T. Oakes: "Airline technical requirements for 1975 STOL and V/STOL systems". SAE Paper No. 700312.

1-58 B.A.M. Botting: "Market research - an economic necessity". Two-day convention on economic factors in aviation, Royal Aero. Soc., 13-14th May 1970.

1-59 G.C. Kenyon and H.M. Drake: "Technological factors in short haul air transportation". AIAA Paper No. 70-1287.

1-60 L. Pazmany: "Future trends in general aviation". AIAA Paper No. 70-1220.

1-61 R.E. Black, D.G. Murphy and J.A. Stern: "The crystal ball focuses on the next generation transport aircraft". SAE Paper No. 710750.

1-62 Various contributors: "Two day convention on aviation's place in transport, 12/13th May 1971, Royal Aero. Soc.

1-63 L.K. Loftin Jr.: "Aeronautical Vehicles - 1970 and beyond". J. of Aircraft, Vol. 8, No. 12, Dec. 1971, pp. 939-951.

1-64 E. Torenbeek and G.H. Berenschot: "Preliminary design of a next-generation short-haul airliner". Delft University of Technology, Dept. of Aeron. Eng., Memorandum M-184, July 1972.

Airworthiness requirements

1-65 Anon.: "Federal Aviation Regulations Parts 1-189". Department of Transportation: Federal Aviation Administration.

1-66 Anon.: British Civil Airworthiness Requirements, Sections D and K.

1-67 W. Tye: "Influence of Recent Civil Airworthiness Requirements on Aircraft Design". Journal of the Royal Aero. Soc. Vol. 52, No. 452, Aug. 1948, pp. 513-522.

1-68 W. Tye: "Modern Trends in Civil Airworthiness Requirements". Journal of the Royal Aero. Soc. Vol. 56, Feb. 1952, pp. 73-108.

1-69 W. Tye: "The Arithmetic of Airworthiness". Journal of the Royal Aero. Soc. Vol. 58, March 1954, pp. 195-200.

1-70 W. Tye: "Philosophy of Airworthiness". AGARD Report 58, Aug. 1956.

1-71 H. Caplan: "The Investigation of Aircraft Accidents and Incidents". Journal of the Royal Aero. Soc. Vol. 59, Jan. 1955, pp. 45-60.

1-72 H. Caplan: "Aircraft Design Philosophy". Journal of the Royal Aero. Soc. Vol. 60, May 1956, pp. 301-312.

1-73 P.P. Baker, S.G. Corps, Capt. D.F. Redup and S.M. Harris: "All-Day Symposium on Flight Testing for the Certification of Civil Transport Aircraft". Journal of the Royal Aero. Soc. Vol. 71, Nov. 1967, pp. 745-772.

1-74 Anon.: "Provisional Acceptable Means of Compliance - Aeroplane Performance". ICAO Circular 58 -AN/53, July-Aug. 1959.

1-75 Anon.: "International Standards - Airworthiness of Aircraft". ICAO-ANNEX 8 to the Convention on International Civil Aviation. Fifth Edition, April 1962, amended up to Dec. 7th, 1972.

1-76 Anon.: Basic Glider Criteria Handbook - 1962 Revision. Federal Aviation Agency - Flight Standards Service, Washington D.C.

1-77 C.C. Jackson: "V_1, V_2 and all that ....". Flight Int. CCJ's Column, 1964-1966.

1-78 H. Slaughter and N.S. Dobi: "Changing Airworthiness Requirements for Air Taxi Operators and their Effect on Manufacturers of small Airplanes". SAE Paper No. 690320.

1-79 J.D. Harris: "Airworthiness Regulations - National and International". The Aeronautical Journal, Vol. 73, May 1969, pp. 453-459.

1-80 J.H.H. Grover: "Beechcraft 99 Performance". Flight Int. 25 Sept. 1969, p. 487.

1-81 W. Tye: "Airworthiness and the Air Registration Board". The Aeronautical Journal Vol. 74, Nov. 1970, pp. 873-887.

1-82 D.J. Wodraska: "The Gates Learjet 24C". Shell Aviation News. 389, 1970, pp. 14-19.

1-83 Anon.: Economic Regulations Part 298. Civil Aeronautics Board.

1-84 D.A. Kier: "Flight Comparison of Several Techniques for Determining the minimum Flying Speed for a Large, Subsonic Jet Transport". NASA TN D-5806, June 1970.

1-85 Anon.: OSTIV Airworthiness Requirements for Sailplanes. Organisation Scientifique et Technique Internationale du Vol à Voile - Sept. 1971.

1-86 Anon.: "Part 298 Contenders". Flight Int.,Vol. 103, No. 3352, 7 June 1973, p. 898.

1-87 Anon.: "FAA Orders Noise Limits". Business & Commercial Aviation, Dec. 1973, pp. 80-82.

1-88 Anon.: "FAA Airworthiness Regulation Review Set". Aviation Week & Space Technology, Vol. 100, No. 8, Feb. 25, 1974, p. 28.

CHAPTER 2.

General arrangement and configuration development

2-1 G.R. Edwards: "Problems in the development of a new aeroplane". J.R.Ae.S., Vol. 53, 1949, pp. 197-252.

2-2 M.L. Pennell: "Evolution of the Boeing jet tanker transport". Aeronautical Eng. Review, Aug. 1954, pp. 32-36.

2-3 Anon.: "Cessna studying radical business aircraft designs". Aviation Week and Space Technology, Nov. 6, 1961.

2-4 H. Wocke: "Ueberlegungen zur Entwicklung eines Geschäftsflugzeugs mit Strahlturbinenantrieb". Jahrbuch 1965 der WGLR, pp. 55-63.

2-5 D. Stinton: "The anatomy of the aeroplane". London, G.T. Foulis Co. Ltd., 1966.

2-6 K. Kens and H.J. Nowarra: "Die deutschen Flugzeuge 1933 - 1945". J.F. Lehmann Verlag, Munich.

2-7 S.D. Davies: "The history of the Avro Vulcan". The Aeronautical Journal, May 1970, pp. 350-364.

2-8 L.T. Goodmanson: "Transonic transports".12th Anglo-American Aeronautical Conference, 7th-9th July, 1971, Calgary.

2-9 W.C. Dietz: "Preliminary design aspects of design-to-cost for the YF-16 prototype fighter". AGARD CP No. 147, Vol 1, Oct. 1973.

2-10 W.C. Swan: "Design evolution of the Boeing 2707-300 supersonic transport". Part 1 - configuration development, aerodynamics, propulsion and structures. AGARD CP No. 147, Vol. 1, Oct. 1973.

Engine location

2-11 G.S. Schairer: "Pod mounting of jet engines". Fourth Anglo-American Aeronautical Conference, London, 1953, pp. 29-46.

2-12 B.T. Salmon: "High speed transport turbojet installation considerations". SAE Paper No. 85, April 1953.

2-13 E.S. Allwright: "Engineering features of rear engine installations in transport aircraft". Lecture for the R.Ae.S., Dec. 17, 1959, London.

2-14 D.J. Lambert: "Design of jet transports with rear-mounted engines". Aerospace Eng., Oct. 1960, pp. 30-35 and 72-74.

2-15 C. Dawson: "British push up performance of rear-engine airliners". Space-Aeronautics, Sept. 1960, pp. 52-54.

2-16 Anon.: "The design of modern pure jet transports with rear-mounted engines. Reviewed from engineering and operational aspects". ESSO World 60/61, page 35.

2-17 Anon.: "The Vickers VC-10". Aircraft Eng., Vol. 34, No. 400, June 1962.

2-18 R. Cabiac: "Voici pourquoi notre Caravelle est unique". Aviation Magazine, Feb. 1963, pp. 20-23.

2-19 J.E. Steiner: "The development of the Boeing 727". J.R.Ae.S., Vol. 67, Feb. 1963, pp. 103-110.

2-20 Anon.: "The Bac 1-11". Aircraft Eng., Vol. 35, No. 5, May 1963.

2-21 G.F. Mahony: "Balance of power". A slight dissertation how many engines and where to put them. Flight Int., June 29, 1967, pp. 1046-1047.

2-22 Laser: "How many engines, and where?". Flight Int., Aug. 8, 1968, pp. 206-207.

2-23 F.S. Hunter: "L-1011 S-duct versus DC-10's straight-through engine inlet" American Aviation, September 2, 1968, pp. 39-40.

2-24 W.C. Swan and A. Sigalla: "The problem of installing a modern high bypass engine on a twinjet transport aircraft". AGARD CP-124, April 1973.

Tailplane configuration

2-25 M. Duguet: "L'Aérodynamique des fuseaux réacteurs à l'arrière du fuselage". Proceedings of the Second European Aeronautical Congress, 1956.

2-26 G.M. Moss: "Some aerodynamic aspects of rear-mounted engines". J. of the Royal Aero. Soc., Dec. 1964, pp. 837-842.

2-27 Anon.: "Erfahrungen mit dem Ueberziehverhalten von Flugzeugen mit T-Leitwerk". Luftfahrttechnik – Raumfahrttechnik, Jan. 1965, pp. 19-22.

2-28 A.L. Byrnes: "Effect of horizontal stabilizer vertical location on the design of large transport aircraft". AIAA Paper No. 65-331.

2-29 D.J. Lambert: "A systematic study of the factors contributing to post-stall longitudinal stability of T-tail transport configuration". AIAA Paper No. 65-737.

2-30 R.S. Shevell and R.D. Schaufele: "Aerodynamic design features of the DC-9". AIAA Paper No. 65-738.

2-31 R.T. Taylor and E.J. Ray: "Factors affecting the stability of T-tail transports". Journal of Aircraft, Vol. 3, No. 4, July-Aug. 1966, pp. 359-364.

2-32 E.B. Trubshaw: "Low speed handling with special reference to the super stall". J. of the Royal Aero. Soc., Vol. 70, No. 667, July 1966, pp. 695-704.

2-33 D.P. Davies: "Handling the big jets". Published by the A.R.B., Second Ed., 1968. "The superstall", pp. 115-128.

2-34 D.A. Lovell: "A low speed wind-tunnel investigation of the tailplane effectiveness of a model representing the airbus type of aircraft". ARC R and M No. 3642, April 1969.

The undercarriage

2-35 R.M. Robbins: "Flight characteristics of the Boeing B-47 Stratojet" Aviation Week, 30 April 1951, pp. 25-31.

2-36 H.G. Conway: "Landing gear design". Textbook Royal Aeronautical Society, Chapmann & Hall Ltd., London, 1958.

Unconventional aircraft configurations

2-37 Anon.: "Konstruktionsbeispiele aus dem Flugzeugbau", Vol. 4: Fahrwerk. Fachbücher für Luft- und Raumfahrt. Luftfahrtverlag Walter Zuerl.

2-38 A. Cameron-Johnson: "The undercarriage in aeroplane project design". Aircraft Eng., Feb. 1969, pp. 6-11

2-39 J.K. Northrop: "The development of all-wing aircraft". J. of the Royal Aero. Soc. 1947, pp. 481-510.

2-40 G.H. Lee: "Tailless aircraft design problems". J. of the Royal Aero. Soc. 1947, pp. 109-131.

2-41 I.L. Ashkenas: "Range performance of turbojet aircraft". J. Aero. Sciences, Vol. 15, Feb. 1948.

2-42 J.H. Stevens: "The shape of the aeroplane", Chapter 13, Hutchinson, London, 1953.

2-43 H. Wocke: "Einige Ueberlegungen über die Anwendbarkeit der Entenanordnung bei Unterschallverkehrs-flugzeugen". Jahrbuch 1959 der WGLR, pp. 129-137.

2-44 D. Küchemann: "Aircraft shapes and their aerodynamics for flight at supersonic speeds". Proc. 2nd Int. Aeron. Congress, ICAS, Zürich, 1960.

2-45 G.H. Lee: "The possibilities of cost reduction with all-wing aircraft". J. of the Royal Aero. Soc., 1965, pp. 744-749.

2-46 H. Behrbohm: "Basic low speed aerodynamics of the short-coupled canard configuration of small aspect ratio". SAAB TN No. 60, July 1965.

CHAPTER 3.

3-1 J. Morris and D.M. Ashford: "Fuselage configuration studies". SAE Paper No. 670370, April 1967.

3-2 E.S. Krauss: "Die Formgebung von Rümpfen neuerer Verkehrsflugzeuge und ihr Einfluss auf die Wirt-schaftlichkeit im Flugbetrieb". Luft- und Raumfahrttechnik, May 1970, pp. 127-132.

3-3 A.A. Badiagin: "Concerning an efficient slenderness ratio for the fuselage of civilian aircraft". In: Methods of selection and approximate calculation of air design parameters. Trudy Inst., Moscow, 1961.

3-4 M.O. Wilmer: "Some aspects of optimisation in the design of civil aircraft". R.Ae.S. Two-day convention on economic factors in aviation, May 1970.

3-5 R.A. McFarland: "Human factors in air transport design". McGraw-Hill Book Company, Inc. 1946.

3-6 B.S. Shenstone: "Why airlines are hard to please". J. of the Royal Aero. Soc., May 1958, pp. 319-336.

3-7 R.A. McFarland: "Human body size and passenger vehicle design". SAE Paper No. 142A, Oct. 1962.

3-8 G. Nason: "Interior design and the airliner". The Architectural Review 140 (1966), pp. 413-422.

3-9 R.G. Mitchell: "Evaluation of economics of passenger comfort standards", SAWE Paper No. 338, 1962.

3-10 R. Maccabee: "Light aircraft design handbook". Loughborough University of Technology, TT 6801 Feb. 1970.

3-11 E.D. Keen: "Freighters - a general survey". J. of the Royal Aero. Soc., March 1959, pp. 135-152.

3-12 Anon.: "New aspects of air freight". Interavia, 1961, pp. 645-650.

3-13 C.A. Hangoe: "World-wide survey of cargo densities". SAWE Technical Paper No. 339, 1962.

3-14 A.H. Stratford: "Air cargo operational and economic problems". Chapter 4 of "Air transport economics in the supersonic era". MacMillan, London 1967.

3-15 M. Heinemann and M.A. Hiatt: "Quick-Change (QC) airplane systems: a prospective". J. of Aircraft. Jan.-Feb. 1967, pp. 42-47.

3-16 Several contributors: "Air logistics". Proceedings of the IAS National Midwestern Meeting, Tulsa, Oaklahoma U.S.A., Oct. 3-5, 1960.

3-17 J. Doetsch: "Neue Verfahren zum Verladen von Luftfrachtgütern". Luftfahrttechnik-Raumfahrttechnik, Feb. 1968, pp. 51-56.

3-18 J.E. Nichols and R.L. Meyers: "Design for quick turnaroud - payload system considerations". ASME
Conf. Proc. of the Annual Aviation and Space Conference, June 1968, pp. 373-377.

3-19 R.C. Hornbug, W.A. Alden and M. Newman: "Preliminary design of all-cargo aircraft". Ref. 3-18, pp.
365-372.

3-20 S.M. Levin: "Uncompromised cargo - the Mach 0.9 box". Space/Aeronautics, Oct. 1969, pp. 34-44.

3-21 J.B. Teeple, H.J. Bond and R.B. Sleight: "How to design a cockpit: from the man out". Aviation Age,
Jan. 1956.

3-22 D.W. Conover: "Cockpit landing visibility". SAE Paper No. 920 B, Oct. 1964.

3-23 Anon.: "Pilot visibility from the flight deck; design objectives for commercial aircraft". SAE AS
580 A.

3-24 Anon.: "Location and actuation of flight deck controls for commercial transport type aircraft". SAE
ARP 268 C, 1952, Rev. 1962.

3-25 T.R. Nettleton: "Handling qualities research in the development of a STOL utility transport aircraft".
Canadian Aero. and Space Journal, March 1966, pp. 93-104.

3-26 W.J. Rainbird, R.S. Crabbe, D.J. Peake and R.F. Meyer: "Some examples of separation in three-dimensional flows". Can. Aero. and Space Journal, Dec. 1966, pp. 409-423.

3-27 D.J. Peake: "Three-dimensional flow separations on up-swept rear fuselages". Can. Aero. and Space
Journal, Dec. 1969, pp. 399-408.

CHAPTER 4.

General literature and textbooks

4-1 Anon.: "Powerplant Review". A survey of the world's turbine engines. The Aeroplane, March 15, 1957,
pp. 360-365.

4-2 W.R. Hawthorne and W.T. Olson: "Design and performance of gas turbine power plants". High speed aerodynamics and jet propulsion, Vol. XI, Princeton Univ. Press, Princeton, 1959.

4-3 O.E. Lancaster (ed.): "Jet propulsion engines". High speed aerodynamics and jet propulsion, Vol. XII,
Princeton Univ. Press, Princeton, 1959.

4-4 M.J. Zucrow: "Aircraft and missile propulsion" (Vol. 1: Thermodynamics of fluid flow and application
to thermal engines; Vol. 2: The gas turbine power plants, turboprop, turbojet, ramjet and rocket engine). Wiley and Sons, New York, 1958.

4-5 H. Cohen, G.F.C. Rogers and H.I.H. Saravanamuttoo: "Gas turbine theory". Second Edition, Longman,
London, 1974.

4-6 W.J. Hesse and N.V.S. Mumford Jr.: "Jet propulsion for aerospace applications". Pitman Publishing
Corporation, 1964.

4-7 P.G. Hill and C.R. Peterson: "Mechanics and thermodynamics of propulsion". Addison-Wesley Publ. Cy.,
Reading, Massachusetts, 1965.

4-8 W. Thomson: "Thrust for flight". Pitman Publishing Cy., 1969.

4-9 I.E. Treager: "Aircraft gas turbine engine technology. McGraw-Hill Book Cy., 1970.

4-10 P.J. McMahon: "Aircraft Propulsion". Pitman Aeronautical Engineering Series, 1971.

RECIPROCATING ENGINES

4-11 A.W. Morley: "Performance of a piston-type aero-engine". London, Pitman and Sons Ltd., 1951.

4-12 J. Liston: "Powerplants for aircraft". McGraw-Hill Book Company, Inc. 1953.

4-13 F. Eisfeld: "Der Kolbenmotor als Antrieb für Leichtflugzeuge und Möglichkeiten seiner Verbesserung".
Jahrbuch 1965 der WGLR, pp. 327-336.

4-14 L. von Bonin and K. Grasmann: "Zukunftaussichten der Kolbenmotoren und Turbotriebwerke für Leichtflugzeuge". Jahrbuch 1965 der WGLR, pp. 337-352.

4-15 L.M. Yanda: "Correction of turbocharged engine performance to standard conditions and prediction of non-standard day performance". SAE Paper No. 690309, March 1969.

4-16 W.A. Wiseman: "Altitude performance with turbosupercharged light aircraft engines". SAE Paper No. 622A, Jan. 1963.

4-17 W.A. Wiseman and E.M. Ounsted: "Tiara light aircraft engines – A new generation". SAE Paper No. 700205.

4-18 J.L. Dooley: "Two-stroke light-aircraft engine potential". SAE Paper No. 670238.

4-19 D.L. Satchwell: "Performance characteristics of aircraft with turbo-supercharged engine and cabin – 1968". SAE Paper No. 680266.

4-20 D.E. Cole: "The Wankel engine". Scientific American, Aug. 1972, pp. 14-23.

4-21 C. Jones and H. Lamping: "Curtiss-Wright's development status of the stratified charge rotating combustion engine". SAE Paper No. 710582.

Turboprop engines

4-22 B. Pinkel and I.M. Karp: "A thermodynamic study of the turbine-propeller engine". NACA TN 2653, March 1952.

4-23 J. Szydlowski: "Bedeutung der Propellerturbine für das leichte 4- bis 6-sitzige Mehrzweckflugzeug". WGLR Jahrbuch 1960, pp. 289-296.

4-24 R.M. Sachs: "Small propeller turbines compared with other power plants in low speed flight applications". AIAA Paper No. 64-799, Oct. 1964.

4-25 E.P. Cockshutt and C.R. Sharp: "Gas turbine cycle calculations: design-point performance of turbopropeller and turboshaft engines". NRC Report LR-449, Nov. 1965.

4-26 E.P. Cockshutt: "Cycle calculations for turboprop engines". Can. Aeron. and Space Journal, Feb. 1968, Vol. 14, pp. 61-71.

4-27 E.J. Kordik: "Kleine Fluggasturbinen für Hubschrauber, Propeller- und Strahlflugzeuge". Jahrbuch 1969 der DGLR, pp. 194-234.

4-28 E.P. Neate and J.J. Petraits: "The Allison Model 250 engine: a case for the small turboprop". SAE Paper No. 7002-6.

Turbojet engines

4-29 M.S. Chapell and E.P. Cockshutt: "A comprehensive method for calculating turbojet and turbofan design-point performance". Can. Aeron. and Space Journal, June 1964, Vol. 10 No. 6.

4-30 S. Boudigues: "Défense et illustration des propulseurs pour avions rapides". Technique et Sciences Aéron. et Spatiales, July-Aug. 1964, pp. 271-285.

4-31 L.R. l'Anson: "The application of the high bypass turbofan for business and executive aircraft". SAE Paper No. 660221, April 1966.

4-32 C.L. Bagby and W.L. Andersen: "Effect of turbofan cycle variables on aircraft cruise performance". Journal of Aircraft, Sept.-Oct. 1966.

4-33 R.E. Neitzel and M.C. Hemsworth: "High-bypass turbofan cycles for long-range subsonic transports". Journal of Aircraft, Vol. 3, No. 4, July-Aug. 1966, pp. 354-358.

4-34 J.P. Armstrong and A.T. Jones: "The advantages of three-shaft turbofan engines for civil transport operation". The Aeronautical Journal, Vol. 73, Jan. 1969, pp. 25-33.

4-35 W.H. Sens and R.M. Meyer: "New generation engines - the engine manufacturers' outlook". SAE Paper No. 680278, April-May 1968.

4-36 E. Torenbeek: "Analytical method for computing turbo-engine performance at design and off-design conditions". Memorandum M-188, Delft University of Technology, Department of Aeronautical Engineering, Delft, Jan. 1973.

4-37 J.H. Horlock and J.F. Norbury: "The aero-thermodynamics of subsonic propulsion". Int. congress on subsonic aerodynamics, New York Acad. of Sciences, 22nd Nov. 1968, Vol. 154, No. 2, pp. 549-575.

4-38 B. Wrigley: "Engine performance considerations for the large subsonic transport". Von Karman Institute for Fluid Dynamics, Lecture Series 16, April 1969.

4-39 W. Dettmering and F. Fett: "Methoden der Schuberhohung und ihre Bewertung". Zeitschrift für Flugwiss., Aug. 1969, Vol. 19, No. 8, pp. 257-267.

4-40 R.H. Weir: "Propulsion prospects". The Aeronautical Journal, Vol. 73, Nov. 1969, pp. 923-934.

4-41 J. Wotton: "Engine development - where now?". Flight Int., Dec. 4, 1969, pp. 888-890.

4-42 R.M. Denning and J.A. Hooper: "Prospects for improvement in efficiency of flight-propulsion systems". AIAA Paper No. 70-873.

4-43 R.L. Cummings and H. Gold: "Low-cost engines for Aircraft". In: NASA SP-259, VII, Nov. 1970.

4-44 L.G. Dawson: "Propulsion". In: "The future of aeronautics", edited by J.E. Allen & J. Bruce, Hutchinson of London, 1970.

4-45 E.S. Taylor: "Evolution of the jet engine". Astronautics and Aeronautics, Nov. 1970, pp. 64-72.

4-46 H. Gevelhoff: "Ein Beitrag zur optimalen Auslegung von Installierten, Luftansangenden Strahltriebwerken". Dissertation, Dec. 1970, T.H. Aachen.

4-47 G. Rosen: "Prop-fan. A high thrust, low noise propulsor". SAE Paper No. 710470.

4-48 M.A. Beheim a.o.: "Subsonic and supersonic propulsion". In: Vehicle technology for civil aviation, NASA SP-292, Nov. 1971, pp. 107-156.

4-49 P. Alesi: "Des hautes températures devant turbine sur turboréacteurs et turbines à gaz". l'Aéronautique et l'Astronautique, No. 33, 1972-1, pp. 69-78.

4-50 R.W. Koenig and L.H. Fischback: "GENENG - A program for calculating design and off-design performance for turbojet and turbofan engines". NASA TN D-6552 and D-6553, Feb. 1972.

4-51 J.F. Dugan Jr.: "Engine selection for transport and combat aircraft". Von Karman Institute of Fluid Dynamics Lecture Series 49, April 1972 (NASA TM-X-68009).

4-52 D.G.M. Davies: "Variable pitch fans". Interavia, March 1972, pp. 241-243.

4-53 W.L. McIntire: "Engine and airplane - will it be a happy marriage?" SAWE Paper No. 910, May 1972.

4-54 M.J. Banzakein and S.B. Kazin: "The NASA/GE Quiet engine A". AIAA Paper No. 72-659.

4-55 E.A. White and G.L. Wilde: "Engines for civil V/STOL". The Aeronautical Journal, Oct. 1972, pp. 627-632.

CHAPTER 5.

General weight and performance "guesstimates"

5-1 I.H. Driggs: "Aircraft design analysis". J. of the Royal Aero. Soc., 1950, Vol. 54, pp. 65-116.

5-2 H.K. Millicer: "The design study". Flight, 17 Aug. 1951, pp. 201-205.

5-3 G. Backhaus: "Grundbeziehungen für den Entwurf optimaler Verkehrsflugzeuge". Jahrbuch der WGLR, 1958, pp. 201-213.

5-4 D. Küchemann and J. Weber: "Analysis of some performance aspects of various types of aircraft, designed to fly at different ranges and speeds". Progress in Aeronautical Sciences, Vol. 9, 1968, pp. 329-456.

5-5 R.A. Werner and G.F. Wislecinus: "Analysis of airplane design by similarity considerations". AIAA Paper No. 68-1017.

5-6 A. Krauss: "Vorhersage des Abfluggewichtes von Verkehrsflugzeugen". Luftfahrttechnik-Raumfahrttechnik, Feb. 1966, pp. 54.58.

5-7 E. Vallerani: "Evaluation of aircraft weight by the method of partial growth factors - case of passenger transport aircraft". Rivista di Ingegneria, Dec. 1967, pp. 989-1002 (in Italian).

5-8 W.E. Caddell: "On the use of aircraft density in preliminary design". SAWE Paper No. 813, May 1969.

5-9 W. Richter et al.: "Luftfahrzeuge". Das Fachwissen des Ingenieurs. Carl Hansen Verlag, München, 1970, section 4.2.4.

5-10 D. Howe: "The prediction of empty weight ratio and cruise performance of very large subsonic jet transport aircraft". Cranfield Report, Aero No. 3, 1971.

Initial drag prediction

5-11 H.G. Sheridan: "Aircraft preliminary design methods used in the weapon systems analysis division". NAVWEPS Report No. R-5-62-13, June 1962.

5-12 S.F. Hoerner: "Fluid-dynamic drag". Published by the author, 1965.

5-13 J.E. Linden and F.J. O'Brinski: "Some procedures for use in performance prediction of proposed aircraft designs". SAE Paper No. 650800.

5-14 B. McCluney and J. Marshall: "Drag development of the Belfast". Aircraft Engineering, Oct. 1967, pp. 33-37.

5-15 A.B. Haines: "Subsonic aircraft drag: an appreciation of present standards". The Aeronautical Journal, Vol. 72, No. 687, March 1968, pp. 253-266.

5-16 D.A. Norton: "Airplane drag prediction". Annals of the New York Academy of Sciences, Nov. 1968, Vol. 154, Art. 2, pp. 306-328.

Evaluation of performance requirements

5-17 B. Göthert: "Einflusz von Flächenbelastung, Flügelstreckung und Spannweitenbelastung auf die Flugleistungen". Luftfahrtforschung, Vol. 6, 1939, pp. 229-250.

5-18 W.B. Oswald: "Designing to the new CAA transport category performance requirements". SAE Quarterly Transactions, Vol. 1, No. 2, April 1947, pp. 321-333.

5-19 H.H. Cherry and A.B. Croshere Jr.: "An approach to the analytical design of aircraft". SAE Quarterly Transactions, Jan. 1948, Vol. 2, No. 1, pp. 12-18.

5-20 I.L. Ashkenas: "Range performance of turbojet airplanes". J. of the Aeronautical Sciences, Vol. 15, Feb. 1948, pp. 97-101.

5-21 D.H. Perry: "Exchange rates between some design variables for an aircraft just satisfying take-off distance and climb requirements". RAE Technical Report No. 69167, Aug. 1969.

5-22 H. Slaughter and N.S. Dobi: "Changing airworthiness requirements for air taxi operators and their effect on manufacturers of small airplanes". SAE Paper No. 690320.

5-23 J. Roskam and D.L. Kohlman: "An assessment of performance, stability and control improvements for general aviation aircraft". SAE Paper No. 700240.

5-24 D.O. Carpenter and P. Gotlieb: "The physics of short take-off and landing". AIAA Paper No. 70-1238.

5-25 R.H. Wild: "Format for flight". Flight Int., 11 March 1971, pp. 338-340, 350.

5-26 J.D. Raisbeck: "Consideration of application of currently available transport-category aerodynamic technology in the optimization of general aviation propeller-driven twin engine design". SAE Paper No. 720337.

5-27 E. Torenbeek: "An analytical expression for the balanced field length". AGARD Lecture Series No. 56, April 1972.

5-28 M.L. Gallay: "Flight of aircraft with partial and unbalanced thrust". NASA TT F-734, April 1973.

Aircraft synthesis and optimization

5-29 K.L. Sanders: "The optimum design of long-range aircraft". Aircraft Eng., April 1957, Vol. 29, No. 338, pp. 98-106.

5-30 H. Kimura, S. Kikuhura and J. Kondo: "Operations research in the basic design of the YS-11 transport airplane". Advances in Aeronautical Sciences, Vol. 3, pp. 557-574.

5-31 K.L. Sanders: "Aircraft Optimization". SAWE Paper No. 289, May 1961.

5-32 J.K. Wimpress and J.M. Swihart: "Influence of aerodynamic research on the performance of supersonic airplanes". J. of Aircraft, Vol. 1, No. 2, March-April 1964.

5-33 H. Pakendorf: "Zur Optimierung der Triebwerksanlage eines Kurzstreckenflugzeuges, unter besonderer Berücksichtigung hoher by-pass Verhältnisse". WGLR Jahrbuch 1966, pp. 261-270.

5-34 C.L. Bagly and W.L. Andersen: "Effect of turbofan cycle variables on aircraft cruise performance". J. of Aircraft, Sept.-Oct. 1966, Vol. 3, No. 5, pp. 385-389 (also AIAA Paper No. 65-796).

5-35 M.L. Olason and D.A. Norton: "Aerodynamic design philosophy of the Boeing 737". Journal of Aircraft, Nov.-Dec. 1966, Vol. 3, No. 6, pp. 524-528.

5-36 G.J. Schott Jr.: "Analytical propulsion research". Luchtvaarttechniek, J. of the Dutch Society of Engineers, Jan. 20, 1967.

5-37 G.H. Lee: "The aerodynamic design philosophy of the Handley Page Jetstream". Aircraft Engineering, Sept. 1967, pp. 10-21.

5-38 V.A. Lee, H.G. Ball and E.A. Wadsworth: "Computerized aircraft synthesis". J. of Aircraft, Sept.-Oct. 1967, Vol. 4, No. 5, pp. 402-408.

5-39 Z.M. v. Krzywoblocki and W.Z. Stepniewski: "Application of optimization techniques to the design and operation of V/STOL aircraft". International Congress on Subsonic Aeronautics, Nov. 1968, Annals of the New York Academy of Sciences, Vol. 154, Art. 2, pp. 982-1013.

5-40 D. Ryle: "Parametric studies in aircraft design", AGARD Lecture Series No. 16, April 1969.

5-41 B.H. Little: "Advanced computer technology". AGARD Lecture Series No. 16, April 1969.

5-42 D.D. Hufford, J.A. Ross and K.W. Hoefs: "The economics of subsonic transport airplane design, evaluation and operation". SAE Paper No. 710423.

5-43 R.E. Wallace: "Parametric and optimisation techniques for airplane design synthesis". AGARD Lecture Series No. 56, April 1972.

5-44 D.L.I. Kirkpatrick: "Review of two methods of optimising aircraft design". AGARD Lecture Series No. 56, April 1972.

5-45 R.E. Wallace: "A computerized system for the preliminary design of commercial airplanes". Document of the Boeing Company, Commercial Airplane Group, Seattle, Washington.

5-46 R.R. Heldenfels: "Integrated, computer-aided design of aircraft". 43rd AGARD Flight Mech. Panel Symposium, Florence, Oct. 1973.

5-47 D.L.I. Kirkpatrick and M.J. Larcombe: "Initial design optimization on civil and military aircraft". 43rd AGARD Flight Mechanics Panel Symposium, Florence, Oct. 1973.

Community noise considerations

5-48 Anon.: "Definitions and procedures for computing the perceived noise level of aircraft noise". Aerospace Recommended Practice 865, SAE, 15th Oct. 1964.

5-49 F.B. Greatrex and R. Bridge: "The evaluation of the engine noise problem". Aircraft Engineering, Feb. 1967, pp. 6-10.

5-50 E.J. Richards: "Aircraft noise-mitigating the nuisance". Aircraft Eng., Feb. 1967, pp. 11-18.

5-51 H. Drell: "Impact of noise on subsonic transport design". SAE Paper No. 700806.

5-52 R.E. Russell and J.D. Kester: "Aircraft noise, its source and reduction". SAE Paper No. 710308.

5-53 J.F. Dugan: "Engine selection for transport and combat aircraft". AGARD Lecture Series No. 56, April 1972 (NASA TM-X-68009).

5-54 L.G. Dawson and T.D. Sills: "An end to aircraft noise?". The Aeronautical Journal, May 1972, pp. 286-297.

5-55 Anon.: "The future of short-haul air transport within Western Europe". Report of the Netherlands V/STOL Working Group, June 1973, NLR, Amsterdam.

5-56 D. Howe: "The weight, economic and noise penalties of short haul transport aircraft resulting from the reduction of balanced field length". Cranfield Institute of Technology Report Aero. No. 24, Jan. 1974.

CHAPTER 6.

Choice of the number and type of engine(s)

6-1 S.J. Moyes and W.A. Pennington: "The influence of size on the performance of turbojet engines". Third Anglo-American Aeronautical Conference, Brighton, 1951, pp. 545-562.

6-2 A.E. Russell: "The choice of power units for civil aeroplanes". Journal of the Royal Aero. Soc., August 1954, pp. 523-539.

6-3 F.H. Keast: "Big engines or small?". The Aeroplane, August 10, 1956, pp. 189-191.

6-4 W.A. Pennington: "Choice of engines for aircraft". Shell Aviation News, Jan. 1959, pp. 14-19.

6-5 N. Scholz and G. Preininger: "Einige Gesichtspunkte zur Frage der günstigsten Schubklasse von TL-Triewerken". Luftfahrttechnik 8, No. 12, December 1962, pp. 327-332.

6-6 D.R. Newman: "The De Havilland 121 Trident". Aircraft Engng., May 1962, pp. 149-152.

6-7 R.W. Higgins: "The choice between one engine or two". Aircraft Engng., Nov. 1962, pp. 19-21.

6-8 J. Karran: "Airbus, 2, 3 or 4 engines?". The Aeroplane, Nov. 3, 1966, pp. 4-5.

6-9 G.F. Mahoney: "Balance of power". Flight Int., June 29, 1967, pp. 1046-1047.

6-10 Laser: "How many engines and where?". Flight Int., August 8, 1968, pp. 206-207.

6-11 Anon.: "Two or three engines?". Flight Int., Sept. 1969, pp. 446-447.

6-12 N.L. Gallay: "Flight of aircraft with partial and unbalanced thrust". NASA TT F-734, April 1973.

Propellers

6-13 H. Glauert: "The elements of aerofoil and airscrew theory". McMillan, New York, 1943.

6-14 G.C.I. Gardiner and J. Mullin: "The design of propellers". Journal of the Royal Aero. Soc., Vol. 53, 1949, pp. 745-762.

6-15 J.L. Crigler and R.E. Jaquis: "Propeller-efficiency charts for light airplanes". NACA TN 1338, 1947.

6-16 Anon.: "SBAC Standard method of propeller performance estimation". Society of British Aircraft Constructors Ltd., April 1950.

6-17 J. Gilman Jr.: "Propeller performance charts for transport airplanes". NACA TN 2966, 1953.

6-18 Anon.: "Aircraft Propeller Handbook". ANC-9 Bulletin, Sept. 1956.

6-19 Anon.: "Choice of propellers for turbine engines in the medium power range". Rotol Ltd. Performance Office Report No. 1104, Issue II, July 1959.

6-20 G. Rosen: "New problem areas in aircraft propeller design". Can. Aero. Journal, June 1960, pp. 213-220.

6-21 Anon.: "Generalized method of propeller performance estimation". Hamilton Standard Publication PDB 6101A, 1963.

6-22 K.D. Wood: "Aerospace vehicle design". Vol. 1: Aircraft Design, 2nd. edition. Johnson Publishing Company, Boulder, Colorado, 1966.

6-23 D.P. Currie: "Propeller design considerations for turbine powered aircraft". SAE Paper No. 680227.

6-24 G. Rosen and W.M. Adamson: "Next generation V/STOL propellers". SAE Paper No. 680281.

6-25 G. Rosen and C. Rohrbach: "The quiet propeller - A new potential". AIAA Paper No. 69-1038.

6-26 R. Worobel: "Computer program user's manual for advanced general aviation propeller study". NASA CR-2066, May 1972.

Powerplant installation

6-27 J. Seddon: "Air intakes for gas turbines". J. Royal Aero. Soc., Oct. 1952, pp. 747-781.

6-28 D. Küchemann and J. Weber: "Aerodynamics of propulsion". McGraw-Hill Book Cy., 1953, New York.

6-29 G.S. Schairer: "Pod mounting of jet engines". Fourth Anglo-American Aeronautical Conference, London, 1953, pp. 29-46.

6-30 G. Schulz: "Aerodynamische Regeln für den Einbau von Strahltriebwerksgondeln". Z. für Flugwiss. 3

(1955), Vol. 5, pp. 119-129.

6-31 J.T. Kutney and S.P. Piszkin: "Reduction of drag rise on the Convair 990 aircraft". J. of Aircraft, January 1964, Vol. 1, p. 8.

6-32 R.S. Shevell and R.D. Schaufele: "Aerodynamic design features of the DC-9". AIAA Paper No. 65-738.

6-33 M.L. Olason and D.A. Norton: "Aerodynamic design philosophy of the Boeing 737". AIAA Paper No. 65-739.

6-34 R.T. Taylor and E.J. Ray: "Factors affecting the stability of T-tail transports". J. of Aircraft, Vol. 3 no. 4, July/August 1966, pp. 359-364.

6-35 W.M. Magruder: "Development of requirement, configuration and design for the Lockheed 1011 jet transport". SAE Paper No. 680688.

6-36 K.S. Lawson: "The influence of the engine on aircraft design". Royal Aero. Soc., Symposium lecture on V/STOL design, London, 1969.

6-37 J.A. Bagley: "Some aerodynamica problems of powerplant installation on swept-winged aircraft". Royal Aeron. Soc., Symposium lecture on V/STOL design, London, 1969.

6-38 D.D. Hufford, J.A. Ross and K.W. Hoefs: "The economics of subsonic transport airplane design, evaluation and operation". SAE Paper No. 710423.

6-39 P.G.R. Williams and D.J. Stewart: "The complex aerodynamic interference pattern due to rear fuselage mounted powerplants". AGARD Conference Proceedings CP-71.

6-40 W.C. Swan and A. Sigalla: "The problem of installing a modern high bypass engine on a twinjet transport aircraft". AGARD CP-124, April 1973.

6-41 J. Barche: "Beitrag zum Interferenzproblem von über den Tragflügeln angeordneten Triebwerken". DGLR Annual Conference, 1969.

6-42 H. Griem, J. Barche, H.J. Beisenherz and G. Krenz: "Some low speed aspects of the twin-engine short haul aircraft VFW 614". AGARD-CP-160, Takeoff and landing, April 1974.

Thrust reversal

6-43 J.C. Pickard: "Thrust reversers for jet aircraft". Aerospace Engineering, Jan. 1961.

6-44 M.J. Green: "Rolls-Royce thrust reversers-compatibility". SAE Paper No. 69040.

6-45 J.S Mount and D.W.R. Lawson: "Developing, qualifying and operating business jet thrust reversers". SAE Paper No. 690311.

6-46 S.K. Wood and P.E. Mc.Coy: "Design and control of the 747 exhaust reverser system". SAE Paper No. 690409.

6-47 D.J. Lennard: "Design features of the CF-6 engine thrust reverser and spoiler". SAE Paper No. 690411.

6-48 Anon.: "Approximate estimation of braking thrust of propellers (piston engines)." R.Ae.S. Data Sheets "Performance", Sheet ED 1/2, Feb. 1963.

Auxiliary Power Units

6-49 J. Wotton: "Integrated auxiliary power". Flight Int., 8 August 1968, pp. 210-212.

6-50 B.H. Nicholls and A.D. Meshew: "Auxiliary power systems for 1975 fighter aircraft". SAE Paper No. 680311.

6-51 W.E. Arndt: "Secondary power requirements for large transport aircraft". SAE Paper No. 680708.

6-52 H.S. Keltto: "Installation of the APU in the 747 airplane". SAE Paper No. 680709.

6-53 P.G. Stein: "The Hamilton Standard APU for Lockheed's Tri Star Airliner". SAE Paper No. 700814.

6-54 D.A. Malohn: "TSCP 700 Auxiliary Power Unit for the DC-10 Aircraft". SAE Paper No. 700815.

6-55 L.H. Allen: "APU selection - An airline viewpoint". SAE Paper No. 700816.

6-56 W.P. Frey and E. Schnell: "Auxiliary power units for secondary power systems". AGARD-CP-104, June 1972.

6-57 C.H. Paul: "The development of small gas turbines for aircraft auxiliary power". The Aeronautical Journal, Oct. 1970, pp. 797-805.

CHAPTER 7.

Textbooks

7-1 A.F. Donovan and H.R. Lawrence: "Aerodynamic components of aircraft at high speeds". Volume VII of "High speed aerodynamics and jet propulsion". Princeton University Press, 1957.

7-2 F.W. Riegels: "Aerodynamische Profile". Oldenburg, 1958.

7-3 H. Schlichting and E. Truckenbrodt: "Aerodynamik des Flugzeuges". (Part II). Springer-Verlag, Berlin, 1960.

7-4 G. Corning: "Supersonic and subsonic airplane design". Published by the author. Second Edition, 1964.

7-5 I.H. Abbott and A.E. von Doenhoff: "Theory of wing sections". Dover Publications Inc., New York, 1958.

7-6 A.C. Kermode: "The aeroplane structure". Pitman Publishing Cy., Second edition, 1964.

7-7 D.O. Dommasch, S.S. Sherbey and T.F. Connolly: "Airplane aerodynamics". Pitman Publishing Corp., New York, Fourth Edition, 1967.

7-8 D.P. Davies: "Handling the big jets". Air Registration Board, Second edition, May 1968.

Wing design philosophy and development

7-9 W.S. Farren: "The aerodynamic art". J. of the Royal Aeron. Society, Vol. 60, July 1956, pp. 431-449.

7-10 J.A. May: "Aerodynamic design of the Vickers VC-10". Aircraft Eng., June 1962, pp. 158-164.

7-11 B.J. Prior: "Aerodynamic design of the BAC 1-11". Aircraft Eng., May 1963, pp. 149-152.

7-12 J.K. Wimpress and J.M. Swihart: "Influence of aerodynamic research on the performance of supersonic airplanes". J. of Aircraft, Vol. 1, No. 2, March-April 1964.

7-13 R.S. Shevell and R.D. Schaufele: "Aerodynamic design features of the DC-9". AIAA Paper No. 65-738.

7-14 M.L. Olason and D.A. Norton: "Aerodynamic design philosophy of the Boeing 737". AIAA Paper No. 65-739.

7-15 J.H.D. Blom: "Fokker F-28 evolution and design philosophy - aerodynamic design and aeroelasticity". Aircraft Eng., June 1967, pp. 17-21.

7-16 R.D. Schaufele and A.W. Ebeling: "Aerodynamic design of the DC-9 wing and high lift system". SAE Paper No. 670846, 1967.

7-17 H. Herb: "Zur Entwicklungsgeschichte der Flugmechanik". DLR Forschungsbericht 67-44, July 1967.

7-18 G.H. Lee: "Aerodynamic considerations of a medium Mach business aeroplane". SAE Paper No. 670244, April 1967.

7-19 J.H. Paterson: "Aerodynamic design features of the C-5A". SAE Paper No. 670847.

7-20 D. Ryle: "Wing design, body design, high lift systems and flying qualities". AGARD-VKI Lecture Series 16, April 21-25, 1969.

7-21 C.H. Hurkamp, W.M. Johnston and J.H. Wilson: "Technology assessment of advanced general aviation aircraft". NASA CR-114339, June 1971.

7-22 D. Küchemann: "Aerodynamic design". The Aeron. Journal, Feb. 1969, pp. 101-110.

7-23 J.M. Swihart (ed.): "Jet transport design". AIAA Selected Reprints/Volume VIII, Nov. 1969.

7-24 R.E. Bates: "Progress on the DC-10 development program". AIAA Paper No. 69-830.

7-25 T.I. Ligum: "Aerodynamics and flight dynamics of turbojet aircraft". NASA Technical Translation TT F-542.

7-26 J.D. Raisbeck: "Consideration of application of currently available transport-category aerodynamic technology in the optimization of general aviation propeller-driven twin design". SAE Paper No. 720337.

7-27 A.L. Byrnes Jr.: "Aerodynamic design and development of the Lockheed S-3A Viking". AIAA Paper No. 72-746.

7-28 D.M. McRae : "The aerodynamic development of the wing of the A-300B". The Aero. Journal, July 1973, pp. 367-379.

7-29 H. Wittenberg: "Some considerations on the design of very large aircraft". Jaarboek 1974 of the Netherlands Association of Aeronautical Engineers (paper No. 3).

Wings for high-speed aircraft

7-30 W.A. Waterton: "Some aspects of high performance jet aircraft". J. of the Royal Aeron. Soc., June 1953, pp. 375-390.

7-31 K.E. Van Every: "An engineering comparison using straight, swept, and delta wings". Interavia, Vol. 8, No. 1, 1953, pp. 23-27. Followed by various other contributions on pp. 27-35.

7-32 C.L. Johnson: "Airplane configurations for high speed flight". Interavia, Vol. 9, No. 1, 1954, pp. 47-51.

7-33 E.W.E. Rogers and J.M. Hall: "An introduction to the flow about plane swept-back wings at transonic speeds". J. of the Royal Aeron. Soc., Vol. 64, August 1960.

7-34 R.C. Lock and E.W.E. Rogers: "Aerodynamic design of swept wings and bodies". Advances in Aeron. Sciences, Vol. 3. Pergamon Press, London.

7-35 J.A. Bagley: "Aerodynamic principals for the design of swept wings". Progress in Aeronautical Sciences, Vol. 3. Pergamon Press, London.

7-36 R.C. Lock: "The aerodynamic design of swept winged aircraft at transonic and supersonic speeds". J. of the Royal Aeron. Soc., Vol. 67, No. 6, June 1963, pp. 325-337.

7-37 K.G. Hecks: "The high-speed shape". Flight Int., Jan. 2 1964, pp. 13-18.

7-38 A.B. Haines: "Recent research into some aerodynamic design problems of subsonic transport aircraft". ICAS Paper No. 68-10, Sept. 1968.

7-39 T.G. Ayers: "Supercritical aerodynamics worthwhile over a range of speeds". Astronautics and Aeronautics, August 1972, pp. 32-36.

Discussions of shape parameters

7-40 H.H. Cherry and A.B. Croshere Jr.: "An approach to the analytical design of aircraft". SAE Quarterly Transactions, Vol. 2, No. 1, Jan. 1948, pp. 12-18.

7-41 I.L. Ashkenas: "Range performance of turbojet airplanes". J. of the Aeron. Sciences, Vol. 15, Feb. 1948, pp. 97-101.

7-42 G. Gabrielli: "A method for determining the wing area and its aspect ratio in aircraft design". Monografie del Laboratorio di Aeron. Politecnico di Torino, No. 294.

7-43 G. Backhaus: "Grunbeziehungen für den Entwurf optimaler Verkehrsflugzeuge". Jahrbuch 1958 der WGL, pp. 201-213.

7-44 W. Lehmann: "Wahl der Profildicke und Flügelpfeilung bei Verkehrsflugzeugen". Luftfahrttechnik, Nov. 1961, pp. 323-326.

7-45 K.L. Sanders: "Aircraft optimization". SAWE Paper No. 289, 1961.

7-46 D. Küchemann and J. Weber: "An analysis of some performance aspects of various types of aircraft designed to fly over different ranges at different speeds". Progress in Aeron. Sciences, Vol. 9, pp. 324-456, 1968.

7-47 J. Roskam and D.L. Kohlmann: "An assessment of performance, stability and control improvements for general aviation aircraft". SAE Paper No. 700240.

7-48 R.A. Cole: "Exploiting AR". Shell Aviation News, 1970, No. 387, pp. 10-15.

7-49 K.H. Bergley: "Debating A.R. (aspect ratio)". Shell Aviation News, 1971, No. 398, pp. 14-18.

7-50 D.L.I. Kirkpatrick: "Review of two methods of optimising aircraft design". AGARD Lecture Series No. 56 (Paper 12), April 1972.

Airfoil sections

7-51 I.H. Abbott, A.E. von Doenhoff and L.S. Stivers: "Summary of airfoil data". NACA Report 824, 1945.

7-52 L.K. Loftin: "Theoretical and experimental data for a number of NACA 6A-Series airfoil sections". NACA TR No. 983, 1948.

7-53 Anon.: "Critical Mach number for high-speed aerofoil sections". Royal Aero. Soc. Data Sheets, Aerodynamics Sub. Series, Vol. 2 "Wings", Sheet 00.03.01, April 1953.

7-54 B.N. Daley and R.S. Dick: "Effect of thickness, camber and thickness distribution on airfoil characteristics at Mach numbers up to 1.0". NACA TN 3607, 1956.

7-55 A.B. Haines: "Wing section design for sweptback wings at transonic speeds". J. of the Royal Aero. Soc., Vol. 61, April 1957, pp. 238-244.

7-56 H.H. Pearcy: "The aerodynamic design of section shapes for swept wings". Second Int. Congress of the Aeron. Sciences, Zürich, 1960. Advances in Aeron. Sciences, Vol. 3, pp. 277-322. Pergamon Press.

7-57 Anon.: "A method of estimating the drag-rise Mach number for two-dimensional airfoil sections". Transonic Data Memorandum 6407, Royal Aero. Soc., 1965.

7-58 D.W. Holder: "Transonic flow past two-dimensional aerofoils". J.R.Ae.S. Vol. 68, August 1964, pp. 501-516.

7-59 J.W. Boerstoel: "A survey of symmetrical transonic potential flow around quasi-elliptical airfoil sections". NLR-TR 136, Jan. 1967.

7-60 L.R. Wootton: "Effect of compressibility on the maximum lift coefficient of airfoils at subsonic speeds". J. of the Royal Aero. Soc., Vol. 71, July 1967.

7-61 G.Y. Nieuwland: "Transonic potential flow around a family of quasi-elliptical airfoil sections". NLR Report TRT 172, 1967.

7-62 E.C. Polhamus: "Subsonic and transonic aerodynamic research". NASA SP-292, 1971.

7-63 R.H. Liebeck and A.I. Ormsbee: "Optimization of airfoils for maximum lift". J. of Aircraft, Vol. 5, Sept./Oct. 1970, pp. 409-415.

7-64 G.J. Bingham and K.W. Noonan: "Low-speed aerodynamic characteristics of NACA 6716 and NACA 4416 airfoils with 35-percent-chord single-slotted flaps". NASA TN X-2623, May 1974.

7-65 R.T. Whitcomb and L.R. Clark: "An airfoil shape for efficient flight at supercritical Mach numbers". NASA TM X-1109, July 1965.

7-66 R.J. McGhee and W.D. Beasly: "Low-speed characteristics of a 17-percent-thick airfoil section designed for general aviation application". NASA TN D-7428.

7-67 W.H. Wentz, Jr.: "New airfoil sections for general aviation aircraft". SAE Paper No. 730876.

7-68 J.J. Kacprzynski: "Drag of supercritical airfoils in transonic flow". AGARD Conference Proceedings No. CP-124, April 1973.

7-69 W.E. Palmer: "Thick-wing flight demonstrations". SAE Paper No. 720320.

Low-speed stalling

7-70 R.F. Anderson: "Determination of the characteristics of tapered wings". NACA Report 572, 1936.

7-71 H.A. Soulé and R.F. Anderson: "Design charts relating to the stalling of tapered wings". NACA Report 703, 1940.

7-72 H.H. Sweberg and R.C. Dingeldein: "Summary of measurements in Langley full-scale tunnel of maximum lift coefficients and stalling characteristics of airplanes". NACA Report No. 829.

7-73 W.H. Philips: "Appreciation and prediction of flying qualities". NACA Report 927, 1949.

7-74 G.B. McCullough and D.E. Gault: "Examples of three representative types of airfoil-section stall at low speed". NACA Technical Note 2502, 1951.

7-75 J.A. Zalovchick: "Summary of stall warning devices". NACA TN 2676, 1952.

7-76 D. Küchemann: "Types of flow on swept wings". J. Roy. Aero. Soc., Vol. 57, pp. 683-699, Nov. 1953.

7-77 A.D. Young and H.B. Squire: "A review of some stalling research", with an appendix on "Wing sections and their stalling characteristics". ARC R and M 2609, 1951.

7-78 H.O. Palme: "Summary of stalling characteristics and maximum lift of wings at low speeds". SAAB TN 15,

April 1953.

7-79 G.G. Brebner: "The design of swept wing planforms to improve tip-stalling characteristics". RAE Report Aero 2520 (ARC 17624), July 1954.

7-80 J. Black: "Flow studies of the leading edge stall on a swept-back wing at high incidence". J. of the Royal Aeron. Soc., Vol. 60, Jan. 1956, pp. 51-60.

7-81 D.E. Gault: "A correlation of low-speed, airfoil-section stalling characteristics with Reynolds number and airfoil geometry". NACA TN 3963, 1957.

7-82 G.C. Furlong and J.G. McHugh: "A summary and analysis of the low speed longitudinal characteristics of swept wings at high Reynolds number". NACA Report 1339, 1957.

7-83 R.L. Maki: "The use of two-dimensional data to estimate the low-speed wing lift coefficient at which section stall first appears on a swept wing". NACA RM A51E15, 1951.

7-84 K.P. Spreeman: "Design guide for pitch-up evaluation and investigation at high-subsonic speeds of possible limitations due to wing-aspect-ratio variations". NACA TM X-26, 1959.

7-85 J.G. Wimpenny: "Low speed stalling characteristics". AGARD Report 356, 1961.

7-86 A. Spence and D. Lean: "Some low-speed problems of high speed aircraft". J. of the Royal Aeron. Soc., Vol. 66 No. 616, April 1962, pp. 211-225.

7-87 J. Fletcher: "The stall game". Shell Aviation News, 1966, Number 341, pp. 16-19.

7-88 C.L. Bore and A.T. Boyd: "Estimation of maximum lift of swept wings at low Mach numbers". J. of the Royal Aeron. Soc., Vol. 67, April 1963, pp. 227-239.

7-89 C.W. Harper and R.L. Maki: "A review of the stall characteristics of swept wings". NASA TN D-2373, July 1964.

7-90 Ph. Poisson-Quinton and E. Erlich: "Analyse de la stabilité et du controle d'un avion au dela de sa portance maximale". AGARD Conference Proceedings No. CP-17, 1966.

7-91 H.B.M. Thomas: "A study of the longitudinal behaviour of an aircraft at near-stall and post-stall conditions". AGARD Conference Proceedings No. CP-17, 1966.

7-92 P.D. Chappell: "Flow separations and stall characteristics of plane, constant-section wings in subcritical flow". J. of the Royal Aeron. Soc., Vol. 72, 1968, pp. 82-90.

7-93 D. Isaacs: "Wind tunnel measurements of the low speed stalling characteristics of a model of the Hawker-Siddeley Trident 1C". ARC R and M Report No. 3608, May 1968.

7-94 B. van den Berg: "Reynolds number and Mach number effects on the maximum lift and the stalling characteristics of wings at low speeds". National Aerospace Laboratory NLR TR 69025 U.

7-95 G.J. Hancock: "Problems of aircraft behaviour at high angles of attack". AGARDograph 136, 1969.

7-96 M.A. McVeigh and E. Kisielowski: "A design summary of stall characteristics of straight wing aircraft". NACA CR-1646, June 1971.

7-97 D.N. Foster: "The low-speed stalling of wings with high-lift devices". AGARD Conference Proceedings CP-102 (Paper 11), April 1972.

7-98 J.K. Wimpress: "Predicting the low speed stall characteristics of the Boeing 747". AGARD Conference Proceedings CP-102 (Paper No. 21), Nov. 1972.

7-99 T. Schuringa: "Aerodynamics of the wing stall of the Fokker F-28". AGARD Conference Proceedings CP-102 (Paper 20), Nov. 1972.

7-100 W.D. Horsfield and G.P. Wilson: "Flight development of the stalling characteristics of a military trainer aircraft". AGARD Conference Proceedings CP-102, Nov. 1972.

7-101 V.E. Lockwood: "Effect of Reynolds number and engine nacelles on the stalling characteristics of a twin-engine light airplane". NASA TN D-7109, Dec. 1972.

7-102 H. Griem, J. Barche, H.J. Beisenherz and G. Krenz: "Some low-speed aspects of the twin-engine short haul transport VFW 614". AGARD CP No. 160 (Paper 11).

7-103 C.R. Taylor: "Aircraft stalling and buffeting - introduction and overview". AGARD Lecture Series No. LS-74, March 1975.

7-104 W. McIntosh and J.K. Wimpress: "Prediction and analysis of the low speed stall characteristics of the Boeing 747". AGARD Lecuture Series No. LS-74 (Paper 3), March 1975.

High-lift devices

7-105 M.A. Garbell: "Wing flaps in light aircraft design". J. of the Aeron. Sciences, Jan. 1945, pp. 14-20.

7-106 H.O. Palme: "Summary of wind tunnel data for high-lift devices on swept wings". SAAB T. Note 16, April 1953.

7-107 J.F. Cahill: "Summary of section data on trailing edge high-lift devices". NACA TR 938, 1949.

7-108 R. Duddy: "High lift devices and their uses". J. Roy. Aero. Soc., Oct. 1949, Vol. 53, pp. 859-900.

7-109 A.D. Young: "The aerodynamic characteristics of flaps". ARC R and M 2622, 1953.

7-110 A.D. Young: "Flaps for landing and take-off". Chapter 14 of: "The principles of the control and stability of aircraft" edited by W.J. Duncan. Cambridge Aeronautical Series.

7-111 G.H. Lee: "High Maximum Lift". The Aeroplane, Oct. 30, 1953.

7-112 T.R.F. Nonweiler: "Flaps, slots and other high-lift aids". Aircraft Engineering, Sept. 1955.

7-113 K.L. Sanders: "High-lift devices, a weight and performance trade-off methodology". SAWE Technical Paper No. 761, May 1963.

7-114 S.T. Harvey and D.A. Norton: "Development of the Model 727 Airplane high lift system". Society of Automotive Engineers S 408, April 1964.

7-115 Anon.: "High lift devices for short field performance". Interavia No. 4/1964, pp. 569-572.

7-116 J.K. Wimpress: "Shortening the take-off and landing distances of high speed aircraft". AGARD Report 501, June 1965.

7-117 R.D. Schaufele and A.W. Ebeling: "Aerodynamic design of the DC-9 wing and high-lift system". SAE Paper No. 670846.

7-118 J.H. Paterson: "Aerodynamic design features of the C-5A". Aircraft Engineering, June 1968, pp. 8-15.

7-119 J.C. Wimpenny: "The design and application of high-lift devices". Annals of the New York Academy of Sciences, Vol. 154, Art. 2, pp. 329-366.

7-120 J.K. Wimpress: "Aerodynamic technology applied to takeoff and landing". Annals of the New York Academy of Sciences, Vol. 154, Art. 2, pp. 962-981.

7-121 A.D. Hammond: "High-lift aerodynamics". Proceedings of Conference on vehicle technology for civil aviation, NASA SP-292, 1971, pp. 15-26.

7-122 M.A. McVeigh and E. Kisielowski: "A design summary of stall characteristics of straight wing aircraft". NACA CR-1646, June 1971.

7-123 J.A. Thain: "Reynolds number effects at low speeds on the maximum lift of two-dimensional aerofoil-sections equipped with mechanical high lift devices". Nat . Res. Council of Can. Quart. Bull. of the Div. of Mech. Eng. and the N.A.E., 30 Sept. 1973, pp. 1-24.

7-124 R.J. Margason and H.L. Morgan Jr.: "High-lift aerodynamics - trends, trades and options". AGARD Conference on Takeoff and Landing, CP No. 160, April 1974.

7-125 A.M.O. Smith: "High-lift aerodynamics". 37th Wright Brothers lecture, AIAA Paper No. 74-939.

Buffeting and high-speed stalling

7-126 C.J. Wood: "Transonic buffeting on aerofoils". J. of the Royal Aeron. Soc., Vol. 64 No. 599, Nov. 1960, pp. 683-686.

7-127 W.R. Burris and D.E. Hutchins: "Effect of wing leading edge geometry on maneuvering boundaries and stall departure". AIAA Paper No. 70-904.

7-128 E.J. Ray and L.W. McKinney: "Maneuver and buffet characteristics of fighter aircraft". AGARD Conference Proceedings CP-102, Nov. 1972 (Paper 24).

7-129 G.F. Moss, A.B. Haines and R. Jordan: "The effect of leading edge geometry on high speed stalling". AGARD Conference Proceedings CP-102, Nov. 1972.

7-130 H.H.B.M. Thomas: "On problems of flight over an extended angle of attack range". The Aero. Journal, Vol. 77, Aug. 1973, pp. 412-423.

7-131 D.G. Mabey: "Beyond the buffet boundary". The Aero. Journal, April 1973, pp. 201-215.

7-132 H. John: "Critical review of methods to predict the buffet penetration capability of aircraft".
    AGARD Lecture Series No. 74, March 1975.

Wing structure

7-133 S.J. Pipitone: "Modern wing structures". Aircraft Production, Jan. 1950, pp. 27-32.

7-134 H.H. Gardner: "Structural problems of advanced aircraft". Journal of the Royal Aeron. Soc., April
    1952, pp. 221-250.

7-135 E.D. Keen: "Integral construction". J. of the Royal Aeron. Soc., April 1953, pp. 215-227.

7-136 O. Ljungström: "Wing structures of future aircraft". Aircraft Eng., May 1953, pp. 128-132.

7-137 G.H. Lee: "The aerodynamic and aeroelastic characteristics of the crescent wing". J. of the Royal
    Aeron. Soc., Vol. 59 No. 529, Jan. 1955, pp. 37-44.

7-138 E.G. Broadbent: "Aeroelastic problems in connection with high-speed flight". J. of the Royal Aeron.
    Soc., Vol. 60, July 1956, pp. 459-475.

7-139 A.F. Newell: "Impressions of the structural design of American civil aircraft". The Aeroplane,
    August 15, 1958, pp. 229-231.

7-140 D.M. McElhinney: "Structural design of the VC-10". Aircraft Eng., June 1962, pp. 165-171, 178.

7-141 K. Bentley: "BAC One-eleven structural design". Aircraft Eng., May 1963, pp. 142-147.

7-142 Anon.: "Wing design-civil transport aircraft". Cranfield Institute of Technology, Sheet DES 430
    (unpublished).

7-143 Anon.: "Potential structural materials and design concepts for light aircraft". NASA CR-1285, March
    1969.

7-144 W.W. Williams: "An advanced extensible wing flap system for modern aeroplanes". AIAA Paper No. 70-
    911.

CHAPTER 8.

Weight subdivision and limitations

8-1 United States Department of Defence: "Weight and balance data reporting forms for aircraft". Military
    Standard, MIL-STD-254 (ASG), Aug. 26, 1954.

8-2 Direction Technique et Industrielle de l'Aéronautique: "Devis de poids (avions)". AIR-2001/C, Edition
    No. 4, Dec. 15, 1959.

8-3 Society of British Aircraft Constructors: "Standard method for the estimation of Direct Operating
    Costs of aircraft". Dec. 1959.

8-4 Bundesminister für Verteidigung: "Normstelle der Luftfahrt". No. LN 9020, Nov. 1962.

8-5 J.R. McCarthy: "Definition and equipment list and standard form for presentation of weight and
    balance data". SAWE Technical Paper No. 354, 1962.

8-6 (British) Ministry of Aviation: "Weight, geometric and design data". AVMIN Form 2492, 1964.

8-7 "Glossary of standard weight terminology for commercial aircraft". Society of Aeronautical Weight
    Engineers, Revision 1964.

8-8 "Proposed Glossary of Standard Weights Terminology". International Air Traffic Association, Specific-
    ation No. 100.

8-9 C. Payton Autry, P.J. Baumgaertner: "The design importance of airplane mile costs versus seat mile
    costs". SAE Paper No. 660277.

8-10 "Standard method of estimating comparative Direct Operating Costs of turbine-powered transport air-
    planes". Air Transport Association of America, Dec. 1967.

8-11 "Recommended standard data format of transport airplane characteristics for aircraft planning".
    National Aerospace Standard NAS 3601", 1968/1970.

General weight considerations and prediction methods

8-12 N.S. Currey: "Structure weight". Interavia, Vol. 4, pp. 89-92, Feb. 1949.

8-13 F. Grinsted: "Aircraft structural weight and design efficiency". Aircraft Eng., July 1949, pp. 214-217.

8-14 L.W. Rosenthal: "The weight aspect in aircraft design". Journal of the Royal Aeron. Soc., Vol. 54, March 1950, pp. 187-210.

8-15 E. Weining: "Design factors in the development of light aircraft". Aeron. Eng. Review, July 1950, pp. 18-19.

8-16 A. Schritt, W. Buckley: "A realistic approach to structural weight estimation". SAWE Paper No. 73, May 1952.

8-17 J. Taylor: "Structure weight". J. of the Royal Aero. Soc., Vol. 57, pp. 646-652, Oct. 1953.

8-18 K. Thalau: "Geschwindigkeit, Konstruktionsgewicht und Nutzlast moderner Verkehrsflugzeuge". Jahrbuch der WGLR, 1953, pp. 110-123.

8-19 F.C. Hopton-Jones: "A practical approach to the problem of structural weight estimation for preliminary design". SAWE Paper No. 127, May 1955.

8-20 R.M. Simonds: "A generalized graphical method of minimum gross weight estimation". SAWE Paper No. 135, May 1956.

8-21 M.G. Heal: "Problems in estimating structure weight". Aeron. Eng. Review, March 1957, pp. 52-56.

8-22 W.E. Caddell: "The development of generalized weight estimating methods". SAWE Paper No. 219, May 1959.

8-23 Anon.: "An introduction into aeronautical weight engineering". SAWE, 1959.

8-24 C.R. Liebermann: "The unity equation and growth factor". SAWE Paper No. 267, May 1960.

8-25 M.E. Burt: "Effects of design speed and normal acceleration on aircraft structure weight", ARC CP 490, 1960.

8-26 H. Hertel: "Grundlagenforschung für Entwurf und Konstruktion von Flugzeugen". Arbeitsgemeinschaft für Forschung, Nordrhein-Westfalen, Vol. 102, 1961.

8-27 M.E. Burt: "Structural weight estimation for novel configurations". J. of the Royal Aeron. Soc., Vol. 66, Jan. 1962, pp. 15-30.

8-28 E.E. Sechler and L.G. Dunn: "Airplane structural analysis and design". Dover Publications New York, 1963.

8-29 J.J. Pugliese: "Gross weight estimation of an attack airplane by generalized graphical solution". SAWE Paper No. 364, 1963.

8-30 R.J. Atkinson: "Structural design". J. of the Royal Aero. Soc., Vol. 67, pp. 692-695, Nov. 1963 (also: RAE TN No. Structures 333).

8-31 A.C. Kermode: "The aeroplane structure". Second Edition, Pitman and Sons, London, 1964.

8-32 W.J. Strickler: "Application of regression and correlation techniques in mass properties engineering". SAWE Paper No. 422, May 1964.

8-33 A.C. Kermode: "The aeroplane structure", Chapter 6: "Weight". Pitman and Sons Ltd., London, Second Edition, 1964.

8-34 R. Riccius: "Untersuchungen über die Gewichte vertikalstartender Flugzeuge'. ILTUB Jahrbuch 1965/1966.

8-35 J.A. Neilson: "Value of a pound". SAWE Paper No. 586, May 1966.

8-36 M.A. Kochegura: "Determination of the weight of an empty aircraft by methods of mathematical statistics". USAF Foreign Technology Division FTD-MT-24-224-68, Aug. 1968.

8-37 H.L. Roland: "General approach to preliminary design weight analysis and structural weight prediction". Short Course in modern theory and practice of weight optimization and control for advanced aeronautical systems, University of Tennessee, Nov. 1968.

8-38 W.H. Ahl: "Rational weight estimation based on statistical data". SAWE Paper No. 791, May 1969.

8-39 W.E. Caddell: "On the use of aircraft density in preliminary design". SAWE Paper No. 813, May 1969.

8-40 R.S. St. John: "The derivation of analytical-statistical weight prediction techniques". SAWE Paper No. 810, May 1969.

8-41 R.N. Staton: "Constrained regression analysis - A new approach to statistical equation development". SAWE Paper No. 762, May 1969.

8-42 C. Vivier and P. Cormier: "Masse d'un avion". AGARD Lecture Series No. 56 on "Aircraft performance prediction and optimization". April 1972.

8-43 D.P. Marsh: "Post-design analysis for structural weight estimation". SAWE Paper No. 936, May 1972.

8-44 W. Schneider: "Die Entwicklung und Bewertung von Gewichtsabschätzungsformeln für Flugzeugentwürfe unter Zuhilfnahme von Methoden der Mathematischen Statistik und Warscheinlichkeitsrechnung". Dissertation, Technical University of Berlin, Feb. 1973.

8-45 A.A. Blythe: The hub of the wheel - A project designer's view of weight". SAWE Paper No. 996, June 1973.

8-46 W. Schneider: "Project weight prediction based on advanced statistical methods". Paper presented at the 43rd AGARD Flight Mechanics Panel Meeting, Symposium on "Aircraft design integration and optimization". Florence, Oct. 1-4, 1973.

Complete weight prediction methods

8-47 I.H. Driggs: "Aircraft design analysis". J. of the Royal Aero. Soc., Vol. 54, Feb. 1950, pp. 65-116.

8-48 M. Vautier and M. Dieudonné: "Le probleme des poids dans l'aviation" (2 Parts), Service de Documentation et d'Information Technique de l'Aéronautique, 1950.

8-49 O. Köhler: "Gewichtsunterlagen für den Flugzeugentwurf". Luftfahrttechnik, Dec. 1955, pp. 134-139 and Jan. 1956, pp. 15-18.

8-50 D. Howe: "Initial aircraft weight prediction". College of Aeronautics Note 77, 1957.

8-51 F.K. Teichmann: "Airplane design manual". Chapter 8: "Preliminary weight estimate". Pitman Publishing Co., New York, fourth edition, 1958.

8-52 H.G. Sheridan: "Aircraft preliminary design methods used in the weapon systems analysis division". US Navy BUWEPS Report No. R-5-62-13, June 1962.

8-53 E. Sechler and L.G. Dunn: "Airplane structural analysis and design", Chapter 1: "The airplane layout". Dover Publications Inc., New York, Jan. 1963.

8-54 W.H. Marr: "Basic weight trends for bomber and transport aircraft". SAWE Paper No. 434, May 1964.

8-55 K.D. Wood: "Aircraft Design". Vol. 1 of "Aerospace vehicle design", second edition, 1966, Johnson Publishing Cy.

8-56 H.L. Roland: "Parametric weight-sizing methods - structure, propulsion, fixed equipment - fighters (USAF & USN)". June 30, 1965, General Dynamics Fort Worth Division, MR-S5-040, Revision, Sept. 30, 1966.

8-57 G. Corning: "Subsonic and supersonic aircraft design". Pp. 2:27 to 2:35: "Weight estimation". College Park, Maryland, second edition, 1966.

8-58 W. Richter et al: "Luftfahrzeuge". Das Fachwissen des Ingenieurs. Carl Hanser Verlag, München, 1970.

8-59 D. Howe: "Structural weight prediction". Cranfield Institute of Technology DES903, 1971 (unpublished).

8-60 D. Howe: "Empty weight and cruise performance of very large subsonic jet transports". Cranfield Institute of Technology, Report Aero No. 3, 1972.

8-61 R.N. Staton: "Weight estimation methods". SAWE Journal, April-May 1972, pp. 7-11.

8-62 H.F. Kooy and H. Rekersdrees: "Weight estimation method for subsonic transport aircraft". Fokker Report H-0-15, June 1972 (unpublished).

8-63 L.M. Nicolai: "Design of airlift vehicles". USAF Academy, Dept. of Aeronautics, Aero 464, 1972.

General wing structure weight considerations

8-64 R.J. Lutz: "Applications of optimum design principles to structural weight estimation". SAWE Paper

No. 205, April 1951.

8-65 D.J. Farrar: "The design of compression structures for minimum weight". J. of the Royal Aero. Soc., Nov. 1949, pp. 1041-1052.

8-66 F. Shanley: "Weight-strength analysis of aircraft structures", McGraw-Hill Book Cy. Inc., New York, 1952.

8-67 O. Ljungström: "Wing structures of future aircraft". Aircraft Eng., May 1953, pp. 128-132.

8-68 A.L. Kolom: "Optimum design considerations for aircraft wing structures". Aero. Eng. Review, Oct. 1953, pp. 31-41.

8-69 C.R. McWorther: "Considerations of bending and torsional stiffness in the design of wings for minimum weight". SAWE Paper No. 84, 1953.

8-70 L.D. Green and J. Mudar: "Estimating structural box weight", Aeron. Eng. Review, Feb, 1958, pp. 48-51.

8-71 S. Sichveland, F.M. de Graan and R.H. Trelease: "The weight engineer's approach to the problem of fatigue in aircraft structures". SAWE Paper No. 172, 1958.

8-72 N.N. Fadeev: "A theoretical formula for the weight of a tapered wing". In: "Methods of selection and approximate calculation of aircraft design parameters". Trudy Institute No. 138, Moscow 1961, pp. 28-52.

8-73 R.J. Taylor: "Weight prediction techniques and trends for composite material structure". SAWE Paper No. 887.

8-74 B. Sealman: "Multitapered wings". J. of Aircraft, July-Aug. 1965, pp. 348-349.

8-75 B. Sealman: "Effect of wing geometry on volume and weight". J. of Aircraft, Vol. 1 No. 5, Sept.-Oct. 1964, p. 305.

8-76 D.H. Emero and L. Spunt: "Wing box optimization under combined shear and bending". J. of Aircraft, Vol. 3, No. 2, March-April 1966, pp. 130-141.

8-77 C.A. Garrocq and J.T. Jackson: "Estimation of wing box weights required to preclude aeroelastic instabilities". SAWE Paper No. 500, May 1966.

8-78 K.L. Sanders: "A review and summary of wing torsional stiffness criteria for predesign and weight estimations". SAWE Paper No. 632, May 1967.

8-79 D.J. Lamorte: "Non-optimum factor and preliminary weight estimation of a boron composite wing structure". SAWE Paper No. 891, May 1971.

Weight prediction of wing and tailplane structure

8-80 W. Tye and P.E. Montangnon: "The estimation of wing structure weight". ARC R and M 2080, 1941.

8-81 F. Grinsted: "Simple formulae for predicting the weights of wing, fuselage and tail unit structures". RAE Report Structures No. 24, 1948.

8-82 F. Grinsted: "Prediction of wing structure weight". RAE Report Structures No. 15, 1948.

8-83 G.K. Gates: "Weight estimation of metal wings". Aircraft Eng., April 1949, p. 116.

8-84 J.F. Carreyette: "Aircraft wing weight estimation". Aircraft Eng., Jan. 1950, pp. 8-11 and April 1950, p. 119.

8-85 E.L. Ripley: "A simple method of tail unit structure weight estimation". RAE Report Structures No. 94, Nov. 1950.

8-86 J. Kelley Jr.: "Wing weight estimation". AAI Technical Report 5161.

8-87 I.H. Driggs: "Aircraft design analysis". J. of the Royal Aero. Soc., Vol. 54, Feb. 1950.

8-88 J. Solvey: "The estimation of wing weight". Aircraft Eng., May 1951, pp. 143-144.

8-89 E.L. Ripley: "A method of wing weight prediction". RAE Report Structures No. 109, May 1951.

8-90 M.E. Burt: "Weight prediction of ailerons and landing flaps". RAE Report Structures No. 116, Sept. 1951.

8-91 A. Hyatt: "A method for estimating wing weight's. J. of the Aero. Sciences, Vol. 21 No. 6, June 1954, pp. 363-372.

8-92 M.E. Burt: "Weight prediction for wings of box construction". RAE Report Structures No. 186, 1955.

8-93 W. v. Nes and O. Köhler: "Das Gewichtsanteil der tragenden Teile am Flügelgewicht". Luftfahrttechnik, Nov. 1056, pp. 206-210.

8-94 K.L. Sanders: "Abschätzung des Flügelgewichtes". Luftfahrttechnik, Oct. 1957, p. 224.

8-95 D. Howe: "Initial aircraft weight prediction". College of Aeronautics Note No. 77, 1957.

8-96 M. Schwartzberg: "Blown flap system for STOL performance - weight considerations". Aerospace Eng., March 1959, pp. 48-52.

8-97 C.R. Ritter: "Rib weight estimation by structural analysis". SAWE Paper No. 259, 1960.

8-98 K.L. Sanders: "High-lift devices; a weight and performance trade-off technology". SAWE Paper No. 761, May 1969.

8-99 W.W. Williams: "An advanced extensible wing flap system for modern aeroplanes". AIAA Paper No. 70-911, July 1970.

8-100 R.L. Gielow: "Performance prediction and evaluation of propulsion-augmented high lift systems". AIAA Paper No. 71-990, Oct. 1971.

8-101 E. Torenbeek: "Prediction of wing group weight for preliminary design". Aircraft Eng., July 1971, pp. 16-21. Summary in Aircraft Eng., Feb. 1972, pp. 18-19.

8-102 F.O. Smetana: "A design study for a simple to fly, constant attitude light aircraft. NASA CR-2208, March 1973.

Wings for high-speed aircraft

8-103 R.E. Lowry: "Problems and solutions of delta wings". SAWE Paper No. 77, 1952.

8-104 W.J. Conway: "Factors affecting the design of thin wings". SAE Preprint No. 357, Oct. 1954.

8-105 R.L. Hammitt: "Structural weight estimation by the weight penalty concept for preliminary design". SAWE Paper No. 141, 1956.

8-106 M.G. Heal: "Structural weights on supersonic aircraft with low aspect ratio unswept wings". SAWE Paper No. 193, 1956.

8-107 A.C. Robinson: "Problems associated with weight estimation and optimization of supersonic aircraft". SAWE Paper No. 234, 1959.

8-108 R.A. Anderson: "Weight-efficiency analysis of thin-wing construction". Transactions of the ASME, Vol. 79, July 1957 (II), pp. 974-979.

8-109 D.J. Johns: "Optimum design of a multicell box to a given bending moment and temperature distribution". College of Aeronautics Note No. 82, April 1958.

8-110 M.E. Burt: "Structural weight estimation for novel configurations". J. of the Royal Aero. Soc., Vol. 66, Jan. 1962, pp. 15-30.

Fuselage structure

8-111 L.W. Rosenthal: "The influence of aircraft gross weight upon the size and weight of hulls and fuselages". J. of the Royal Aero. Soc., Vol. 51, Nov. 1947, pp. 874-883.

8-112 F. Grinsted: "Simple formulae for predicting the weights of wing, fuselage and tail unit structures". RAE Report Structures No. 24, May 1948.

8-113 E.L. Ripley: "A method of fuselage structure weight prediction". RAE Report Structures No. 93, 1950.

8-114 W.R. Micks: "Structural weight analysis. Fuselage and shell structures". The Rand Corporation, Report No. R-172. 1950.

8-115 M.E. Burt and J. Philips: "Prediction of fuselage and hull structure weight". RAE Report Structures No. 122, April 1952.

8-116 L.D. Green: "Fuselage weight prediction". SAWE Paper No. 126, May 1955.

8-117 R.L. Hammitt: "Structural weight estimation by the weight penalty concept for preliminary design". SAWE Paper No. 141, May 1956.

8-118 E.W. Tobin Jr.: "A method for estimating optimum fuselage structural weight". SAWE Paper No. 152, May 1957.

8-119 A.A. Badiagin: "Concerning an efficient slenderness ratio for the fuselage of civilian aircraft". In: "Methods of selection and approximate calculation of air design parameters". Trudy Institute No. 138, Moscow, 1961, pp. 19-27.

8-120 J. Morris and D.M. Ashford: "Fuselage configuration studies". SAE Paper No. 670370, April 1967.

8-121 A.R. Di Pierro: "Minimum weight analysis of fuselage frames". SAWE Paper No. 826, May 1970.

8-122 D.E. Poggio: "Theoretical and real weight of shell fuselages". Ingegneria, Jan. 1971, pp. 1-12.

8-123 D.M. Simpson: "Fuselage structure weight prediction". SAWE Paper No. 981, June 1973.

Alighting gear

8-124 J. Philips: "A method of undercarriage weight estimation". RAE Report Structures No. 198, March 1956.

8-125 C.R. Liebermann: "Rolling type alighting gear weight estimation". SAWE Paper No. 210, May. 1959.

8-126 M.E. Burt and E.L. Ripley: "Prediction of undercarriage weights". RAE Report Structures No. 80, June 1950.

8-127 P.R. Kraus: "An analytical approach to landing gear weight estimation". SAWE Paper No. 829, May 1970.

The powerplant

8-129 G.R. Holzmeier: "A rational method for estimating fuel system weight in preliminary design". SAWE Paper No. 147, 1957.

8-130 W.C. Crooker: "Aircraft fuel system weight estimation for the tri-sonic era". SAWE Paper No. 232, May 1959.

8-131 G. Rosen: "New problem areas in aircraft propeller design". Canadian Aero. Journal, Vol. 6, June 1960, p. 219.

8-132 Anon.: "Hamilton Standard propeller and gear box weight generalization". Figs. 1, 2 and 3 of Publication PDB 6101, 1963.

8-133 R.C. Engle: "Jet engine weight and thrust trends including future development promises by the engine manufacturers". SAWE Paper No. 682, May 1968.

8-128 I.H. Driggs and O.E. Lancaster: "Engine weights". In: "Gasturbines for aircraft". Section 8.9, 1955, Ronald Press, New York.

8-134 M.L. Yaffee: "Propeller research gains emphasis". Aviation Week and Space Technology, Nov. 1969, pp. 56-65.

8-135 R.P. Gerend and J.P. Roundhill: "Correlation of gasturbine engine weights and dimensions". AIAA Paper No. 70-669.

8-136 J.F. Dugan Jr.: "Engine selection for transport and combat aircraft". NASA TMX-68009, April 1972. (Also in AGARD Lecture Series 49).

Airframe services and equipment, operational items, payload

8-137 C.K. McBaine: "Weight estimation of aircraft hydraulic systems". SAWE Paper No. 128, 1955.

8-138 J.R. McCarty: "A review and revised approach to the average passenger weight". SAWE Paper No. 223, May 1959.

8-139 C.A. Hangoe: "Comparison of passenger service equipment". SAWE Paper No. 220, May 1959.

8-140 G.R. Williams: "Optimization of fluid lines". SAWE Paper No. 291, May 1961.

8-141 Anon.: "The use of standard baggage weights". European Civil Aviation Conference, Strassbourg, July 1961, Vol. II, Section 2, Doc. 8185, ECAC/4-2, pp. 355-356.

8-142 R.G. Mitchell: "Evaluation of economics of passenger comfort standards". SAWE Paper No. 338, May 1962.

8-143 C.A. Hangoe: "World-wide survey of cargo densities". SAWE Paper No. 339, May 1962.

8-144 J.R. McCarty: "Airline new aircraft evaluations". SAWE Paper No. 619, May 1967.

8-145 J. Morris and D.M. Ashford: "The use of standard baggage weights". SAWE Preprint.

8-146 R.J. Taylor and K. Smith: "Advanced aircraft parametric weight analysis". SAWE Paper No. 637, May 1967.

8-147 B.H. Nicholls and A.D. Meshew: "Auxiliary power systems for 1975 fighter aircraft". SAE Paper No. 680311, April-May 1968.

8-148 T.P. Clemmons: "Systems design for weight optimization". SAWE Paper No. 757, May 1969.

8-149 H.L. Roland: "Advanced design weight analysis and systems and equipment weight prediction". SAWE Paper No. 790, May 1969.

8-150 D.M. Cate: "A parametric approach to estimate weights of surface control systems of combat and transport aircraft". SAWE Paper No. 812, May 1969.

8-151 P.A. Ward and W.G. Lydiard: "Aircraft auxiliary power system and their influence on power plant design". Lecture presented at Symposium of the Royal Aero. Soc., London, 1969.

8-152 R.S. Kaneshiro: "Weight estimation of hydraulic secondary power system". SAWE Paper No. 935, May 1972.

Some recent publications on weight prediction

8-153 J. Banks: "Preliminary weight estimation of canard configured aircraft". SAWE Paper No. 1015, May 1974.

8-154 A. Krzyzanowski: "A method for weight/cost trade-offs in preliminary vehicle design". SAWE Paper No. 1017, May 1974.

8-155 R.N. Staton: "Fuselage basic shell weight prediction". SAWE Paper No. 1019, May 1974.

8-156 W. Schneider: "A procedure for calculating the weight of wing structures with increased service life". SAWE Paper No. 1021, May 1974 (summary in SAWE Journal), Vol. 34 No. 1, Jan. 1975, pp. 1-12 and 40-41.

8-157 J.L. Anderson: "A parametric analysis of transport aircraft system weights and costs". SAWE Paper No. 1024, May 1974.

8-158 C.R. Glatt: "WAATS - a computer program for weights analysis of advanced transportation systems" NASA CR-2420, Sept. 1974.

8-159 B. Saelman: "Methods for better prediction of gross weight". SAWE Paper No. 1041, May 1975.

8-160 R. St. John: "Weight effects of structural material variation". SAWE Paper No. 1044, May 1975.

Balance and loadability

8-161 K.L. Sanders: "Simpler wing location for a specified longitudinal stability". Space/Aeronautics, March 1960, page 67-70.

8-162 J.R. McCarty: "Passenger seating pattern. A statistically based cabin load or passenger seating assumption applicable to airline operation". SAWE Paper No. 250, May 1960.

8-163 G.H. Hopper: "The influence of balance and loadability on the design of commercial passenger transports". SAWE Paper No. 269, May 1960.

8-164 D.J. Lambert: "Design of jet transport with rear-mounted engines". Aerospace Engineering, Oct. 1960, page 30-35, 72, 74.

8-165 G.W. Benedict: "Methods of evaluating aircraft loadability". SAWE Paper No. 334, May 1962.

8-166 K.L. Sanders and D.O. Nevinger: "Balancing options in aircraft configuration design". SAWE Paper No. 840, May 1970.

8-167 H. Waldon: "Theoretical vs. actual seating patterns variations in wide body airvraft". SAWE Paper

No. 1080, May 1975.

8-168 R. Maswell: "A loadability comparison: L1011/DC-10-10". SAWE Paper No. 1094, May 1975.

CHAPTER 9.

General aspects of airplanes stability and control

9-1 C.D. Perkins and R.E. Hage: "Airplane performance, stability and control", New York, John Wiley and Sons, Inc., 1949.

9-2 C.F. Joy: "Power controls for aircraft". J. of the Royal Aero. Soc., Jan. 1952, pp. 7-24.

9-3 C.S. Draper: "Flight control". J. of the Royal Aero. Soc., Vol. 59, July 1955, pp.451-478.

9-4 A.W. Babister: "Aircraft stability and control". International Series of Monographs in Aeronautics and Astronautics, Pergamon Press, 1961.

9-5 O.H. Gerlach: "Vliegeigenschappen I". Lecture notes VTH-D10, Delft University of Technology, Dept. of Aeron. Eng., Dec. 1967 (in Dutch).

9-6 S.B. Dickinson: "Aircraft stability and control for pilots and engineers". Pitman and Sons Ltd., 1968.

9-7 R.B. Holloway, P.M. Burris and R.P. Johannes: "Aircraft performance benefits from modern control systems technology". J. of Aircraft, Vol. 7 No. 6, Nov.-Dec. 1970, pp. 550-553.

9-8 F. O'Hara: "Stability augmentation in aircraft design". The Aeron. Journal of the Royal Aeron. Soc., Vol. 75 No. 724, April 1971, pp. 293-304.

9-9 D.P. Davies: "Handling the big jets". Air Registration Board, third edition, Dec. 1971.

9-10 B. Etkin: "Dynamics of atmospheric flight". J. Wiley Publ. Corp., New York, 1972.

9-11 F.O. Smetana, D.C. Summey and W.D. Johnson: "Riding and handling qualities of light aircraft - a review and analysis". NASA CR-1975, March 1972.

9-12 J. Roskam: "Flight dynamics of rigid and elastic airplanes". University of Kansas, 1972.

Flying qualities criteria

9-13 Anon.: "Design objections for flying qualities of civil transport aircraft", SAE ARP 842 B.

9-14 C. Leyman and E.R. Nuttall: "A survey of aircraft handling criteria". ARC Current Paper No. 833.

9-15 H.A. Shomber and W.M. Gertsen: "Longitudinal handling qualities criteria: an evaluation". AIAA Paper No. 65-780.

9-16 F. O'Hara: "Handling criteria". J. of the Royal Aeron. Soc., Vol. 71, April 1967, pp. 271-291.

9-17 E.E. Larrabee and J.P. Tymczyszyn: "The effect of flying qualities requirements on the design of general aviation aircraft in the 1980's". AIAA Paper No. 68-189.

9-18 Anon.: "Flying qualities of piloted airplanes". MIL-F-8785 B (ASG), August 1969.

9-19 C.R. Chalk et.al.: "Background information and user guide for MIL-F-8786 B (ASG), Military specification - flying qualities of piloted airplanes". T. Report AFFDL-TR-69-72, August 1969.

9-20 Anon.: "Flying qualities of piloted V/STOL aircraft". MIL-F-83300, Dec. 1970.

Tailplane and control surface design

9-21 W.S. Brown: "Spring tab controls". ARC R and M 1979, 1941.

9-22 W.H. Philips: "Application of spring tabs to elevator controls". NACA TR 797, 1944.

9-23 M.B. Morgan and H.H.B.M. Thomas: "Control surface design". J. of the Royal Aeron. Soc., Vol. 49, 1945, pp. 431-510.

9-24 L.E. Root: "The effective use of aerodynamic balance on control surfaces". J. of the Aeron. Sciences, April 1945, pp. 149-163.

9-25 W.M. Phillips: "Appreciation and prediction of flying qualities". NACA TR 927, 1949.

9-26 F.B. Baker: "Choice of fin area and dihedral". Aircraft Eng., March 1948, pp. 87-88.

9-27 O.R. Dunn: "Aerodynamically boosted surface controls and their application to the D.C. 6 transport". Second International Aeronautical Conference, New York, 1949.

9-28 D.E. Morris: "Designing to avoid dangerous behaviour of an aircraft due to the effects on control hinge moments of ice on the leading edge of the fixed surface". ARC Current Paper 66, 1952.

9-29 J.C. Wimpenny: "Stability and Control in Aircraft Design". J. of the Royal Aeron. Soc., May 1954, pp. 329-360.

9-30 B.R.A. Burns: "Design considerations for the satisfactory stability and control of military aircraft". AGARD CP-119 Stability and Control.

9-31 W.D. Thompson: "Improvements in flying qualities of modern light planes". SAE Paper No. 622D, Jan. 1963.

9-32 E. Obert: "Low-speed stability and control characteristics of transport aircraft with particular reference to tailplane design". AGARD CP-160 (Paper 10), April 1974.

Determination of aerodynamic characteristics

9-33 D.J. Lyons and P.L. Bisgood: "An analysis of the lift slope of aerofoils of small aspect ratio wings including fins, with design charts for aerofoils and control surfaces". ARC R and M 2308, 1945. Also: Aircraft Eng., Sept. 1947, pp. 278-286.

9-34 D.E. Morris and J.C. Morrall: "Effect of slipstream on longitudinal stability of multi-engined aircraft". ARC R and M 2701, Nov. 1948.

9-35 A. Silverstein and S. Katzoff: "Design charts for predicting downwash angles and wake characteristics behind plain flaps an flapped wings". NACA TR 648.

9-36 J. Weil and W.C. Sleeman Jr.: "Prediction of the effects of propeller operation on the static longitudinal stability of single-engine tractor monoplanes with flaps retracted". NACA TR 941.

9-37 S. Katzoff and H. Sweberg: "Ground effect on downwash angle and wake location". NACA TR 738.

9-38 D.E. Hoak: "USAF Stability and Control Datcom". Wright Patterson Air Force Base. Oct. 1960 (Revision August 1968).

9-39 D.E. Ellison and D.E. Hoak: "Stability derivative estimation at subsonic speeds". Annals of the New York Academy of Sciences, Vol. 154, Part 2, pp. 367-396.

9-40 G.M. Moss: "Some aerodynamic aspects of rear mounted engines". J. of the Royal Aeron. Soc., Vol. 68, Dec. 1964, pp. 837-842.

9-41 Anon.: Data Sheets, Aerodynamics, Vols. I, II and III, Royal Aeronautical Society.

9-42 J. Roskam: "Methods of estimating stability and control derivatives of conventional subsonic airplanes". The University of Kansas, 1971.

9-43 C.H. Wolowicz: "Longitudinal aerodynamic characteristics of light, twin-engine, propeller driven airplanes". NASA TN D-6800, 1972.

9-44 C.H. Wolowicz and R.B. Yancey: "Lateral-directional aerodynamic characteristics of light, twin-engine, propeller-driven airplanes". NASA TN D-6946, Oct. 1972.

9-45 R.K. Heffly and W.F. Jewell: "Aircraft handling qualities data". NASA CR-2144, Dec. 1972.

9-46 C.G.B. Mitchell: "A computer programme to predict the stability and control characteristics of subsonic aircraft". RAE Technical Report No. 73079, Sept. 1973.

9-47 M.I. Goldhammer and N.F. Wasson: "Methods for predicting the aerodynamic and stability and control characteristics of STOL aircraft". Technical Report AFFDL-TR-73-146, Dec. 1973.

9-48 F.O. Smetana, D.C. Summey, N.S. Smith and R.K. Carden: "Light aircraft lift, drag and pitching moment prediction - a review and analysis". NASA CR-2523, May 1975.

Engine failure considerations

9-49 A.H. Yates: "Control in flight under asymmetric power". Aircraft Eng., Sept. 1947, pp. 287-290.

9-50 L.E. Wright: "Flight on asymmetric engine power". Aircraft Eng., Dec. 1950, pp. 350-355.

9-51 A. Hammer: "Die analytische und experimentelle Ermittlung der Mindestkontrollgeschwindigkeiten von Flugzeugen (minimum control speeds) unter Berücksichtigung von Flugeigenschaftsforderungen". Dissertation Berlin, 1971.

9-52 E.J.N. Archbold and K.T. McKenzie: "Response in yaw". J. of the Royal Aeron. Soc., Vol. 50, 1946, pp. 275-285.

9-53 M.E. Kirchner: "Turboprop airplane control problems associated with engine failure considerations". SAE Paper No. 613, 1955.

9-54 J. Mannée: "Windtunnelinvestigation of the influence of the aircraft configuration on the yawing and rolling moment of a twin-engined, propeller-driven aircraft with one engine inoperative". NLL Report A 1508 B, 1963.

Airplane spinning

9-55 A.I. Neihouse: "Tail-design requirements for satisfactory spin recovery for personal-owner-type light airplanes". NACA TN-1329, June 1947.

9-56 L.J. Gale and I.P. Jones Jr.: "effects of antispin fillets and dorsal fins on the spin recovery characteristics of airplanes as determined from free-spinning-tunnel tests". NACA TN-1779, Dec. 1948.

9-57 J.S. Bowman: "Airplane spinning". NASA SP-83, May 1965.

9-58 A.I. Neihouse, I. Anshal, W.J. Klinar and S.H. Scher: "Status of spin research for recent airplane designs". NASA TR R-57, 1960.

CHAPTER 10.

10-1 H.G. Conway: "Landing gear design". Textbook Royal Aeronautical Society, Chapman & Hall Ltd., London, 1958.

10-2 G. Bock: "Operations from unprepared and semiprepared airfields". Agardograph 45, Sept. 1960.

10-3 Anon.: "ICAO Aerodrome Manual, Part 2". Doc. 7920-AN/865/2, Second edition, 1965.

10-4 A. Cameron-Johnson: "The undercarriage in aeroplane project design". Aircraft Eng., Feb. 1969, pp. 6-11.

10-5 R.C. Cussons: "Bogie undercarriages", J. Royal Aeron. Soc., July 1952.

10-6 J.A. Skinner: "Testing runway foundations and pavements". Airport Paper No. 17, Proc. of the Institution of Civil Engineers, 1951.

10-7 L.S. Bialkowski: "The basic problem of undercarriage geometry". Aircraft Eng., August 1953.

10-8 E.G. Collinson: "A note on the load transference on multi-wheel bogie undercarriages". J. of the Royal Aero. Soc., Oct. 1949.

10-9 Anon.: "Konstruktionsbeispiele aus dem Flugzeugbau". Band 4 Fahrwerk, Fachbücher für Luft- und Raumfahrt. Luftfahrtverlag Walter Zuerl.

10-10 R. Hadekel: "The mechanical characteristics of pneumatic tyres". S & T, Memo No. 10/52, Ministry of Supply, Nov. 1952.

10-11 R.O. Dickinson: "A fresh approach to aircraft landing gear design". ASME Paper No. 57-SA-30.

10-12 W.E. Eldred: "Landing gear design as applied to modern aircraft". SAE Paper No. 146, Sept./Oct. 1953.

10-13 E. Schumacher: "Die Fahrwerke der heutigen Flugzeuge". Luftfahrttechnik, Nov./Dec. 1958, pp. 295-306 and 320-338, Jan. 1959, pp. 22-30.

10-14 E. Overesch: "The problems of exact calculation of takeoff and landing characteristics of conventional transport aircraft ". AGARD Report 417, Jan. 1963.

10-15 V.K. Karrask: "The trim of an aircraft with a landing gear of the tricycle type in a crosswind". Trudy Institute, Moscow, 1961.

10-16 R. Lucien: "Military requirements and undercarriage design". Interavia 6/1961, pp. 839-842.

10-17 G. Bruner: "L'attérrisseur à amortisseur horizontal "jockey" de la société Messier pour avions cargos". DOCAERO, No. 66, Jan. 1961, pp. 5-16.

10-18 E. Schumacher: "Die Fahrwerke der heutigen Flugzeuge". Luftfahrttechnik 5 (1959) No. 1, Jan 15, pp. 22-30, No. 4, March, pp. 89-92.

10-19 W.W. Williams, G.K. Williams and W.C.J. Garrard: "Soft and rough field landing gears". SAE Paper No. 650844, Oct. 1965.

10-20 K.S. Carter: "The landing gear of the Lockheed SST". SAE Paper No. 650224, April 1965.

10-21 Anon.: "Some current types of landing gear". Aircraft Engineering, Jan. 1968, pp. 26-31.

10-22 S.W.H. Wood: "Problems of undercarriage design for V/STOL aircraft". The Aeronautical Journal, Feb. 1969, pp. 157-168.

10-23 R.A. Werner and G.F. Wislecinus: "Analysis of airplane design by similarity considerations". AIAA-Paper No. 68-1017, Oct. 1968.

10-24 L.G. Hoare: "Aircraft landing gear". Aircraft Eng., Jan. 1968, pp. 6-8.

10-25 W.J.G. Pinsker: "The dynamics of aircraft rotation and liftoff and its implication for tail clearance requirements, especially with large aircraft". ARC R and M No. 3560, 1969.

10-26 J.F. O'Hara: "Aircraft crosswind performance". AGARD Report 492, Oct. 1964.

10-27 W.B. Horne, T.J. Yager and G.R. Taylor: "Recent research on ways to improve tire traction on water, slush or ice". Luchtvaarttechniek 5, Sept. 1966.

10-28 J.W.H. Thomas: "Design for runway conditions". J. of the Royal Aeron. Soc., Sept. 1965, pp. 571-576.

10-29 H.S.D. Yang: "C-5A main landing gear bogie pitch control". AIAA Paper No. 70-914.

10-30 Anon.: Draft International Standard ISO/DIS 3324/1: "Aircraft tyres and rims". 1974/1975.

CHAPTER 11.

Prediction of aerodynamic characteristics

11-1 Anon.: "Report of the definitions panel on definitions to be used in the description and analysis of drag". ARC CP No. 369, 1957.

11-2 G. Gabrielli: "On the subdivision in different "forms" of the aircraft drag at the maximum speed". Troisième Congrès Aéronautique Européen, 1958, pp. 398-407 and 979-984.

11-3 H. Schlichting and E. Truckenbrodt: "Aerodynamik des Flugzeuges", Band 1 und 2, Springer Verlag, Berlin, 1962.

11-4 D.E. Hoak and D.E. Ellison: "USAF Stability and Control DATCOM". McDonnell Douglas Airplane Cy., Oct. 1960, Revised Aug. 1968.

11-5 Anon.: "Subsonic lift-dependent drag due to wing trailing vortices". R.Ae.S. Data Sheets, "Wings" 02.01.02, 1965.

11-6 B.W. McCormick: "Aerodynamics of V/STOL flight". Academic Press, New York, London, 1967.

11-7 J. Williams and A.J. Ross: "Some airframe aerodynamic problems at low speeds". Annals of the New York Academy of Sciences, Vol. 154, Art. 2, Nov. 22, 1968, pp. 264-305.

11-8 D.A. Norton: "Airplane drag prediction". Annals of the New York Academy of Sciences, Vol. 154, Part 2, pp. 306-328, 1968.

11-9 J.C. Wimpenny: "The design and application of high lift devices". Annals of the New York Academy of Sciences, Vol. 154, Art. 2, Nov. 22, 1968, pp. 329-366.

11-10 D.E. Ellison and D.E. Hoak: "Stability derivative estimation at subsonic speeds". Annals of the New York Academy of Sciences, Vol. 154, Part 2, Nov. 22, 1968, pp. 367-396.

11-11 B.H. Little: "Scaling effects on shock-induced separations". AGARD-VKI Lecture Series No. 16, April 21-25, 1969.

11-12 G.J. Hancock: "Problems of aircraft behaviour at high angles of attack". AGARDograph 136, 1969.

11-13 J.H. Patterson: "A survey of drag prediction techniques applicable to subsonic transport aircraft design". AGARD Conference on aerodynamic drag, CP-124, 1973.

11-14 S.F.J. Butler: "Aircraft drag prediction for project appraisal and performance estimation". AGARD Conference Proc. No. 124, April 1973.

11-15 J.G. Callaghan: "Aerodynamic prediction methods for aircraft at low speeds with mechanical high lift devices". AGARD Lecture Series No. 67, May 1974.

11-16 G.M. Bowes: "Aircraft lift and drag prediction and measurement". AGARD Lecture Series No. 67, May 1974.

11-17 J.K. Wimpress: "Predicting the low speed stall characteristics of the Boeing 747". AGARD CP-102 (Paper 21), April 1972.

Performance prediction - general literature

11-18 C.D. Perkins and R.E. Hage: "Airplane performance, stability and control". John Wiley, 1949.

11-19 J. Rotta: "Ueber die Flugleistungsmechanik des Flugzeuges mit Turbinenstrahltriebwerk". Zeitschrift für Flugwiss., Vol. 4, 1953.

11-20 Various contributors: "Flight Test Manual". Vol. I, Performance, AGARD, 1954.

11-21 G.S. Schairer and M.L. Olason: "Some performance considerations of a jet transport airplane". First Turbine Powered Air Transportation Meeting, IAS, August 1954.

11-22 R. Ludwig: "Ein Beitrag zur dimensionslosen Methode der Leistungsberechnungen von Flugzeugen mit Strahlantrieb". Zeitschrift für Flugwissenschaften, Vol. 6, 1955.

11-23 Anon.: Data Sheets "Performance". Vols. 1, 2 and 3. Royal Aeronautical Society and Engineering Sciences Data Unit.

11-24 A. Miele: "Flight mechanics". Vol. 1, Theory of flight paths, Pergamon Press, London, 1962.

11-25 D. Fiecke: "Stationär betrachtete Flugleistungen". Luftfahrttechnik, October 1962, pp. 266-283.

11-26 D.O. Dommasch, S.S. Sherby and T.F. Connally: "Airplane aerodynamics". Pitman Publishing Corp., New York, 4th Edition, 1967.

11-27 Boeing Cy.: "Jet transport performance methods". The Boeing Company, Commercial Airplane Group, Seattle, Washington, Boeing Document No. D6-1420, 6th Edition, May 1969.

11-28 R.L. Schultz and N.R. Zagalsky: "Aircraft performance optimization". J. of Aircraft, Vol. 9, Feb. 1972, pp. 108-114.

11-29 H. Wittenberg: "Prestatieleer van vliegtuigen". Dictaat VTH-D14, Delft University of Technology, Dept. of Aeron. Eng., July 1971 (in Dutch).

11-30 R.F. Creasy: "Propulsion/aircraft matching experience". AGARD Lecture Series No. 65, May 1974.

11-31 F.O. Smetana, D.C. Summery and W.D. Johnson: "Point and path performance of light aircraft". NASA CR-2272, June 1973.

11-32 W.J. Moran: "Performance methods for aircraft synthesis". SAWE Paper No. 909, May 1972.

Climb, cruise and descent performance

11-33 K.J. Lush: "Total energy methods". AGARD Flight Test Manual, Vol. I (Performance), 1954.

11-34 Anon.: "Estimation of cruise range and endurance". ESDU Data Sheets 73018 and 73019, Oct. 1973.

11-35 H. Friedel: "Flight-manoeuvre and climb-performance prediction". AGARD Lecture Series No. 56, April 1972.

11-36 R.K. Page: "Range and radius-of-action performance prediction for transport and combat aircraft". AGARD Lecture Series 56, April 1972.

11-37 J.J. Spillman: "Climb and descent techniques". Short course: "Aircraft performance estimation". Cranfield Institute of Technology, Feb, 1973.

11-38 J.J. Spillman: "Cruise characteristics". Short course: "Aircraft performance estimation". Cranfield Institute of Technology, Feb. 1973.

Takeoff performance, general

11-39 F. Handley Page: "Towards slower and safer flying, improved take-off and landing and cheaper
airports". J.R.Ae.S., Dec. 1950, pp. 721-739.

11-40 R.E. Gillman: "Performance standards in practice". The Aeroplane, May 13, 1955, pp. 634-638.

11-41 G.S. Alias: "Notes on the ground-run of jet-propelled aircraft during take-off and landing". AGARD
Report 82, 1956.

11-42 G. Mathias: "On the optimum utilization of an airplane high-lift device for minimum take-off run
and climb distance". Zeitschrift für Flugwissenschaften, 1961, pp. 276-284.

11-43 E. Overesch: "The problems of exact calculation of take-off and landing characteristics of conventional transport aircraft". AGARD Report 417, 1963.

11-44 B.N. Tomlinson and M. Judd: "Some calculations of the take-off behaviour of a slender wing SST
design constrained to follow a specific pitch-attitude time history". ARC R & M 3493, RAE TR 65174,
1965.

11-45 J.K. Wimpress: "Shortening the take-off and landing for high-speed aircraft". AGARD Report 501 and
Aircraft Engineering, June 1966, pp. 14-19.

11-46 W.J.G. Pinsker: "The dynamics of aircraft rotation and lift-off and its implication for tail
clearance requirements, especially with large aircraft". ARC R & M 3560, 1967.

11-47 D.P. Davies: "Handling the big jets". Second edition, Air Registration Board, 1968.

11-48 J.K. Wimpress: "Technology of take-off and landing". Annals of the New York Academy of Sciences,
Vol. 154, Part 2, Nov. 22, 1968, pp. 962-981.

11-49 T.G. Foxworth and H.F. Mathinsen: "Another look at accelerate-stop criteria". AIAA Paper No. 69-772.

11-50 D.H. Perry: "An analysis of some major factors involved in normal take-off performance". ARC CP No.
1034, 1969.

11-51 D.H. Perry: "Exchange rates between some design variables for an aircraft just satisfying take-off
distance and climb requirements". RAE Technical Report 69197, 1969.

11-52 J. Williams: "Airfield performance prediction methods for transport and combat aircraft". AGARD
Lecture Series No. 56, April 1972.

Landing performance

11-53 G.S. Alias: "Notes on the ground-run of jet-propelled aircraft during landing and take-off".
AGARD Flight Test Panel Meeting, August 1956.

11-54 Anon.: " A first approximation to the landing distance from 50 ft". R.Ae.S. Data Sheet "Performance"
EG 6/1, 1960 (See also Sheets EG 6/3 and 6/4).

11-55 R. Staubenfiel: "Computation of the shortest landing distance". DVL Bericht 130, July 1960.

11-56 J.M.N. Willis: "Effects of water and ice on landing". Shell Aviation News, Number 296, 1963, pp.
16-20.

11-57 W.J.G. Pinsker: "The landing flare of large transport aircraft". ARC R & M No. 3602, Nov. 1967.

11-58 M.D. White:"Proposed analytical model for the final stages of a landing of a transport airplane".
NASA TN D-4438, 1968.

11-59 D.H. Perry: "A first-order analysis of landing performance based on current British Civil Airworthiness Requirements". RAE Technical Memorandum Aero 132 (1970).

11-60 A.J. Walton: "Landing performance of conventional aircraft". Short course: "Aircraft performance
estimation". Cranfield Institute of Technolgy, Feb, 1933.

Economics

11-61 W.C. Mentzer and H.E. Nourse: "Some economic aspects of transport airplane performance". J. of the
Aeron. Sciences, April-May 1940.

11-62 P.G. Masefield: "Some problems and prospects in civil air transport". J.R.Ae.S., April 1955, pp. 235-248.

11-63 R.G. Thorne: "The influence of range, speed and lift-drag ratio on the operating costs of a civil aircraft ". RAE Tech. Note Aero 2487, Nov. 1956.

11-64 L.B. Aschenbeck: "Passenger air-line economics". Aeron. Eng. Review, Dec. 1966, pp. 39-43.

11-65 R.G. Thorne: "The estimation of civil aircraft direct operating costs". Aircraft Eng., Feb. 1957, pp. 56-57.

11-66 F.H. Robertsen: "Note on an improved short-cut method for estimating aircraft d.o.c.". J. of the Royal Aeron. Soc., Jan. 1957, pp. 52-53.

11-67 N.E. Rowe: "Complexity and progress in transport aircraft". J. of the Royal Aeron. Soc., Nov. 1958, pp. 787-795.

11-68 Anon.: "Standard method for the estimation of direct operating costs of transport airplanes". Society of British Aircraft Constructors, Issue No. 4, 1959.

11-69 G. Backhaus: "Einflusz der ökonomischen Forderungen auf die Verkehrsflugzeugentwicklung". Deutsche Flugtechnik, 1961, Vol. 2, pp. 48-57.

11-70 M.H. Smith and H.P. Schmidt: "A rational method for selecting business aircraft". SAE Paper No. 797C, 1964.

11-71 A.F. Watts: "Aircraft turbine engines – development and procurement costs". The Rand Corporation, Memorandum RM-4670-PR, Nov. 1965.

11-72 E.L. Courtney: "Some factors affecting fares". J. of the Royal Aeron. Soc., Nov. 1965, pp. 727-732.

11-73 R.H. Whitberg: "An airline view on cheap short-range air transport". J. of the Royal Aeron. Soc., Nov. 1965, pp. 732-739.

11-74 C.J. Hamshaw Thomas: "Steps towards lower operational costs with conventional jet transport". J. of the Royal Aeron. Soc., Nov. 1965, pp. 737-743.

11-75 A.H. Stratford: "The prospects of lower airline and airport costs". J. of the Royal Aeron. Soc., Nov. 1965, pp. 749-755.

11-76 E.H. Yates: "Cost analysis as an aid to aircraft design". J. of Aircraft, March-April 1965, pp. 100-107 (also: AIAA Paper No. 64-178).

11-77 E.C. Wells: "Some economic aspects of air transport design". Fifth Dr. Albert Plesman Memorial Lecture, Delft, September 5, 1966.

11-78 E.L. Thomas: "ATA Direct Operating Cost formula for transport aircraft". SAE Paper No. 660280.

11-79 G.S. Levenson and S.M. Barro: "Cost-estimating relationships for aircraft airframes". The Rand Corporation, Memorandum RM-4845-PR, May 1966.

11-80 C. Peyton Autry and P.J. Baumgaertner: "The design importance of airplane mile costs versus seat mile costs". SAE Paper No. 660277.

11-81 W.G. Kaldahl: "Factors affecting utilization". SAE Paper No. 660279.

11-82 Anon.: "Operating costs, the intractable other half". The Aeroplane, Dec. 22, 1966, pp. 14-15.

11-83 Anon.: "Standard method of estimating direct operating costs of turbine powered transport airplanes". Air Transport Association of America, Dec. 1967.

11-84 A.M. Jackes: "The influence of performance characteristics on the economic effectiveness of transport vehicles". AIAA Paper No. 67-802, Oct. 1967.

11-85 M. Besinger: "Einige Wirtschaftliche Aspekte beim Entwurf von Verkehrsflugzeuge". WGLR Jahrbuch 1967, pp. 68-77.

11-86 D.J. Lloyd-Jones: "Airline equipment planning". AIAA Paper No. 67-392.

11-87 A.H. Stratford: "Air transport economics in the supersonic era". Macmillan/London, Melbourne, Toronto, St. Martin's Press/New York, 1967.

11-88 R.B. Ormsby: "Total airline profit program". SAE Paper No. 690413.

11-89 J.E. Gorham: "Long term trends in airlines economics". SAE Paper No. 690414.

11-90 H.D. Kysoz: "A value analysis approach to evaluating business aircraft". Bus. & Comm. Av., May 1969, pp. 90-94, 162.

11-91 H. Wittenberg: "De vervoersprestatie en de exploitatiekosten van subsone en supersone verkeersvlieg-
tuigen voor de lange afstand". Tijdschrift voor vervoerswetenschap (Magazine for transport science),
1970, Nos. 1, 2 (in Dutch).

11-92 D.D. Hufford, J.A. Ross and K.W. Hoefs: "The economics of subsonic transport airplane design,
evaluation and operation". SAE Paper No. 710423.

11-93 R.H. Wild: "The state of the art in light aircraft design". Interavia, April 1973, pp. 346-348.

11-94 R. Jensen: "The weight/performance interface - an argument for weight control". SAWE Technical
Paper No. 967, June 1973.

11-95 G.P. Sallee: "Aircraft economics and its effects on propulsion system design". AIAA Paper No.
73-808.

CHAPTER 12.

Structural design of aircraft

12-1 A.E. Russell: "Some factors affecting large transport airplanes with turboprop engines". J.R.Ae.S.
Feb. 1950, pp. 67-106.

12-2 G. de Vries: "Safeguards against flutter of airplanes". NACA TM1423, Translated from La Recherche
Aéronautique, No. 12, 1949 and No. 13, 1950.

12-3 E.D. Keen: "Integral construction, application to aircraft design and effect on production methods".
J.R.Ae.S., Vol. 57, April 1953, pp. 215-227.

12-4 O. Ljungström: "Wing structures of future aircraft". Aircraft Eng., Vol. 25, No. 251, May 1953, pp.
128-132.

12-5 E.F. Bruhn and A.F. Schmitt: "Analysis and design of aircraft structures". Tri-State Offset Company,
Ohio, 1958.

12-6 A.F. Newell: "Impressions of the structural design of American civil aircraft". The Aeroplane, Aug.
15 and 22, 1958.

12-7 A.J. Troughton: "Relationship between theory and practice in aircraft structural problems". J.R.Ae.S.,
Nov. 1960, pp. 653-667.

12-8 A.F. Newell: "Recent British progress in aeronautics. Part One: Structural design". Aircraft Eng.,
Sept. 1961, pp. 248-254.

12-9 A.F. Newell and D. Howe: "Aircraft design trends". Aircraft Eng.,Vol. 34, No. 399, May 1962, pp. 131-
139.

12-10 D.M. Mc.Elhinney: "Structural design of the Vickers VC-10". Aircraft Eng., June 1962, pp. 165-171,
178.

12-11 K. Bentley: "Structural design of the BAC 1-11". Aircraft Eng., May 1963, pp. 142-147.

12-12 E.E. Sechler and L.G. Dunn: "Airplane structural analysis and design". Dover Publications, New York,
1963.

12-13 A.C. Kermode: "The aeroplane structure". Pitman & Sons, London, second edition, 1964.

12-14 Anon.: "Trident structural design". Aircraft Eng., June 1964, pp. 166-171.

12-15 P.L. Sandoz: "Structural considerations for long-haul transport aircraft". AIAA Paper No. 66-882.

12-16 W.T. Shuler: "Large cargo airplane structural considerations". SAE Paper No. 660669.

12-17 G. Gerard: "Structural guidelines for materials development; some vehicle performance and design
generalizations". AIAA Paper No. 68-331.

12-18 K.L. Sanders: "High-lift devices, a weight and performance trade-off methodology". SAWE Paper No.
761, May 1969.

12-19 Anon.: "Potential structural materials and design concepts for light aircraft". NASA CR-1285, March
1969.

12-20 W.M. Laurence: "Special structural considerations for wide-body commercial jet transports". AIAA
Paper No. 70-845.

12-21 D.P. Marsh: "Post-design analysis for structural weight estimation". SAWE Technical Paper No. 936, May 1972.

12-22 P.L. Sandoz: "Structural design of future commercial transports". AIAA Paper No. 73-20.

12-23 A.I. Gudkov and P.S. Leshakov: "External loads and aircraft strength". NASA TT F-753, July 1973.

Appendix A. Definitions relating to the geometry and aerodynamic characteristics of airfoils

SUMMARY

This appendix deals with the most common definitions of some major geometric and aerodynamic characteristics of wings with rounded noses and sharp trailing edges. Formulas
and graphs are presented for calculating the Mean Aerodynamic Chord and the mean quarter-
chord point for straight-tapered wings, with or without prismoidal center section.

Nomenclature

A   - aspect ratio
a.c.   - aerodynamic center
b   - span
c   - chord
$\bar{c}$ - chord length of the MAC
c_g - chord length of the SMC
c.p.   - center of pressure
e   - fraction of the chord
i   - wing angle of incidence
MAC   - Mean Aerodynamic Chord
O   - origin of system of axis (wing apex)
S   - gross wing area
S_{net} - net wing area
S_{wet} - wetted (exposed) wing area
SMC   - Standard Mean Chord
t   - maximum airfoil section thickness
X   - wing axis, coinciding with root chord
x   - coordinate in X-direction
$\bar{x}$ - x-coordinate of mean quarter-chord point
Y   - lateral axis
y   - ordinate of section, measured from chord line; lateral coordinate of a wing section, measured from X-axis
$\bar{y}$ - y-coordinate of mean quarter-chord point of a half wing
y_c - mean-line ordinate of a section
Δy - leading edge sharpness parameter
Z   - axis perpendicular to XOY-plane
z   - coordinate in Z-direction
$\bar{z}$ - z-coordinate of mean quarter-chord point
α - angle of attack
α_{L_o} - wing zero-lift angle
α_{ℓ_o} - section zero-lift angle
Γ - dihedral angle
ε - twist angle
ε_a - aerodynamic twist angle
ε_g - geometric twist angle
η - non-dimensional lateral coordinate
Λ - angle of wing sweep
λ - taper ratio
ϕ - semi-span (non-dimensional) of prismoidal center section; tail angle

Subscripts

r   - root
t   - tip

A-1. GENERAL

This appendix deals with definitions of the geometry and some aerodynamic properties of symmetrical wings. Only wing sections with rounded noses and sharp trailing edges are considered. Slender wings such as low aspect ratio delta wings are not treated. Most definitions are also applicable to horizontal tailplanes and subscripts w or h have therefore been omitted.
Various formulas will be presented without proof. More details and background can be found in the publications mentioned in the list of references.

A-2. WING SECTIONS

A-2.1. Geometric definitions

A WING SECTION is formed by the external contour of a wing cross section with a plane parallel to the plane of symmetry of the wing, or a plane perpendicular to the quarter-chord line. The cambered wing sections of most conventional families of airfoils are obtained by the combination of a MEAN LINE and a THICKNESS DISTRIBUTION (Fig. A-1).

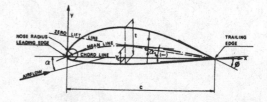

Fig. A-1. Geometric definitions of a wing section

The LEADING and TRAILING EDGE are defined, respectively, as the forward and rearward extremities of the mean line.

The CHORD LINE is defined as the straight
line connecting the leading and trailing
edges. Its length is defined as the CHORD
LENGTH or simply THE CHORD (c).
SECTION THICKNESS (t) is the maximum distance between corresponding points on the
upper and lower section surface (Fig. A-1).
It is usually expressed as a thickness/
chord ratio (t/c).
The NOSE RADIUS defines the sharpness of
the section at the leading edge. Its center is located on the tangent to the mean
line, drawn at .5% of the chord.
CAMBER is the distance between corresponding points on the mean line and the chord
line; mean line ordinate: y_c. The maximum
camber is frequently referred to as "the
camber".

The following geometric parameters have also proved to be useful in the prediction of
certain aerodynamic characteristics of wing
sections:
LEADING EDGE SHARPNESS PARAMETER:

$$\frac{\Delta y}{c} = \frac{y_{6\%} - y_{.15\%}}{c} \times 100\% \qquad (A-1)$$

Values for Δy for several NACA sections are
presented in Fig. A-2.
TAIL ANGLE (ϕ) is the angle between the
tangents to the upper and lower surfaces
at the trailing edge (Figs. A-1 and A-3).

A-2.2. Aerodynamic definitions

ANGLE OF ATTACK (α): angle between the
chord line and the direction of the undisturbed flow. (Positive α: nose-up, see
Fig. A-1).
THE ZERO LIFT ANGLE (α_l): angle of attack
for zero lift. Positive α_{l_o}: nose up; for
sections with positive camber α_{l_o} is negative, as in Fig. A-1.
ZERO-LIFT LINE: line through the trailing
edge parallel to the direction of the undisturbed airflow at α_{l_o}.
CENTER OF PRESSURE (c.p.): the intersection
of the vector, representing the resulting
aerodynamic force, with the chord line. The

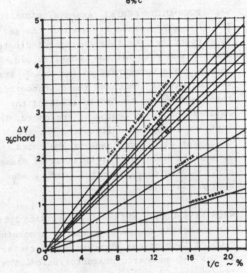

Fig. A-2. Leading edge sharpness parameter
of NACA sections (Ref. A-6)

c.p. varies with α for cambered sections.
AERODYNAMIC CENTER: a point about which the
pitching moment for given dynamic pressure
is essentially independent of the angle of
attack up to maximum lift in subcritical
flow. For design purposes it is generally
acceptable to assume the a.c. as .25 c

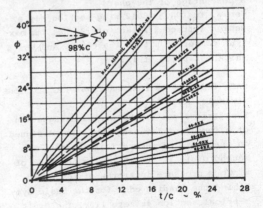

Fig. A-3. Tail angle of NACA sections
(Ref. A-6)

(low-subsonic speeds).

A-2.3. Nomenclature for some NACA sections

NACA FOUR-DIGIT SERIES (example: NACA 2415).
The numbering system is based on the section geometry. The first integer indicates
the maximum value of the mean-line ordinate y_c as a percentage of the chord. The
second integer indicates the distance from
the leading edge to the location of the
maximum camber in tenths of the chord. The
last two integers indicate the section
thickness as a percentage of the chord.
Hence, the NACA 2415 section has 2% camber
at .4 of the chord from the leading edge
and is 15% thick.

NACA FIVE-DIGIT SERIES (example: NACA 23012)
The numbering system is based on a combination of theoretical aerodynamic characteristics and geometric characteristics. The
first integer indicates the amount of camber in terms of the relative magnitude of
the design lift coefficient*; the design
lift coefficient is three halves of the
first integer. The second and third integers together indicate the distance from
the leading edge to the location of the
maximum camber; this distance as a percentage of the chord is one-half the number represented by these integers. The last two
integers indicate the section thickness as
a percentage of the chord.
Hence, the NACA 23012 wing section has a
design lift coefficient of .3, has its maximum camber at .15 of the chord, and has a
thickness ratio of 12%.

NACA 6-SERIES OR LAMINAR-FLOW WING SECTIONS
(example: NACA 65_3-218)
The first digit is the series designation
of a family of low-drag airfoils, designed
on the basis of theoretical methods. The
second digit is the chordwise position of
minimum pressure in tenths of the chord
behind the leading edge for the basic symmetrical section at zero lift. The digit

*Definition in Ref. A-1.

438

following the comma or written as a subscript gives the range of lift coefficient
in tenths above and below the design lift
coefficient in which pressure gradients
favorable for obtaining low drag exist on
both surfaces. The digit following the dash
gives the design lift coefficient in tenths.
The last two digits represent the thickness
of the wing section as a percentage of the
chord.
Hence, NACA 65_3-218 is a 6-series section.
The point of minimum pressure at zero lift
for the 65-018 section is at .50 of the
chord. The favorable range of lift coefficients is between -.1 and .5, the design
lift coefficient is .2 and the maximum
thickness 18% of the chord.
Some modifications of the 6-series sections
are designated by replacing the dash by a
capital A. These sections are substantial-
ly straight on both surfaces from about .8
of the chord to the trailing edge. The 6-
series airfoils without this designation
have cusped trailing edges.

A-3. WINGS

A-3.1. Wing planform

A system of axes is generally used as indicated in Fig. A-4. The origin is located

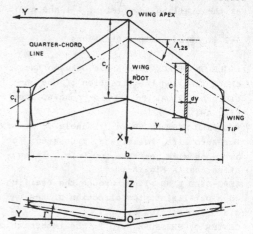

Fig. A-4. System of axis in relation to
the wing planform

at the WING APEX 0, which is the intersection of the leading edges of both wing halves. The X axis coincides with the chord of the root section, positive backwards. The Y axis is perpendicular to the plane of symmetry, positive to port side. The Z axis is positive upwards.

LEADING and TRAILING EDGE: the lines through the leading and trailing edges of the wing chords.

ROOT CHORD and TIP CHORD (c_r, c_t): the chord length of the wing sections in the plane of symmetry and at the outer extremity of the wing, respectively. For rounded tips the definition of the tip chord is given in Fig. A-4.

TAPER RATIO (λ) is the ratio of the tip chord to the root chord:

$$\lambda = c_t/c_r \tag{A-2}$$

Straight-tapered wings have straight leading and trailing edges.

GROSS or DESIGN WING AREA (S): the area enclosed by the wing outline, including wing flaps in the retracted position and ailerons, but excluding fillets or fairings, projected on the XOY plane. The leading and trailing edge are assumed to be extended through the nacelles and fuselage to the XOZ plane in any reasonable manner (example in Fig. A-5)

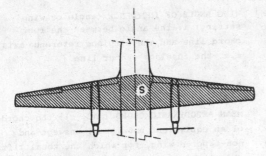

Fig. A-5. Definition of gross wing area

$$S = 2 \int_0^{b/2} c\, dy \tag{A-3}$$

NET WING AREA (S_{net}): the gross wing area minus the projection of the central wing part, inside the fuselage.

EXPOSED or WETTED WING AREA * (S_{wet}): the net external wing surface area, which is exposed to the airflow.

SPAN (b): the distance between the wing tips, measured perpendicular to the XOZ plane, navigation lights excluded.

GEOMETRIC or STANDARD MEAN CHORD (SMC): the length of this chord (c_g) is equal to the gross area, divided by the span:

$$c_g = S/b \tag{A-4}$$

For straight-tapered wings:

$$c_g = c_r \frac{1 + \lambda}{2} \tag{A-5}$$

ASPECT RATIO (A): the span divided by the geometric mean chord:

$$A = b/c_g = b^2/S. \tag{A-6}$$

For a straight tapered wing the root chord can be calculated for given gross area, aspect ratio and taper ratio from:

$$c_r = \frac{2}{1 + \lambda} \sqrt{S/A} \tag{A-7}$$

QUARTER-CHORD LINE: the line through all points at .25 c of the sections.

(ANGLE OF) DIHEDRAL ($\Gamma_{.25}$): the angle between the projection of the quarter-chord line on the YOZ plane and the Y axis (positive upwards, Fig. A-4). Negative angle: ANHEDRAL.

SWEEP ANGLE of the quarter-chord line ($\Lambda_{.25}$): the angle between the projection of the quarter-chord line on the XOY plane and the Y axis. Positive angle backwards (sweepback), negative forwards (sweep forward).

Sweep angles of other characteristic lines (trailing edge, leading edge, mid-chord line) have a similar definition and can be calculated for a straight-tapered wing from the following relationship:

*In some publications the term "exposed wing area" has the same meaning as the net wing area defined above, i.e. a projected area.

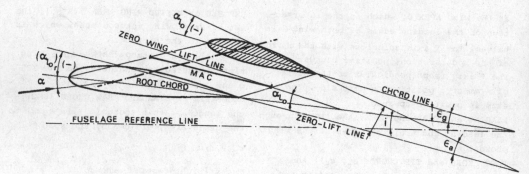

Fig. A-6. Definition of wing twist

$$\tan \Lambda_2 = \tan \Lambda_1 + \frac{4}{A} \frac{1-\lambda}{1+\lambda} (e_1 - e_2)$$

where e_1 and e_2 denote the fraction of the chord according to the definition of the sweep angles Λ_1 and Λ_2, respectively. For example, the trailing edge sweep angle can be calculated from $\Lambda_{.25}$ by substitution of $e_1 = .25$ and $e_2 = 1.0$. For a wing (or tailplane) with straight trailing edge ($\Lambda_2 = 0$) we have:

$$\tan \Lambda_{.25} = \frac{3(1-\lambda)}{A(1+\lambda)} \qquad (A-8)$$

A-3.2. (Wing) twist and incidence

(WING) TWIST (ε) is the angle of incidence of a wing section relative to that of the root section, measured in a plane parallel to the XOZ plane (Fig. A-6). Positive twist: nose rotated upwards, WASH-IN. Negative twist: nose rotated downwards, WASH-OUT.

Geometric twist (ε_g) is the twist of the chord line of a section relative to the chord line of the root section.

Aerodynamic twist (ε_a) is the twist of the zero-lift line of a section relative to the zero-lift line of the root section. For an arbitrary section the geometric and aerodynamic twist are interrelated as follows:

$$\varepsilon_a = \varepsilon_g + \left(\alpha_{\ell_o}\right)_r - \alpha_{\ell_o} \qquad (A-9)$$

The WING TWIST ANGLE usually refers to the twist angle at the tip chord:

$$\left(\varepsilon_a\right)_t = \left(\varepsilon_g\right)_t + \left(\alpha_{\ell_o}\right)_r - \left(\alpha_{\ell_o}\right)_t \qquad (A-10)$$

For a wing with LINEAR TWIST, the aerodynamic twist angle increases in proportion to the lateral coordinate:

$$\varepsilon_a = \eta \left(\varepsilon_a\right)_t \qquad (A-11)$$

where

$$\eta = \frac{y}{b/2} \qquad (A-12)$$

A LINEAR LOFTED (GEOMETRIC) TWIST is obtained on a wing where the intermediate sections are formed by linear lofting between the root and tip sections:

$$\varepsilon_g = \left(\varepsilon_g\right)_t \frac{\lambda\eta}{1 - (1 - \lambda)\eta} \qquad (A-13)$$

WING ANGLE OF INCIDENCE (angle of wing setting) i: the angle between the root chord line and the airplane reference axis, e.g. the fuselage center line.

A-3.3. Aerodynamic definitions

MEAN AERODYNAMIC CHORD (MAC, $\bar{c}$): the chord of an equivalent untwisted, unswept and non-tapered wing, for which the total lift and pitching moment are essentially equal to the lift and pitching moment on the actual wing. The derivation of $\bar{c}$ can be found in Ref. A-9:

$$\bar{c} = \frac{2}{S} \int_0^{b/2} c^2 \, dy \qquad (A-14)$$

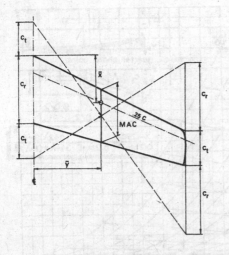

Fig. A-7. Graphical construction of the MAC

For straight-tapered wings

$$\bar{c} = \frac{2}{3} c_r \frac{1 + \lambda + \lambda^2}{1 + \lambda} \qquad (A-15)$$

Alternatively, the simple graphic construction depicted in Fig. A-7 can be used to find the length and location of the MAC of a wing half. For wings with prismoidal inboard sections and tapered outboard sections, Fig. A-8 may be used.

AERODYNAMIC CENTER (a.c.): a point in the XOZ plane about which the aerodynamic pitching moment coefficient of the wing is essentially constant up to maximum lift in subcritical flow.

For moderate sweep angle a reasonable approximation for the a.c. is the MEAN QUARTER-CHORD POINT. For a half wing, this point is located at .25 $\bar{c}$, with coordinates:

$$\bar{x} = .25 \, c_r + \frac{2 \tan \Lambda}{S} .25 \int_0^{b/2} cy \, dy \quad (A-16)$$

$$\bar{y} = \frac{2}{S} \int_0^{b/2} cy \, dy \qquad \text{(half wing)} \qquad (A-17)$$

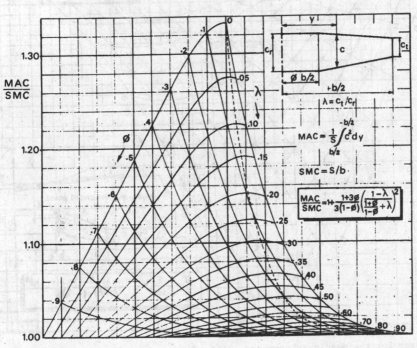

Fig. A-8. Diagram for the Mean Aerodynamic Chord of straight-tapered wings with or without prismoidal inboard section

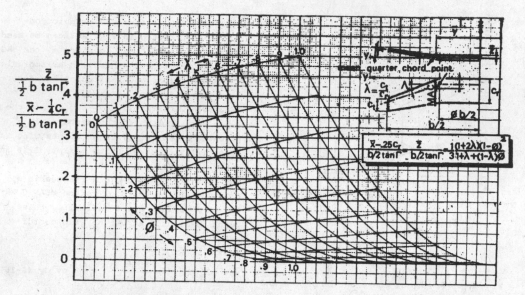

Fig. A-9. Diagram for the mean quarter-chord point for straight-tapered wings with or without prismoidal center section

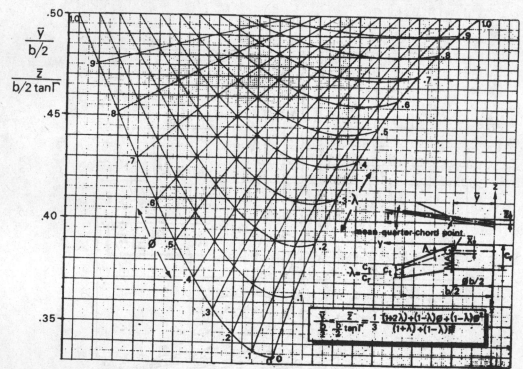

Fig. A-10. Diagram for the mean quarter-chord point of straight-tapered wings with or without prismoidal center section

$$\bar{z} = \frac{2 \tan \Lambda_{.25}}{S} \int_{o}^{b/2} cy \, dy \qquad (A-18)$$

For straight-tapered wings the MAC is located at the lateral coordinate:

$$\bar{y} = \frac{b}{2} \frac{1 + 2\lambda}{3(1 + \lambda)} \qquad \text{(half wing)} \qquad (A-19)$$

while the quarter-chord point of this chord is identical to the mean quarter-chord point. Figs. A-9 and A-10 can be used for more complex wings shapes.

WING ANGLE OF ATTACK (α): the angle between the root chord and the direction of the undisturbed airflow (Fig. A-6).

THE ZERO-LIFT LINE is drawn through the trailing edge of the MAC, parallel to the undisturbed airflow, in a position at which the wing lift is zero (Fig. A-6).

ZERO-LIFT ANGLE (α_{L_o}) is the angle between the root chord and the zero-lift line; α_{L_o} is positive when the root chord nose is directed upwards relative to the airflow (Fig. A-6).

REFERENCES

A-1. I.H. Abbott, A.E. von Doenhoff: "Theory of Wing Sections". Dover Publications, Inc., New York, 1958.

A-2. F.W. Riegels: "Aerodynamische Profile". Published by R. Oldenbourg, Munich, 1958.

A-3. R.C. Pankhurst: NPL Aerofoil Catalogue and Bibliography ARC R & M No. 3311, 1963.

A-4. T. Nonweiler: "The Design of Wing Sections". Aircraft Engng., July 1956, pp. 216-227.

A-5. Anon.: R.Ae.S. Engineering Sciences DATA Sheet WINGS 01.01.05, Oct. 1958.

A-6. D.E. Hoak and J.W. Carlson (red): "USAF Stability and Control Handbook", prepared by Douglas Aircraft Cy., Oct. 1960. Rev. 1968.

A-7. Anon.: "Dictionary of Technical Terms for Aerospace Use", NASA SP-7, First Edition, 1965.

A-8. W.S. Diehl: "The Mean Aerodynamic Chord and the Aerodynamic Centre of a Tapered Wing". NACA Report 751, June 1942.

A-9. A.H. Yates: "Notes on the Mean Aerodynamic Chord and the Mean Aerodynamic Centre of a Wing". Journal of the Royal Aeron. Soc., June 1952, pp. 461-474.

Appendix B. The computation of circumferences, areas and volumes of curves, sections and bodies

SUMMARY

Methods are presented for the calculation of circumferences, projected areas, wetted areas and volumes of sections, fuselages, wings and tailplanes, fuel tanks and engine nacelles.
The methods presented are simplified results derived from Ref. B-1; they can be readily applied in the preliminary stage of an aircraft design in cases where the shape and dimensions are not known in great detail.

NOMENCLATURE

A_c — cross-sectional area; area of a wing section

b — span (no index: wing span); width

C — circumferential length

c — chord length of an airfoil section

D — diameter

h — height

k — factor for calculating areas, volumes, etc.

l — length

S — area (no index: gross wing area)

S_{net} — net wing area

t — maximum thickness of an airfoil section

β — (fan) nacelle forebody length/total length

λ — wing taper ratio; fineness ratio

τ — ratio of t/c ratios at tip and root

φ — shape parameter

Subscripts

A — area enclosed by a curve

C — circumference

c — cross section; cylindrical mid-section

ef — fan exhaust opening

eg — gas generator exhaust opening

f — fuselage

g — gas generator

h — highlight of an intake

n — nacelle; nose section of fuselage

p — plug

t — tail section of fuselage; fuel tank

V — volume

W — external area

wet — wetted area

B-1. FUSELAGE

The present methods refer to the gross wetted area and volume of streamline bodies by which most fuselages can be approximated. Cockpit hoods, air scoops, fillets and the like, as well as non-exposed parts at the junction of wing and empennage, must be calculated separately, e.g. with the data on wing sections (Section B-2), and subtracted.

B-1.1. General method

The generalized curves in Fig. B-1 are derived from approximations of the external lines by polynomials with fractional exponents (Ref. B-1). In most cases the fuselage can be divided into a nose section, a cylindrical mid-section and a tail section. The shape parameter φ is measured from the plan view and the factors k_A, k_C, k_V and k_W are then determined in order to calculate the characteristic areas and volume as follows (symbol in Fig. B-2):

Cross-sectional area* (frontal area):

$$A_c = k_A \, b_{f_{max}} \, h_{f_{max}} \qquad (B-1)$$

Circumferential length of the cross section:

$$C_f = 2 \, k_C \, (b_{f_{max}} + h_{f_{max}}) \qquad (B-2)$$

where k_C is obtained from Fig. B-1. If necessary, the cross section is subdivided into several parts.
Fuselage volume:

$$V_f = A_c \, (l_c + k_{V_n} \, l_n + k_{V_t} \, l_t) \qquad (B-3)$$

where k_V is found by measuring φ in the plan view for the nose section (n) and the tail section (t) and using Fig. B-1.
Fuselage wetted area:

$$S_{f_{wet}} = C_f \, (l_c + k_{W_n} \, l_n + k_{W_t} \, l_t) \qquad (B-4)$$

*at the fuselage station where the width and height are maximum

446

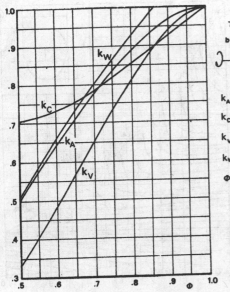

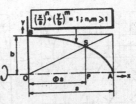

$k_A = \dfrac{\text{AREA OF SECTION OBSA}}{a \cdot b}$

$k_C = \dfrac{\text{CIRCUMFERENCE BSA}}{a + b}$

$k_V = \dfrac{\text{VOLUME OF BODY OF REVOLUTION}}{\pi a b^2}$

$k_W = \dfrac{\text{WETTED AREA OF BODY OF REV.}}{2\pi a b}$

$\Phi = \dfrac{OP}{OA}$

Fig. B-1. Factors for calculating the area, circumference, volume and wetted area of sections and bodies

FRONT VIEW   PLANVIEW

Fig. B-2. Definition of streamline body geometry

where C_f is given by (B-2) and the factor k_W is obtained from Fig. B-1.

Projected areas of plan view and side view can be calculated in the same way as cross-sectional areas by subdividing them into several parts and using the appropriate k_A factors given in Fig. B-1.

For a cross section built up from two circular segments, the frontal area, volume and wetted area are first calculated from the plan view, assuming the fuselage to be a body of revolution. Correction factors given in Fig. B-3 are then applied.

B-1.2. Quick method for bodies of revolution

The following approximations apply to fuselages with cylindrical mid-sections:

$$\text{volume} = \frac{\pi}{4} D_f^2 \, l_f \left(1 - \frac{2}{\lambda_f}\right) \qquad (B-5)$$

$$\left.\text{wetted area} = \pi D l_f \left(1 - \frac{2}{\lambda_f}\right)^{2/3} \left(1 + \frac{1}{\lambda_f^2}\right)\right\}_{\lambda_f \geqslant 4.5} \qquad (B-6)$$

where λ_f is the fuselage fineness ratio: $\lambda_f = l_f / D_f$.

For fully streamlined shapes without cylindrical mid-section the following expressions apply:

$$\text{volume} = \frac{\pi}{4} D_f^2 \, l_f \left(.50 + .135 \frac{l_n}{l_f}\right) \qquad (B-7)$$

$$\text{wetted area} = \pi D l_f \left(.50 + .135 \frac{l_n}{l_f}\right)^{2/3} \times$$

$$\left(1.015 + \frac{.3}{\lambda_f^{1.5}}\right) \qquad (B-8)$$

where l_n is the length of the nose section in front of the maximum cross section.

B-2. WINGS AND TAILPLANES

For most subsonic airfoil sections the following simple expressions are reasonably accurate:

$$C = 2c \, (1 + .25 \, t/c) \qquad (B-9)$$

447

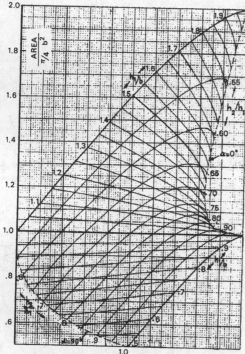

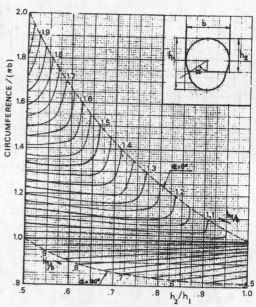

Fig. B-3. Area and circumference of double-bubble and flattened fuselage cross sections

$$A = .68 \, tc \qquad (B-10)$$

A more accurate result can be obtained with Fig. B-1.

The wetted area of a linear lofted wing without nacelles is derived by integration of the circumferential length in a spanwise direction. A simple approximation for the result of this procedure is:

$$S_{wet} = 2S_{net} \left| 1 + .25 \, (t/c)_r \, \frac{1 + \tau\lambda}{1 + \lambda} \right| \qquad (B-11)$$

where $\lambda = c_t/c_r$ and $\tau = (t/c)_t / (t/c)_r$. In this case the root section is not taken at the wing centerline, as usual, but at the wing-fuselage intersection. For a wing with nacelles the wetted area should be reduced by an amount equal to the total area of the wing inside the nacelle structure.

B-3. FUEL TANK VOLUME

a. Most bladder tanks can be compared with a geometric body having parallel end faces.

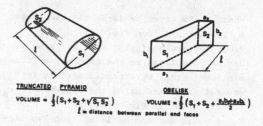

Fig. B-4. Volume of bodies with parallel end faces

Their volume is obtained from the appropriate formula given in Fig. B-4.

b. For integral tanks the external shape of the structure enclosing the tank may be used to apply Fig. B-4. In this case the volume of the structure, which is equivalent to about 4% of the tank volume, should be subtracted to obtain the net tank volume.

c. External streamlined fuel tanks can be treated in the same way as the fuselage. The net tank volume is 4% less than the volume enclosed by the skin.

In project design a check on the available tank capacity is usually made in order to check the wing size or thickness/chord ratio. A first guess for the total tank volume available in a wing with linear lofted intermediate airfoils is:

$$V_t = .54 \frac{S^2}{b} (t/c)_r \frac{1+\lambda\sqrt{\tau}+\lambda^2\tau}{(1+\lambda)^2} \qquad \text{(B-12)}$$

where

S	= gross wing area
b	= wing span
$(t/c)_r$	= thickness/chord ratio at the wing root
λ	= taper ratio
τ	= $(t/c)_t / (t/c)_r$

Statistical data were used to calculate the constant .54 in (B-12) but the accuracy is not very high. Thus, if the available tank volume appears to be critical, a more precise calculation is necessary to account for the actual section shape and the wing structural layout.

Equation B-12 may also be used for calculating the tank capacity of a part of the wing. In that case the geometric definitions of S, b, etc. apply only to the wing part containing fuel.

Note that the Usable Fuel Capacity* is some 5% less than the tank volume to allow for expansion of the fuel.

B-4. ENGINE NACELLES AND AIR DUCTS

In the most general case, the engine nacelle group may consist of a fan cowling, a gas generator cowling and a plug in the hot flow (Fig. B-5). The wetted areas of these components may be computed as follows:

External wetted area of fan cowling:

$$l_n D_n \left[2 + .35\beta + .8\beta \frac{D_h}{D_n} + 1.15(1-\beta) \frac{D_{ef}}{D_n} \right] \qquad \text{(B-13)}$$

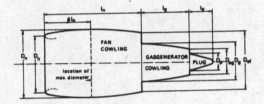

Fig. B-5. Geometry of a turbofan engine pod

Wetted area of gas generator cowling:

$$\pi l_g D_g \left[1 - \frac{1}{3} (1 - \frac{D_{eg}}{D_g}) \left\{ 1 - .18 \left(\frac{D_g}{l_g} \right)^{\frac{5}{3}} \right\} \right] \qquad \text{(B-14)}$$

Wetted area of plug:

$$.7 \pi l_p D_p \qquad \text{(B-15)}$$

REFERENCES

B-1. E. Torenbeek: "The computation of characteristic areas and volumes of major aircraft components in project design". Delft University of Technology, Dept. of Aeron. Engng., Memorandum M-188, Feb. 1973.

B-2. A.H. Schmidt: "A simplified method for estimating the wetted area of aircraft fuselages". SAWE Technical Paper No. 308, 1962.

B-3. R.T. Bullis: "Geometric analysis". SAWE Technical Paper No. 1025, May 1974.

*defined in Section 8.2.2.

Appendix C. Prediction of wing structural weight

SUMMARY

The derivation of the present method (Ref. C-1) is based on a generalized expression for
the material required to resist the root bending moment due to wing lift in a specific
operational condition. The average stress is related to the loading index of the compression structure. A correction is made for the extra weight to provide the torsional
stiffness required to withstand wing flutter. Separate contributions for high-lift devices, spoilers and speed brakes are given. Statistical analysis of the wing weight of
many aircraft types actually built has yielded the necessary factors of proportionality.
The method is applicable to light and transport category aircraft with wing-mounted engines in front of the elastic axis or engines not mounted to the wing.

b - wing span

b_{fs} - structural flap span (see Fig. C-1)

b_s - structural wing span
($b_s = b \cos^{-1} \Lambda_{1/2}$)

C_{L_α} - lift-curve slope of the wing
($C_{L_\alpha} = dC_L/d\alpha$)

c - chord length

K_g - gust alleviation factor

k - correction factors for taper ratio, non-optimum weight, etc.

M_c - design cruising Mach number

N_e - number of wing-mounted engines

n_{ult} - ultimate load factor (1.5 times limit load factor)

S - gross projected wing area

S_f - projected area of flaps

t - maximum thickness of wing section or flap section

t/c - thickness/chord ratio

V_C - Design Cruising speed (EAS)

V_D - Design Diving speed (EAS)

V_{lf} - design speed for flaps in landing configuration (EAS)

W - weight

W_{des} - design All-Up Weight of aircraft

W_w - wing group weight as defined in AN 9103-D

$\Lambda_{1/2}$ - sweep angle of mid-chord line (Fig. C-1)

Λ_f - average sweep angle of flap structure (Fig. C-1)

δ_f - maximum flap deflection angle at V_{lf}, measured streamwise

η_s - distance of the strut mounting on a braced wing from the wing root, divided by the wing semi-span (η defined by equation A-12)

λ - taper ratio ($\lambda = c_t/c_r$)

SUBSCRIPTS

b - basic

e - engines

f - flap

hld - high-lift devices

lef - leading-edge devices

r - wing root

sp - spoilers; speed brakes

t - wing tip

tef - trailing-edge devices

w - wing

C-1 INTRODUCTION

The method presented in this Appendix is based on Ref. C-1, where the primary wing structure weight is derived from the requirement that in a specified critical flight condition the bending moment due to wing lift must be resisted. The weight of high-lift devices is based on a critical loading condition at the flap design speed. Application of the method to several high-subsonic short-haul airliners led to the conclusion that for this category the original method in Ref. C-1 results in an underestimation of the wing weight. It was assumed that the main reasons for this are the extra weight required to provide adequate stiffness against wing flutter and the weight penalty due to the long service life required, resulting in a rather low level flight stress level. Only the first of these aspects has resulted in a modifi-

cation of the original formula. Other minor modifications have been introduced in the engine relief factor and in the weight of the high-lift devices. A weight penalty for spoilers and speed brakes has been introduced.

C-2. BASIC WING STRUCTURE WEIGHT

The weight of the basic wing structure (i.e. the wing group weight less the weight of high-lift devices, spoilers and speed brakes) is given by the equation:

$$W_{w_{basic}} = \text{constant} \times k_{no} k_\lambda k_e k_{uc} k_{st} \times$$
$$\left[k_b n_{ult} (W_{des} - .8 W_w) \right]^{.55} \times$$
$$b^{1.675} (t/c)_r^{-.45} (\cos\Lambda_{1/2})^{-1.325}$$

(C-1)

where the constant is:
8.94x10^{-4}, W_w and W_{des} in lb, b in ft, or
4.58x10^{-3}, W_w and W_{des} in kg, b in m.
For geometrical definitions see Fig. C-1.

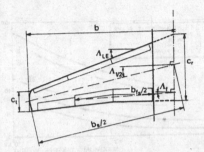

Fig. C-1. Geometric definitions

The correction factors are defined as
follows:

$$k_{no} = 1 + \sqrt{\frac{b_{ref}}{b_s}} \qquad (C-2)$$

where b_{ref} = 6.25 ft (1.905 m). The factor
k_{no} represents the weight penalties due to
skin joints, non-tapered skin, minimum
gauge, etc.

$$k_\lambda = (1 + \lambda)^{.4} \qquad (C-3)$$

where λ is the wing taper ratio, $\lambda = c_t/c_r$.
The bending moment relief factor due to
the engine and nacelle installation is:

k_e = 1.0 engines not wing-mounted
    .95 two wing-mounted engines in front
       of the elastic axis
    .90 four wing-mounted engines in front
       of the elastic axis

The correction factor for undercarriage
suspension is:

k_{uc} = 1.0 wing-mounted undercarriage
     .95 undercarriage not mounted to wing

For the extra weight required to provide
stiffness against flutter the following
correction is proposed for high-subsonic
jet aircraft with engines not mounted to
the wing or two engines wing-mounted in
front of the elastic axis of the wing:

$$k_{st}=1+\text{constant} \times \frac{(b \cos\Lambda_{LE})^3}{W_{des}} \left\{\frac{V_D/100}{(t/c)_r}\right\}^2 \cos\Lambda_{1/2}$$
$$(C-4)$$

where the constant is 1.50x10^{-5} when b is
in ft, W_{des} in lb and V_D in kts, or 9.06x
10^{-4} when b is in m, W_{des} in kg and V_D in
m/s. For low-subsonic aircraft and high-
subsonic aircraft with four wing-mounted
jet engines k_{st} = 1.0.
The correction factor for strut location
on braced wings (k_b = 1.0 for cantilever
wings) is:

$$k_b = 1 - \eta_s^2 \qquad (C-5)$$

The maximum weight with wing tanks empty
should be used for the design weight W_{des}
in eq. C-1, although the Maximum Zero Fuel
Weight is acceptable for most preliminary
designs.
The ultimate load factor n_{ult}, equal to
1.5 times the limit load factor, is the
higher of the gust and the maneuver load
factor. The gust load is based on the design cruising speed V_c at an altitude of
20,000 ft for pressurized cabins and sea
level for non-pressurized cabins.
The load factors are determined in accordance with the airworthiness regulations
FAR 23.341, FAR 25.341 or BCAR Section D
ch. 3.1.4., as appropriate. The original
formula in Ref. C-1 is based on the simplified gust alleviation factor found in
earlier regulations:

$$K_g=.8-\frac{1.6}{(W_{des}/S)^{3/4}} \quad \text{for } W_{des}/S>16 \text{ lb/sq.ft}$$

$$K_g=.3\times(W_{des}/S)^{1/4} \quad \text{for } W_{des}/S \leq 16 \text{ lb/sq.ft}$$
$$(C-6)$$

In this equation W_{des}/S is in lb/sq.ft.
The gust load is determined from the
following expression for the wing lift-
curve slope:

$$C_{L_\alpha} = \frac{2\pi}{2/A+\sqrt{\frac{1}{\cos^2\Lambda_{1/2}}-M_c^2+(2/A)^2}} \quad (\text{rad}^{-1}) (C-7)$$

453

where M_c corresponds to V_c in the operational condition mentioned previously.

C-3. HIGH-LIFT DEVICES, SPOILERS AND SPEED BRAKES

The weight of high-lift devices can be broken down as follows:

$$W_{hld} = W_{tef} + W_{lef} \qquad (C-8)$$

The trailing-edge flap weight is calculated from:

$$\frac{W_{tef}}{S_f} = \text{constant} \times k_f \, (S_f b_{fs})^{3/16} \times$$
$$\left[\left(\frac{V_{lf}}{100} \right)^2 \frac{\sin\delta_f \, \cos\Lambda_f}{(t/c)_f} \right]^{3/4} \qquad (C-9)$$

where the constant is equal to .105 when W_{tef} is in lb, S_f in sq.ft, b_{fs} in ft and V_{lf} in kts, or 2.706 when W_{tef} is in kg, S_f in m^2, b_{fs} in m and V_{lf} in m/s. The flap deflection angle δ_f and the flap thickness/chord ratio $(t/c)_f$ are measured streamwise. Other geometric definitions are given in Fig. C-1.

The factor k_f represents the effect of the flap configuration:

$$k_f = k_{f1} \, k_{f2} \qquad (C-10)$$

where

$k_{f1} = 1.0$: single slotted; double slotted, fixed hinge

1.15: double slotted, 4-bar movement; single slotted Fowler

1.30: double slotted Fowler

1.45: triple slotted Fowler

$k_{f2} = 1.0$: slotted flaps with fixed vane

1.25: double slotted flaps with "variable geometry", i.e. extending flaps with separately moving vanes or auxiliary flaps

For variable geometry flaps the $(t/c)_f$ ratio in (C-9) refers to the retracted position. If the flap speed V_{lf} is not known, 1.8 times the stalling speed in the landing configuration may be assumed as a first guess.

In the absence of the various data required to compute the flap weight according to (C-9), the data of Fig. C-2 may be

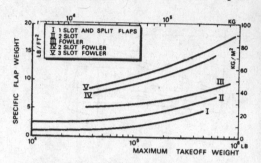

Fig. C-2. Specific weight of trailing-edge high-lift devices (Ref. C-2)

used. The specific weight of leading edge high lift devices can be read from Fig. C-3.

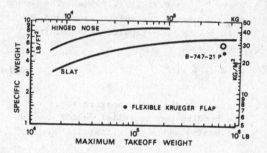

Fig. C-3. Specific weight of leading-edge high-lift devices (Ref. C-2)

The weight of spoilers and speed brakes W_{sp} may be taken either as 2.5 lb (sq.ft (12.2 kg/m^2) or 1.5% of the wing weight.

C-4. WING GROUP WEIGHT

The structural weight of the wing group, as defined in accordance with AN-9103D, is given by:

$$W_w = W_{w_{basic}} + 1.2 (W_{hld} + W_{sp}) \qquad (C-11)$$

where the basic wing weight is defined by (C-1), W_{hld} by (C-8) and the spoiler and speed brake weight by the data given in Section C-3.
A first estimate of the wing weight must be available for substitution in (C-1). Equation 8-12 in Section 8.4.1 is sufficiently accurate to make a second iteration superfluous.

The standard error of prediction of the present method is 9.64%, but this figure can be considerably improved by adaptation of the constants of proportionality of (C-1) and (C-9) to accurately known weights of similar airplane wings. In this case the formula obviously applies to a very restricted category of aircraft only.

REFERENCES

C-1. E. Torenbeek: "Prediction of wing group weight for preliminary design". Aircraft Engng., July 1971, pp. 16-21.

C-2. W. Schneider: "Die Entwicklung und Bewertung von Gewichtsabschätzungsformeln für den Flugzeugentwurf unter Zuhilfnahme von Methoden der mathematischen Statistik und Wahrscheinlichkeitsrechnung". Dissertation, University of Berlin, 23-2-1973.

Appendix D. The weight penalty method for fuselage structural weight prediction

SUMMARY

The basic weight penalty method for fuselage structural weight prediction was introduced for the first time by Burt (Ref. D-1) and a number of other publications based on the same approach have been published since (Refs. D-2 through D-5). The present method represents a synthesis of these publications, updated for modern pressurized fuselages with or without engines mounted to the side of the rear fuselage section.

457

NOMENCLATURE

b_f - maximum width of fuselage

c_r - theoretical wing root chord (see Fig. C-1)

D_f - diameter of main fuselage lobe

f - design hoop stress in fuselage skin in level flight

h_f - maximum depth of fuselage

K_f - ratio of fuselage structure weight to gross shell weight

k_λ - correction factor on shell weight for fuselage slenderness

k_{bh} - proportionality factor for bulkhead weight

k_{fl} - proportionality factor for floor weight

l_e - distance from engine support to the rear wing center section spar

l_t - distance between quarter-chord points of wing root and horizontal

tail root (Fig. D-2)

MTOW - Maximum Takeoff Weight

n_{ult} - ultimate load factor corresponding to W_{to} (1.5 times limit load factor)

P_{fl} - maximum floor loading

Δp - maximum operational differential pressure in cabin or pressurized fuselage section

S_G - gross shell area (all openings faired over, no excrescences, fairings or blisters)

S_{bh} - bulkhead area

V_D - Design Diving speed (EAS)

W_f - fuselage structure weight

W_{fl} - floor weight

W_{fr} - gross frame weight

W_G - gross shell weight

W_{sk} - gross skin weight

W_{str} - gross stringer and longeron weight

W_{to} - Maximum Takeoff Weight

D-1. SURVEY OF THE METHODOLOGY

A detailed subdivision of all weight contributions is given in Table D-1. Not all items mentioned are applicable to each configuration. For example: main landing gear doors may be counted as part of the wing group if the main landing gear is retracted into the wing.

Computation takes place in four stages.

STAGE 1: Calculation of the weight of the fuselage shell, which carries the primary loads and contributes approximately a third to one half of the fuselage weight (gross shell weight).

STAGE 2: The weight of material removed for cutouts and openings is subtracted from the gross shell weight (net shell weight).

STAGE 3: Weight is added for the materials used to fill the holes and for the surround structure required to recover strength (modified shell weight).

STAGE 4: Weight contributions and penalties are added for floors, bulkheads, support structure and various additional items, resulting in the total fuselage group weight.

D-2. GROSS SHELL WEIGHT

The weight of the fuselage shell structure is divided into the amounts contributed by the skin, stringers and frames:

$$W_G = W_{sk} + W_{str} + W_{fr} \qquad (D-1)$$

D-2.1. Gross skin weight

The gross skin weight W_{sk} is the greatest of the values given by (D-2), (D-4) and (D-5).

$$W_{sk} = \text{constant} \times k_\lambda \, S_G^{1.07} \, V_D^{.743} \qquad (D-2)$$

where the constant is equal to .00575 when W_{sk} is in lb, S_G in sq.ft and V_D in knots, or .05428 when W_{sk} is in kg, S_G is in m^2 and V_D in m/s. Allowing for the influence of the fuselage slenderness ratio, the factor k_λ is approximated as follows:

$$k_\lambda = .56 \left(\frac{l_t}{b_f + h_f} \right)^{3/4} \qquad (D-3)$$

for $l_t / (b_f + h_f)$ up to 2.61, while for higher

GROSS SHELL		
SKIN		
STRINGERS AND LONGERONS		
FRAMES		
GROSS SHELL MODIFICATIONS		
MATERIAL REMOVED	−	
PASSENGER AND CREW DOORS		
CARGO HOLD DOORS		
(LARGE) FREIGHT HOLD DOORS & RAMPS		
ESCAPE HATCHES		
ENCLOSURES AND WINDSHIELDS		
WINDOWS & PORTS		
LANDING GEAR DOORS		
EQUIPMENT ACCESS DOORS		
SPEEDBRAKES		
MODIFIED SHELL		( )
FLOORING		
PASSENGER CABIN FLOOR, BEAMS AND RAILS		
FREIGHT COMPARTMENT FLOOR & LOADING SYST.		
CARGO/BAGGAGE HOLD FLOOR		
FLIGHT DECK FLOOR		
EQUIPMENT BAY FLOOR		
BULKHEADS AND PRESSURE FLOORS		
FRONT PRESSURE BULKHEAD		
REAR PRESSURE BULKHEAD		
LANDING GEAR WHEELBAYS		
COCKPIT BULKHEAD		
SPECIAL (MOUNTING) FRAMES		
SUPPORT STRUCTURE		
WING-FUSELAGE INTERCONNECTION		
TAILPLANE SUPPORT		
ENGINE(S) SUPPORT		
LANDING GEAR SUPPORT		
FUSELAGE TANKS SUPPORT		
ADDITIONAL WEIGHT ITEMS		
FAIRINGS & FILLETS		
AIRSCOOPS		
STAIRS		
PAINT, SEALING, REDUX		
FASTENERS, JOINTS		
MISCELLANEOUS		
TOTAL BODY GROUP WEIGHT		

Table D-1: Weight breakdown of the fuselage group (applicable to tail booms as well)

values $k_\lambda = 1.15$. The definition of l_t is slightly modified as compared with that given in Ref. D-1, in order to account for the effect of wing sweep on the fuselage bending moment (Fig. D-2).

The gross skin weight based on cabin pressure, for constant skin thickness over the complete fuselage shell, is:

$$W_{sk} = \text{constant} \times \Delta p \, D_f \, S_G \, \frac{f_{ref}}{f} \qquad (D-4)$$

where the constant of proportionality is .007 when W_{sk} is in lb, Δp in lb/sq.in. and S_G in sq.ft., or 1.595 when W_{sk} is in kg, Δp in kg/cm² and S_G in m². Equation D-4 represents the skin weight required to resist the design cabin pressure differen-

tial Δp, based on a mean hoop stress level of $f_{ref} = 12,000$ lb/sq.in. (843.7 kg/cm²). The actual value for f to be anticipated may be substituted in (D-4), but in the absence of this information assume $f_{ref}/f=1$.

The minimum value for W_{sk}, based on a minimum gauge of .8 mm (.315 in.), is equal to:

$$W_{sk} = \frac{W_{sk}}{S_G} \, S_G \qquad (D-5)$$

where $W_{sk}/S_G = .445$ lb/sq.ft (2.173 kg/m²).

In all equations the gross shell area S_G is defined as the area of the entire outer surface of the fuselage, assuming that all holes for doors, windows, cutouts, etc. are faired over, and all local excrescences such as blisters, wheel fairings and canopies are removed and faired over. The gross shell area can be estimated with the methods given in Appendix B.

D-2.2. Gross stringer and longeron weight

The graphs presented in Ref. D-1 can be approximated as follows:

$$W_{str}=\text{constant}\times k_\lambda \, S_G^{1.45} \, V_D^{.39} \, n_{ult}^{.316} \quad (D-6)$$

where the constant of proportionality is .000635 when W_{str} is in lb, S_G in sq.ft and V_D in knots, or .0117 when W_{str} is in kg, S_G in m² and V_D in m/s.

D-2.3. Gross standard frame weight

The curves in Ref. D-1 are approximated as follows:

Freighters:

$$W_{fr} = .32 \, (W_{sk} + W_{str}) \qquad (D-7)$$

All other types:

$$W_{sk}+W_{str}>630 \text{ lb (286 kg): } W_{fr}=.19 \, (W_{sk}+W_{str}) \qquad (D-8)$$

$$W_{sk}+W_{str}<630 \text{ lb (286 kg):}$$

$$W_{fr} = \text{const} \times (W_{sk}+W_{str})^{1.13} \qquad (D-9)$$

459

where the constant is .0822 when W_{fr}, W_{sk} and W_{str} are in lb, or .0911 when all weights are in kg.

D-3. GROSS SHELL MODIFICATIONS

Computation of the total net weight penalty relative to the gross shell weight is a lengthy task. In the conceptual design phase the required information is not always present and available data may be inaccurate and/or ill defined. The alternative way of allowing for the departure of the fuselage design from the ideal is to use a simple overall correction factor for the gross shell weight, as suggested in Ref. D-1. This can be done with reasonable accuracy where the design is conventional. Defining a correction factor K_f as the ratio of the total fuselage group weight to the gross shell weight, we find:

$$W_f = K_f W_G \qquad (D-10)$$

Values of K_f have been calculated for several aircraft types, with the following result:
Hawker Siddeley HS-125 : K_f = 2.40
Fokker F-27/100 : K_f = 1.82
Fokker F-28/1000 : K_f = 1.83
McDonnell Douglas DC-9/10: K_f = 2.20
McDonnell Douglas DC-8/55: K_f = 1.88
Lockheed C-5A (freighter): K_f = 2.51
Except in the case of the C-5A, a heavy freighter with fuselage-mounted undercarriage, a systematic explanation of the variation in K_f cannot be given. Instead of the rough estimate given by (D-10), the more detailed computation method can be followed, using Table D-1 for collecting the various contributions.

D-3.1. Removed material

Fig. D-1 lists the openings to be considered for a typical passenger transport. Allowance must be made for cutouts of appreciable size, while for small openings it is reasonable to assume that the weight of skin removed is equal to the reinforce-

ment added.
The weight of removed material is obtained by multiplying the specific gross shell weight (W_G/S_G) by the "wetted" area of each opening. This procedure is performed for all items listed in Table D-1 under the heading "gross skin modifications", as far as they are applicable to the airplane under consideration. The bay required at the wing-fuselage connection will be treated in a different way by equation D-21 or D-22.

D-3.2. Doors, hatches, windows and enclosures

The weight of these items is taken into account by addition of
a. the actual weight of the doors, etc., including any operating mechanism ("fillings"), and
b. the surround structure weight, i.e. door landings, frames, etc.
The data presented in Table D-2 are not applicable to the (large) freighthold doors of freighter airplanes. They have been obtained by comparing the results of various references and those of detailed fuselage weight breakdowns.
If data relating to the number and dimensions of access doors are not available, their net weight penalty (i.e. removed material included) may be assumed at 1.5 to 2% of the gross shell weight.
Note that landing gear wheelbays are generally not pressurized and that the "unpressurized" data must be used for landing gear doors. Nose landing gear doors are included in the weight penalty given in Section D-5.2 and may therefore be omitted here. Main landing gear doors are frequently regarded as a wing weight contribution. For freighter airplanes the following net weight penalty (i.e. removed material and surround structure are taken into account) is suggested for (large) freighthold doors:
Side doors - Weight = $4 \sqrt{\Delta p}$ lb/sq.ft of height × width, with Δp in lb/sq.in.
 - Weight = $73.65 \sqrt{\Delta p}$ kg/m^2 of height × width, with Δp in kg/cm^2.

WEIGHT CONTRIBUTION	FILLINGS*		SURROUNDS	
	PRESSURIZED	UNPRESSURIZED	PRESSURIZED	UNPRESSURIZED
PASSENGER AND CREW DOORS	$2.40\sqrt{\Delta p}\ bh$	2bh	$15\sqrt{A_{ap}}$ FRONT, AFT $20\sqrt{A_{ap}}$ ABOVE WING	$10\sqrt{A_{ap}}$
CARGO HOLD DOORS (BELLY)	$2.65\sqrt{\Delta p}\ bh$	2bh	$24\sqrt{A_{ap}}$ FRONT $34\sqrt{A_{ap}}$ REAR	
ESCAPE HATCHES	$1.75\sqrt{\Delta p}\ bh$	2bh	$18\sqrt{A_{ap}}$	$7.5\sqrt{A_{ap}}$
COCKPIT WINDOW GLAZING	$.18A_{ws}\Delta p^{.25}\sqrt{b_f V_D}$	$(3.25+.011V_D)A_{ws}$	FRAME INCLUDED IN FILLING	$2\sqrt{A_{ap}}$
CANOPIES SLIDING	$7A_{ws}^{.8}\ \Delta p^{.25}$	$(.20+.007V_D)A_{ws}$	$8.4\sqrt{A_{ap}}$	
CANOPIES HINGED	$(7A_{ws}^{.8}-20)\Delta p^{.25}$	$.0046V_D\ A_{ws}$		
CANOPIES FIXED	-	$2.5\ A_{ws}$		
WINDOWS AND PORTS	$2.7\ A_{ap}\sqrt{b_f}$	2.5 bh	FRAME INCLUDED IN FILLING	
EQUIPMENT BAY/ACCESS DOORS, LANDING GEAR DOORS	4.5 bh	3.3 bh	$6.7\sqrt{A_{ap}}$	
SPEEDBRAKES	TRANSPORT A/C: 2 TO 3 lb/ft² JET TRAINERS : 5 TO 7 lb/ft²			

WEIGHT IN LB

WEIGHT CONTRIBUTION	FILLINGS*		SURROUNDS	
	PRESSURIZED	UNPRESSURIZED	PRESSURIZED	UNPRESSURIZED
PASSENGER AND CREW DOORS	$44.2\sqrt{\Delta p}\ bh$	9.765 bh	$22.3\sqrt{A_{ap}}$ – FRONT, AFT $29.8\sqrt{A_{ap}}$ – ABOVE WING	$14.9\sqrt{A_{ap}}$
CARGO HOLD DOORS (BELLY)	$48.8\sqrt{\Delta p}\ bh$	9.765 bh	$35.7\sqrt{A_{ap}}$ – FRONT $50.6\sqrt{A_{ap}}$ – REAR	
ESCAPE HATCHES	$32.2\sqrt{\Delta p}\ bh$	9.765 bh	$26.8\sqrt{A_{ap}}$	$11.2\sqrt{A_{ap}}$
COCKPIT WINDOW GLAZING	$4.31\ A_{ws}\ \Delta p^{.25}\sqrt{b_f V_D}$	$(15.9+.104V_D)\ A_{ws}$	FRAME INCLUDED IN FILLING	$2.98\sqrt{A_{ap}}$
CANOPIES SLIDING	$41.3\ A_{ws}^{.8}\ \Delta p^{.25}$	$(.98+.0664V_D)\ A_{ws}$	$12.5\sqrt{A_{ap}}$	
CANOPIES HINGED	$(41.3\ A_{ws}^{.8}-17.6)\Delta p^{.25}$	$.0436\ V_D\ A_{ws}$		
CANOPIES FIXED	-	$12.2\ A_{ws}$		
WINDOWS AND PORTS	$23.9\ A_{ap}\sqrt{b_f}$	12.2 bh	FRAME INCLUDED IN FILLING	
EQUIPMENT BAY/ACCESS DOORS, LANDING GEAR DOORS	22 bh	16.1 bh	$9.97\sqrt{A_{ap}}$	
SPEEDBRAKES	TRANSPORT A/C: 10 TO 15 KG/M² JET TRAINERS : 25 TO 35 KG/M²			

WEIGHT IN KG

*ALL ITEMS QUOTED INCLUDE OPERATING MECHANISM

A_{ws} = windshield wetted area, frame included (sq.ft or m²); A_{ap} = "wetted" area of aperture or opening (sq.ft or m²); b = width (ft or m); b_f = fuselage width (ft or m); V_D = Design Diving speed (knots or m/s EAS); Δp = maximum operational pressure differential of relevant fuselage section (lb/sq. inch or kg/cm²)

Table D-2. Gross weight penalties of fillings and surrounds (References D-1 through D-7 and various detail weight data)

461

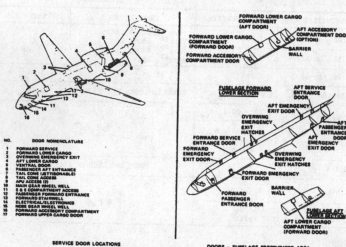

Fig. D-1. Fuselage doors of the Douglas DC-8 and DC-9 aircraft (Ref. D-6)

Rear loading doors, ramp and headroom door included –

Weight = 3 $\sqrt{\Delta p}$ lb/sq.ft of total wetted area, with Δp in lb/sq.in.

Weight = 55,24 $\sqrt{\Delta p}$ kg/m^2 of wetted area, with Δp in kg/cm^2.

Nose loading door weight – 15 lb/sq.ft (73.236 kg/m^2) of frontal area.

D-4. FLOORING

Typical floor weights are:
for passenger transports – 1 to 2 lb/sq.ft (5 to 10 kg/m^2)
for heavy freighters   – 4 to 5 lb/sq.ft (20 to 25 kg/m^2)
For a more detailed calculation it is necessary to allow for the floor support configuration and the floor loading.

D-4.1. Passenger cabin and freight hold floors

The diagram in Ref. D-1 is approximated as follows:

$$W_{fl} = k_{fl} (S_{fl})^{1.045} \qquad (D-11)$$

Floors are categorized as follows:

TYPE A: Passenger floors with beams resting directly on the fuselage frames, normally used in flat-bottomed fuselages:
$k_{fl} = .85$ when W_{fl} is in lb and S_{fl} in sq. ft, or 4.62 when W_{fl} is in kg and S_{fl} in m^2.

TYPE B: Passenger floors with beams attached to the fuselage sides and possibly braced to the bottom by struts, normally used in pressurized fuselages with under-floor cargo holds:

$$k_{fl} = const \times \sqrt{P_{fl}} \qquad (D-12)$$

where the constant of proportionality is equal to .125 when W_{fl} is in lb, S_{fl} in sq.ft and P_{fl} in lb/sq.ft, or .3074 when W_{fl} is in kg, S_{fl} in m^2 and P_{fl} in kg/m^2. The factor k_{fl} includes floor panels, lateral and longitudinal stiffeners, support struts and seat tracks.

TYPE C: Freight compartment floors designed for the carriage of containerized freight –
$k_{fl} = 2.0$ when W_{fl} is in lb and S_{fl} in sq. ft, or 10.867 when W_{fl} is in kg and S_{fl} in m^2.

TYPE D: Freight compartment floors designed for the carriage of bulk freight – $k_{fl} = 2.70$ when W_{fl} is in lb and S_{fl} in sq.ft, or 14.67 when W_{fl} is in kg and S_{fl} in m^2.

TYPE E: Freight compartment floors for

462

heavy freighters, intended for all-vehicle operation - k_{fl} = 3.85 when W_{fl} is in lb and S_{fl} in sq.ft, or 20.92 when W_{fl} is in kg and S_{fl} in m^2.

D-4.2. Various other floors

The weight of floors in (belly) cargo and baggage holds is related to the total cargo or baggage hold volume and the permissible floor loading as follows:

$$W_{fl} = \text{constant} \times (\text{volume})^{.7} \sqrt{P_{fl}} \qquad (D-13)$$

where the constant is .4 when the volume is in cu.ft and the floor loading in lb/sq.ft, or 1.0 when the volume is in m^3 and the floor loading in kg/m^2.

Flight deck floors have a specific weight of approximately 80% of a passenger cabin floor weight per unit of area.

Equipment bay floors have to be considered only if a separate equipment bay is provided. Their weight is approximately 1.5% of the gross shell weight for transport aircraft and .5 lb/cu.ft (8 kg/m^3) of fuselage volume (Ref. D-3) for other types.

D-5. PRESSURE BULKHEADS AND FRAMES

D-5.1. Pressure cabin bulkheads

Flat bulkheads in front or at the rear of the pressurized section or cabin have a weight of:

$$W_{bh} = 20 + .19 \, \Delta p^{.8} \, S_{bh}^{1.2} \quad (lb)$$
$$\qquad (D-14)$$
$$W_{bh} = 9.1 + 12.48 \, \Delta p^{.8} \, S_{bh}^{1.2} \quad (kg)$$

where S_{bh} is the actual area of each bulkhead in sq.ft (m^2) and Δp the design cabin pressure differential in lb/sq.in (kg/cm^2). Due to their more favorable loading condition, spherical bulkheads are considerably lighter (Ref. D-6). The following approximations may be used:

$$W_{bh} = 20 + .11 \, \Delta p^{.8} \, S_{bh}^{1.2} \quad (lb)$$

$$W_{bh} = 9.1 + 7.225 \, \Delta p^{.8} \, S_{bh}^{1.2} \quad (kg) \quad (D-15)$$

where S_{bh} is the $\frac{\pi}{4}$ (diameter)^2 in sq.ft or m^2 and Δp is in lb/sq.in or kg/cm^2 respectively.

D-5.2. Wheelbays for retractable undercarriages

If the bays are surrounded by a pressurized section, the sidewalls, frames and roofs are pressure bulkheads; (D-14) may be used for weight estimation. Alternative method (Ref. D-5):

Nose landing gear bay weight - .282% of the MTOW plus 7 lb (3.18 kg). Nosewheel bay doors are included in this figure and should be subtracted if they have been counted previously as a "gross shell modification" (Section D-3.2).

For main landing gear wheelbays inside a pressurized fuselage (section), we have:

$$\text{Weight} = .015 \, \Delta p^{.8} \, W_G \quad (\Delta p \text{ in lb/sq.in})$$
$$\qquad (D-16)$$
$$\text{Weight} = .125 \, \Delta p^{.8} \, W_G \quad (\Delta p \text{ in } kg/cm^2)$$

For other aircraft types - generally with non-pressurized fuselages - Ref. D-3 gives:

nose landing gear bay weight =
$$.26 \times 10^{-3} \, n_{ult} \, W_{to} \qquad (D-17)$$
main landing gear bay weight =
$$10^{-3} \, n_{ult} \, W_{to} \qquad (D-18)$$

On several designs, for example high-wing airplanes, the main landing gear is suspended from the fuselage and retracts into a separate fairing outside the fuselage external lines. The weight per fairing is approximately:

$$\text{Weight} = .03 \, \sqrt{V_D} \, (\text{wetted area})^{1.2} \quad (lb)$$
$$\qquad (D-19)$$
$$\text{Weight} = .328 \, \sqrt{V_D} \, (\text{wetted area})^{1.2} \quad (kg)$$

Landing gear doors are included. The wetted area is in ft^2 (m^2) and the Design Diving speed in kts (m/s) respectively.

Cockpit bulkheads typically have a specific

weight of .75 lb/sq.ft (3.66 kg/m^2).

Special (mounting) frames are required at any abrupt change in cross-section where loads have to be redistributed - e.g. near intake or exhaust openings - and for the support of engine pods mounted to the sides of the rear fuselage section:

$$\text{Weight (lb)} = 1.2 \ (\text{area})^{1.2} - \text{area in sq.ft}$$
$$\text{(D-20)}$$
$$\text{Weight (kg)} = 9.42 (\text{area})^{1.2} - \text{area in m}^2$$

D-6. SUPPORT STRUCTURE

D-6.1. Wing/fuselage connection

For low-wing transport aircraft with a continuous torque box across the fuselage, Ref. D-5 states:

$$\text{Weight} = 45 + .907 \times 10^{-3} \ n_{ult} \ W_{to} \ \text{(lb)}$$
$$\text{(D-21)}$$
$$\text{Weight} = 20.4 + .907 \times 10^{-3} \ n_{ult} \ W_{to} \ \text{(kg)}$$

The expression given in Ref. D-1 generally results in a rather lower weight and is probably based on data for unpressurized fuselages:

$$\text{Weight(lb)} = .345 \times 10^{-3} \ (n_{ult} \ W_{to})^{1.185} -$$
$$W_{to} \ \text{in lb}$$
$$\text{(D-22)}$$
$$\text{Weight(kg)} = .4 \times 10^{-3} \ (n_{ult} \ W_{to})^{1.185} -$$
$$W_{to} \ \text{in kg}$$

For high-wing configurations with continuous torque box across the fuselage, the penalty is roughly two-thirds of these values.

If the wing torsion box is not continuous, the weight penalty may be three times greater than that given by (D-22). However, the wing weight is reduced accordingly (Ref. D-1).

D-6.2. Engine support structure*

The direct weight penalty is approximately

*For fuselage-mounted engines only

2 to 3% of the bare engine weight. An indirect, but by no means negligible effect is caused by the increased fuselage bending moment on touchdown. A calculation method is not available to the author; the following expression applying to rear fuselage-mounted engines is therefore presented as a tentative suggestion.

$$\text{Weight} = \text{constant} \times \frac{l_e^2}{b_f} W_{to} \qquad \text{(D-23)}$$

where the constant of proportionality is equal to 2×10^{-4} when l_e and b_f are in ft, or 6.56×10^{-4} when l_e and b_f are in m. Definitions of l_e and b_f are given in Fig. D-2.

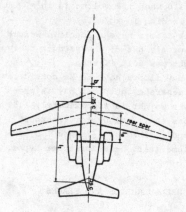

Fig. D-2. Geometrical definitions

D-6.3. Other support structures

The tailplane support structure weighs approximately 10% of the tailplane structure group weight, or .25% of the MTOW in the case of central fin tailplane configurations.

For nose landing gears the support structure weight is included in the figure given in Section D-5.2.

For wing-mounted main landing gears, but fuselage-mounted sidestays, the fuselage weight penalty is typically 5% of the main landing gear weight, or .15% of the MTOW. For fuselage-mounted main landing gears:

$$\text{Weight} = 3 \times 10^{-3} \, n_{ult} \, W_{to} \qquad (D-24)$$

The tank support structure weighs approximately .2 lb per U.S. gal. (.024 kg/liter) of tank volume for fuselage-mounted tanks.

D-7. ADDITIONAL WEIGHT ITEMS

Wing/fuselage fairings may be quite heavy - sometimes of the order of 5% of the wing structure weight. They are usually considered as part of the wing group weight.

Fairings for a double-bubble fuselage weigh about 1.5% of the gross shell weight.
For air intakes (buried engines) the duct skin is regarded as part of the engine section weight. An allowance may be required for the modification of frames and holes in the shell.
Passenger stairs weigh approximately 20 lb per ft (30 kg/m) extended length.
For paint, sealing and redux a weight penalty of 1 to 2% of W_G may be assumed, while for production joints in the fuselage of a transport aircraft the weight penalty is about 2 to 3% of W_G.

REFERENCES

D-1. M.E. Burt and J. Philips: "Prediction of fuselage and hull structure weight". RAE Report Structures 122, April 1952.
D-2. D. Howe: "Structural weight prediction". Cranfield Institute cf Technology, DES 903.
D-3. R.L. Hammit: "Structural weight estimation by the weight penalty concept for preliminary design". SAWE Technical Paper No. 141, 1956.
D-4. B.B. Coker: "Problems in airframe development associated with weight and balance control in heavy logistics transport vehicles such as the C-5A transport". Short Course, University of Tennessee, Nov. 1968.
D-5. D.M. Simpson: "Fuselage structure weight prediction". SAWE Paper No. 481, June 1973.
D-6. D.P. Marsh: "Post-design analysis for structural weight estimation". SAWE Paper No. 936, May 1972.
D-7. R.J. Atkinson: "Structural design". RAE Tech. Note Structures 133, May 1963.

Appendix E. Prediction methods for lift and pitching moment of aircraft in the en route configuration

SUMMARY

A concise collection of formulas, generalized data and methods is presented for estimating the lift and pitching moment coefficients at subcritical flight speeds in the cruise configuration. All methods are readily applicable in the preliminary design stage of conventional aircraft.

Some effects of wing-fuselage interference and the trim load on the tailplane are included and conditions are derived for choosing the wing and horizontal stabilizer angles of incidence relative to the fuselage.

NOMENCLATURE

A	- aspect ratio; $A = b^2/S$ (no index: wing aspect ratio)
a.c.	- aerodynamic center
B_p	- number of blades per propeller
b	- span; width (no index: wing span)
C_1/C_4	- coefficients in Diederich's method
C_L	- 3-dimensional lift coefficient; $C_L = L/qS$
C_{L_α}	- lift-curve slope; $C_{L_\alpha} = dC_L/d\alpha_f$
C_{L_o}	- C_L when the fuselage is horizontal ($\alpha_f = 0$)
C_m	- 3-dimensional pitching moment coefficient; $C_m = M/qS\bar{c}$
$C_{m_{ac}}$	- C_m about an axis through the aerodynamic center
c	- chord
$\bar{c}$	- mean aerodynamic chord (MAC) defined in Appendix A, Section A-3.3
c_g	- mean geometric chord (SMC), defined in Appendix A, Section A-3.1
c.g.	- center of gravity
c_ℓ	- 2-dimensional lift coefficient; $c_\ell = $ lift/(unit span x qc)
c_{ℓ_α}	- lift-curve slope of an airfoil section
c_{ℓ_a}	- "additional" lift coefficient
c_{ℓ_b}	- "basic" lift coefficient
c_{ℓ_i}	- design lift coefficient
c_m	- 2-dimensional pitching moment coefficient; $c_m = $ pitching moment/ (unit span x qc^2)
$c_{m_{ac}}$	- c_m about the airfoil aerodynamic center
$c_{m_{1/4}}$	- c_m about the airfoil quarter-chord point
D_p	- propeller diameter
E	- Jone's edge-velocity correction; $E = $ semiperimeter/span
f	- Anderson's correction factor for C_{L_α}; Diederich's lift distribution function
G	- Anderson's factor for computing

$C_{m_{ac}}$	due to twist
h_f	- fuselage height
i	- angle of wing/fuselage, tailplane/ fuselage or tailplane/wing setting
J	- Anderson's factor for computing α_{L_o}
K_I, K_{II}	- factors for calculating the lift on the wing plus body
k	- ratio of βc_{ℓ_α} to 2π
k_n, k_p	- factor for the effect of nacelles and propellers on x_{ac}
k_s	- correction factor for $C_{L_{max}}$
L	- lift
L_a, L_b	- Anderson's lift functions
l	- length; moment arm
M	- pitching moment; Mach number
m	- geometric parameter defining the vertical position of the horizontal stabilizer
q	- dynamic pressure; $q = \frac{1}{2}\rho V_o^2$
R	- Reynolds number ; $R = V\,l/\nu$
r	- geometric parameter defining the horizontal stabilizer longitudinal position
S	- area (no index: gross wing area)
t	- airfoil thickness, defined in Appendix A, Section A-3.1
V_o	- flight speed
x_o	- coordinate measured from the MAC leading edge, measured in the direction of zero lift, positive to the rear
y	- spanwise coordinate, measured from the airplane centerline, positive to port
α	- angle of attack (no index: measured relative to the wing zero-lift line)
$(\alpha_{L_o})_f$	- zero-lift angle of attack of the wing relative to the fuselage datum line
α_{ℓ_o}	- zero-lift angle of an airfoil section, defined in Appendix A, Section A-2.2
β	- Prandtl's compressibility correction: $\beta = \sqrt{1 - M^2}$
Δ	- increment; for example: $\Delta_h \ldots = $

increment due to horizontal tail

Δy – leading edge sharpness parameter of a section, defined in Fig. A-2 of Appendix A

ϵ – aerodynamic twist; angle of downwash

η – non-dimensional spanwise station; $\eta = y/\frac{b}{2}$

$\Lambda_{1/4}$ – sweep angle of quarter-chord line, defined in Appendix A, Section A-3.1

Λ_{β} – corrected sweep angle; $\tan \Lambda_{\beta} = \tan \Lambda_{1/4}/\beta$

λ – taper ratio (no index: wing taper ratio, defined in Appendix A, Section A-3.1).

ν – kinematic viscosity

ρ – air density

ϕ'_{TE} – airfoil section trailing-edge angle, defined in Appendix A, Section A-2.1, and Fig. A-3

INDICES

A-h   – aircraft minus horizontal tail
a   – additional lift
ac   – aerodynamic center
b   – basic lift
cg   – center of gravity
ex   – exposed
f   – fuselage
fn   – nose section of the fuselage
h   – horizontal tailplane
LE   – leading edge
n   – nacelle(s)
net   – net wing area
n   – neutral point
p   – propeller
r   – root
TE   – trailing edge
t   – tip
z   – vertical displacement of wing

E-1. APPLICABILITY OF THE METHODS

Apart from the simplifications occasionally made in the derivation of the methods presented, some general restrictions on the validity of the methods must be mentioned.

a. Flight speeds are subcritical, i.e. shock waves are absent and compressibility effects are restricted to those that can be analyzed with subsonic potential flow theory.

b. Angles of attack are relatively small, so that the flow is predominantly non-separated.

c. Wing aspect ratios exceed $4/\cos \Lambda_{1/4}$ and wing sweep angles are less than 35°, approximately.

d. Only power-off conditions are considered.

e. The effects of aero-elasticity are ignored.

f. Ground effect is not considered; see Appendix G, Section G-7.

The above conditions are generally met by "conventional" subsonic aircraft in most flight conditions. Other types of aircraft and types of flow can be analyzed with the USAF Stability and Control DATCOM (Ref.E-3) and the R.Ae.S. Data Sheets (Ref. E-5). In the sections which follow it is assumed that a copy of Abbott and Von Doenhoff's textbook, "Theory of wing sections" (Ref. E-14), is available to the reader.

E-2. CONTRIBUTIONS TO THE LIFT

The airplane lift is thought of as being composed of wing lift, fuselage lift, horizontal tailplane lift, nacelle lift and lift due to the powerplant installation. In this Appendix only the "clean" configuration will be considered, i.e. flaps, undercarriage, spoilers, etc. are assumed retracted.

In the preliminary design stage the airplane lift is frequently approximated by the lift on the isolated gross wing[*].

[*] Definition in Appendix A, Section A-3.1, and Fig. A-5.

469

It is therefore assumed that the fuselage lift is roughly equal to the lift on the wing center section in the absence of the fuselage. Other contributions are ignored. However, if the fuselage width is relatively large, wing/fuselage interference effects on lift should be taken into account. Particularly in the case of a forward c.g. location the horizontal tail download may be appreciable. Methods for predicting these effects are therefore included in the present survey. The effect of nacelles on lift is generally small and difficult to predict and is frequently ignored, but the shift of the aerodynamic center due to nacelles should be taken into account. Power effects on lift and pitching moment are of vital importance to the performance and operation of propeller-driven V/STOL aircraft, but the subject is considered to be outside the scope of this appendix. It is covered thoroughly in the DATCOM method and in Refs. E-39 through E-43.

For a given Mach number the lift is a linear function of the angle of attack (see Fig. E-1):

$$C_L = C_{L_\alpha} \left\{ \alpha_f - \left(\alpha_{L_o} \right)_f \right\} \qquad (E-1)$$

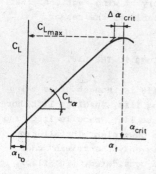

Fig. E-1. General shape of the lift curve

This expression is valid up to angles of attack approaching the critical angle of attack. Except insofar as it may give him some idea of certain stalling properties (Réf. E-7), the C_L vs. α curve in the stalling region is generally of little

concern to the designer in the pre-design stage. It is usually found that $\Delta\alpha_{crit}$ is between 1° and 3° (see Fig. E-1). The power-off airplane lift is

$$C_L = (C_L)_{A-h} + C_{L_h} \frac{S_h}{S} \frac{q_h}{q} \qquad (E-2)$$

where the first contribution is due to the wing, fuselage, nacelles and (windmilling or feathered) propeller(s), i.e. the aircraft minus the horizontal tailplane. The second term is a correction for the tailplane lift.

E-3. LIFTING PROPERTIES OF AIRFOIL SECTIONS

Section properties may form a basis for estimating the wing characteristics with a reasonable degree of accuracy. For several standard airfoil sections these characteristics may be obtained from Refs. E-12 through E-16 and many other NACA publications. An example is given in Fig. E-2. For non-standard airfoils the DATCOM presents a generalized method based on the leading edge sharpness parameter Δy defined in Appendix A, Fig. A-2.

E-3.1. The zero-lift angle

The zero-lift angle may be computed from potential flow methods or empirical data:
four-digit NACA airfoils: $\alpha_{l_o} = -(\%$ wing camber) (deg.)
five-digit NACA airfoils: $\alpha_{l_o} = -4\ c_{l_i}$ (deg.)
six-digit NACA airfoils: $\alpha_{l_o} = -6.6\ c_{l_i}$ (deg.)

E-3.2. Lift-curve slope

According to the DATCOM method, the lift-curve slope is given by:

$$c_{l_\alpha} = \frac{1.05}{\beta} \left[\frac{c_{l_{\alpha_i}}}{\left(c_{l_\alpha} \right)_{theory}} \right] \left(c_{l_\alpha} \right)_{theory} \qquad (E-3)$$

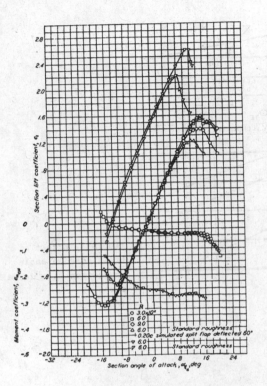

Fig. E-2. Experimental data on the NACA 64-
215 airfoil section (Ref. E-14)

$$(c_{\ell_\alpha})_{theory} = 2\pi + 4.7t/c(1 + .00375\phi'_{TE})$$

$$(rad^{-1}) \qquad (E-4)$$

ϕ'_{TE} = section trailing-edge angle (degrees),
defined in Fig. E-3.
See also Fig. A-3 of Appendix A.

For the ratio of experimental to theoreti-
cal c_{ℓ_α} for boundary layer transition near
the leading edge, see Fig. E-3. For thick-
ness/chord ratios between 10% and 20%,
c_{ℓ_α} = 6.1 per radian is a representative
value. Ref. E-5 gives a method for transition at 50% chord.

E-3.3. Maximum lift

The best approach is to use experimental

data. Instead of presenting a generalized
method, Fig. E-4 is proposed as a guideline to the designer who is interested in
the highest value of c_{ℓ}-max that can possibly be achieved for a given t/c and chord
Reynolds number, provided the section is
suitably cambered and an optimum nose shape
is chosen. An envelope of c_{ℓ}-max values for
NACA sections is shown on which the following comments may be made:
a. For t/c<10% the c_{ℓ}-max envelope is not
very sensitive to Reynolds effects.
b. For t/c<12% leading edge stall predomi-
nates. The leading edge sharpness parameter
Δy, defined in Fig. A-2 of Appendix A, may
be used to predict c_{ℓ}-max using the DATCOM
method.
c. For t/c>12% trailing edge stall is predominant and c_{ℓ}-max is sensitive to
Reynolds number effects.
d. High c_{ℓ}-max values are obtained with 5-
digit series airfoils. Laminar flow air-
foils with $c_{\ell_i} \sim .4$ and the maximum thickness
not too far aft have c_{ℓ}-max values some
.08-.12 lower.
e. The ultimate c_{ℓ}-max for airfoils of the
standard NACA series appears to be approx.
1.8 for the NACA 23012 airfoil at $R=9\times10^6$.
Higher values are possible with special
airfoil designs, e.g. recent experiments
with a 17% thick supercritical rear loading
section indicate that even values slightly
above 2.0 are possible (Ref. E-18).

E-4. WING LIFT AND LIFT DISTRIBUTION

E-4.1. Lift-curve slope

a. For straight wings ($\Lambda_{1/4}=0°$) in incompressible flow, according to Ref. E-20:

$$C_{L_{W_\alpha}} = f \frac{c_{\ell_\alpha}}{E + c_{\ell_\alpha}/\pi A} \qquad (E-5)$$

where f is a correction factor for wing
taper, shown in Fig. 8 of Ref. E-14. For
.2<λ<1.0 f may be assumed equal to .995.
Jone's edge velocity factor E is equal to

471

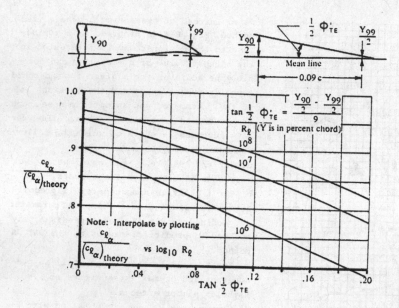

Fig. E-3. Effect of
trailing-edge angle
on section lift curve
slope
(Ref. E-3)

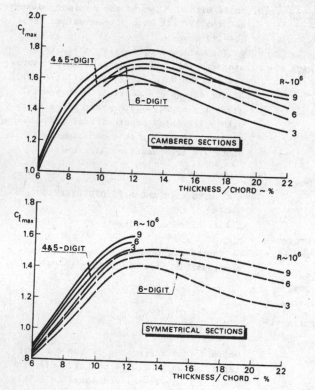

Fig. E-4. Envelopes of c_ℓ-max
values obtainable with NACA standard airfoils
(composed from experimental data
in Ref. E-14)

the ratio of the planform semiperimeter to the span of the wing. For straight-tapered wings a reasonably accurate approximation is:

$$E = 1 + \frac{2\lambda}{A(1 + \lambda)} \qquad (E-6)$$

For a 2-dimensional lift-curve slope of 2π/radian:

$$C_{L_{w_\alpha}} = \frac{2\pi}{1 + \frac{2}{A}\frac{1+2\lambda}{1+\lambda}} \qquad (rad^{-1}) \qquad (E-7)$$

The effect of compressibility may be incorporated by replacing A bij βA and $C_{L_{w_\alpha}}$ by $\beta C_{L_{w_\alpha}}$.

b. Swept wings, compressible flow (DATCOM method).

$$\beta C_{L_{w_\alpha}} = \frac{2\pi}{\frac{2}{\beta A} + \sqrt{\frac{1}{k^2 \cos^2\Lambda_\beta} + (\frac{2}{\beta A})^2}} \qquad (rad^{-1}) \quad (E-8)$$

where $\tan\Lambda_\beta = \frac{\tan\Lambda_{\frac{1}{2}}}{\beta}$

and $k = \frac{\beta c_{\ell_\alpha}}{2\pi}$ $(c_{\ell_\alpha}$ $rad^{-1})$

As compared to results from lifting surface theory (Ref. E-31), (E-8) yields good results for $\Lambda_\beta > 30°$; $C_{L_{w_\alpha}}$ is overestimated by $\sim 4\%$ for $\Lambda_\beta = 0°$ and by $\sim 2\%$ for $\Lambda_\beta = 20°$.

E-4.2 Spanwise lift distribution

The lift may be divided into additional and basic lift:

$$c_\ell = c_{\ell_a} + c_{\ell_b} \qquad (E-9)$$

In terms of the non-dimensional parameters

L_a and L_b used by Anderson:

$$c_\ell \frac{c}{c_\alpha} = L_a C_L + \frac{\varepsilon_t c_{\ell_\alpha}}{E} L_b \qquad (E-10)$$

$$L_a = \frac{c_{\ell_a}}{C_L}\frac{c}{c_g} \qquad (E-11)$$

$$L_b = \frac{c_{\ell_b}}{c_g}\frac{c}{\varepsilon_t c_{\ell_\alpha}} E \qquad (E-12)$$

The definition of wing twist ε_t is explained in Appendix A, Section A-3.2. Anderson presents tables for L_a and L_b for straight-tapered wings with linear twist in incompressible flow. The following more general semi-empirical method by Diederich (Ref. E-21) yields satisfactory results for pre-design purposes. It is valid for wings with arbitrary planform and lift distribution, provided the quarter-chord line of a wing half is approximately straight. This method can thus be used for straight and swept wings in compressible, subcritical flow.

a. Additional lift distribution.

$$L_a = C_1 \frac{c}{c_g} + C_2 \frac{4}{\pi} \sqrt{1-\eta^2} + C_3 f \qquad (E-13)$$

For coefficients C_1, C_2 and C_3, see Fig. E-5; for lift distribution function f, see Fig. E-6. For straight wings f is elliptical and (E-13) can be simplified to:

$$L_a = C_1 \frac{c}{c_g} + (C_2+C_3) \frac{4}{\pi} \sqrt{1-\eta^2}$$
$$\text{for } \Lambda_{1/4} = 0° \qquad (E-14)$$

This result is similar to the well-known approximation by Schrenk (Ref. E-19), provided $C_1 = C_2 + C_3 = .5$.

473

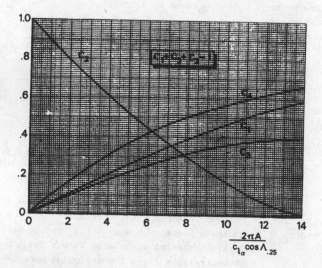

Fig. E-5. Factors in
Diederich's method (Ref. E-21)

b. Basic lift distribution.
The formula derived by Diederich in Ref.
E-21 may be modified to a form similar to
that used by Anderson:

$$\frac{L_b}{\beta E} = L_a\, C_4\, \cos\Lambda_\beta \left(\frac{\varepsilon}{\varepsilon_t} + \alpha_{o_1} \right) \qquad (E-15)$$

where $\alpha_{o_1} = - \int_o^1 \frac{\varepsilon}{\varepsilon_t} L_a\, d\eta$ $\qquad$ (E-16)

and the factor C_4 (Fig. E-5) is identical
to the product of k_o and k_1 used in Ref.
E-21. The factor α_{o_1} is equal to the local
aerodynamic twist at the spanwise station
for which $c_{\ell_b} = 0$, assuming a wingtip twist
angle of one degree relative to the root.
For the case of a linear twist[*] distri-
bution ($\varepsilon = \eta\varepsilon_t$) and elliptic L_a, $\alpha_{o_1} = 4/3\pi$.
For straight-tapered unswept wings with
linear twist distribution, (E-16) can be
evaluated with L_a in (E-14) to yield:

$$-\alpha_{o_1} = C_1 \frac{1+2\lambda}{3(1+\lambda)} + (C_2 + C_3) \frac{4}{3\pi} \qquad (E-17)$$

[*] Definition in Section A-3.2 of Appendix
A.

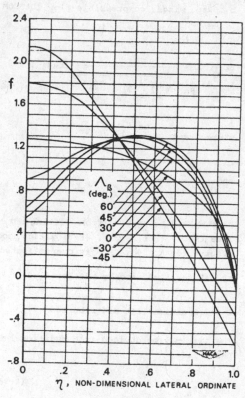

Fig. E-6. The lift distribution function f
(Ref. E-21)

474

The factor α_{o_1} is very similar to Anderson's factor J for straight wings with rounded tips (Ref. E-20, but see also Ref. E-14).

For straight wings with linear lofted twist [*] Fig. E-7 can be used. This diagram is based on results of lifting surface theory, reported in Ref. E-31, from which it can also be concluded that relative to Fig. E-7, α_{o_1} for swept wings should be reduced by approximately .0006 per degree of Λ_β. For arbitrary twist distributions α_{o_1} must be computed numerically.

E-4.3. Zero-lift angle

The lift coefficient in the linear range is

$$C_{L_w} = C_{L_{w_\alpha}} \left\{ \alpha_r - \left(\alpha_{\ell_{o_r}}\right) - \alpha_{o_1} \, \varepsilon_t \right\} \tag{E-18}$$

The angle of attack α_r is defined relative to the root chord, while the zero-lift angle of attack is:

[*] Definition in Section A-3.2 of Appendix A.

$$\left(\alpha_{L_{o_r}}\right) = \left(\alpha_{\ell_{o_r}}\right) + \alpha_{o_1} \, \varepsilon_t \tag{E-19}$$

For the root section α_{ℓ_o} can be obtained from airfoil characteristics (Section E-3). Note that for washout ε_t is negative.

E-4.4. Maximum lift

The basic procedure for estimating the maximum lift of high aspect-ratio straight wings is explained in, for example, Refs. E-14 and E-32. Abbott and Von Doenhoff's summary is quoted here:

"The maximum lift coefficient of the wing may be estimated from the assumption that this coefficient is reached when the local section lift coefficient at any position along the span is equal to the local c_ℓ-max for the corresponding section. This value may be found conveniently by the process indicated in Fig. E-8. Spanwise variations of the local c_ℓ-max and of the additional c_{ℓ_a} for $C_L=1$ and c_{ℓ_b} distributions are plotted. The spanwise variation of c_ℓ-max minus c_{ℓ_b} is plotted and

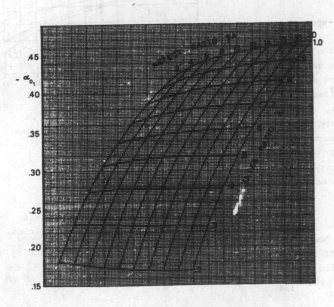

Fig. E-7. Zero-lift angle of attack per unit of twist for straight wings (Ref. E-31)

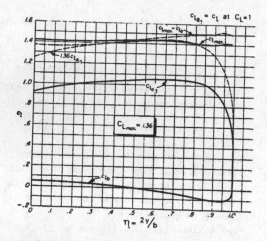

$c_{l_{a_1}} = c_l$ at $C_L = 1$

$c_{l_{max}} - c_{l_b}$

$c_{l_{max}}$

$1.36 c_{l_{a_1}}$

$c_{l_{a_1}}$

$C_{L_{max}} = 1.36$

c_{l_b}

$\eta = {}^{2y}/_b$

Fig. E-8. Example of lift distribution for a straight wing (Ref. E-14).

the minimum value of the ratio of $(c_{l_{max}} - c_{l_b})$ to c_{l_a} at $C_L = 1$ is then found.
This ratio is considered to be the maximum lift coefficient C_L-max of the wing."
A simpler but less accurate approach is:

$$C_{L_{max}} = k_s \frac{\left(c_{l_{max}}\right)_r + \left(c_{l_{max}}\right)_t}{2} \qquad (E-20)$$

where $k_s = .88$ for $\lambda = 1$ and $.95$, approximately, for tapered wings.
This expression does not take the effects of wing twist into account. The procedure described loses validity as wing sweep is increased, mainly due to the effects of the spanwise pressure gradients, which cause crossflows in the boundary layer. In spite of this, Callaghan reports in Ref. E-11 that the same approach works quite well for moderate sweep angles. The following procedures are suggested.
a. Twisted swept wings, airfoil - section variation along the span.
The basic and additional lift variations are computed by any suitable method, e.g. (E-13) and (E-15). The total lift coefficient for each section is compared with the corresponding value of $c_{l_{max}} \cos \Lambda_{1/4}$

and the procedure described previously for straight wings is applied to estimate C_L-max. Alternatively, (E-20) may be corrected by multiplication by $\cos \Lambda_{1/4}$.
b. Untwisted, constant airfoil-section swept wings.
C_L-max is derived from the section c_l-max with Fig. E-9. The parameter Δy may be obtained from Fig. A-2 of Appendix A. Alternatively:

$$C_{L_{max}} = 0.9 c_{l_{max}} \cos \Lambda_{1/4} \text{ for } t/c > 0.12$$

$$(E-21)$$

Compressibility affects C_L-max at Mach numbers above .2 approximately. The DATCOM method (Fig. 4.1.3.4-15) or Ref. E-29 can be used to estimate the order of magnitude of this effect.

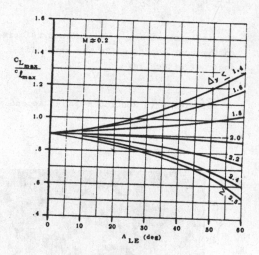

$M \approx 0.2$

$\frac{C_{L_{max}}}{c_{l_{max}}}$

Δy

1.4

1.6

1.8

2.0

2.2

2.4

2.6

Λ_{LE} (deg)

Fig. E-9. Effect of sweep angle on maximum lift (DATCOM method, Ref. E-3)

E-5. PITCHING MOMENT OF THE WING

It follows from the definition of the aerodynamic center of the wing in Section A-3.3 of Appendix A that:

$$C_{m_w} = \left(C_{m_{ac}}\right)_w + C_{L_w} \frac{x_{cg} - (x_{ac})_w}{\bar{c}} \qquad (E-22)$$

476

E-5.1. Aerodynamic center

Location of the aerodynamic center, relative to the MAC leading edge: see Fig.E-10. This diagram has been compiled from the results of lifting surface theory. The MAC can be computed with the data of Appendix A, Section A-3.3.

E-5.2. Pitching moment $\left(C_{m_{ac}}\right)_w$

$$\left(C_{m_{ac}}\right)_w = \left(C_{m_{ac}}\right)_{basic} + \Delta_\varepsilon C_{m_{ac}} \qquad (E-23)$$

The first term is the contribution of the spanwise airfoil-section camber distribution:

$$\left(C_{m_{ac}}\right)_{basic} = \frac{2}{s\bar{c}} \int_o^{b/2} c_{m_{ac}} c^2 \, dy \qquad (E-24)$$

The value of $c_{m_{ac}}$ can be obtained from potential flow theory or from experimental data on airfoils. It is primarily sensitive to the shape of the mean line. For a constant airfoil-section wing $\left(C_{m_{ac}}\right)$-basic $= c_{m_{ac}}$.

The last term in (E-23) is the contribution of the basic lift distribution due to twist:

$$\Delta_\varepsilon C_{m_{ac}} = -\frac{2}{s\bar{c}} \int_o^{b/2} c_{\ell_b} c \, y \tan\Lambda_{ac} \, dy \qquad (E-25)$$

Assuming that all section aerodynamic centers are located on a straight quarter-chord line:

$$\Delta_\varepsilon C_{m_{ac}} = -\frac{1}{2} A \frac{c_g}{\bar{c}} \tan \Lambda_{1/4} \int_o^1 c_{\ell_b} \frac{c}{c_g} \eta \, d\eta \qquad (E-26)$$

The ratio $c_g/\bar{c}$=SMC/MAC may be obtained from Fig. A-8 of Appendix A.

a. For arbitrary twist distribution, the integral must be solved numerically, using the basic lift distribution obtained from any suitable method, e.g. Diederich's.

b. For linear twist distribution, straight-tapered wings with rounded tips in incompressible flow, Anderson has evolved (E-25) to:

$$\Delta_\varepsilon C_{m_{ac}} = -G \frac{c_g}{\bar{c}} \frac{\varepsilon_t c_{\ell_\alpha}}{E} A \tan\Lambda_{1/4} \qquad (E-27)$$

A diagram to compute G can be found on page 20 of Ref. E-14. This result applies to small sweep angles only, since the effect of wing sweep on the lift distribution is not considered.

c. For linear lofted straight-tapered wings Kapteyn presents diagrams in Ref. E-31, based on lifting-surface theory (Ref. E-30). The result can be generalized as follows:

$$\frac{-\beta\left(\Delta_\varepsilon C_{m_{ac}}\right)}{\varepsilon_t C_{L_{w_\alpha}} (\beta A) \tan\Lambda_\beta} =$$

$$.0066 + .029\lambda - .03\lambda^2 + .00273(\lambda - .095)(\beta A) \quad (E-28)$$

E-6. WING/FUSELAGE INTERFERENCE EFFECTS ON LIFT

The major aerodynamic interference effects between the wing and the non-cambered fuselage are (Fig. E-11):

a. In inviscid flow the resultant force on a closed body is zero. In viscous flow and in the presence of the lifting wing, however, the lift on the body nose, which is in the upwash field, is counteracted only partly by the download on the rear fuselage which is in the downwash behind the wing.

b. The flow component normal to the fuselage axis (crossflow) induces increased effective angles of attack of the wing sections, particularly near the wing/fuselage junction.

c. There is a wing lift carry-over into the body, although this lift component is less than the lift that would be produced by the wing section replacing the fuselage if the latter were absent.

d. Vertical displacement of the wing relative to the fuselage centerline alters the

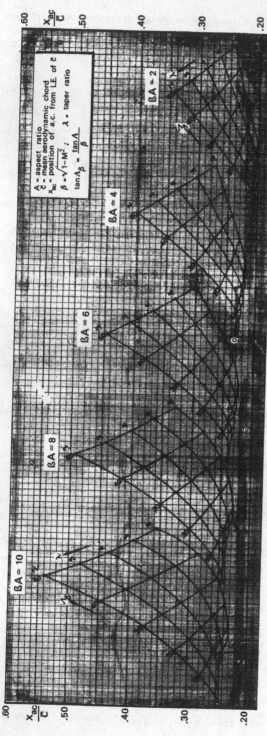

Fig. E-10. Aerodynamic center of lifting surfaces at subsonic speeds (Ref. E-31).

Within the chart:

A = aspect ratio
$\bar{c}$ = mean aerodynamic chord
x_{ac} = position of a.c. from L.E. of $\bar{c}$
$\beta = \sqrt{1-M^2}$; λ = taper ratio
$\tan\Lambda_\beta = \dfrac{\tan\Lambda}{\beta}$

$\beta A = 2$

$\beta A = 4$

$\beta A = 6$

$\beta A = 8$

$\beta A = 10$

$\dfrac{x_{ac}}{\bar{c}}$

.60 .50 .40 .30 .20

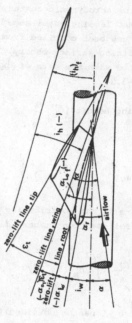

Fig. E-12. Geometric definitions of the wing/body/tail combination.

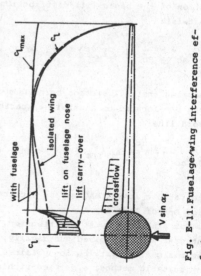

Fig. E-11. Fuselage/wing interference effects on the spanwise lift distribution.

flow pattern, as explained by Hoerner (Ref. E-38, p. 8-17). This results in a lift reduction for high-wing and a lift increment for low-wing configurations ($\Delta_z C_L$).

The lift components can be added together to obtain the total wing fuselage lift:

$$C_{L_{wf}} = \left(C_{L_\alpha}\right)_{wf} \left[\left(\alpha_f - \alpha_{0_1} \varepsilon_t\right) + \frac{K_{II}}{K_I}\left\{i_w - \left(\alpha_{\ell_0}\right)_r\right\}\right] +$$

$$+ \Delta_z C_L \qquad (E-29)$$

The various angles are defined in Fig. E-12. The wing/body incidence i_w is treated in Section E-9, while the lift-curve slope of the wing/body combination is:

$$\left(C_{L_\alpha}\right)_{wf} = K_I C_{L_{w_\alpha}} \qquad (E-30)$$

$$K_I = \frac{\left(\partial C_L / \partial_\alpha\right)_{wf} \text{ for } i_w = \text{const.}}{C_{L_{w_\alpha}}} \qquad (E-31)$$

$$K_{II} = \frac{\left(\partial C_L / \partial i_w\right)_{wf} \text{ for } \alpha_f = \text{const.}}{C_{L_{w_\alpha}}} \qquad (E-32)$$

For $\Delta_z C_L$ Hoerner states:

$$\Delta_z C_L\, S/c_r b_f = \begin{array}{l} -.1: \text{high wing} \\ 0 : \text{mid wing} \\ +.1: \text{low wing} \end{array}$$

The net wing area is defined in Appendix A, Section A-3.1. A method to estimate K_I and K_{II} can be found in Ref. E-37. For fuselages with near-circular cross-section, with $b_f/b < .2$, this method can be approximated by:

$$K_I = \left(1 + 2.15 \frac{b_f}{b}\right) \frac{S_{net}}{S} + \frac{\pi}{2C_{L_{w_\alpha}}} \frac{b_f^2}{S} \qquad (E-33)$$

and

$$K_{II} = \left(1 + .7 \frac{b_f}{b}\right) \frac{S_{net}}{S} \qquad (E-34)$$

The net wing area S_{net} is defined as the projection of the part of the wing outside the fuselage, assuming a mid-wing configuration with the same gross area (cf. Appendix A, Section A-3.1). For fuselages with swept-up tails, the flow pattern is considerably more complicated (cf. Section 3.5.1) and the previous methods have very limited value.

All interference effects treated thus far are potential-flow effects and must therefore be considered as "minimum effects" valid for small angles of attack. At high angles of attack separation near the wing/fuselage junction may result in a reduction of c_L-max. These effects may be minimized by suitable filleting, adaptation of the planform shape (e.g. a locally extended root chord), suitable twist and airfoil section modification. From Fig. E-11 it can be concluded that fuselage effects may be small, provided the separation on the isolated wing occurs on the outer wing first. The presence of the fuselage may improve the stalling characteristics by shifting the separation more to the inboard wing. No general rule for estimating the body effect on C_L-max can be given. In preliminary design, therefore, the wing/fuselage C_L-max is frequently assumed equal to that of the wing. The critical angle of attack is modified accordingly. This assumption may be somewhat conservative in the case of a high-wing location and optimistic for mid-wing configurations.

E-7. WING/FUSELAGE PITCHING MOMENT

The pitching moment equation for the wing/fuselage combination is written as follows:

$$\left(C_m\right)_{wf} = \left(C_{m_{ac}}\right)_{wf} + C_{L_{wf}} \frac{x_{cg} - \left(x_{ac}\right)_{wf}}{\bar{c}} \qquad (E-35)$$

E-7.1. Aerodynamic center

The a.c. location of the wing/body combination is sensitive to the pressure distribution and not to the integrated lift

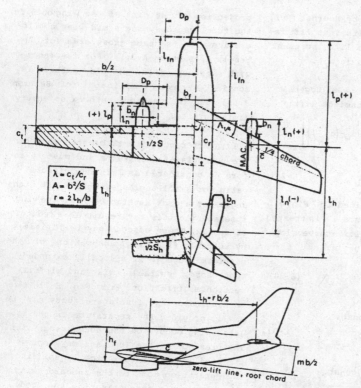

Fig. E-13. Wing/body/tail geometry

$$\lambda = c_t/c_r$$
$$A = b^2/S$$
$$r = 2l_h/b$$

force and it is therefore difficult to predict accurately. Only two corrections to the wing a.c. location will be considered here:

$$\left(\frac{x_{ac}}{\bar{c}}\right)_{wf} = \left(\frac{x_{ac}}{\bar{c}}\right)_w + \frac{\Delta_{f1}x_{ac}}{\bar{c}} + \frac{\Delta_{f2}x_{ac}}{\bar{c}} \qquad (E-36)$$

The correction $\Delta_{f1}x_{ac}$ represents the forward shift due to the fuselage sections forward and aft of the wing. The nose in particular contributes to this shift. Based on experimental data in Ref. E-6:

$$\frac{\Delta_{f1}x_{ac}}{\bar{c}} = -\frac{1.8}{\left(C_{L_\alpha}\right)_{wf}} \frac{b_f h_f l_{fn}}{S\bar{c}} \qquad (E-37)$$

For geometric definitions see Fig. E-13. The correction $\Delta_{f2}x_{ac}$ accounts for the lift loss in the region where the wing/fuselage lift carry-over is concentrated and is derived from the DATCOM method:

$$\frac{\Delta_{f2}x_{ac}}{\bar{c}} = \frac{.273}{1 + \lambda} \frac{b_f\, c_g\, (b-b_f)}{(\bar{c})^2 (b+2.15b_f)} \tan\Lambda_{1/4}$$

$$(\text{for } \frac{b_f}{b} < .2) \qquad (E-38)$$

E-7.2 Pitching moment $\left(C_{m_{ac}}\right)_{wf}$

$$\left(C_{m_{ac}}\right)_{wf} = \left(C_{m_{ac}}\right)_w + \Delta_f C_{m_{ac}} \qquad (E-39)$$

The fuselage contribution $\Delta_f C_{m_{ac}}$ can be obtained from Munk's theory (Ref. E-33). For fuselages with near-circular cross-sections:

$$\Delta_f C_{m_{ac}} = -1.8 \left(1 - \frac{2.5b_f}{l_f}\right) \cdot \frac{\pi\, b_f h_f l_f}{4\, S\bar{c}} \frac{C_{L_0}}{\left(C_{L_\alpha}\right)_{wf}}$$

$$(E-40)$$

where C_{L_0} is C_L for $\alpha_f = 0$ (cf. Section E-9.3) and C_{L_α} is per radian. If the cross-section is not near-circular, $\Delta_f C_{m_{ac}}$ should be multiplied by the ratio of the actual cross-section area to $(\pi/4) b_f h_f$.

E-8. NACELLE AND PROPELLER CONTRIBUTIONS

Some aerodynamic effects of engine nacelle installation are discussed qualitatively in Section 6.5. Accurate prediction of these effects is generally not possible in the preliminary design stage. It is therefore suggested that only the effects in the aerodynamic center location be estimated for each nacelle separately and then added:

$$\Delta_n \frac{x_{ac}}{\bar{c}} = \sum k_n \frac{b_n^2 l_n}{S\bar{c} \left(C_{L_\alpha}\right)_{wf}} \qquad (C_{L_\alpha} \text{ in rad}^{-1}) \quad (E-41)$$

where $k_n \sim -4.0$ for nacelles mounted in front of the LE of the wing

$k_n \sim -2.5$ for jet engine pods mounted to the sides of the rear fuselage.

The geometry is defined in Fig. E-13. The distance l_n is positive when the nacelle is in front of, negative when behind 1/4 MAC.

Propeller effects can be very considerable in the power-on configuration; power-off effects on lift are small. However, lift on a windmilling propeller at incidence causes a shift in the a.c. For windmilling tractor propellers the data given by Perkins and Hage (Ref. E-1) are approximated by:

$$\Delta_p \frac{x_{ac}}{\bar{c}} = -.05 \sum \frac{B_p D_p^2 l_p}{S\bar{c} \left(C_{L_\alpha}\right)_{wf}} \qquad (C_{L_\alpha} \text{ in rad}^{-1})$$

$$(E-42)$$

where B_p is the number of blades per propeller. Other definitions: see Fig. E-13. The effects of nacelles and propellers on x_{ac} are appreciable in the case of wing-mounted engines with horizontally opposed cylinders, relatively high-powered airplanes with a single tractor propeller in the

fuselage nose and jet airplanes with podded engines mounted to the sides of the rear fuselage.

E-9. LIFT OF THE COMPLETE AIRCRAFT

In this paragraph the location of the aerodynamic center and moment about the a.c. of the complete aircraft minus horizontal tail, as obtained from the previous sections, will be referred to as x_{ac} and $C_{m_{ac}}$. The moment coefficient is:

$$C_{m_{A-h}} = C_{m_{ac}} + C_L \frac{x_{cg} - x_{ac}}{\bar{c}} \qquad (E-43)$$

E-9.1. Tailplane lift

The horizontal tailplane lift to trim is obtained from the conditions of longitudinal equilibrium and lift = weight:

$$\Delta_h C_L = C_{L_h} \frac{S_h}{S} \frac{q_h}{q} = \frac{\bar{c}}{l_h} \left(C_{m_{ac}} + C_L \frac{x_{cg} - x_{ac}}{\bar{c}}\right) \qquad (E-44)$$

E-9.2. Total trimmed airplane lift

By combining (E-2), (E-29) and (E-44), neglecting nacelles and lift due to the propulsion system, we find:

$$C_L = C_{L_{wf}} \left(1 + \frac{x_{cg} - x_{ac}}{l_h}\right) + \frac{\bar{c}}{l_h} C_{m_{ac}} \qquad (E-45)$$

E-9.3. Wing/body incidence

The wing incidence (relative to the fuselage reference line), required to obtain a specified C_{L_o} with the fuselage reference axis[*] horizontal ($\alpha_f = 0$), is obtained from (E-29), (E-30) and (E-45):

$$i_w = \frac{C_{L_{wf}}^* - \Delta_z C_L}{K_{II} C_{L_{w_\alpha}}} + \frac{K_I}{K_{II}} \alpha_{o_1} \varepsilon_t + \left(\alpha_{\ell_o}\right)_r \qquad (E-46)$$

[*] Strictly speaking, the fuselage floor line should be used as the reference for defining $\alpha_f = 0$.

where

$$C_{L_{wf}}^* = \left(C_{L_o} - \frac{\bar{c}}{l_h} C_{m_{ac}}\right) \bigg/ \left(1 + \frac{x_{cg} - x_{ac}}{l_h}\right) \qquad (E-47)$$

The values of C_{L_o} and x_{cg} frequently correspond to mean values of C_L and airplane weight[*] at the design cruising altitude, but there may also be other factors to consider (cf. Section 7.7.2).

E-9.4. Trimmed lift curve

The data generated may now be used to obtain the lift curve of the trimmed airplane in the linear range, with the angle attack defined relative to the fuselage datum line:

$$C_L = C_{L_\alpha} \left\{ \alpha_f - \left(\alpha_{L_o}\right)_f \right\} \qquad (E-1)$$

where

$$C_{L_\alpha} = K_I \left(1 + \frac{x_{cg} - x_{ac}}{l_h}\right) C_{L_{w_\alpha}} \qquad (E-48)$$

and

$$\left(\alpha_{L_o}\right)_f = - \frac{C_{L_o}}{C_{L_\alpha}} \qquad (E-49)$$

provided i_w is chosen in accordance with (E-46) and (E-47).

The tailplane effect on C_L-max can be found from the C_L-α curve, assuming that the critical angle of attack is not affected by the presence of the horizontal tailplane.

E-10. AIRPLANE PITCHING MOMENT AND NEUTRAL POINT (STICK FIXED)

E-10.1. The stick-fixed neutral point

According to Section 9.2.1 of Chapter 9:

[*] e.g. a payload equal to 50-60% of the maximum payload and half the trip fuel burnt.

$$\frac{x_n}{\bar{c}} = \frac{x_{ac}}{\bar{c}} + \frac{C_{L_{h_\alpha}}}{C_{L_\alpha}} \left(1 - \frac{d\varepsilon_h}{d\alpha}\right) \frac{S_h l_h}{s\bar{c}} \frac{q_h}{q} \qquad (E-50)$$

Contrary to the previous section, C_{L_α} in this formula refers to the untrimmed (stick fixed) condition:

$$C_{L_\alpha} = \left(C_{L_\alpha}\right)_{wf} + C_{L_{h_\alpha}} \left(1 - \frac{d\varepsilon_h}{d\alpha}\right) \frac{S_h}{S} \frac{q_h}{q} \qquad (E-51)$$

and x_{ac} refers to the airplane less horizontal tail.

In the absence of better information, q_h/q may be assumed equal to .85 for a fuselage-mounted stabilizer and .95 for a fin-mounted stabilizer, except in the case of a T-tail (with $q_h/q \sim 1$).

The lift-curve slope of the tailplane can be computed with the method presented for wings in Sections E-3 and E-4, with the following typical corrections (where applicable):
- a reduction of 8% for an unsealed full-span gap between the elevator and the stabilizer,
- a reduction of 5% if the elevator is beveled over its full root chord to accommodate rudder deflection.

These data refer to conventional, fuselage-mounted fins (class A in Fig. 2-23); for other configurations Ref. E-48 may be consulted. In the case of twin fins mounted as endplates to the stabilizer, the aerodynamic aspect ratio of the horizontal tailplane is approximately 1.5 times the geometric aspect ratio.

The downwash gradient in unpowered flight is approximately:

$$\frac{d\varepsilon_h}{d\alpha} = 1.75 \frac{C_{L_{w_\alpha}}}{\pi A (\lambda r)^{.25} (1 + |m|)} \qquad (E-52)$$

The geometric parameters r and m are defined in Fig. E-13. Pylon-mounted jet engine nacelles at the sides of the rear fuselage reduce the factor $(1 - d\varepsilon_h/d\alpha)$ by approximately 10%. Due to the propeller-fin effect at zero thrust $(1-d\varepsilon_h/d\alpha)$ is decreased by approximately $.012\ B_p$ (approximation of data in Ref. E-1).

E-10.2. Horizontal stabilizer incidence

For zero elevator and trim tab deflection, the tailplane lift for a symmetrical section is:

$$C_{L_h}=C_{L_{h_\alpha}}\left|\alpha(1-\frac{d\varepsilon_h}{d\alpha})+i_h\right| \qquad (E-53)$$

where α and i_h are measured relative to the wing zero-lift line (Fig. E-12). For a fixed stabilizer the condition may be imposed that the airplane is trimmed at $C_L=C_{L_o}$ with the elevator neutral in order to minimize the tailplane drag.

Provided that:

$\varepsilon_h=0$ for $C_L=0$,

ε_h is a linear function of α, and

$C_{L_w}=C_{L_o}$,

the required incidence relative to the fuselage datum line is:

$$(i_h)_f=\frac{C_{m_{ac}}+C_{L_o}\dfrac{x_{cg}-x_{ac}}{\bar{c}}}{C_{L_{h_\alpha}}\dfrac{S_h l_h}{S\bar{c}}\dfrac{q_h}{q}}+\frac{d\varepsilon_h/d\alpha}{C_{L_{w_\alpha}}}C_{L_o} \qquad (E-54)$$

and that relative to the wing zero-lift line is:

$$i_h=(i_h)_f-\frac{C_{L_o}}{C_{L_\alpha}} \qquad (E-55)$$

C_{L_o} being the lift coefficient when the fuselage is at zero angle of attack (Section E-9.3).

E-10.3. Pitching moment curve

For zero elevator deflection and angles of attack in the linear range:

$$C_m=C_{m_o}+\frac{dC_m}{dC_L}C_L \qquad (E-56)$$

where

$$\frac{dC_m}{dC_L}=-\frac{x_n-x_{cg}}{\bar{c}} \qquad (E-57)$$

and

$$C_{m_o}=C_{m_{ac}}-C_{L_{h_\alpha}}i_h\frac{S_h l_h}{S\bar{c}}\frac{q_h}{q} \qquad (E-58)$$

A summary of the various contributions to C_m is given in Fig. E-14.

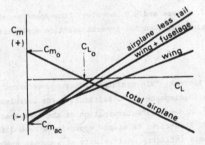

Fig. E-14. Pitching moment curve in the linear angle of attack range for zero elevator deflection.

REFERENCES

E-1. C.D. Perkins and R.E. Hage: "Airplane performance, stability and control". New York, John Wiley and Sons, Inc., 1949.

E-2. H. Schlichting and E. Truckenbrodt: "Aerodynamik des Flugzeuges". Vol. II. Springer Verlag, 1960.

E-3. D.E. Hoak and D.E. Ellison: "USAF Stability and Control Datcom". Wright Patterson Air Force Base, October 1960 (rev. August 1968).

E-4. D.E. Ellison and D.E. Hoak: "Stability derivative estimation at subsonic speeds". Annals of the New York Academy of Sciences, Vol. 154, Part. 2, pp. 367-396.

E-5. Anon.: Data Sheets, Aerodynamics, Vols. I, II and III, Royal Aeronautical Society.

E-6. W. Just: "Flugmechanik, Steuerung und Stabilität von Flugzeugen". Verlag Flugtechnik, Stuttgart, 1966.

E-7. G.J. Hancock: "Problems of aircraft behaviour at high angles of attack". AGARDograph 136, April 1969.

E-8. J. Roskam: "Methods of estimating stability and control derivatives of conventional subsonic airplanes". The University of Kansas, 1971.

E-9. C.H. Wolowicz: "Longitudinal aerodynamic characteristics of light, twin-engine, propeller driven airplanes". NASA TN D-6800, 1972.

E-10. R.K. Heffley and W.F. Jewell: "Aircraft handling qualities data". NASA CR-2144, December 1972.

E-11. J.G. Callaghan: "Aerodynamic prediction methods at low speeds with mechaninal high lift devices". AGARD Lecture Series No. 67, May 1974.

Airfoil sections

E-12. I.H. Abbott, A.E. Von Doenhoff and L.S. Stivers: "Summary of Airfoil Data". NACA Report 824, 1945.

E-13. F.W. Riegels: "Aerodynamische Profile". R. Oldenburg, Munich, 1958.

E-14. I.H. Abbott and A.E. Von Doenhoff: "Theory of wing sections". McGraw-Hill Book Comp. Inc., New York; also: Dover Publications Inc., New York, 1959.

E-15. L.K. Loftin: "The effects of variations in Reynolds number between 3.0×10^6 and 25.0×10^6 upon the aerodynamic characteristics of NACA 6-series airfoil sections". NACA TN 1773, 1948.

E-16. L.K. Loftin: "Theoretical and experimental data for a number of NACA 6A-series airfoil sections". NACA Report 903, 1948.

E-17. D.M. McRae: "Aerodynamics of mechanical high-lift devices". AGARD Lecture Series No. 43, April 20-24, 1970.

E-18. R.J. McGhee and W.D. Beasley: "Low-speed aerodynamic characteristics of a 17 percent-thick airfoil section designed for general aviation applications", NASA TN D-7428, December 1973.

Wings

E-19. O. Schrenk: "Ein einfaches Näherungsverfahren zur Ermittlung von Auftriebsverteilungen längs der Tragflügelspannweite". Zeitschrift für Luftwissenschaften, Vol. 7, No. 4, pp. 118-120.

E-20. R.F. Anderson: "Determination of the characteristics of tapered wings". NACA Rep. 572, 1936.

E-21. F.W. Diederich: "A simple approximate method for calculating spanwise lift distributions and aerodynamic influence coefficients at subsonic speeds". NACA TN 2751, 1952.

E-22. J. de Young and C.W. Harper: "Theoretical symmetric span loading at subsonic speeds for wings having arbitrary planform". NACA Rep. 921, 1948.

E-23. R. Stanton Jones: "An empirical method for rapidly determining the loading distribution on swept back wings". College of Aeronautics Report No. 32, 1950.

E-24. G.H. Lee: "High maximum lift". The Aeroplane, October 30 and November 6, 1953.

E-25. V. Holmboe: "Charts for the position of the aerodynamic centre at low speeds and small angles of attack for a large family of tapered wings". SAAB TN 27, 1954 (reproduced in "Airplane Design", by K.D. Wood).

E-26. Anon.: "Method for the rapid estimation of theoretical spanwise loading due to a change of incidence". R.Ae.S. Transonic Data Memorandum 6208, 1962.

E-27. Anon.: "Graphical method for estimating the spanwise distribution of aerodynamic centre on wings in subsonic flow". R.Ae.S. Transonic Data Memorandum 6309, Sept. 1963.

E-28. C.L. Bore and A.T. Boyd: "Estimation of maximum lift of swept wings at low Mach numbers". J. of the R.Ae.S., Vol. 67, April 1963, pp. 227-239.

E-29. L.R. Wootton: "The effect of compressibility on the maximum lift coefficient of aerofoils at subsonic airspeeds". J. of the R.Ae.S., Vol. 71, July 1967.

E-30. T.E. Labruyere and J.G. Wouters: "Computer application of subsonic lifting surface theory". NLR Report TR 70088.

E-31. P. Kapteyn: "Design charts for the aerodynamic characteristics of straight and swept, tapered, twisted wings at subsonic speeds". Delft University of Technology, Dept. of Aeronautical Engineering, M-180, May 1972.

E-32. M.A. McVeigh and E. Kisielowski: "A design summary of stall characteristics of straight wing aircraft". NASA CR-1646, June 1971.

Bodies and wing/body combinations

E-33. M.M. Munk: "The aerodynamic forces on airship hulls". NACA T. Rep. 184, 1924.

E-34. J.H. Allen: "Estimation of the forces and moments acting on inclined bodies of revolution". NACA RM A 9126, 1949.

E-35. H. Schlichting: "Calculation of the influence of a body on the position of the aerodynamic centre of aircraft with sweptback wings". ARC R and M 2582, 1947.

E-36. A.H. Flax and H.R. Lawrence: "The aerodynamics of low aspect -ratio wings and wing-body combinations". Proceedings of the Third Anglo-American Aeronautical Conference 1951, pp. 363-398.

E-37. W.C. Pitts, J.N. Nielsen and G.E. Kaattari: "Lift and center of pressure of wing-body-tail combinations at subsonic, transonic and supersonic speeds". NACA T. Rep. 1307, 1957.

E-38. S.F. Hoerner: "Fluid dynamic drag". Published by the author, 1965.

Propeller effects

E-39. R. Smelt and H. Davies: "Estimation of increase in lift due to slipstream". ARC R and M 1788, 1937.

E-40. D.E. Morris and J.C. Morrall: "Effect of slipstream on longitudinal stability of multi-engined aircraft". ARC R and M 2701, November 1948.

E-41. J. Weil and W.C. Sleeman, Jr.: "Prediction of the effects of propeller operation on the static longitudinal stability of single-engine tractor monoplanes with flaps retracted", NACA T. Rep. 941.

E-42. R.E. Kuhn: "Semi-empirical procedure for estimating lift and drag characteristics of propeller-wing-flap configurations for vertical and short take-off and landing aeroplanes". NASA Memorandum 1-16-59L, 1959.

E-43. B.W. McCormick: "Aerodynamics of V/STOL Flight". Academic Press, New York/London, 1967.

Downwash

E-44. A. Silverstein and S. Katzoff: "Design charts for predicting downwash angles and wake characteristics behind plain and flapped wings". NACA Rep. 648, 1939.

E-45. J. De Young and W.H. Barling: "Prediction of downwash behind swept-wing airplanes at subsonic speeds". NACA TN 3346.

E-46. F.W. Diederich: "Charts and tables for use in calculations of downwash of wings of arbitrary plan form". NACA TN 2353, 1951.

E-47. J.L. Decker: "Prediction of downwash at various angles of attack for arbitrary tail locations". Aero. Eng. Review, August 1956, pp. 22-61.

Tailplane characteristics

E-48. D.J. Lyons and P.L. Bisgood: "An analysis of the lift slope of aerofoils of small aspect ratio wings including fins, with design charts for aerofoils and control surfaces". R and M 2308, 1945. Also in Aircraft Eng., September 1947, pp. 278-286.

Appendix F. Prediction of the airplane polar at subcritical speeds in the en route configuration

SUMMARY

A concise collection of formulas, generalized data and methods is presented for estimating the airplane drag at subcritical speeds in the en route configuration. All methods are applicable in the preliminary design stage of conventional aircraft.

Drag associated with the trailing vortices and profile drag are computed for the isolated, smooth airplane parts. Some corrections are given to account for the interaction of the flow fields of parts placed in close proximity. Finally, a number of data and methods are given to account for the effects of surface imperfections, powerplant installation, protuberances and other extras.

NOMENCLATURE

A	–	aspect ratio (no index: wing aspect ratio); constant in drag equation; planform area
A_c	–	cross-sectional area of a streamline body
B	–	factor of C_L in drag equation
b	–	span; width
c	–	coefficient of proportionality,
C_c	–	conversion constant
c	–	chord
$\bar{c}$	–	Mean Aerodynamic Chord
C_1,C_2,C_3	–	coefficients in Diederich's method (cf. Appendix E)
C_{0_1},C_{1_1}	–	coefficients determining the vortex-induced drag due to twist
C_{ci}	–	total circumferential length of wing/fuselage intersection
c_d,C_D	–	two- and three-dimensional drag coefficient
C_{D_π}	–	C_D based on frontal area
C_{D_o}	–	zero-lift drag coefficient
c_{d_p},C_{D_p}	–	two- and three-dimensional profile drag coefficient
C_{D_v}	–	vortex-induced drag coefficient
$C_{D_{ref}}$	–	reference drag coefficient at $C_L = C_{L_{ref}}$
C_F	–	skin friction drag coefficient of smooth flat plate
c_g	–	geometric mean chord
c_ℓ,C_L	–	two- and three-dimensional lift coefficient
C_{L_o}	–	C_L at $\alpha_f = 0$
$c_{\ell_\alpha},C_{L_\alpha}$	–	lift-curve slope of section and aircraft respectively
c_{ℓ_i},C_{L_i}	–	design lift coefficient of section and wing respectively
$C_{L_{ref}}$	–	C_L for minimum (reference) C_D
$C_{m_{ac}}$	–	aerodynamic pitching moment about the aerodynamic center of aircraft less tail
D	–	drag; diameter; factor of C_L^2 term in drag equation
D_o	–	diameter of capture area of engine air
D_1,D_2	–	factors in fuselage drag equation
D_c	–	drag due to cross flow component of V_∞
d_m	–	diameter of cylindrical forebody of boat tail
E	–	Jones' edge velocity factor (cf. Appendix E)
e	–	Oswald's factor for the induced drag coefficient
$\overline{FR}$	–	boat tail fineness ratio
f	–	drag area; Diederich's sweep function (cf. Appendix E)
h	–	height
k	–	equivalent sand grain size
L	–	boat tail length
l	–	length
M_∞	–	flight Mach number
M_f,M_g	–	Mach number of fan and gas generator flows respectively
$\dot{m}_a,\dot{m}_c,\dot{m}_j$	–	mass flow per unit time of engine air, cooling air and engine exhaust flow respectively
NPR	–	Nozzle Pressure Ratio; i.e. (total pressure at nozzle exit)$/P_\infty$
N_e	–	number of engines
P_b	–	brake power
P_∞	–	ambient pressure (static)
q_∞	–	dynamic head ($q_\infty=\frac{1}{2}\rho_\infty V_\infty^2=\frac{1}{2}\gamma P_\infty M_\infty^2$)
R	–	Reynolds number
r	–	radius of curvature
S	–	area; no subscript: gross wing area
S_{net}	–	net wing area (cf. Appendix A, Section A-3.1.)
T	–	thrust
T_∞	–	ambient temperature (static)
T_c	–	propeller thrust coefficient
t	–	thickness; canopy nose or tail section length
u,v,w	–	Anderson's factors in vortex-induced drag coefficient of wing
V	–	volume; velocity of flow
V_∞	–	flight speed
v_e,v_i	–	exhaust and inlet velocity respectively
$\dot{W}_a,\dot{W}_j$	–	weight flow per second of engine air and jet flow respectively
x	–	longitudinal coordinate
x_T	–	distance of transition region downstream from the nose or leading edge

y - spanwise coordinate, measured from wing centerline

α - angle of attack

α_f - angle of attack relative to fuselage centerline

α_f' - α_f for minimum fuselage drag coefficient

α_i - induced angle of attack

β - Prandtl-Glauert compressibility factor ($\beta = \sqrt{1-M^2}$); fuselage tail upsweep angle; length of cowl forebody/total cowl length; boat tail angle

γ - ratio of specific heats (for air: $\gamma = 1.4$)

Δ - increment; e.g. $\Delta_i C_D$ = increment in C_D due to interference

δ - increment of plane wing vortex-induced drag coefficient due to additional lift

ε - angle of twist; downwash angle

η - non-dimensional spanwise coordinate ($\eta = 2y/b$)

η_p - propeller efficiency

θ - relative atmospheric temperature $T_\infty / (T_\infty$ at sea level, standard conditions)

θ_j - deflection angle of jet flow relative to free stream

Λ - sweepback angle of quarter-chord line (unless otherwise stated); no index: wing sweepback angle

Λ_β - corrected sweepback angle (tan $\Lambda_\beta =$ (tan $\Lambda)/\beta$)

λ - taper ratio; slenderness ratio

ν - coefficient of kinematic viscosity

ρ_∞ - density of ambient air

σ - relative density $= \rho_\infty/(\rho_\infty$ at sea level, standard conditions)

φ - shape factor in profile drag coefficient equation

F-1. DRAG COMPONENTS

Airplane drag can be subdivided in a number of ways. The terminology used in several schemes is discussed in Section 11.2 of Chapter 11. The usual breakdown for the purpose of performance analysis is as follows:

Subscripts

A - afterbody section

a - engine airflow

ac - aerodynamic center, aircraft less horizontal tailplane

b - base; brake

C - mid-section of body

c - cross flow; cooling; coefficient

cg - center of gravity

cp - center of pressure

e - excrescence(s); engine; exhaust

F - forebody section

f - fuselage; fan

g - gas generator; gross

h - highlight; horizontal tailplane

i - interference; inlet; design condition

j - jet flow

LE - leading edge

l - lift

N - nose section

n - nacelle

P - pressure

p - profile; propeller, plug

py - pylon

r - root

s - slipstream

ss - slats and steps

st - strut

t - tip

uc - undercarriage

v - vortex; vertical tailplane

w - wing

wet - wetted by airflow

β - boat tail

¼ - quarter-chord line

½ - mid-chord line

$\alpha\beta$ - denotes effects of cross flow and fuselage upsweep

$$C_D = C_{D_o} + \frac{C_L^2}{\pi Ae} \qquad (F-1)$$

Section 5.3.1. gives an explanation of these terms and an initial estimation method for C_{D_o} and the Oswald factor e. At the stage where the layout and principal geometric properties have already been defined, a more detailed analysis becomes appropriate. The information required for this is given in this Appendix.

Drag can be broken down into the following components (see Table F-1):

a. Vortex-induced drag: pressure drag associated with the kinetic energy required to generate trailing vortices and downwash. Provided all main airplane parts are considered as isolated bodies, most components of vortex-induced drag can be calculated fairly accurately, using classical potential flow methods.

b. Profile drag: drag due to the boundary layer and regions of separated flows around the main airplane parts placed in isolation. For well-streamlined, smooth aircraft components, the skin friction drag dominates at small angles of attack, while pressure drag is only a fraction of the total drag. Profile drag can be estimated reasonably accurately for the most common airplane shapes.

c. Interference effects: corrections to allow for the interaction of the flow field around the airplane parts. Although for a well-designed airplane the interference drag is not more than 5-10% of the total zero-lift drag, and may even be negative, prediction methods are very unreliable for most components.

d. Drag due to protuberances, surface imperfections and other extras: calculation must be done partly on the basis of known design geometry (cockpit drag, intake scoop drag, etc.) and partly on a statistical basis (drag of rivets, joints, gaps, leaks, etc.).

The vortex-induced drag of a plane wing (i.e. the wing is not twisted) is zero for $C_L = 0$, but profile drag, interference drag and several vortex drag components are minimum at some positive reference (datum) value for C_L, e.g. in cruising flight. Hence, the condition for minimum C_D will be found for $0 < C_L < C_{L_{cr}}$ and (F-1) therefore approximates the actual polar curve only in a limited region of C_L (cf. Fig. 5-4).

The drag polar at subcritical speeds can be obtained by computation of all items mentioned in Tables F-1. The drag components are presented in the following form:

$$C_{D_j} = A_j + B_j C_L + D_j C_L^2 \qquad (F-2)$$

To avoid confusion as to the reference area to which these components are related, most data will be presented in the form of a drag area $f = (C_D S)_j^*$. Components can be related to the gross wing area as follows:

$$C_{D_j} = \frac{(C_D S)_j}{S} \qquad (F-3)$$

All drag areas can be added and the total drag coefficient is:

$$C_D = \frac{\Sigma (C_D S)_j}{S} \qquad (F-4)$$

Drag components are very often expressed as "counts"; 1 count is $\Delta C_D = .0001$, based on the wing area. Addition of all components in Table F-1[**] results in the final expression:

$$C_D = A + B C_L + D C_L^2 \qquad (F-5)$$

Alternatively, the drag polar may be expressed as follows:

$$C_D = C_{D_{ref}} + \beta (C_L - C_{L_{ref}})^2$$

where

$$C_{L_{ref}} = -B/2D; \quad C_{D_{ref}} = A - D C_{L_{ref}}^2; \quad \beta = D \qquad (F-6)$$

A polar of the shape given by (F-1) can be obtained by plotting calculated values of C_D vs. C_L^2 and making a straight line approximation in the C_L region between the high-speed cruise and approximately 70% of

[*] An exception is made for wing drag components, related to the gross wing area S.
[**] To avoid mistakes it is useful to fill out Table F-1 in counts, by multiplying all components by 10^4. The values of A, B and D are then divided by 10^4.

A. VORTEX DRAG	A (ΣA_v)	B (ΣB_v)	D (ΣD_v)
DRAG OF PLANE WING			
DRAG DUE TO WING TWIST			
WING TIP CORRECTION			
DRAG DUE TO FUSELAGE LIFT			
HORIZONTAL TAILPLANE DRAG (TRIM)			
B. PROFILE DRAG (SMOOTH CONDITION)	(ΣA_p)	(ΣB_p)	(ΣD_p)
WING – MINIMUM DRAG			
– INCREMENTAL DRAG			
FUSELAGE – BASIC DRAG			
– BOATTAIL AND BASE DRAG			
– UPSWEEP AND INCIDENCE EFFECTS			
TAILBOOMS			
NACELLES – EXTERNAL (FAN) COWLING			
– GAS GENERATOR			
– PLUG			
– PYLONS			
HORIZONTAL TAILPLANE – MINIMUM DRAG			
– DRAG DUE TO δ_e AND α_h			
VERTICAL TAILPLANE			
C. INTERFERENCE CORRECTIONS	(ΣA_i)	(ΣB_i)	(ΣD_i)
WING/FUSELAGE – VORTEX DRAG			
– VISCOUS INTERFERENCE DRAG			
– LIFT EFFECT ON FUSELAGE DRAG			
NACELLE/AIRFRAME – VORTEX DRAG			
– PROFILE DRAG			
TAILPLANE/TAILPLANE			
TAILPLANE/AIRFRAME			
SUB-TOTAL			

D. PROTUBERANCES, SURFACE IMPERFECTIONS, ETC.	A (ΣA_e)	B (ΣB_e)	D (ΣD_e)
FIXED UNDERCARRIAGE	( )	( )	( )
MAIN UNDERCARRIAGE			
NOSE UNDERCARRIAGE			
TAILWHEEL/TAILSKID			
PARTLY RETRACTED LANDING GEARS			
CANOPIES AND CABIN WINDOWS	( )	( )	( )
PROTRUDING COCKPIT ENCLOSURES			
FLIGHT DECK WINDOWS			
FAIRINGS AND BLISTERS	( )	( )	( )
WHEEL-WELL FAIRINGS			
BLISTERS			
EXTERNAL FUEL TANKS	( )	( )	( )
BASIC PROFILE DRAG			
INSTALLATION AND INTERFERENCE CORRECTIONS			
STRUTS AND WIRES	( )	( )	( )
ENGINE INSTALLATION			
EXTERNAL DRAG – ENGINE AIR INTAKE(S)/SPILLAGE			
– ENGINE AIR OUTLET(S)/REVERSERS			
– OIL COOLER(S)/RADIATOR(S)			
INTERNAL DRAG – ENGINE COOLING			
– OIL COOLER(S)/RADIATOR(S)			
SLIPSTREAM EFFECTS – PROFILE DRAG			
– VORTEX-INDUCED DRAG			
JET INTERFERENCE	( )	( )	( )
EXCRESCENCES, SURFACE IMPERFECTIONS, OTHER EXTRAS			
SURFACE IMPERFECTIONS			
HIGH-LIFT SYSTEM			
FLIGHT CONTROL SYSTEM			
VORTEX GENERATORS, FENCES, STRAKES			
AIRFRAME SYSTEMS INSTALLATION			
MISCELLANEOUS, UNACCOUNTABLE ALLOWANCE			
TOTAL	(ΣA)	(ΣB)	(ΣD)

NOTE: $C_D = \Sigma A + \Sigma B \cdot C_L + \Sigma D \cdot C_L^2$

Table F-1. Breakdown of the drag for the purpose of drag estimation

$C_{L_{max}}$

The most important restrictions to most of the methods presented are the same as thos stated in Appendix E for the lift prediction. In terms of wing geometry:

$t/c > 9\%$

$A > 4/\cos \Lambda_{\frac{1}{2}}$

$\Lambda_{\frac{1}{2}} < 35^{\circ}$

These approximate limits generally define wings with attached flow at small angles of attack and full leading edge suction, provided they have adequate camber and washout and the angle of attack is in the normal operating regime. Flight speeds are limited to subcritical Mach numbers. An estimation of the drag-critical Mach number can be found in Section 7.5.2.

The data presented in this Appendix are sufficiently complete to make a drag analysis for the most common aircraft shapes. However, the designer should be on the alert for special drag components to be anticipated in his particular design. The well-known book by Hoerner (Ref. F-18) and other publications listed in the references can be consulted in such cases, and if a more detailed analysis is desirable.

F-2. PRIMARY COMPONENTS OF VORTEX-INDUCED DRAG

The lift-dependent drag due to trailing vortices is associated with the kinetic energy lost in the system of trailing vortices behind the aircraft. For the wing it is given by the equation

$$C_{D_V} = \int_{0}^{1} \frac{c_{\ell} c}{c_g} \alpha_i \, d\eta \qquad (F-7)$$

where

$$\alpha_i = \frac{1}{4\pi A} \int_{-1}^{+1} \frac{d}{d\eta'} \left(\frac{c_{\ell} c}{c_g} \right) \frac{d\eta'}{\eta - \eta'} \qquad (F-8)$$

These equations are valid only for symmetrical lift distributions.

F-2.1. Untwisted plane wings

Assuming for the moment* that all lift of the airplane is generated by an untwisted plane wing, the vortex drag coefficient, based on S, can be defined as:

$$C_{D_V} = (1 + \delta) \frac{C_L^2}{\pi A} \qquad (F-9)$$

Equations F-7 and F-9 can be combined:

$$1 + \delta = \pi A \frac{\int_{0}^{1} \frac{c_{\ell} c}{c_g} \alpha_i \, d\eta}{\left[\int_{0}^{1} \frac{c_{\ell} c}{c_g} d\eta \right]^2} \qquad (F-10)$$

with $\delta = 0$ for an elliptical lift distribution.

The solution of (F-8) and (F-10) requires a knowledge of the lift distribution along the span, which may be obtained for potential flow using any classical lifting line or lifting surface theory. Although in many institutes these are available in the form of standard computer programs, it may be desirable to use a simpler approach in the preliminary design stage. Several possibilities will be given from which a choice can be made:

METHOD A. Garner has demonstrated (Ref. F-39) that there is a correlation between δ and the spanwise center of pressure $\overset{*}{\eta}_{cp}$:

$$\delta = 46.264 \left(\eta_{cp} - \frac{4}{3\pi} \right)^2 \qquad (F-11)$$

where

$$\eta_{cp} = \int_{0}^{1} \frac{c_{\ell} c}{C_L c_g} \eta \, d\eta \qquad (F-12)$$

For a known lift distribution this integral is simpler to evolve than (F-8) and (F-10). For example, the expression for the

*Corrections will be dealt with in Section F-4.

additional lift distribution given by Diederich (Section 4.2. of Appendix E) can be used:

$$\eta_{cp} = C_1 \int_0^1 \frac{c}{c_g} \eta \, d\eta + C_2 \frac{4}{3\pi} + C_3 \int_0^1 f \eta \, d\eta \qquad (F-13)$$

Although Diederich's approximation for the lift distribution is not very accurate, it has the advantage that arbitrary wing planforms can be dealt with. For the case of a straight-tapered wing, we have

$$\int_0^1 \frac{c}{c_g} \eta \, d\eta = \frac{1 + 2\lambda}{3(1 + \lambda)} \qquad (F-14)$$

while the last integral in (F-13) can be approximated by:

$$\int_0^1 f \eta \, d\eta = \frac{4}{3\pi} + .001 \Lambda_\beta \quad (\Lambda_\beta \text{ deg.}) \qquad (F-15)$$

Hence, for straight-tapered wings:

$$\eta_{cp} = C_1 \frac{1+2\lambda}{3(1+\lambda)} + \left(C_2 + C_3\right) \frac{4}{3\pi} + .001 \Lambda_\beta C_3 \quad (F-16)$$

where C_1, C_2 and C_3 are factors given in Fig. E-5 of Appendix E and Λ_β is in degrees. The author's experience is that (F-11) gives accurate results for corrected aspect ratios (βA) up to 10 when compared with lifting surface theory.

METHOD B. Anderson has generalized the results of Prandtl's lifting line theory for straight-tapered wings with rounded tips in incompressible flow (Ref. F-29). A summary is given in Fig. 10 of Ref. F-48 in the form of a diagram for u,

$$u \equiv \frac{1}{1+\delta} \qquad (F-17)$$

Lifting surface theory (e.g. Refs. F-40 and F-41) gives very similar results. The results in Ref. F-41 can be represented by:

$$\delta = \left\{ .0015 + .016 \ (\lambda - .4)^2 \right\} (\beta A - 4.5) \qquad (F-18)$$

for $6 < \beta A < 30$; $.3 < \lambda < 1.0$; $\Lambda_{\frac{1}{2}} = 0$

METHOD C. Results of an NLR computer program (Refs. F-40 and F-41) are presented in Fig. F-1, for $.2 < \lambda < .5$ and Λ_β up to $60°$. Similar results may also be obtained with Ref. F-36.

F-2.2. Drag due to twist

The shape of the lift distribution of twisted wings is no longer similar for different values of C_L. For zero lift the induced drag of a plane wing equals zero, but the basic lift distribution of a twisted wing causes a vortex-induced drag at zero lift. For straight wings with linear twist (low speeds) Anderson (Ref. F-29) represents the drag coefficient due to twist, related to the gross wing area, as follows:

$$\Delta_\varepsilon C_{D_v} = C_L \left(\frac{\varepsilon_t \, c_{\ell_\alpha}}{E} \right) v + \left(\frac{\varepsilon_t \, c_{\ell_\alpha}}{E} \right)^2 w \quad (F-19)$$

Diagrams for determining v and w can be found in Ref. F-48, Figs. 11 and 12. For a near-elliptical additional lift distribution a first approximation is:

$$\Delta_\varepsilon C_{D_v} = 3.7 \times 10^{-5} \varepsilon_t^2 \quad (\varepsilon_t \text{ deg.}) \quad (F-20)$$

For straight and swept wings with linear lofted twist (subcritical flow) Ref. F-41 states:

$$\Delta_\varepsilon C_{D_v} = \varepsilon_t^2 C_{o_1} + \varepsilon_t C_L C_{1_1} \qquad (F-21)$$

The vortex drag factors are computed from lifting surface theory and depicted in Figs. F-2 and F-3.

Some comments can be made on the results of the methods summarized above:

1. For a representative straight wing (A = 8, λ = .4) the induced drag due to a linear washout of 4 degrees at the tip is roughly twice as high when compared with a linear lofted[*] wing with the same twist angle.

2. An "optimum twist angle" can be defined for which $\Delta_\varepsilon C_{D_v}$ reaches an extreme value. In terms of

[*]linear and linear lofted twist are defined in Appendix A, Section A-3.2.

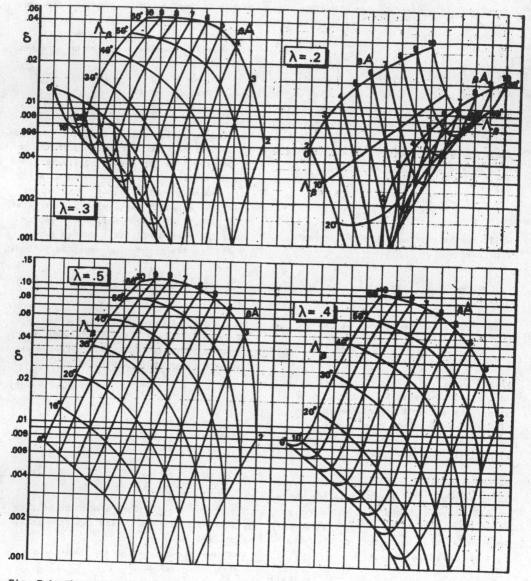

Fig. F-1. The vortex-induced drag factor δ (Ref. F-41)

Anderson's factors:

$$\varepsilon_{t_{opt}} = -\frac{1}{2}\left(\frac{E\,C_L}{c_{l_\alpha}}\right)\frac{v}{w} \tag{F-22}$$

corresponding to:

$$\left(\Delta_\varepsilon\,C_{D_v}\right)_{opt} = -\frac{1}{4}\,C_L^{\,2}\,\frac{v^2}{w} \tag{F-23}$$

Since w is positive, the induced drag is reduced when the twist is chosen according to (F-22) as compared with the plane, untwisted wing. This subject is treated more thoroughly in Ref. F-38 for swept wings. It should be noted, however, that for sweptback wings more aerodynamic twist than this optimum is usually required in order to lessen the tendency towards tip stalling.

494

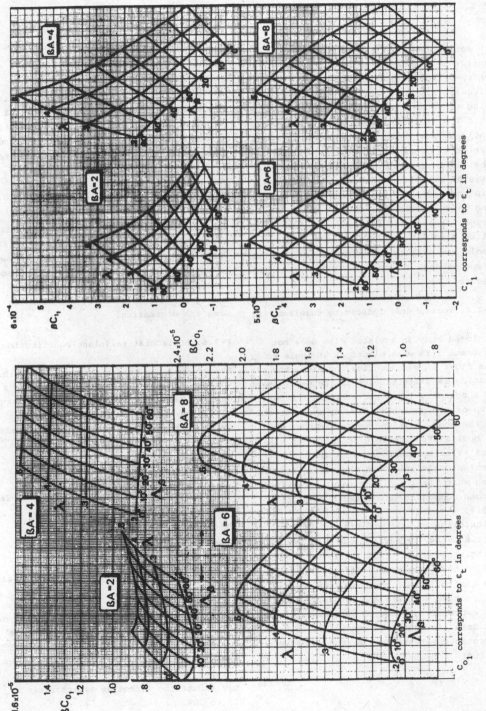

Fig. F-2. The factor C_0 for the computation of vortex-induced drag due to twist (Ref. F-41)

C_0 corresponds to ε_t in degrees

Fig. F-3. The factor C_1 for the computation of vortex-induced drag due to twist (Ref. F-41)

C_1 corresponds to ε_t in degrees

F-2.3. Wing tip correction on vortex-induced drag

In some potential flow methods the wing tip vortex is assumed to be concentrated at the extreme tip of the wing. Anderson takes a circular tip shape, others assume square wing tips. As shown by Hoerner (Ref. F-18, page 7-5) the tip vortex location may vary noticeably with the tip shape. This observation is confirmed by Schaufele and Ebeling in SAE Paper No. 670846 where they give an example of a minor change in the tip shape of a transport aircraft, resulting in an appreciable gain in takeoff weight in the case of a second segment climb gradient limitation. As no analytical tools are available on this subject, no attempt is made to present a calculation procedure here.

F-2.4. Vortex drag induced by fuselage lift

A closed body in inviscid flow does not generate lift when placed at an angle to the flow. In viscous flow, however, the situation is different due to the development of a boundary layer and subsequent breakdown of the potential flow at the rear end of the body. Lift and vortex-induced drag of bodies are much more sensitive to viscous effects than those of a wing. This section deals only with the vortex-induced drag which is the direct result of lift; a second component is discussed in Section F-3.4.

Calculation of lift and drag on isolated fuselages may be carried out with the methods given in Refs. F-57 and F-60. However, due to the interference effects caused by the airflow around the wing, the result will not be accurate and the following simple expression, based on various experimental data, is therefore considered to be an acceptable alternative for bodies of revolution:

$$C_D S = .15 \ \alpha_f^2 \ V_f^{2/3} \qquad (F-24)$$

where V_f is the volume of the fuselage and

α_f is, in radians:

$$\alpha_f = \frac{C_L - C_{L_o}}{C_{L_\alpha}} \qquad (F-25)$$

for C_{L_α} in rad^{-1}.

For bodies with rectangular cross-sections the vortex drag due to lift is at least twice as high as the result of (F-24). The effect of rounding off the corners is not known quantitatively, but is expected to be appreciable.

F-2.5. Nacelle contribution

The effect of nacelles on vortex drag is appreciable if they are located in the flow field of the wing. It does not make sense to estimate the vortex-induced drag associated with the lift on an isolated nacelle, as interference effects with the wing are dominating.

F-2.6. Horizontal tailplane contribution

For an isolated horizontal tailplane, the vortex-induced drag can be obtained in a similar manner as for the wing. To reduce this drag, the tailplane may also be twisted. Ignoring this effect and assuming a deviation from the elliptical drag we have:

$$C_D S = 1.02 \ \frac{C_{L_h}^2 \ S_h}{\pi A_h} \qquad (F-26)$$

where the tailplane lift is derived in Appendix E as:

$$C_{L_h} = \frac{C_{m_{ac}} + C_L \ (x_{cg} - x_{ac})/\bar{c}}{S_h \ell_h / (S\bar{c})} \qquad (E-44)$$

When dual fins are mounted as endplates on the horizontal stabilizer, the effective aspect ratio is approximately 50% higher than the geometric (see Fig. 2-26). A gap between the stabilizer and the elevator may reduce the effective aspect ratio by 15 to 20%.

It is noted that (F-26) gives only part of the trim

drag; profile drag due to elevator deflection (Section F-3.6.) and interference with the wing (Section F-4.4.) alter the result appreciably.
Data required to compute C_{L_h} are given in Appendix E. Alternatively, a simple approximation can be obtained by assuming the center of gravity to coincide with the aerodynamic center of the aircraft less tail. Hence,

$$\Delta C_D = 1.02 \frac{C_{m_{ac}}^2}{\pi A_h \ S_h/S \ (\ell_h/\bar{c})^2}$$

F-3. PROFILE DRAG OF SMOOTH, ISOLATED MAJOR COMPONENTS

Skin friction and pressure drag due to the boundary layer and limited regions of separation are referred to as profile drag. The wing, fuselage, nacelles and empennage are considered to be smooth streamline bodies; corrections are made to allow for the interaction of the flow fields, protuberances and surface roughness, etc. There are two exceptions to this:
a. Roughness is only taken into account insofar it affects boundary layer transition.
b. The wetted area does not include those parts of the wing and fuselage, etc. that are not actually exposed to the flow (cf. Section E-4.1.).

F-3.1. The flat plate analogy

Provided the pressure distribution in potential flow over the airplane part considered is known, the boundary layer may be analyzed, at least theoretically. The pressure distribution must then be corrected for the effect of the boundary layer and a second iteration may be started. In practice, this leads to considerable difficulties, but the reader who is in a position to avail himself of such methods may use them to advantage.
In most pre-design applications, the airplane shape is not yet fully defined and for this reason the "flat plate analogy"

is frequently used. This method assumes that the actual shape of the major parts can be compared to smooth streamline bodies, and excrescences and surface irregularities are ignored by fairing over the body to achieve smooth contours. Provided that the component is well streamlined, a close relation is found between the body profile drag and that of a flat plate. As the skin friction drag of a flat plate (Fig. F-4)[*] can be computed very accurately, this provides a good basis for estimating the profile drag of the component considered.
The practical application of this flat plate analogy is as follows:
1. The wetted area of an airfoil section, wing or body is computed (Appendix B). Corrections are applied to account for the interconnection of the parts, e.g. the wing/body junction (cf. Section F-4.1.).
2. The skin friction coefficient C_F is determined for a smooth flat plate (Fig. F-4) with the same wetted area and projected length, the transition being located at the same position downstream from the leading edge or nose. The Reynolds number is based on the (chord) length.
3. A shape correction factor φ is determined to account for the following factors:
a. The boundary layer around a body or airfoil section develops in a different way from the flat plate flow. Boundary layer velocity profiles and local skin friction coefficients are therefore different.
b. Due to the thickness of the body, the average velocity just outside the boundary layer is higher than the free stream velocity. Local values of the dynamic head and the skin friction per unit area are therefore increased.
c. Unlike a flat plate, a body has a frontal area and undergoes a noticeable pressure drag, particularly at high thickness/chord or diameter/length ratios. This pressure drag is generally of the order of 5 to

[*]The curves shown in this figure are valid for incompressible flow. At subsonic speeds the effect of compressibility is small.

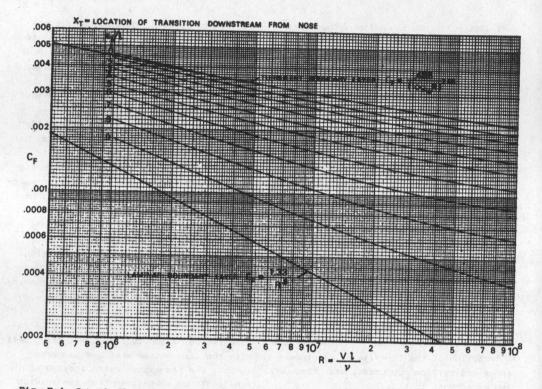

X_T = LOCATION OF TRANSITION DOWNSTREAM FROM NOSE

C_F

$R = \dfrac{V\ell}{\nu}$

Fig. F-4. Smooth flat plate skin friction coefficient (incompressible)

10% of the skin friction'drag.

4. The drag area of the body is finally computed as:

$$C_D S = C_F \, (1 + \varphi) \, S_{wet} \qquad (F-27)$$

The validity of the flat plate analogy is subject to a number of restrictions:

a. Airfoil sections should have a thickness/chord ratio of not more than approximately 25% and body diameter/length ratios should not be more than .25 to .35.

b. The parts should have smooth surface contours, free from kinks or steps. Conical or cylindrical body sections are acceptable, provided the transitions to the adjacent sections are gradual.

c. Lift, or angle of incidence, is zero or small. Experimental data (e.g. in Ref. F-44) are usually consulted wherever these conditions are not fulfilled and appropriate corrections are applied for lifting airfoils.

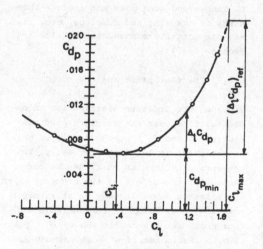

Fig. F-5. Profile drag of an airfoil section in two-dimensional flow

F-3.2. Wing sections

Fig. F-5 depicts a typical experimentally determined drag polar of a wing section. The profile drag coefficient, related to the chord, is split for the purpose of drag estimation as follows:

$$c_{d_p} = c_{d_{p_{min}}} + \Delta_\ell \, c_{d_p} \qquad (F-28)$$

For most sections the condition for minimum profile drag is the design lift coefficient c_{ℓ_i}, while $\Delta_\ell \, c_{d_p}$ is the drag increment for lift coefficients different from c_{ℓ_i}. For most sections c_{ℓ_i} is approximately equal to 1/10 of the camber as a percentage of the chord.

a. The minimum profile drag coefficient is:

$$c_{d_{p_{min}}} = 2 \, C_F \, (1 + \varphi_w) \qquad (F-29)$$

The factor 2 is necessary because c_{d_p} is based on the chord length and C_F on the wetted area, both sides being exposed to the flow. The thickness correction factor φ_w, given in Fig. F-6, not only accounts for the aerodynamic effect of thickness, but also for the difference between the circumference and twice the projected chord length. It is found that moderate section camber has only a very slight effect on the profile drag at c_{ℓ_i}. If only the thickness/chord ratio is known, the following approximation may be used:

$$\varphi_w = 2.7 \, t/c + 100 \, (t/c)^4 \quad \text{for } t/c \leqslant .21 \quad (F-30)$$

b. Boundary layer transition is frequently assumed to occur at the leading edge, but for straight high aspect ratio wings the result is conservative, particularly at Reynolds numbers less than 10^7. On the other hand, wind tunnel measurements with free transition yield erroneous results due to scale effects and to the effects of the unavoidable surface irregularities of practical wing constructions. In particular, the low-drag "bucket" on laminar flow sections will seldom be realized in practice, except on carefully polished high performance sailplanes. Wind tunnel experiments are therefore usually carried out with roughness strips at the position where the transition is likely to occur on the aircraft in flight. For Reynolds numbers up to 10^7 a transition region 15 to 20%* of the chord from

*Mean value for the upper and lower surface.

I: "Conventional" airfoils (NACA 4- and 5- digit series):

$$\varphi_w = \left(\frac{t/c}{.20}\right)^{1.38} \varphi_{20}$$

II: Cusped trailing edge, max. thickness at 40% chord (NACA 6-digit series):

$$\varphi_w = \left(\frac{t/c}{.20}\right)^{1.10} \varphi_{20}$$

III: NACA 6A- and similar series with straight trailing edge:

$$\varphi_w = \frac{t/c}{.20} \varphi_{20}$$

NOTE: these curves have been derived by approximating the various curves in R.Ae.S. Data Sheets "Wings" 02.04.02 and 02.04.03.

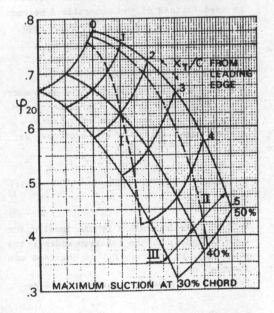

Fig. F-6. Shape factor for some airfoils

the leading edge will be representative of
good average straight wing constructions.
Near the wing/fuselage and wing/nacelle
junctions and in the propeller slipstream,
however, transition to turbulent flow is
to be expected at about 5 to 10% of the
·chord. This is also generally the case with
sweptback wings, where spanwise pressure
gradients cause cross flows at the leading
edge, provoking early transition. At
Reynolds numbers above 3×10^7 the transition can be assumed to occur at 5% of the
chord for all section shapes.

c. The lift correction to the profile drag
can be related to the lift increment relative to c_{ℓ_i}:

$$\frac{\Delta_\ell \, c_{d_p}}{\left(\Delta_\ell \, c_{d_p}\right)_{ref}} = f\left(\frac{c_\ell - c_{\ell_i}}{c_{\ell_{max}} - c_{\ell_i}}\right)^2 \qquad (F-31)$$

where $\left(\Delta_\ell \, c_{d_p}\right)_{ref}$ is representative of the
(extrapolated) profile drag increment at
the stalling angle of attack (cf. Fig. F-5).
The following expressions have been derived
from experimental data on NACA 4 and 5 digit sections:

$$\text{for } R \leqslant 10^7: \left(\Delta_\ell \, c_{d_p}\right)_{ref} = \frac{67 \, c_{\ell_{max}}}{(\log_{10} R)^{4.5}} -$$

$$- .0046 \, (1 + 2.75 \, t/c)$$
$$(F-32)$$

$$\text{for } R > 10^7: \left(\Delta_\ell \, c_{d_p}\right)_{ref} = .01 \, c_{\ell_{max}} -$$

$$- .0046 \, (1 + 2.75 \, t/c)$$

The Reynolds number in this expression
should be based on a condition at high lift,
for example at 70% to 80% of c_ℓ-max, as the
flight speed is lower than the condition
for which the minimum profile drag is com-
puted. The ratio $\Delta_\ell \, c_{d_p} / \left(\Delta_\ell \, c_{d_p}\right)_{ref}$ is
plotted in Fig. F-7 as a mean line; most
conventional sections are within ± 10% from
this line. The figure suggests that for
$c_\ell - c_{\ell_i}$ up to 85% of $c_{\ell_{max}} - c_{\ell_i}$ the following straight line approximation is ac-
ceptable:

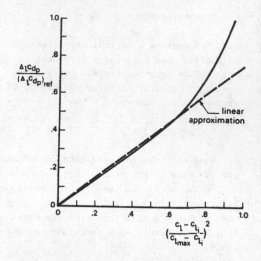

Fig. F-7. Generalized profile drag increment with lift

$$\Delta_\ell \, c_{d_p} = .75 \left(\Delta_\ell \, c_{d_p}\right)_{ref} \left(\frac{c_\ell - c_{\ell_i}}{c_{\ell_{max}} - c_{\ell_i}}\right)^2 \quad (F-33)$$

This expression may also be used for laminar flow sections, where the region of low
drag is not obtained for practical constructions. If low drag can be achieved,
experimental data with free transition must
be used instead of the generalized method
presented above; cf. Refs. F-44 through
F-46 and F-48.

F-3.3. Wings

The drag coefficient is obtained from section characteristics. On the basis of strip
theory, i.e. assuming that there is no
spanwise interaction of the sections, we
have:

$$C_{D_p} = \frac{2}{S} \int_{b_f/2}^{b/2} c_{d_p} \, c \, dy \qquad (F-34)$$

where c_{d_p} is determined by the local c_ℓ for
each C_L and b_f is the fuselage width at the
wing root. In this way C_{D_p} is obtained as a
function of C_L.
Provided the geometry of the wing is de-

fined and the sectional shape is known for a number of spanwise stations, the profile drag coefficient can be obtained from (F-29), corrected for the sweepback angle:

$$c_{d_{P_{min}}} = 2 C_F (1 + \varphi_w \cos^2 \Lambda_{\frac{1}{2}}) \qquad \text{(F-35)}$$

The shape factor φ_w, based on the streamwise section, is given by Fig. F-6 or (F-30).

Any variation in operational conditions results in variations of C_L and the Reynolds number. As it is not customary to calculate a drag polar for each operational value of R, the minimum profile drag is determined for an average cruise condition. However, the profile drag increment for $C_L > C_{\ell_i}$ should be estimated for a representative value of R at the lower end of the operational flight envelope (flaps up), for example, at a Reynolds number equal to half the value in cruising flight.

The integral in (F-34) may be evaluated as follows:
a. Assume a combination of R per unit length and C_L, corresponding to the flight condition to be considered.
b. Compute the spanwise lift distribution using any suitable method available, for example, Diederich's method (Section E-4.2. of Appendix E).
c. Compute the local value of R, related to the chord, at a number of spanwise stations and calculate the minimum profile drag coefficient with (F-35).
d. Estimate c_{ℓ}-max from experimental data, or Section E-3.3. of Appendix E, for each spanwise station.
e. Compute $\Delta_{\ell} c_{d_p}$ for each station using experimental data or (F-32) and (F-33).
f. For each station plot $c_{d_p}\cdot c$ versus y, draw a line through the calculated points and integrate numerically according to (F-34).
g. Repeat the process for several values of C_L, to obtain $C_{D_p} = f(C_L)$.
For straight wings this procedure, based on strip theory, yields good results, but for sweptback wings the method is open to criticism in view of the effects of crossflows in the boundary layer.

Instead of the method explained previously,

a much simpler procedure will usually be acceptable:
a. For a representative station midway between the wing/fuselage junction and the wing tip, the local section properties are estimated: $\bar{R}$, $\bar{c}_{\ell_i}$, $\bar{c}_{\ell_{max}}$, $\bar{c}_{d_{P_{min}}}$ with (F-35) and $\Delta_{\ell} \left(\bar{c}_{d_p}\right)_{ref}$ with (F-32).
b. The three-dimensional profile drag coefficient is derived from these data:

$$C_{D_p} = \bar{c}_{d_{P_{min}}} \frac{S_{net}}{S} + .75 \left(\Delta_{\ell} \bar{c}_{d_p}\right)_{ref} \left(\frac{C_L - C_{L_i}}{C_{L_{max}} - C_{L_i}}\right)^2$$

$$\text{(F-36)}$$

where C_{L_i} is assumed equal to $\bar{c}_{\ell_i}$ and $C_{L_{max}}$ can be estimated with Section E-4.2. of Appendix E. The net wing area, defined in Appendix A, Section A-3.1., is obtained from the gross wing area by subtracting the projected area of the wing section covered by the fuselage.
Equation F-36 can readily be rewritten as an equation expressing C_{D_p} in terms of a constant, a term proportional to C_L and a term proportional to C_L^2. Typical minimum profile drag coefficients range between .006 and .010.

F-3.4. Fuselages and tail booms

The aerodynamic ideal of a smooth streamlined shape, as defined in Section F-3.1., is not approximated with certain low-speed aircraft, such as agricultural aircraft, light private aircraft and small transports of simple construction. In those cases the designer must rely on experimental and statistical data of similar shapes, as presented in Ref F-20, for example

An impression of the drag coefficient of streamlined fuselages can be gained from Fig. F-8. As these data are valid for rather outdated fuselages, measured at low Reynolds number, a more detailed calculation is usually made, using the flat plate analogy. The outer contour is considered to be a streamlined shape, assuming that

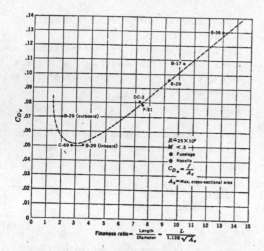

Fig. F-8. Effect of fineness ratio on fuselage and nacelle frontal drag coefficient
(Ref. F-4)

all excrescences such as wheel well fairings, canopy enclosures and airscoops,
etc. have been removed or faired over.
For streamlined fuselages with pointed tail
sections with a length/diameter ratio of at
least 2 and without a jet efflux, we may
write:

$$C_D S = (C_D S)_{basic} + \Delta_{\alpha\beta} (C_D S) \qquad (F-37)$$

a. Streamlined fuselages without blunt nose
or tail sections, no jet efflux.
The basic fuselage drag is the profile drag
of an uncambered, smooth streamline body at
zero angle of attack:

$$(C_D S)_{basic} = C_F S_{f_{wet}} (1 + \varphi_f) \qquad (F-38)$$

The wetted area is the gross wetted area,
reduced by those areas of the fuselage that
are not exposed to the flow (cf. Section
F-4.1.). The value of C_F is read off from
Fig. F-4 for R based on the fuselage length.
Transition is assumed to occur at or very
near the nose. The shape factor φ_f is mainly a function of the effective fuselage
slenderness ratio:

$$\lambda_{eff} = \frac{\ell_f}{D_{f_{eff}}}$$

or

$$\lambda_{eff} = \frac{\ell_N + \ell_A}{D_{f_{eff}}} + 2$$

whichever is less (F-39)

where

$$D_{f_{eff}} = \sqrt{\frac{4}{\pi} A_c} \qquad (F-40)$$

and the fuselage nose and tail section
lengths ℓ_N and ℓ_A are defined in Fig. F-10.
Fig. F-9 shows the shape factor φ_f vs. the
reciprocal value of the slenderness ratio,
according to several prediction methods.
Experiments are also shown, although several points refer to fuselages with swept-up
tail sections, having approximately 5% more
drag than the axisymmetric fuselage.
The following relation is suggested for fuselages without a blunt nose and with an
optimum pointed tail section:

$$\varphi_f = \frac{2.2}{\lambda_{eff}^{1.5}} + \frac{3.8}{\lambda_{eff}^3} \qquad \text{for } \frac{\ell_A}{D_{f_{eff}}} \geq 2 \qquad (F-41)$$

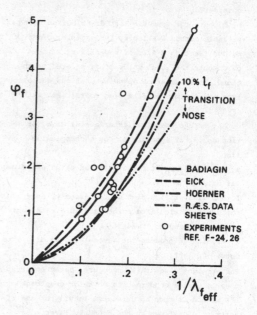

Fig. F-9. Fuselage shape factor

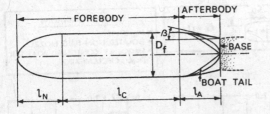

Fig. F-10. Nomenclature for determining the drag of fuselages with short afterbody or a jet efflux

b. Fuselages with blunt and/or short afterbodies, no jet efflux.

Small areas normal to the flow at the rear end of the fuselage may be present, e.g. in the case of an APU installation or a blunt tail fairing. Provided the base area is less than approximately 10% of the fuselage cross-sectional area and there is no jet efflux, it is sufficiently accurate to increase the basic fuselage drag area by an amount which can be written as:

$$\Delta C_D S = .13 \times \text{base projected area} \qquad (F-42)$$

No method is available to give an accurate prediction of the drag due to a large base area and/or a short afterbody, such as is used on some freighter aircraft (Figs. 3-3b, 3-4 and 3-21d). The following procedure will at least yield a qualitatively correct result for short afterbodies and boat tails, with or without jet efflux.

1. Calculate the basic fuselage drag coefficient with the method given in Section a above, assuming the afterbody length/diameter ratio equal to 2, hence:

$$\lambda_{eff} = \frac{\ell_N + \ell_C}{D_{f_{eff}}} + 2$$

or

$$\lambda_{eff} = 4 + \frac{\ell_N}{D_{f_{eff}}}$$

whichever is less (F-43)

2. Calculate the base drag that would occur in the absence of an afterbody, i.e. with the fuselage truncated immediately behind the cylindrical section. Using Hoerner's data (Ref. F-58):

$$\left(C_{D_b}\right)_{ref} = .015 \sqrt{\frac{D_{f_{eff}}}{C_F(\ell_N + \ell_C)}} \qquad (F-44)$$

Alternatively, this coefficient may be assumed equal to .13.

3. A pressure drag increment is added to the basic fuselage drag coefficient to account for actual fuselage afterbody length and shape:

$$\Delta\left(C_D S\right)_A = \left(C_{D_b}\right)_{ref} \frac{\Delta C_{D_A}}{\left(C_{D_b}\right)_{ref}} \frac{\pi}{4} D_{f_{eff}}^2 \quad (F-45)$$

The afterbody pressure drag increment is given in Fig. F-11 as a fraction of the reference base drag.

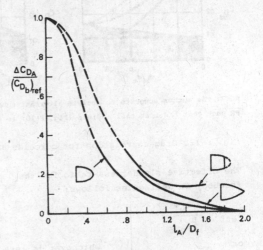

NOTE: ℓ_A and D_f are defined in Fig. F-10

Fig. F-11. Pressure drag increment vs. fineness ratio of short afterbodies without jet efflux (Refs.: F-18, F-65 and F-98)

c. Fuselages with propulsive jets.
The basic fuselage drag is subdivided into contributions of the forebody and the afterbody boat tail drag:

$$\left(C_D S\right)_{basic} = \left(C_D S\right)_F + C_{D_\beta} \frac{\pi}{4} D_{f_{eff}}^2 \quad (F-46)$$

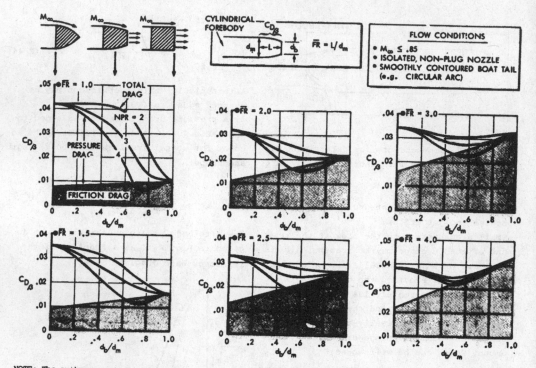

NOTE: The author suggests to analyze alternative boat tail shapes (e.g. parabolic) by taking an equivalent $\overline{FR}$ such that the boat tail angle β (Fig. F-10) is equal to β for the comparable circular arc.

Fig. F-12. Drag correlation for circular arc boat tails (Ref. F-98)

The effective slenderness ratio for the forebody is defined as follows:

$$\lambda_{eff} = \frac{2 \ell_N + \ell_C}{D_{f_{eff}}}$$

or

$$\lambda_{eff} = 2\left(1 + \frac{\ell_N}{D_{f_{eff}}}\right) \qquad (F-47)$$

whichever is less

The forebody drag area $(c_D S)_F$ is calculated with (F-38), using the forebody wetted area as the reference and taking the shape correction factor equal to half the value for normal fuselages:

$$\varphi_f = \frac{1.1}{\lambda_{eff}^{1.5}} + \frac{1.9}{\lambda_{eff}^{3}} \qquad (F-48)$$

The boat tail drag coefficient C_{D_β} is read off from Fig. F-12 for the appropriate

Nozzle Pressure Ratio (NPR) which should be obtained from the engine type specification. Fig. F-12 does not apply to boat tails with a base area; the reader is referred to Ref. F-86 for this case.

d. Correction for angle of attack and upsweep effects.

A generally accepted method is not available, but a reasonable approximation can be made on the basis of the cross flow concept (Ref. F-62). For an isolated fuselage in uniform flow the drag on a fuselage element due to the cross flow (Fig. F-13) is:

$$d D_c = c_{d_c} \cdot \tfrac{1}{2}\rho_\infty \left\{V_\infty \sin(\alpha_f - \beta)\right\}^2 \cdot \frac{b \, dx}{\cos\beta} \sin|\alpha_f - \beta| \qquad (F-49)$$

The effective drag coefficient c_{d_c} is that of an infinitely long cylinder, with cross

504

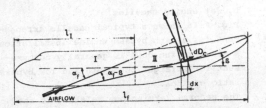

Fig. F-13. Derivation of fuselage cross flow drag

section equal to the local fuselage cross section, perpendicular to an airflow. This coefficient must be determined for a Reynolds number, based on $V \sin(\alpha_f - \beta)$, and corrected for the finite length of the fuselage. It can be obtained from Ref. F-18 or F-57. For fuselages with a circular cross section $c_{d_c} \approx 1.0$ and for a rectangular cross section with rounded corners $c_{d_c} = 1.5$ to 2.0.

Integration of (F-49) yields:

$$\frac{D_c}{\frac{1}{2}\rho_\infty V_\infty^2} = \Delta_{\alpha\beta}(C_D S) = \int_0^{\ell_f} |\sin^3(\alpha_f - \beta)| c_{d_c} \times \frac{b \, dx}{\cos\beta} \qquad \text{(F-50)}$$

This integral can be computed numerically as a function of α_f. For the simplified case of an axisymmetrical forebody (Section I: $\beta = 0$) and a constant upsweep angle of the afterbody (Section II) and assuming $c_{d_c} = 1.0$:

$$\Delta_{\alpha\beta}(C_D S) = A_I |\sin^3\alpha_f| + A_{II} \frac{|\sin^3(\alpha_f - \beta)|}{\cos\beta} \qquad \text{(F-51)}$$

where A_I and A_{II} are the planform areas of sections I and II (Fig. F-13). By way of example, the result is plotted in Fig. F-14. In spite of the crude assumptions made, the result appears to be realistic when compared with the experimental data reported in Ref. F-62. For α_f or β in excess of $5°$ the extra drag is a significant fraction of the basic fuselage drag coefficient. The minimum of the curves can be shown to occur at:

$$\alpha_f' = \beta \frac{\sqrt{A_I/A_{II}} - 1}{A_I/A_{II} - 1} \qquad \text{(F-52)}$$

In order to fit the curves given by (F-50) or (F-51) into the usual concept of a parabolic polar, they may be approximated by

$$\Delta_{\alpha\beta}(C_D S) = D_1 + D_2 \left(\alpha_f - \alpha_f'\right)^2 \qquad \text{(F-53)}$$

where the constants D_1 and D_2 are obtained by plotting $\Delta_{\alpha\beta}(C_D S)_f$ versus $(\alpha_f - \alpha_f')^2$, and replacing the result by a straight line. In order to express $\Delta_{\alpha\beta}(C_D S)_f$ in terms of C_L, it is noted that:

$$\alpha_f = \frac{C_L - C_{L_o}}{C_{L_\alpha}} \qquad \text{(F-25)}$$

See Section E-9.4. of Appendix E.

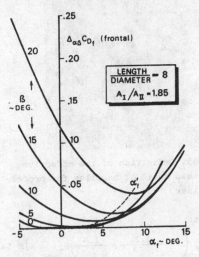

Fig. F-14. Estimated fuselage profile drag increment due to angle of attack and rear fuselage upsweep

F-3.5. Engine nacelles

Although in preliminary design nacelle drag is frequently assumed to be independent of C_L, interference effects with the wing, for example, can be quite appreciable (cf. Section F-4.3.).

a. Propeller aircraft.

The calculation procedure is analogous to that for the fuselage basic drag, explained in Section F-3.4. The nacelle (spinner included) is considered to be a smooth streamline body, generally with low slenderness ratio and fully turbulent boundary layer (Fig. F-15). The drag area per nacelle is:

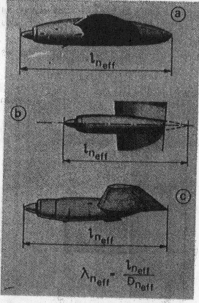

Fig. F-15. Definition of the effective slenderness ratio of nacelles for propeller engines

celle is:

$$C_D S = C_{F_,} \left(1 + \frac{2.2}{\lambda_{n_{eff}}^{1.5}} + \frac{3.8}{\lambda_{n_{eff}}^{3}}\right) S_{n_{wet}} \quad \text{(F-54)}$$

In order to define $\lambda_{n_{eff}}$, a rational choice of the nacelle length must be made. If the centerline of the nacelle is approximately lined up with the chord, as in Fig. F-15c, the presence of the wing will lead to a reduction in pressure drag. The effective nacelle slenderness ratio should be increased accordingly. Instead of, or in addition to, using the method of (F-54), a rough figure can be obtained from Fig. F-8.

b. Jet engine nacelles.

Fig. F-16 shows a typical nacelle arrangement for a high bypass engine installation. A separate fan cowl is generally not present for straight turbojets and low bypass ratio engines.

The general terminology relating to the subdivision of a powered jet engine nacelle drag can be found in the literature, e.g. Ref. F-78. Engine thrust is assumed to be defined as "standard gross" and "standard net thrust", as in the ESDU Data Sheet 69006. Internal friction losses of the inlet diffuser are not considered; these are accounted for in the (installed) engine thrust determination.

The total drag per nacelle less interference is subdivided into drag contributions of the fan and gas generator cowlings, the plug in the primary flow and the pylon. Reference will be made to the geometric definitions of Fig. B-5 of Appendix B.

The (fan) cowling drag area is written as the sum of a friction drag and an afterbody pressure drag component:

$$C_D S = C_F \left\{ \beta (1 + \varphi_n)^{5/3} + (1-\beta) \right\} S_{n_{wet}} + \left(C_{D_\beta}\right)_p \frac{\pi}{4} D_n^2 \quad \text{(F-55)}$$

where β denotes the cowling forebody length divided by the cowling length.

The flat plate skin friction coefficient C_F (Fig. F-4) is determined for R based on the length of the cowl or nacelle and fully turbulent flow. The (fan) cowling wetted area can be calculated with Section B-4 of

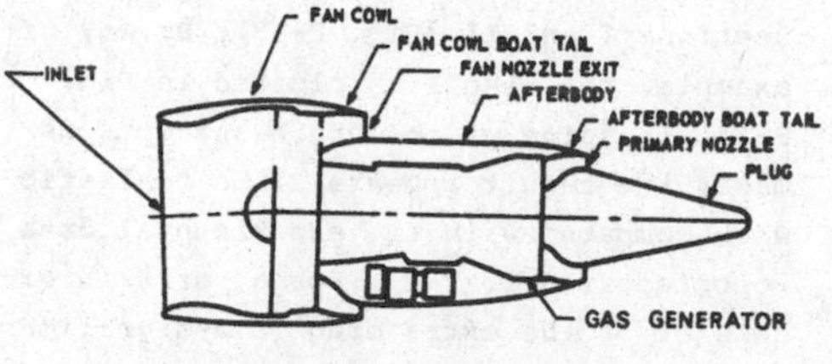

Fig. F-16. Typical high bypass engine nacelle

506

Appendix B.

In (F-55) the terms in the brackets represent the skin friction drag of the forebody, corrected by means of φ_n for excess velocities associated with the cowl curvature and angle of attack and the skin friction drag of the afterbody (boat tail). According to Ref. F-95 the following correlation applies to the NACA-1 Series of inlets:

$$\varphi_n = .33 \frac{D_n - D_h}{\beta \ell_n} \left(1 + 1.75 \frac{D_h^2 - D_o^2}{D_n^2 - D_h^2}\right) \qquad (F-56)$$

where D_o is the diameter of the capture area, i.e. the internal stream tube of the engine flow at large distance in front of the engine. It can be shown that:

$$\frac{D_h^2 - D_o^2}{D_n^2 - D_h^2} = \frac{1 - 31.55 \dot{W}_a \sqrt{\theta}/(p_\infty M_\infty D_h^2)}{(D_n/D_h)^2 - 1} \qquad (F-57)$$

The last term of (F-55) is the boat tail pressure drag area. The pressure drag coefficient $(C_{D_\beta})_p$ can be obtained from Fig. F-12 by subtracting the skin friction drag from the total boat tail drag, as indicated, or from Ref. F-86.

The fan cowl drag is sensitive to the geometries of the cowling and the fan nozzle, and to the engine thrust setting. If (F-56) cannot be used because the cowl design is not yet completely defined, a typical figure may be assumed for a well-designed medium to high bypass ratio turbofan cowling, in cruising flight:

$$C_D S \approx 1.25 C_F S_{n_{wet}} \qquad (F-58)$$

The expressions given for the engine cowl drag are valid only if the intake lip is well designed and operates at approximately the datum engine mass flow. In that case there is no appreciable loss in lip suction and the pre-entry drag is fully balanced by the suction forces on the front cowl. No additive drag is present in this condition. The validity of this assumption must be verified in later aerodynamic development.

The gas generator cowling drag component is frequently quoted by the manufacturer in the engine brochure as it is usually considered to constitute an effective thrust loss. In the absence of this information the following expression can be used:

$$C_D S = C_F \left(\frac{M_f}{M_\infty}\right)^{11/6} \left(\frac{1 + .116 M_\infty^2}{1 + .116 M_f^2}\right)^{2/3} S_{g_{wet}} +$$

$$+ \left(C_{D_\beta}\right)_p \frac{\pi}{4} D_g^2 \qquad (F-59)$$

The first term is the scrubbing drag, associated with the high velocity jet to which the gas generator cowling is exposed (Ref. F-93). The skin friction drag coefficient for fully turbulent flow is related to free stream conditions and the length of the gas generator cowling plus the fan nozzle length. The Mach number of the fully expanded fan flow M_f can be obtained from the engine manufacturer's brochure or (H-21) in Appendix H. The gas generator cowling wetted area can be computed with Section B-4 of Appendix B. The part of the pylon which is immersed in the fan exhaust flow must be included in this wetted area.

The second term of (F-59) is the gas generator boat tail pressure drag area. Generalized data for calculating $\left(C_{D_\beta}\right)_p$ are not available, but Fig. F-17 is thought to be fairly representative.

Scrubbing drag of the plug is usually considered as a loss in engine gross thrust. If no data are quoted in the engine brochure its order of magnitude can be estimated as follows:

$$C_D S = C_F S_{p_{wet}} \left(\frac{M_g}{M_\infty}\right)^{11/6} \left(\frac{1 + .116 M_\infty^2}{1 + .116 M_g^2}\right)^{2/3} \qquad (F-60)$$

The plug wetted area is given in Section B-4 of Appendix B, the fully expanded hot flow Mach number M_g being derived from the engine type specification or (H-20) of Appendix H. The plug drag amounts to roughly 5% of the total engine pod drag in cruising flight.

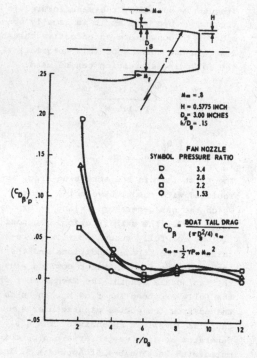

Fig. F-17. Boat tail pressure drag coefficient vs. radius of curvature (Ref. F-89)

Pylon drag is similar to wing profile drag
and is determined from:

$$C_D S = C_F \left\{ 1 + 2.75 \ (t/c)_{py} \cos^2 \Lambda_{py} \right\} S_{py_{wet}}$$

$$(F-61)$$

The relative thickness of the pylon is defined in streamwise direction and the
sweepback angle applies to the quarter-
chord line of the pylon.

F-3.6. Tailplane profile drag

For the horizontal tailplane the profile
drag can be subdivided into basic profile
drag and profile drag due to lift associated with incidence and elevator deflection:

$$C_D S = (C_D S)_{h_{basic}} + \Delta_\ell (C_D S)_h \qquad (F-62)$$

The basic or minimum profile drag is similar to the wing profile drag:

$$(C_D S)_{h_{basic}} = 2 \ C_F \left\{ 1 + 2.75 (t/c)_h \cos^2 \Lambda_{\frac{1}{2} h} \right\} S_h$$

$$(F-63)$$

where C_F can be read off from Fig. F-4.
The reference length for R and t/c is the
mean geometric tailplane chord, and it is
usual to assume transition at the leading
edge. To account for interference with the
airplane, S_h may be assumed as the gross
horizontal tailplane area, including parts
covered by the fuselage or vertical tailplane (cf. Section F-4.4.).
Drag increment due to incidence and elevator deflection is difficult to compute accurately. From the very scanty data available it was concluded that this term is
mainly dependent upon C_{L_h} irrespective of
the tailplane angle of attack and the elevator deflection. This drag contribution
can therefore be related to the vortex-induced drag of the tailplane by introducing
an effective Oswald factor of the order of
$.75 \cos^2 \Lambda_h$ for the horizontal tailplane.
Hence:

$$\Delta_\ell (C_D S)_h = \frac{.33 \ C_{L_h}^2}{\cos^2 \Lambda_h \ \pi A_h} \ S_h \qquad (F-64)$$

Some comments on the calculation of C_{L_h} are
given in Section F-2.6.

Vertical tailplane drag is given by:

$$(C_D S)_v = 2 \ C_F \left\{ 1 + 2.75 \ (t/c)_v \cos^2 \Lambda_{\frac{1}{2} v} \right\} S_v$$

$$(F-63)$$

where C_F is read off from Fig. F-4. The
mean geometric chord of the vertical tailplane is the reference length for R and
t/c. It is usual to assume transition at
the leading edge. The gross vertical tailplane area is used to account for interference (cf. Section F-4.4.). Drag due to
sideload on the vertical tailplane is not
present in symmetrical flight. The case of
engine failure will be treated in Appendix
G, Section G-8.3.

508

F-4. SUBCRITICAL INTERFERENCE EFFECTS AND CORRECTIONS

For a well designed configuration the effects of interference will be limited to a bare minimum. This can be obtained by suitable filleting, fairings and wing/fuselage blending, etc. If insufficient care is taken in the aerodynamic design, however, interference effects will be considerable, especially at high angles of attack and high-subsonic speeds.

Most interference effects cannot be calculated accurately in preliminary design, and the approximations presented here must be considered as "minimum effects".

F-4.1. Wetted area corrections

Viscous interference drag is sometimes approximated by first calculating the drag of the isolated gross wing and fuselage, etc., and then simply adding these components. It is thus assumed that interference may be accounted for simply by ignoring the fact that the intersections of airframe components are not actually exposed to the flow, which implies that all wetted areas mentioned in the previous section are gross areas. For example: the wing wetted area includes the center section and is therefore equal to twice the gross area. The only justification for such a procedure is its simplicity, which may be acceptable for small contributions such as tailplane/fuselage interference, but for larger contributions, like wing/fuselage interference, an alternative method is desirable. For other components - e.g. nacelle/airframe interference - the isolated drag will simply be factored.

A comprehensive survey of the physical aspects of interference and many experimental data are given by Hoerner (Ref. F-18).

F-4.2. Wing/fuselage interference

a. Vortex-induced drag
The lifting wing induces an upwash in front and a downwash aft of the wing/fuselage

junction. Hence, the body nose will experience lift increment and the afterbody a reduction in lift. In addition, the fuselage cross flow component at angle of attack leads to increased effective angles of attack near the wing/body junction. Combination of both effects will generally result in an increment of the lift coefficient for a given angle of attack and of the vortex-induced drag for a given lift. On the other hand, for a given lift of the complete configuration, the fuselage lift reduces the wing lift and consequently vortex-induced drag. It seems fair to assume that for a well designed configuration these effects are of the same order of magnitude. The only interference correction suggested here is the effect of the wing lift carry-over by the fuselage at zero fuselage angle of attack.

The results obtained by Lennertz (Ref. F-100) for constant lift along the span and by Marx (Ref. F-101) for an elliptical lift distribution can be brought into reasonable agreement by the following expression for mid-wing configurations with circular fuselage cross-sections:

$$\Delta_i C_{D_V} = \frac{.55 \, \eta_f}{1 + \lambda} (2 - \pi \eta_f) \frac{C_{L_o}^2}{\pi A} \qquad (F-65)$$

where η_f = fuselage diameter/wing span. For η_f up to .15 this result compares reasonably well with the method based on experimental data given by Hoerner (Ref. F-18, page 8-18). No attempt is made to present corrections for the effect of low-wing or high-wing positions.

b. Viscous interference
Near the intersections of the wing and fuselage contours there is a thickening of the boundary layers and an increase in the local flow velocity; both effects increase the profile drag. At small angles of attack the result is approximately:

$$\Delta_i (C_D S)_p = 1.5 \, C_F \, t_r \, C_{ci} \cos^2 \Lambda_{\frac{1}{2}} \qquad (F-66)$$

where C_{ci} is the total circumferential

length for both wing halves of the wing/
fuselage intersection line at which the
boundary layers interact. For a mid-wing
configuration C_{ci} is approximately $4\frac{1}{2}$
times the root chord. Equation F-66 is
based on an interpretation of various results in Ref. F-18.

Due to lift effects, the pressure gradients
and section forces increase at the upper
side of the wing and decrease at the lower
side. This may lead to premature separation near the wing/fuselage junction, especially with low-wing configurations, but
this tendency may be suppressed by suitable filleting. However, the profile drag
increment of the fuselage section above
the wing due to increased velocities cannot be avoided, while on high-wing configurations the fuselage drag is reduced at
high lift. A simple correction based on a
theoretical analysis of typical configurations is suggested here:

$$\frac{\Delta_i (C_D S)_p}{C_F C_L c_r D_f} = \begin{array}{l} + .88 - \text{ low wing} \\ - .81 - \text{ high wing} \end{array} \qquad (F\text{-}67)$$

where C_F is the local skin friction coefficient of the fuselage near the wing/body
interconnection. No correction is required
for mid-wing configurations.

Finally, the cross flow effects on fuselage drag (cf. Section F-3.4.) are altered
as a consequence of the circulation, inducing a change in the effective angle of
attack along the fuselage. Equation F-50
may therefore be modified into:

$$\Delta_{\alpha\beta} (C_D S) = \int_o^{\ell_f} |\sin^3 (\alpha_f - \epsilon - \beta)| \, c_{d_c} \, b \, \frac{dx}{\cos\beta}$$

$$(F\text{-}68)$$

Note that in front of the wing the downwash angle ϵ is negative (i.e. upwash)
while at the wing/fuselage intersection
there is no cross flow component due to
the straightening effect of the wing.
Equation F-60 may be solved if detailed
data are available for calculating the
flow field around the wing, but in view
of the very approximate character of the

cross flow component the labor involved
may not be justified. It is suggested that
this form of interference should be taken
into account by reducing the angle α_f' in
(F-52) by an amount equal to $\beta \cos\Lambda/A$,
while D_2 in (F-53) remains unchanged. Realizing that there is no induced upwash or
downwash for $C_L = 0$, we can derive:

$$\Delta_i (C_D S)_p = \frac{2 \beta \cos\Lambda}{A} D_2 \frac{C_L}{C_{L_\alpha}} \qquad (F\text{-}69)$$

where β is the rear fuselage upsweep angle.

F-4.3. Nacelle/airframe interference

a. Propeller aircraft.

The interference effects due to nacelle
installation are comparable to the wing/
fuselage interference effects, but they
are more difficult to calculate. A thorough but outdated investigation can be
found in Ref. F-71. The drag appears to
be a function of the relative fore- and
aft-position, the height and angle of inclination of the nacelle centerline relative to the wing chord, the shape of the
rear end and the amount of filleting.
The most important contribution at low angles of attack is the vortex-induced drag
caused by the local change in wing lift
due to the nacelle. Typical figures (from
Ref. F-18, page 8-18) are:

$$\left. \begin{array}{l} \text{low wing: } \Delta_i \, C_D S = .004 \times \\ \text{nacelle frontal area} \\ \text{high wing: } \Delta_i \, C_D S = .008 \times \\ \text{nacelle frontal area} \end{array} \right\} \begin{array}{l} \text{per} \\ \text{nacelle} \end{array} \quad (F\text{-}70)$$

Negligible drag is experienced if the nacelle centerline coincides with the local
chord.

Drag due to interference from the boundary
layers of the nacelle and the wing can be
taken into account by ignoring the wetted
area correction required according to Section F-4.1. and by adding the profile drag
of the wing area that is not actually exposed to the flow (Ref. F-18, page 8-18).

At high angles of attack the interference drag may increase considerably when vortices are formed at the nacelle/wing junctions, especially when the nacelle is high on the wing. No method of analysis is available for this effect.

b. Jet aircraft.
Jet engine pods and the jet efflux interact with the flow around the airframe in a number of complex ways that are still the subject of investigation (cf. Section 6.5.2.). For wing-mounted nacelles the determining factors are the nacelle fore- and aft position, the distance between the nacelle centerline and the wing chord and jet efflux effects. It is generally found that an interference drag penalty of approximately 20% of the nacelle-plus-pylon drag must be accepted for low bypass engines with long ducts. High bypass engines with short fan ducts exhibit a favorable interference drag (Refs. F-109 and F-24) which is of the order of 5 to 10 counts in cruise conditions. For fuselage-mounted nacelles the shape of the pylon and the fuselage contour are important. The inter-ference drag penalty for straight or low bypass jet engines is of the order of 50% of the profile drag of the isolated nacelle plus pylon. No data for high bypass engine pods are available to the author.

F-4.4. Tailplane/airframe interference

a. Wing/tailplane interference, vortex-induced drag.
The tailplane lift acts normal to the local airflow. Due to the downwash behind the wing a positive tail load (up-load) has a drag component. On the other hand, this tail load partly offsets the wing lift for a given total lift and hence the vortex-induced drag of the wing is reduced. It can readily be shown that these effects result in the following contribution, ignoring a second order term:

$$\Delta_i \left(C_D S\right)_v = C_{L_h} C_L \left(\frac{d\varepsilon}{dC_L} - \frac{2}{\pi A}\right) S_h \qquad (F-71)$$

where C_{L_h} is defined by (E-44), see Section F-2.6. An estimate of $d\varepsilon/dC_L$ can be made with the data in Section E-10.1. of Appendix E. It is noted that (F-71) may

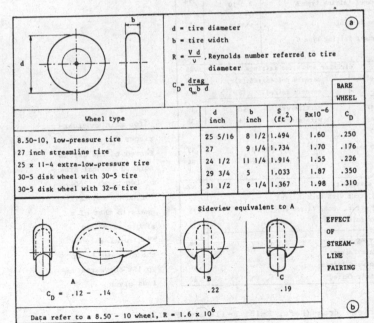

Fig. F-18. Drag of wheels with tires, with and without streamlined fairings (fenders) (Refs. F-110, F-111)

well represent a negative term, i.e. a drag reduction.

b. Viscous interference (tailplane/tailplane and tailplane/airframe).
As the effect of filleting is usually not known and this drag is a small term anyhow - a few percent of the tailplane profile drag - it is justifiable to assume that no extra drag need be accounted for, provided the gross area of the tailplanes is used in Section F-3.6. instead of their net wetted area.

F-5. PROTUBERANCES, SURFACE IMPERFECTIONS AND OTHER EXTRA'S

A large collection of detailed information such as that given in Hoerner's book (Ref. F-18) is required to analyze a particular design in detail, but in project design all required details are usually not yet known. What follows is therefore partly a statistical indication of the most relevant drag penalties that can be foreseen, provided close attention will be paid to detail design.

CONFIGURATION	REMARKS			C_{D_π} *
 **wheel type 8.5-10 ****	no streamline members, no fairings			1.28
	with stream-line members	junctions not faired		.56
		junctions A and B faired		.47
		junctions A, B and C faired		.43
		with wheel fairing type C (Fig. F-17)		.36
 **streamline member**	no fairing	27-inch streamline wheels		.28
				.29
	wheel fairing	type B	8.5-10 wheels	.27
		type C		.25
	no fairing	27-inch streamline wheels		.25
				.31
	wheel fairing type A		8.5-10 wheels	.23
	no fairing			.51
	wheel fairing type C			.34
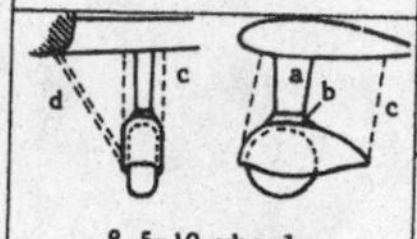 **8.5-10 wheels**	circular strut, no fairings			.05
	streamline strut	corners not faired (a)		.26
		corners faired (b)		.17
	trouser fairing	cantilever (c)		.17
		with sidestay (d)		.38
	8.5-10 wheels	not faired		.53
		with fairing c		.34
NOSE GEAR	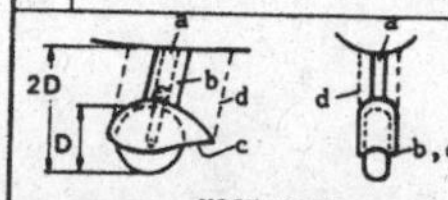 round strut with fork (a)			.64
	faired strut with fork (b)			.42
	faired strut, wheel faired (c)			.15
	trouser fairing (d)			.29
TAILWHEEL	 no fairing			.58
	with rear fairing			.49
	with forward fairing			.41
	completely faired			.27

MAIN UNDERCARRIAGE

*for main undercarriages C_{D_π} is referred to the circumscribed frontal area of two tires (2bD), for nose- and tailgears to that of a single tire (bD)
**with other types the drag can be up to 15% above the values given

Fig. F-19. Fixed undercarriage drag (Refs. F-18, F-111 and F-112)

512

F-5.1. Fixed undercarriages

Fig. F-18a presents the drag coefficients of several types of isolated wheels. As shown by Fig. F-18b, drag can be appreciably reduced by various types of streamlined fairings.

The drag coefficient of complete main undercarriages is usually related to the circumferential area of two wheels. Fig. F-19 indicates that the effect of fairings and streamlining is noticeable. Data are also given for nose gears and tail wheels, both with and without streamline caps. Tail skids have a drag area between .043 and .12 sq.ft (.004 and .011 m^2) typically. All data on main undercarriages in Fig. F-19 refer to wing- or fuselage-mounted gears. The drag of nacelle-mounted cantilever fixed undercarriages is approximately comparable to the data of the fifth configuration. The drag of a partly retracted landing gear can be found in Ref. F-111.

F-5.2. Canopies and windshields

a. Protruding cockpit enclosures.
Systematic drag measurements on a variety of shapes have been reported in Ref. F-115. Some results are summarized in Figs. F-20 and F-21.

The following observations are made.
1. A minimum value of c_{D_π} = .04 can be achieved for a well-streamlined windshield without cylindrical mid-section, and approximately .045 with cylindrical mid-section, provided optimum nose and tail sections are used.
2. Below a length/diameter ratio of 2½ to 3 for the nose and tail sections, drag increases sharply.
3. For a short conical nose section (t ≈ h) and a long tail section, drag can be appreciably decreased by properly rounding the windshield/hood junction.
4. Rounding the hood-tail section junction for a short tail (t ≈ 2h) is not very effective.
5. In order to obtain a low windshield drag it appears imperative to have a radius at the windshield/hood junction of at least 20% and a tail length of at least 3½ times the windshield height.

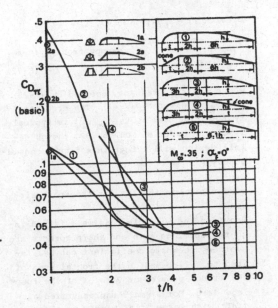

Fig. F-20. Effect of nose and tail length on the drag of windshields (derived from Ref. F-115)

A similar conclusion is found in Ref. F-66 for helicopter canopies.

6. The drag coefficient may amount to approximately .47 for the aerodynamically unfavorable shapes that are used on some agricultural aircraft.

The data of Ref. F-115 have been used to derive the following expression:

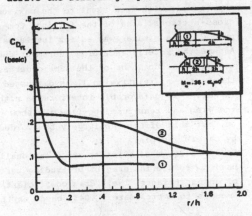

Fig. F-21. Effect of radius at windshield/hood and hood/tail junctions on windshield drag (derived from Ref. F-115)

$$C_D S \approx .85 \ (1 + \varphi_f) \ (1 + 4.5\alpha_f) \times$$

$$\left| \left(C_{D_\pi}\right)_{basic} + \Delta_{ss} C_{D_\pi} \right| \times canopy \ frontal \ area$$
$$(F-72)$$

where α_f is in radians, defined by (F-25), and φ_f is given by (F-41). The drag penalty for steps and slots $\Delta_{ss} C_{D_\pi}$ is typically of the order of .01 to .02, though detail design will affect this figure considerably.

b. Cockpit windshields for transport aircraft

Provided the windshield is well faired into the fuselage nose and sharp corners are avoided, the drag coefficient can be limited to .02, based on the frontal area of the window panels, or 2 to 3% of the fuselage drag. For a small unpressurized aircraft with large, flat window panels this coefficient may be as large as .08; more details can be found in Ref. F-115.

F-5.3. Wheel-well fairings* and blisters

Two methods are suggested to calculate the drag of wheel-well fairings.
Method A: use the data on cockpit canopies in Fig. F-20.
Method B: reduce each fairing to a body of revolution with equivalent diameter based on a circle with area equal to the maximum cross-sectional area of the fairing. Compute C_D in a similar way as for fuselages. Assume the interference drag increment due to installation to be of the same order as the factor φ_f relative to 1. As discussed in Ref. F-24 unfavorable interference with the wing and rear fuselage flow may show up, but this effect can largely be avoided by careful shaping.
Small fairings or blisters are occasionally used to reduce the drag of protruding parts of undercarriages. Their drag coefficient amounts to approximately .045, based on the

*Examples: Transall C-160, Lockheed C-130, C-141 and C-5A, Nord 262, Bréguet 941, Short Belfast and also the design in Fig. 1-4.

frontal area. For more details see Ref. F-18.

F-5.4. External fuel tanks

The basic profile drag of the isolated tank can be calculated in the same way as the fuselage drag. According to Fig. 3-2, a slenderness ratio of 4 to 5 results in minimum drag for a given volume. The corresponding drag coefficient is .025, based on (volume)$^{2/3}$.
Interference effects with the airframe are usually accounted for by multiplying the basic profile drag by an installation factor. Typical factors are:
1.20 - mounted symmetrically to the wing tip
1.30 - mounted fairly close to the wing, e.g. on a pylon
1.50 - mounted flush to the wing or fuselage.
Wing tip tanks will also affect the induced drag of the wing. This can be taken into account by assuming a percentage increment of the effective wing aspect ratio which is equal to half the percentage increment in span caused by the addition of the tanks (Ref. F-18, page 7-7).

F-5.5. Streamlined struts

According to Ref. F-18 (page 6-5) the profile drag coefficient based on the chord length is:

$$c_{d_{st}} = .015 \ (1 + t/c) + (t/c)^2 \qquad (F-73)$$

This equation is valid for subcritical Reynolds numbers i.e. R up to 7×10^4 for $t/c = .4$ and 3×10^4 for $t/c = .3$. At high Reynolds numbers the method used for wing sections (Section F-3.2.) can be employed. The usual t/c ratio is approximately .30. Ignoring interference, the strut drag area is:

$$\frac{D_{st}}{q_\infty} = c_{d_{st}} \times chord \times total \ length \qquad (F-74)$$

The interference drag with the aircraft can

be limited to a typical penalty of 10% of the profile drag by proper filleting.

F-5.6. Powerplant installation drag

The forces exerted on the airframe, both externally and internally, by the airflow passing through a jet engine, are included in the definitions of thrust and nacelle drag and do not appear under this heading. Other engine installation drag components may generally be classified as follows.

a. External drag: the sum of all streamwise forces associated with the external flow around the installed powerplant, e.g. the cooling system intake scoops.

b. Internal drag: the forces exerted by the internal flow on the surface which bounds it, e.g. the drag due to pressure losses in the oil cooling systems.

c. Slipstream effects for propeller aircraft: the increase in profile drag of aircraft parts in the slipstream and the increment of vortex-induced drag due to the change in lift distribution.

d. Jet interference, of importance with wing-mounted engines (cf. Section F-4.3.b.). In the case of propeller engines the effects of slipstream interference are directly related to the engine operating condition. As the drag polar is usually defined in the power-off or low power condition, this contribution can be considered as an effective reduction in propeller efficiency (cf. Section 6.3.2.).

a. Reciprocating engines.
For air cooled engines the sources of installation drag are engine cooling drag, air intake and exhaust external drag, oil cooler internal and external drag and engine air intake momentum drag.
Cooling drag is given by:

$$D_c = \dot{m}_c (V_\infty - v_e) \qquad (F-73)$$

For a given engine type the cooling air mass flow $\dot{m}_c$ is determined by the ambient temperature, the engine rating and the cylinder head temperature. The exit velocity is mainly determined by the pressure loss

in the inlet and the cylinder baffles; the exhaust opening is normally designed to accommodate the desired airflow in cruising flight. These characteristics should be decided in cooperation with the engine manufacturer.

Cooling drag can be considered as an equivalent loss in shaft power:

$$\Delta P_b = D_c V_\infty / \eta_p \qquad (F-74)$$

This term can be dealt with as a separate contribution in performance calculations for each engine working condition and flight speed. If, on the other hand, it is desirable to include a first estimate in the drag polar for cruising flight, it may be assumed that:

$$(C_D S)_c = C_c \times \frac{P_b T_\infty^2}{\sigma V_\infty} \qquad (F-75)$$

$C_c = 4.9 \times 10^{-7}$; $C_D S$ in sq.ft, P_b in hp, T_∞ in R and V_∞ in ft/s
$C_c = 5.9 \times 10^{-10}$; $C_D S$ in m^2, P_b in kgm/s, T_∞ in K and V_∞ in m/s

The increase in wing lift due to slipstream entails an increase in vortex-induced drag, which is less than the vortex-induced drag increment of the wing for the same lift increment. From Ref. F-18 the following expression can be derived for the apparent increase in the wing aspect ratio to which this effect is equivalent:

$$\frac{\Delta A}{A} = .5 \frac{8}{\pi} T_c \frac{S_s}{S} \qquad (F-76)$$

where

$$T_c = \frac{\text{thrust per propeller}}{\rho V_\infty^2 D_p^2} \qquad (F-77)$$

and S_s is the total projected area of the wing part immersed in the propeller slipstreams.

For steady cruising flight (F-76) can be modified to represent an effective increment in engine power:

$$\frac{\Delta P_b}{\eta_p P_b} = \frac{-\Delta C_D}{C_D} = .5 \frac{S_s / N_e}{\frac{\pi}{4} D_p^2} \frac{C_L^2}{\pi A} \quad \text{(for T=D) (F-78)}$$

The effect of the slipstream on profile drag is accounted for in the form of a reduction in effective propeller efficiency (cf. Section 6.3.2.).

Other installation drag increments are not amenable to analysis before the inlet, exhaust and oil cooling systems have been designed in detail. In preliminary design it may be assumed that $C_DS = .02$ sq.ft $(.0019 \text{ m}^2)$ per 100 takeoff bhp.

b. Turboprop engines.

The external drag of the engine air inlets is included in the nacelle drag - see Section F-3.5. - unless protruding intake scoops are used; in this case assume $C_DS = .10$ x the scoop frontal area at datum airflow. The oil cooler drag area is approximately .20 x the cooler inlet scoop frontal area, or .0065 sq.ft $(.0006 \text{ m}^2)$ per 100 takeoff eshp. This figure includes internal drag.

Occasionally, protruding exhaust pipes are used, with their axis at a large angle to the flow. A drag coefficient of .5 may be assumed on the basis of their projected frontal area.

Engine air intake momentum drag is accounted for in the calculation of the net thrust of the engine gas flow:

$$T_{net} = T_g \cos\theta_j - \dot{m}_j V_\infty \qquad (F-79)$$

where θ_j is the mean deflection angle of the exhaust gases relative to the airflow. Note that even without deflection ($\theta_j = 0$) T_{net} need not be positive, and furthermore that for

$$\theta_j > \cos^{-1} \frac{\dot{m}_j V_\infty}{T_g} \qquad (F-80)$$

there can be no positive net thrust component. In both cases there is, in effect, a momentum drag. In the extreme case of $\theta_j = 90^\circ$:

$$C_DS = C_j \times \frac{\dot{w}_j}{\sigma V_\infty} \qquad (F-81)$$

where $C_j = 24$; C_DS in sq.ft, $\dot{w}_j$ in lb/s, V_∞ in ft/s or $C_j = 1.6$; C_DS in m^2, $\dot{w}_j$ in kg/s, V_∞ in m/s This contribution is of the order of 15% of the total airplane zero lift drag, which is generally unacceptable.

Slipstream effects are similar to those occurring with reciprocating engines.

c. Turbojet engines.

For a daisy-type ejector nozzle/noise suppressor, assume $C_DS = .025$ x nozzle area. The drag of a thrust reverser on turbofan engines amounts to between 3 and 5% of the nacelle drag, although this figure depends on the type of installation.

Cooling drag is caused by the ventilation/ cooling of the space between the hot engine sections and the surrounding structure; it amounts to about 5% of the nacelle drag.

The drag of air inlets for buried engines may be included in the fuselage or wing drag in the case of a fully integrated inlet, by properly accounting for the increase in wetted area. Scoop type inlets (cf. Fig. 2-21c) must be treated separately. Their drag is not only dependent upon the shape of the scoop, but also on the inlet velocity ratio v_i/V_∞. The external drag coefficient is as low as $C_{D_\pi} = .06-.07$ for $v_i/V_\infty = 1$, but for a typical cruising value of $v_i/V_\infty = .6$ to $.8$, a figure of $C_{D_\pi} = .10$ may be obtained. Assume $C_{D_\pi} = .25$ for boundary layer diverters.

F-5.7. Excrescences, surface imperfections and other extras

A representative breakdown of this drag contribution for a subsonic jet transport is depicted in Fig. F-22. For this aircraft the profile drag penalties are summarized as follows:

wing : 6% of the wing profile drag

fuselage + empennage : 7% of the fuselage drag

engine installation :15% of the nacelle drag

systems : 3% of the zero-lift drag

For project development a detailed assessment can be made of all contributions, using the data given in Ref. F-18, for example. A target can then be set to provide a guideline for detail design, aerodynamic

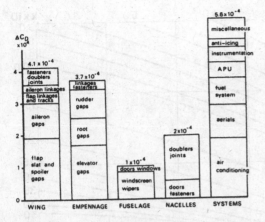

Fig. F-22. Typical breakdown of excrescence drag at cruise (Ref. F-25)

development and manufacturing tolerances. In this survey a statistical approach will be presented instead.

a. Surface imperfections must be accepted in the manufacturing process: doublers, skin joints, steps, gaps, fasteners, rivets, screws, holes, doors, scratches, indentations, waviness. If the size of these roughnesses is less than a critical size, the wing or body can be regarded as aerodynamically smooth and there is no drag penalty. For a roughness immersed in a turbulent boundary layer, this condition is derived from Ref. F-128:

$$\left(\frac{k}{\ell}\right)_{crit} = \frac{39.5}{R^{.94}} \qquad (F-82)$$

where k is the equivalent sand grain size of the roughness, ℓ the body or chord length, and R is based on ℓ.
For transport aircraft the surface imperfections appear to be equivalent to a sand grain size of .001 inch (25 microns) for average surfaces and about half this value for very carefully treated and smooth surfaces. Using this figure, the profile drag increment for all individual major items can be computed from Fig. F-23. Light aircraft have a roughness drag of the order of 10-15 counts or more.
Painted surfaces can be treated in similar

way, using Fig. F-23, provided the equivalent grain size is known. There is a considerable variation in this quantity (cf. Ref. F-129).

b. Retracted high-lift systems and flight control surfaces.
Surface imperfections such as steps, gaps and discontinuities are unavoidable with retracted flaps and slats, ailerons, elevators, rudder and spoilers. Exposed flap hinges, linkages and tracks are usually present. Some examples are presented in Fig. F-24. For slotted flaps the drag penalty may easily amount to 10-12 counts if no attempt is made to cover the gap properly. The penalty may be reduced to 2-3 counts, based on the projected area of the wing or tailplane with flaps or controls. For light aircraft the control surfaces entail an appreciable gap drag of the order of 25% of the tailplane profile drag. Leading-edge devices such as slats or plain leading-edge flaps not only provoke boundary layer transition – to be accounted for in the profile drag estimation – but also cause pressure drag at the various discontinuities and steps. The drag area is roughly .007 sq.ft per ft of total slat span in front view (.002 m^2 per m).

c. Airframe installations contribute to the drag. These include air conditioning system inlets and outlets and momentum drag, antennas, lights, fuselage skin waves and leaks due to pressurization, fuel system (dumping provisions), APU installation, instrumentation, anti-icing devices (rubber boots) and windscreen wipers. This drag is of the order of 6 to 8% of the fuselage drag.

d. Vortex generators, wing fences, fuselage strakes: the drag is obviously very much dependent upon the installation.* The or-

*Strictly speaking, their drag will be negative, relative to the condition where these devices are not present, for flight conditions where they improve the flow.

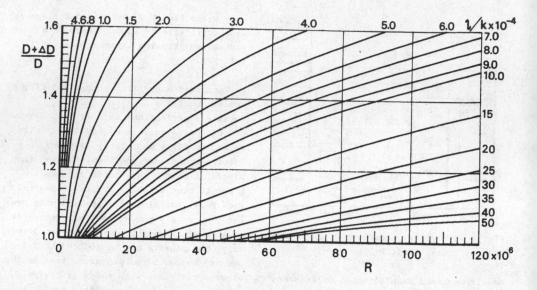

D = profile drag of an aerodynamically smooth surface

ΔD = drag increment due to surface roughness

ℓ = characteristic length, wing chord or body length

k = equivalent sand grain size

R = Reynolds number

More complete information for higher Reynolds number, taking into account compressibility effects, can be found in ESDU Data Sheet Item No. 73016, dtd. July 1973

Fig. F-23. Wing or body drag due to surface roughness (Ref. F-128)

der of magnitude is typically a few counts (e.g. Ref. F-134).

e. Miscellaneous sources of lift-dependent drag.

For a number of components and drag items such as engine nacelle drag and various interference effects, only the drag in cruising conditions has been considered. In addition, the major sources of profile drag increment with lift have been given for a smooth surface, whereas it is recognized that the profile drag of a rough body will increase more rapidly with incidence. Published methods for estimating this contribution are not available and the author therefore suggests that 20% of the profile drag increment should be assumed for $C_L \neq C_{L_o}$, or approximately:

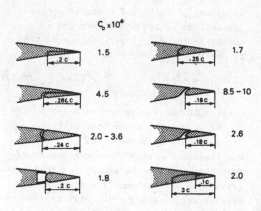

C_D referred to projected area of wing or tailplane with flaps or controls

Fig. F-24. Drag due to gaps caused by retracted flaps and control surfaces (Ref. F-8)

518

$$\Delta c_D = \frac{.0015}{\cos^2 \Lambda} \left(c_L - c_{L_0} \right)^2 \qquad (F-83)$$

It is, of course, obvious that this assump-

tion is open to criticism. The best method is to derive the required value from the difference between the actual and estimated drag polars of existing aircraft.

REFERENCES

Drag prediction methodology and general considerations

F-1. B.M. Jones: "The streamline aeroplane". J.R.Ae.S., May 1929.

F-2. W.B. Oswald: "General formulas and charts for the calculation of airplane performance". NACA Report No. 408, 1932.

F-3. H. Eick: "Der Mindestwiderstand von Schnellflugzeugen". Luftfahrtforschung 1938, Band XV, pp. 445-461.

F-4. C.D. Perkins and R.E. Hage: "Airplane performance, stability and control". Wiley, New York, 1949. Chapter 2: "Drag estimation".

F-5. I.H. Driggs: "Aircraft design analysis". J.R.Ae.S., Vol. 54, Feb. 1950, pp. 65-116.

F-6. E.J. Richards: "A review of aerodynamic cleanness". J.R.Ae.S., Vol. 54, 1950, pp. 137-186.

F-7. H.K. Millicer: "The design study". Flight, Aug. 17, 1951.

F-8. D. Fiecke: "Die Bestimmung der Flugzeugpolaren für Entwurfszwecke". I. Teil: "Unterlagen". Deutsche Versuchsanstalt für Luftfahrt E.V., June 1956.

F-9. Anonymus: "The estimation of the drag of an aeroplane". Cranfield Institute of Technology, Aerodynamics Lecture Supplement No. 11, April 1956.

F-10. Anonymus: "Report of the definitions panel on definitions to be used in the description and analysis of drag". ARC C.P. 369, 1957.

F-11. D.D. Dommasch: "Airplane aerodynamics", Pitman Publishing Corp., 1958.

F-12. G. Gabrielli: "On the subdivision in different "forms" of drag of the aircraft at the maximum speed". Troisième Congrès Aéronautique Européen, 1958. Vol. II and III, pp. 398, 407, 981-984.

F-13. D. Fiecke: "Der Einfluss des Widerstandes auf die Flugzeugentwicklung, Teil II". Flug-Revue, June 1962, pp. 40-42, 44, 46, 48.

F-14. D.E. Hoak and D.E. Ellison: "USAF Stability and Control Datcom". Wright Patterson Air Force Base, Oct. 1960, Rev. 1968.

F-15. H.G. Sheridan: "Aircraft preliminary design methods used in the weapon systems analysis". U.S. Navy BuWeps. Report R-5-62-13, June 1962.

F-16. W.H. Cook, C.S. Howell and J.K. Wimpress: "Aerodynamic performance". Air, Space and Instruments, Edited by S. Lees, Draper Anniversary Volume, 1963, pp. 238-256.

F-17. J.E. Linden and F.J. O'Brimski: "Some procedures for use in performance prediction of proposed aircraft designs". SAE Preprint No. 650800, Oct. 1965.

F-18. S.F. Hoerner: "Fluid-dynamic drag". Published by the author, 1965.

F-19. A.K. Martynov: "Practical aerodynamics". Pergamon Press, 1965, Chapter 10: "The aerodynamic characteristics of aircraft".

F-20. K.D. Wood: "Airplane design". Second edition, 1966. Johnson Publishing Company, Boulder, Colorado.

F-21. A.B. Haines: "Subsonic aircraft drag". The Aeron. Journal of the R.Ae.S. Vol. 72, No. 687, March 1968, pp. 253-266.

F-22. D.A. Norton: "Airplane drag prediction". Annals of the New York Academy of Sciences, Vol. 154 part 2, pp. 306-328, 1968.

F-23. D.C. Leyland and B.R.A. Burns: "Methods of estimating the drag characteristics, manoeuvrability and performance of strike/fighter aircraft". Short Course on Aircraft Performance estimation, Cranfield Inst. of Techn., Feb. 1973.

F-24. J.H. Paterson, D.G. MacWilkinson and W.T. Blackerby: "A survey of drag prediction techniques applicable to subsonic and transport aircraft design". AGARD Conference on aerodynamic drag CCP-124, 1973.

F-25. S.F. Butler: "Aircraft drag prediction for project appraisal and performance estimation". AGARD CP-124, 1973.

F-26. D.G. MacWilkinson, W.T. Blackerby and J.H. Paterson: "Correlation of full-scale drag predictions with flight measurements on the C-141 aircraft". NASA CR-2333 and CR-2334, Feb. 1974.

F-27. J.G. Callaghan: "Aerodynamic prediction methods for aircraft at low speeds with mechanical high-lift devices". AGARD Lecture Series No. 67, May 1974.

F-28. G.M. Bowes: "Aircraft lift and drag prediction and measurement". AGARD Lecture Series No. 67, May 1974.

Lift-induced drag of wings

F-29. R.F. Andersen: "Determination of the characteristics of tapered wings". NACA Report No. 572, 1936.

F-30. J. De Young and C.W. Harper: "Theoretical symmetrical span loading at subsonic speeds for wings having arbitrary planforms". NACA Report No. 921, 1948.

F-31. T.R.F. Nonweiler: "Lift curve slope and induced drag factors of large aspect ratio straight-tapered wings". J.R.Ae.S. Vol. 64 No. 592, April 1960, pp. 224-225.

F-32. K.L. Sanders: "Subsonic induced drag". J. Aircraft, Vol. 2 No. 4, July-Aug. 1965, pp. 347-348.

F-33. R.C. Frost and R. Rutherford: "Subsonic wing span efficiency". AIAA Journal April 1963, pp. 931-933.

F-34. D. Gardner and J. Weir: "The drag due to lift of plane wings at subsonic speeds". J.R.Ae.S., May 1966.

F-35. D. Gardner: "On the value of C_L-crit associated with the drag due to lift of plane wings in subsonic, subcritical flow". British Aircraft Corporation (Preston) Report Ae. 250, 1966.

F-36. Anonymus: "Subsonic lift-dependent drag due to wing trailing vortices". R.Ae.S. Data Sheets Wings 02.01.02.

F-37. J.L. Lundry: "Minimum swept-wing induced drag with constraints on lift and pitching moment". J.Aircraft, Vol. 4, No. 1, Jan.-Feb. 1967, pp. 73-74.

F-38. B.G. Gilman and K.P. Burdges: "Rapid estimation of wing aerodynamic characteristics for minimum induced drag". J. Aircraft, Vol. 4 No. 6, Nov.-Dec. 1967, pp. 563-565.

F-39. H.C. Garner: "Some remarks on vortex drag and its spanwise distribution in incompressible flow". The Aeron. Journal of the R.Ae.S., July 1968, pp. 623-625.

F-40. T.E. Labrujere and J.G. Wouters: "Computer application of subsonic lifting surface theory". N.L.R. Technical Report No. 70088.

F-41. P. Kapteijn: "Design charts for the aerodynamic characteristics of straight and swept, tapered, twisted wings at subsonic speeds". Delft University of Technology, Dept. of Aeron. Engng., Memorandum M-180, May 1972.

Profile drag of wings and airfoils

F-42. H.B. Squire and A.D. Young: "The calculation of the profile drag of airfoils". A.R.C. R & M 1818, Nov. 1937.

F-43. Anonymus: "Profile drag of smooth wings". R.Ae.S. Data Sheets Wings 02.04.02/03.

F-44. I.H. Abbott, A.E. von Doenhoff and L.S. Stivers Jr.: "Summary of airfoil data". NACA Report No. 824, 1945.

F-45. L.K. Loftin Jr.: "Theoretical and experimental data for a number of NACA 6A-series airfoil sections". NACA Report No. 903, 1948.

F-46. L.K. Loftin Jr. and W.J. Bursnall: "The effects of variations in Reynolds number between 3.0×10^6 and 25.0×10^6 upon the aerodynamic characteristics of a number of NACA 6-series airfoil sections". NACA Report No. 964, 1950.

F-47. J. Weber and G.G. Brebner: "A simple estimate of the profile drag of swept wings". RAE TN Aero 2168, June 1952.

F-48. I.H. Abbott and A.E. von Doenhoff: "Theory of wing sections". Dover Publications 1960.

F-49. J.F. Nash, T.H. Moulden and J. Osborne: "On the variation of profile drag coefficient below the critical Mach number". ARC CP No. 758, Nov. 1963.

F-50. J.C. Cooke: "The drag of infinite swept wings with an addendum". ARC CP No. 1040, June 1964.

F-51. H.H. Pearcey and J. Osborne: "On estimating two-dimensional section drag". ARC 27872, 28th. March 1966.

F-52. T. Cebeci and A.M.O. Smith: "On the calculation of profile drag of aerofoils at low Mach numbers". J. Aircraft, Vol. 5, No. 6, Nov.-Dec. 1968.

F-53. A. Barkhem: "Skin-friction formula for tapered and delta wings". J. Aircraft, Vol. 6, No. 3, 1969, page 284.

Fuselages

F-54. M.M. Munk: "The aerodynamic forces on airship hulls". NACA Report No. 184, 1928.

F-55. A.D. Young: "The calculation of the total and skin friction drags of bodies of revolution at zero incidence". ARC R & M 1874, April 1939.

F-56. Anonymus: "Drag of streamline bodies". R.Ae.S. Data Sheets Bodies 02.04.01/02.

F-57. H.J. Allen: "Estimation of the forces and moments acting on inclined bodies of revolution of high fineness ratio". NACA R & M A9I26, Nov. 1949.

F-58. S.F. Hoerner: "Base drag and thick trailing edges". J. Aeron. Sciences, Oct. 1950.

F-59. H.R. Kelly: "The estimation of normal force, drag and pitching moment coefficients for blunt-based bodies at large indidences". J. Aeron. Sciences Vol. 21, Aug. 1954, pp. 549-555.

F-60. E.J. Hopkins: "A semi-empirical method for calculating the pitching moment of bodies of revolution at low Mach numbers". NACA RM A51 C14, 1955.

F-61. A.A. Badiagin: "Concerning an efficient slenderness-ratio for the fuselage of civilian aircraft". In: "Methods of selection and approximate calculation of air design parameters". Moscow Aviation Institute Transactions, Sept. 27th, 1962.

F-62. T.R. Nettleton: "A method of estimating the effects of rear fuselage up sweep and fuselage cross sectional shape on fuselage drag". Unpublished Report of De Havilland Canada, Nov. 1964.

F-63. H. Hertel: "Full integration of VTOL power plants in the aircraft fuselage" AGARD CP-9, Vol. I, 1966.

F-64. D. Gyorgyfalvy: "Effect of pressurization on airplane fuselage drag". J. of Aircraft, Vol. II, Nov.-Dec. 1965, pp. 531-537.

F-65. W.A. Mair: "Reduction of base drag by boat-tailed afterbodies". Aeron. Quarterly, Vol. 20, Nov. 1969, pp. 307-320.

F-66. C.N. Keys and R. Wiesner: "Guidelines for reducing helicopter parasite drag". J. of the American Helicopter Soc., Jan. 1975, pp. 31-40.

Nacelles and engine installations, propeller aircraft

F-67. D.H. Wood: "Tests of nacelle-propeller combinations in various positions with reference to the wing". NACA Reports No. 415, 436, 462, 1932/1933.

F-68. J.G. Lee: "Air-cooled vs. liquid cooled aircraft". Journal of the Aeron. Sciences, Vol. 8, No. 6, April 1941, pp. 219-229.

F-69. J.V. Becker: "High speed tests of radial engine nacelles". NACA Wartime Report No. L-229.

F-70. Davies: "A review of windtunnel tests at R.A.E. on typical engine nacelle installations". RAE Report BA 1475.

F-71. R. Smelt, A.G. Smith and B. Davison: "The installation of an engine nacelle on a wing". ARC R. & M. 2406, 1950.

F-72. A.L. Courtney: "The estimation of powerplant drag of radial air-cooled engine installations at high speed and in temperature conditions". RAE Aero TN 1776.

F-73. Squire: "Calculation of the effect of slipstream on lift and drag". Part I, II and III R.A.E. Reports Aero 2083 A, B and C.

F-74. J.V. Becker: "Windtunnel investigation of air inlet and outlet openings on a streamline body". NACA Report No. 1038, 1951.

Turbojet engine nacelles and installation drag

F-75. J. Seddon: "Air intakes for gasturbines". J.R.Ae.S. Oct. 1952, pp. 747-781.

F-76. D. Küchemann and J. Weber: "Aerodynamics of propulsion". McGraw-Hill Book Comp., Inc., New York, 1953.

F-77. M. Sibulkin: "Theoretical and experimental investigations of additive drag". NACA Report No. 1187, 1954.

F-78. A.R.C. Panel: "Definitions to be used in the description and analysis of drag". J.R.Ae.S. Nov. 1958, pp. 796-801. See also: J.R.Ae.S. Aug. 1955.

F-79. G. Schulz: "Aerodynamische Regeln für den Einbau von Strahltriebwerksgondeln". Zeitschrift für Flugwissenschaften, Vol. 3, 1955, pp. 119-129.

F-80. J.S. Dennard: "The total pressure recovery and drag characteristics of several auxiliary inlets at transonic speeds". NASA Memo 12-21-58 L.

F-81. G. Schulz: "Der Kraftangriff bei Strahltriebwerken und ihren Verkleidungen". Z. Flugwissenschaften, Vol. 9, Sept. 1956, pp. 285-290.

F-82. E.E. Honeywell: "Compilation of power-off base drag data and empirical methods for predicting power-off base drag". Convair Report TM 334-337, 1959.

F-83. J.E. Steiner: "The development of the Boeing 727". J.R.Ae.S. Vol. 67, Feb. 1963, pp. 103-110.

F-84. J. Bogdanovic: "A method for determining propulsion system requirements for long-range, long-endurance aircraft". AIAA Paper No. 64-783.

F-85. C.R. Palmer: "Engine pod drag". Rolls-Royce (Hucknall) Brochure HK 15, March, 1964.

F-86. H.Mc. Donald and P.F. Hughes: "A correlation of high subsonic afterbody drag in the presence of a propulsive jet or support sting". J. of Aircraft, Vol. 2, No. 3, May-June 1965, pp. 202-207.

F-87. R.L. Lawrence: "Afterbody flow fields and skin friction on short duct fan nacelles". J. Aircraft, Vol. 2, No. 4, July-Aug. 1965.

F-88. J.S. Mount: "Effect of inlet additive drag on aircraft performance". J. Aircraft, Vol. 2, No. 5, Sept.-Oct. 1965, pp. 374-378.

F-89. W.C. Swan: "A discussion of selected aerodynamic problems on integration of propulsion systems with airframe on transport aircraft". AGARDograph 103, Oct. 1965.

F-90. L.R. l'Anson: "The application of the high bypass turbofan for business and executive aircraft". SAE Preprint No. 660221.

F-91. J.T. Kutney: "High-bypass vs. low-bypass engine installation considerations". SAE Paper No. 660775.

F-92. W.S. Viall: "Aerodynamic considerations for engine inlet design for subsonic high-bypass fan engines". SAE Paper No. 660733.

F-93. J.E. Green: "Short cowl front fan turbojets; friction drag and wall-jet effects on cylindrical afterbodies". RAE Technical Report, No. 67144.

F-94. P. Taylor: "Inlet and fan aerodynamics". SAE Paper No. 680711.

F-95. R. Hetherington: "Engine component design problems associated with large subsonic transports". Lecture given at the Von Karman Institute, Brussels, 23rd. April, 1969.

F-96. W. Tabakoff and H. Sowers: "Drag analysis of powered nacelle fan jet engine model tests". Zeitschrift für Flugwiss., Vol. 4, April 1969, pp. 134-144.

F-97. C.E. Swavely: "Twin jet aircraft aft- fuselage performance prediction". Proceedings of the Airforce airframe compatability symposium 24/26th. June, 1969. T. Report AFAPL-69-103.

F-98. D. Bergman: "Implementing the design of airplane engine exhaust systems". AIAA Paper No. 72-112.

F-99. W. Swann and A. Sigalla: "The problem of installing a modern high by-pass engine on a twin jet transport aircraft". AGARD Conference on aerodynamic drag, CP-124, 1973.

Interference effects

F-100. J. Lennertz: "Beitrag zum theoretischen Behandlung des gegenseitigen Einflusses von Tragfläche und Rumpf". Abhandlungen aus dem Institut der Technische Hochschule, Aachen, Heft 8, 1928, pp. 1-30.

F-101. A.J. Marx: "Korte beschouwing over de invloed van de romp op de geïnduceerde weerstand van de vleugel". NLL Report V 1299, 1943 (in Dutch).

F-102. A.A. Nikolski: "On lift properties and induced drag of wing-fuselage combinations". NASA RE 5-I-59W.

F-103. H.R. Lawrence and A.H. Flax: "Wing-body interference at subsonic and supersonic speeds, survey and new developments". J. Aeron. Sciences, Vol. 21 No. 5, May 1954.

F-104. J.T. Keetney and S.P. Piszkin: "Reduction of drag rise on the Convair 990 airplane". J. of Aircraft, Jan.-Feb. 1964, (AIAA Preprint No. 63-276).

F-105. S. Neumark: "Lift due to interference between an aerofoil and an external non-lifting body". ARC R & M No. 3411, May 1964.

F-106. J. Seddon: "Factors determining engine installation drag of subsonic and supersonic long-range aircraft". AGARD CP-9, Part I, 1966.

F-107. D.J. Raney, A.G. Kurn and J.A. Bagley: "Windtunnel investigation of jet interference for underwing installation of high bypass ratio engines". ARC CP No. 1044, March 1968.

F-108. P. Williams and D. Stewart: "The complex aerodynamic interference pattern due to rear fuselage mounted power-plants". AGARD CP-71.

F-109. J.C. Patterson Jr.: "A wind-tunnel investigation of jet-wake effect of a high-bypass engine on wing-nacelle interference drag of a subsonic transport". NACA TN No. D-4693, 1968 (see also NACA TN No. D-6067, 1970).

Undercarriages

F-110. F.B. Bradfield: "Wheels, fairings and mudguards". ARC R & M 1479, 1932.

F-111. Herrnstein and Biermann: "The drag of airplane wheels, wheel fairings and landing gears". NACA Reports No. 485, 1934 and 518, 522, 1935 (Parts I, II and III).

F-112. H. Harmon: "Drag determination of the forward component of a tricycle landing gear". NACA TN No. 788, 1940.

F-113. P.A. Hufton and J.R. Edwards: "Note on a method of calculating the take-off distance". RAE Departmental Note B.A. Performance No. 20, Aug. 1940.

Canopies and windshields

F-114. J.H. Hartley, D. Cameron and W.H. Curtis: "Note on windtunnel tests on the design of cabins". ARC R & M No. 1811, 1937.

F-115. R.G. Robinson and J.B. Delano: "An investigation of the drag of windshields in the 8-foot high-speed windtunnel". NACA Report No. 730, 1942.

F-116. T.V. Somerville and N. Sharp: "Note on the drag and pressure distribution on two types of cabin". RAE Report No. B.A. 1654.

F-117. Anonymus: "The drag of fighter-type canopies at subcritical Mach-numbers". Engineering Sciences Data Sheets Item No. 67041.

External stores

F-118. J.B. Berry: "Examples of airframe-stores interference". AGARD CP-71, Sept. 1970.

F-119. P.G. Pugh and P.G. Hutton: "The drag of externally carried stores - its prediction and alleviation". AGARD CP-124, 1973.

F-120. J.B. Berry: "External store aerodynamics for aircraft performance prediction". AGARD LS-67, 1974.

Trim drag

F-121. C.H. Naylor: "Notes on the induced drag of a wing-tail combination". ARC R & M 2528, 1954.

F-122. H. Behrbohm: "Basic low-speed aerodynamics of short-coupled canard configurations of small aspect ratio". SAAB TN No. 60, July 1966.

F-123. V.L. Marshall: "Aircraft trim drag". Thesis, Cranfield Institute of Technology, 1970.

F-124. L.W. McKinney and S.M. Dollyhigh: "Some trim drag considerations for manoeuvring aircraft". AIAA Paper No. 70-932.

Rougnness

F-125. A.D. Young: "Surface finish and performance". Aircraft Engng. Sept. 1939.

F-126. M.J. Wood: "The effect of some common surface irregularities on a wing". NACA TN No. 695, March 1939.

F-127. J.H. Quin Jr.: "Summary of drag characteristics of practical-construction wing sections". NACA Report No. 910, 1948.

F-128. Anonymus: "Wing or body drag due to surface roughness". R.Ae.S. Data Sheets Wings 02.04.08. and 02.04.10.

F-129. A.D. Young: "The drag effects of roughness at high subcritical speeds". J.R.Ae.S. Vol. 54, 1950, pp. 534-540.

F-130. E.A. Horton and N. Tetervin: "Measured surface defects on typical transonic airplanes and analysis of their drag contribution". NASA TN D-1024, Oct. 1962.

F-131. J.F. Nash and P. Bradshaw: "The magnification of roughness drag by pressure gradients". J.R.Ae.S., Vol. 71, No. 673, Jan. 1967.

F-132. T.A. Cook: "The effects of ridge excrescences and trailing-edge control gaps on twodimensional aerofoil characteristics". RAE Technical Report 71080, April 1971.

F-133. J.I. Simper and P.G. Hutton: "Results of a series of wind-tunnel model breakdown tests on the Trident 1 aircraft and a comparison with drag estimates and full scale flight data". ARC CP No. 1170, 1971.

F-134. J.I. Simper: "Results of a series of wind tunnel tests on the Victor B.Mk. 2 aircraft and a comparison with drag estimates and full scale flight". ARC CP No. 1283, 1974.

Appendix G. Prediction of lift and drag in the low-speed configuration

SUMMARY

A collection of generalized data and methods for estimating the lift curve and drag polar in the configuration for low-speed flight is presented. Passive* trailing-edge and leading-edge high-lift devices are considered. Glauert's linear theory for thin airfoils with deflected flaps forms the basis for most prediction methods; correction factors are given for taking nonlinearity and flow separation into account.
Prediction methods are also added for estimating the drag due to extension of a retractable undercarriage, the effects of ground proximity on lift and drag, and the increase in drag associated with the failure of an outboard engine.

*no action is taken to augment the external flow by means of blowing or suction.

NOMENCLATURE

Symbols with a ' (prime) refer to quantities defined based on the extended chord or to the lift-curve slope with flaps deflected.

A - aspect ratio (no index: wing aspect ratio)

A_N - area of a nozzle

a.c.   - aerodynamic center

B_p - number of blades per propeller

b - span; width

c - chord

$\bar{c}$ - mean aerodynamic chord (MAC)

c_g - mean geometric chord ($c_g = S/b$)

c_d, C_D - two- and three-dimensional drag coefficient, respectively

C_{D_o} - zero-lift drag coefficient

c_{d_p}, C_{D_p} - two- and three-dimensional profile drag coefficient, respectively

C_{D_v} - vortex-induced drag coefficient

$C_{D_{trim}}$ - trim drag coefficient

c_ℓ, C_L - two- and three-dimensional lift coefficient, respectively

c_{ℓ_o}, C_{L_o} - lift coefficient at zero wing angle of attack

$c_{\ell_\alpha}, C_{L_\alpha}$ - two- and three-dimensional lift-curve slopes, for constant δ_f

c_{ℓ_δ} - $\partial c_\ell / \partial \delta_f$ for constant α

C_{L_∞} - lift coefficient out of ground effect

c_m, C_m - two- and three-dimensional pitching moment coefficient, respectively

C_{y_v} - side force coefficient of vertical tailplane ($C_{y_v} = Y_v/q_v S_v$)

D_i - engine inlet diameter

D_p - propeller diameter

e - Oswald factor (no index: wing Oswald factor)

$F(\delta)$ - function for determining the profile drag increment due to split flaps

F_{uc} - function defining lift effect on undercarriage drag

h - height of aerodynamic center above ground plane

h_{eff} - effective height of MAC above ground plane

K_{as}, K_σ - factors for determining drag due to asymmetric flight condition

K_b - flap-span factor on lift

K_c - flap-chord factor on lift

K_{ff} - flap/fuselage lift interference factor

K_{2-} - correction factor for profile drag due to flap

k_c - flap-chord factor for split flap

k_d - profile drag factor due to flap

k_ℓ - factor characterizing profile drag increment with lift

k_s - factor allowing for nonlinearity in lift curve

k_δ - flap angle correction factor for split flap

ℓ_h, ℓ_v - moment arm of horizontal and vertical tailplane, respectively (see Figs. E-13 and 9-21)

ℓ_{uc} - length of undercarriage

M - Mach number

$\dot{m}$ - engine air mass flow (windmilling engine)

q - dynamic head ($q = \frac{1}{2}\rho V^2$)

R_c - Reynolds number based on chord

S - area (no index: gross wing area)

T - thrust

ΔT - net thrust loss of inoperative engine plus windmilling and propeller drag

t - absolute profile thickness

V - flight velocity

V_N - average velocity of engine nozzle flow

u.c.   - undercarriage

v, w - induced drag factors

W_{to} - Maximum Takeoff Weight

x - longitudinal coordinate (origin: MAC leading edge)

Y_v - sideforce on vertical tailplane

y - lateral coordinate

Y_e - yawing moment arm of inoperative engine

Δy - leading edge sharpness parameter

z - vortex-induced drag factor for fuselage lift carry-over effect

z_h - distance from flap hinge to chord

α - angle of attack

α_i - induced angle due to trailing vortices

α_δ — theoretical flap lift factor $(\alpha_\delta = c_{\ell_\delta}/c_{\ell_\alpha})$

β — span factor for ground effect

δ — deflection angle of flap or slat, normal to hinge line

Δ — increment or decrement (example: Δ_f = increment due to flap)

ε — downwash angle

η_δ — flap lift effectiveness

θ — angle characterizing relative flap or slat chord

$\Lambda_{\frac{1}{4}}$ — sweep angle of quarter-chord line (no index: wing sweep angle)

$\Lambda_{\frac{1}{2}}$ — sweep angle of mid-chord line (no index: wing sweep angle)

Λ_h, Λ_v — sweep angle of horizontal and vertical tailplane quarter-chord lines, respectively

λ — taper ratio (no index: wing taper ratio)

μ — factor determining the wing pitching moment

ρ — density of ambient air

σ — ground effect function; side-wash angle

Subscripts

A — auxiliary flap section

a.c. — aerodynamic center

as — asymmetric flight condition

crit — critical (stalling) condition

e — engine

ewm — engine windmilling

F — forward flap section

f — flap

ff — flap/fuselage interference

ft — front tire(s)

h — horizontal tailplane

i — inboard end of flap

LE — leading edge

mg — main gear

N — nozzle

n — nacelle

ng — nose gear

o — outboard end of flap

p — profile

prop — propeller

r — rudder

ref — reference configuration

rt — rear tires (bogie u.c.)

s — leading-edge high-lift device (slat)

TE — trailing edge

uc — undercarriage

v — vertical tailplane; vortex-induced

w — wing

wf,ws — part of the wing with flaps or slats

∞ — undisturbed flow; out of ground effect

$\frac{1}{4}$ — quarter-chord (line)

$\frac{1}{2}$ — mid-chord (line)

.2;60 — split flap with c_f/c = .20 at $\delta_f = 60^\circ$

G-1. INTRODUCTION

The low-speed performance of aircraft, especially for transport aircraft, has received considerable added emphasis during recent years. It is therefore essential in the preliminary design stage to be able to predict the aerodynamic characteristics (lift and drag, trimmed condition) which can be used with a certain amount of confidence to compute low-speed performance and handling qualities and to provide a realistic goal for further aerodynamic development.

The present compilation of methods and data will enable the designer to make a good estimate of the lift curve and the drag polar for configurations with leading- and trailing-edge high-lift devices deflected. Split flaps, plain flaps, and single slotted, double slotted, triple slotted and Fowler flaps are the most commonly used trailing-edge devices to be dealt with, and in addition some data are also given on plain leading-edge flaps, slats and Krueger flaps. A method for estimating the pitching moment change due to flap deflection is presented for the purpose of calculating the tailplane load required to trim the aircraft.

The two-dimensional (sectional) lift increment at small angle of attack due to
trailing-edge flap deflection will be calculated on the basis of Glauert's theory
for thin airfoils with plain flaps, deflected over small angles. Efficiency factors representing the ratio of experimental to theoretical lift increment are given
for the various types of flaps to be considered. The prediction method for the maximum section lift coefficient is based on
McRae's approach (Ref. G-23), which proved
useful for all common types of flaps except split flaps. The conversion from two-
dimensional to three-dimensional lift coefficients is based on classical methods,
supplemented by approximate corrections
for the presence of the fuselage.

The prediction method presented for the
profile drag increment due to flap deflection again uses the lift increment according to Glauert's theory, assuming that,
owing to viscous effects, a certain percentage of the lift increment will act
perpendicular to the flap chord. Corrections for vortex-induced drag and trim drag
are based on classical theory.

Reliable methods for determining the effect of leading-edge devices on lift and
drag are not available in the open literature. A first-order approximation is
therefore presented which is based on the
observation that for airfoil and wings
with slats deflected the stalling angle of
attack is the most characteristic parameter.

A separate paragraph is devoted to the effects of undercarriage extension, ground
proximity, and engine failure on lift and
drag. Although these items are of vital
importance to the prediction of transport
aircraft performance, in particular takeoff (climb) and landing (climb) performance, little useful information is available in the literature and the methods
proposed must be considered only as first-
order approximations.

G-2. EFFECT OF TRAILING-EDGE FLAP DEFLEC-
TION ON AIRFOIL SECTION LIFT

G-2.1. General aspects

The traditional approach to the calculation of wing lift is based on the assumption that the lift generated by a two-
dimensional wing (airfoil section) can be
considered as a starting point. A suitable
conversion to the three-dimensional wing
is made by means of semi-empirical corrections for part span, fuselage effects, etc.
This procedure is acceptable when spanwise
flows and interference effects are or of
minor importance or completely absent. It
is therefore frequently used in the preliminary design stage of aircraft categories that are in present use in civil aviation, in the realization that during the
configuration development an adequate aerodynamic development program will be initiated in order to optimize the wing and
flap system.

Theoretical methods of calculating the aerodynamic characteristics of flapped sections have recently come to a stage of development where lift and pitching moment
can be predicted with reasonable accuracy.
A survey of the state of the art is given
in Ref. G-31. Particularly for the most
effective flap systems (multiple element
flaps, slats), the flow is very sensitive
to the details of the configuration: the
flap and slat shape, their relative position, shroud shape, etc. This would require
an optimization program which is a complex
subject in itself. Instead of this, we will
present generalized semi-empirical methods
based on thin airfoil theory and experimental data. Where possible, an indication
will be given of the sensitivity of flap
performance to design details (average
"good" design, best design, poor design).
The basic effect of trailing-edge flap deflection on the lift curve of an airfoil

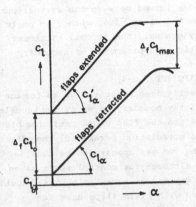

Fig. G-1. Effect of trailing-edge flap deflection on section lift

section is depicted in Fig. G-1. It will be assumed that for conventional wing shapes the lift curves are linear for moderate angles of attack. It can be observed that even if the lift curve of the basic section is fairly nonlinear due to trailing edge stall, the flapped section exhibits a more linear behaviour when the flow near the trailing edge is cleaned up by the flap.
The determination of the lift curve will be subdivided into prediction of the increment in lift at zero angle of attack ($\Delta_f c_{\ell_o}$), the (increment in) maximum lift coefficient and the lift-curve slope at small angles of attack, c_{ℓ_α}.

G-2.2. Lift increment at zero angle of attack

For small flap deflections the rate of change of lift with flap deflection at constant angle of attack is frequently used as a measure of flap effectiveness:

$$c_{\ell_\delta} = \left(\frac{\partial c_\ell}{\partial \delta_f}\right)_\alpha \qquad (G-1)$$

Another convention used is the rate of change of zero-lift angle of attack with flap deflection:

$$\frac{\partial \alpha}{\partial \delta_f} = \alpha_\delta = \frac{\left(c_{\ell_\delta}\right)_\alpha}{\left(c_{\ell_\alpha}\right)_{\delta_f}} \qquad (G-2)$$

The basis for several estimation methods is formed by Glauert's linearized theory for thin airfoils with flaps (Ref. G-57). A result obtained from this theory for the lift due to flap deflection is as follows:

$$\alpha_\delta = 1 - \frac{\theta_f - \sin\theta_f}{\pi} \qquad (G-3)$$

where

$$\theta_f = \cos^{-1}\left(2\,\frac{c_f}{c} - 1\right) \qquad (G-4)$$

Fig. G-2 gives a plot of α_δ as a function

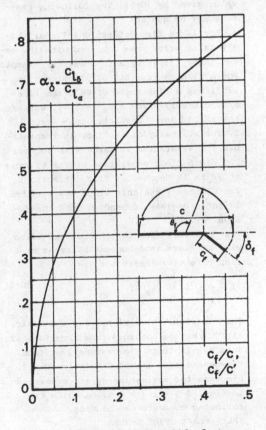

Fig. G-2. Theoretical flap lift factor

529

of the relative flap chord. The theoretical $\Delta_f c_{\ell_o}$ is thus:

$$\Delta_f c_{\ell_o} = \left(c_{\ell_\delta}\right)_\alpha \delta_f = \frac{\left(c_{\ell_\delta}\right)_\alpha}{\left(c_{\ell_\alpha}\right)_{\delta_f}} c_{\ell_\alpha} \delta_f = \alpha_\delta \, c_{\ell_\alpha} \, \delta_f$$

$$(G-5)$$

It is found that the theoretical lift effectiveness cannot be realized in practice. For small flap deflections approximately 70 to 85% of the theoretical value is possible, depending upon the type of flap system, while for large flap angles, e.g. those used in the landing configuration, the lift effectiveness may go down to approximately 50 percent of the theoretical value given by (G-5). The following reasons may be quoted:

1. For large flap deflections linear theory is in error when compared with the exact theory given by Ref. G-66, for example. For a flap chord ratio of 35 percent (α_δ = .707) and a flap angle of 60°, the linearized theory gives $\Delta_f c_{\ell_o}$ = 4.44, whereas the exact theory predicts $\Delta_f c_{\ell_o}$ = 4.0.

2. The viscosity of the flow is responsible for separation at large flap angles. For plain flaps separation starts to occur at 10 to 15 degrees of flap deflection; for slotted flaps this value may vary from 20 to 35 degrees, depending upon the design details.

The departure from the theoretical value will be taken into account by means of a flap effectiveness factor η_δ:

$$\Delta_f c_{\ell_o} = \eta_\delta \, \alpha_\delta \, c_{\ell_\alpha} \, \delta_f \qquad (G-6)$$

The numerical value of c_{ℓ_α} for the basic airfoil may be obtained from experimental data available to the designer, or from the generalized method in Section E-3.2. Diagrams for estimating η_δ are presented in Figs. G-3 through G-6, on which the following comments can be made.

SPLIT FLAPS (Fig. G-3)
The effectiveness factor does not exceed 70 percent, even at small deflections, due to the wake formed between the airfoil and the flap. The decay of η_δ with flap angle is fairly gradual. The airfoil thickness/chord ratio is a parameter of secondary importance.

PLAIN FLAPS (Fig. G-4)
Flap effectiveness is sensitive to the condition of the boundary layer at the knuckle and therefore the flap chord ratio has been used as a correlation parameter. The effect of sealing the gap appears to be considerable and although for control surfaces an unsealed gap may be acceptable, this should be avoided with plain flaps used for increasing the lift.

SINGLE SLOTTED FLAPS (Fig. G-5)
For efficiently designed slotted flaps the lift increment due to flap deflection is not affected by the boundary layer of the basic airfoil, since a new boundary layer forms over the flap surface. The lift effectiveness, however, is very sensitive to the flap and slot geometry. Most of the early systematic measurements by NACA were made on flaps with optimum slot shape for each deflection, although for reasons of structural simplicity a single slotted flap is frequently supported by means of a fixed hinge. In this latter case performance is sensitive to the hinge location. Generally speaking, the lower the hinge, the better the lift effectiveness will be, provided a good shroud and gap shape are present.

Slotted flaps require a rearward flap motion in order to ensure a good slot. The airfoil chord is thus extended effectively and this in itself contributes to the lift. This can be taken into account by referring the section lift to the extended chord, as defined in Fig. G-7, and then converting the result to the original chord as follows:

$$c_\ell = \left(c_{\ell_o}' + \Delta_f c_{\ell_o}'\right) \frac{c'}{c} \qquad (G-7)$$

where c_{ℓ_o}' and $\Delta_f c_{\ell_o}'$ are based on the extended chord c'. Assuming that for the basic section c_{ℓ_o} is not altered after chord extension $\left(c_{\ell_o}' = c_{\ell_o}\right)$, we have:

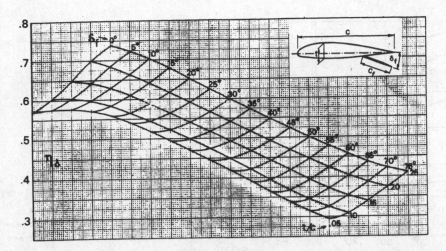

Fig. G-3. Lift-effectiveness factor for split flaps (derived from experimental data in Ref. G-34 and the USAF Datcom, Table 6.1.1.1.-24)

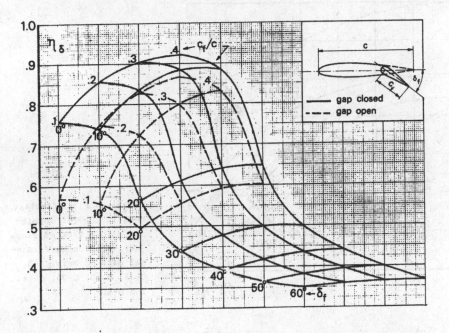

Fig. G-4. Lift effectiveness factor for plain flaps (derived from experimental data in the USAF Datcom, Table 6.1.1.1.-A and Ref. G-64)

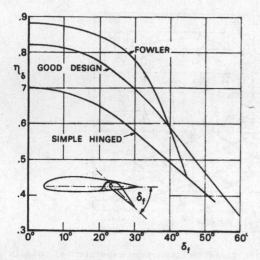

Fig. G-5. Lift effectiveness of single slotted (Fowler) flaps (derived from experimental data in Refs. G-33, G-35, G-37 and G-19)

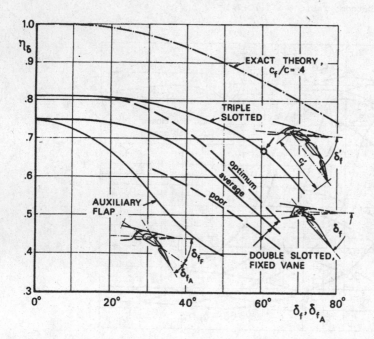

Fig. G-6. Flap effectiveness factor for double and triple slotted flaps (derived from experimental data in Refs. G-3, G-19, G-29, G-39, G-44 and G-48)

$$\Delta_f c_{\ell_o} = \Delta_f c_{\ell_o}' \frac{c'}{c} + c_{\ell_o} \left(\frac{c'}{c} - 1 \right) \qquad \text{(G-8)}$$

where $\Delta_f c_{\ell_o}'$ is defined by (G-6), noting that α_δ is also based on the extended chord (c_f/c'). Replacing the basic section lift gradient by 2π rad^{-1}, we have:

$$\Delta_f c_{\ell_o}' = 2\pi \, \eta_\delta \, \alpha_\delta' \, \delta_f \qquad \text{(G-9)}$$

where δ_f is in radians and α_δ' can be obtained from (G-3) and (G-4) or from Fig. G-2.

For a given flap geometry and type of support, the amount of chord extension can be

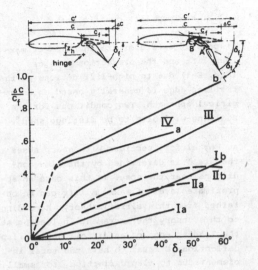

I: Fixed hinge - a: $z_h/c_f = .2$
            b: $z_h/c_f = .4$

II: Typical optimum flap position
            a: single slotted
            b: double slotted, fixed vane

III: Double slotted, variable geometry, with flap extension

IV: Fowler   - a: single slotted, double slotted with fixed vane
            b: double and triple slotted, with flap extension

Fig. G-7. The extended chord - its definition and some typical values for practical configurations

calculated; in the absence of information of this type Fig. G-7 may be used. In the case of a fixed hinge position the chord extension is given by:

$$\frac{c'}{c} = 1 + 2 \frac{z_h}{c} \tan \tfrac{1}{2} \delta_f \qquad (G-10)$$

FOWLER FLAPS (Fig. G-5)

Aerodynamically, the Fowler flap acts in a similar manner to a single slotted flap and the same basic method can be used for both. The effect of chord extension is much larger (Fig. G-7), however, although this is partly offset by a reduction in c_f/c' and a lower value of α_δ'. For flap angles up to 30 - 35 degrees the flap ef-

fectiveness is found to be superior to that for single slotted flaps.

DOUBLE SLOTTED TRAILING-EDGE FLAPS

The single slotted flap loses its lift effectiveness when the deflection angle exceeds about 40 degrees. The double slotted flap may be considered as a single slotted flap with a turning vane in the slot to recover the flow turning effectiveness. Consequently, the important parameters for the 2-slot flap are more complicated than for the 1-slot flap and so is the prediction of its lift effectiveness. Two main categories will be distinguished here.

a. The flap with fixed vane. The prediction method for $\Delta_f c_{\ell_o}$ is similar to that for the single slotted flap, except that the factor η_δ is plotted in Fig. G-6. Typical curves are given for a good, average design with fixed hinge, together with a poor design and an indication of what can be expected when the flap support mechanism is such that an optimum gap is realized for each deflection. The maximum useful deflection is about 50 to 55 degrees.

b. The double slotted flap with variable geometry. The forward flap is deflected to a maximum angle of 30 degrees, typically, and the aft section up to 30 or 40 degrees relative to the forward flap. There may also be a backward movement of the auxiliary flap. The total lift increment of this configuration may be calculated as follows:

$$\Delta_f c_{\ell_o} = \Delta_1 c_{\ell_o} + \Delta_2 c_{\ell_o} \qquad (G-11)$$

where $\Delta_1 c_{\ell_o}$ is the increment in c_{ℓ_o} due to the combined flap at a deflection equal to that of the forward flap section, assuming that the flap chords of the aft and forward flap coincide and no second slot is present. This contribution can be calculated with the method given above for single slotted flaps (Fig. G-5).
$\Delta_2 c_{\ell_o}$ is the increment in c_{ℓ_o} due to deflection of the auxiliary flap relative to the forward section. Again the method for single slotted flaps is used, except that a reduced lift effectiveness factor η_δ must be substituted in (G-9), given in

Fig. G-6, because the presence of the forward flap reduces the effectiveness of the auxiliary flap.

The factor $\alpha_\delta{}'$ in (G-9) must be related to the ratio of the flap chord to the extended chord, which is defined in Fig. G-7 and obtained as follows:

a. The auxiliary flap is rotated from its deflected position about point A, until the two flap chords coincide.

b. Both flap sections are then rotated from the deflected position of the forward flap about point B until they coincide with the wing chord.

c. The distance from the leading edge of the airfoil to the trailing edge thus obtained is the extended airfoil chord c'. Using the typical figures given in Fig. G-7, the extended chord is calculated as follows:

$$\frac{c'}{c} = 1 + \frac{\Delta c}{c_f} \frac{c_f}{c} \qquad (G-12)$$

It should be noted that larger values of $\Delta c/c_f$ than those given in Fig. G-7 are feasible, although at the expense of more structural complications and weight.

TRIPLE SLOTTED TRAILING EDGE FLAPS

Insufficient data have been published on triple slotted flaps to form the basis of a generalized prediction method. The data given in Ref. G-29 have been transformed into a single point for η_δ in Fig. G-6. An equivalent flap deflection angle $\delta_f{}'$ was defined to characterize the combined effect of the complex flap system, while the factor $\alpha_\delta{}'$ was determined on the basis of the equivalent flap chord ratio $c_f{}'/c'$ to derive $\alpha_\delta{}'$ from (G-3) and (G-4). In spite of this simplified approach the point is in fair agreement with the statement in Ref. G-29 that 81 percent of the potential flow value can be obtained at a total flap angle of 60 degrees. The suggested curve for η_δ is therefore assumed at 81 percent of the potential flow value (exact theory of Ref. G-66) up to 50 degrees of flap angle.

G-2.3. Maximum lift coefficient

Viscous effects dominate the flow at maximum lift and the prediction of $\Delta_f c_{\ell_{max}}$ (Fig. G-1) due to high-lift devices at the trailing edge is generally based on an empirical approach. Two conditions for the stalling will have to be distinguished[*].

a. For airfoil sections with sharp noses the stall is determined by the flow conditions near the nose. In this case an approximate level of maximum lift can be obtained from thin airfoil theory. According to this theory the increment in loading at the leading edge of an airfoil due to flap deflection is equal to half the total increment due to flap deflection, for small flap chord ratios. For airfoils whose maximum lift is controlled by leading edge stall, the maximum lift increment due to flap deflection can be shown (Ref. G-16) to be equal to:

$$\Delta_f c_{\ell_{max}} = \Delta_f c_{\ell_o} \frac{\sin\theta_f}{\pi - (\theta_f - \sin\theta_f)} \qquad (G-13)$$

where θ_f is defined by (G-4). The theoretical ratio of $\Delta_f c_{\ell_{max}}/\Delta_f c_{\ell_o}$ varies between .4 and .5 for practical flap chord ratios. Combination of (G-13) with (G-3) and (G-5) yields a simple expression for the theoretical $\Delta_f c_{\ell_{max}}$ of sections with sharp leading edges. Assuming $c_{\ell_\alpha} = 2\pi$, we find:

$$\Delta_f c_{\ell_{max}} = 2\delta_f \sin\theta_f \qquad (G-14)$$

b. When the basic section displays a separation associated with the pressure gradient at the rear part of the airfoil, this trailing edge stall will be delayed by the local suction produced by deflection of an effective trailing-edge flap. In this case the gain in maximum lift will be of the same magnitude as the lift increment at small angles of attack or may even be

[*]A more complete explanation of this subject is given in Ref. G-23.

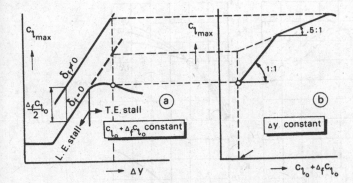

Fig. G-8. Effect of leading
edge sharpness and flap deflection on maximum lift

slightly higher.

Conclusions that can be drawn from the above are
illustrated in Fig. G-8. In Fig. 8a the sharpness
of the airfoil nose is characterized by the leading-edge sharpness parameter Δy (see Appendix A,
Fig. A-2), which has proved to be a useful correlation parameter for predicting the maximum lift
associated with stall at the leading edge. For Δy
up to 1.2 - 1.5 percent chord, the maximum lift
associated with leading edge stall (long bubble)
is roughly constant. For larger Δy values $c_{l_{max}}$
increases approximately proportionally with Δy
(short bubble) up to a point where trailing edge
stall begins to dominate. Thus for relatively thin
airfoils, deflection of a trailing-edge flap results in a maximum lift increment which is predicted theoretically by (G-13), while thicker airfoils with a trailing edge stall on the basic airfoil have a maximum lift increment of the order of
50 to 100 percent of the lift increment at small
angles of attack[*].

Consider now an airfoil with a trailing edge stall
when the flap is retracted (Fig. G-8b). Small flap
deflections clean up the flow near the trailing
edge, resulting in a maximum lift increment which
is of the same order of magnitude as the lift increment at small angles of attack. At a certain
flap angle the load induced at the airfoil nose
will increase up to a point at which leading edge
stall will occur first, as dictated by the criter-
ion in Fig. G-8a. The slope of the $c_{l_{max}}$ vs. $\Delta_f c_{l_o}$
curve will be approximately .5 : 1 for this region

*Ref. G-16 suggests an average of $\Delta_f c_{l_{max}}$
$= \frac{2}{3} \Delta_f c_{l_o}$

of flap angles.

The above observations form the basis of
the following simple, unified prediction
method for the maximum lift of flapped
airfoils.

a. Plain and slotted (Fowler) trailing-
edge flaps.

The lowest value of the following items
determines the maximum lift coefficient:

$$c_{l_{max}}' = .533 \frac{\Delta y}{c} \left(\frac{R_c}{3 \times 10^6} \right)^{.08} + \tfrac{1}{2} \left(c_{l_o} + \Delta_f c_{l_o}' \right)$$

for $\Delta y/c$ (in %) $\geqslant 1.5$, or:

$$c_{l_{max}}' = \left(c_{l_{max}} \right)_{\delta_f = 0} + \Delta_f c_{l_o}' \qquad (G-15b)$$

(G-15a)

where R_c is the Reynolds number based on
the original chord length, and the maximum
lift coefficient based on this chord is
given by:

$$c_{l_{max}} = \frac{c'}{c} c_{l_{max}} \qquad (G-16)$$

Equation G-15 has been derived from experimental data in the literature quoted in
the references. The method gives acceptable results not only for plain and single
slotted flaps, but also for multiple-ele-
ment flaps with or without Fowler motion.

b. Split flaps.

Airfoil sections with split flaps cannot
be dealt with by the previous method, prob-

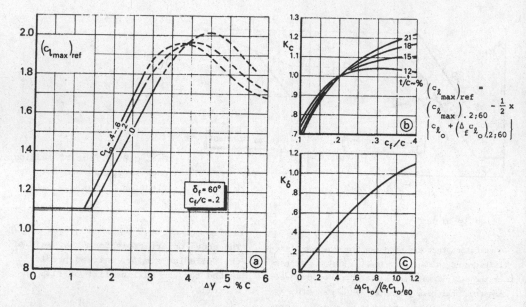

Fig. G-9. Charts for estimating the increment in maximum lift due to split flaps. For Δy: see Fig. A-1 of Appendix A. (derived from Ref. G-34)

ably because there is always a wake at the trailing edge when the flap is deflected, and leading edge and trailing edge stall will occur simultaneously. For split flaps the curve of Fig. G-8b has, typically, a mean slope of .75 : 1 and is slightly curved. A different prediction method has therefore been devised, consisting of the following steps:

1. Calculate $\Delta_f c_l$ vs. δ_f with the method given in Section G-2.2. for split flaps (Fig. G-3).

2. Calculate $\Delta_f c_{l_0}$ for a flap chord ratio $c_f/c = .20$ and a flap angle $\delta_f = 60°$. This is denoted by $\left(\Delta_f c_{l_0}\right)_{.2;60}$

3. Determine $c_{l_{max}}$ for $c_f/c = .2$ and $\delta_f = 60°$, as follows:

$$\left(\Delta c_{l_{max}}\right)_{.2;60} = \left(c_{l_{max}}\right)_{ref} - \left(c_{l_{max}}\right)_{\delta_f=0} +$$
$$+.5 \left\{ c_{l_0} + \left(\Delta_f c_{l_0}\right)_{.2;60} \right\} \qquad (G-17)$$

where $\left(c_{l_{max}}\right)_{ref}$ is obtained from Fig. G-9a.

4. Correct for flap chord ratio and flap angle as follows:

$$\Delta c_{l_{max}} = k_c \, k_\delta \left(\Delta_f c_{l_{max}}\right)_{.2;60} \qquad (G-18)$$

k_c and k_δ are correction factors for the flap chord ratio and flap angle given in Figs. G-9b and G-9c, respectively. The values for $\Delta_f c_{l_0}$ to be used here have been calculated in step 1.

G-2.4. Lift-curve slope

The lift gradient is affected by flap deflection in a number of ways:

1. The chord extension increases c_{l_α}. This can be allowed for by the multiplication factor c'/c.

2. The potential flow effect of flap deflection on the lift-curve slope given by the exact theory in Ref. G-66 indicates that $\Delta_f c_l$ is reduced with increasing α. Hence c_{l_α}' (based on the extended chord) decreases not only with the flap deflection angle, but also with the angle of attack, and the result is a nonlinearity in the lift curve which is particularly pronounced for large flap angles.

3. The effect of viscosity on the lift ef-

fectiveness of a flap increases with the angle of attack, thus reducing $\Delta_f c_\ell$ with increasing α.

Fig. G-10 shows that for small flap angles

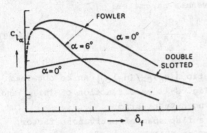

Fig. G-10. Variation of section lift-curve slope with flap angle

the effect of the chord extension dominates, but that this effect is cancelled at large flap angles by the other effects. The reduction in lift-curve slope is more pronounced at angles of attack approaching the stall.

No method for calculating the lift-curve slope is available in the literature. The relationship presented here,

$$\frac{c_{\ell_\alpha}' \text{ (flaps down)}}{c_{\ell_\alpha} \text{ (flaps up)}} = \frac{c'}{c}\left(1 - \frac{c_f}{c'} \sin^2 \delta_f\right)$$

$$\text{(for } \alpha = 0 \text{ to } 5^\circ\text{)} \qquad \text{(G-19)}$$

approximates the results of the exact theory fairly accurately and is in qualitative agreement with experimental data. In individual cases the reduction of c_{ℓ_α} with δ_f may be considerably more than that indicated by (G-19), particularly when the shape of the slot is not optimized, and when the stalling condition occurs at small angles of attack.

G-3. LIFT OF AIRCRAFT WITH DEFLECTED TRAILING-EDGE FLAPS

The prediction of aircraft lift is based on a build-up of various contributing components: the aircraft in the clean configuration (see Appendix E), trailing-edge flaps, the horizontal tailplane and corrections for the presence of the fuselage, powerplant installation, etc. The contribution of leading-edge devices will be dealt with separately in Section G-5. The lift curve may thus be expressed in the form of a lift increment relative to the en route configuration:

$$\Delta C_L = \Delta_f C_L + \sum \Delta C_L + \Delta_h C_L \qquad \text{(G-20)}$$

The direct contribution of the trailing-edge flap $\Delta_f C_L$ is dealt with in Section G-3.1., the various corrections $\sum \Delta C_L$ are discussed briefly in Section G-3.2., and the tailplane contribution $\Delta_h C_L$ is finally given in Section G-3.3.

G-3.1. Wing lift

In much the same way as was done for airfoil sections, the wing lift curve will be computed from the following characteristics:

$\Delta_f C_{L_o}$: the lift increment at zero angle of attack;

$\Delta_f C_{L_{max}}$: the increment in maximum lift coefficient;

C_{L_α}' : the lift-curve slope with flaps deflected.

A result similar to Fig. G-1 is found, except that c_ℓ is replaced by C_L.

The reader's attention is drawn to the fact that the maximum lift defined in this appendix refers to the top of the lift curve. As explained in Section 5.4.4., this figure cannot be used directly to determine the stalling speed (see also Ref. G-32).

a. Increment in lift for $\alpha = 0^\circ$.

Lifting surface theory can be employed to determine $\Delta_f C_{L_o}$ for moderate flap deflections. The method of Ref. G-64 relies on a knowledge of the section flap effectiveness and is summarized as follows:

$$\Delta_f C_{L_o} = \Delta_f c_{\ell_o}\left(\frac{C_{L_\alpha}}{c_{\ell_\alpha}}\right)\left[\frac{(\alpha_\delta) C_L}{(\alpha_\delta) c_\ell}\right] K_b \qquad \text{(G-21)}$$

$\Delta_f c_{\ell_o}$ is the section lift increment for

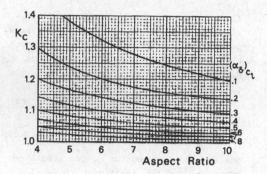

Fig. G-11. Variation of flap chord factor with aspect ratio (Ref. G-64)

$\alpha = 0$ due to flap deflection for a representative section, e.g. halfway along the semi-flap span. Experimental results should preferably be used, but when these are not available, the method of Section G-2.2. or any other suitable method may be used (e.g. Ref. G-19 or Ref. G-65). C_{L_α} and c_{ℓ_α} are the lift-curve slopes of the basic wing (see Section E-4.1.) and the basic airfoil (see Section E-3.2.), respectively, while $(\alpha_\delta)_{C_L}$ is the three-dimensional flap effectiveness parameter:

$$(\alpha_\delta)_{C_L} = \frac{C_{L_\delta}}{C_{L_\alpha}} \qquad (G-22)$$

and $(\alpha_\delta)_{c_\ell}$ is the two-dimensional flap effectiveness parameter:

$$(\alpha_\delta)_{c_\ell} = \frac{\Delta_f c_{\ell_0}}{c_{\ell_\alpha} \delta_f} = \eta_\delta \, \alpha_\delta \qquad (G-23)$$

The ratio $\left[(\alpha_\delta)_{C_L} / (\alpha_\delta)_{c_\ell}\right]$ can be obtained from Fig. G-11 as a function of $(\alpha_\delta)_{c_\ell}$ and the wing aspect ratio.

K_b is a flap span effectiveness factor, defined as follows:

$$K_b = \frac{\Delta_f C_{L_0} \text{ (partial span)}}{\Delta_f C_{L_0} \text{ (full span)}} \qquad (G-24)$$

The curve of K_b versus flap span and taper ratio, as derived from empirical data, is plotted in Fig. G-12. For flaps other than inboard the value of K_b is obtained by superposition of the flaps, as shown schematically in Fig. G-13 for the case of a flap interrupted by the fuselage.

In the case of inboard flaps which are augmented by ailerons acting as flaps ("flaperons"), the follow-

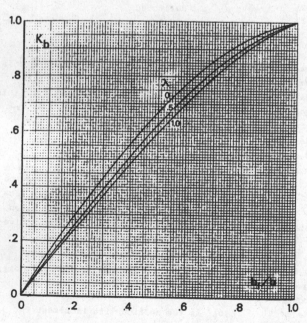

Fig. G-12. Variation of span factor K_b with flap span for inboard flaps (Ref. G-64)

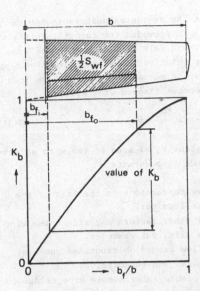

Fig. G-13. Flap span factor for flaps other than inboard flaps and definition of S_{wf}

ing procedure is proposed:

1. Estimate the $\Delta_1 c_{\ell_o}$ due to deflection of the aileron for a representative local section.

2. Compute $\Delta_1 C_{L_o}$, assuming that $\Delta_1 c_{\ell_o}$ acts full-span of the combined inboard flaps and the deflected ailerons.

3. Estimate the $\Delta_2 c_{\ell_o}$ due to inboard flap deflection for a representative inboard section.

4. Compute $\Delta_2 C_{L_o}$ due to $\Delta_2 c_{\ell_o} - \Delta_1 c_{\ell_o}$ over the inboard flap span with (G-21).

5. Add $\Delta_1 C_{L_o}$ and $\Delta_2 C_{L_o}$ to obtain $\Delta_f C_{L_o}$.

b. Maximum lift increment.

The maximum lift increment of a representative section is used for predicting $\Delta_f C_{L_{max}}$. Section data should preferably be derived from experimental data; if these are not available, the method explained in Section G-2.3. may be used.

The two-dimensional data for efficient slotted flap configurations* are converted into a three-dimensional lift increment as

*For plain flaps the effect of sweepback is much more pronounced, e.g. proportional to $\cos^3 \Lambda$ (Ref. G-22).

follows:

$$\Delta_f C_{L_{max}} = .92 \, \Delta_f c_{\ell_{max}} \frac{S_{wf}}{S} \cos\Lambda_{\frac{1}{4}} \qquad (G-25)$$

S_{wf}/S is the ratio of the wing area affected by the trailing-edge flaps to the total wing area. For straight-tapered wings we have:

$$\frac{S_{wf}}{S} = \frac{b_{f_o} - b_{f_i}}{b} \left\{ 1 + \frac{1-\lambda}{1+\lambda} \left(1 - \frac{b_{f_o} + b_{f_i}}{b} \right) \right\} \quad (G-26)$$

The nomenclature is defined in Figs. G-13 and G-14. For moderately tapered wings the numerical value of S_{wf}/S is approximately equal to the factor K_b defined previously. The factor .92 in (G-25) takes into account the loss of lift near the flap tips, as illustrated by Fig. G-14, where the shaded area represents the increment in maximum lift in the hypothetical case that all sections of the flapped part of the wing stall simultaneously.

c. Lift-curve slope.

The lift-curve slope is corrected for the effects referred to in Section G-2.4. a. above, as follows:

$$\frac{C_{L_\alpha}' \text{ (flaps down)}}{C_{L_\alpha} \text{ (flaps up)}} = 1 + \frac{\Delta_f C_{L_o}}{\Delta_f c_{\ell_o}} \times$$

$$\left| \frac{c'}{c} \left(1 - \frac{c_f}{c'} \sin^2 \delta_f \right) - 1 \right| \qquad (G-27)$$

where $\Delta_f C_{L_o} / \Delta_f c_{\ell_o}$ is the ratio of the three-dimensional to two-dimensional lift increments at $\alpha = 0$, given by (G-21).

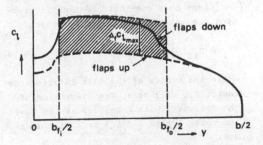

Fig. G-14. Lift distribution of a wing-fuselage combination with deflected flaps.

G-3.2. Various contributions

The corrections on the lift curve ($\sum \Delta C_L$) to be discussed will simply be expressed in the form of ΔC_L values, as the absence of reliable methods precludes systematic treatment of these effects on $\Delta_f C_{L_0}$, $\Delta_f C_{L_{max}}$ and C_{L_α}' cannot be presented in view. We shall therefore assume that the complete lift curve will be shifted up or down by the stated value of ΔC_L.

a. Wing/fuselage interference.

Some interference effects between the flows around the wing and the fuselage have been discussed in Section E-6 of Appendix E. Similar, but more complicated, flow phenomena will be observed on wing/flap/fuselage combinations. Analytical treatment of this effect being ruled out, the designer must rely on rules of thumb in the preliminary design stage.

The most conservative approach is to completely ignore the lift carry-over by the fuselage. Though this may be justified for large flap deflections, when there is a large gap between the fuselage and the flaps, the carry-over effect cannot be ignored altogether in all cases. Potential flow theory indicates that the lift carry-over by the fuselage is between one-half and two-thirds of the lift generated by the wing center section in the absence of the fuselage, assuming the flaps to be extended to the centerline. The experiments reported by Hoerner (Ref. G-10, page 8-18) indicate that in the case of wings without flaps only one-third remains in practice. We may therefore conclude that

$$\Delta C_L = K_{ff} \frac{2}{1+\lambda} \frac{b_{fi}}{b} \Delta_f c_{\ell_0} \qquad (G-30)$$

where the choice of the lift interference factor K_{ff} will be at the discretion of the designer, with 0 and 2/3 as the lower and upper limits and 1/3 as a good average.

b. Flap cutouts.

Flap cutouts are sometimes required to prevent the engine exhaust impinging on the flaps. Provided the cutout is not too wide, it is fair to assume that the loss in lift relative to the uninterrupted flap is about 50 percent of the relative loss in effective flapped wing area:

$$\Delta C_L = - .5 \frac{\Delta S_{wf}}{S} \Delta c_\ell \qquad (G-31)$$

ΔS_{wf} is the total area of the wing affected by the flap cutouts.

c. Other corrections to the lift of the aircraft less tail.

Several other factors may affect the wing maximum lift, for example:
- the flow around obstructions such as flap tracks and supports,
- unfavorable interference effects due to engine nacelles or engine intake scoops,
- losses incurred by devices needed to provide acceptable handling and control at the stall, and
- the effect of aeroelastic deformation on the lift, particularly on large aircraft (Ref. G-32).

Each of these may affect $C_{L_{max}}$ by as much as .1 to .2, but prediction of their magnitude is generally impossible in the preliminary design stage.

G-3.3. Contribution of the horizontal tailplane

The horizontal tail load required to trim out the nose-down pitching moment due to flap deflection reduces the lift. This contribution can be derived from semi-empirical data on pitching moment variations due to flap deflection. A relatively simple approach is thought to be acceptable for the purpose of determining the trim load.

a. Section pitching moment.

When experimental data are not available, the generalized expression quoted in Ref. G-5 may form a useful starting point:

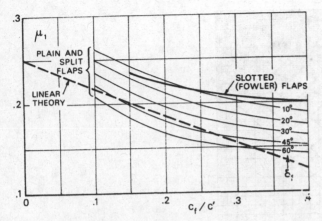

Fig. G-15. The pitching moment function μ_1

$$\Delta_f c_{m_{\frac{c}{4}}} = -\mu_1 \; \Delta_f c_{\ell}' \; \left(\frac{c'}{c}\right)^2 - \frac{c_{\ell}}{4} \; \frac{c'}{c}\left(\frac{c'}{c} - 1\right) +$$

$$+ \left(c_{m_{\frac{c}{4}}}\right)_{\delta_f = 0} \; \left[\left(\frac{c'}{c}\right)^2 - 1\right] \qquad (G-32)$$

The extended chord c' (cf. Fig. G-7) allows for the effect of the backward movement of the flap while it is being extended.

The first contribution in (G-32) is due to the increased section camber. The factor μ_1 is defined as follows:

$$\mu_1 = -\frac{\Delta_f c_m'}{\Delta_f c_{\ell}'} \qquad (G-33)$$

According to Glauert's linear theory for small flap deflections, we have

$$\mu_1 = \frac{1}{2} \left(1 - \frac{c_f}{c}\right) \frac{\sin \theta_f}{\pi - (\theta_f - \sin\theta_f)} \qquad (G-34)$$

where θ_f is defined by (G-4). Fig. G-15 shows that the theoretical value of μ_1 generally underpredicts the pitching moment coefficient. It has been found that for slotted flaps, with or without Fowler movement, most data are on a single line, provided the second term in (G-32), representing the theoretical rearward shift of the airfoil aerodynamic center, is halved. In the case of split and plain flaps the flap angle is observed to exert a pronounced influence on μ_1. The last term of (G-32), being generally of a low order,

can be ignored and the following practical equation is proposed:

$$\Delta_f c_{m_{\frac{c}{4}}} = -\mu_1 \; \Delta_f c_{\ell} \; \frac{c'}{c} - \frac{c_{\ell}}{8} \; \frac{c'}{c} \; \left(\frac{c'}{c} - 1\right) \qquad (G-35)$$

where μ_1 can be obtained from Fig. G-15.

b. Wing pitching moment.
References G-5 and G-60 convert the two-dimensional effect into a three-dimensional pitching moment change approximately in the following manner:

$$\Delta_f C_{m_{\frac{c}{4}}} = \mu_2 \; \Delta_f c_{m_{\frac{c}{4}}} + .7 \; \frac{A}{1 + 2/A} \; \mu_3 \; \Delta_f c_{\ell} \; \tan\Lambda_{\frac{c}{4}} \qquad (G-36)$$

where $\Delta_f c_{m_{\frac{c}{4}}}$ can be obtained from (G-35) by substituting:

$$c_{\ell} \approx C_L + \Delta_f c_{\ell} \left(1 - \frac{S_{wf}}{S}\right) \qquad (G-37)$$

The correction factor μ_2 (Fig. G-16) takes into account the part-span effect on a straight wing, while μ_3 is a function determining the effect of wing sweep (Fig. G-17).

c. Pitching moment, aircraft less tail.
Most measurements of flapped wing sections have been related to the .25-chord point or to the a.c. of the basic wing section. For this reason we take the aerodynamic pitching moment about the mean quarter chord point as:

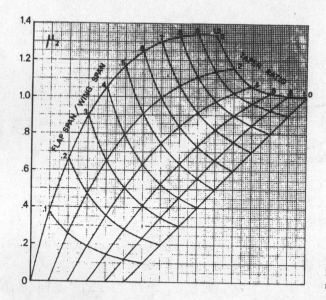

Fig. G-16. Correction factor for part-span effect on pitching moment (derived from Ref. G-5)

$$C_{m_{\frac{1}{4}}} = \left(C_{m_{\frac{1}{4}}}\right)_{\delta_f=o} + \Delta_f C_{m_{\frac{1}{4}}} + \sum \Delta C_{m_{\frac{1}{4}}} \qquad (G-38)$$

of Appendix E, replacing C_{L_o} by $C_{L_o} + \Delta_f C_{L_o}$. The second contribution may be approximated by ignoring

The pitching moment coefficient with flaps retracted can be obtained from (E-35) of Appendix E:

$$\left(C_{m_{\frac{1}{4}}}\right)_{\delta_f=o} = C_{m_{ac}} + \left(C_L - \Delta_f C_L\right)\left(.25 - \frac{x_{ac}}{\bar{c}}\right) \quad (G-39)$$

where the aerodynamic center refers to the aircraft less tail with flaps retracted (See Appendix E, Section E-7). The term $\Delta_f C_{m_{\frac{1}{4}}}$ in (G-38) can be calculated as a function of C_L using (G-36) and (G-37).

The term $\sum \Delta C_{m_{\frac{1}{4}}}$ is a formal statement for various other effects:

1. The effect of lift carry-over by the fuselage, relative to the theoretical scheme in which the flaps extend to the plane of symmetry.
2. The effect of flap deflection on fuselage forebody and afterbody contributions to the pitching moment.
3. Other effects, e.g. contributions of flap cutouts, nacelle interference, etc.

The first effect may be estimated by using (E-40)

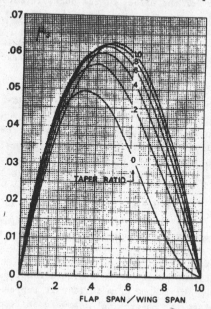

Fig. G-17. The function μ_3 for determining the pitching moment increment of flaps on sweptback wings. (Ref. G-5).

altogether the lift carry-over by the fuselage. In view of absence of data a verification for this approach cannot be given. It should be noted that the first two effects mentioned are of opposite sign.

d. Tailplane lift.

The tail load change required to trim the aircraft after flap deflection is, ignoring $\sum \Delta C_{m_{\frac{1}{4}}}$ in (G-38), approximately:

$$\Delta_h C_L = \Delta C_{L_h} \frac{S_h}{S} = \left\{ \Delta_f C_L \left(\frac{x_{cg} - x_{\frac{1}{4}}}{\bar{c}} \right) + \Delta_f C_{m_{\frac{1}{4}}} \right\} \frac{\bar{c}}{\ell_h}$$

(G-40)

where $\Delta_f C_L$ and $\Delta_f C_{m_{\frac{1}{4}}}$ can be computed by means of the methods presented previously.

G-4. PREDICTION OF THE LOW-SPEED DRAG POLAR

The drag of the aircraft with flaps down (undercarriage retracted) can be obtained from the drag polar in the clean configuration as follows:

$$C_D = C_{D_o} + \frac{C_L^2}{\pi A e} + \Delta_f C_{D_p} + \Delta_f C_{D_v} + \Delta_{trim} C_D \quad (G-41)$$

where $\Delta_f C_{D_p}$, $\Delta_f C_{D_v}$, and $\Delta_{trim} C_D$ are the increments in profile, vortex-induced, and trim drag coefficients. In (G-41) the basic polar is written as a simple parabolic approximation, but other relationships may be used as well. Its derivation is discussed in Appendix F. Although, in principle, all drag contributions discussed in that appendix are slightly altered when the flaps are deflected and the operational conditions are changed, we shall discuss only the most pertinent corrections required.

As illustrated by Fig. G-18, the approach adopted here is artificial in the sense that, for the large values of C_L to be achieved with deflected flaps, the basic drag coefficient of the aircraft (flaps up) is assumed to be extrapolated beyond

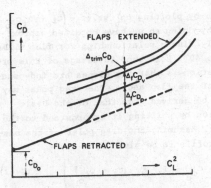

Fig. G-18. Drag polars with trailing-edge flaps retracted and extended

maximum lift values.

G-4.1. Profile drag

a. Drag of airfoil sections.

Fig. G-19a shows an example of drag polars

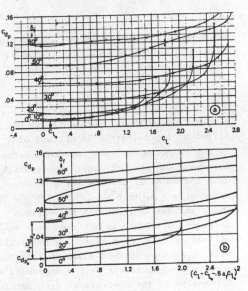

Fig. G-19. Section drag polars of NACA 23012 airfoil with a single slotted flap (Ref. G-33)

for several deflection angles of an airfoil section with a single slotted flap. These polars have been converted in Fig.

G-19b by plotting c_{d_p} vs. $\{c_\ell - (c_{\ell_0} + \frac{1}{2}\Delta_f c_{\ell_0})\}^2$ and by doing so we have obtained approximately linear relationships for flap angles up to 40°. A further advantage of this presentation is that the slopes are independent of the flap angle. The drag polar may thus be derived from that of the basic section by shifting it upwards and to the right. Assuming the drag polar of the basic profile to be given by:

$$c_{d_p} = c_{d_{p_0}} \left\{ 1 + k_\ell \left(c_\ell - c_{\ell_0} \right)^2 \right\} \qquad (G-42)$$

we obtain the polar with deflected flap:

$$c_{d_p} = c_{d_{p_0}} + \Delta_f c_{d_{p_0}} + c_{d_{p_0}} k_\ell \left\{ c_\ell - \left(c_{\ell_0} + \frac{1}{2}\Delta_f c_\ell \right) \right\}^2 \qquad (G-43)$$

Consequently, the drag increment due to flap deflection is:

$$\Delta_f c_{d_p} = \Delta_f c_{d_{p_0}} - c_{d_{p_0}} k_\ell \Delta_f c_\ell \left\{ c_\ell - \left(c_{\ell_0} + \frac{1}{2}\Delta_f c_{\ell_0} \right) \right\} \qquad (G-44)$$

where $\Delta_f c_{d_{p_0}}$ is the increment in the minimum profile drag coefficient. In (G-44) this increment will be defined for:

$$c_\ell = 1 + \frac{1}{2}\left(c_{\ell_0} + \Delta_f c_{\ell_0} \right)$$

a condition which is representative of flight at approximately 70 percent of $c_{L_{max}}$. It may be noted that most handbook methods are based on Ref. G-5, where the profile drag increment is defined for a constant value of $\alpha = \alpha_{\ell_0} + 6°$.
For practical applications the various terms in (G-44) may be obtained from experimental data. When these are not available, k_ℓ can be assumed equal to 1.0. The profile drag increment may be expressed as a fraction of the theoretical drag obtained, assuming that $\Delta_f c_\ell$ acts normal to the flap chord:

$$\Delta_f c_{d_{p_0}} = \text{constant} \frac{c_f}{c} \Delta_f c_\ell \sin\delta_f \qquad (G-45)$$

Taking $\Delta_f c_\ell$ in accordance with Glauert's linear thin airfoil theory and allowing for chord extension, a generalized result is:

$$\Delta_f c_{d_{p_0}} = k_d c_{\ell_\alpha} \alpha_\delta' \frac{c_f}{c} \delta_f \sin\delta_f + c_{d_{p_0}} \left(\frac{c'}{c} - 1 \right) \qquad (G-46)$$

The factor α_δ' is based on the ratio c_f/c', as explained in Section G-2.2. The factor k_d is plotted in Fig. G-20 for various flap

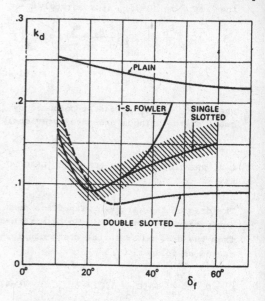

Note: the shaded area indicates a typical variation due to shroud and slot geometry differences for single slotted flaps.

Fig. G-20. The factor k_d for determining the increment in profile drag caused by flap deflection

configurations. This figure shows that for plain flaps the drag increment is of the order of 20 to 25 percent of the "theoretical" value; for single and double slotted flaps these figures are 10 to 15 and 8 to 10 percent, respectively. The relatively high value of k_d for slotted flaps at small deflection angles is caused by unfavorable slot flow and is affected to a large extent by the shape of the shroud.
Split flaps do not lend themselves to this approach. The following relationship has

been derived from systematic measurements in Ref. G-34:

$$\Delta_f c_{d_{p_o}} = .55 \frac{c_f}{c} \left[\frac{c_f/c}{(t/c)^{3/2}} \right]^{2/9} F(\delta) \quad (G-47)$$

where $F(\delta)$ is given in Fig. G-21. Alterna-

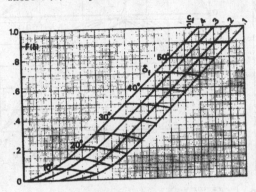

Fig. G-21. Profile drag function k_δ for split flaps (derived from Ref. G-34)

tively, the generalized and more detailed method in Ref. G-71 may be used.

b. Wing profile drag increment.
The profile drag increment of the wing due to flap deflection is obtained from:

$$\Delta_f C_{D_p} = K_{2-3} \frac{S_{wf}}{S} \Delta_f c_{d_{p_o}} \cos\Lambda_{\frac{1}{4}} - k_\ell \; c_{d_{p_o}} \Delta_f C_{L_o} \times$$
$$\left\{ C_L - (C_{L_o} + \tfrac{1}{2} \Delta_f C_{L_o}) \right\} \quad (G-48)$$

where $\Delta_f c_{d_{p_o}}$ is the two-dimensional drag increment, given by (G-46) or (G-47), and K_{2-3} is a correction factor allowing for wing/fuselage interference, sweep effects, suspension effects, etc. A figure of $K_{2-3} = 1.15$ is proposed, based on the case reported in Ref. G-24.

G-4.2. Vortex-induced drag

The change in vortex-induced drag is caused by a change in the spanwise lift distribution (see Fig. G-14). The following expression is a synthesis of several existing

methods in the literature (Refs. G-5, G-60 and G-65):

$$\Delta_f C_{D_v} = (w+z) \left(\Delta_f c_\ell \right)^2 + v \, C_L \, \Delta_f c_\ell \quad (G-49)$$

The term $w \left(\Delta_f c_\ell \right)^2$ represents the induced drag increment due to flaps extending to the wing centerline, for a wing with elliptic loading when the flaps are retracted. The variation of w with wing geometry is shown in Fig. G-22b. The last term is a correction for the nonellipticity of the spanwise lift distribution of the basic wing. The variation of v with wing geometry is shown in Fig. G-22a.
It will be noted that for certain taper ratios the v and w factors are of opposite sign and their contributions counteract each other. In fact, for nontapered wings the elliptic loading may be approached when the flaps are deflected and the induced drag increment may be small or even negative. The term $z \left(\Delta_f c_\ell \right)^2$ in (G-49) takes into account the effect of the flap cutout and lift carry-over by the fuselage (c.f. Section G-3.2.). The simple drag increment given by Hoerner on page 8-18 of Ref. G-10 is found to be in fair agreement with other data (Refs. G-5 and G-60) and can be converted into:

$$z = \frac{.07}{1+\lambda} \left(1 - K_{ff} \right)^2 \frac{b_{f_i}}{b} \quad \left(\text{for } \frac{b_{f_i}}{b} < .2 \right) (G-50)$$

where b_{f_i} is defined in Fig. G-13 and the choice of K_{ff} was discussed in Section G-3.2. An average value $K_{ff} = 1/3$ appears to be acceptable, provided the flap edge is as close to the fuselage as possible.

G-4.3. Trim drag

On condition that the airplane vortex-induced drag is calculated by assuming that all lift is provided by the aircraft less tail, the trim drag can be taken into account as follows, ignoring a second-order term:

$$\Delta_{trim}(C_D S) = \left(\frac{C_{L_h}^2}{\pi A_h e_h} + C_{L_h} \sin\epsilon_h - 2 \frac{C_L}{\pi A} C_{L_h} \right) S_h$$

$$(G-51)$$

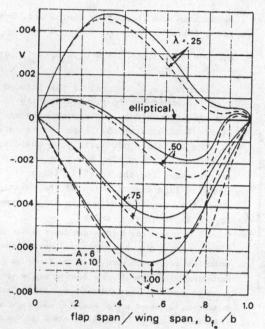

a. Factor v

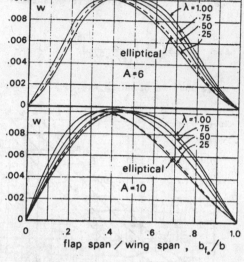

b. Factor w

Fig. G-22. Factors of induced drag of a wing with flaps (Ref. G-60)

The first contribution to trim drag is the basic vortex-induced drag due to tailplane lift, based on free-stream conditions. The second contribution is the component of the tailplane lift in the direction of the free-stream as a consequence of the local downwash. The last term corrects the vortex-induced drag of the wing to account for the tailplane lift.

The tailplane lift coefficient is defined by the condition that the aircraft will be trimmed, while the downwash angle is given by:

$$\varepsilon_h = \left(\frac{\partial \varepsilon_h}{\partial C_L}\right)_{\delta_f} C_L + \left(\frac{\partial \varepsilon_h}{\partial \delta_f}\right)_{C_L} \delta_f \qquad (G-52)$$

The estimation of the partial derivatives in this equation is usually very difficult in the preliminary design stage and for this reason we will neglect the net effect of the second and third contribution in

(G-51)*. The only term remaining can be evolved with (G-40) into an approximate trim drag coefficient (based on S) relative to the basic polar:

$$\Delta_{trim}C_D = \frac{\Delta_f C_{m_{\frac{1}{4}}} \left(\Delta_f C_{m_{\frac{1}{4}}} + 2 C_{m_{ac}}\right)}{\pi A_h e_h \ (\ell_h/\bar{c})^2 \ S_h/S} \qquad (G-53)$$

where $\Delta_f C_{m_{\frac{1}{4}}}$ is defined by (G-36) and $C_{m_{ac}}$ refers to the aircraft less tail with flaps up (Appendix E, Section E-7.2.). The Oswald factor e_h is approximated by:

$$e_h = 1 - \frac{.25}{\cos^2 \Lambda_h} \qquad (G-54)$$

and takes into account the profile drag increment due to elevator deflection.** The

*Note that for an elliptic lift distribution ε_h at infinity behind the wing is equal to $2C_L/\pi A$ and the two contributions cancel each other out.

**Cf. Appendix F, Section F-3.6.

simplified calculation procedure presented
here may in some cases lead to considerable errors. A more detailed prediction is
required as soon as more accurate data become available, particularly with respect
to the downwash at the tail.

G-5. LEADING-EDGE HIGH-LIFT DEVICES

Leading-edge flaps and slats* level off the
high peak suction pressures near the airfoil nose, thereby delaying leading edge
separation to a higher angle of attack
(Fig. G-23). Since leading-edge flaps do

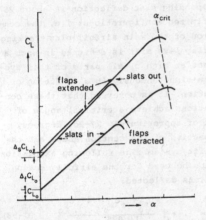

Fig. G-23. The lift curve with and without
slats

not significantly alter the trailing edge
flow about an unflapped section, they are
most effective on thin airfoil sections.
Airfoils with a deflected, high-efficiency
trailing-edge flap will frequently exhibit
a leading edge stall, even on relatively
thick sections. Leading-edge slats or flaps
suppress this stall, thereby increasing the
stalling angle of attack and hence maximum
lift. However, a slat wake may cause undesirable interference with the flow around
trailing-edge flaps and the increment in
$c_{\ell_{max}}$ due to the slat with trailing-edge

*The terminology is explained in Section
7.6.3.

flaps deflected is typically some 15 percent lower than the $c_{\ell_{max}}$ increment of the
unflapped airfoil.

Reliable generalized methods for predicting the effect of leading-edge devices are
not known to the author, with the exception of the (maximum) lift increment due
to plain leading-edge flaps (Ref. G-19).
The following data, indicating an order of
magnitude of the effect of leading-edge
devices on lift and drag, are the author's
provisional conclusions from the very
scanty literature on the subject. In view
of the many factors involved in the aerodynamic design of leading-edge devices
these data must only be considered indica-
tive.

G-5.1. Sections with plain leading-edge flaps

The leading-edge droop causes a lift loss
at zero angle of attack which can be derived from Glauert's linear thin airfoil
theory:

$$\Delta_s c_{\ell_o} = - \frac{\theta_s - \sin\theta_s}{\pi} \delta_s c_{\ell_\alpha} \qquad (G-55)$$

where

$$\theta_s = \cos^{-1} \left(1 - 2 \frac{c_s}{c}\right) \qquad (G-56)$$

The increment in $c_{\ell_{max}}$ may be estimated on
the assumption that this increment is entirely due to the change in the ideal angle of attack. The ideal angle of attack
is the angle at which the flow comes
smoothly onto the airfoil, causing no sin-
gularity at the nose. Using the data from
Ref. G-19, we may approximate the result
by:

$$\frac{\Delta\alpha_{crit}}{\delta_s} = .58 \sqrt{\frac{c_s}{c}} \qquad (G-57)$$

Experimental data are in fair agreement
with this expression for δ_s up to 20 to 25
degrees. Larger deflection angles cause
only a slight increase in α_{crit} up to δ_s =
30 degrees, further deflection resulting in

a decreasing α_{crit}.
Leading-edge flap deflection has no effect
on c_{ℓ_α}, provided the angle of attack is
positive and the flap is deflected not
much more than 20 degrees. Pronounced non-
linearities in the lift curve will be observed for negative and small positive angles of attack when the flap is deflected
more than 25 degrees.

G-5.2. Sections with slats and Krueger flaps

As in the case of the plain leading-edge
flap, extension of a slat results in a
slight decrement in lift at zero angle of
attack, caused by the nose droop. However,
the slat increases the effective chord
length and this causes a lift increment
when c_ℓ is based on the original chord.
Both effects are generally of the same order of magnitude and it is therefore reasonable to assume that $\Delta_s c_{\ell_o} = 0$, while
the lift gradient is not affected to any
considerable degree by slat deflection.
The estimation of $c_{\ell_{max}}$ with a deflected
slat is complicated by a number of factors:

1. For a given configuration and deflection of the trailing-edge flap system, the
slat will be more effective in terms of
$\Delta_s c_{\ell_{max}}$ on an airfoil with a sharp nose
than on a well-rounded airfoil nose. For
an optimized slat configuration the stall
may be of the trailing-edge type whereas
without a slat it is of the leading-edge
type.
2. The slat position relative to the airfoil (slat deflection and gap between the
airfoil and the slat) has a very pronounced effect on the lift and profile
drag increment. An elucidating discussion
can be found in Ref. G-29.
3. The design of the trailing-edge flap
system and the slat should be matched in
order to obtain the highest performance.
4. Compressibility effects may set an upper limit to the performance of complex
high-lift systems. In fact, Callaghan suggests in Ref. G-31 that the prediction of

maximum lift increment due to slats should
be based on the condition that sonic flow
sets a limit to the pressure peaks.
Refs. G-27 and G-29 demonstrate that the
details of a slat design optimization
scheme are of paramount importance: for a
particular configuration $\Delta_s c_{\ell_{max}}$ due to
the slat may vary between .5 and 1.2. We
therefore propose to differentiate between
simple (nonoptimized) high-lift systems
and fully optimized configurations.
a. Simple slat configurations: the maximum
lift increment is about 2.2 $\sqrt{c_s/c}$ with
trailing-edge flaps retracted and 1.9
$\sqrt{c_s/c}$ with trailing-edge flaps out. The
corresponding slat deflection is about 25°.
b. Optimized configurations: i.e. the combination of the main airfoil plus trailing-
edge flap plus slat is designed in such a
way that at high c_ℓ all parts of this system are simultaneously in a condition of
separation. It is observed that these configurations achieve a critical angle of
attack of approximately $27-2c_\ell$ degrees,
irrespective of the section shape and the
high-lift system. The following expression
may thus be used for the airfoil section
with slats deflected:

$$c_{\ell_{max}} = (1-k_s)\,\frac{c_{\ell_o}+\Delta_f c_{\ell_o}+.47\,c_{\ell_\alpha}}{1+.035\,c_{\ell_\alpha}} \qquad (G-58)$$

where c_{ℓ_α} is in rad^{-1} and k_s is a factor
which takes into account the nonlinearity
of the lift curve at high lift. It is generally between .03 and .15, with $k_s = .07$
as a good average. Factors c_{ℓ_o} and c_{ℓ_α} in
(G-58) are the lift at zero angle of at-
tack and the lift-curve slope for the airfoil with or without flaps deflected, to
be obtained from experimental data or from
Sections G-2.2. and G-2.4. of this appendix.

The maximum lift capability of Krueger
flaps is very similar to that of slats. In
view of the pitching moment behavior at
large lift coefficients Krueger flaps are
generally used inboard only on swept wings,
while slats may be used both full span and

outboard.

G-5.3. Wing lift with leading-edge devices

For plain leading-edge flaps the increment
in the critical angle of attack given by
(G-57) can be converted into a maximum
lift increment:

$$\Delta_s C_{L_{max}} = .58 \sqrt{\frac{c_s}{c}}\, \delta_s\, c_{\ell_\alpha}\, \frac{S_{ws}}{S}\, \cos^2\Lambda_{\frac{1}{4}}$$

(G-59)

where S_{ws} is the projected wing area e-
quipped with slats.
For full-span optimum slat configurations
it was found that (G-58) can be used in a
slightly modified form:

$$C_{L_{max}} = .93\, \frac{C_{L_o} + \Delta_f C_{L_o} + .47\, C_{L_\alpha} \cos\Lambda_{\frac{1}{4}}}{1 + .035\, C_{L_\alpha} \cos\Lambda_{\frac{1}{4}}} \quad (G-60)$$

In this expression C_{L_o} is defined at $\alpha=0$
for the wing.
For part-span slats the $C_{L_{max}}$ increment is
given by:

$$\Delta_s C_{L_{max}} = \Delta_s c_{\ell_{max}}\, \frac{S_{ws}}{S}\, \cos^2\Lambda_{\frac{1}{4}} \quad (G-61)$$

where $\Delta_s c_{\ell_{max}}$ is obtained by subtracting
the maximum lift coefficient for the basic
airfoil (slats in) from that with slats
out. The sweep effect in (G-61) is an approximation of the curve given in Ref.
G-31.

G-5.4. Drag due to leading-edge devices

Leading-edge devices have little influence
on the profile camber and when deflected
they cause negligible aerodynamic twist.
Their effect on the vortex-induced drag
may therefore be neglected.

An example of profile drag polars of an
airfoil with and without slat is depicted
in Fig. G-24, which shows that
a. the drag increment at 70 percent of
$c_{\ell_{max}}$ is approximately 70 counts;
b. when the polar with slat out is based

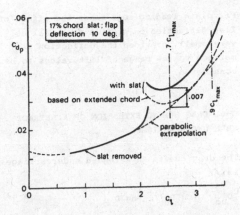

Fig. G-24. Drag polars of an airfoil with
single slotted flap, with and without flap
(Ref. G-53)

on the extended chord, there is virtually
no drag increase between 70 and 90 percent
of $c_{\ell_{max}}$;
c. the profile drag increases sharply at
lower lift coefficients, probably due to
flow separation on the lower surface of
the slat.
The slat has a large influence on the airfoil pressure distribution, resulting in
decreased flow velocities near the leading
edge of the basic airfoil and consequently
a profile drag reduction. On the other
hand, the slat itself not only increases
the wetted area, but it carries a relatively high load and the local friction
will be high, due to the high flow velocities. Slat gap and deflection angle are
major parameters contributing to these
flow phenomena.
The following provisional rule of thumb
may be suggested on the basis of Fig. G-24:

$$\Delta_s C_{D_p} = \left(C_{D_p}\right)_{basic}\, \frac{S_{ws}}{S}\, \frac{c_s}{c}\, \cos\Lambda_{\frac{1}{4}} \quad (G-62)$$

where $\left(C_{D_p}\right)_{basic}$ is the profile drag of
the airfoil with slats and trailing edge
flaps retracted. The reservation must be
made that (G-62) is valid only in the
range of lift values where the slat position is optimum.

For plain leading-edge flaps the effect of flap deflection on profile drag will be very small, provided the deflection is matched to the range of lift values to be considered.

G-6. DRAG DUE TO EXTENSION OF A RETRACTABLE UNDERCARRIAGE

The drag due to an extended undercarriage is:

$$C_{D_{uc}} = \Delta_{mg} C_{D_{uc}} + \Delta_{ng} C_{D_{uc}} \qquad (G-63)$$

where Δ_{mg} and Δ_{ng} denote the contributions of the main and nose gear, respectively. The undercarriage drag is determined not only by its drag area, but also by the local flow conditions. Consequently, the angle of attack (or lift coefficient) and the position of the high-lift devices have to be taken into account:

$$C_{D_{uc}} = F_{uc}(\alpha, \delta_f) \left(C_{D_{uc}} \right)_{basic} \qquad (G-64)$$

The following methods for estimating the basic undercarriage drag coefficient are proposed.

METHOD 1:

$$\left(C_{D_{uc/basic}} \right) = \frac{1.5 \sum S_{ft} + .75 \sum S_{rt}}{S} \qquad (G-65)$$

where in the case of non-bogie type gears ΣS_{ft} denotes the total frontal area of all tires, and in the case of bogie-type gears ΣS_{ft} is the frontal area of the front tires only. ΣS_{rt} is the total frontal area of the rear tires in the case of bogie-type gears and is equal to zero for non-bogie type gears.

Equation G-65 represents the total basic undercarriage drag coefficient and not just the tire drag coefficient.

METHOD 2:

When the tire size and the undercarriage configuration are not (yet) known, a purely statistical expression can be used:

$$\left(C_{D_{uc}} \right)_{basic} = constant \frac{W_{to}^{.785}}{S} \qquad (G-66)$$

where the constant is equal to 4.05×10^{-3} when W_{to} is in lb and S in sq.ft, or 7×10^{-4} when W_{to} is in kg and S in m^2.

Factor $F_{uc}(\alpha, \delta_f)$ depends on various factors.

a. The nose gear drag will not be affected by α or δ_f to any appreciable extent, nor will it affect the airframe drag.

b. The main gear is placed below the wing and will thus be sensitive to the local flow conditions.

c. For a given aircraft general arrangement and geometry, the induced effects of wing and flap lift reduce the undercarriage drag.

d. For given flight conditions (incidence and flap angle), the local position of the undercarriage relative to the wing, the wing thickness and camber, and the spanwise lift distribution all affect the undercarriage drag.

e. The undercarriage will have an effect on wing lift and hence on the vortex-induced drag. In the clean configuration the circulation will be slightly increased by the blocking effect of the airflow due to the main gear. When the trailing-edge flaps are deflected, the wake generated by the undercarriage may hit the flaps and in that case the lift will be reduced.

f. Part of the undercarriage of propeller aircraft will be in the slipstream and experiences an effective drag increment, which will depend upon the flight condition.

A theoretical analysis of the function $F_{uc}(\alpha, \delta_f)$ cannot (and need not) be made in the preliminary design stage. Instead, a reasonable approximation may be used:

$$F_{uc}(\alpha, \delta_f) = \left\{ 1 - .04 \frac{C_L + \Delta_f C_{L_o} (1.5 S / S_{wf} - 1)}{\ell_{uc}/c_g} \right\}^2 \qquad (G-67)$$

where ℓ_{uc} is the length of the main gear legs, i.e. the distance between the local wing chord and the wheel axis. Equation G-67 is based on Helmholtz' theorem, applied to the lift of the basic wing section and the flap separately.

It should be noted that retraction of the undercarriage during takeoff will initially cause the drag to increase some 20 to 30 percent, mainly due to the extension of the wheelbay doors and the drag of the wheelbays.

G-7. GROUND EFFECTS

The basic nature of the changes in the flow field due to ground effect can be analyzed with simple mathematical models, using the Prandtl lifting-line concept. Wieselsberger has derived some of the results given below (Ref. G-74) in a slightly modified form, which are still satisfactory for present-day conventional wing shapes, provided the distance to the ground is not too small and the lift coefficient is not too high. Wieselsberger's results can be applied readily in preliminary design to wings with flaps retracted, but unfortunately the results are not very good for the condition with flaps deflected at high lift. Some modifications were found to be necessary.

G-7.1. Ground effect on lift

a. In two-dimensional analysis the image vortex of the airfoil below the ground plane induces a velocity distribution at the airfoil in the opposite direction to the free-stream velocity, thus reducing the lift. In addition, the camber and incidence of the airfoil are effectively increased. For low lift coefficients and moderate wing height these opposing effects are approximately equal in magnitude, and the ground effect on lift can be calculated with sufficient accuracy by ignoring them. At low wing heights and high lift coefficients, however, particularly with flaps deflected, the induced horizontal velocity is dominant, causing a decrease in airfoil lift from its free-stream value.

When the airfoil section is sufficiently high above the ground, it can be replaced by a single vortex and the velocity induced by the image vortex can be derived from Helmholtz' law:

$$\frac{\Delta V}{V_\infty} = - \frac{c_\ell}{8\pi \ h/c} \qquad (G-68)$$

where h is the height of the airfoil a.c. above the ground plane. Provided $\Delta V/V_\infty \ll 1$, the reduction in effective velocity must be regained by increasing the angle of attack in order to maintain c_ℓ constant:

$$\Delta\alpha = \frac{c_\ell^2}{4\pi \ h/c \ c_{\ell_\alpha}} \qquad (G-69)$$

The effective increase in camber is proportional to $\Delta V/V_\infty$ and the chord length, and inversely proportional to the a.c. height. Assuming that the average upwash is equal to the upwash at the mid-chord point, we have for constant c_ℓ:

$$\Delta\alpha = - \ .25 \ \frac{\Delta V}{V_\infty} \frac{c}{2h} = - \frac{c_\ell}{64\pi \ (h/c)^2} \qquad (G-70)$$

For finite wings the effects described above are less due to the finite length of the bound vortex. The correction factor β proposed in Ref. G-95 takes this into account. Thus we find for the total effect due to the bound vortex:

$$\Delta_1\alpha = \beta \ \frac{C_L}{4\pi \ h/c_g} \left(\frac{C_L}{C_{L_\alpha}} - \frac{1}{16 \ h/c_g} \right) \qquad (G-71)$$

For several heights of the trailing edge above the ground, $\Delta_1\alpha$ is shown in Fig. G-25a, and β is given in Fig. G-26.

b. The induced upwash due to the images of the trailing vortices was first analyzed by Wieselsberger. It is usually expressed as a reduction of the angle of attack required to achieve a given C_L:

$$\Delta_2\alpha = - \ \sigma \ \alpha_i \qquad (G-72)$$

According to the classical lifting line theory for high aspect ratio straight wings with elliptic lift distribution:

$$\alpha_i = \frac{C_L}{\pi A}$$

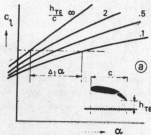

a. Airfoil section with flap ; effect of trailing edge height on the lift curve

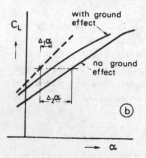

b. Ground effect on the lift curve of a wing

Fig. G-25. Ground effect on airfoil lift

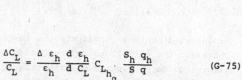

c. Effect of wing height and lift coefficient on ground effect

and hence

$$\Delta_2 \alpha = - \frac{\sigma \, C_L}{\pi A} \qquad (G-73)$$

A more generally valid expression for swept wings of arbitrary aspect ratio is:

$$\Delta_2 \alpha = -\sigma \, C_L \left(\frac{1}{C_{L_\alpha}} - \frac{1}{c_{\ell_\alpha}} \right) \qquad (G-74)$$

The ground effect factor σ can be obtained from Fig. G-26.
The combined effects of $\Delta_1 \alpha$ en $\Delta_2 \alpha$ on the lift-curve of a wing are shown in Fig. 25b.

c. The image vortices of the wing trailing vortex system induce an upwash at the horizontal tailplane. For a given angle of attack and tailplane setting the associated lift increment is:

$$\frac{\Delta C_L}{C_L} = \frac{\Delta \, \varepsilon_h}{\varepsilon_h} \frac{d \, \varepsilon_h}{d \, C_L} C_{L_{h_\alpha}} \frac{S_h}{S} \frac{q_h}{q} \qquad (G-75)$$

The position of the tailplane relative to the wing is of paramount importance. Methods for computing $\Delta \, \varepsilon_h / \varepsilon_h$ and $d \, \varepsilon_h / d \, C_L$ can be found in Refs. G-19 and G-95.
The induced tailplane lift must be trimmed out by elevator deflection. Moreover, ground effect causes a shift in the wing center of pressure and this effect must also be counteracted by elevator deflection. A secondary effect is the induced downwash or upwash at the horizontal tail due to the image vortices of the tailplane itself.
The various effects of ground proximity on tailplane lift are difficult to calculate and due to their relatively small magni-

tude they are frequently neglected.

d. The pressure distribution around an airfoil changes considerably when it is placed near the ground, in particular when effective trailing-edge flaps are deflected. A change occurs in the spanwise load distribution, leading to an increased root-stalling tendency on swept wings. The induced adverse pressure gradient at the leading edge may result in early separation, while blocking or even reversal of the flow can occasionally be observed below the wing. As a rule a reduction in $C_{L_{max}}$ will be observed. This effect must not be ignored, particularly for STOL configurations (Ref. G-96), and an assessment by means of wind tunnel experiments must be made in the early stages of aerodynamic development. The effect has an important bearing on the minimum unstick speed V_{MU} (see Appendix K, Section K-2).

The total effect on lift may be found by adding the contributions discussed in a. and b. above. Defining the lift coefficient at a given incidence out of ground effect as C_{L_∞}, and substituting the expression for C_{L_α} given in Section E-4.1. of Appendix E, we find the following approximation for the lift in ground effect at the same angle of attack:

$$\frac{C_L}{C_{L_\infty}} = 1+\sigma - \frac{\sigma A \cos\Lambda_{\frac{1}{2}}}{2\cos\Lambda_{\frac{1}{2}}+\sqrt{A^2+(2\cos\Lambda_{\frac{1}{2}})^2}} - \frac{\beta}{4\pi\, h/c_g} \times$$
$$\left(C_{L_\infty} - \frac{C_{L_\alpha}}{16\, h/c_g}\right) \qquad (G-76)$$

This expression has been plotted in Fig. G-25c for a straight aspect ratio 7 wing, as a function of wing height for several lift coefficients. The figure shows that for lift coefficients up to two the effect of the induced upwash and camber dominate, while for higher values the effective lift is reduced due to the decrement in effective flow velocity.

G-7.2. Ground effect on drag

The aerodynamic phenomena referred to in

the previous section result in the following expression for the vortex-induced drag in ground effect of the wing:

$$C_{D_V} = \frac{1-\sigma}{1 - \frac{\beta\, C_L}{4\pi\, h/c_g}} \left(C_{D_V}\right)_\infty \qquad (G-77)$$

where $\left(C_{D_V}\right)_\infty$ denotes the vortex-induced drag in the absence of ground effect. There is also a reduction in the profile drag caused by the reduced effective flow velocity. Assuming that only the wing profile drag is affected, the following expression can be derived for the total drag reduction due to ground effect:

$$\Delta C_D = - \frac{4\pi\,\sigma\, h/c_g - \beta\, C_L}{4\pi\, h/c_g - \beta\, C_L} \left(C_{D_V}\right)_\infty - \frac{\beta\, C_L}{4\pi\, h/c_g} \left(C_{D_{p/w}}\right)$$
$$(G-78)$$

where the vortex-induced and profile drag coefficients of the wing (with or without flaps extended) may be obtained from Sections G-4.1. and G-4.2., respectively. The functions σ and β are plotted in Fig. G-26. The effect of ground proximity on trim drag can be calculated in principle, but the amount of work involved is probably not justified in view of the unavoidable inaccuracy of the calculation.

G-8. DRAG DUE TO ENGINE FAILURE

The drag increment in steady flight following engine failure is composed of engine windmilling drag, propeller drag, and drag due to the asymmetric flight condition:

$$\Delta C_D = \Delta C_{D_{ewm}} + \Delta C_{D_{prop}} + \Delta C_{D_{as}} \qquad (G-79)$$

G-8.1. Engine windmilling drag

The drag of a windmilling gas-turbine engine is composed of external drag due to spillage of the inlet and internal drag associated with pressure losses in the flow through the windmilling engine. Very little is known about external drag;

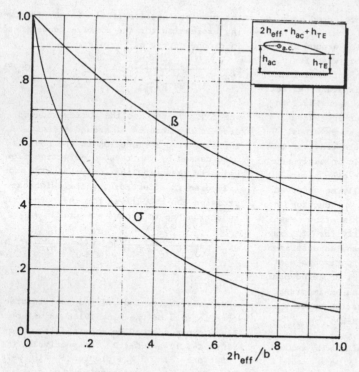

$$\sigma = \exp\left\{-2.48\ (2\ h_{eff}/b)^{.768}\right\}$$

$$\beta = \sqrt{1+(2\ h_{eff}/b)^2} - 2\ h_{eff}/b$$

Fig. G-26. The ground effect functions σ and β (Ref. G-95)

a reasonable figure is probably:

$$C_D S = .1 \times \frac{\pi}{4}\ D_i^2 \qquad (G-80)$$

where D_i is the engine inlet diameter. The internal drag can be obtained from the momentum theorem:

$$D = \dot{m}(V - V_N) \qquad (G-81)$$

where V_N is the mean flow velocity in the nozzle exit and $\dot{m}$ the windmilling mass flow. Equation G-81 may be further expanded by assuming that the nozzle exit temperature is equal to the static temperature plus 80 percent of the temperature rise due to the stagnation effect. Taking the static pressure of the exit flow equal to the ambient pressure, we obtain for the total drag area due to a windmilling engine:

$$\Delta (C_D S)_{wm} = .0785 D_i^2 + \frac{2}{1+.16M^2}\ A_N\ \frac{V_N}{V}\ (1 - \frac{V_N}{V})$$

$$(G-82)$$

Typical values for V_N/V are:
straight turbojet and turboprop engines: .25
low bypass ratio engines, mixed flow: .42
primary airflow of high bypass engines: .12
fan airflow of high bypass engines: .92
For bypass engines with separate nozzles for the hot and cold flow, the contributions of both flows to the internal drag must be added.

G-8.2. Propeller drag

A detailed procedure for calculating propeller windmilling drag can be found in Ref. G-100. For a feathered, stopped propeller the following approximation can be derived from Ref. G-98:

$$\Delta (C_D S)_{prop} = .00125\ B_p\ D_p^2 \qquad (G-83)$$

554

where B_p is the number of blades per propeller and D_p the propeller diameter.

G-8.3. Drag due to the asymmetric flight condition

Engine failure causes yawing and rolling moments when, as is usual, the inoperative engine is located outside the XOZ plane of the aircraft. These moments must be trimmed out by rudder and aileron/spoiler deflection. The following drag contributions result from this asymmetric condition:

a. induced drag due to the normal force on the vertical tailplane,
b. profile drag increment due to rudder and aileron (spoiler) deflection,
c. airframe drag due to sideslip,
d. vortex-induced drag due to the change in wing lift distribution, associated with sideslip and the asymmetric slipstream behind propellers, and
e. a change in the profile drag of the parts of the airplane immersed in the propeller slipstream.

The pilot may choose the angle of bank or sideslip after engine failure, within certain limits.* The various drag components are depicted in Fig. 48 of Ref. G-31 as a function of the sideslip angle. It is generally found that the total asymmetric drag after failure of a starboard engine is minimum for a small negative slip angle, i.e. the airplane sideslips in the direction of the operative engine. However, there is very little drag increment in a flight without sideslip. In that condition the most important drag contribution is caused by the vertical tailplane. The lift-induced drag of the vertical tailplane for zero slip is given by:

$$\Delta(C_DS) \approx \frac{C_{y_v}^2}{\pi A_{v_{eff}}} S_v + C_{y_v} S_v \sin \sigma_v \qquad (G-84)$$

where

*Cf. Section 9.6.1.

$$C_{y_v} = \frac{\Delta T}{q_v S_v} \frac{y_e}{\ell_v} \qquad (G-85)$$

and $A_{v_{eff}}$ is the effective (aerodynamic) aspect ratio of the vertical tailplane (cf. Fig. 2-26).

σ_v is the mean sidewash angle at the vertical tailplane associated with the asymmetric downwash behind the wing due to the asymmetric lift distribution.

ΔT is the net thrust loss of the failed engine plus the engine windmilling and propeller drag.

y_e is the yawing moment arm of the failed engine. Note that in the case of rear-mounted engines this distance can be reduced to a certain extent by suitable choice of the direction of the engine exhaust flow (cf. Fig. 6-18).

The profile drag increment due to rudder deflection may be estimated by using the generalized data on plain flap effectiveness and drag increment presented in the relevant sections of this Appendix and Appendix E. The result is approximated by:

$$\Delta(C_DS) = \frac{2.3}{\pi} \sqrt{S_r S_v} \left(A_{v_{eff}}\right)^{-4/3} (\cos\Lambda_v)^{1/3} \times$$
$$C_{y_v}^2 \qquad (G-86)$$

The drag area due to engine failure is found by adding (G-84) and (G-86) and substituting (G-85). The result can be written in the following form:

$$\Delta(C_DS)_{as} = K_{as} \left| \frac{\text{thrust per engine}}{q} + \Delta(C_DS)_{wm} + \Delta(C_DS)_{prop} \right|^2 + K_\sigma \frac{\text{thrust per engine}}{W} \qquad (G-87)$$

where

$$K_{as} = \left(\frac{y_e}{\ell_v}\right)^2 \frac{1}{S_v} \frac{1}{\pi A_{v_{eff}}} \left| 1 + 2.3 \sqrt{\frac{S_r}{S_v}} \right.$$
$$\left. \left(A_{v_{eff}} / \cos\Lambda_v\right)^{-1/3} \right| \qquad (G-88)$$

$$K_\sigma = \frac{y_e}{\ell_v} \frac{d\sigma_v}{dC_L} S \qquad (G-89)$$

For jet aircraft it is reasonable to as-

sume $d\sigma_v/dC_L = 0$, hence $K_\sigma = 0$. For propeller aircraft the order of magnitude of
$d\sigma_v/dC_L$ is such that the associated asymmetric drag accounts for a very large contribution. The author does not have at his
disposal any data for estimating $d\sigma_v/dC_L$,
although Ref. G-101 presents a qualitative
assessment of the effect. It is found that
in some cases, particularly on high-wing

aircraft, the effect of σ_v is equivalent
to an increase of 100 to 200 percent in the
yawing moment, resulting in a very large
increase in asymmetric drag. It is therefore recommended that the expression presented above be substantiated at the earliest possible moment by appropriate wind-
tunnel experiments.

REFERENCES

General literature, high-lift devices

G-1. H.A. Wilson and L.K. Loftin: "Landing characteristics of high-speed wings". NACA Res. Mem. L8A28e, 1948.

G-2. R. Duddy: "High lift devices and their uses". J. Royal Aero. Soc., Oct. 1949, Vol. 53, pp. 859-900.

G-3. J.F. Cahill: "Summary of section data on trailing-edge high-lift devices". NACA TR 938, 1949.

G-4. F.E. Weick, L.E. Flanagan Jr. and H.H. Cherry: "An analytical investigation of effects of high-lift flaps on takeoff of light airplanes". NACA TN 2404, 1951.

G-5. A.D. Young: "The aerodynamic characteristics of flaps". ARC R and M 2622, 1953.

G-6. A.D. Young: "Flaps for landing and takeoff". Chapter 14 of: "The principles of the control and stability of aircraft", edited by W.J. Duncan. Cambridge Aeronautical Series.

G-7. G.H. Lee: "High Maximum Lift". The aeroplane, October 30, 1953.

G-8. T.R.F. Nonweiler: "Flaps, slots and other high-lift aids". Aircraft Eng., Sept. 1955.

G-9. G. Chester Furlong and J.G. McHugh: "A summary and analysis of the low-speed longitudinal characteristics of swept wings at high Reynolds number". NACA Report 1339, 1957.

G-10. S.F. Hoerner: "Fluid dynamic drag". Published by the author, 1958.

G-11. I.H. Abbott and A.E. von Doenhoff: "Theory of wing sections". Dover Publications Inc., New York.

G-12. J. Williams and S.F.J. Butler: "Aerodynamic aspects of boundary layer control for high lift at low speeds". AGARD Report 414, 1963.

G-13. W.H. Kuhlman: "The Douglas double-slotted flap". Part II of: "Boundary layer and flow control".
Volume 1, edited by G.V. Lachmann, Pergamon Press Inc.

G-14. S.T. Harvey and D.A. Norton: "Development of the Model 727 Airplane high lift system". Society of Automotive Engineers S 408, April 1964.

G-15. J.K. Wimpress: "Shortening the takeoff and landing distances of high speed aircraft". AGARD Report 501, June 1965.

G-16. B.W. McCormick: "Aerodynamics of V/STOL Flight". Chapter 6: "Unpowered flaps". Academic Press, New York/London, 1967.

G-17. R.D. Schaufele and A.W. Ebeling: "Aerodynamic design of the DC-9 wing and high-lift system". SAE Paper No. 670846.

G-18. J.H. Paterson: "Aerodynamic design features of the C-5A". Aircraft Eng., June 1968, pp. 8-15.

G-19. D.E. Hoak and J.W. Carlson: "USAF stability and control DATCOM". Douglas Aircraft Company, Rev. 1968.

G-20. G.J. Hancock: "Problems of aircraft behaviour at high angles of attack". AGARDograph 136, April 1969.
Appendices AI.2 and AII.2.

G-21. J.C. Wimpenny: "The design and application of high-lift devices". Annals of the New York Academy of Sciences, Vol. 154, Art. 2, pp. 329-366.

G-22. J.K. Wimpress: "Aerodynamic technology applied to takeoff and landing". Annals of the New York Academy of Sciences, Vol. 154, Art. 2, pp. 962-981.

G-23. D.M. McRae: "General description and comments on $C_{L_{max}}$ and stalling behaviour". J. Royal Aero. Soc.

(534), Vol. 73, June 1969, pp. 535-541. Also in AGARD Lecture Series No. 23.

G-24. D.N. Foster: "Some aspects of the RAE high-lift research programme". J. Royal Aero. Soc. (534), Vol. 73, June 1969, pp. 541-546.

G-25. K.L. Sanders: "High-lift devices, a weight and performance trade-off methodology". SAWE Technical Paper No. 761, May 1963.

G-26. A.D. Hammond: "High-lift aerodynamics". Proceedings of Conference on vehicle technology for civil aviation, NASA SP-292, 1971, pp. 15-26.

G-27. F. Mavriplis: "Aerodynamic research on high-lift systems". Canadian Aeronautics and Space Journal, May 1971, pp. 175-183.

G-28. J.A. Thain: "Reynolds number effects at low speeds on the maximum lift of two-dimensional aerofoil sections equipped with mechanical high lift devices". Quarterly Bulletin of the Div. of Mech. Eng. and the NAE, Canada, July 1-Sept. 30, 1973, pp. 1-24.

G-29. B.L.G. Ljungström: "Experimental high lift optimization of multiple element airfoils". AGARD-CP-143, April 1974.

G-30. R.J. Margason: "High-lift aerodynamics - trends, trades and options". AGARD Conference on Takeoff and Landing, CP-160, April 1974.

G-31. J.G. Callaghan: "Aerodynamic prediction methods for aircraft at low speeds with mechanical high lift aids", AGARD Lecture Series 67, May 1974.

G-32. W.McIntosh: "Prediction and analysis of the low speed stall characteristics of the Boeing 747". AGARD Lecture Series 74, March 1975.

Experimental results, high-lift devices

G-33. C.J. Wenzinger and T.A. Harris: "Wind-tunnel investigation of an NACA 23012 airfoil with various arrangements of slotted flaps". NACA TR 664, 1939.

G-34. C.J. Wenzinger and T.A. Harris: "Wind-tunnel investigation of NACA 23012, 23021 and 23030 airfoils with various sizes of split flap". NACA TR No. 668, 1939.

G-35. C.J. Wenzinger and T.A. Harris: "Wind-tunnel investigation of an NACA 23021 airfoil with various arrangements of slotted flaps". NACA TR No. 677, 1939.

G-37. J.G. Lowry: "Wind-tunnel investigation of an NACA 23012 airfoil with several arrangements of slotted flaps with extended lips". NACA TN No. 808, May 1941.

G-38. I.H. Abbott, A.E. von Doenhoff and L.S. Stivers, Jr.: "Summary of airfoil data". NACA TR No. 824, 1945.

G-39. A.L. Braslow and L.K. Loftin: "Two-dimensional wind-tunnel investigation of an approximately 14-percent-thick NACA 66-series-type airfoil section with a double-slotted flap". NACA TN 1110, 1946.

G-40. G.M. McCormack and V.I. Stevens: "An investigation of the low-speed stability and control characteristics of swept-forward and swept-back wings in the Ames 40- by 80-foot wind tunnel". NACA RM No. A6K15, 1947.

G-41. F.F. Fullmer, Jr.: "Two-dimensional wind-tunnel investigation of the NACA 64-012 airfoil equipped with two types of leading-edge flap". NACA TN, No. 1277, May 1947.

G-42. F.F. Fullmer, Jr.: "Two-dimensional wind-tunnel investigation of an NACA 64-009 airfoil equipment with two types of leading-edge flap". NACA TN No. 1624.

G-43. J.F. Cahill: "Two-dimensional wind-tunnel investigation of four types of high-lift flap on an NACA 65-210 Airfoil Section". NACA TN 1191, 1947.

G-44. J.F. Cahill and R.M. Racisz: "Wind-tunnel investigations of seven thin NACA airfoil sections to determine optimum double-slotted flap configurations". NACA TN 1545, 1948.

G-45. J.C. Sivells and S.H. Spooner: "Investigation in the Langley 19-foot pressure tunnel of two wings of NACA 65-210 and 64-210 airfoil sections with various type flaps". NACA TR No. 942, 1949.

G-46. R. Hills, R.E.W. Harland and R.H. Whitbey: "Wind tunnel tests on lateral control with high-lift flaps made on the S24/37". ARC R and M No. 2452, 1950.

G-47. H. Davies, J.E. Adamson and J. Seddon: "Wind tunnel tests on the Supermarine S.24/37, a high-wing monoplane with a variable-incidence wing". ARC R and M No. 2451, 1951.

G-48. J.A. Kelly and N.F. Hayter: "Lift and pitching moment at low speeds of the NACA 64A010 airfoil section equipped with various combinations of a leading-edge slat, a leading-edge flap, split flap and double-slotted flap". NACA TN 3007, Sept. 1953.

G-49. H.O. Palme: "Summary of wind tunnel data for high-lift devices on swept wings". SAAB TN 16, 1953.

G-50. B.J. Gambucci: "Section characteristics of the NACA 0006 airfoil with leading-edge and trailing-edge flaps". NACA TN No. 3797, Dec. 1956.

G-51. C.C. Furlong and J.C. McHugh: "A summary and analysis of the low speed longitudinal characteristics of swept wings at high Reynolds numbers". NACA TR 1339, 1957.

G-52. R.L. Naeseth and E.E. Davenport: "Investigation of double slotted flaps on a swept-wing transport model", NACA TN D-103, Oct. 1959.

G-53. D.N. Foster, H.P.A.H. Irwin and B.R. Williams: "The two-dimensional flow around a slotted flap". ARC R and M No. 3681, 1971.

G-54. W.H. Wentz, Jr.: "New airfoil sections for general aviation aircraft". SAE Paper No. 730876.

G-55. G.J. Bingham and K.W. Nooman: "Low-speed aerodynamic characteristics of NACA 6716 and NACA 4416 airfoils with 35-percent-chord single-slotted flaps". NASA TM X-2623, May 1974.

G-56. C.J. Wenzinger and W.E. Gauvain: "Wind-tunnel investigation of an NACA airfoil with a slotted flap and three types of auxiliary flap". NACA TR No. 679, 1939.

Calculation of aerodynamic characteristics with high-lift devices

G-57. H. Glauert: "Theoretical relationships for an airfoil with hinged flap". ARC R and M No. 1095, 1927.

G-58. H.A. Pearson: "Span load distribution for tapered wings with partial-span flaps". NACA Report 585, 1936.

G-59. A. Silverstein and S. Katzoff: "Design charts for predicting downwash angles and wake characteristics behind plain and flapped wings". NACA Report 648.

G-60. H.A. Pearson and R.F. Anderson: "Calculation of the aerodynamic characteristics of tapered wings with partial-span flaps". NACA Report 665, 1939.

G-61. J. de Young: "Theoretical symmetric loading due to flap deflection for wings of arbitrary planform at subsonic speeds". NACA TR No. 1071, 1952.

G-62. D. Küchemann: "A simple method of calculating the span and chordwise loading on straight and swept wings of any given aspect ratio at subsonic speeds". ARC R and M 2935, 1952.

G-63. D. Fiecke: "Die Bestimmung der Flugzeugpolaren für Entwurfszwecke". I. Teil: Unterlagen. Deutsche Versuchsanstalt für Luftfahrt, Bericht Nr. 16, 1956.

G-64. J.G. Lawry and E.C. Polhamus: "A method for predicting lift increment due to flap deflection at low angles of attack". NACA TN 3911, 1957.

G-65. Anon.: Royal Aeronautical Society DATA Sheets, "Aerodynamics", Volume 2 "Wings" and Vol. 4 "Flaps".

G-66. J.A. Hay and W.J. Eggington: "An exact theory of a thin aerofoil with large flap deflection". J. Royal Aero. Soc. Vol. 60 (551), Nov. 1956, pp. 753-757.

G-67. A. Roshko: "Computation of the increment of maximum lift due to flaps". Douglas Aircraft Rep. SM-23626, 1959.

G-68. C.L. Bore and A.T. Boyd: "Estimation of maximum lift of swept wings at low Mach numbers". J. Royal Aero. Soc., April 1963.

G-69. J. Mc.Kie: "The estimation of the loading on swept wings with extending chord flaps at subsonic speeds". ARC CP No. 1110, 1970.

G-70. Anon.: "Lift coefficient increment at low speeds due to full-span split flaps". ESDU Item Number 74009, May 1974.

G-71. Anon.: "Low-speed drag coefficient increment at zero lift due to full-span split flaps". ESDU Item Number 74010, July 1974.

G-72. Anon.: "Rate of change of lift coefficient with control deflection for full-span plain controls". ESDU Item Number 74011, July 1974.

G-73. Anon.: "Conversion of lift coefficient increment due to flaps from full span to part span". ESDU Item Number 74012, July 1974.

Aerodynamic effects of ground proximity

G-74. C. Wieselsberger: "Über den Flügelwiderstand in der Nähe des Bodens". Z. Flugtechn. u. Motorluft-schiffahrt 12, 1921, pp. 145-154 (English translation: NACA TM 77, 1922).

G-75. I. Tani, M. Taima and S. Simidu: "The effect of ground on the aerodynamic characteristics of a monoplane wing". Rep. Aero. Res. Inst., Tokyo Imperial University No. 156, 1937 (Also: ARC R and M 3376, 1938).

G-76. I. Tani, H. Itokawa and M. Taima: "Further studies of the ground effect". Rep. Aero. Res. Inst., Tokyo Imperial University No. 158, 1937.

G-77. W.S. Brown: "Windtunnel corrections on ground effect". ARC R and M No. 1865, July 1938.

G-78. J. Wetmore and L. Turner: "Determination of ground effect from tests of a glider in towed flight". NACA Report No. 695, 1940.

G-79. S. Katzoff and H.N. Sweberg: "Ground effect on downwash angles and wake location". NACA Report 738, 1943.

G-80. Y. Hamal: "Modification des propriétés aérodynamiques d'une aile au voisinage du sol". Centre National d'Etudes et de Recherches Aéronautiques (Bruxelles). Mémoire No. 4, 1935.

G-81. G. Chester Furlong and T.V. Bollech: "Effect of ground interference on the aerodynamic and flow characteristics of a 42° sweptback wing at Reynolds numbers up to 6.8×10^6". NACA Report 1218, 1955.

G-82. R.M. Licher: "Increase in lift for two- and three-dimensional wings near the ground". Douglas, Santa Monica Division, Rep. SM-22615, 1956.

G-83. K. Gersten: "Über die Berechnung des induzierten Geschwindigkeitsfeldes von Tragflügeln". Jahrbuch 1957 der WGL, pp. 172-190.

G-84. D. Kohlman: "A theoretical method of determining the ground effect on lift and pitching moment for wings of arbitrary planform". Boeing Document D3-1861, 1958.

G-85. F. Thomas: "Aerodynamische Eigenschaften von Pfeil- und Deltaflügeln in Bodennähe". Jahrbuch 1958 der WGL, pp. 53-61.

G-86. K. Gersten: "Berechnung der aerodynamischen Beiwerte von Tragflügeln endlicher Spannweite in Boden-nähe". Abhandlungen der Braunschweigischen Wissenschaftlichen Gesellschaft, Vol. XII, 1960, pp. 95-115.

G-87. J.A. Bagley: "The pressure distribution on two-dimensional wings near the ground". ARC R and M No. 3238, 1960.

G-88. M.P. Fink and J.L. Lastinger: "Aerodynamic characteristics of low-aspect-ratio wings in close prox-imity to the ground". NASA TN D-926, July 1961.

G-89. U. Ackermann: "Ein Doppeltraglinienverfahren zur Untersuchung des Flügels in Bodennähe". Jahrbuch 1962 der WGLR, pp. 104-109.

G-90. P.R. Owen and H. Hogg: "Ground effect on downwash with slipstream". ARC R and M No. 2449, 1952.

G-91. G.H. Saunders: "Aerodynamic characteristics of wings in ground proximity". Canadian Aeronautic and Space Journal, June 1965, pp. 185-192.

G-92. U. Ackermann: "Zur Berechnung der aerodynamischen Beiwerte und des Strömungsfeldes von Tragflügeln in Bodennähe unter Berücksichtigung von Nichtlinearitäten". Doctor's Thesis, Technological University of Darmstadt, 1966.

G-93. D.J. Willig: "A method of computing Indicated Airspeed in ground effect". J. of Aircraft, Vol. 5 No. 4, July/August 1968, pp. 412-414.

G-94. K. Gersten and J. von der Decken: "Aerodynamische Eigenschaften schlanker Flügel in Bodennähe". Jahrbuch 1966 der WGLR, pp. 108-125.

G-95. Anon.: "Low-speed longitudinal aerodynamic characteristics of aircraft in ground effect". ESDU
Item Number 72023.

G-96. L.B. Gratzer and A.S. Mahal: "Ground effects in STOL Operations". AIAA Paper No. 71-579 (also: J.
of Aircraft, Vol. 9 No. 3, March 1972, pp. 236-242).

G-97. D. Hummel: "Nichtlineare Tragflügeltheorie in Bodennähe". Z. für Flugwiss., Dec. 1973, pp. 425-442.

Effects of engine failure on aerodynamic characteristics

G-98. H.K. Millicer: "The design study". Flight, August 17, 1951, pp. 201-205.

G-99. B. Wrigley: "Engine performance considerations for the large subsonic transport". Lecture given at
the Von Karman Institute, Brussels, 23rd April 1969.

G-100. Anon.: "Approximate estimation of drag of wind-milling propellers". Royal Aero. Soc. Data Sheets
"Performance", Sheet ED 1/1, April 1962.

G-101. J. Mannée: "Wind tunnel investigation of the influence of the aircraft configuration on the yawing
and rolling moment of a twin-engined, propeller-driven aircraft with one engine inoperative". NLL
Report A 1508 B, 1963.

Appendix H. Procedures for computing turbo-engine performance for aircraft project design work

SUMMARY

This appendix contains a survey of analytical equations for computing the aerothermodynamic performance of gas turbine engines: straight turbojets, bypass engines (turbofans) and turboprop engines. Both the gas generator and the overall engine performance are dealt with, on the assumption that the engine operates under its design conditions. An approximate expression is also presented for relating the cruise and climb thrust (at altitude) for a turbojet engine to the takeoff thrust.

The method can be used to perform parametric design studies of gas turbine engines for the purpose of investigating the optimum aircraft / engine combination. For this reason the Turbine Entry Temperature, the Overall Pressure Ratio and the Bypass Ratio can be dealt with as explicit variables.

NOMENCLATURE

A_d – fan inlet duct cross-sectional area

A_e – total exhaust nozzle area of fan and gas generator flows

A_i – intake area of total engine flow

a_{oo} – speed of sound at sea level ISA

C_F – skin friction coefficient for smooth duct flow, turbulent b.l.

C_p – specific fuel consumption of turbo-prop engine

C_T – specific fuel consumption of turbojet engine

c_p – specific heat of engine air at constant pressure

G – gas generator power function

H – heating value of fuel

K_d – duct pressure loss factor

M_o – flight Mach number

M_i – Mach number at engine inlet face

M_o^* – design flight Mach number for turboprop engine

$\dot{m}$ – mass flow per unit time (no index: total engine mass flow)

N – engine h.p. compressor rpm

P_{br} – brake horsepower

P_{eq} – equivalent horsepower

$P_{g_{is}}$ – convertible energy generated by gasifier

p – static pressure

p_o – ambient pressure

p_{oo} – p_o at sea level

p_t – total (stagnation) pressure

R_e – Reynolds number

T – thrust; (static) temperature

T_o – ambient temperature

T_{oo} – T_o at sea level

T_j – net jet thrust of turboprop engine

T_t – total (stagnation) temperature

T_{to} – takeoff thrust (static, sea level)

V_o – flight speed

v – velocity of fully expanded exhaust flow

$\dot{W}$ – weight flow per unit time of engine air

$\dot{W}_F$ – fuel weight flow per unit time

γ – ratio of specific heats (for ambient air: $\gamma = 1.4$)

δ – relative ambient pressure ($\delta = p_o/p_{oo}$)

ε – total pressure ratio of compressor or fan

η – efficiency

η_B – combustion efficiency

η_c – isentropic compressor efficiency

η_d – isentropic fan intake duct efficiency

η_f – isentropic fan efficiency

η_i – gas generator intake stagnation pressure ratio

η_{mech} – efficiency of mechanical transmission (gearbox)

η_n – isentropic efficiency of expansion process in nozzle

η_p – propulsive efficiency

η_{prop} – propeller efficiency

η_t – isentropic turbine efficiency

η_{tf} – product of η_f and η_t

η_{th} – thermal efficiency of gas generator

η_{tot} – overall engine efficiency

θ – relative ambient temperature ($\theta = T_o/T_{oo}$)

κ – temperature function of compression process

λ – bypass ratio

μ – ratio of stagnation to static temperature of ambient air

σ – jet velocity coefficient

ϕ – nondimensional Turbine Entry Temperature

φ – gross thrust parameter

ψ – corrected specific thrust

Subscripts

B – combustion chamber

c – high-pressure compressor

d – fan intake duct

e – nozzle exhaust

F – fuel

f – fan

g – gas generator

i – intake of gas generator

j – exhaust jet of turboprop engine

n – nozzle

t – turbine

tf – combination of turbine and fan

to – takeoff, sea level static condition

The station numbering system used is in accordance with Fig. 4-16: 0 - ambient; 1 - nacelle leading edge; 2 - (l.p.) compressor entry; 3 - (h.p.) compressor exit; 4 - turbine entry; 5 - turbine exit

H-1. SCOPE OF THE METHOD

Most of the equations presented in this appendix have been derived in Ref. H-1 within the framework of propulsion system analysis for aircraft with Laminar Flow Control (LFC). They are presented here in a slightly modified form to make them readily usable for takeoff, climb and cruise conditions for normal subsonic engines without special facilities for LFC. In addition, the result of Ref. H-6 has been reproduced, yielding a simple method for computing the jet engine thrust lapse with altitude.

The classical procedure for analyzing engine performance is to carry out a cycle analysis for the various thermodynamic processes in the engine. Many examples of this procedure can be found in the literature, e.g. Ref. H-7. In Ref. H-1 it is shown that engine performance can be written in closed form if the following simplifications are considered acceptable:
a. The thrust is referred to fully expanded conditions. Thus instead of the usual standard net thrust definition, we use the ideal thrust:

$$T = \dot{m}_g \ (v_g - V_o) + \dot{m}_f \ (v_f - V_o) \qquad (H-1)$$

for fully expanded exhaust flows.
b. The fuel mass flow is neglected relative to the engine air mass flow:

$$\dot{m}_F << \dot{m}_g + \dot{m}_f \qquad (H-2)$$

c. The ratio of specific heats is assumed constant throughout all engine processes ($\gamma = 1.4$).
d. The fan and exhaust flows are unmixed.
e. Power extraction and/or bleed air take-off for airframe services are ignored.

Although assumptions b. and c. will result in considerable errors in individual thermodynamic processes, it is experienced that several effects cancel each other out when the overall engine performance is considered and for most conventional engine configurations the results are found to be reasonably accurate.

Several equations similar to those derived in Ref. H-1 can also be found in Refs. H-2 through H-4. The background to the cycle analysis and the various assumptions and approximations from which the simple analytical expressions have been derived are discussed in Refs. H-1, H-4 and H-6. The present method is useful for application in aircraft project design work, in which errors of a few percent are acceptable. The results of the present method are not, however, accurate for low engine ratings, particularly when the hot flow nozzle is not choked. An alternative method for turbojets can be found in Ref. H-8.

Finally, it should be noted that various definitions used in this appendix are dealt with in greater detail in Section 4.3.

H-2. THE GAS GENERATOR

The following major engine parameters are used in the analysis.
TURBINE ENTRY TEMPERATURE T_{t_4}, a quantity which particularly affects the (specific) thrust, as discussed in Section 4.4.2. High values of specific thrust are obtainable when turbine blade cooling is used, but the effect of the cooling on engine performance will not be taken into account here. Instead, the data in Ref. H-5 suggest that a correction for cooling can be made by substituting in the present equations an equivalent value of T_{t_4} which is about 30K (54R) below the actual T_{t_4} of the (cooled) blades.
OVERALL PRESSURE RATIO ε_c, an engine characteristic which is of vital importance to the engine (specific) fuel consumption, as

discussed in Section 4.4.1. For the usual engine configuration with a fan in front of the gas generator, ε_c is equal to the combined effect of the fan and the remaining portion of the engine compressor.

The analysis is simplified by introducing the following characteristic functions: the non-dimensional Turbine Entry Temperature:

$$\phi = \frac{T_{t_4}}{T_o} \tag{H-3}$$

and the temperature function of the compression process:

$$\kappa = \mu(\varepsilon_c^{\frac{\gamma-1}{\gamma}} - 1) = \mu(\varepsilon_c^{.2857} - 1) \tag{H-4}$$

where

$$\mu = 1 + \frac{\gamma-1}{2} M_o^2 = 1 + .2 M_o^2 \tag{H-5}$$

Gas generator performance may be expressed in nondimensional form in terms of a GAS GENERATOR FUNCTION,

$$G = \frac{P_{g_{is}}}{\dot{m}_g c_p T_o} \tag{H-6}$$

and a THERMAL EFFICIENCY,

$$\eta_{th} = \frac{P_{g_{is}} - \frac{1}{2} \dot{m}_g V_o^2}{\dot{m}_g c_p (T_{t_4} - T_{t_3})} \tag{H-7}$$

where $P_{g_{is}}$ is the gas generator CONVERTIBLE ENERGY, defined as the portion of the gas generator hot-gas energy that can be converted into useful propulsive power. The convertible energy is equivalent to the kinetic energy of the gases for the hypothetical case that they would expand to ambient pressure in an isentropic process. The rest of the energy is lost in the form of heat added to the atmosphere.
The gas generator function G can be calculated as follows:

$$G = \left(\phi - \frac{\kappa}{\eta_c}\right)\left[1 - \frac{1}{\eta_i} .2857 \frac{1.01}{(\kappa+\mu)(1 - \frac{\kappa/\phi}{\eta_c \eta_t})}\right] \tag{H-8}$$

where a combustion chamber pressure loss of 3.5 percent has been assumed. The gas generator intake stagnation pressure ratio is defined as follows:

$$\eta_i = P_{t_2}/P_{t_o} \tag{H-9}$$

The isentropic compressor efficiency η_c is the ratio of the total temperature rise in an isentropic compression to the actual temperature rise in a (polytropic) compression with the same pressure ratio. The isentropic turbine efficiency η_t is the ratio of the actual total temperature drop in the (polytropic) expansion process to the temperature drop in an isentropic expansion with the same pressure ratio. Some guidelines with respect to the choice of η_i, η_c and η_t are given in Section H-7.

The thermal efficiency is computed as follows:

$$\eta_{th} = \frac{G - .2 M_o^2}{\phi - \mu - \frac{\kappa}{\eta_c}} \tag{H-10}$$

Instead of G, a velocity coefficient σ may be used to characterize the gas generator performance:

$$\sigma = \left(\frac{\frac{1}{2} \dot{m}_g V_o^2}{P_{g_{is}}}\right)^{\frac{1}{2}} = \frac{M_o}{\sqrt{\frac{2}{\gamma-1} G}} = \frac{M_o}{\sqrt{5G}} \tag{H-11}$$

Unlike G, the velocity ratio is characterized by its large variation with flight speed.

H-3. SPECIFIC PERFORMANCE OF STRAIGHT JET ENGINES

The PROPULSIVE EFFICIENCY is defined as follows:

$$\eta_p = \frac{T V_o}{P_{g_{is}} - \frac{1}{2} \dot{m}_g V_o^2} \tag{H-12}$$

and can be computed from:

$$\eta_p = \frac{2\sigma \ (\sqrt{\eta_n}-\sigma)}{1-\sigma^2} \qquad\qquad (H-13)$$

where σ is given by (H-11) and the isentropic nozzle efficiency is defined in a manner similar to the turbine efficiency η_t.

The OVERALL EFFICIENCY is defined as follows:

$$\eta_{tot} = \frac{T \ V_o}{\dot{W}_F \ H} \qquad\qquad (H-14)$$

and can be computed from:

$$\eta_{tot} = \eta_B \ \eta_{th} \ \eta_p = \frac{.4 \ \eta_B \ M_o^2}{\phi - \mu - \frac{\kappa}{\eta_c}} \left(\frac{1}{\sigma} \ \sqrt{\eta_n} - 1\right) \qquad (H-15)$$

The SPECIFIC FUEL CONSUMPTION is related to the overall efficiency by:

$$C_T = \frac{\dot{W}_F}{T} = .2788 \ \frac{M_o \ \sqrt{\theta}}{\eta_{tot}} \qquad (h^{-1}) \qquad (H-16)$$

This quantity may be obtained from the CORRECTED SPECIFIC FUEL CONSUMPTION:

$$\frac{C_T}{\sqrt{\theta}} = .711 \ \frac{\phi - \mu - \kappa/\eta_c}{\sqrt{5\eta_n G} - M_o} \qquad (H-17)$$

for a combustion efficiency of 98 percent. The SPECIFIC THRUST is the engine thrust divided by the intake mass flow per unit time. The following applies to the CORRECTED SPECIFIC THRUST:

$$\psi = \frac{T}{\dot{W} \ \sqrt{\theta}} = 34.714 \ (\sqrt{5\eta_n G} - M_o) \qquad (sec) \qquad (H-18)$$

H-4. SPECIFIC PERFORMANCE OF TURBOFAN ENGINES

The BYPASS RATIO λ, defined as the ratio of the mass flows per unit time through the fan and the gas generator, respectively, is a fundamental parameter of turbofan engines, affecting engine layout, specific thrust, specific fuel consumption, weight, drag and noise. Its significance is dealt with thoroughly in Section 4.

An "optimum" FAN PRESSURE RATIO may be defined, resulting in maximum propulsive efficiency and thrust and minimum C_T for a given gas generator. Its value can be obtained from:

$$\varepsilon_{f_{opt}} = \left\{ 1 + \frac{\eta_{tf}^2 \ G - .2 \ \eta_i \ M_o^2}{\mu \ (1 + \eta_{tf} \ \lambda)} \right\}^{3.5} \qquad (H-19)$$

This pressure ratio corresponds to the "optimum" jet velocities of the fully expanded hot and cold flows:

$$\frac{v_{g_{opt}}}{\sqrt{\theta}} = a_{oo} \sqrt{\frac{5\eta_n \ G + \lambda \ M_o^2 \ \eta_i/\eta_{tf}}{1 + \eta_{tf} \ \lambda}} \qquad (H-20)$$

where a_{oo} is the speed of sound at sea level ISA (1116.9 ft/s = 340.4 m/s).

$$v_{f_{opt}} = \eta_t \ \eta_f \ v_{g_{opt}} = \eta_{tf} \ v_{g_{opt}} \qquad (H-21)$$

The following equations are valid on condition that the fan pressure ratio is optimized according to (H-19).

Propulsive efficiency, defined by (H-12):

$$\eta_p = \frac{2\sigma^2}{1-\sigma^2} \left| \sqrt{\eta_n (1+\eta_{tf}\lambda) \left(\frac{1}{\sigma^2} + \frac{\eta_d}{\eta_{tf}}\lambda\right)} - (1+\lambda) \right| \qquad (H-22)$$

overall efficiency, defined by (H-14):

$$\eta_{tot} = .4 \ \frac{\eta_B \ M_o^2}{\phi - \mu - \kappa/\eta_c} \left| \sqrt{\eta_n (1+\eta_{tf}\lambda) \left(\frac{1}{\sigma^2} + \frac{\eta_d}{\eta_{tf}}\lambda\right)} - (1+\lambda) \right| \qquad (H-23)$$

corrected specific fuel consumption, for a combustion efficiency of 98 percent:

$$\frac{C_T}{\sqrt{\theta}} = \frac{.711 \ (\phi - \mu - \kappa/\eta_c)}{\sqrt{5\eta_n (1+\eta_{tf}\lambda) (G + .2M_o^2 \frac{\eta_d}{\eta_{tf}}\lambda)} - (1+\lambda) M_o}$$

$$(h^{-1}) \qquad (H-24)$$

and the corrected specific thrust:

$$\psi = \frac{T}{\dot{W}\sqrt{\theta}} = 34.714 \ \left| \frac{1}{1+\lambda} \ \sqrt{5\eta_n (1+\eta_{tf}\lambda)} \times \right.$$

$$\left. \sqrt{(G + .2M_o^2 \frac{\eta_d}{\eta_{tf}}\lambda)} - M_o \right| \qquad (sec) \qquad (H-25)$$

The ISENTROPIC FAN INLET DUCT EFFICIENCY η_d in these equations is the total temperature rise resulting from an isentropic compression from ambient to fan total intake pressure, divided by the actual temperature rise in an adiabatic compression to the same inlet pressure. Provided the intake of the gas generator and the fan have similar performance, the relationship between η_d and η_i is as follows:

$$\eta_i^{\frac{\gamma-1}{\gamma}} = \frac{1 + \eta_d \frac{\gamma-1}{2} M_o^2}{1 + \frac{\gamma-1}{2} M_o^2}$$

or: (H-26)

$$\eta_i = \left(\frac{1 + .2\, \eta_d\, M_o^2}{1 + .2\, M_o^2}\right)^{3.5}$$

A convenient approximation is:

$$1 - \eta_i = (1 - \eta_d) \frac{.7\, M_o^2}{1 + .2\, M_o^2} \qquad \text{(H-27)}$$

H-5. THRUST LAPSE RATES, INTAKE AND EXHAUST AREAS OF TURBOJET AND TURBOFAN ENGINES

The engine thrust can be derived from the specific thrust, provided the engine mass flow is known. To this end an estimation must be made of the total nozzle area A_e, which is assumed to be fixed.
A correlation of $T_{to}/(A_e\, P_{oo})$ vs. the specific thrust in static takeoff conditions is presented in Fig. H-1. The same figure can also be used to find the fan intake area A_i:

$$\frac{A_i P_{oo}}{T_{to}} = \frac{1 + .6\, M_i^2}{.04\, M_i\, \psi_{to}} \qquad \text{(H-28)}$$

where M_i is the mean Mach number at the fan face and ψ_{to} the corrected specific static takeoff thrust at sea level. An average value $M_i = .5$ appears to be representative for sizing the fan inlet area.

THE GROSS THRUST PARAMETER φ is now defined as:

$$\varphi = \frac{1 + \dfrac{\text{gross thrust}}{A_e\, P_o}}{\mu^{3.5}} \qquad \text{(H-29)}$$

This parameter is frequently used in the presentation of nondimensional engine performance. Its meaning can be demonstrated by taking the sealevel static condition for a straight turbojet with choked nozzle. For that condition the following relation can be derived:

$$\varphi_{to} = (1+\gamma_e)\, \frac{\text{static exhaust nozzle pressure}}{\text{ambient pressure}}$$

(H-30)

where $\gamma_e \approx 1.33$

The general case of a turbofan engine (static, takeoff) yields:

$$\varphi_{to} = 1 + \frac{T_{to}}{A_e P_{oo}} \qquad \text{(H-31)}$$

The corrected intake weight flow is $\dot{W}\sqrt{T_{t_2}}/P_{t_2}$, where T_{t_2} and P_{t_2} are the total intake temperature and pressure, respectively. Both the gross thrust parameter and the nondimensional weight flow are related to the corrected engine rpm $(N/\sqrt{T_{t_2}})$. The compressor speed is not a convenient parameter for a generalized method as it is greatly dependent upon the particular compressor design, and should therefore be eliminated. The following approximate relationship can be deduced from consideration of several nondimensional graphs obtained from engine performance brochures:

$$\frac{\dot{W}\sqrt{\theta}}{\delta \dot{W}_{to}} = \mu^3 \sqrt{\frac{\varphi}{\varphi_{to}}} \qquad \text{(H-32)}$$

The square-root relationship has no physical meaning, but it gives a reasonable approximation (see Fig. H-2) and is convenient for solving the equations analytically. For normal operational conditions in takeoff, climb and cruising flight, the nondimensional mass flow is within $\pm 5\%$ of the value at takeoff and the approximation given by (H-32) is acceptable. Equation H-32 can be combined with:

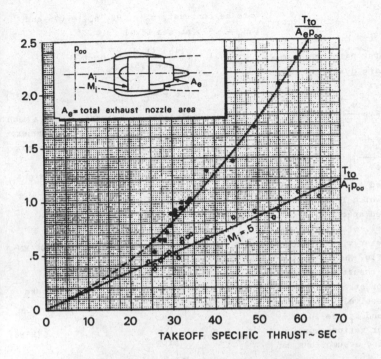

Fig. H-1. Intake and exhaust areas as determined by the takeoff specific thrust (civil jet engines, fixed nozzle)

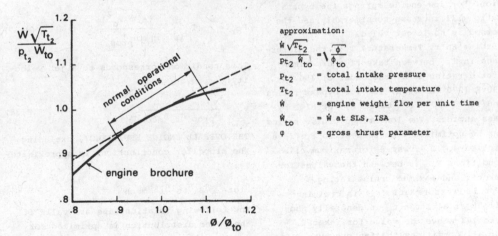

approximation:

$$\frac{\dot{W}\sqrt{T_{t_2}}}{p_{t_2}\hat{W}_{to}} = \sqrt{\frac{\phi}{\phi_{to}}}$$

p_{t_2} = total intake pressure

T_{t_2} = total intake temperature

$\dot{W}$ = engine weight flow per unit time

$\hat{W}_{to}$ = $\dot{W}$ at SLS, ISA

ϕ = gross thrust parameter

Fig. H-2. Nondimensional engine air weight
flow and gross thrust parameter for the
Rolls-Royce RB183 turbofan engine

$$\frac{\dot{W}\sqrt{\theta}}{\delta \dot{W}_{to}} = \frac{T}{\delta T_{to}} \cdot \frac{\psi_{to}}{\psi} \qquad (H-33)$$

to yield the THRUST LAPSE RATE T/T_{to}. The solution of the resulting quadratic equation is approximated by:

$$\frac{T}{\delta T_{to}} = \frac{\psi}{\psi_{to}} \mu^{1.5} \left| \frac{1}{\sqrt{\varphi_{to}}} + .6 \frac{\varphi_{to}-1}{\varphi_{to}} \mu^{1.5} \times \right.$$

$$\left. \left(\frac{\psi}{\psi_{to}} + \frac{34.714 \, M_o}{\psi_{to}} \right) \right| \qquad (H-34)$$

where ψ is the corrected specific thrust for the condition for which T is required, and ψ_{to} is the corrected specific thrust in the takeoff (sealevel, static).

The gross thrust parameter φ_{to} is calculated with (H-31) and Fig. H-1. Expressions for the specific thrust are given by (H-18) and (H-25) for straight jet and turbofan engines, respectively. The engine parameters λ, T_{t_4}, ε_c and ε_f are in principle not constant under various working conditions. The following assumptions can be made.

Bypass Ratio λ: assumed constant and equal to the value in the takeoff condition. For low engine ratings the bypass ratio will increase considerably and the method is no longer valid.

Turbine Entry Temperature T_{t_4}: the difference in T_{t_4} between takeoff and maximum cruise rating is of the order of 150 to 200 K (270 to 360 R) for low bypass ratios and 50 to 100 K (90 to 180 R) for high bypass engines. The long-range cruise rating can be optimized with respect to T_{t_4} (see Section 4.4.2.), while for maximum climb conditions T_{t_4} is between the values for takeoff and maximum cruise rating.

Overall Pressure Ratio ε_c: in cruising flight at altitude ε_c is generally about 5 to 10% above the value for takeoff. For all conditions the fan pressure ratio ε_f is assumed optimum for minimum fuel consumption.

A simple approximation can be derived for the thrust during takeoff from series expansion of (H-25) and (H-32), assuming φ

to be constant, $\eta_d = .98$, $\eta_{tf} = .75$ and $\eta_n = .97$, at sea level

$$\frac{T}{T_{to}} = 1 - \frac{.454 \, (1 + \lambda)}{\sqrt{(1 + .75\lambda) \, G}} M_o + \left(.6 + \frac{.13\lambda}{G} \right) M_o^2 \qquad (H-35)$$

This approximation is accurate up to a Mach number of approximately .3. The gas generator function G is given by (H-8). Typical values are:

$G = .9$, low-bypass engines and straight jets

$G = 1.1$, high-bypass engines

H-6. SPECIFIC PERFORMANCE OF TURBOPROP ENGINES

The EQUIVALENT POWER is the sum of brake horse power and jet power:

$$P_{eq} = P_{br} + \frac{T_j V_o}{\eta_{prop}} \qquad (H-36)$$

Power distribution between the propeller shaft and exhaust gases is "optimum" (for maximum overall efficiency) if:

$$P_{br} = \eta_t \, P_{g_{is}} - \frac{\frac{1}{2} \dot{m} V_o^2}{\eta_t \, (\eta_{prop})^2} \qquad (H-37)$$

This condition corresponds to a jet velocity:

$$v_j = \frac{V_o}{\eta_{prop} \, \eta_t \, \eta_{mech}} \qquad (H-38)$$

THE OVERALL ENGINE EFFICIENCY, excluding the propeller contribution, is approximately:

$$\eta_{tot} = \eta_B \, \eta_{th} \, \eta_t \, \eta_{mech} \qquad (H-39)$$

The following relationships are valid if the power distribution is optimized for $\eta_{prop} = .85$ and $\eta_B \, \eta_{mech} = .95$, at a flight Mach number equal to M_o^*.

(BRAKE) SPECIFIC FUEL CONSUMPTION:

$$C_p = \frac{\text{fuel consumption per hour}}{\text{brake horsepower}} \qquad (H-40)$$

$$C_p = \text{constant} \ \frac{\phi - \mu - \kappa/\eta_c}{\eta_t \ G - .28 \ (M_o^*)^2/\eta_t} \qquad \text{(H-40)}$$

wnere the constant is equal to .1426 if C_p is in lb/hp/h, or .0647 if C_p is in kg/hp/h.

SPECIFIC SHAFT POWER:

$$\frac{P_{br}}{\dot{W}\theta} = \text{constant} \ \left| \eta_t \ G - .28 \ (M_o^*)^2/\eta_t \right| \qquad \text{(H-41)}$$

SPECIFIC EQUIVALENT POWER:

$$\frac{P_{eq}}{\dot{W}\theta} = \text{constant} \ \left| \eta_t \ G - (M_o^*)^2 (.4 - .193/\eta_t) \right| \qquad \text{(H-42)}$$

where the constant in (H-41) and (H-42) is equal to 173 if P_{br} is in hp and $\dot{W}$ in lb/s, or 382 if $\dot{W}$ is in kg/s.

The Turbine Entry Temperature in cruising conditions is about 50 to 100 K (90 to 180 R) below the value for takeoff rating.

H-7. CYCLE EFFICIENCIES AND PRESSURE LOSSES

The gas generator intake and fan inlet duct efficiencies η_i and η_d may be calculated with the method in Ref. H-9. A simpler approach will be permissible in most cases if the inlet duct is of regular shape, without sharp bends or divergences. In that case we may use the following equation, derived from Ref. H-10:

$$1 - \eta_d = 4 K_d C_F \left(\frac{M_d}{M_o} \right)^2 \frac{\text{duct length}}{\text{inlet diameter}} \qquad \text{(H-43)}$$

K_d is the ratio of the total duct pressure loss to that due to skin friction only, in a constant-ar   .uct. For a good design $K_d \approx 1.3$. C_F is the skin friction coefficient in a straight duct, approximated by:

$$C_F = \frac{.05}{(R_e)^{1/5}} \qquad \text{(H-44)}$$

where the Reynolds number is based on the average conditions in the duct and the mean duct diameter.

M_d = mean Mach number in the intake duct:

$$M_d = \frac{24.77}{A_d P_{oo}} \frac{\dot{W}\sqrt{\theta}}{\delta} \qquad \text{(H-45)}$$

The air weight flow $\dot{W}$ is determined from the specific thrust and the thrust. For a generalized study of engines for high-subsonic aircraft it is reasonable to assume $M_d = .5$.

The stagnation pressure ratio η_i is obtained from (H-27).

Other engine component efficiency factors may be assumed as follows:

η_f = .85 to .87 - takeoff
 .82 to .85 - cruising
η_c = .84 to .86
η_t = .87 to .89
η_n = .96 to .98

The lower values are applicable to small engines, the higher ones to large civil aircraft engines. Engines can be designed in such a way that high fan efficiencies are obtained in cruising flight and lower values under takeoff conditions. This may be desirable for high-bypass engines in view of their unfavorable thrust lapse with altitude and flight speed.

REFERENCES

H-1. E. Torenbeek: "The propulsion of aircraft with laminar flow control". Report VTH-150, Delft University of Technology, Dept. of Aeron. Eng., 1968.

H-2. S. Boudigues: "Défense et illustration des propulseurs pour avions rapides". Techniques et Science Aéronautiques et Spatiales, July - Aug. 1964 pp. 271-285.

H-3. W. Dettmering and F. Fett: "Methoden der Schuberhöhung und ihre Bewertung". Zeitschrift für Flugwissenschaften. August 1969, pp. 257-267.

H-4. D.G. Shepherd: "Aerospace propulsion". New York, American Elsevier, 1972.

H-5. R.E. Neitzel and H.C. Hemsworth: "High-bypass turbofan cycles for long-range subsonic transports", Journal of Aircraft, Vol. 3 No. 4, July - August 1966, pp. 354-358.

H-6. E. Torenbeek: "Analytical method for computing turbo engine performance at design and off-design conditions". Memorandum M-188, Delft University of Technology, Dept. of Aeron. Eng. Jan. 1973.

H-7. H. Cohen, G.F.C. Rogers and H.I.H. Saravanamuttoo: "Gas turbine theory". Longman Group Ltd., London, Second Edition, 1974.

H-8. B. Ahren: "General performance investigation of turbojet engines". SAAB Technical Note No. 52.

H-9. E.W. Dunlap: "Flight Test Handbook - Performance". Chapter 1, Part 1. AFFTC-TR-59-47, Jan. 1960.

H-10. W.J. Hesse and N.V.S. Mumford Jr.: "Jet propulsion for aerospace applications". Second Edition, 1964, Pitman Publ. Corp., New York.

Appendix J. Principal data of the US and ICAO standard atmospheres

SUMMARY

The Standard Atmosphere presented in this appendix is the US Standard Atmosphere 1962, prepared under the sponsorship of the NASA, USAF and USWB (United States Weather Bureau). This atmosphere is in agreement with the ICAO Standard Atmosphere over their common altitude range. For quick reference standard data for sea level are given, together with some of relative values at various practical flying altitudes.

Altitude m	Pressure ratio δ	Temperature ratio θ	$\sqrt{\theta}$	Density ratio σ	Kin.Viscosity ratio ν/ν_o
0	1	1	1	1	1
1,500	.83450	.96616	.98294	.86373	1.1271
3,000	.69192	.93233	.96557	.74214	1.2754
4,500	.56973	.89849	.94789	.63410	1.4495
6,000	.46564	.86465	.92987	.53853	1.6549
7,500	.37751	.83082	.91149	.45439	1.8989
9,000	.30340	.79698	.89274	.38069	2.1905
10,500	.24154	.76314	.87358	.31651	2.5415
11,000 *	.22336	.75187	.86710	.29708	2.6743
12,000	.19078	.75187	.86710	.25374	3.1311
13,500	.15059	.75187	.86710	.20029	3.9666
15,000	.11887	.75187	.86710	.15810	5.0251
16,500	.09383	.75187	.86710	.12480	6.3660
18,000	.07407	.75187	.86710	.09851	8.0648
19,500	.05847	.75187	.86710	.07776	10.2169

ENGLISH UNITS

Altitude ft	Pressure ratio δ	Temperature ratio θ	$\sqrt{\theta}$	Density ratio σ	Kin.Viscosity ratio ν/ν_o
0	1	1	1	1	1
5,000	.83205	.96562	.98266	.86167	1.1293
10,000	.68770	.93124	.96501	.73848	1.2806
15,000	.56434	.89687	.94703	.62924	1.4586
20,000	.45954	.86249	.92870	.53281	1.6693
25,000	.37109	.82811	.91001	.44812	1.9203
30,000	.29696	.79373	.89092	.37413	2.2214
35,000	.23531	.75935	.87141	.30987	2.5852
36,089 *	.22336	.75187	.86710	.29708	2.6743
40,000	.18509	.75187	.86710	.24617	3.2273
45,000	.14555	.75187	.86710	.19358	4.1040
50,000	.11446	.75187	.86710	.15223	5.2189
55,000	.09001	.75187	.86710	.11971	6.6367
60,000	.07078	.75187	.86710	.09414	8.4395
65,000	.05566	.75187	.86710	.07403	10.7321

* tropopause

Table J-1. Atmospheric properties for several geopotential altitudes

STANDARD SEA LEVEL DATA	Symbols	TECHNICAL UNITS *		SI UNITS
		ENGLISH	METRIC	METRIC
PRESSURE	P_o	2116.22 lb$_f$/sq.ft (14.696 lb$_f$/sq.in)	10332.27 kg$_f$/m^2	101325.0 Newtons/m^2 (1013.25 millibar)
TEMPERATURE	t_o T_o	59 F 518.67 R	15 C 288.15 K	15 C 288.15 K
DENSITY	ρ_o	.0023769 slugs/ft^3	.12492 kg$_f$ s^2/m^4	1.2250 kg/m^3
VELOCITY OF SOUND	a_o	1116.45 ft/s 661.48 knots	340.294 m/s 1225.06 km/h	340.294 m/s
KINEMATIC VISCOSITY	ν_o	1.5723×10^{-4} sq.ft/s	1.4607×10^{-5} m^2/s	1.4607×10^{-5} m^2/s
ACCELERATION DUE TO GRAVITY	g_o	32.1741 ft/s^2	9.80665 m/s^2	9.80665 m/s^2

* Contrary to the convention used in this book, the subscript f for "force" has been retained here in order to avoid confusion with the SI system.

Table J-2. Standard sea level data

Appendix K. The definition and calculation of the takeoff field length required for civil transport aircraft

SUMMARY

A summary and explanation are given of the various definitions associated with the take-off of civil transport aircraft, taking into account the event of engine failure. The schedule to be set up for determining the takeoff reference speeds is discussed and a suggested generalized procedure for analyzing the takeoff performance is presented. A survey of the methods and data required to actually perform the calculations for zero wind and a level runway concludes this appendix, which is intended to give adequate information for evaluating the takeoff performance in the preliminary design stage.

The principles of takeoff field length determination apply to civil transport aircraft (weight ⩾ 12,500 lb = 5,670 kg) only, but the methods of analysis and data presented may well be used for other aircraft categories, e.g. light aircraft.

NOMENCLATURE

A — aspect ratio; factor in the deceleration

a — acceleration or deceleration

BFL — Balanced Field Length

C_D — (aircraft) drag coefficient

C_{D_O} — zero-lift drag coefficient

C_L — (aircraft) lift coefficient

c.g. — center of gravity

D — drag

e — Oswald factor

F(h) — non-dimensional height function

$F(\gamma)$ — non-dimensional climb angle function

g — acceleration due to gravity

h — height

L — lift

l_b — wheelbase

N — normal ground reaction force

n — load factor

n_α — rate of change of load factor with α

Q_{max} — total maximum torque of the wheelbrakes

r_t — (deflected) tire radius

S — wing area; distance

$T, \bar{T}$ — thrust, mean thrust respectively

t — time

u.c. — undercarriage

V — speed (relative to the ground and the air)

W — weight

α — angle of attack

γ — angle of climb

Δ — increment

n_b — braking efficiency

θ — angle of pitch

μ — friction coefficient

ρ — air density

Subscripts

A — airborne phase

b — braking

cg — center of gravity

cp — center of pressure

eff — effective

G — ground roll

LOF — liftoff

MC — Minimum Control (Speed)

MU — Minimum Unstick (Speed)

N — all engines operating

N-1 — one engine inoperative

ROT — rotation

S — stalling

STOP — stopping

T — thrust line

to — takeoff

uc — undercarriage

x — moment of engine failure

O — standstill at beginning of runway

1 — decision point; main u.c.

2 — 35 ft height point (one engine inoperative); nose u.c.

3 — 35 ft height point (all engines operating)

K-1. REFERENCE DISTANCE DEFINITIONS

Although the FAR and BCAR codes differ in detail, both attempt to ensure a similar level of safety in the takeoff with all engines operating, and in the event of an engine failure occurring at any point. Consequently, the basic procedures for takeoff calculations are identical. Definitions associated with the takeoff are introduced by reference to Fig. K-1. Although the example applies to a twin-engine airplane, the requirements for three- and four-engine airplanes differ only in detail. The following principal distances* determine the takeoff field length required for a fixed weight, altitude, temperature and airplane configuration (e.g. flap setting).

a. The TAKEOFF RUN required is derived from calculations by taking the distance needed to accelerate to the moment of liftoff, plus a proportion of the airborne dis-

*The definitions given here are interpretations of the formal definitions and procedures in FAR 25.103 through 25.121 and BCAR Section D Ch. D2-3.

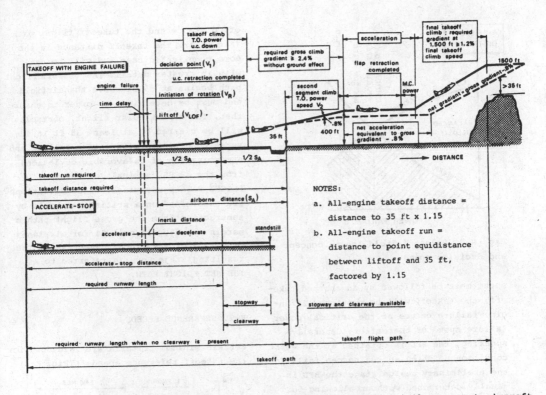

Fig. K-1. Takeoff procedures and requirements for a twin-engine civil transport aircraft

tance to a height of 35 ft (FAR: half, BCAR: a third). In the all-engine case, a safety margin of 15% is added. The greater of these distances must not exceed the takeoff ground run available - i.e. the length of the runway, having a hard prepared surface compatible with the aircraft weight and undercarriage design.

b. The EMERGENCY or ACCELERATE-STOP DISTANCE required is the distance to accelerate to the speed at which the critical engine fails (V_x), plus the distance to come to a standstill by means of braking, when the pilot decides to abort the takeoff. This distance must not exceed the length of the takeoff ground run plus the stopway (if present). The latter has a surface capable of supporting the aircraft with little damage and is suitable for braking.

c. The TAKEOFF DISTANCE required is the distance needed to accelerate, rotate to flying attitude and climb to a height of 35 ft. In the all-engines case a safety margin of 15% is added. The greater of these distances must not exceed the length of the takeoff runway plus clearway* (if present). The latter is essentially free of large protruding obstacles, but does not necessarily have a solid surface. For example, water may form a clearway.

d. The BALANCED FIELD LENGTH (BFL) is the distance required in the situation where the emergency distance becomes identical with the takeoff distance to 35 ft with engine failure. The corresponding value of V_x is the CRITICAL POWER FAILURE SPEED (Fig. K-2). Engine failure prior to this

*Fig. K-1 assumes the stopway to be included in the clearway, but there is no unanimity about this.

575

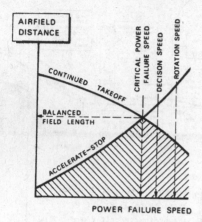

Fig. K-2. Balanced field length concept
and decision speed

speed must be followed by an aborted takeoff; the takeoff will be continued if engine failure occurs at the critical power
failure speed or thereafter. Generally
speaking, the BFL is obtained at the worst
condition at which an engine can fail. In
the preliminary design stage the BFL is
usually determined without allowing for
the existence of a stopway or clearway.

e. The takeoff field length required is
the greatest of the emergency distance,
the takeoff distance to 35 ft with engine
failure and the all-engine takeoff distance to 35 ft, in the last case multiplied by 1.15.
Contrary to the all-engine case, the unfactored BFL is accepted by the airworthiness authorities as the reference engine
failure case, the reason being that engine
failure at the critical power failure
speed is considered as an extremely unlikely combination of unfavorable conditions, to which no extra safety margin
need be added.

f. The TAKEOFF PATH from rest to at least
1,500 ft (450 m)* is divided into the take-

*More precisely: the altitude where the
flaps are retracted and the airspeed is
increased to the operational climb speed.

off distance and the takeoff flight path
(Fig. K-1). The takeoff distance is the
more complicated case and will be dealt
with in greater detail. The takeoff flight
path begins at 35 ft above the airfield
and must be determined in order to ensure
that, for a particular flight, obstacles
will be cleared by at least 35 ft in adverse conditions, with engine failure and
a combination of unfavorable deviations
from the GROSS (nominal, calculated)
FLIGHT PATH performance. These deviations
are determined on a statistical basis by
subtracting from the gross flight path a
margin of .8%, .9% or 1.0% for airplanes
with 2, 3 or 4 engines respectively. The
resulting trajectory is referred to as
THE NET FLIGHT PATH.

K-2. REFERENCE SPEEDS

Certain minimum criteria are laid down for
the takeoff reference speeds* (Fig. K-3)

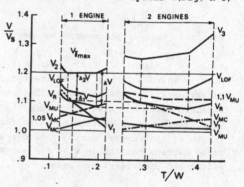

Fig. K-3. Example of a takeoff reference
speed schedule for a twin-engine aircraft

to ensure adequate safety and handling,
particularly in the case of engine failure,
minor mis-handling on the part of the pilot and atmospheric turbulence.

*The nomenclature of several reference
airspeeds in this section is simplified;
for full details the reader should refer
to the airworthiness rules, particularly
as regards V_s.

As already mentioned in Section 5.4.4., THE STALLING SPEED V_S can be approximated as follows:

$$V_S = \sqrt{\frac{W}{S} \frac{2}{\rho} \frac{1}{(1.13) \, C_{L_{max}}}} \qquad (K-1)$$

where C_L-max is the "physical" maximum lift coefficient, i.e. C_L at the top of the C_L vs. α curve, while the statistical factor 1.13 takes account of the speed loss in the FAA stall maneuver.

THE MINIMUM CONTROL SPEED (in the air) V_{MC} is the lowest airspeed at which it has proved to be possible to recover control of the airplane after engine failure with a bank angle of up to five deg. and take-off thrust (power). The broad requirement is that V_{MC} must not exceed 1.2 V_S with MTOW. For a specified pressure altitude and flap angle there is only one V_{MC}, regardless of the weight. Hence, the ratio V_{MC}/V_S increases when the T/W ratio increases because of decreasing takeoff weight.

In the project design phase an estimate of V_{MC} may be based on wind tunnel measurements or statistical evidence (cf. Section 9.6.1. and Table 9-3).

There is also a minimum control speed on the ground below which the takeoff must always be aborted when engine failure occurs. This characteristic speed is generally of minor importance for performance estimates in the preliminary design stage, provided the vertical tailplane size and rudder capacity are adequate.

THE MINIMUM UNSTICK SPEED V_{MU} is the airspeed at and above which it can be demonstrated by means of flight tests that the aircraft can safely leave the ground and continue the takeoff; V_{MU} is usually very close to V_S. In view of the required positive gradient it may be a function of the T/W ratio.

Alternatively, V_{MU} may be determined by the geometry of the aircraft as the airspeed for which the fuselage tail scrapes the ground prior to liftoff. In view of the great influence of the ground proximity effects, V_{MU} is very difficult to es-

timate in the preliminary design stage. An attempt may be made to estimate the geometry-limited V_{MU}, using the C_L-α curve with ground effect included.

THE TAKEOFF SAFETY SPEED V_2 is the airspeed obtained at the 35 ft height point. The broad requirement stipulates:

$$V_2 \geqslant V_{2_{min}} \qquad (K-2)$$

$$V_2 \geqslant 1.1 \, V_{MC}$$

where $V_{2_{min}}$ must not be less than 1.2 V_S, except in the case of aircraft with four power units where the application of power results in a significant reduction of the one-engine-inoperative power-on stalling speed (in this case: $V_{2_{min}} = 1.15 \, V_S$). This requirement is intended to ensure an adequate safe climbout with the critical engine inoperative. V_2 may be increased relative to the minimum values in order to improve the climb gradient with a failed engine ("overspeed"). It should be noted that during the airborne phase dV/dt must not be negative at any point.

THE ALL-ENGINES SCREEN SPEED V_3 is the airspeed attained at the 35 ft height point with all engines operating. Since rotation of the airplane is initiated at V_R, generally determined by the engine-out case, V_3 is greater than V_2 by an amount depending on the T/W ratio. For instance, $V_3 = V_2 + 10$ kts ($V_2 + 5$ m/s) is a typical value (not a requirement).

THE LIFTOFF SPEED (or touchoff speed) V_{LOF} is the airspeed at which the aircraft first becomes airborne. This speed is determined by the rotation speed V_R and the piloting technique during rotation. A positive climb gradient potential (out of ground effect) must be present at V_{LOF}.

THE ROTATION SPEED V_R is the reference speed for the pilot at which he raises the nosewheel. The broad requirement is:

$$V_R \geqslant V_1$$

$$\qquad (K-3)$$

$$V_R \geqslant 1.05 \, V_{MC}$$

In addition, V_R must be chosen such that

V_2 is reached at 35 ft, taking into account the speed increment between V_R and V_2 (ΔV). If the airplane is rotated at its maximum practicable rate of rotation, V_{LOF} must not be less than $1.10\ V_{MU}$ in the all-engine operating condition, or $1.05\ V_{MU}$ in the one-engine-out case.

At very low T/W ratios, V_R and V_{LOF} may be increased in order to obtain a positive climb gradient potential at V_{LOF}*. However, when we have

$$\Delta C_{D_{uc}} < \gamma_{2_{min}}\ C_{L_2} \quad \text{(approximately)} \quad \text{(K-4)}$$

it can be shown that the second segment climb requirement* is more critical. This is the usual condition.

THE DECISION SPEED V_1. The procedure explained in Section K-1, leading to the BFL, results in a critical power failure speed V_x. The pilot will need time to recognize the failure and to decide whether to abort or continue the takeoff. During this decision time (typically 1 second for a test pilot) the airplane accelerates to a speed, called the decision speed V_1. However, once the rotation is initiated, the pilot must continue the takeoff when an engine fails, hence: $V_1 \leqslant V_R$, while V_1 must also be at or above V_{MC} on the ground.

For the aircraft in Fig. K-3 the interrelation between the various characteristic speeds is depicted as a function of the T/W ratio, both for the all-engine case and the one-engine-out case. The following observations apply fairly generally to aircraft with adequate one-engine-out directional control properties:

a. For moderate T/W, V_2 is usually equal to $V_{2_{min}}$. The margin relative to V_{MC} is generally not critical, provided the vertical tailplane design is adequate.

b. For high T/W ratios (low weight) V_2 is determined by the required margin relative to V_{MC}, and consequently V_2/V_S, V_3/V_S and the field length increase. This situation

is typical for a twin-engine aircraft with wing-mounted engines.

c. For very low T/W ratios (hot and high airfields) V_R, V_{LOF} and V_2 may be increased up to the point where V_2 equals the speed for maximum climb gradient, in order to meet the climb gradients required.

d. At a particular T/W ratio V_1 becomes equal to V_R; for T/W ratios below this value, the one-engine-out takeoff is always critical and thus the field length is not balanced.

e. In practical takeoff calculations the engine-out case determines the reference speeds, particularly V_R which is equal for both the all-engines and the one-engine-out case. It is therefore appropriate to analyze the engine failure case prior to the all-engines takeoff.

K-3. PROCEDURE FOR DETERMINING THE TAKEOFF FIELD LENGTH

In what follows it is assumed that the ambient conditions (temperature, pressure altitude), aircraft weight and position of the high-lift devices are specified. Aerodynamic data (lift curve, drag polar) are assumed to be known, as well as the effects of ground proximity on these characteristics. Runway gradients and wind effects are usually ignored in pre-design calculations.

Steps 1 through 17 of the following procedure apply to the engine failure case, except when otherwise specified. Various methods of analysis are given in Section K-4.

1. Calculate V_S, $V_{2_{min}}$ and V_{MC} and the minimum allowable V_2.

2. Check the second segment climb gradient for $V = V_{2_{min}}$ by means of WAT curves (Section 11.6.2.). If necessary: increase V_2 to the airspeed at which this requirement is satisfied.

3. If even the climb gradient at the optimum speed for maximum climb gradient $V_{\gamma_{max}}$ is too low, reduce the flap deflection angle, if possible, or the takeoff

*An explanation and summary of the climb requirements can be found in Section 11.6.2.

578

weight.

4. Calculate the airborne distance $(S_A)_{N-1}$ and speed increment $(\Delta_2 V)_{N-1}$ for several values of V_{LOF}. Ground effect, undercarriage retraction initiated 3 seconds after liftoff, and extra drag due to engine failure must be taken into account (see Appendix G).

5. Determine V_{LOF} for which $V_{LOF} + (\Delta_2 V)_{N-1}$ $= V_2$ and the corresponding air distance.

6. Check the first segment climb potential (Section 11.6.2.) and increase V_{LOF} and V_2 if necessary. Alternatively, the flap angle may be reduced or the takeoff weight decreased, and the procedure is then started afresh.

7. Calculate the speed increment $(\Delta_1 V)_{N-1}$ and distance travelled $(S_R)_{N-1}$ during rotation prior to liftoff and $V_R = V_{LOF} - (\Delta_1 V)_{N-1}$ A normal rate of rotation is assumed. Make sure that V_R has an adequate margin to V_{MC}. Increase V_R to 1.05 times V_{MC} if necessary.

8. Analyze the rotation phase for the maximum practicable rate of rotation, both for the all-engines-operating and the engine failure case and make sure that V_{LOF} has an adequate margin relative to V_{MU}. If this is not so, increase V_R.

9. After the final choice of V_R, the (normal) rotation phase and airborne distances may be analyzed, resulting in final values of $(S_R)_{N-1}$, V_{LOF}, $(S_A)_{N-1}$ and V_2.

10. Choose several values for $V_x \geqslant V_{MC}$ on the ground, e.g. V_x = 90%, 95% and 100% of V_R.

11. Determine the ground run from standstill to V_x (S_{o-x}) for all values of V_x with all engines operating.

12. Determine the ground run from V_x to V_R (S_{x-R}) for all values of V_x with one engine inoperative. A realistic assumption should be made concerning the decay of thrust with time after engine failure. It is preferable to use the engine manufacturer's data, if available.

13. Determine the takeoff distance with engine failure at V_x:

$$S_{N-1} = S_{o-x} + S_{x-R} + (S_R)_{N-1} + (S_A)_{N-1} \qquad (K-5)$$

for the chosen values of V_x and plot S_{N-1} vs. $(V_x/V_S)^2$ *.

14. Determine the distance to come to a standstill from V_x: S_{STOP}.

15. The accelerate-stop (emergency) distance,

$$S_{AS} = S_{o-x} + S_{STOP} \qquad (K-6)$$

is computed for the chosen values of V_x and plotted vs. $(V_x/V_S)^2$ *.

16. The intersection of S_{N-1} and S_{AS} defines the critical power failure speed and the BFL (Fig. K-2).

17. Calculate V_1 and make sure that $V_1 \leqslant V_R$. If this is not so, the condition of engine failure one second prior to V_R determines the field length required with engine failure.

18. Analyse the rotation phase in the all-engine takeoff, assuming V_R equal to the one-engine-out case. The distance travelled is $(S_R)_N$, and V_{LOF} follows from the speed increment $(\Delta_1 V)_N$.

19. Analyze the airborne phase in the all-engine takeoff. The distance travelled is $(S_A)_N$ and the speed increment $(\Delta_2 V)_N$.

20. The factored takeoff distance is:

$$S_N = 1.15 \left[S_{o-R} + (S_R)_N + (S_A)_N \right] \qquad (K-7)$$

21. The takeoff field length required is the greater of the BFL and S_N.

K-4. METHODS AND DATA FOR THE ANALYSIS OF THE TAKEOFF

The following methods are generally used in the preliminary design stage, when the data required are not sufficiently complete to warrant a detailed analysis.

K-4.1. The ground run from standstill to V_x

The distance required to accelerate the aircraft from standstill to V_x is:

*S_{N-1} and S_{AS} vs. $(V_x/V_S)^2$ are almost linear relationships

$$S_{o-x} = \int_{0}^{V_x} \frac{V}{a}\, dV = \frac{1}{2g} \int_{0}^{V_x^2} \frac{dV^2}{a/g} \qquad (K-8)$$

Assuming the thrust vector and the runway to be horizontal, it can be shown that:

$$\frac{a}{g} = \frac{dV/dt}{g} = \frac{T}{W} - \mu - \left(C_{D_G} - \mu\, C_{L_G}\right) \frac{\frac{1}{2}\rho V^2 S}{W} \qquad (K-9)$$

For an arbitrary relationship of engine thrust with speed, the integral in (K-8) may be solved numerically. Numerous methods have been developed in the literature to simplify this process (e.g. Refs. K-1 through K-13). A well-known approximation is: $T = \text{constant} = \bar{T}$. In that case substitution of (K-9) into (K-8) results in an integral, which can be solved analytically:

$$S_{o-x} = \frac{V_x^2}{2g\,(\bar{T}/W - \mu)} \cdot \frac{1}{a}\, \ln\!\left(\frac{1}{1-a}\right) \qquad (K-10)$$

$$\text{where } a = \frac{C_{D_G} - \mu\, C_{L_G}}{(1.13)\,C_{L_{max}}\,(\bar{T}/W - \mu)}\left(\frac{V_x}{V_S}\right)^2 \qquad (K-11)$$

Coefficients of rolling friction are fundamentally a function of tire pressure and ground speed; good average values are:

hard runways (concrete, tarmac):	$\mu = .02$
hard turf, gravel	$: \mu = .04$
short, dry grass	$: \mu = .05$
long grass	$: \mu = .10$
soft ground	$: \mu = .10$
	to .30

The mean thrust $\bar{T}$ is defined at a mean velocity given by Fig. K-4. When the numerical procedure is used instead, the thrust lapse with speed is obtained from the engine manufacturer's brochure supplemented by a propeller diagram in the case of propeller-powered aircraft (cf. Section 6.3.2.). Alternatively, Fig. 4-35 may be used for jet engines as a first approximation. The angle of incidence is assumed invariable during the ground roll for nosewheel aircraft and is specified by the three-point ground attitude. Note that during the roll C_L and C_D are affected by the ground effect. For tailwheel aircraft an optimum attitude is defined by the condition that the accel-

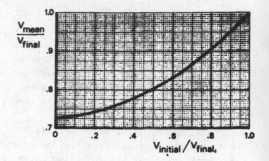

Fig. K-4. Mean speed for estimating the distance required to accelerate from a given initial speed to a given final speed

eration has a maximum when:

$$C_{D_G} - \mu\, C_{L_G} = C_{D_o} - \frac{1}{4}\, \mu^2\, \pi Ae \qquad (K-12)$$

$$\text{for } C_{L_G} = \frac{1}{2}\, \mu\, \pi Ae$$

The ground effect can be estimated from the data given in Appendix G, Section G-7.

K-4.2. The ground run from V_x to V_R

A precise analysis of the motion of the aircraft after engine failure is complicated by several factors which affect the variation of external forces with time (see Fig. K-5).

a. Immediately after engine failure, the thrust decays in a finite time (typically 4 seconds) to zero or idling thrust.

b. Engine failure causes windmilling drag of the dead engine and extra drag due to the asymmetric flight condition. Additional drag is also created in the rotation and flare maneuvers, and this is considerably affected by the piloting technique. The download on the tailplane must be compensated for by extra wing lift, resulting in increased induced drag. The ground effect gradually decreases after liftoff; hence the induced drag increases. Retraction of

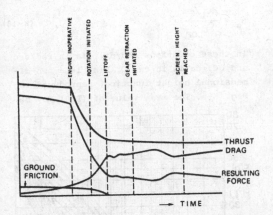

Fig. K-5. Forces on the airplane during the takeoff with engine failure

the undercarriage, initiated 3 seconds after liftoff, results in another drag variation.

c. Ground friction drag is related to the lift and vanishes at liftoff. The calculated total force component in the direction of the flight path has an irregular shape, but the variation may not be observed in practice due to the dynamic character of the motion.

d. Another complication is that, especially on large aircraft, the motion of the lowest point, and not just the center of gravity, must be observed. During the rotation phase prior to liftoff, and immediately after, a check should be made to see whether the fuselage tail has adequate clearance to the ground. When the screen is passed, the total height gain of the c.g. during the airborne phase may be considerably more than 35 ft for large aircraft.

Equation (K-8) applies to the acceleration phase from V_x to V_R, provided the integration is carried out from V_x to V_R. The following analytical expression can be used instead of a numerical procedure:

$$S_{x-R} = \frac{V_x^2}{2g(\bar{T}/W-\mu)} \cdot \frac{1}{a} \ln\left[\frac{1-a}{1-a(V_R/V_x)^2}\right] \quad (K-13)$$

It is noted that $\bar{T}$ and "a" have different values from the previous case, due to the thrust reduction and drag increment after engine failure. The velocity for which $\bar{T}$ is to be determined can be obtained from Fig. K-4.

K-4.3. The rotation phase

Assuming the mean rate of rotation about the lateral axis equal to $(d\theta/dt)_R$ and the mean acceleration along the runway conservatively equal to the value at liftoff, we have:

$$S_R = \frac{1}{2}(V_R + V_{LOF})\frac{\alpha_{LOF} - \alpha_G}{(d\theta/dt)_R} \quad (K-14)$$

and

$$V_{LOF} = V_R + \Delta_1 V = V_R + g\left(\frac{T-D}{W}\right)_{LOF}\frac{\alpha_{LOF} - \alpha_G}{(d\theta/dt)_R} \quad (K-15)$$

In view of the short duration of the rotation phase (2 to 4 seconds), this simple approximation will be acceptable in the pre-design stage. Typical values for $(d\theta/dt)_R$ are quoted in Section 10.3.1. The liftoff angle of attack is found from the C_L-α curve in proximity to the ground.

K-4.4. The airborne phase

A detailed study of a particular takeoff problem will usually involve a numerical calculation of the step-by-step type, assuming a control law for the lift coefficient, the angle of attack or pitch, or the elevator deflection. The problem with these methods is to define an adequate control law, which usually has to be based on previous experience with similar types of aircraft.

Many attempts have been made to develop analytical methods for simple control laws, such as taking C_L = constant or assuming the flare-up to be a circular arc with n = L/W = constant. We shall quote two representative examples here:

a. The AGARD Flight Test Handbook method (Ref. K-9), where the equations of motion are linearized and solved analytically for

581

$C_L = C_{L-LOF} + \Delta C_L$ = constant and (T-D) = constant. The flare becomes a sector of a phugoid for this particular case. The method gives the height, speed and distance travelled vs. time after liftoff. However, the guidelines for choosing ΔC_L are not very practical for civil aircraft. The author's experience is that the airborne distance is reasonably well predicted for $\Delta C_L/C_{L-LOF}$ = 0 to .05 for the N-1 takeoff, and for $\Delta C_L/C_{L-LOF}$ = .10 to .15 when all engines are operating.

b. The method developed by Perry (Ref. K-12) uses a constant rate of pitch after liftoff: $(d\theta/dt)_A$. The equations of motion are linearized by assuming V = constant and (T-D) = constant. The advantage of this method is that the control law is representative for practical civil aircraft piloting techniques, while - contrary to C_L - the pitch angle can be directly observed. As the speed increment after liftoff is ignored in solving the equation of motion normal to the flight path, the shape of the flare-up may not be realistic for high T/W ratios.

The method devised by Perry may be summarized as follows:

1. The airborne distance is composed of a flare-up, during which the flight path angle γ is increased from zero* to γ_2 at V_2 (or γ_3 at V_3) and a phase with constant climb angle γ_2 or γ_3.

2. The height gain after liftoff is given by:

$$h = \frac{V_{LOF}^2}{g} \frac{T-D}{W} F(\dot{\theta}) F(h) \qquad \text{(K-16)}$$

where

$$F(\dot{\theta}) = 1 + \frac{V_{LOF}}{2g} \frac{W}{T-D} n_\alpha \left(\frac{d\theta}{dt}\right)_A \qquad \text{(K-17)}$$

and

*The actual climb angle of the c.g. at liftoff is of the order of a half to one degree, due to the extension of the undercarriage and rotation.

582

$$n_\alpha = \frac{dC_L/d\alpha}{C_{L_{LOF}}} \qquad \text{(K-18)}$$

In these equations $(d\theta/dt)_A$ is in rad/sec and, $dC_L/d\alpha$ is in rad^{-1}. $F(h)$ is a non-dimensional height function, reproduced in Fig. K-6. The mean value of (T-D) may be

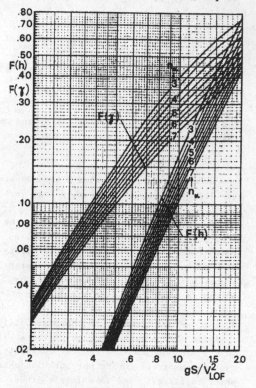

Fig. K-6. The functions F(h) and F(γ) used in Perry's method for the analysis of the airborne path (Derived from Ref. K-12)

taken as being halfway the airborne distance. A provisional assumption may be made for $(d\theta/dt)_A$, for instance 1°/sec ($d\theta/dt$ = .018) for the case with engine failure, or 2°/sec ($d\theta/dt$ = .035) for the all-engine case.

3. The climb angle during the transition is given by:

$$\gamma = \frac{dh}{dS} = \frac{T-D}{W} F(\dot{\theta}) F(\gamma) \qquad \text{(K-19)}$$

where F(γ) is a non-dimensional flight path

angle function depicted in Fig. K-6.

4. The end of the transition is defined by

$$F(\gamma) = \frac{\gamma_2}{\frac{T-D}{W} F(\dot\theta)} \qquad (K-20)$$

In the all-engine case γ_3 is used instead of γ_2. Using Fig. K-6 both the height and distance travelled at the end of the flare-up are found from this condition. If this height is less than h_{to} = 35 ft (10.7 m), an additional phase must be added with climb angle γ_2 or γ_3. If the takeoff height is reached before the end of the transition, $F(h)$ is found by substitution of h_{to} into (K-16) and the airborne distance is obtained from Fig. K-6.

5. A check on $(d\theta/dt)_A$ can be made by assuming that during the flare-up the pitch angle increases linearly from α_{LOF} to the pitch angle at V_2 in stationary flight:

$$\left(\frac{d\theta}{dt}\right)_A = \frac{V_{LOF}\left[\alpha_{V_2} - \alpha_{LOF} + \gamma_2\right]}{\text{transition distance}} \qquad (K-21)$$

For the all-engine case V_3 and γ_3 are used instead of V_2 and γ_2.

6. The speed increment after liftoff is obtained from the energy equation:

$$\Delta_2 V = \frac{gS}{V_{LOF}}\left(\frac{T-D}{W} - \frac{h}{S}\right) \qquad (K-22)$$

where h and S are the height and distance travelled after liftoff at the end of the flare-up, or at the 35 ft (10.7 m) height point, whichever is less. It should be noted that γ_2 and γ_3 can only be calculated when V_2 and V_3 are known. The calculation is therefore iterative.

K-4.5. The stopping distance

The velocity after engine failure is depicted in Fig. K-7 as a function of time. Initially an appreciable overshoot is observed due to the still considerable thrust of the inoperative engine immediately after failure. Time delays are necessary to allow for failure recognition and decision (1 second) and subsequent operation of the

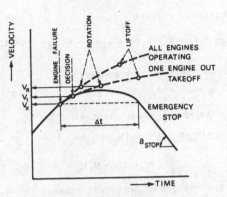

Fig. K-7. Time history of the airspeed during the takeoff

wheelbrakes, throttle closure, lift dumpers and airbrakes.

Integration of the velocity yields the emergency brake distance

$$S_{STOP} = V_x \Delta t + \frac{1}{2g} \int_{V_x}^{o} \frac{dV}{a/g} \qquad (K-23)$$

where Δt is approximately 3 to 4 seconds. In Fig. K-1 $V_x \Delta t$ is referred to as the inertia distance.

In principle, the equation of motion during braking is similar to (K-9) for the acceleration phase. However, consideration must be given to the fact that the retarding ground force mainly acts at the main undercarriage, as the nosewheel is usually not provided with brakes.

The deceleration during the steady braked roll can be derived from the conditions of equilibrium (Fig. K-8):

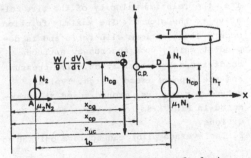

Fig. K-8. Forces on the aircraft during the braked roll

horizontal forces:

$$\frac{W}{g}\frac{dV}{dt} + D - T + \mu_1 N_1 + \mu_2 N_2 = 0 \qquad (K-24)$$

forces normal to the runway:

$$L - W + N_1 + N_2 = 0 \qquad (K-25)$$

and the moment about point A:

$$W\left\{l_b - (x_{uc} - x_{cg})\right\} - L\left\{l_b - (x_{uc} - x_{cp})\right\} -$$

$$\frac{W}{g}\left(-\frac{dV}{dt}\right)h_{cg} + D h_{cp} - T h_T - N_1 l_b = 0 \qquad (K-27)$$

In the absence of nosewheel braking we take $\mu_1 = \mu$ and $\mu_2 = 0$, resulting in the following expression for the deceleration:

$$\frac{a}{g} = \frac{-dV/dt}{g} = A_1 \mu + \frac{A_2 C_{D_G} - A_3 \mu C_{L_G}}{(1.13) C_{L_{max}}}\left(\frac{V}{V_S}\right)^2 - A_4 \frac{T}{W}$$

$$(K-28)$$

where

$$A_1 = \frac{l_b - (x_{uc} - x_{cg})}{l_b + \mu h_{cg}} \quad ; \quad A_2 = \frac{l_b + \mu h_{cp}}{l_b + \mu h_{cg}} \approx 1 \quad ;$$

$$A_3 = \frac{l_b - (x_{uc} - x_{cp})}{l_b + \mu h_{cg}} \quad ; \quad A_4 = \frac{l_b + \mu h_T}{l_b + \mu h_{cg}}$$

If brakes are also present on the nose-wheel, having the same braking effectiveness as those on the main u.c., A_1 through A_4 are equal to 1. Some data for calculating the stopping distance are given below.

a. The friction coefficient μ.
1. The tyre-to-runway friction during braking is a function of the slip ratio, i.e. the relative velocity of the tire relative to the ground[*]. The maximum friction occurs at an optimum slip ratio and is dependent mainly upon the runway surface condition and contamination, tire pressure and the type of tread design. Typical curves for μ_{max} vs. ground speed are presented in Fig. K-9a for several runway conditions.
2. The average pilot can achieve only 30 to

[*]A precise definition and background information can be found in Ref. K-31

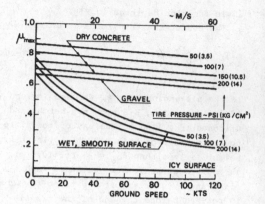

a. Typical braking coefficients of friction

Fig. K-9. Some data on braking performance

50 percent of μ_{max} (hence $\mu \approx .35$) in view of his inability to maintain the optimum slip ratio. For this reason, anti-skid devices based on mechanical or electronic pressure control systems have been developed. Their effectiveness may be expressed in the form of an efficiency factor:

$$\eta_b = \frac{\mu}{\mu_{max}} \qquad (K-29)$$

Fig. K-9b presents typical values for η_b,

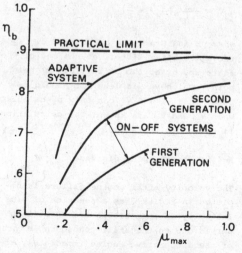

b. Typical anti-skid system efficiency

Fig. K-9 (continued)

although in practice wide variations will be found. The simple, first-generation anti-skid systems based on the on-off principle achieve low values of η_b, of the order of .5 to .6 at low μ_{max} (wet runways). Adaptive brake pressure control systems boost this performance considerably, and the most recent systems have a value of up to η_b = .9, almost independent of the runway condition.

3. Brake effectiveness is reduced by various dynamic effects such as normal load variation, undercarriage vibration and suspension effects. These may reduce η_b by up to 20% (Ref. K-31).

4. The total torque obtainable from all brakes (Q_{max}) forms a limiting factor at low ground speeds on dry runways, when μ is very high. To ensure passenger comfort during normal braking the maximum torque is usually so designed that decelerations will not exceed .5 to .6 g during a maximum braking effort. On light aircraft and small passenger transports a typical value for Q_{max} results in a wheel braking force of .35 W at zero ground speed.

5. A very large amount of heat is developed during braking and the brakes and tires may become very hot. A limit is set to the maximum heat sink capacity, which may have different values for various operating conditions (e.g. normal braking or emergency braking).

The effective friction coefficient may be obtained by taking the least of the following:

$$\mu = .85 \, \eta_b \, \mu_{max}$$

and

$$\mu = \frac{Q_{max}}{N_1 \, r_t} \qquad \text{(K-30)}$$

where r_t is the radius of the deflected tire, which is a function of the tire load. In view of these characteristics, it may be generally observed that at low ground speed μ increases with speed - N_1 decreases and r_t increases, but less rapidly - while at high ground speeds the available friction is critical and decreases with speed,

particularly on wet runways.

b. Engine thrust during braking.
The operational program for manipulating the thrust must be chosen in accordance with the airworthiness rules. The FAR 25 requirements do not allow thrust reversal, and idle forward thrust is assumed in this case. British rules allow the use of reverse thrust under certain conditions, but the failed engine will cause asymmetry and in the pre-design stage it is wise to be conservative and neglect the reverse thrust in performance estimation.

c. Aerodynamic properties.
The deflection and effectiveness of the high-lift devices, the effect of ground proximity, the design and operation of lift dumpers (ground spoilers) and reverse thrust - particularly for propeller aircraft - are all factors affecting the lift and drag coefficients. In the absence of aerodynamic data on spoilers, their effect may be assumed to be equivalent to an increment in the mean deceleration of .08 to .10 g.

A simple solution of (K-23) can be obtained by taking $\mu = \mu_{eff}$ = constant and T = 0. The result is as follows:

$$S_{STOP} = \frac{v_x^2}{2g \, A_1 \, \mu_{eff}} \cdot \frac{1}{a} \ln\left(\frac{1}{1-a}\right) + v_x \, \Delta t \qquad \text{(K-31)}$$

where $a = \dfrac{A_3 \, \mu_{eff} \, C_{L_G} - A_2 \, C_{D_G}}{(1.13) \, A_1 \, \mu_{eff} \, C_{L_{max}}} \left(\dfrac{v_x}{v_S}\right)^2$ (K-32)

An even simpler solution is found when the deceleration during braking is assumed constant:

$$S_{STOP} = \frac{v_x^2}{2a_{STOP}} + v_x \, \Delta t \qquad \text{(K-33)}$$

The highest values obtainable for a_{STOP} are typically:

.55g - dry runway, maximum effort, ignoring passenger tolerance,

.35g - wet runway, modern braking with anti-skid, lift dumpers, reverse thrust,

.15g - wet runway, simple braking; or flooded runway, reverse thrust,

while Δt is approximately $3\frac{1}{2}$ seconds.

REFERENCES

Analysis of takeoff performance

K-1. W.S. Drehl: "The calculation of takeoff run". NACA Technical Report No. 450, 1932.

K-2. B. Göthert: "Der Abflug von Landflugzeugen mit besonderer Berücksichtigung des Uebergangsbogens". Jahrbuch 1937 der Deutschen Luftfahrtforschung.

K-3. J.R. Ewans and P.A. Hufton: "Note on a method of calculating takeoff distances". RAE Technical Note Aero 880 (ARC 4783), 1940.

K-4. W.R. Buckingham: "A theoretical analysis of the airborne path during takeoff". Aircraft Eng., Jan. 1958, pp. 5-8.

K-5. D.J. Kettle: "Ground performance at takeoff and landing". Aircraft Eng., Jan. 1958, pp. 2-4.

K-6. A.D. Edwards: "Performance estimation of civil jet aircraft". Aircraft Eng. 1950, pp. 70-75.

K-7. Anon.: "Estimation of takeoff distance". R.Ae.S. Data Sheet "Performance" EG 5/1, 1952.

K-8. G.E. Rogerson: "Estimation of takeoff and landing airborne paths". Aircraft Eng., Nov. 1960, pp. 328-330.

K-9. F.E. Douwes Dekker and D. Lean: "Takeoff and landing performance". AGARD Flight Test Manual, Vol. 1, Chapter 8. Pergamon Press, 1962.

K-10. L. Bournet: "Estimation de la longueur de roulement au décollage, un problème simple parfois méconnu Technique et Science Aéronautique, 1967, pp. 213-222.

K-11. A.R. Krenkel and A. Salzmann: "Takeoff performance of jet-propelled conventional and vectored-thrust aircraft". J. of Aircraft, 1968, pp. 429-436.

K-12. D.H. Perry: "The airborne path during takeoff for constant rate-of-pitch manoeuvres". ARC CP No. 1042, 1969.

K-13. Anon.: "Estimation of takeoff distance". ESDU Data Sheets "Performance" EG 5/1, 1971/1972.

K-14. J. Collingbourne: "A digital computer program (EMA) for estimating aircraft takeoff and accelerate-stop distances". RAE Tech. Memo Aero, 1252, 1970.

K-15. D.H. Perry: "A review of methods for estimating the airfield performance of conventional fixed wing aircraft". RAE Tech. Memo Aero 1264, 1970.

K-16. J. Williams: "Airfield performance prediction methods for transport and combat aircraft". AGARD Lecture Series No. 56, April 1972.

K-17. R.N. Harrison: "Takeoff and climb characteristics". Short course on aircraft performance estimation, Cranfield Institute of Technology, Feb. 1973.

Braking friction, water, snow, slush and ice on runways

K-18. J.W. Wetmore: "The rolling friction of several airplane wheels and tires and the effect of rolling friction on takeoff". NACA Technical Report No. 583, 1973.

K-19. E.C. Pike: "Coefficients of friction". J. of the Royal Aero. Soc., December 1949.

K-20. M.N. Gough, R.H. Sawyer and J.P. Trant: "Tire-runway braking coefficients". AGARD Report 51, Feb. 1956.

K-21. W.B. Horne, U.T. Joyner and T.J.W. Leland: "Studies of retardation force developed on an aircraft tire rolling in slush or water". NASA TN D-552, Sept. 1960.

K-22. J.A. Zabovchik: "Ground deceleration and stopping of large aircraft". AGARD Report 231, Oct. 1958.

K-23. W.B. Horne and J.W.L. Tafford: "Influence of tire tread pattern and runway surface conditions on braking friction and rolling resistance". NASA TN D-1376, Sept. 1962.

K-24. G. Ciampolini: "A method of evaluating runway friction for the prediction of actual takeoff runs".

AGARD Report 418, 1963.

K-25. H.R. Herb: "Problems associated with the presence of water, slush, snow and ice on runways". AGARD Report 500, 1966.

K-26. R.L. Maltby and H.W. Chinn: "Effects of slush on takeoff". Shell Aviation News, No. 296-1963, pp. 8-11.

K-27. T.G. Foxworth and H.F. Marthinsen: "Another look at accelerate-stop criteria". AIAA Paper No. 69-772.

K-28. L.S. McBee: "Effective braking - a key to air transportation progress". SAE Paper No. 640376.

K-29. E.G. Wilkinson: "Lift-dump system shortens landing roll on ice runways". Space/Aeronautics, Oct. 1969, pp. 73-75.

K-30. J.T. Yager: "A comparison of aircraft and ground vehicle stopping performance on dry, wet, flooded, slush, snow and ice-covered runways". NASA TN D-6098, 1970.

K-31. Anon.: "Frictional and retardation forces on aircraft tyres". ESDU Data Sheets No. 71025, 71026 and 72008.

K-32. J.T. Yager, W. Pelham Phillips and P.L. Deal: "Evaluation of braking performance of a light, twin-engine airplane on grooved and ungrooved pavements". NASA TN D-6444, 1971.

K-33. J.L. McCarthy: "Wear and related characteristics of an aircraft tire during braking". NASA TN D-6963, Nov. 1972.

STOL - takeoff performance

K-34. P.L. Sutcliffe, V.K. Merrick and A.R. Howell: "Aerodynamics and propulsion of minimum field aircraft". Proc. 8th Anglo-American Aeron. Conf., 181-232, London, 1961.

K-35. J. Hamann: "Contribution à la définition d'un avion léger STOL". Jahrbuch 1965 der WGLR, pp. 133-140.

K-36. F.H. Schmitz: "Takeoff trajectory optimization of a theoretical model of a STOL aircraft". AIAA Paper No. 69-935.

K-37. D.O. Carpenter and P. Gotlieb: "The physics of short takeoff and landing (STOL)", AIAA Paper No. 70-1238.

K-38. R.K. Ransone: "STOL definition and field length criteria". AIAA Paper No. 70-1240.

Index